HANDBOOK OF AGING AND THE SOCIAL SCIENCES

EIGHTH EDITION

THE HANDBOOKS OF AGING

Consisting of Three Volumes

Critical comprehensive reviews of research knowledge,
theories, concepts, and issues

Editors-in-Chief
Laura L. Carstensen

and

Thomas A. Rando

Handbook of the Biology of Aging, 8th Edition
Edited by Matt Kaeberlein and George M. Martin

Handbook of the Psychology of Aging, 8th Edition
Edited by K. Warner Schaie and Sherry L. Willis

Handbook of Aging and the Social Sciences, 8th Edition
Edited by Linda K. George and Kenneth F. Ferraro

HANDBOOK OF AGING AND THE SOCIAL SCIENCES

EIGHTH EDITION

Edited by

LINDA K. GEORGE AND KENNETH F. FERRARO

Associate Editors

DEBORAH CARR, JANET M. WILMOTH, AND DOUGLAS A. WOLF

AMSTERDAM • BOSTON • HEIDELBERG • LONDON
NEW YORK • OXFORD • PARIS • SAN DIEGO
SAN FRANCISCO • SINGAPORE • SYDNEY • TOKYO

Academic Press is an imprint of Elsevier

Academic Press is an imprint of Elsevier
32 Jamestown Road, London NW1 7BY, UK
525 B Street, Suite 1800, San Diego, CA 92101-4495, USA
225 Wyman Street, Waltham, MA 02451, USA
The Boulevard, Langford Lane, Kidlington, Oxford OX5 1GB, UK

Seventh edition 2011
Eighth edition 2016

Copyright © 2016, 2011 Elsevier Inc. All rights reserved

No part of this publication may be reproduced or transmitted in any form or by any means, electronic or mechanical, including photocopying, recording, or any information storage and retrieval system, without permission in writing from the Publisher. Details on how to seek permission, further information about the Publisher's permissions policies and our arrangements with organizations such as the Copyright Clearance Center and the Copyright Licensing Agency, can be found at our website: www.elsevier.com/permissions.

This book and the individual contributions contained in it are protected under copyright by the Publisher (other than as may be noted herein).

Notices

Knowledge and best practice in this field are constantly changing. As new research and experience broaden our understanding, changes in research methods, professional practices, or medical treatment may become necessary.

Practitioners and researchers may always rely on their own experience and knowledge in evaluating and using any information, methods, compounds, or experiments described herein. In using such information or methods they should be mindful of their own safety and the safety of others, including parties for whom they have a professional responsibility.

To the fullest extent of the law, neither the Publisher nor the authors, contributors, or editors, assume any liability for any injury and/or damage to persons or property as a matter of products liability, negligence or otherwise, or from any use or operation of any methods, products, instructions, or ideas contained in the material herein.

Library of Congress Cataloging-in-Publication Data
A catalog record for this book is available from the Library of Congress.

British Library Cataloguing-in-Publication Data
A catalogue record for this book is available from the British Library.

ISBN: 978-0-12-417235-7

For information on all Academic Press publications
visit our website at http://store.elsevier.com

Publisher: Nikki Levy
Acquisition Editor: Emily Ekle
Editorial Project Manager: Barbara Makinster
Production Project Manager: Melissa Read
Designer: Matthew Limbert

Printed and bound in the United States of America

Dedications

To my awesome siblings, Kathie, Kent, and Karen,
who have known and loved me longer than anyone.
—LKG

To Linda C. Ferraro, my gracious wife: *Many women do noble things,
but you surpass them all* (Proverbs 31.29).
—KFF

Contents

Foreword xi
Preface xiii
About the Editors xv
List of Contributors xvii

I
THEORY AND METHODS

1. Aging and the Social Sciences: Progress and Prospects
LINDA K. GEORGE AND KENNETH F. FERRARO

Theoretical and Conceptual Developments 4
Methods and Data 6
Emerging Themes in Aging Research 9
What Aging Research Contributes to the Social Sciences: The Big Picture 16
References 20

2. Trajectory Models for Aging Research
SCOTT M. LYNCH AND MILES G. TAYLOR

Growth Modeling in a Nutshell 25
Latent Class Modeling in a Nutshell 31
Latent Class Growth Analysis 38
Growth Mixture Modeling 39
Important Issues in the Implementation of Trajectory Methods 43
Conclusion 49
References 50

II
SOCIAL STRUCTURES AND PROCESSES

3. Biodemography: Adding Biological Insight into Social, Economic, and Psychological Models of Population and Individual Health Change with Age
EILEEN M. CRIMMINS AND SARINNAPHA M. VASUNILASHORN

Introduction 56
Expansion of the Demographic Approach: The Process of Health Change 56
The Expanded Biodemographic Model of Health 57
Measuring Biomarkers in Population Studies 59
Use of Biomarkers in Assessing Population Health and Health Care Use and Its Effectiveness 60
Summary Indices of Biological Risk 63
Genetic Markers as a New Frontier 68
Summary 69
Appendix: Information on Biomarkers Often Used in Social Science Research 69
References 72

4. Late-Life Disability Trends and Trajectories
DOUGLAS WOLF

Introduction 77
The Conceptualization of Disability 79
Measuring Disability 81
Evidence on Trends in Disability Prevalence 84

CONTENTS

Individual-Level Disability Trajectories 89
Conclusion 94
References 95

5. Early Life Origins of Adult Health and Aging
DIANA KUH AND YOAV BEN-SHLOMO

Introduction 101
Historical Overview 103
Early Origins of Adult Disease: From FOAD Through DOHaD to a Life Course Perspective 104
Early Life Origins of Functional Aging in a Life Course Perspective 109
Conclusions and Future Directions 116
References 118

6. Racial and Ethnic Inequalities in Health
JACQUELINE L. ANGEL, STIPICA MUDRAZIJA, AND REBECCA BENSON

Racial and Ethnic Inequalities in Health 123
Theoretical Perspectives 125
Research Across Minority Groups 130
Summary and Conclusion 133
References 137

7. Immigration, Aging, and the Life Course
JUDITH TREAS AND ZOYA GUBERNSKAYA

Introduction 143
Immigration as a Life-Course Experience 144
The Principle of Life-Span Development 146
The Principle of Agency 147
The Principle of Time and Place 147
The Principle of Timing 148
The Principle of Linked Lives 149
Immigrants and Families 150
Socioeconomic Outcomes of Older Immigrants 152
The Health of Older Immigrants 154
Conclusion 155
References 157

8. Gender, Time Use, and Aging
LIANA C. SAYER, VICKI A. FREEDMAN, AND SUZANNE M. BIANCHI

Introduction 163
Measuring Time Allocation in Later Life 165
"A Day in the Life" of Older Adults 167
The Social versus Solitary Dimension of Time 173
Caregiving, Time Use, and Well-Being 174
Future Directions 176
Acknowledgments 178
References 178

9. Social Networks in Later Life
BENJAMIN CORNWELL AND MARKUS H. SCHAFER

Introduction 181
Network Concepts and Definitions 182
Why and How Social Networks Matter 185
Aging and Social Network Change 189
Social Networks and Stratification 191
Emerging Topics in Network-Gerontology 193
Conclusions 196
References 197

III
SOCIAL FACTORS AND SOCIAL INSTITUTIONS

10. Stability, Change, and Complexity in Later-Life Families
J. JILL SUITOR, MEGAN GILLIGAN, AND KARL PILLEMER

Introduction 206
Theoretical Roots and Conceptual Advances 207
Substantive Advances 209
Relationship Quality Between Older Parents and Adult Children 212
Sibling Relations 215
Grandparent–Grandchild Relations 217
Marriage in the Later Years 219
Conclusion 220
References 221

11. The Influence of Military Service on Aging
JANET M. WILMOTH AND ANDREW S. LONDON

Introduction 227
Cohort Flow, Periods of War, and the Composition of the US Older Adult Population 228
Military Service as a "Hidden Variable" in Aging Research 230

Mechanisms Through Which Military Service Influences Aging 232
An Overview of Military Service and Aging Among Specific War Cohorts 234
Studying Military Service and Aging 242
References 246

12. Religion, Health, and Aging
NEAL KRAUSE AND R. DAVID HAYWARD

Introduction 251
Religious Involvement over the Life Course 252
Religion, Health, and Well-Being 254
From Correlation to Explanation: Identifying the Health-Related Dimensions of Religion 256
Spiritual Struggles: Assessing the Dark Side of Religion 264
Race/Ethnicity, Religion, and Health 265
Conclusions 266
Acknowledgment 267
References 267

13. Evolving Patterns of Work and Retirement
KEVIN E. CAHILL, MICHAEL D. GIANDREA, AND JOSEPH F. QUINN

Introduction 271
The Beginning and End of Earlier and Earlier Retirement 273
A Closer Look at the Retirement Process in the Modern Era 275
Changes to the Traditional Pillars of Retirement Income and How They Relate to Labor Force Participation 277
The Increasing Importance of Macroeconomic Influences 282
The Potential Benefits of Continued Work Later in Life 284
Disclaimer and Acknowledgments 287
References 287

14. Productive Engagement in Later Life
NANCY MORROW-HOWELL AND EMILY A. GREENFIELD

Introduction 293
Conceptual Issues 294
Relevance of Productive Engagement in Later Life 296

Scholarship on the Antecedents and Outcomes of Productive Engagement 299
Challenges and Future Directions 306
References 309

15. Aging, Neighborhoods, and the Built Environment
CAROL S. ANESHENSEL, FREDERICK HARIG, AND RICHARD G. WIGHT

Introduction 315
Theoretical Models of Neighborhood 316
Neighborhood Structure and the Health of Older Persons 321
The Built Environment and the Health of Older Persons 326
Discussion and Directions for Future Research 330
References 332

16. Abusive Relationships in Late Life
KAREN A. ROBERTO

Introduction 337
A Socioecological Framework for Understanding Elder Abuse 340
Vulnerabilities and Risk for Elder Abuse 342
Perpetrators of Elder Abuse 345
Responses to Elder Abuse 348
Future Research 350
References 351

17. The Impact of Disasters: Implications for the Well-Being of Older Adults
LISA M. BROWN AND KATHRYN A. FRAHM

Introduction 357
Types and Definitions of Disasters 358
Influence of Residential Environment on Disaster-Related Activities and Outcomes 359
Group Evacuation versus Individual/Independent Evacuation 360
Temporary Evacuation, Transfer, or Permanent Relocation 361
Age and Vulnerability 362
Stress and Coping 363
Age and Resilience 363
Disaster-Related Physical and Mental Health Issues 365
Social Factors and Disaster Response Outcomes 366
Role of Formal and Informal Social Support and Social Networks 367

Future Directions 369
Conclusion 370
References 370

18. End-of-Life Planning and Health Care
DEBORAH CARR AND ELIZABETH LUTH

Introduction 375
Death and Dying in the United States 376
Advance Care Planning 380
Public Policy Innovations 385
Conclusion and Future Directions 388
References 390

IV
AGING AND SOCIETY

19. Organization and Financing of Health Care
MARILYN MOON

Introduction 397
A Brief History of Medicare and Medicaid 399
Moving into an Era of Health Reform 403
Supplementing Medicare with Medicaid and other Insurance 405
The Affordability and Sustainability Questions about the Future of Medicare and Medicaid 410
Major Options for Reform 411
The Issues of Financing Medicare and Medicaid 415
References 416

20. Innovations in Long-Term Care
JOSEPH E. GAUGLER

Introduction 419
Defining Long-Term Care 421
A Brief Historical Overview of Long-Term Care in the United States 422
Selected Innovations in Long-Term Care 424
Looking Toward the Future of Long-Term Care 435
References 436

21. Politics and Policies of Aging in the United States
ROBERT B. HUDSON

Introduction 441
The Altered Political Perceptions of Older Americans 442
The Transformation of Seniors' Political Environment 446
Accounting for Old-Age Policy Enactments 451
Emerging Issues 455
References 457

22. The Future of Retirement Security in Comparative Perspective
JOHN B. WILLIAMSON AND DANIEL BÉLAND

Introduction 462
Social Security in the United States 462
Employer-Sponsored Pensions in the United States 465
International Developments and Lessons 467
Conclusion 476
Acknowledgments 478
References 478

23. Health Inequalities Among Older Adults in Developed Countries: Reconciling Theories and Policy Approaches
AMÉLIE QUESNEL-VALLÉE, ANDREA WILLSON, AND SANDRA REITER-CAMPEAU

Introduction 483
Theories of Health Inequality in Older Age 484
Welfare States and the Interplay of Social Solidarity and Equity 489
WHO Age-friendly Environments Programme 494
Promising Avenues for Sociological Research 496
References 498

Author Index 503
Subject Index 521

Foreword

The near-doubling of life expectancy in the 20th century represents extraordinary opportunities for societies and individuals. Just as sure, it presents extraordinary challenges. In the years since the last edition of the *Handbook of Aging* series was published, the United States joined the growing list of "aging societies" alongside developed nations in Western Europe and parts of Asia; that is, the U.S. population has come to include more people over the age of 60 than under 15 years of age. This unprecedented reshaping of age in the population will continue on a global scale and will fundamentally alter all aspects of life as we know it.

Science is responsible for the extension of life-expectancy and science is now needed more than ever to ensure that added years are high quality. Fortunately, the scientific understanding of aging is growing faster than ever across social and biological sciences. Along with the phenomenal advances in the genetic determinants of longevity and susceptibility to age-related diseases has come the awareness of the critical importance of environmental and psychological factors that modulate and even supersede genetic predispositions. The *Handbooks of Aging* series, comprised of three separate volumes, the *Handbook of the Biology of Aging*, the *Handbook of the Psychology of Aging*, and the *Handbook of Aging and the Social Sciences*, is now in its eighth edition and continues to provide foundational knowledge that fosters continued advances in the understanding of aging at the individual and societal levels.

Attention to the science of aging involves a concomitant increase in the number of college and university courses and programs focused on aging and longevity. With this expansion of knowledge, the *Handbooks* play an increasingly important role for students, teachers and scientists who are regularly called upon to synthesize and update their comprehension of the broader field in which they work. The *Handbook of Aging* series provides knowledge bases for instruction in these continually changing fields, both through reviews of core and newly emerging areas, historical syntheses, methodological and conceptual advances. Moreover, the interdisciplinary nature of aging research is exemplified by the overlap in concepts illuminated across the *Handbooks*, such as the profound interactions between social worlds and biological processes. By continually featuring new topics and involving new authors, the series has pushed innovation and fostered new ideas.

One of the greatest strengths of the chapters in the *Handbooks* is the synthesis afforded by preeminent authors who are at the forefront of research and thus provide expert perspectives on the issues that current define and challenge each field. We express our deepest thanks to the editors of the individual volumes for their incredible dedication and contributions to the series. It is their efforts to which the excellence of the products is largely credited. We thank Drs. Matt Kaeberlein and George M. Martin editors of the *Handbook of the Biology of Aging*;

Drs. K. Warner Schaie and Sherry L. Willis, editors of the *Handbook of the Psychology of Aging*; and Drs. Linda K. George and Kenneth F. Ferraro, editors of the *Handbook of Aging and the Social Sciences*. We would also like to express our appreciation to our publishers at Elsevier, whose profound interest and dedication has facilitated the publication of the *Handbooks* through their many editions. And we continue to extend our deepest gratitude to James Birren for establishing and shepherding the series through the first six editions.

Thomas A. Rando
Laura L. Carstensen
Stanford Center on Longevity,
Stanford University

Preface

Social science scholarship on aging is alive and well. Although the last edition of the *Handbook of Aging and the Social Sciences* appeared only 5 years ago, the growth of research since then on aging individuals, populations, and as a dynamic culmination of the life course has been extraordinary. There are many reasons for this stunning growth in the quantity and quality of aging research. Infrastructure and the methodological tools necessary for rigorous, sophisticated research have increased and become widely available to the scientific community. In the social sciences, the increase in data sources covering significant portions of the life course from a broad spectrum of societies, the increased coupling of social and biological data, and statistical advances have been especially important. New, energetic cohorts of scholars have posed fresh, innovative research questions to the field and demonstrated the importance of those questions for a deeper understanding of aging. And, of course, the complexities of population dynamics, cohort succession, and policy changes modify the world and its inhabitants in ways that must be vigilantly monitored so that aging research remains relevant and accurate.

This is the eighth edition of the *Handbook of Aging and the Social Sciences* and we have endeavored to do justice to the research topics and questions that, in our judgment, represent both foundational, classic, and ever-important topics critical to aging research in the social sciences and emerging and timely topics that expand the field in exciting ways. This edition of the *Handbook* includes 23 chapters. Seventeen of these chapters address topics that did not appear in the seventh edition; most of them address topics that did not appear in any previous edition of the *Handbook*. Of the six topics covered in this edition that also appeared in the seventh edition, four were written by different authors.

Because each edition of the *Handbook* includes chapters that differ from the previous edition, each edition is a stand-alone volume. Thus, chapters in the seventh edition, as well as even earlier ones, remain important compilations of aging research.

Just as the majority of chapters in this edition of the *Handbook* address new topics, most of the chapter authors also are new to this edition. Specifically, of the 47 chapter authors who contributed to this edition, 40 of them did not participate in the previous edition. Most of the new chapter authors are well-established scholars, but they are also relatively young. Without question, they will be among the premier scholars of aging for decades to come and it has been a great pleasure to include their impressive contributions to this edition of the *Handbook*.

Unlike the *Handbook of the Biology of Aging* and the *Handbook of the Psychology of Aging*, this *Handbook* is intended to cover a spectrum of disciplines. As a result, the chapters in this volume were written by scholars that include demographers, economists, epidemiologists, gerontologists, political scientists, psychologists, social workers, sociologists, and statisticians. Likewise, chapters address topics at the micro- and macro-levels, as well as topics that address the intersection of individual and aggregate factors. The result is a rich array of topics and perspectives that cover much, though by no means all, of the landscape of aging research in the social sciences.

Chapter authors were asked to contribute scholarly reviews of their topics, devoting special attention to what is new and exciting (theoretically, methodologically, and substantively) and to priority issues for future research. They meticulously crafted chapters that stand as exemplary reviews of the state-of-the-science and point the way to exciting ways to advance the field. We found reading these chapters and corresponding with the authors to be enlightening and we stand in awe of the knowledge and insights that the authors generously shared.

We also owe huge debts of gratitude to our Associate Editors: Deborah Carr, Janet Wilmoth, and Doug Wolf. They were involved in every part of this *Handbook*, including selection of topics and authors, providing feedback to chapter authors, and writing superb chapters themselves. As a group, they beautifully buttressed us in areas where our knowledge was less extensive than theirs and provided insights and suggestions that improved the volume in multiple ways.

Cohort succession signifies the entrance of fresh, new generations, but also brings the exit of those who created the world that new cohorts enter. Bob Binstock was the senior editor of the seven previous editions of the *Handbook of Aging and the Social Sciences*. Bob died in 2011. Bob was, without question, the guiding spirit and the meticulous work-horse of the *Handbook of Aging and the Social Sciences* series. And this was but one of his monumental contributions to aging research and policy. Linda George had the privilege of co-editing the third through seventh editions of the *Handbook* with Bob. They had a wonderful working partnership that she will always treasure. She agreed to serve as senior editor of this edition only if she succeeded in recruiting a co-editor in whose intelligence, research contributions, service to the field, and judgment she had total trust. Ken Ferraro was her choice and he graciously accepted the call. Her choice could not have been better. How fortunate she's been to work with two remarkable men who did their share of the work and more.

This *Handbook* is intended to serve as a useful resource, an inspiration to those searching for ways to contribute to the aging enterprise, and a tribute to the rich bodies of scholarship that comprise aging research in the social sciences.

Linda K. George and Kenneth F. Ferraro

About the Editors

Linda K. George is professor of Sociology at Duke University where she also serves as associate director of the Duke University Center for the Study of Aging and Human Development. She is a fellow and past president of the Gerontological Society of America (GSA). She is former chair of the Aging and Life Course Section and the Sociology of Mental Health Section of the American Sociological Association (ASA). She is former editor of the *Journal of Gerontology, Social Sciences*. She is currently associate editor of *Social Psychology Quarterly* and former associate editor of *Demography*. She is the author or editor of 8 books and author of more than 250 journal articles and book chapters. She co-edited the third, fourth, fifth, sixth, and seventh editions of the *Handbook of Aging and the Social Sciences*. Her major research interests include social factors and illness, stress and social support, and mental health and well-being across the life course. Among the honors she has received are Phi Beta Kappa, the Duke University Distinguished Teaching Award, the Mentorship Award from the Behavioral and Social Sciences Section of GSA, the Dean's Mentoring Award from the Graduate School of Duke University, the Kleemeier Award from the GSA, and the Matilda White Riley Award from the ASA.

Kenneth F. Ferraro is distinguished professor of Sociology and founding director of the Center on Aging and the Life Course at Purdue University. He is the author of over 100 peer-reviewed articles and 2 books and has edited 4 editions of *Gerontology: Perspectives and Issues*. His recent research focuses on health inequality over the life course. Current projects examine the early origins of adult health, health disparities, and the health consequences of obesity. With interests in how stratification processes unfold over the life course, he has developed cumulative inequality theory for the study of human development, aging, and health. A fellow of the Gerontological Society of America (GSA), he formerly edited *Journal of Gerontology: Social Sciences* and chaired the Behavioral and Social Sciences section of GSA. He also is a member of the honorary Sociological Research Association and former chair of the Section on Aging and Life Course of the American Sociological Association (ASA). GSA has honored him with the Distinguished Mentor Award, Richard Kalish Innovation Publication Award, and the Best Paper Award for Theoretical Developments in Social Gerontology. ASA honors from the Section on Aging and the Life Course include Outstanding Publication Award and Matilda White Riley Distinguished Scholar Award.

Deborah Carr is professor of Sociology at Rutgers University where she also is a faculty member at the Institute for Health, Health Care Policy & Aging Research, and holds a secondary appointment at the School of Social Work. She is a fellow of the Gerontological Society of America (GSA) and a member of the honorary Sociological Research Association. She is the 2014–15 chair of the Aging and Life Course Section of the American Sociological Association (ASA). She is editor of the *Journal of Gerontology: Social Sciences* for the 2015–18 term and formerly served as deputy editor of *Journal of Marriage and Family*, and *Social Psychology Quarterly*. She is the author or editor of five books including the *Encyclopedia of the Life*

Course and Human Development (Cengage, 2009). She has authored 70 journal articles and more than 2 dozen book and encyclopedia chapters. She is an investigator on several major studies of aging and the life course including the Midlife in the United States (MIDUS) and Wisconsin Longitudinal Study (WLS), and is chair of the Board of Overseers of the General Social Survey (GSS). Her major research interests include stress, health and well-being over the life course. Her specific research projects focus on death, dying and bereavement; families and health; and the psychosocial consequences of body weight over the life course.

Janet M. Wilmoth received a Ph.D. in Sociology and Demography, with a minor in Gerontology, from the Pennsylvania State University. She is professor of Sociology at Syracuse University where she also serves as the director of the Aging Studies Institute, senior research affiliate in the Center for Policy Research, and senior fellow in the Institute for Veterans and Military Families. She is a fellow of the Gerontological Society of America (GSA), current secretary/treasurer of GSA's Behavioral and Social Science Section, and past secretary/treasurer of the Section on Aging and the Life Course of the American Sociological Association (ASA). She has authored of over 50 articles and book chapters, and co-edited *Gerontology: Perspectives and Issues* (third and fourth editions) and *Life Course Perspectives on Military Service*. Her research examines older adult migration and living arrangements, health status, and financial security, and explores how military service shapes various life course outcomes related to marriage and family, economic well-being, health conditions, and disability. She has received several teaching awards, including the School of Liberal Arts Excellence in Education Award at Purdue University and the Chancellor's Award for Public Engagement and Scholarship-Faculty and Staff Inspiration at Syracuse University.

Douglas Wolf is the Gerald B. Cramer Professor of Aging Studies and a professor of Public Administration and International Affairs at the Maxwell School of Citizenship and Public Affairs at Syracuse University. Previously he was a senior research associate and director of the Population Studies Center at the Urban Institute, and was a research scientist at the International Institute of Applied Systems Analysis in Laxenburg, Austria. His research focusses mainly on family and household demography, late-life disability, and informal care and its consequences for care providers and care receivers. His research has been published in demography, gerontology, public policy, economics, health, and evaluation journals. He has served on the editorial boards of *Demography*, *Journal of Gerontology: Social Sciences*, *Journal of Marriage and the Family*, *Journal of Population Aging*, *Population Research and Policy Review*, and *Demographic Research*, and as director of the Center for Aging and Policy Studies at Syracuse University. At present he is a co-investigator for the National Health and Aging Trends Study (NHATS), a longitudinal study that collects data on a sample of Medicare beneficiaries ages 65 and older.

List of Contributors

Carol S. Aneshensel Department of Community Health Sciences, University of California, Los Angeles, Los Angeles, CA, USA

Jacqueline L. Angel The University of Texas at Austin, Austin, TX, USA

Daniel Béland Johnson-Shoyama Graduate School of Public Policy, University of Saskatchewan, Saskatoon, Saskatchewan, Canada

Rebecca Benson The University of Texas at Austin, Austin, TX, USA

Yoav Ben-Shlomo School of Social and Community Medicine, University of Bristol, Bristol, UK

Suzanne M. Bianchi Department of Sociology and California Center for Population Research, University of California-Los Angeles, Los Angeles, CA, USA

Lisa M. Brown Palo Alto University, Palo Alto, CA, USA

Kevin E. Cahill Sloan Center on Aging & Work at Boston College, Chestnut Hill, MA, USA

Deborah Carr Department of Sociology and Institute for Health, Health Care Policy and Aging Research, Rutgers University, New Brunswick, NJ, USA

Benjamin Cornwell Department of Sociology, Cornell University, Ithaca, NY, USA

Eileen M. Crimmins Davis School of Gerontology, University of Southern California, Los Angeles, CA, USA

Kenneth F. Ferraro Center on Aging and the Life Course, Purdue University, West Lafayette, IN, USA

Kathryn A. Frahm School of Aging Studies, University of South Florida, Tampa, FL, USA

Vicki A. Freedman Institute for Social Research, University of Michigan, Ann Arbor, MI, USA

Joseph E. Gaugler Center on Aging, School of Nursing, University of Minnesota-Twin Cities, Minneapolis, MN, USA

Linda K. George Center for the Study of Aging and Human Development, Duke University, Durham, NC, USA

Michael D. Giandrea US Bureau of Labor Statistics, Office of Productivity and Technology, Washington, DC, USA

Megan Gilligan Human Development and Family Studies, Iowa State University, Ames, IA, USA

Emily A. Greenfield School of Social Work Affiliate of the Institute for Health, Health Care Policy, & Aging Research Rutgers, The State University of New Jersey, New Brunswick, NJ, USA

Zoya Gubernskaya Department of Sociology, University at Albany, State University of New York, Albany, NY, USA

Frederick Harig Department of Community Health Sciences, University of California, Los Angeles, Los Angeles, CA, USA

R. David Hayward School of Public Health, University of Michigan, Ann Arbor, MI, USA

Robert B. Hudson School of Social Work, Boston University, Boston, MA, USA

Neal Krause School of Public Health, University of Michigan, Ann Arbor, MI, USA

Diana Kuh MRC Unit for Lifelong Health and Ageing and MRC National Survey of Health and Development, University College London, London, UK

Andrew S. London Department of Sociology, Aging Studies Institute, Center for Policy Research, and Institute for Veterans and Military Families, Syracuse University, Syracuse, NY, USA

Elizabeth Luth Department of Sociology and Institute for Health, Health Care Policy and Aging Research, Rutgers University, New Brunswick, NJ, USA

Scott M. Lynch Department of Sociology, Duke University, Durham, NC, USA

Marilyn Moon American Institutes for Research, Washington, DC, USA

Nancy Morrow-Howell George Warren Brown School of Social Work, Washington University, St. Louis, MO, USA

Stipica Mudrazija University of Southern California, Los Angeles, CA, USA

Karl Pillemer Department of Human Development, Cornell University, Ithaca, NY, USA

Amélie Quesnel-Vallée Department of Epidemiology, Biostatistics, and Occupational Health; Department of Sociology; Centre for Population Dynamics, McGill University, Montréal, QC, Canada

Joseph F. Quinn Department of Economics, Boston College, Chestnut Hill, MA, USA

Sandra Reiter-Campeau Faculty of Medicine, Université de Montréal, Montréal, QC, Canada

Karen A. Roberto Center for Gerontology and The Institute for Society, Culture and Environment, Virginia Tech, Blacksburg, VA, USA

Liana C. Sayer Sociology Department and Maryland Population Research Center, University of Maryland, College Park, MD, USA

Markus H. Schafer Department of Sociology, University of Toronto, Toronto, ON, Canada

J. Jill Suitor Department of Sociology, Center on Aging and the Life Course, Purdue University, West Lafayette, IN, USA

Miles G. Taylor Pepper Institute on Aging and Public Policy, Florida State University, Tallahassee, FL, USA

Judith Treas Department of Sociology, Center for Demographic & Social Analysis, University of California, Irvine, Irvine, CA, USA

Sarinnapha M. Vasunilashorn Beth Israel Deaconess Medical Center, Harvard Medical School, Brookline, MA, USA

Richard G. Wight Department of Community Health Sciences, University of California, Los Angeles, Los Angeles, CA, USA

John B. Williamson Department of Sociology, Boston College, Chestnut Hill, MA, USA

Andrea Willson Department of Sociology, Social Science Centre, The University of Western Ontario, London, ON, Canada

Janet M. Wilmoth Department of Sociology, Aging Studies Institute, Center for Policy Research, and Institute for Veterans and Military Families, Syracuse University, Syracuse, NY, USA

Douglas Wolf Aging Studies Institute, Syracuse University, Syracuse, NY, USA

PART I

THEORY AND METHODS

1 *Aging and the Social Sciences: Progress and Prospects* 3
2 *Trajectory Models for Aging Research* 23

CHAPTER 1

Aging and the Social Sciences: Progress and Prospects

Linda K. George[1] and Kenneth F. Ferraro[2]

[1]Center for the Study of Aging and Human Development, Duke University, Durham, NC, USA
[2]Center on Aging and the Life Course, Purdue University, West Lafayette, IN, USA

OUTLINE

Theoretical and Conceptual Developments	4
Cumulative Advantage/Disadvantage Theory	5
Cumulative Inequality Theory	6
Methods and Data	6
Data Developments	6
Statistical Sophistication	8
Emerging Themes in Aging Research	9
Increased Attention to Cohort Analysis	9
The Effects of Social and Economic Disruptions on Aging	10
Gradual, Incremental Cultural Change	12
What Aging Research Contributes to the Social Sciences: The Big Picture	16
References	20

"The only constant is change." This quote, heard frequently today, is attributed to Heraclitis of Ephesus, a Greek philosopher who lived from approximately 535 BC to 475 BC. One wonders what it was about life at about 500 years before the birth of Christ that led Heraclitis to that conclusion. Was the pace of social change so rapid that it led to this inference? Was it the rhythms of nature that triggered this observation? Or, perhaps, was it the flow of everyday life that convinced Heraclitis that he was not the same person today that he was yesterday or would be tomorrow? At any rate, it is clear that humans have long been aware that change is ubiquitous.

Scholars of aging arguably devote more of their intellectual activity to studying and understanding change than those in any other field. Aging itself is change – some of it easily observable; some of it occurring at the cellular and molecular levels and requiring years or even decades to be measurable and the fodder for scientific inquiry. Aging individuals are embedded in macro-, meso-, and micro-environments in which change also is omnipresent. And a fundamental assumption of the social sciences is that those constantly changing environments affect the ways in which people age. Thinking seriously about the complexity of change leads to the conclusion that considerable audacity and fortitude are required to study aging and lay claim to understanding or explaining its dynamics. And yet that is precisely what aging researchers do.

Audacity and fortitude also are required in any attempt to summarize the state-of-the-science with regard to social science aging research. Yet, the goal of this chapter is to provide a partial summary of the state-of-the-field. More specifically, the purpose of this chapter is to review, in broad brush, recent theoretical, methodological, and selected substantive developments in aging research in the social sciences. We used the approximate dates of 1996–2015 as the focus of this review. This is an arbitrary window of time, but we believe that it is a reasonable temporal scope for summarizing current significant issues in aging research.

The chapter is organized into four sections. The first section reviews theoretical and conceptual developments in the field; the second provides an update of advances in data, methods, and statistical techniques that have become central in aging research. The third and longest section reviews three thematic topics that have emerged as cutting-edge issues in social research on aging and the life course. In the concluding section, we briefly comment upon the broader issue of how aging research contributes to major issues and assumptions in the social sciences.

Considerable subjectivity was employed in developing this chapter, especially in identifying emerging substantive issues. It is possible to produce a veritable "laundry list" of recent and emerging themes in aging research. We selected only three, with the unifying theme being "big picture" influences on aging. Undoubtedly, other scholars would have selected other developments in the field. Other scholars may disagree with our labeling these research topics as "recent" or "new." This is inevitable. Nonetheless, we hope that this chapter captures much of the theoretical, methodological, and substantive "action" of the past two decades in social science research on aging.

THEORETICAL AND CONCEPTUAL DEVELOPMENTS

Arguably, the biggest "story" in aging research for the past several decades has been developments in, advances in, and the greatly increased volume of research that incorporates the life course perspective. The life course perspective is not a theory per se; rather, it is a set of five principles that contextualize individual lives in a number of ways (Elder, Johnson, & Crosnoe, 2003). The principle of *life span development* states that human development and aging are lifelong processes – that patterns observable over time link distal and proximal events and experiences across the life course. The principle of *agency* focuses on the ways that individuals construct their own lives by the choices they make within the opportunities and constraints of their environments. The principle of *time and place* states that human lives develop in historical and geographic contexts that strongly affect the opportunities and constraints available. The principle of *timing* states that the effects of events and other experiences vary, depending on the individuals' ages or life stages. Finally, the principle of *linked lives* focuses on the social networks and relationships that also structure the opportunities and constraints available to individuals. Although temporality, especially biographical and historical time, is widely viewed as the hallmark of the life course perspective, context is its major foundation.

Questions arise at times about the relationships between life course research and gerontological research, especially whether gerontological theory and research will be or have been eclipsed by the life course perspective. In order to document its strengths, life course scholars sometimes critique gerontological research that does not incorporate one or more principles of the life course perspective. Nonetheless, multiple research questions appropriately focus on late life and need not incorporate explicit life course principles (e.g., studies of variability within the older population, studies that examine the effects of interventions or policies on older adults). Virtually all studies of older adults, however, should recognize that research participants are members of cohorts measured at specific historical times – and therefore it cannot be assumed that the findings will generalize to other cohorts and historical contexts.

Because the life course perspective is not a theory, its principles need to be incorporated and tested in conjunction with established theories. This cross-fertilization of life course principles with mainstream social science theories has expanded rapidly. Several examples provide illustrations of this cross-fertilization but do not comprise a comprehensive inventory of relevant topics. Life course principles of life span development, agency, timing, and linked lives have been incorporated in stress process theory. This research has provided important knowledge about the persistent effects of early severe trauma on the mental health and well-being of older adults (e.g., Danese & McEwen, 2012; Shaw & Krause, 2002). Another profitable area of research focuses on the ways in which educational achievements and occupational choices in young adulthood affect financial security in later life (Cahill, Giandrea, & Quinn, 2006). And perhaps no topic has been more thoroughly investigated than the effects of childhood conditions (traumatic events, persistent poverty, and poor health) on morbidity and mortality in middle and late life (for a review, see Chapter 5, this volume).

Cumulative Advantage/Disadvantage Theory

If there has been a *bona fide* theory based on the life course perspective, specifically the principle of life span development, it is cumulative advantage/disadvantage theory (CA/DT). The major hypothesis of CA/DT was developed by Robert Merton (1968), who called it the Matthew Effect, based on a verse in the Gospel of Matthew (13:12). The Matthew Effect refers to a pattern in which those who begin with advantage accumulate more advantage over time and those who begin with disadvantage become more disadvantaged over time (Dannefer, 1987; O'Rand, 1996). The result is ever-widening differences between the advantaged and disadvantaged. This simple theory has been supported in many domains of life (Rigney, 2010). When applied to trajectories of advantage and disadvantage over long periods of time, CA/DT is obviously compatible with the life course perspective. And research on life course patterns often finds support for CA/DT.

Nonetheless, as Rigney's review of research (2010) documents, CA/DT does not always apply. A key example is late life health. Individuals who begin adulthood in excellent health do not become healthier over time with physical and mental well-being peaking at the end of life. These early advantaged individuals are likely, on average, to have better health than persons who entered adulthood with poor health or experienced health problems as young adults. But their trajectory of health is not monotonic improvement over the life course. As a consequence, aging researchers often label their theoretical foundation as cumulative disadvantage theory.

There is increasing and appropriate recognition that CA/DT is oversimplified. CA/DT posits two trajectories when, in fact, phenomena of interest are typically characterized by multiple trajectories. Depending on the phenomenon under investigation, two of the trajectories may resemble straightforward cumulative advantage and disadvantage, but there will be other meaningful trajectories as well. Despite its shortcomings as a universally applicable theory, CA/DT has been tremendously useful in emphasizing the importance of early social status and cohort membership on life course trajectories and has generated a large volume of important research.

Cumulative Inequality Theory

To capture more of the contingencies involved in how status and life experience influence the aging process, cumulative inequality theory (CIT) integrates elements from multiple conceptual approaches, most notably but not limited to: life course perspective (Elder, 1998), CA/DT, and stress process theory (Pearlin, Schieman, Fazio, & Meersman, 2005). Formulated in five axioms and 19 propositions, the theory builds upon but is distinctive from prior approaches in several ways (Ferraro, Shippee, & Schafer, 2009).

First, CIT prioritizes perceptions of the aging experience while juxtaposing the systemic generation of inequality with human agency (Schafer, Ferraro, & Mustillo, 2011). Social structures constrain choices, and both influence aging. Second, rather than assume inexorable effects of early disadvantage, CIT specifies that exposures to risks and resources also shape life trajectories. Indeed, the timely activation of resources may nullify or compensate for the effects of negative exposures. Third, the influence of family lineage is emphasized in the theory, noting the roles that genes and environment have on status differentiation. It calls for more attention to the intergenerational transmission of risks and resources. Finally, the theory integrates selection processes into the study of inequality. Given that inequality itself is an engine of mortality and other forms of nonrandom selection, failure to consider selection processes may lead to misrepresenting inequality in later life (i.e., typically underestimating inequality).

Several longitudinal studies testing elements of the theory reveal the importance of intergenerational influences on health outcomes – ranging from adult depression (Goosby, 2013) to myocardial infarction (Morton, Mustillo, & Ferraro, 2014) – but also how those health risks may be amplified or diminished by resources and lifestyle choices. Indeed, in a study of racial disparities in health, Kail and Taylor (2014, p. 805) reported that "mobilizing financial resources into insurance coverage is protective" against functional limitations. Other studies testing elements of the theory reveal that both psychosocial resources and how one interprets life experiences are consequential to status attainment and health (Wickrama & O'Neal, 2013; Wilkinson, Shippee, & Ferraro, 2012). The emerging picture from empirical tests of the theory is that there are powerful systemic influences on exposure to risk, opportunity, and inequality but that these influences on well-being in later life are often contingent on how the exposures are interpreted and whether resources can be activated to address them.

METHODS AND DATA

Data Developments

One of the greatest boons to aging research has been the proliferation of longitudinal data sets covering long periods of time. The increased availability of high-quality data sets in the past two decades or so has transformed aging research. Space limitations preclude a description of all the valuable longitudinal data sets available. Several major differences in data sources, however, will be reviewed. With few exceptions, we focus on data sets with three or more times of measurement, which is the minimum number of data points for modeling trajectories.

Age Ranges and Times of Measurement. Some studies were designed to focus on the dynamics of late life; others followed samples from young adulthood to late life; and still others recruit age-heterogeneous samples at baseline and follow them for significant periods of time. The Health and Retirement Study (HRS), for example, was originally designed to follow individuals from late middle-age until very old age or death. Additional cohorts have been added during the past two decades, however, resulting in some cohorts entering the study relatively early in adulthood (Institute for Social Research, 2014a). The Wisconsin Longitudinal Study (WLS), in contrast, recruited participants during their senior year of high school in 1957 and continues to collect data. Last surveyed in 2011, study participants were approximately 72 years old (University of Wisconsin, 2014). The Americans Changing Lives (ACL) study began in 1986 and recruited a sample of adults age 25 and older (Institute for Social Research, 2014b).

The intervals between measurements also vary across data sources. The HRS began data collection of its original cohort in 1992 and interviews participants every 2 years. The WLS includes seven times of measurement to date and the intervals between them range from 7 to 17 years. The ACL has four times of measurement at intervals ranging from 8 to 10 years. Some studies also oversample specific subgroups of interest, which can enhance opportunities for analyses based on middle and late life. The ACL, for example, oversampled both African Americans and adults age 60 and older.

Academic versus Government Sponsorship. Virtually all large-scale longitudinal studies are funded by government agencies. The distinction here is between studies that were funded via grants to academic institutions and studies carried out by government agencies. The HRS, WLS, and ACL are examples of studies designed and conducted by universities and funded by federal grants. Examples of government-conducted longitudinal studies include the Second Longitudinal Study of Aging (LSOA-II) (CDC, 2014b) and the Medicare Current Beneficiary Survey (MCBS) (CMS, 2014). Virtually all federally funded longitudinal studies focus on health. The data sets also include, to varying degrees, information about social, economic, and psychological characteristics of study participants. In general, data sources that are funded by grants to academic institutions include richer social science content than those conducted by government agencies.

National versus Regional/Local Samples. All the longitudinal data sources mentioned above were designed to be based on nationally representative samples. Data sets based on regional or local samples also offer important research opportunities. The Established Populations for Epidemiologic Studies of the Elderly (EPESE) is an example of data collected at the local or regional level that has made important contributions to aging research (ICPSR, 2014). The EPESE Program included local/regional data collected from adults age 65 and older at four sites: East Boston, MA, USA; New Haven, CT, USA; Iowa and Washington Counties, IA, USA; and central North Carolina, USA. The research design included four in-person interviews over a 10-year period, with brief telephone interviews administered in the years between the in-person interviews. A common set of survey questions were asked at each site, supplemented with site-specific interview content. A few years later, the Hispanic EPESE was added, with the same basic research design (Sociometrics, 2014). The sample included Hispanic older adults, both native and foreign-born, living in five southwestern states. Hundreds of scientific articles have been published using data from one or more EPESE sites, testifying to the value of non-national samples. Other longitudinal studies based on local/regional samples yielded important findings as well (e.g., Alameda County Study).

Merging Survey and Administrative Data. Another trend during the past two decades has been merging survey data from older adults with federal and, occasionally, state administrative data. Merging data from these sources greatly expand the research questions that can be addressed. The most frequently used administrative data base is the National Death Index (NDI), which includes data from death certificates in all 50 US states. Investigators routinely use the NDI to determine study participants' mortality status and date of death and can use the NDI Plus for cause-of-death data. Although not every name submitted to the NDI can be definitively matched, the overall accuracy of the NDI is excellent (e.g., Lash & Silliman, 2001). The other major administrative data set often merged with longitudinal survey data is Medicare claims data, which include detailed information about the use and costs of inpatient and outpatient health care (CDC, 2014a). Most major longitudinal studies use the NDI and many (e.g., the HRS and EPESE) also obtain Medicare claims data.

Biomarker, Genetic, and Physical Performance Data. Another important trend in longitudinal studies of aging is the collection of biological and physical performance data. Advances in data collection methods now allow biological data to be easily obtained via non- or minimally invasive methods, including buccal swabs for DNA and urine and saliva samples for selected biomarkers. Highly trained interviewers often collect blood samples; measure height, weight, waist circumference, and blood pressure; and/or administer physical performance tests. To date, the genetic and biomarker data typically have been collected at a single point in time. An exception is the National Social Life, Health, and Aging Project (NSHAP; NORC, 2014). To date, NSHAP has collected two waves of data and biomeasures were collected at both test dates, permitting longitudinal analyses spanning about 5 years. This trend will undoubtedly continue in other longitudinal studies, resulting in multiple waves of biological and genetic data that are linked to rich survey and administrative data.

Non-US Databases. An important and relatively new resource for research on aging is the availability of large-scale longitudinal studies conducted in countries other than the United States. Especially rich data are available from Europe and the Pacific Rim. European examples include the English Longitudinal Study of Ageing (ELSA) and the Survey of Health, Ageing, and Retirement in Europe (SHARE). The ELSA began in 2002, conducts interviews biannually, and has completed six waves of data, with a seventh in progress (ELSA, 2014). SHARE also interviews participants biannually; it began in 2004 and five waves are complete. SHARE's baseline sample included older adults from 11 countries. By Wave 5, 15 countries had participated (SHARE, 2014). Both the ELSA and SHARE are modeled on the HRS in design and content. Two studies from the Pacific Rim are especially rich in times of measurement. The Australian Longitudinal Study of Aging (ALSA) began in 1992 and completed 12 times of measurement (Luszcz et al., 2014). The Chinese Longitudinal Healthy Longevity Study (CLHLS) has conducted six waves to date (1998, 2000, 2002, 2005, 2008–2009, and 2014) and focuses on the oldest-old (Chinese Longitudinal Healthy Longevity Survey, 2014). All four of these data sets include biomarker and physical performance tests at one or more times of measurement.

Statistical Sophistication

The statistical armamentarium for analysis of three or more waves of longitudinal data has grown in volume and sophistication over the past two decades. The concept of trajectory – a distinct temporal pattern observed over multiple times of measurement – has become a staple of aging research. Some studies include multiple times of measurement over a relatively short period of time, permitting estimation of fine-grained trajectories (e.g., patterns of onset, stability, and recovery of disability).

Other studies examine long-term trajectories of stability and change, such as those hypothesized in CA/DT theory. A variety of statistical techniques can be used to model trajectories, as reviewed in Chapter 2 of this volume.

Structural equation modeling remains an important analytic tool for analysis of longitudinal data. Its unique characteristics include the option of estimating reciprocal relationships between variables over time, production of distinct measurement and explanatory models, the ability to correct for unreliability of measurement, and estimation of direct and indirect effects of explanatory variables on the outcome of interest.

Multilevel modeling is now frequently used to jointly examine the effects of individual-level and aggregate contextual variables on outcomes of interest. These models have proven especially useful in studies examining the effects of environmental characteristics on outcomes of interest. In many of these studies the research question focuses on whether environmental characteristics are related to outcomes of interest after the effects of individual characteristics are statistically controlled (e.g., is living in a high-poverty neighborhood associated with mortality after individual poverty status is taken into account?). It also is possible to estimate interactions between individual-level and contextual variables. In aging research, most multilevel studies examine the effects of neighborhood characteristics on health and quality of life. Chapter 16 reviews this research.

EMERGING THEMES IN AGING RESEARCH

In this section, three topical areas of aging research are briefly reviewed. Although there have been notable advances on dozens of research topics, we focus on emerging themes not covered in detail in one of the chapters in this volume. In addition, we selected three research themes that focus on the effects of macro-level characteristics of social structure that have potentially important implications for aging and/or older adults. Of course, some of the topics are mentioned in the chapters that follow, but we think these emerging themes nonetheless merit additional consideration.

Increased Attention to Cohort Analysis

The term "cohort," of course, refers to a set of people who experience the same event at the same time. Although any event can define a cohort, social scientists typically use the term to refer to birth cohorts – to people born at the same or approximately the same time – and that is how the term is used here. Norman Ryder's classic article (1965) was the first systematic consideration of cohort as a social, rather than simply actuarial, phenomenon. For Ryder, cohort differences are evidence of social change. In his words, "cohorts do not cause change; they permit it. If change does occur, it differentiates cohorts from one another, and the comparison of their careers becomes a way to study change" (p. 844). Ryder hypothesized that four types of circumstances were most likely to differentiate cohorts. First, cohort size is important – very large and very small cohorts experience different structural opportunities and constraints from each other and from cohorts of more usual sizes. Second, major social and historical events cause significant differences across cohorts. Ryder posited that societal disruptions had the strongest and most lasting effects on cohorts who were adolescents or young adults at the time of the events. Thus, although all cohorts experience the disruptive event, it is the young who are permanently changed by it (a line of reasoning compatible with the life course principle of timing). Third, wide variations in the influx of migrants to a society can change the character of a cohort and differentiate it from those before and after it. A variant of this, with the same result, is widespread migration from rural to urban areas. Fourth, Ryder believed that technological innovation was a primary trigger for cohort differentiation – and argued that technological advances were most targeted at and welcomed by adolescents and young adults.

To support the claim that cohort analysis has become increasingly popular in aging research, we conducted an informal analysis of journal articles published between 1970 and the first half of 2014. We used the Web of Science core collection and narrowed the search to journal articles categorized as falling under at least one of three topics: gerontology, geriatrics, and sociology (aging was not a topic offered). Using these criteria, Web of Science identified 7635 articles in which the word "cohort" appeared in the title or abstract. Examining the distribution of these articles by date of publication is illuminating. Less than half a percent of the titles appeared between 1970 and 1979 and slightly more than 1% were published between 1980 and 1989. About 16.5% were published in the 1990s and approximately 39.5% were published between 2000 and 2009. In the interval between 2010 and June 30, 2014, 41.8% of the articles were published. Even we were surprised to find that the largest percentage of articles appeared in the most recent four and a half years. There are obvious limitations to this analysis (e.g., we cannot know whether investigators simply began to use the term "cohort" more frequently in article titles and abstracts). Nonetheless, if the trend observed in this highly unsophisticated analysis is generally accurate, explicit attention to cohorts is increasingly common in research on aging.

Most studies of cohort differences in later life examine health outcomes. Examples include cohort differences in the relationship between education and health (Lynch, 2006), in depression during late life (Yang, 2007), in the extent to which segregated southern schools partially account for Black–White health disparities in late life (Frisvold & Golberstein, 2013), and in the relationship between women's labor market participation and health (Pavalko, Gong, & Long, 2007). Cohort analysis is valuable for outcomes other than health as well, such as the discrepancy between chronological and "felt" age (Choi, DiNitto, & Kim, 2014) and patterns of gradual retirement (Giandrea, Cahill, & Quinn, 2009).

Although there are significant exceptions, few studies either empirically test or even speculate about the specific social changes that trigger cohort differences. As a result, cohort analysis often appears simply descriptive. But the best cohort studies are those that not only describe cohort differences, but also attempt to explain the reasons for them. Frisvold and Golberstein's (2013) study of how segregated schools and their subsequent demise are associated with cohort differences in race disparities in health is an example of a study that aims to explain cohort differences and not simply describe them. The increased attention to cohort differences is an important contribution to aging and life course research. The contributions of cohort analysis could be even greater if this research routinely addressed potential explanations for cohort differences.

The Effects of Social and Economic Disruptions on Aging

Social and economic disruptions have long been of interest in the social sciences. Major shocks to social structure provide a rare opportunity to not only study the consequences of and responses to significant disruptions, but also to highlight social arrangements before the disruptions that were not fully understood. Large-scale events typically receive substantial attention by both scientists and the general public. Much less attention has been paid to the differential implications of these disruptions for population subgroups, including older adults. Recently, however, the implications of large-scale social and economic changes for older adults have received increased attention.

The Great Recession

Perhaps no social and economic disruption in the past quarter century generated more scholarly and public attention than the Great Recession that began in 2008 and continues to shape the lives of the citizens of many countries, including the United States. No age group has been unaffected by the consequences of this major disruption, but there are reasons to believe that older adults are suffering at least as much as their younger counterparts. As a *New York Times* headline stated, "In Hard Economy for All Ages, Older Isn't Better.... It's Brutal" (*New York Times*, 2013). A growing body of research addresses the effects of this cataclysm on older adults. Several topics on the consequences of the Great Recession have received empirical attention. First and of obvious concern is whether the increased rates of unemployment in the population at large affected older adults. There appears to be both good and bad news on that front. On the positive side, older adults (variously defined as those age 55, 60, and 65 and older) have lower rates of unemployment than any age group – indeed, rates of unemployment are strongly and inversely related to age (US Bureau of Labor Statistics, 2013). On the other hand, between 2008 and 2010, the unemployment rate of older adults roughly quadrupled and has declined little since then. In addition, the length of time between job loss and reemployment is significantly longer for older adults than their younger counterparts and many older adults opt out of job seeking after a relatively short period of unemployment. A second important issue is whether the Great Recession altered the plans of those nearing retirement. It is too early to know definitively the extent to which persons at or nearing conventional retirement ages are postponing retirement, but there is strong evidence that these individuals report that they plan to retire later than they had intended prior to the Great Recession (e.g., McFall, 2011). Chapter 14 describes the impact of long-term macroeconomic trends on the labor force participation of older adults and reviews in detail the economic and labor force consequences of the Great Recession.

The consequences of the Great Recession are not limited to purely economic issues. A small base of research on the mental health consequences of the Great Recession is emerging, almost none of which focuses on older adults. Cagney and colleagues, however, report that increases in neighborhood foreclosures are associated with increases in depressive symptoms among older adults in NSHAP, controlling on demographic characteristics, socioeconomic status (SES), and physical functioning (Cagney, Browning, Iveniuk, & English, 2014). The spillover from the Great Recession also may affect the family lives of older adults. Livingston and Parker (2010) report that between 2007 and 2009 the number of older adults with custodial care of grandchildren increased by nearly 20%, although these grandparents are a small proportion of the older population.

Hurricane Katrina and Other Disasters

Natural disasters are another form of disruption and have received increased attention in aging research. According to the Centers for Disease Control and Prevention, "In Louisiana during Hurricane Katrina, roughly 71% of the victims were older than 60 and 47% were over the age of 75" (CDC, 2013, p. 1). Given these large percentages, it would be logical to assume that older victims received a significant portion of the publicity, aid, and health monitoring in the aftermath of Katrina. There is little evidence, however, to support that assumption. Public discourse about Katrina gave little attention to age groups other than displaced children. No local or regional disaster plans included procedures for transferring residents out of nursing homes – and residents of those homes fared especially badly (CDC, 2013). Some research examined the effects of Katrina on mental health, the coping strategies used by older victims, and the ways that family support did or did not ease the trauma of older victims (e.g., Cherry et al., 2010; Henderson, Roberto, & Kamo, 2010; Kamo, Henderson, & Roberto, 2011; and for a review of research on the effects of disasters on older adults, see Chapter 18).

The examples above illustrate the increased attention paid to events that threaten preexisting structural arrangements and their consequences for older adults. We applaud this trend and encourage broader attention to major social disruptions – for the United States as a whole, such as the Great Recession, and for specific regions or cities, such as Hurricane Katrina.

Gradual, Incremental Cultural Change

Not all consequential social changes take the form of sudden social disruptions; gradual and/or incremental cultural changes also can have important implications for older adults. In fact, social scientists are probably more likely to miss or understudy the effects of more gradual social change than sudden disruptions. The history of aging research reveals numerous gradual changes, the significance of which was not recognized until a critical mass of older adults was affected. Family care for impaired older adults has occurred at least since the beginning of recorded history. Nonetheless, it was not until the vast majority of adults lived until late life and gradual social changes (e.g., women's labor force participation, intergenerational geographic mobility) made family caregiving difficult for a significant proportion of older spouses and adult children that the concept of caregiver burden became a topic of scientific interest. Indeed, the term "caregiver" did not appear in public discourse until the 1980s. Similarly, the transition from defined benefit to defined contribution pension plans was underway for a decade or so before the implications of this transition for the financial security of retired adults became an issue in aging research. A gradual societal change that has received significant recent attention is the health effects of income inequality.

Income Inequality and Health

Income inequality refers to the size of the gap between the richest and poorest members of society – the wider the gap, the greater the inequality. Although there are gaps between the bottom and top of the income ladder in all societies, the size of the gap varies widely across countries and over time. The United States has higher income inequality than any other developed country in the world and the gap between the richest and the poorest has widened substantially over a relatively short period of time in the United States, with no apparent end in sight (*The Economist*, 2013). The implications of income inequality for economic growth, social cohesion, and health are now "hot topics" in the social sciences, politics, and public discourse.

Conceptual and Methodological Issues. The outcomes of income inequality for which there has been substantial research include rates of labor force participation, workers' earnings, economic growth, general trust, civic engagement, life expectancy, and other health indicators. For decades, most economists argued that the net effects of economic inequality (both income and wealth inequality) are beneficial. Mainstream economic theory posited that income inequality motivates workers to increase their job skills and productivity in order to climb the economic ladder. In turn, more productive workers not only increase their own incomes, but also spur economic growth for the society as a whole. In contrast, Marxian theorists and other social scientists argued that because income inequality concentrates capital in their control, the very rich are motivated to cut labor costs as much as possible. As a result, increasing income inequality depresses workers' wages and increases unemployment. Quite recently, economists have found, using data from the United States, that high levels of income inequality suppress rather than facilitate economic development (Stiglitz, 2012), necessitating that economic theory recognize that there is a threshold beyond which high income inequality has negative effects on nations' economic growth.

The potential link between income inequality and health is especially important for older adults. Research examining the relationships between income inequality and health is voluminous and inconsistent. Some studies report significant correlations between income inequality and a variety of health outcomes. Other studies, however, report nonsignificant relationships.

Four aspects of research design may account for much of the inconsistency in previous studies of income inequality and health. These methodological issues largely result from theoretical ambiguity about the expected relationships between income inequality and health.

First, is the selection of control variables. In order to isolate the effects of income inequality on health, researchers have controlled on a variety of other structural characteristics. The most important of these is economic growth, typically measured as Gross Domestic Product (GDP). Strong and robust relationships between economic growth and multiple indicators of population health have been observed for decades (e.g., Easterlin, 1974). If income inequality is to continue to receive scientific attention, it must be significantly related to health with GDP taken into account. Research findings demonstrate that relationships between income inequality and multiple outcomes are substantially reduced if GDP is controlled. In most instances, coefficients for income inequality remain statistically significant; in others, they do not. Other frequently used control variables include proportion of GDP spent on social and health programs, political regime, and women's rights. The general hypothesis underlying inclusion of these characteristics is that income inequality may be compensated for by government policies that redistribute resources from the rich to the poor. In general, the more control variables measuring economic growth and welfare state benefits that are included in analyses, the more that the net effects of income inequality are reduced. Theoretically, it is not clear whether these structural characteristics should be conceptualized as control variables, included to test whether the relationships between income inequality and health are spurious, or as mediators of those relationships.

A second important decision in research on income inequality is whether to model the effects of income inequality on health (or other outcomes) solely at the aggregate level or to use multilevel models that incorporate both individual and aggregate predictors of health. Multilevel models are generally viewed as superior to aggregate-only models because the former allow researchers to determine if income inequality is significantly related to health once individual-level predictors are taken into account. Most attention in multilevel models has focused on whether coefficients for income equality are significant net of individual-level income. Again, research findings have been inconsistent. Another, less recognized advantage of multilevel models is the ability to test whether income inequality interacts with individual-level characteristics to exacerbate or reduce disparities across population subgroups. Although arguments favoring multilevel models are strong, the underlying issue is theoretical. Do we expect high levels of income inequality to harm individual health, population health, or both? It is possible that high levels of income inequality harm population health, but not individual-level health (or vice versa). This could happen if elevated income inequality has small effects on multiple risk factors for mortality and morbidity and it is the cumulative or aggregated strength of these multiple small effects that links high income inequality to poorer population health. Choice of the level of the health outcome should be based on theoretical grounds. If the outcome of interest is an indicator of population health, aggregate-only models are appropriate.

A third decision that investigators face is the choice of a unit of analysis. Most early studies compared the relationships between income inequality and potential outcomes using the nation state as the unit of analysis. An increasing number of studies, however, use units of analysis that are smaller than countries, including states or provinces, metropolitan areas, and neighborhoods. The choice of a unit of analysis is undoubtedly determined in part by data availability (e.g., if county-level data are not available for important variables, another unit of analysis for which data are available must be used). There are countervailing advantages and disadvantages to country versus smaller units of analysis.

The disadvantage of country-level variables is that they include a great deal of unmeasured heterogeneity both within and across countries. Economic conditions and public policies often differ substantially across geographic units within a country, and ignoring that variability may mask relationships that would be observed with smaller units of analysis. Unmeasured heterogeneity is undoubtedly even greater across countries, with cultural preferences and unique aspects of national history ignored. The advantages of country-level analyses are that results presumably apply to the population as a whole and many structural characteristics are, by definition, nationally homogeneous (e.g., GDP, political structure). The advantage of using smaller units of analysis in the same country is that some important structural characteristics are national and, thus, constants that need not be included in predictive models, thus permitting fine-grained analyses of other structural characteristics. This also is the primary disadvantage of small units of analysis – if the effects of income inequality differ primarily across countries but are homogeneous within countries, important information about the effects of income inequality will be missed. Again, theory should provide guidance about the most appropriate unit of analysis, but there is little evidence of that in extant research.

Fourth and finally is the question of whether geographic units are the optimal basis for studying the effects of income inequality on health and other outcomes. Although the vast majority of income equality research is based on comparisons across geographic units, other strategies are available. Zheng and George (2012) argue that the best way to study income inequality is to relate time-based trajectories of inequality to health. Time-based analyses permit investigators to determine whether patterns of increasing (or decreasing, although to our knowledge, decreasing levels of income inequality have never been observed) income inequality are associated with worse health. Using time-based trajectories, the temporal order between changes in income inequality and changes in health and the lag times between changes in income inequality and changes in health can be observed. Lag time is important, but theory to date has not addressed this issue. Cross-sectional studies, while plentiful, are of dubious value. It is highly unlikely that increases in income inequality trigger immediate changes in health. Because the lag times between changes in income inequality and changes in health are unknown, trajectory analyses could shed light on that dynamic.

Income Inequality, Aging, and Health. The vast majority of research in this field focuses on the relationship between income inequality and mortality. Other studies examine self-rated health, physical functioning/disability, and mental health. Because older adults have higher rates of death and disability, and are more likely than their younger counterparts to rate their health as fair or poor, studies based on age-heterogeneous samples are clearly relevant to the older population. Mental disorders are less common in later life than in middle or young adulthood, however, and are not reviewed here. Space limitations preclude an extensive review of the voluminous literature linking income inequality and health.

Mortality is the health outcome most frequently related to income inequality. Most studies report a positive and significant relationship between income inequality and mortality rates at the aggregate level, especially in the United States (e.g., Kaplan, Pamuk, Lynch, Cohen, & Belfour, 1996 – a study of US states; Ross et al., 2000 – a study of US states and Canadian provinces, with a significant relationship observed only in the United States). Some aggregate studies report that income inequality is associated with mortality rates with median income and/or poverty rates also controlled (e.g., Kawachi & Kennedy, 1997 – a study of US states; Wilkinson & Pickett, 2008 – a study of US states). Other studies report no significant relationship between income inequality and mortality at the aggregate level (e.g., Beckfield, 2004 – a study of 115 countries). As noted above, an important issue in multilevel studies is whether this relationship remains robust with individual-level income statistically controlled. Results are inconsistent, with some studies reporting that the income inequality–mortality link remains strong and significant (e.g., Lochner, Pamuk, Maduc, Kennedy, & Kawachi, 2001 – a study of US states; Shi & Starfield, 2001 – a study of US metropolitan areas) and others reporting that the association is rendered nonsignificant (e.g., Fiscella & Franks, 1997 – a study of US communities).

Because fewer studies examine health outcomes other than mortality, it is difficult to summarize the pattern of results. Studies of the relationship between income inequality and self-rated health report inconsistent relationships. This is especially true of multilevel studies that include individual-level as well as aggregate predictors. The single multilevel study of physical functioning reported a significant and positive relationship between increasing income inequality and both physical functioning and activity limitations (Zheng & George, 2012 – a temporal study of the US population between 1984 and 2007).

Few multilevel studies examined interactions between income inequality and individual-level predictors, but this appears to be worth additional effort. Diez-Roux, Link, and Northridge (2000) examined the relationship between income inequality and cardiovascular disease in a multilevel study. The unit of analyses was US states. The direct effect of income inequality was significant and in the expected direction for women, but not men. The interaction between income inequality and individual-level income was strong and significant. As expected, the combination of high income inequality and low personal income predicted cardiovascular disease. In contrast, Sturm and Gresenz (2002), in a multilevel study of income inequality and number of chronic physical illnesses, tested the same interaction and it was not significant. To evaluate whether rising income inequality contributes to status-based health disparities, Zheng and George (2012) examined interactions between income inequality and family income, education, employment, marital status, gender, and race-ethnicity. Coefficients for the first four variables were statistically significant and in the predicted direction. That is, the *protective* effects of individual income, education, employment, and marriage strengthened as income inequality increased. The interaction between income inequality and gender also was significant, with rising income inequality having stronger negative effects on physical functioning for men than for women. These results suggest that increasing income quality may exacerbate SES- and gender-based health disparities, although there was no evidence of elevated risk for racial and ethnic minorities once SES indicators were taken into account.

Two relatively recent reviews reach similar conclusions to those referenced above. The first is a literature review based on 168 published analyses of the relationships between income inequality and multiple health outcomes (Wilkinson & Pickett, 2006). The authors conclude that 70% of the studies totally or partially support the hypothesis that high income inequality is associated with poorer health. They also found that studies based on larger geographic units of analysis were more likely to support the hypothesis that income inequality is positively related to worse health than those based on small areas. They suggest that studies that sample small areas are "too small to reflect the scale of social class differences in a society" (p. 1768). The second review reports the results of a meta-analysis based on 28 studies that cumulatively included more than 61 million respondents (Kondo et al., 2009). The health outcomes examined were mortality and self-rated health. The results suggested that income inequality is significantly related to both mortality and self-rated fair/poor health in the expected direction, although the size of the coefficient is modest. Kondo et al. also observed significant relationships between specific study characteristics and the odds of a negative association of income inequality with health. Specifically, results were stronger and larger in studies characterized by higher levels of income inequality, longer duration of follow-up, that used data from 1990 and later, and explicitly modeled time lags. In line with the conclusions of both reviews, other studies have empirically examined the effects of size of the geographic unit of analysis, time lags, and income inequality thresholds (Blakely, Kennedy, Glass, & Kawachi, 2000; Kondo, van Dam, Sembajwe, Kawachi, & Yamagata, 2012), demonstrating that these study characteristics strongly affect the size and significance of relationships between income inequality and health outcomes.

Income inequality is only one example of a wide range of patterns of gradual social change that may affect population health as a whole and the health of older adults in particular. Examples of other gradual social trends that may be worth examining include the increasing age at first marriage in the United States, which has implications for the aging of those cohorts and their parents (US Census Bureau, 2012), the steadily increasing proportion of the population, including the older population, living in near poverty (Heggeness & Hokayem, 2013), and the increasing income residential segregation in the United States (Fry & Taylor, 2012). The effects of these cumulative small social changes are easily overlooked. And yet a core premise of social science research is that the larger environment substantially determines the opportunities and constraints within which societal members live their lives. We suggest that these kinds of structural changes merit closer scrutiny from social scientists.

WHAT AGING RESEARCH CONTRIBUTES TO THE SOCIAL SCIENCES: THE BIG PICTURE

Opportunities for innovative and rigorous aging research have never been better. A proliferation of data sets in which large numbers of individuals are followed over long periods of time became available in the past two decades. The ability to merge survey data and other data sources (including, but not limited to Medicare files, Census data, and the NDI) also has broadened the range of research questions that can be addressed. High-quality data sources are available for a growing number of countries other than the United States. Statistical techniques designed for multiple times of measurement and multiple levels of analysis are now readily available in standard statistical packages. Obviously, advances in these and other components of the infrastructure on which aging research rests will continue in the future. But plenty of exciting research questions can be addressed with the resources available now.

Aging research is important for many reasons, ranging from answering basic questions about relatively regular patterns of human development across adulthood to understanding the importance of age structures for social institutions to providing data that guide the development of social/health interventions and public policy. Recently and across many disciplines, the term "big questions" has enjoyed considerable popularity. The phrase "big questions" appears to have originated in philosophy as shorthand for describing the discipline's content and scope (Solomon & Higgins, 2013). Now, however, multiple disciplines are asking their practitioners what big questions they want to answer and how much progress has been made in answering them (e.g., Keeler, 2010; Sussman, 2010). Like other research fields, it may be worthwhile to ask ourselves whether research on aging can or does address big questions. This is not the venue for a comprehensive list of the big questions that aging research can or does address. We will, however, suggest one big question to which aging research makes significant contributions.

For most social science disciplines, a core question has always been: What mechanisms allow societies to survive? Alternatively, what mechanisms convince societal members to create and sustain societies, even when those mechanisms require members to sacrifice some of their own resources, gains, and autonomy? This is a big question that social scientists have tried to answer for more than a century.

Many of the founding fathers of sociology and anthropology looked for answers to the big question of societal order and stability. Durkheim's comparison of mechanical and organic solidarity outlined two modes of sustaining societal order and stability (1997). Similarly, Weber's writings on rationalization and bureaucracy focused on forms of organization that yielded stability and order in increasingly diverse societies (Ritzer & Stepinsky, 2012). In anthropology, Levi-Strauss (1966) posited that although societies vary widely in the structural arrangements that they use to achieve solidarity and stability, all humans share the same underlying patterns of thought. Consequently, no matter how much structural arrangements appear to differ across societies, the functions that they serve are the same.

During the first half of the twentieth century, issues of stability and order were generally studied under the theoretical umbrella of structural functionalism. It is not surprising that structural functionalism focused on the cultural tools and practices that promoted order and stability – that was precisely the purpose of inquiries in that tradition. Nonetheless, structural functionalism was heavily criticized for neglecting conflict, innovation, inequality, and social change. These criticisms led to countervailing theories and research and by the middle of the twentieth century research based on structural functionalist theory had declined substantially.

But social solidarity and order are issues that are too fundamental to lay fallow forever. And no scholar played a larger role in bringing these issues back to the forefront of the social sciences than Pierre Bourdieu and his theory of social reproduction. Bourdieu's theory and research is far-ranging and discussion here will focus on his contribution to understanding the stability of social systems. More specifically, he studied the reproduction of social stratification or, as he preferred to call them, *status hierarchies*. Bourdieu acknowledged the importance of economic capital in reproducing status hierarchies, but argued that a focus solely on material resources is incomplete. He argued for the importance of social capital, cultural capital, and even physical/bodily capital in sustaining stratification systems over long periods of time (Bourdieu, 1977). Much of his research focused on the role of education in social reproduction of social classes. Buttressing his belief that much more than financial capital is involved in this process, his research focused on cultural capital, especially the arts, in reproducing social stratification (Bourdieu, 1984).

Literally hundreds of studies using the social reproduction framework have been published in the last several decades. As might be expected, a disproportionate number of them focus on education and the failure of schools to generate as much upward mobility for disadvantaged students as would be desirable (Aschaffenburg & Maas, 1997; Collins, 2009). The scope of social reproduction studies, however, is quite large and ranges from research on the effects of economic growth in developing countries (Boughey, 2007) to the failure of politics to break patterns of social reproduction (Ruckert, 2010) to feminist critiques of the persistence of traditional family roles (Chodorow, 1978).

It is interesting to note that the valence of social scientists' views of social reproduction changed quite dramatically over time. Durkheim and later scholars who relied on structural functionalism began with the premise that stability and order are problems that societies must resolve to survive. Identifying social processes that promoted stability and social integration was viewed as a testimony to the power of "social facts" to create order out of potential chaos. In short, structural functionalists generally view social institutions and the social arrangements that sustain them favorably. Scholars using the framework of social reproduction tend to take the opposite view. Many studies purport to demonstrate that schools perpetuate social hierarchies rather than reduce them through upward mobility. The social reproduction of inequality is viewed as problematic. Scholars in this tradition clearly favor social institutions that do not reproduce established social hierarchies.

A case also can be made that social reproduction is at the heart of several major theories of aging. Continuity theory, which emerged as a response to and critique of disengagement theory, is essentially a theory of social reproduction (Atchley, 1999). It posits that as older adults experience age-related transitions such as retirement and widowhood, they sustain as many roles and activities that they valued prior to the transitions as possible – in essence, they reproduce the same parameters of their lives that they had previously. Socio-emotional selectivity theory (SST) is similar, in some ways, to continuity theory. According to SST, as aging adults experience declines in their capacities, they express the highest levels of life satisfaction if they release less important roles, relationships, and activities and invest mainly in those that are most meaningful for them (Lockenhoff & Carstensen, 2004). Again, the emphasis of this theory is on continuity or stability of meaningful engagement. Cumulative advantage/disadvantage theory (C/AD) also focuses on stability across individuals over time and, thus, the social reproduction of social stratification. For economic outcomes, C/AD hypothesizes that the rich literally get richer and the poor get poorer. For health outcomes there is no expectation that health improves throughout adulthood, but those who begin adulthood in better health are expected to maintain better health over time than their less advantaged peers. This is a social reproduction scenario.

The valence of aging and life course scholars who document social reproduction is more mixed than was the case historically. Both the disengagement/continuity theory debate and SST focused on identifying the conditions under which older adults are satisfied with their lives. Research based on continuity theory and SST suggests that forms of social reproduction are associated with subjective perceptions of life quality. Given the research questions asked, there is no reason to view these social reproduction processes as anything but positive. Researchers documenting the cumulative effects of advantage and disadvantage are generally less happy with findings that point to the maintenance and the accumulation of resources and deficits due to stratification. Despite the assumption of objectivity in science, it is clear that social scientists (with the possible exception of economists) dislike inequality and the social arrangements that reproduce or increase it.

Whether one views it as a necessary requirement for societal survival or a means of perpetuating inequality (or both), evidence leaves no doubt that social reproduction exists and operates in many areas of life. We suggest, however, that excessive attention to social reproduction and stability promotes an unrealistic and incomplete view of the dynamics of time and aging. This chapter began with reference to the adage that "the only constant is change." At the same time that we are persuaded by rigorous evidence of social reproduction, most of us believe that adage. Everything that we see, hear, and experience tells us that change is a frequent, if not constant, dynamic in the world and in our lives. Thus, science needs to focus on change as well as stability.

We hope to make a case for increased attention to individual and social change. The study of individual change is well established in aging research. There may be a tendency to interpret findings from the perspective of stability rather than change – to focus on the stability of physical, emotional, and cognitive capacities across the life course and to miss the processes that permit stability in outcomes (such as life satisfaction) despite substantial change in objective circumstances. Overall, however, aging research in the social sciences is attentive to individual-level change and, as a result of recent statistical advances, trajectories of change over substantial periods of time.

The study of social change and its relationships with aging is much less explored. Social scientists are cognizant of cohort changes, but are too satisfied with labeling them, rather than explaining them. The three topics that we chose to discuss as recent advances in aging research all focus on social change – in the forms of cohort differences; sudden large-scale changes in the basic infrastructure of society or generated by natural or man-made disasters; and gradual changes that creep up and culminate in changes that no one "saw coming." Social scientists should be heavily invested in studying social change, as well as social stability. Aging researchers should examine how age and age structures affect broad social changes and the consequences of social change for aging adults.

The antecedents and consequences of social change are a "big question" for social scientists. They subsume multiple more specific research questions such as "What kinds and degrees of social change trigger meaningful cohort differences?" and "How and under what conditions does social change alter the well-being of older adults?" Aging and life course researchers are arguably in one of the best positions of all social scientists to tackle questions about social change because they already study individual change and know the tools needed to study change at all levels of aggregation. Advancing our understanding of social change also would contribute to the social sciences more broadly because social change has widespread implications for individuals, social institutions, and societies. Most importantly, aging research has the potential to balance the current emphasis on social reproduction with recognition of the prevalence and importance of social change.

Yes, aging and life course research is alive and well. Important contributions to our understanding of aging and the social contexts within which it unfolds have been impressive over the past two decades or so. There are also, however, important research questions yet to be addressed. Opportunities to generate new understandings of aging, older adults, and an aging population are plentiful. We invite fresh attention to research opportunities that have the potential to optimize the aging experience in spite of current inequalities.

References

Aschaffenburg, A., & Maas, I. (1997). Cultural and educational careers. *American Sociological Review, 62,* 573–587.

Atchley, R. C. (1999). *Continuity and adaptation in aging.* Baltimore, MD: Johns Hopkins University Press.

Beckfield, J. (2004). Does income inequality harm health? New cross-national evidence. *Journal of Health and Social Behavior, 45,* 231–248.

Blakely, T. A., Kennedy, B. P., Glass, R., & Kawachi, I. (2000). What is the lag time between income inequality and health status? *Journal of Epidemiology & Community Health, 54,* 318–319.

Boughey, C. (2007). Educational development in South Africa: From social reproduction to capitalist expansion? *Higher Education Policy, 20,* 5–18.

Bourdieu, P. (1977). *Outline of a theory of practice.* Cambridge: Cambridge University Press.

Bourdieu, P. (1984). *Distinction: A social critique of the judgment of taste.* Cambridge, MA: Harvard University Press.

Cagney, K. A., Browning, C. R., Iveniuk, J., & English, N. (2014). The onset of depression during the Great Recession: Foreclosures and older adult mental health. *American Journal of Public Health, 104,* 498–505.

Cahill, K. E., Giandrea, M. D., & Quinn, J. F. (2006). Retirement patterns from career employment. *The Gerontologist, 46,* 514–523.

CDC (Centers for Disease Control and Prevention). (2013). *CDC's disaster planning goal: Protect vulnerable older adults.* Retrieved from: <http://www.cdc.gov/aging/pdf/disaster_planning_goal.pdf> Accessed 11.06.14.

CDC (Centers for Disease Control and Prevention). (2014a). *Medicare administrative data (Health Indicators Warehouse).* Retrieved from: <http://www.healthindicators.gov/Resources/DataSources/Medicare-Administrative-Data_68/Profile> Accessed 12.08.14.

CDC (Centers for Disease Control and Prevention). (2014b). *The second longitudinal study of aging II.* Retrieved from: <http://www.cdc.gov/nchs/lsoa/lsoa2.htm> Accessed 20.08.14.

Cherry, K. E., Galea, S., Su, L. J., Welsh, D. A., Jazwinski, S. M., Silva, J. L., et al. (2010). Cognitive and psychosocial consequences of Hurricanes Katrina and Rita among middle-aged, older and oldest-old adults in the Louisiana Health Aging Study (LHAS). *Journal of Applied Social Psychology, 40,* 2463–2487.

Chinese Longitudinal Healthy Longevity Survey. (2014). *Chinese longitudinal healthy longevity survey (CLHLS).* Retrieved from: <http://www.geri.duke.edu/chinese-longitudinal-healthy-longevity-survey> Accessed 8.07.14.

Chodorow, N. J. (1978). *The reproduction of mothering: Psychoanalysis and the sociology of gender.* Berkeley, CA: University of California Press.

Choi, N. G., DiNitto, D. M., & Kim, J. (2014). Discrepancy between chronological age and felt age. *Journal of Aging and Health, 26,* 458–473.

CMS (Center for Medicare and Medicaid Services). (2014). *Medicare current beneficiary survey (MCBS).* Retrieved from: <http://www.cms.gov/Research-Statistics-Data-and-Systems/Research/MCBS/index.html?redirect=/mcbs/> Accessed 20.08.14.

Collins, J. (2009). Social reproduction in classrooms and schools. *Annual Review of Anthropology, 38,* 33–48.

Danese, A., & McEwen, B. S. (2012). Adverse childhood experiences, allostasis, allostatic load, and age-related disease. *Physiology & Behavior, 106,* 29–39.

Dannefer, D. (1987). Aging as intracohort differentiation: Accentuation, the Matthew effect, and the life course. *Sociological Forum, 2,* 211–236.

Diez-Roux, A. V., Link, B. G., & Northridge, M. E. (2000). A multi-level analysis of income inequality and cardiovascular disease. *Social Science & Medicine, 50,* 673–687.

Durkheim, E. (1997). *The division of labor in society* (L. Coser, Trans.). Glencoe, IL: The Free Press.

Easterlin, R. A. (1974). Does economic growth improve the human lot? Some empirical evidence. In P. A. David & W. R. Levin (Eds.), *Nations and households in economic growth* (pp. 98–125). Stanford, CA: Stanford University Press.

Elder, G. H., Jr. (1998). The life course as developmental theory. *Child Development, 69*(1), 1–12.

Elder, G. H., Jr., Johnson, M. K., & Crosnoe, R. (2003). The emergence and development of life course theory. In J. T. Mortimer & M. J. Shanahan (Eds.), *Handbook of the life course* (pp. 3–22). New York, NY: Kluver Academic/Plenum Publishers.

ELSA (2014). *ELSA: English longitudinal study of ageing.* Retrieved from: <http://www.elsa-project.ac.uk/> Accessed 12.08.14.

Ferraro, K. F., Shippee, T. P., & Schafer, M. H. (2009). Cumulative inequality theory for research on aging and the life course. In V. L. Bengtson, D. Gans, N. M. Putney, & M. Silverstein (Eds.), *Handbook of theories of aging* (2nd ed., pp. 413–433). New York, NY: Springer.

Fiscella, K., & Franks, P. (1997). Poverty or income inequality as predictor of mortality: Longitudinal cohort study. *British Medical Journal, 314,* 1724.

Frisvold, D., & Golberstein, E. (2013). The effect of school quality on black-white health differences: Evidence from segregated southern schools. *Demography, 50,* 1989–2012.

Fry, R., & Taylor, P. (2012). The rise of residential segregation by income: *Social & Demographic Trends.* Washington, DC: Pew Research Center.

Giandrea, M. D., Cahill, K. E., & Quinn, J. F. (2009). Bridge jobs: A comparison across cohorts. *Research on Aging, 31,* 549–576.

REFERENCES

Goosby, B. J. (2013). Early life course pathways of adult depression and chronic pain. *Journal of Health and Social Behavior, 54*(1), 75–91.

Heggeness, M.L., & Hokayem, C. (2013). *Life on the edge: Living near poverty in the United States, 1966–2011.* U.S. Bureau of the Census. Retrieved from: <www.census.gov/hhes/www/poverty/publications/WP2013-02.pdf> Accessed 19.06.14.

Henderson, T. L., Roberto, K. A., & Kamo, Y. (2010). Older adults' responses to Hurricane Katrina: Daily hassles and coping strategies. *Journal of Applied Gerontology, 29,* 48–69.

ICPSR (Inter-university Consortium for Political and Social Research). (2014). *Established populations for epidemiological studies of the elderly, 1981–1993.* Retrieved from: <http://www.icpsr.umich.edu/icpsrweb/ICPSR/studies/09915> Accessed 18.07.14.

Institute for Social Research. (2014a). *HRS – Health and retirement study.* Retrieved from: <http://hrsonline.isr.umich.edu/> Accessed 12.08.14.

Institute for Social Research. (2014b). *Understanding social disparities in health and aging – The Americans Changing Lives Study.* Retrieved from: <http://www.isr.umich.edu/acl/> Accessed 12.08.14.

Kail, B. L., & Taylor, M. G. (2014). Cumulative inequality and racial disparities in health: Private insurance coverage and black/white differences in functional limitations. *Journal of Gerontology: Social Sciences, 69B*(5), 798–808.

Kamo, Y., Henderson, T. L., & Roberto, K. A. (2011). Displaced older adults' reactions to and coping with the aftermath of Hurricane Katrina. *Journal of Family Issues, 32,* 1346–1370.

Kaplan, G. A., Pamuk, E. R., Lynch, J. W., Cohen, R. D., & Belfour, J. L. (1996). Inequality in income and mortality in the United States: Analysis of mortality and potential pathways. *British Medical Journal, 312,* 999–1003.

Kawachi, I., & Kennedy, B. P. (1997). The relationship of income inequality to mortality: Does the choice of indicator matter? *Social Science & Medicine, 45,* 1121–1127.

Keeler, D. R. (Ed.). (2010). *Environmental ethics: The big questions.* New York, NY: John Wiley & Sons.

Kondo, N., Sembajwe, G., Kawachi, I., van Dam, R. M., Subrmanian, S. V., & Yamagata, Z. (2009). Income inequality, mortality, and self-rated health: Meta-analysis of multilevel studies. *British Medical Journal, 339,* 1–9.

Kondo, N., van Dam, R. M., Sembajwe, G., Kawachi, I., & Yamagata, Z. (2012). Income inequality and health: The role of population size, inequality threshold, period effects, and lag effects. *Journal of Epidemiology & Community Health, 66,* 966–972.

Lash, T. L., & Silliman, R. A. (2001). A comparison of the National Death Index and Social Security Administration databases to ascertain vital statistics. *Epidemiology, 12,* 259–261.

Levi-Strauss, C. (1966). *The savage mind.* Chicago, IL: University of Chicago Press.

Livingston, G., & Parker, K. (2010). Since the start of the Great Recession, more children raised by grandparents: *Social & Demographic Trends.* Washington, DC: Pew Research Center.

Lochner, K., Pamuk, E., Maduc, D., Kennedy, B. P., & Kawachi, I. (2001). State-level income inequality and individual mortality risk: A prospective, multilevel study. *American Journal of Public Health, 91,* 385–391.

Lockenhoff, C. E., & Carstensen, L. L. (2004). Socioemotional selectivity theory, aging, and health: The increasingly delicate balance between regulating emotions and making tough choices. *Journal of Personality, 72,* 1395–1424.

Luszcz, M. A., Giles, L. C., Anstey, K. J., Browne-Yung, K. C., Walker, R. A., & Windsor, T. D. (2014). Cohort profile: The Australian longitudinal study of ageing (ALSA). *International Journal of Epidemiology.* <http://dx.doi.org/10.1093/ije/dyu196>.

Lynch, S. M. (2006). Explaining life course and cohort variation in the relationship between education and health: The role of income. *Journal of Health and Social Behavior, 47,* 324–338.

McFall, B. H. (2011). Crash and wait? The impact of the Great Recession on retirement planning of older Americans. *American Economic Review, 101,* 40–44.

Merton, R. K. (1968). The Matthew effect in science. *Science, 159,* 56–63.

Morton, P. M., Mustillo, S. A., & Ferraro, K. F. (2014). Does childhood misfortune raise the risk of acute myocardial infarction in adulthood? *Social Science & Medicine, 104,* 133–141.

New York Times. (2013). *In hard economy for all ages, older isn't better…It's brutal.* Retrieved from: <www.nytimes.com/2013/02/03/business/americans-closest-to-retirement-were-hardest-hit-by-recession.html> Accessed 4.04.14.

NORC (National Opinion Research Center). (2014). *National social life, health, and aging project.* Retrieved from: <http://www.norc.org/Research/Projects/Pages/national-social-life-health-and-aging-project.aspx> Accessed 30.08.14.

O'Rand, A. M. (1996). The precious and the precocious: Understanding cumulative disadvantage and cumulative advantage over the life course. *The Gerontologist, 36*(2), 230–238.

Pavalko, E. K., Gong, F., & Long, J. S. (2007). Women's work, cohort change, and health. *Journal of Health and Social Behavior, 48,* 352–368.

Pearlin, L. I., Schieman, S., Fazio, E. M., & Meersman, S. C. (2005). Stress, health, and the life course: Some conceptual perspectives. *Journal of Health and Social Behavior, 46,* 205–219.

Rigney, D. (2010). *The Matthew effect: How advantage begets further advantage*. New York, NY: Columbia University Press.

Ritzer, G., & Stepinsky, J. (2012). *Contemporary sociological theory and its classical roots: The basics* (4th ed.). New York, NY: McGraw Hill.

Ross, N. A., Wolfson, M. C., Dunn, J. R., Berthelot, J., Kaplan, G. A., & Lynch, J. W. (2000). Relation between income inequality and mortality in Canada and in the United States: Cross-sectional assessment using census data and vital statistics. *British Medical Journal, 320*, 898.

Ruckert, A. (2010). The forgotten dimension of social reproduction: The Word Bank and the poverty reduction strategy paradigm. *Review of International Political Economy, 17*, 816–839.

Ryder, N. B. (1965). The cohort as a concept in the study of social change. *American Sociological Review, 6*, 843–861.

Schafer, M. H., Ferraro, K. F., & Mustillo, S. A. (2011). Children of misfortune: Early adversity and cumulative inequality in perceived life trajectories. *American Journal of Sociology, 116*(4), 1053–1091.

SHARE. (2014). *SHARE: The study of health, aging, and retirement in Europe*. Retrieved from: <http://www.share-project.org/> Accessed 12.08.14.

Shaw, B. A., & Krause, N. (2002). Exposure to physical violence during childhood, aging, and health. *Journal of Aging and Health, 14*, 467–494.

Shi, L., & Starfield, B. (2001). The effect of primary care physician supply and income inequality on mortality among blacks and whites in U.S. metropolitan areas. *American Journal of Public Health, 91*, 1246–1250.

Sociometrics. (2014). *Hispanic established populations for epidemiological studies of the elderly (HEPESE)*. Retrieved from: <http://www.socio.com/cam3031.php> Accessed 20.08.14.

Solomon, R. C., & Higgins, K. M. (2013). *The big questions* (9th ed.). Independence, KY: Cengage Learning.

Stiglitz, J. E. (2012). *The price of inequality*. New York, NY: W. W. Norton.

Sturm, R., & Gresenz, C. R. (2002). Relations of income inequality and family income to chronic medical conditions and mental health disorders: National survey. *British Medical Journal, 324*, 20.

Sussman, M. D. (2010). The randomized controlled trial: An excellent design, but can it address the big questions in neurodisability? *Developmental Medicine & Child Neurology, 52*, 1066–1067.

The Economist. (2013). *Growing apart: America's income inequality is growing again*. Retrieved from: <http://www.economist.com/news/leaders/21586578-americas-income-inequality-growing-again-time-cut-subsidies-rich-and-invest> Accessed 29.08.14.

University of Wisconsin. (2014). *Wisconsin longitudinal study*. Retrieved from: <http://www.ssc.wisc.edu/wlsresearch/> Accessed 12.08.14.

US Bureau of Labor Statistics. (2013). *Record unemployment among older workers does not keep them out of the job market*. Retrieved from: <http://www.bls.gov/opub/ils/summary_10_04/older_workers.htm> Accessed 17.06.14.

US Census Bureau. (2012). *Figure 1. Median age at first marriage by sex: 1890 to 2010*. Retrieved from: <www.census.gov/hhes/socdemo/marriage/data/acs/ElliottetalPAA2012figs.pdf> Accessed 4.08.13.

Wickrama, K. A. S., & O'Neal, C. W. (2013). Family of origin, race/ethnicity, and socioeconomic attainment: Genotype and intraindividual processes. *Journal of Marriage and Family, 75*(1), 75–90.

Wilkinson, L. R., Shippee, T. P., & Ferraro, K. F. (2012). Does occupational mobility influence health among working women? Comparing objective and subjective measures of work trajectories. *Journal of Health and Social Behavior, 53*(4), 432–447.

Wilkinson, R. G., & Pickett, K. E. (2006). Income inequality and population health: A review and explanation of the evidence. *Social Science & Medicine, 62*, 1768–1784.

Wilkinson, R. G., & Pickett, K. E. (2008). Income inequality and socioeconomic gradients in mortality. *American Journal of Public Health, 98*, 699–704.

Yang, Y. (2007). Is old age depressing? Growth trajectories and cohort variations in late-life depression. *Journal of Health and Social Behavior, 48*, 16–32.

Zheng, H., & George, L. K. (2012). Rising U.S. income inequality and the changing gradient of socioeconomic status on physical functioning and activity limitations, 1984–2007. *Social Science & Medicine, 75*, 2170–2183.

CHAPTER 2

Trajectory Models for Aging Research

Scott M. Lynch[1] and Miles G. Taylor[2]

[1]Department of Sociology, Duke University, Durham, NC, USA [2]Pepper Institute on Aging and Public Policy, Florida State University, Tallahassee, FL, USA

OUTLINE

Growth Modeling in a Nutshell	25	Data Structure and Method	43
Latent Class Modeling in a Nutshell	31	Measurement of Time	44
		Importance of Assumptions	46
Latent Class Growth Analysis	38	Extraction of Classes and Inclusion of Covariates	47
Growth Mixture Modeling	39	Conclusion	49
Important Issues in the Implementation of Trajectory Methods	43	References	50

Life course research investigates how human lives and events unfold over time, at both the individual level and larger levels, such as within families or nations (Elder, 1985). At the individual level, life course research is concerned with the development of individuals as they age, as well as with between-person differences in development. Such differences often exist across sexes, races, socioeconomic classes, and other characteristics. Importantly, the birth cohort to which an individual belongs plays an important role in shaping development, as do period events (e.g., economic depression). Trajectory methods have emerged over the last several decades as important tools for investigating life course dynamics, including between-person differences in development (George, 2009).

Trajectories are simply patterns in values of variables across time. Based on this broad definition, the term "trajectory modeling" refers to a number of qualitatively different methods used to model an even greater number of social phenomena. For example, one may be interested in trajectories of unemployment rates, stock market closing values, or other macro-level phenomena, either for a single case (e.g., the United States) or for multiple cases (e.g., states or nations). Alternatively, one may be interested in trajectories of income or health at the individual level. In some cases, one may use the term "trajectory" to refer to the timing and ordering of life events: school completion, employment, marriage, childbearing, retirement, and death.

In this chapter, we will restrict our discussion of trajectory models to two broad classes – growth curve models (Bollen & Curran, 2006; Meredith & Tisak, 1990) and latent class models (Clogg, 1995; Goodman, 1974; Lazarsfeld & Henry, 1968) – and two of their generalizations, including latent class growth models (Nagin & Odgers, 2010) and growth mixture models (Muthén, 2004; Muthén & Muthén, 2000). While our discussion of these models can easily extend to macro-level units like states and countries, we will focus our discussion on individual-level characteristics. Following our exposition of each of these methods, we discuss a variety of issues relevant to consider when using them in the face of untestable assumptions.

Due to space constraints, we will exclude the latter type of trajectory models mentioned above involving sequences of different types of life events. That is, we will focus only on models of repeated measures (i.e., *levels*) of the same phenomenon, not the timing and pattern of multiple and qualitatively distinct events like school completion, employment, and marriage (i.e., *transitions*). Recent extensions of the software and methods we will discuss can allow transitions between states, but handling multiple, distinct types of transitions is still difficult within the framework we discuss. Models that handle sequences of different events are more commonly called "sequence analyses" (see Barban & Billari, 2012) and historically have required specialized software.

For the purpose of illustrating trajectory methods, we rely on a subset of data from the Health and Retirement Study (HRS), a panel study of adults over age 50 in the United States. Details about the study design can be found elsewhere (RAND HRS Data, Version M., 2013). Our key outcome measure is the body mass index (BMI), a commonly used measure of weight per height (kg/m^2) that is currently the focus of much attention in public health research and the media (National Institutes of Health, 1998). BMI is a good measure with which to illustrate trajectory methods for two reasons. First, BMI is not highly volatile across adulthood for most individuals. While it tends to increase or decrease over age, it does not often change in a dramatic or erratic way. Second, BMI can be treated as both continuous and categorical, with well-established categories. Specifically, a BMI under 20 is considered underweight; a BMI between 20 and 25 is considered normal weight; a BMI between 25 and 30 is considered overweight; and a BMI above 30 is considered obese, with BMIs over 30 often further subdivided into two or more obesity classes. The continuous version of BMI is ideally suited for growth modeling, while the categorical version of BMI is most amenable to latent class methods, as we will discuss.

We restrict the sample to members of the arbitrarily chosen 1951 birth cohort who were interviewed in the 2004, 2006, 2008, and 2010 waves of the study. We further restrict the sample to blacks and whites, to those who survived the entire time period of observation, and to those with complete information on BMI, sex, race, region of birth (south versus elsewhere), and years of schooling. There are two main approaches to trajectory modeling – multivariate methods and hierarchical methods – and both can handle missing data, albeit in somewhat different fashions. However, the main focus of this chapter is not on the intricacies of the methodology. Thus, we simplify our discussion by eliminating data missing due to both item nonresponse and attrition, although we discuss missing data handling, including attrition, briefly later in the chapter. The resulting analytic sample size was $n = 353$, a sample large enough to illustrate all ideas in the chapter but small enough for the construction of readable figures. We note at the outset that we do not focus on any substantive aspect of BMI. BMI tends to decline substantially for some at the very oldest ages as bone density and muscle mass decline; however, our sample is restricted to ages younger than 60.

GROWTH MODELING IN A NUTSHELL

Figure 2.1A shows BMI measures for four sample members, each measured on four occasions. A natural first effort at modeling these data might involve an ordinary least squares (OLS) regression model. Figure 2.1B shows the prediction line. The line appears to fit the data fairly well, with an intercept of 26.8 and a slope of 1.18 BMI units. Note that time is measured as 0, 1, 2, and 3, reflecting waves since baseline. Thus, the average sample member gained 3*1.18 = 3.54 BMI units across the survey period. In other words, the average sample member started the survey period overweight and became slightly obese over the 6 years.

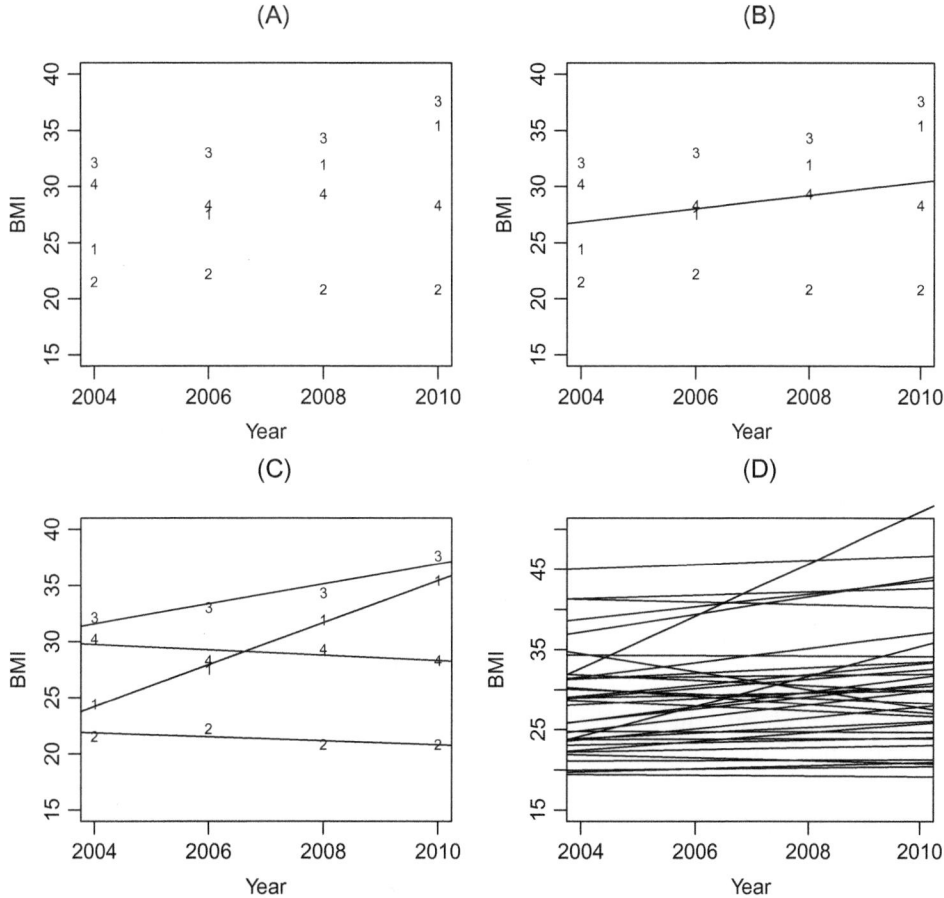

FIGURE 2.1 Plots of individual patterns in BMI across time. Panel A shows measures for four persons (1–4); panel B shows the best-fitting single regression line for these four cases; panel C shows individual-specific regression lines for these four cases; panel D shows regression lines for 35 sample members.

There are at least two, now widely recognized, shortcomings of this modeling strategy. First, given that the data come from a panel, the OLS regression model is inappropriate because the errors are not independent within individuals. Thus, standard errors of the parameter estimates are biased downward, rendering t-tests invalid. Second, the OLS model estimates the average age pattern for the four individuals instead of the age pattern for any single individual, but the average age pattern may not reflect the actual experience of any real person.

Example 2.1

An alternative approach to modeling these data might be to estimate a separate OLS regression model for each person (Bollen & Curran, 2006). Figure 2.1C illustrates the results of this strategy. As the figure shows, while the single regression line for the sample had a positive slope, the slopes of the individual lines are not uniformly positive. Instead, two individuals have regression lines with steeper positive slopes than the average, and two have lines with negative slopes. The intercepts vary as well. Thus, of the four persons, one was obese at baseline and became heavier over the study period; one was obese at baseline but lost weight across time to end slightly overweight (BMI > 25); one began as normal weight but gained considerable weight and was obese by the end of the period; and one began as normal weight but lost weight to become nearly underweight by the end of the period. In short, the single OLS regression model missed considerable heterogeneity across the sample. Figure 2.1D expands Figure 2.1C in showing estimated OLS regression lines for 35 sample members (a 10% subsample). As the figure shows, there is substantial variation in the intercepts and slopes – that is, trajectories.

Figure 2.2 shows a scatterplot of the intercepts and slopes obtained from estimating OLS regression models for all sample members.

The mean intercept and mean slope for the sample are indicated by dashed reference lines. The mean BMI at baseline was just under 30, and the mean rate of change in BMI was just above 0. The histogram above the scatterplot shows the distribution of intercepts for the sample, while the histogram to the right of the scatterplot shows the distribution of slopes for the sample. The figure reveals the considerable heterogeneity that a single OLS regression summary would fail to capture. Furthermore, the correlation between intercepts and slopes is negative and moderate, as the dotted reference line in the scatterplot shows: those with higher baseline BMIs tend to experience less growth, or even decline, in BMI over time, while those with lower baseline BMIs tend to experience greater growth in BMI over time.

This process of estimating an OLS regression model to capture individual-level patterns over time illustrates the key concept underlying growth modeling. Whereas the OLS regression model posits a single value for the intercept and slope, a growth model (GM) posits a unique intercept and slope for each individual as shown in Eqs. (2.1) and (2.2), respectively:

$$\text{OLS: } y_{it} = b_0 + b_1 t_{it} + e_{it} \qquad (2.1)$$

$$\text{GM: } y_{it} = b_{0i} + b_{1i} t_{it} + e_{it} \qquad (2.2)$$

In Eq. (2.1), y_{it} is the outcome for individual i at time t, b_0 is the intercept (the value of y when $t = 0$), b_1 is the slope across time, t_{it} is the time of measurement of y, and e_{it} is an error term that is assumed to follow the usual assumptions that it is normally distributed, homoscedastic, and independent across observations. Note that the subscripting of t implies that individuals do not need to be measured at the same time nor on the same number of occasions.

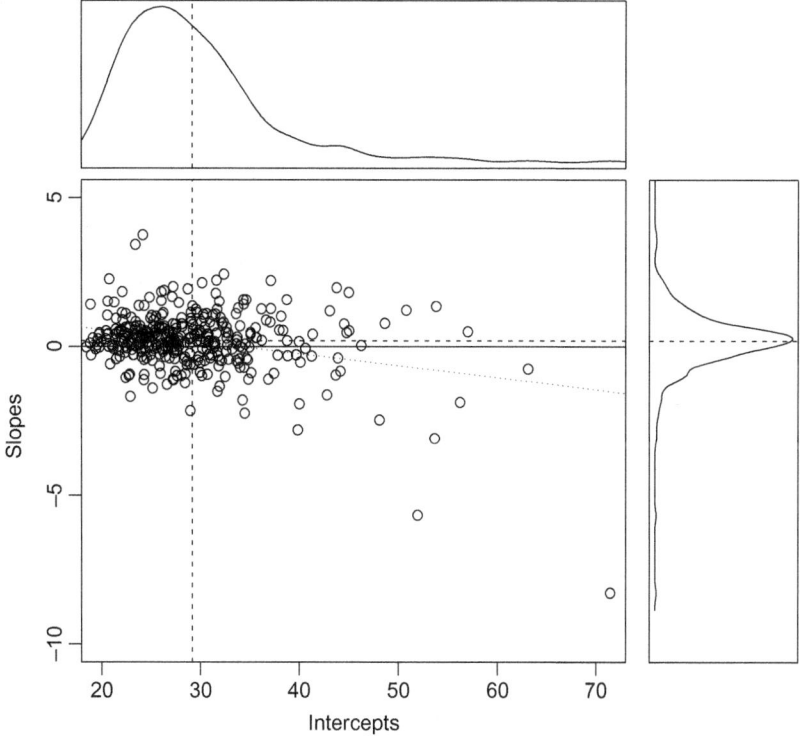

FIGURE 2.2 Scatterplot and histograms of unique individual intercepts and slopes estimated via OLS regression for all sample members. Histogram at top shows distribution of intercepts; histogram at side shows distribution of slopes. Dashed vertical and horizontal lines represent means of intercepts and slopes (respectively). Diagonal dotted reference line reflects correlation between intercepts and slopes.

Equation (2.2) shows the extension of the model to include individual-specific intercepts (b_{0i}) and slopes (b_{1i}). Equation (2.3) shows that this model can be rewritten as an OLS regression model with a common intercept and slope (denoted here as b_{00} and b_{10} – the second subscript of 0 reflects a common intercept for the sample), but with unit-specific "random effects" (u_i and v_i) that allow individual deviations from the average:

$$y_{it} = (b_{00} + u_i) + (b_{10} + v_i)t_{it} + e_{it} \quad (2.3)$$

Estimating this model via OLS is problematic, however, because OLS can estimate an average intercept and slope only; thus, u_i and v_i are relegated to the error term as shown in Eq. (2.4), making it both autocorrelated (due to cross-time commonality reflected in u_i) and heteroscedastic across time (because part of the error is a function of time: $v_i t_{it}$):

$$y_{it} = b_{00} + b_{10}t_{it} + (e_{it} + u_i + v_i t_{it}) \quad (2.4)$$

The model is therefore generally estimated as a hierarchical model, with probability distributions assigned to the random effects at a second level (i.e., "Level 2"). Specifically, u and v are usually assumed to be multivariate normally distributed with a mean vector of 0 and a covariance matrix of Σ, which contains both the variances of u and v and the covariance of u with v.

The normality assumption is a crucial one. While the assumption reduces the number of parameters to be estimated compared to the individual-specific OLS regression model approach, it does so by imposing a specific form for the collection of intercepts and slopes. At the same time, the specification of a specific distribution for the random effects – and the estimation of the associated parameters – enables the random effects approach to make out-of-sample inferences regarding the population. Implications of this assumption are discussed subsequently.

Given that each individual has a unique intercept and slope via the random effects u and v, we can evaluate whether fixed individual-level characteristics, like sex, race, birth region, and education, "explain" some of the variance in them. Figure 2.3 illustrates this idea. The figure replicates the scatterplot from Figure 2.2 but limits the points to black females and white females all of whom were born in the south. As the figure shows via the horizontal and vertical reference lines for the mean intercepts and slopes for each group, black women tend to have much higher baseline BMI, while white women tend to have a much larger growth rate. In fact, white women have positive average growth in BMI over

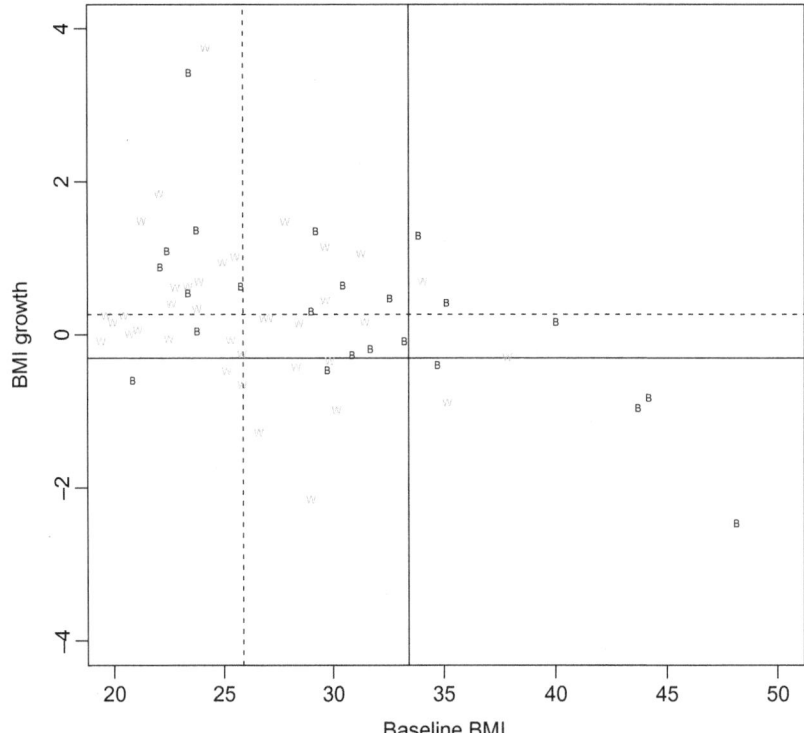

FIGURE 2.3 Scatterplot of intercepts and slopes from individual-level OLS regression models for black (B) and white (W) women born in the south. Vertical reference lines show the mean intercepts for blacks (solid line) and whites (dashed line); horizontal reference lines show the mean slopes.

time, while black women have negative average growth (i.e., they experience decline on average).

In order to capture covariate differences in intercepts and slopes, we can conduct a second-stage (or level) OLS model by regressing the unique intercepts (b_{0i}) and slopes (b_{1i}) on (time invariant) covariates X, as shown in Eq. (2.5):

$$b_{0i}=b_{00}+X_i\gamma+u_i \quad (2.5a)$$

$$b_{1i}=b_{10}+X_i\delta+v_i \quad (2.5b)$$

In this equation, the u_i and v_i are not the same as before: they have become residual terms that reflect unobserved heterogeneity that remains in individual intercepts and slopes after extracting similarities that exist among those who share values of X (e.g., race, sex). u_i and v_i remain as random effects because they are assumed to follow a probability distribution, similar to the typical OLS error term, e_i. In contrast, X are fixed covariates, and so b_{00}, b_{10}, γ, and δ are called "fixed effects." That is, these coefficients are assumed not to vary. For this reason, the model is sometimes referred to as a "mixed model," or a "random coefficient model" (Raudenbush, 2001). Furthermore, the model is often considered a special type of "hierarchical model," with Eq. (2.2) representing level 1, and Eqs. (2.5a) and (2.5b) representing level 2 (Raudenbush & Bryk, 2002). Finally, it is important to note that, because there are two levels of error terms, or "variance components," all growth models are hierarchical models by common statistical terminology. Thus, the terms "growth model" and "hierarchical growth model" are interchangeable, with the latter containing some redundancy.

Table 2.1 presents the results of two sets of models. The upper half of the table shows the results of following the strategy of estimating the two OLS regression models. The lower half of the table shows the results of estimating the

TABLE 2.1 Results of Growth Modeling Via Two Strategies: Separate OLS Regression Models Versus One SEM-Based Multivariate Model

	Intercepts (baseline)	Slopes (growth)
OLS MODELING APPROACH		
Stage 1		
Mean	29.16	0.195
Variance	53.7	1.15
Correlation (b_{0i},b_{1i})		−0.28
Stage 2		
Intercept	34.86 (1.98)***	0.16 (0.30)
Male	0.72 (0.78)	−0.06 (0.12)
Black	2.28 (1.02)*	−0.05 (0.15)
South	−1.16 (0.87)	−0.13 (0.13)
Education	−0.46 (0.14)**	0.01 (0.02)
Correlation (b_{0i},b_{1i})		−0.28
R^2	0.046	0.010
MULTIVARIATE GROWTH APPROACH		
Model 1 (unconditional)		
Mean	29.16 (0.39)	0.196 (0.057)
Variance	50.5 (4.03)	0.40 (0.13)
Correlation (b_{0i},b_{1i})		−0.21
Model 2 (conditional)		
Intercept	34.90 (1.96)***	0.13 (0.29)
Male	0.71 (0.77)	−0.07 (0.12)
Black	2.24 (1.02)*	0.003 (0.15)
South	−1.18 (0.86)	−0.13 (0.13)
Education	−0.46 (0.14)***	0.01 (0.02)
Correlation (b_{0i},b_{1i})		−0.21

Note: Standard errors shown in parentheses. Growth modeling approach estimates standard errors for all parameters, including variances.
# $p < 0.1$, * $p < 0.05$, ** $p < 0.01$, *** $p < 0.001$.

model as a single, two-level hierarchical model using structural equation modeling (SEM) software, *Mplus* (Geiser, 2013; Muthén & Muthén, 1998–2012).

In the Stage 1 OLS analyses, regression models were estimated for each sample member as described above. The mean intercept obtained from these regressions was 29.16, with a variance of 53.7. The mean slope was 0.195, with a variance of 1.15. The correlation between intercepts and slopes was −0.28.

In the Stage 2 OLS analyses (i.e., a *separate*, second set of models), these intercepts and slopes were regressed on covariates in two regression models (one each). In the first model, the intercept for the intercept (b_{00}) – that is, the intercept for the baseline BMI – was 34.86. Males have higher BMIs at baseline than females ($\gamma_1 = 0.72$; $p > 0.05$), blacks have higher baseline BMIs than whites ($\gamma_2 = 2.28$; $p < 0.05$), and persons born in the south have lower baseline BMIs than those born elsewhere ($\gamma_3 = -1.16$; $p > 0.05$). Finally, those with greater schooling have lower BMIs at baseline than those with less, with each year of schooling reducing baseline BMI by $\gamma_4 = 0.46$ units. Race and education were the only significant predictors of baseline BMI. In short, if we wish to predict an individual's BMI, we would use Eq. (2.5a) tailored to this set of covariates. All in all, these covariates explain 4.6% of the variance in the collection of BMI intercepts, indicating that the residual variance, var(u_i), is 95.6% of the total variance of 53.7.

In the second Stage 2 model, the intercept for the slope (b_{10}) – that is, the value of the slope for those with all covariate values set to 0 – was 0.195, indicating that the average sample member saw an increase in BMI of 0.195 units per study wave. No covariates had significant effects on the BMI slope, but the coefficients (δ) for males, blacks, and persons born in the south were negative, while the coefficient for education was positive. The *R*-square for the model predicting BMI slopes was small at 1%. In short, growth in BMI was not predicted well by this set of covariates.

The bottom half of the table replicates the top half but in a growth modeling framework. Model 1 was an "unconditional" model, meaning that covariates were not included; this model corresponds to the Stage 1 model in the top half of the table. The estimated mean intercept and slope were almost identical to those obtained via the OLS approach. The variances, however, were substantially smaller (e.g., 50.5 for the SEM-based approach versus 53.7 for OLS-based approach for the intercept parameter). The reason for this difference is that the OLS modeling approach fits the individual trajectories better, because it does not assume any distribution for the collection of intercepts and slopes together. That is, since the intercepts and slopes are estimated as *n* separate regression models, there is no assumption regarding the combined distribution of intercepts and slopes. In contrast, because the growth modeling approach simultaneously estimates all intercepts and slopes under the assumption that the distribution of them is normal, there are larger, time-specific individual deviations of the observed measures from the individual trajectories. Thus, there is greater measurement error at level 1 and less variance in the random effects.

The bottom quarter of the table shows the results of the growth modeling approach for the conditional growth model, that is, the model in which covariate influences on the growth parameters are simultaneously estimated with the variance in those parameters themselves. These results are very similar to those obtained via the second-stage OLS regression model. However, the coefficients and standard errors differ slightly, with the estimates obtained via growth being better in a statistical sense, because simultaneous estimation of the level 1 and 2 equations is more efficient, and the standard errors do not suffer from heteroscedasticity and autocorrelation implicit under the OLS approach.

LATENT CLASS MODELING IN A NUTSHELL

As shown in the previous section, growth modeling assumes a particular parametric shape for all individuals' trajectories over time. In the example above, while each individual had his/her own unique intercept and slope, all trajectories were assumed to be linear. Deviations of time-specific values of BMI for individuals are assumed to be either measurement error in BMI, as captured by e_{it}, or fluctuations from the trajectory because of time-specific "shocks" that "bump" an individual off his/her linear trajectory. Equation (2.2) can be modified to include such shocks, which account for some of the individual time-specific error represented by e_{it}:

$$y_{it} = b_{0i} + b_{1i}t_{it} + (Z_{it}\phi + e_{it}) \quad (2.6)$$

In Eq. (2.6), Z_{it} is a time-specific variable (or vector) that has an effect (ϕ) on y_{it} (see Bollen & Curran, 2006, p. 192). As an example of such a time-specific "shock," consider that an individual could have surgery that results in significant weight loss that is reflected in a single survey wave but which does not alter his/her fundamental, long-term BMI trajectory. That is, once the individual recovers, s/he regains the lost weight and continues on his/her trajectory established by prior and subsequent BMI measures.

A traditional latent class model applied to repeated measures does not assume a parametric (e.g., polynomial) trajectory shape (Collins & Lanza, 2010). Instead of assuming that individual deviations from a parametric trajectory are attributable to shocks or measurement error, a latent class approach assumes that such deviations may be meaningful, at least insofar as enough sample members experience similar such deviations that, together, they may constitute a separate "class" of individuals.

Example 2.2

Before advancing to more sophisticated latent class methods, consider the BMI data for the first study wave. Figure 2.4A shows a histogram of the data (solid line). We could assume that BMI follows a normal distribution; the figure shows the histogram for the best-fitting normal distribution superimposed over the observed data (dashed line). As the figure shows, the fit is not very good.

A better fit might be obtained by assuming that the observed BMI distribution has arisen from two types of individuals whose BMIs come from two different normal distributions. Perhaps some people have a propensity for heaviness and some have a propensity for normal weight. Thus, the observed distribution of BMI is a "mixture" of a normal distribution with a smaller mean and one with a greater mean, with both distributions also possibly having different variances. Figure 2.4B shows the best-fitting set of two normal distributions. The mean of the heavier BMI distribution was 38.86, and its variance was 97.85. The mean of the lighter BMI distribution was 26.93, and its variance was 18.51. The two distributions do not initially appear to fit the data particularly well, but this is only because the distributions have not been adjusted for the relative proportions of individuals in the population who come from each group. In fact, an estimated 82% of the population belongs to the lighter distribution, while 18% belongs to the heavier distribution.

Figure 2.4C shows the combined mixture distribution; that is, the single distribution implied by the two normal distributions shown in Figure 2.4B when the component distributions are scaled for their relative proportions of members in the population. Figure 2.4D shows the results of a model that assumes there are three BMI classes in the population. As the figure suggests, and various measures of fit (not

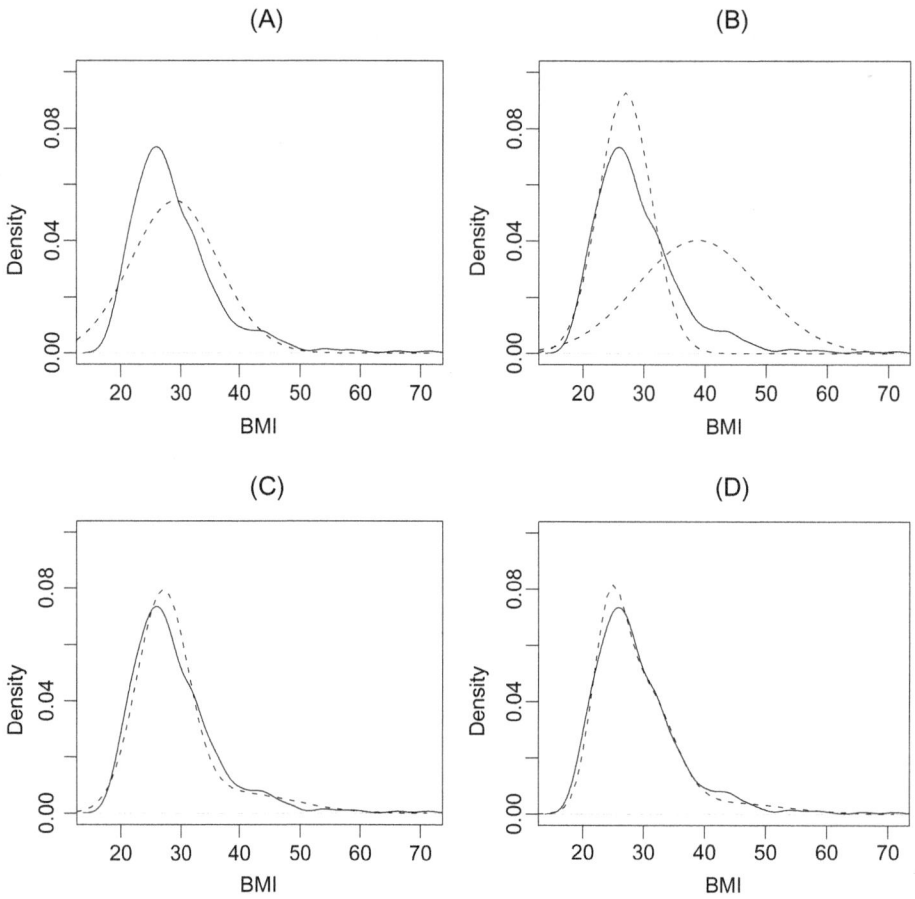

FIGURE 2.4 Histograms of observed wave 1 BMI (solid lines) with additional histograms superimposed (dashed lines). Panel A shows the best-fitting normal distribution, based on the mean and variance of BMI. Panel B shows the best-fitting set of two normal distributions. Panel C shows the best-fitting two-component mixture distribution based on the distributions in panel B. Panel D shows the best-fitting three-component mixture distribution.

shown) indicate, the fit is not substantially better than the two-class model. This mixing of multiple distributions is the key concept underlying latent class models.

Latent class models exploit the law of total probability, such that the probability for an individual's value on a variable of interest is conditional on the latent class to which s/he belongs. Thus, the generic likelihood function for a latent class model is:

$$L(Y) = \prod_{i=1}^{n}\left(\sum_{k=1}^{K} f(y_i | c_k) f(c_k)\right) \quad (2.7)$$

In Eq. (2.7), Y is the complete vector of observed responses $(y_1, \ldots, y_n)$, and the likelihood function is simply the product over individuals, as usual. Each individual's contribution to the likelihood is the sum in parentheses: it is the probability density function for y_i conditional on the membership in each class, c_k, $f(y_i|c_k)$, multiplied by the probability of class membership, $f(c_k)$. This probability of class membership is what differentiates Figure 2.4B from 2.4C: it represents the proportion of individuals in the population that belong in each class.

The conditional density, $f(y_i|c_k)$, may be continuous, as in the example above, or discrete, as we will discuss. The density $f(c_k)$ is generally discrete in latent class modeling, meaning that the number of classes, $k = 1, ..., K$, is distinct and finite. In statistics, this type of model is called a "finite mixture" model, with $f(c_k)$ being the "mixing" distribution, and $f(y_i|c_k)$ being the "mixture component" distributions (Land, 2001). The parameters for the component distributions are unique within a class. In other words, what distinguishes the classes are the values of the parameters – like the mean and variance in the example above – in $f(y_i|c_k)$. Thus, $f(y_i|c_k)$ is often generically denoted: $f(y_i|c_k, \theta_k)$, where θ_k is the unique parameter vector associated with class c_k.

Membership of individuals in each class is generally unknown, but probabilities of an individual's (i) membership in each class (c_k) can be computed once the parameters of the mixture component distributions and the overall sample proportions in each class have been estimated, by using Bayes' Rule (see Lynch, 2007):

$$p(i \in c_k) \equiv p(c_k | y_i) = \frac{f(y_i|c_k)f(c_k)}{\sum_{k=1}^{K} f(y_i|c_k)f(c_k)} \quad (2.8)$$

These probabilities are commonly referred to as "posterior probabilities of class membership" and can be used to assign an individual to a class deterministically by simply assigning an individual to a class based on the class for which s/he has the highest posterior probability of being a member. This assignment process embodies one of the two key assumptions of latent class models: that there is no variation of observations within a latent class. In other words, although individuals have varying probabilities of being in their most likely class, the characteristics of each class are considered identical across the individuals within the class. We discuss both the deterministic assignment to class and variability in y within classes subsequently.

Once class membership has been established, researchers usually engage in a second-stage analysis in which a multinomial logit model is estimated to determine whether covariates predict class membership. In the growth modeling example from the previous section (a one-class model), we found that males, whites, those from the south, and those with greater education had lower estimated baseline values of BMI. Here we found that there were two latent classes for wave 1 BMI, with 82% of the sample in the lighter class and 18% in the heavier class. After assigning class membership deterministically as described above, we estimated a logistic regression model with these covariates predicting membership in the two latent classes. The results (not shown in a table) were similar to those obtained via the growth model: men, southerners, and those with greater education were less likely to be in the heavier class (OR = 0.80, $p > 0.05$; OR = 0.48, $p < 0.1$; OR = 0.69, $p < 0.1$, respectively), and blacks were more likely to be in the heavier class (OR = 1.96, $p < 0.1$).

Extending the latent class model to handle more than one outcome variable is straightforward, involving simply expanding the likelihood function shown in Eq. (2.7) by incorporating additional product terms:

$$L(Y) = \prod_{i=1}^{n} \prod_{j=1}^{J} \left(\sum_{k=1}^{K} f(y_{ij}|c_k) f(c_k) \right) \quad (2.9)$$

With this extension, there are still K latent classes, but now each of the n sample members is measured on J variables, with y_{ij} being the ith person's response on the jth variable.

The second key assumption that underlies latent class analysis is apparent from this set of products: individual responses to items are considered independent, conditional on latent class membership. In other words, once an individual's class membership is established, his response to variable a (i.e., y_{ia}) is unrelated to his response to variable b (i.e., y_{ib}). This is called the "conditional independence assumption" and can be relaxed but generally is not (Vermunt & Magidson, 2002).

The collection of J variables could be multiple measures of a single theoretical construct, such as happiness. In that case, latent class analysis can be viewed as an *alternative to factor analysis* that clusters individuals with similar patterns of response, rather than clustering variables based on their intercorrelations. Thus, latent class is akin to K-means clustering but has a stronger statistical justification underlying it, given that its foundation is based on probability theory (Magidson & Vermunt, 2002).

The collection of J variables could, alternatively, be repeated measures of a single item, like BMI. In that case, the latent classes that emerge from the analysis will represent the common patterns observed in the variable over time: trajectories. Unlike linear or other polynomial growth models, which assume a common average trajectory shape for all individuals, latent class models of repeated measures allow for very different, non-smooth shapes across classes. Thus, latent class modeling is sometimes referred to as a "nonparametric" method. Furthermore, the data may be continuous or fundamentally dichotomous or categorical, unlike in the growth model, which assumes either (1) that the observed data are continuous or (2) that the observed categorical/dichotomous data are simply limited measures of a continuous latent variable (Muthén & Asparouhov, 2006).

Example 2.3

To illustrate this repeated measure latent class model, suppose our BMI data were dichotomized at each wave so that individuals were observed to be obese or not. In that case, there would be $2^4 = 16$ possible trajectories of BMI, ranging from stable-obese to stable nonobese. We may hypothesize that these stable trajectories are the only two that exist in the population, and we may estimate a series of latent class models in order to evaluate that hypothesis.

We estimated three latent class (LC) models – with two, three, and four latent classes – and used the Bayesian Information Criterion (BIC) to determine the best-fitting model. Although all of the analyses discussed here rely on multiple test statistics to determine overall and relative model fit (Geiser, 2013), the BIC is the most commonly used measure to compare LC models, with the smallest BIC indicating the "best" model (Nylund, Asparouhov, & Muthén, 2007). Here a three-class model was found to fit the data best. Given the dichotomous nature of the data in this example, the key model parameters are thresholds on a latent logistic distribution that assign probabilities of obesity to each wave of measurement for those belonging to the class. Table 2.2 presents the results of the analyses and clarifies these ideas.

The upper half of the table presents the probabilities that an individual in a given class is obese at each wave of the study. Class 1 is characterized by having members with low probabilities of obesity at each wave: the probabilities that a member is obese at each wave are 0.018, 0.009, 0, and 0.037, respectively. We might therefore call this class a stable nonobese class. In contrast, for members of class 3, the probabilities exceed 0.98 that they are obese at every wave. Thus, we might call this class a stable-obese class. Members of class 2 have a modest probability of being obese at wave 1 (0.321), but relatively high probabilities of being obese at subsequent waves (>0.6). We might be inclined

TABLE 2.2 Results of Latent Class Model of Dichotomized BMI at Four Time Points

Class	p (o1)	p (o2)	p (o3)	p (o4)	Percentage in class
1	0.018	0.009	0.000	0.037	57
2	0.321	0.605	0.655	0.642	13
3	0.990	0.986	1.000	0.982	30

n	Sequence	$p(c_1)$	$p(c_2)$	$p(c_3)$	Assigned class
"Stable nonobese" (n = 204)					
190	0000	0.992	0.008	0.000	1
10	0001	0.727	0.273	0.000	1
4	1000	0.824	0.176	0.000	1
"Variable weight" (n = 44)					
2	0010	0.000	1.000	0.000	2
7	0011	0.000	0.997	0.003	2
4	0100	0.422	0.578	0.000	2
3	0101	0.016	0.984	0.000	2
5	0110	0.000	0.996	0.004	2
9	0111	0.000	0.883	0.117	2
2	1001	0.092	0.908	0.000	2
4	1011	0.000	0.625	0.375	2
3	1100	0.027	0.973	0.000	2
1	1101	0.001	0.999	0.000	2
4	1110	0.000	0.535	0.465	2
"Stable-obese" (n = 105)					
105	1111	0.000	0.036	0.964	3

Note: In top half of table, p(o*t*) is the probability of obesity at wave *t*. In bottom half of table, $p(c_k)$ is the probability an individual is in class *k*. Binary sequences indicate obesity at each wave; thus, 0000 is nonobese at all four waves.

to call this class a weight-gaining class, but the bottom half of the table shows that members of this class have considerable variability in their obesity patterns over time. The far-right column of the upper half of the table shows the proportion of the population in each class. Fifty-seven percent are in class 1, 13% are in class 2, and 30% are in class 3. It is important to note that these proportions are for the population and not the sample: the assignment of sample members to classes based on posterior probabilities may yield sample proportions that differ from the estimated population proportions.

The bottom half of the table shows the breakdown of the sample by obesity pattern and class assignment. The first column of the table shows the number of sample members with the obesity pattern shown in the second column. The obesity pattern is represented as a binary sequence of four digits (one for each wave) in which a "1" represents obese and "0" represents nonobese, and the position of the digit indicates the wave. As stated earlier, there are 16 possible patterns, given the four waves of measurement. Subsequent columns show the posterior probabilities a sample member with the given obesity pattern is in each class, and the last column shows the assigned class based on these posterior probabilities. As the table shows, 15 of the 16 possible patterns exist in the data. There are three patterns that are associated with class 1, including those who are nonobese at all four waves and those who are obese only at either wave 1 or wave 4. Overall 204 sample members were assigned to the stable nonobese class; just under 58% of the sample.

The last row of the table shows that there is only one pattern associated with the stable-obese class: obesity at all four waves of the study. The middle rows of the table show a number of patterns associated with class 2. What each of the patterns with the highest posterior probabilities for class 2 have in common is that almost all individuals in this class were obese for two study waves. However, the timing of obesity varies considerably. We might therefore call this pattern a variable weight class rather than a weight-gaining one, as mentioned above.

Example 2.4

We replicated the analyses using the original, continuous version of BMI for all four waves. This example is similar to that for the initial latent class example: the goal is to determine how many latent classes (i.e., normal mixture components with unique means and variances) are needed to explain patterns in BMI across the four waves. We estimated a series of models with numbers of latent classes ranging from 1 to 8. Although BIC statistic declined across all models, there was very little improvement beyond five classes. Furthermore, increasing the number of classes beyond five reduced some classes to trivial proportions of the sample, suggesting that they may simply be outliers and not representative of true classes in the population.

Figure 2.5 shows the patterns in mean BMI for the five-class model. The circles represent the estimated means for each class at each time point. The vertical dashed lines are intervals constructed around the estimated means based on the estimated variance in BMI for each class at each time point. The intervals are 68% intervals; they are constructed by adding/subtracting one standard deviation from the mean at the given wave. They can be interpreted such that 68% of all population members belonging in the class should have BMIs within the interval for the given time period. We use 68% intervals in this and subsequent figures for illustration purposes: the distinctions between latent classes are clearer with narrower intervals.

The figure shows that all five classes have relatively linear and flat trajectories across time.

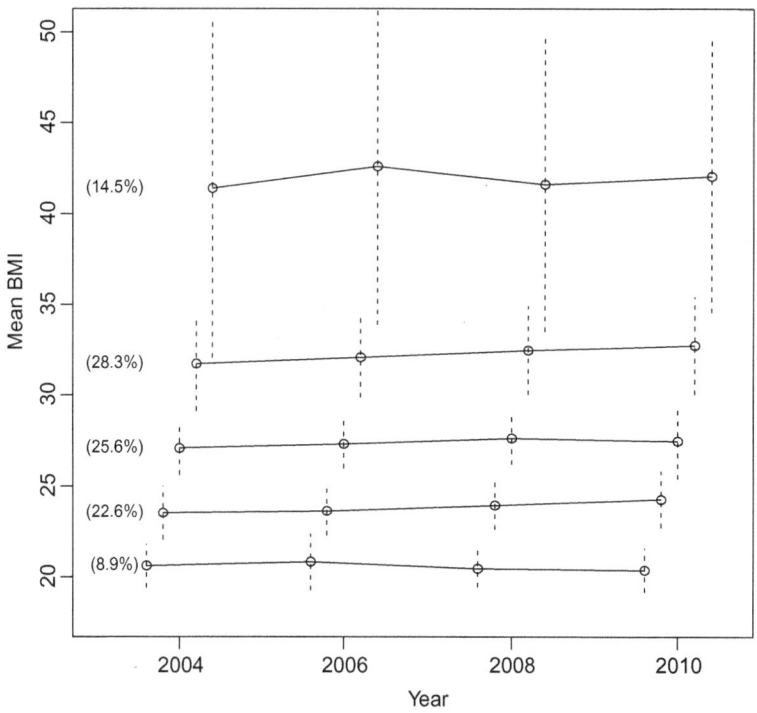

FIGURE 2.5 Nonparametric latent class model results. Circles represent model-predicted means for each class at each time point. Horizontal lines between points represent latent class patterns. Vertical dashed lines are 68% interval estimates for values of y within each class at each wave. Proportions on the left are the proportion of the population in each BMI class.

The classes also appear to follow the standard categorizations for BMI closely. Class 1 has a BMI of about 20, bordering on underweight. Class 2 has a BMI of about 23, centered within the normal weight category. Class 3 has a BMI of about 27, centered within the overweight category. Class 4 has a BMI of about 32, just above the level for obesity. Finally, class 5 has a BMI of about 42, a level that is sometimes called morbidly obese.

The figure also shows the proportion of the population in each class. Just under 10% are in the near-underweight class. Roughly one-quarter of the population are in each of the normal weight, overweight, and obese classes. Finally, about 15% are in the morbidly obese class.

We estimated a multinomial logit model with sex, race, birth region, and education predicting class membership after classes were assigned deterministically. Those results are shown in the top half of Table 2.3. As the results indicate, there are very few significant influences of covariates on class membership, although the signs of the coefficients are consistent. Men are more likely than women to be in classes 2 and 3 (versus 1). Blacks are more likely than whites to be in the heaviest two classes. Finally, those born in the south and those with more education are less likely to be in the heaviest class (versus the lightest class).

We conclude this section by noting that latent class analysis with repeated, continuous measures is incredibly flexible, but simultaneously difficult to implement successfully, in part because of its flexibility. For example, in the analyses just presented, $f(y_{ij}|c_k)$ from Eq. (2.9) is assumed to be a univariate normal distribution for each variable y_j and each class c_k. Each distribution may have its own mean and variance, and although the equation specifies that the normal distributions are independent via the second product term (over J), the J variables could be assumed to come from a multivariate normal distribution with a non-diagonal covariance matrix. In other words, we can essentially relax the conditional independence assumption. However, the number of parameters can get large rather quickly, making estimation difficult, especially given the lack of restriction on the parametric shape of trajectories over time. Imposing a parametric shape reduces the number of parameters to be estimated and can make estimation easier.

TABLE 2.3 Results of Multinomial Logistic Regression Model Predicting Class Membership in 5 Class Nonparametric Latent Class Model, and Results of Multinomial Logistic Regression Model Predicting Class Membership in 3 Class GMM

Nonparametric latent class model (class 1 is the reference)

Variable	Class 2	Class 3	Class 4	Class 5
Intercept	1.01 (1.23)	2.02 (1.19)#	2.21 (1.19)#	2.95 (1.25)*
Male	0.75 (0.50)	1.45 (0.49)**	1.57 (0.49)***	0.73 (0.54)
Black	1.07 (0.70)	0.81 (0.71)	1.53 (0.69)*	1.60 (0.73)*
South	−0.36 (0.46)	−0.63 (0.46)	−0.60 (0.46)	−1.10 (0.53)*
Education	−0.02 (0.09)	−0.10 (0.09)	−0.12 (0.08)	−0.20 (0.09)*

Psuedo $R^2 = 0.04$

Growth mixture model (class 1 is the reference)

Variable	Class 2	Class 3
Intercept	0.73 (0.95)	−2.85 (1.38)*
Male	1.23 (0.52)*	0.63 (0.68)
Black	−0.61 (0.47)	−1.32 (0.63)*
South	0.23 (0.53)	0.32 (0.53)
Education	0.05 (00.06)	0.28 (0.09)**

Note: For the nonparametric latent class model, the higher class number indicates heavier members. For the growth mixture model, class 1 is the heaviest, and class 3 is the lightest.
# $p < 0.1$, * $p < 0.05$, ** $p < 0.01$, *** $p < 0.001$.

LATENT CLASS GROWTH ANALYSIS

It may sometimes be reasonable to impose a parametric shape for trajectories, as we did in the growth modeling example, but we may also believe that the population is partitioned into relatively homogeneous subgroups that are not represented by single trajectory with extensive heterogeneity around its mean and variance. Consider the previous analysis: all of the BMI classes had trajectories that were very linear, and the trajectories were fairly well separated. Thus, it may be reasonable to assume each class has linear, but distinct, trajectories over time. This approach is the basis of Latent Class Growth Analysis (LCGA; Muthén, 2004).

The basic LCGA model assumes that the population consists of homogeneous subgroups that are characterized by different parametric trajectories. Trajectories are assumed to follow a polynomial shape (e.g., linear, quadratic), and the parameters of the random effects representing these trajectories are assumed to cluster. In other words, whereas the latent class analysis described above clusters individuals based on their values on the variables, y_{ij}, LCGA clusters individual growth model parameters (b_{0i} and b_{1i}) that link variables across time. For example, recall the scatterplot in Figure 2.2 showing the collection of intercepts and slopes obtained from separate OLS regressions for each sample member's values of BMI over time. The vertical and horizontal reference lines at the means for the intercepts and slopes, respectively, effectively divide the distribution of intercepts and slopes into four quadrants, and therefore, into four classes of individuals: those with large intercepts and steep slopes, those with large intercepts and shallow slopes, those with small intercepts and steep slopes, and those with small intercepts and shallow slopes. Rather than dividing the intercepts and slopes based on the means for each, which is an arbitrary division and limits the number of classes to four, LCGA determines the best values for the intercept and slope for each class, based on the best-fitting mixture distribution as described above. The key assumption of LCGA is that there is no variance in intercepts and slopes within a class (Kreuter & Muthén, 2007). Thus, a one-class LCGA model would find that the one class has a mean trajectory with parameters equal to the estimated mean of b_{0i} and mean of b_{1i}. In terms of Figure 2.2, this is the point at which the vertical and horizontal reference lines intersect. Individual, time-specific deviations from this single trajectory are considered time-specific individual error, rather than an indication that individuals within the class have differing trajectories as implied by the growth model. Equation (2.2) can therefore be modified as:

$$y_{itk} = b_{0k} + b_{1k} t_{itk} + e_{itk} \quad (2.10)$$

where y_{itk} is the value of the outcome for person i at time t who belongs in class k. b_{0i} has been replaced by b_{0k}, and b_{1i} has been replaced by b_{1k}, indicating that intercepts and slopes are specific to *classes* and not individuals. This equation can be substituted into the likelihood function in Eq. (2.9). Since classes are qualitatively distinct, Eqs. (2.5a) and (2.5b) no longer hold: there is no second-stage linear model predicting class-specific intercepts and slopes. Instead, individuals are assigned (deterministically or otherwise) to classes based on their posterior probabilities of class membership, and a second-stage multinomial logit model is generally used, as in the previous example, to predict class membership using covariates.

Example 2.5

We estimated a series of LCGA models with 1–8 classes. As in the previous analyses, we found that five classes provided the best fit to the data. Adding classes beyond five continued to reduce the BIC statistic, but the returns

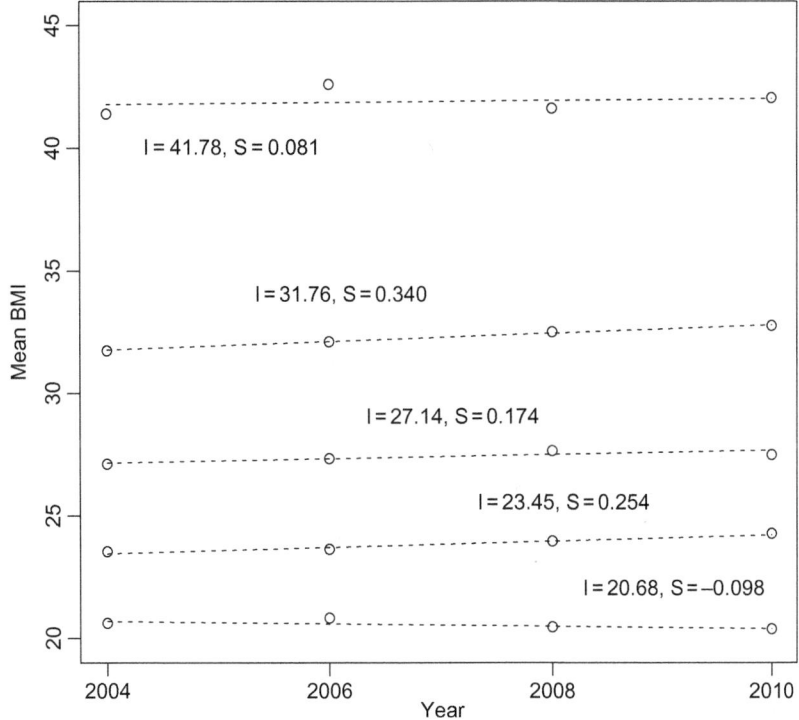

FIGURE 2.6 LCGA results (dashed lines) superimposed over nonparametric latent class estimates of means of BMI at each wave (circles). Figure shows strong linearity in means within classes over time, and so the more restrictive LCGA model fits nearly as well as the less restrictive nonparametric latent class model.

were minimal, and the proportion of the population allocated to each additional class was very small. Figure 2.6 shows the results of the LCGA with five classes. The figure looks remarkably like Figure 2.5, because the nonparametric patterns estimated in the previous latent class model were very linear. Thus the LCGA model, which assumed linear trajectories in our specification, essentially reproduced the nonparametric results. In Figure 2.6, the circles are the estimated means from the previous model, while the dashed lines represent the implied linear trajectories obtained from the LCGA model. The intercepts and slopes are reported in the figure next to each estimated trajectory. The fit is nearly identical under the two models, and the latent class proportions are identical (hence, they are not shown in this figure). Consequently, there is no need to estimate a second-stage multinomial logit model with covariates: the results would be the same as those reported in Table 2.3.

GROWTH MIXTURE MODELING

The LCGA assumption that all individual, time-specific measures' deviations from the individual's class trajectory are error, rather than an indication of within-class variability in intercepts and slopes, is a difficult one to swallow. Growth mixture modeling (GMM) relaxes this assumption and is therefore a generalization of both the growth model and the LCGA (Jung & Wickrama, 2008; Muthén, 2004). The GMM generalizes the growth model by recognizing that the distribution of individual-specific intercepts and slopes may arise not from a single population but from multiple subpopulations. It therefore assumes a mixture distribution for the collection of intercepts and slopes, generalizing the growth model to mixture distributions as LCGA does. The GMM generalizes LCGA by allowing variation of individual trajectories within classes, as growth modeling does for its single class.

Example 2.6

In our final example, we estimated GMMs using our data. Figure 2.7 shows the logical progression to the GMM. Panel A of the figure shows our original OLS-estimated, individual-level intercepts and slopes as first displayed in Figure 2.2, but with the x-axis span reduced to the range of 20–45. Instead of the original unfilled circles showing each person's intercept/slope pair, panel A displays numbers indicating the latent class for each person as determined by the nonparametric latent class model from Example 2.4. The alternating black and gray coloring is displayed simply to aid in visualization of the class separation.

Panel B replicates panel A but shows large filled-in circles for the intercept and slope values from the LCGA model in Example 2.5. These circles appear to be centered over the scatters for each latent class; however, there is considerable variability in intercepts and slopes around each latent class mean. Panel C provides a heuristic illustration of the generalization of the LCGA to the GMM. The ellipses shown in the panel are 68% intervals around the latent class means. The ellipses are based on the within-class covariance matrix for the OLS-estimated intercepts and slopes (from Example 2.1) and provide some indication of the extent of variability in the intercepts and slopes around the means for each latent class. In general, it appears that the data are well represented by these ellipses.

This variability is precisely what GMMs attempt to capture. Specifically, the GMM generalizes Eq. (2.9) to:

$$y_{itk} = b_{0ik} + b_{1ik} t_{itk} + e_{itk} \quad (2.11)$$

Note that the only change in this equation is that the growth parameters b_0 and b_1 have an additional subscript, i, reflecting that, within classes, intercepts, and slopes are allowed to vary across members of the class. Because individuals are allowed to have unique intercepts and slopes, it is reasonable to extend Eqs. (2.5a) and (2.5b) as:

$$b_{0ik} = b_{00k} + X_i \gamma_k + u_i \quad (2.12a)$$

$$b_{1ik} = b_{10k} + X_i \delta_k + v_i \quad (2.12b)$$

These equations show that each individual's unique intercepts and slopes are a function of a class-specific mean intercept and slopes and contributions of covariates, which themselves are allowed to have effects that vary by class (γ_k and δ_k). As with LCGA, class membership can be estimated as a function of covariates via multinomial logistic regression, and so, in the GMM, *both* class membership *and* individual deviations from class mean intercepts and slopes can be influenced by covariates.

Although there is ongoing debate in the literature regarding whether the multinomial logistic regression predicting class membership as a function of covariates should be performed as a separate, second-stage analysis as we did in the LCGA, current research suggests there are advantages to estimating the full model jointly in a single stage (Muthén, 2004). We take that approach here, but discuss this issue in greater depth in the next section. We note at the outset that we estimated models using both the

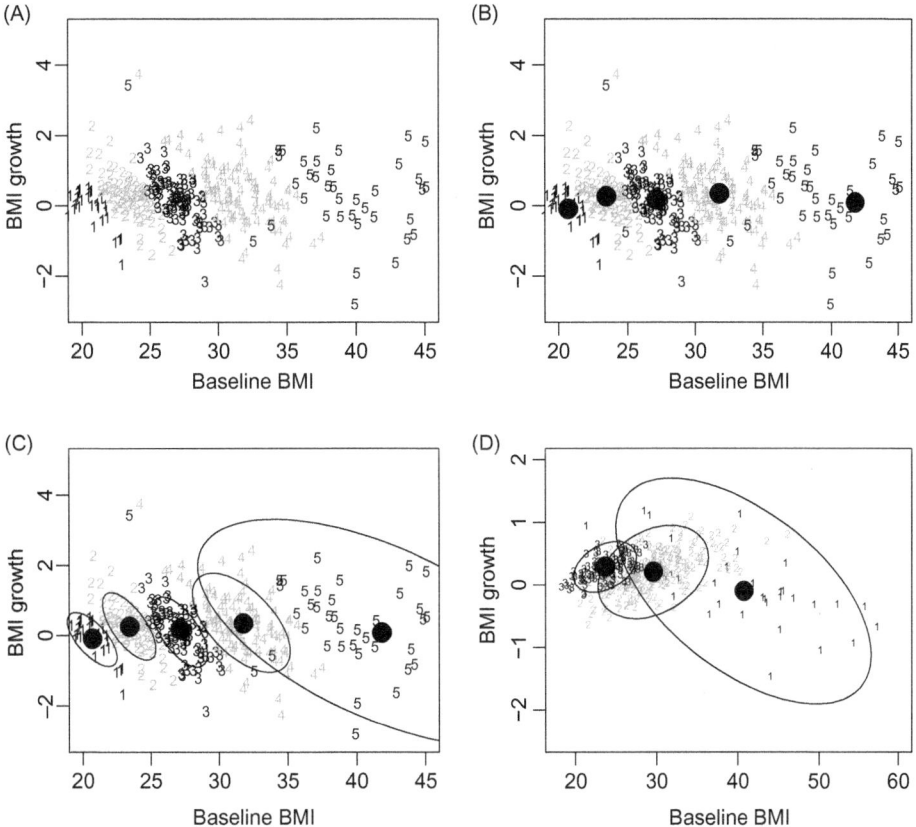

FIGURE 2.7 Results of various latent class modeling strategies. Panel A shows OLS-estimated intercept and slope pairs from Example 2.1, with numbers indicating assignment of individual to latent classes from the nonparametric latent class model in Example 2.4 (coloration is for ease of viewing). Panel B superimposes mean intercepts and slopes for the LCGA with five classes (dark circles). Panel C is heuristic: it shows that the OLS-estimated intercepts and slopes vary considerably around the LCGA-estimated mean intercepts and slopes. The ellipses show that 68% intervals around the means capture much of the within-class variability in intercepts and slopes, motivating the key contribution of the GMM. Panel D shows the actual results of the best-fitting GMM, which is a three-class (rather than five-class) model. OLS intercepts and slopes have been replaced by GMM-estimated intercepts and slopes with much less between-person variance.

original BMI measures and logged-BMI, given that BMI is decidedly right-skewed, and some research indicates that skewness can be problematic in extracting GMMs (Bauer & Curran, 2003). The results using the original BMI measures and logged version did not differ, and so, to keep consistent with the previous examples, we report the results obtained using the original BMI metric.

We estimated GMMs for 1–5 classes, with the covariates allowed to influence both the intercepts and slopes within classes and class membership. We were unable to get any models to converge to reasonable estimates if at all. Thus, in a second round of modeling, we restricted the covariates to influence only class membership, consistent with the approach used in the previous examples (but, again, with covariate influences estimated simultaneously with the extraction of classes). Neither the five-class nor the four-class models converged to reasonable estimates, and the BIC statistics for them were much higher than for two- and three-class models. BIC values indicated a three-class model fit best.

Panel D shows the results of the three-class model. The model-predicted intercepts and slopes for each person are plotted as either "1," "2," or "3" based on class membership. The large, filled-in circles show the mean intercept and slope for each of the latent classes. In class 1, the mean intercept was 40.76, and the mean slope was −0.10 and contains 10% of the population. In latent class 2 (61% of the population), the mean intercept was 29.56, and the mean slope was 0.20. Finally, latent class 3 constituted 29% of the population, with a mean intercept of 23.61 and a mean slope of 0.29. Thus, the first class consists of morbidly obese persons who tend to lose weight over the survey period, while the latter two classes consist of persons who gain weight. The second class consists of persons who are almost obese at baseline and edge into the obese category, and the third class consists of persons who are normal weight at baseline but tend to become overweight and then obese over the survey period.

It is important to notice that the collection of intercepts and slopes in this panel is different from the collection of intercepts and slopes shown in the previous panels. This difference is the key reason that the GMM found three classes, while the other methods found five. The OLS-based approach described in Example 2.1 involves estimating a separate OLS model for each individual. This approach imposes no constraints on the distribution of intercepts and slopes for the entire sample; as a consequence, the OLS models can fit the data very well, and the variance of the intercepts and slopes is maximized, making the scatterplot of them much more spread out. In fact, the R^2s for the observed measures are around 96% for the OLS-based approach.

The growth modeling approach, in contrast, assumes some variance in y is due to measurement error or noise in the measures, and some is due to variance in the level-2 parameters – the random effects. Because it imposes a distributional assumption on the random effects (usually bivariate normal), and a normal distribution on the first-level errors, the growth model does not fit as well as the OLS models, and the random effects are consequently less spread out. Indeed, the R^2s for y from the growth model are in the 0.92 range. In a GMM, both the lack of spread in the level-2 random effects, and the fact that within-class variation in intercepts and slopes is allowed, may lead to the estimation of fewer classes than in the LCGA in which all observed variance in individual trajectories is assumed to be between classes (with deviations of individual measures from the observed trajectory within a class assumed to be noise).

In this example, a three-class model makes sense via an admittedly *post hoc* interpretation. The results suggest there are three types of people in the population: those who struggle with serious weight issues, those who are heavy and become heavier as they age, and those who are normal weight and become heavier as they age. Those who struggle with weight issues tend to gain and/or lose tremendous amounts of weight potentially because of severely restrictive and unsustainable, intermittent diets. Thus, the negative relationship between intercepts and slopes arises via negative feedback: when an individual's weight has become too great, s/he diets; when the weight is lower, the diet ends. For the other two classes, aging is associated with steady weight gain from middle adulthood into young-old age.

The (simultaneously estimated) second-stage logistic regression predicted class membership as a function of the covariates. Those results are shown in the bottom half of Table 2.3. We found that men were more likely to be in the overweight class (2) than the morbidly obese class (1), but no other covariates influenced membership in class 2 versus class 1. We found that blacks were less likely to be in the normal weight class (3) than the obese class, and that those with more education were more likely than those with less to be in the normal weight class than the obese class. These results are somewhat different from those obtained in the previous examples, and some explanations can be found in the next section.

IMPORTANT ISSUES IN THE IMPLEMENTATION OF TRAJECTORY METHODS

There are a number of issues to consider when engaging in trajectory modeling, some theoretical or substantive and some statistical. In this section, we discuss several key issues, including: (1) the data structure and method for estimating trajectory models; (2) the measurement of time and the differentiation of age, cohort, and period patterns; (3) the importance of distributional assumptions; and (4) the method for determining the number of latent classes that exist and the inclusion of covariates as predictors of membership in them.

Data Structure and Method

Trajectory models may be estimated via a variety of both general and specialized software packages, and packages require one of two data structures for implementation: a "long" format or a "wide" format. In the long format, the data follow a person-time structure, so that each person contributes one record for each time period observed. Thus, if we have n persons, all of whom are measured at T time points, the outcome would be represented as an nT-by-1 column vector. Time-varying covariates are included in the data structure simply by changing the value of x_{it} in the appropriate row (time point, t). This data structure arises from the hierarchical and mixed-modeling tradition, in which the outcome is univariate (Raudenbush & Bryk, 2002).

The wide format treats each time-specific measure as a separate variable. Thus, the outcome constitutes an n-by-T matrix. Time-varying covariates are included by adding more columns to the data set. This data structure arises from the multivariate modeling tradition most often associated with structural equation modeling (Bollen & Curran, 2006).

Each data structure – which corresponds to a specific modeling strategy – has its relative (dis)advantages, and the two produce identical results in many circumstances. An advantage of approaches involving the long format is that attrition is simpler to handle *from a mechanical perspective*. Individuals who are missing for a given study wave simply contribute fewer rows to the data set. In contrast, in the wide format, a person missing on a wave is missing on a variable; consequently, missing data methods are required to handle the missingness. It is important to note that both of these strategies rely on the same fundamental assumption about the missing information: it is missing at random (MAR; Little & Rubin, 2002). Therefore, *neither is inherently better at handling missingness*, despite the fact that we have seen occasional claims to the contrary. The MAR assumption is simply explicit in the wide format but implicit in the long format.

Because of the multivariate nature of the modeling procedures, methods utilizing the wide format are more flexible than methods involving the long format. For example, long-format methods are limited to handling a trajectory for only a single variable, whereas wide-format methods can simultaneously model trajectories of multiple variables, as well as multiple processes such as mortality (e.g., Taylor & Lynch, 2004, 2011). Wide-format methods also allow greater flexibility in variance specification. Consider Eq. (2.2) and the text following it. The long-format specification for the error variance of e_{it} is that it is normal with a variance of σ_e^2. Because the wide-format treats time-varying variables as separate variables, error variances can vary over time, rather than assumed constant; thus, σ_e^2 may become $\sigma_{e(t)}^2$. Finally, covariate effects are generally assumed to be constant over time in methods involving the long format. While the effects can be made to vary over time by creating interactions between time and X, such interactions imply a monotonic pattern in the change of effects across time; no such requirement exists for wide-format methods.

Such flexibility, however, comes at a price: namely, there are often numerous parameters that are either freely estimated or constrained by default. If a researcher is ignorant of the defaults, s/he may estimate a model that is, in fact, contradictory to his/her theory.

Measurement of Time

The measurement of time in trajectory models is often a consequence of the data structure used in estimation, and the differentiation of maturation and cohort patterns tends to depend on the measurement of time. Because these models require panel data, it seems that differentiating these two types of temporal patterns – and even including period effects via "shocks" in the error term (see Yang & Land, 2013) – should be fairly straightforward. However, how time enters the model is crucial.

Time is measured in one of two fashions as the basis for trajectories: as the wave of the survey or the age of the respondent at each time of observation. Reconsider Eq. (2.2) in which time is specified as t_{it}. This specification is the most general, representing individual and time-specific measures of time. When data are structured in the long format, t_{it} is usually measured by the age of the respondent in the given wave of the survey. Birth cohort then becomes a time-invariant level-2 measure: an X_i measure in Eqs. (2.5a) and (2.5b). When data are structured in the wide format, however, t_{it} is commonly measured by the wave of the study. The baseline age of the respondent – which simply reflects birth cohort – is often then entered as a level-2 measure (X_i). Under approaches using both data structures, wave-to-wave change reflects maturation. However, the wave-based approach common under the wide data structure format leads to an inflation in the random effects variance for at least the intercept in unconditional models, because members of different birth cohorts are at different stages of maturation at each wave, and this variance is commonly misinterpreted. Mehta and West (2000) discuss this issue in great detail, and so we will not present detailed equations showing the problem. Instead, we will present some simple figures that illustrate this problem and its extension to difficulties in disentangling age and cohort patterns.

Panel A of Figure 2.8 shows trajectories of an arbitrary outcome, y, for three people (p1, p2, p3) each measured on three occasions 5 years apart, but each of whom was a different age at the baseline of the study. p1 was 30 years old at baseline, p2 was 40, and p3 was 50. With the x-axis representing age, it seems that there is a linear trajectory for y across age, but that there may be a cohort difference in y as evidenced by the jumps at the ages at which the cohorts overlap (40 and 50). Thus, if we were to extend each person's trajectory across the entire age range 30–60, there would be variation in the intercepts at age 30, but no variation in the slopes across the cohorts.

Panel B shows the same data but changes the x-axis to reflect study wave rather than age. Comparing this panel to the previous one reveals two consequences. First, variation in values at baseline, which would be captured via the random effects variance for the intercepts, is much greater than that implied by panel A. If we extended the trajectories in panel A back to age 30 for the earlier two cohorts, none of the trajectories would have an intercept greater

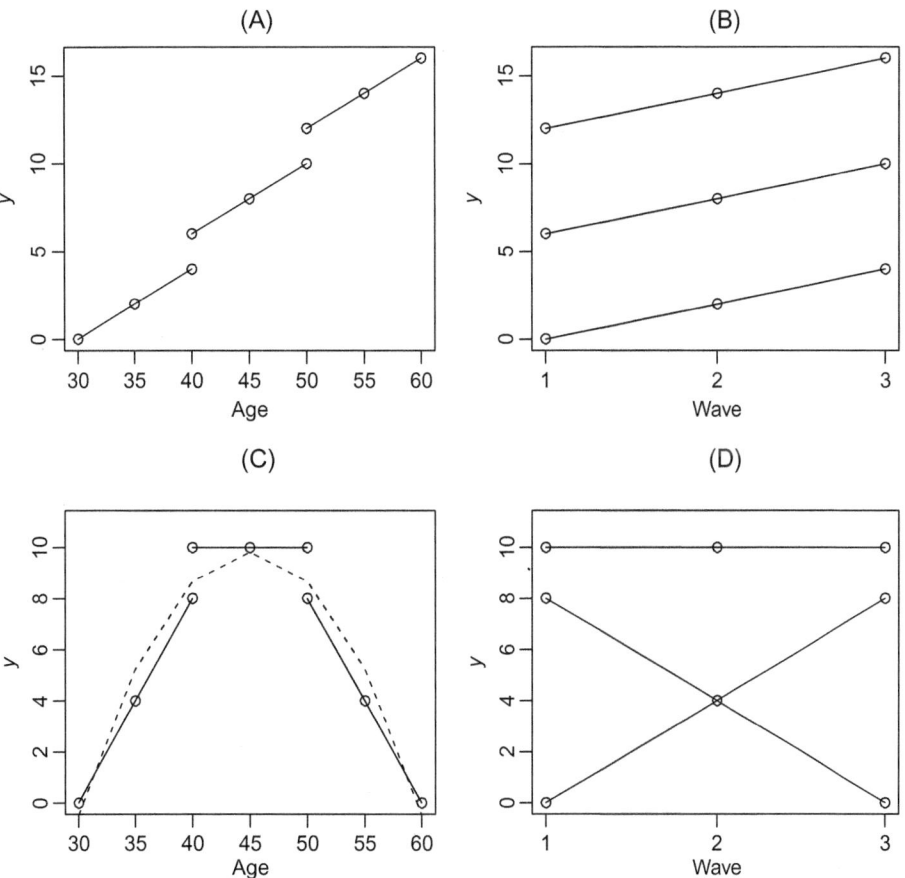

FIGURE 2.8 Illustration of age-based versus wave-based specification of trajectories and the difficulties of disentangling age and cohort patterns. Panel A shows three waves of measurement for three persons at different starting ages at baseline with age as the key measure of time. Panel B shows these measures with wave as the measure of time. Panel C shows three waves of measurement for three persons whose observed measures follow differing, seemingly linear trajectories; however, it is unclear whether the age pattern is specified incorrectly (i.e., it should be quadratic) or whether there are genuine cohort differences in trajectories. Panel D shows these same data with study wave as the key time measure.

than 5. In contrast, in panel B, the intercept for the earliest cohort member is 12, implying much greater variance in the estimated intercepts. Second, the sheer gap between persons is much greater at any given wave in panel B than the gap implied by panel A, were we to extend trajectories for each person across the age range. This result occurs because, while the slopes are identical for all three persons, the variance in the intercept is extended across all study waves. In short, the unconditional wave-based specification confounds age and cohort differences.

Panel C reveals another issue that makes distinguishing age and cohort patterns difficult, even when one uses an age-based model specification. Reconsider the three persons from panels A and B. Suppose that the pattern for p1 is increasing as shown in the left-hand portion of panel C, while the pattern for p2 is flat, and the pattern for p3 is decreasing. Here, it is unclear whether there are cohort differences in both intercepts and slopes, or whether the age pattern has been improperly assumed to be linear, when it is actually quadratic and identical across cohorts as represented by the dashed line. The dashed line deviates somewhat from the solid lines, but in real data it would be impossible to know whether such deviations are due to cohort effects, age effects, misspecification, or measurement error.

In fact, real data are much messier than that shown in these contrived examples, and so assumptions must be made to distinguish the sources of variation. A relatively recent strategy for distinguishing between age and cohort effects in growth models, for example, is "aging-vector analysis" (see Mirowsky & Kim, 2007). The basic idea underlying this methodology is that gaps between cohorts at ages at which cohort measurements overlap, as shown at ages 40 and 50 in panel A, reflect cohort effects/differences. However, such gaps may very well arise for other reasons, including the misspecification of the shape of the age trajectory, nonlinear cohort change in both intercepts and slopes, and misspecification of the error variances as normal and/or constant (or changing) over time.

Finally, panel D shows the data from panel C in a wave-based specification. It is unclear from this view that one would even suspect an inverted quadratic age pattern. Instead, the wave-based specification here would suggest there is tremendous variation in both intercepts and slopes, when in fact there is no variation in intercepts and a single quadratic age pattern rather than multiple linear ones. It is important to note here that the wide format is most often used in latent class analysis, and so the combination of different cohorts could very likely lead to the entirely erroneous conclusion that there are multiple latent classes. This panel therefore illustrates why the outcome may need to be conditioned on covariates (like age or cohort) *before* extraction of latent classes begins, a topic we discuss in greater depth below.

Importance of Assumptions

A number of assumptions must be made when engaging in trajectory modeling, and these assumptions can be critical in influencing results. Therefore, they need explicit consideration before engaging in data analysis. The first assumption that should be considered regards the distribution of the variable constituting a trajectory. The most common assumption is that the variable is normally distributed (or that, conditional on X, it is; this is NOT the same assumption!). While this may seem to be a relatively innocuous assumption, it is a crucial one, because it may very well influence the number of classes one finds (Bauer & Curran, 2003). Figure 2.2 showed the scatterplot of OLS-estimated intercepts and slopes for BMI trajectories in the sample. This scatterplot, and the histograms in the margins, suggest that the distribution of intercepts and slopes is not bivariate normal, despite the implicit assumption under the growth modeling approach that it is. Instead, the distribution of intercepts is right-skewed, and the distribution of slopes is both highly kurtotic (peaked) and right-skewed. A consequence was that the GMM approach found three classes, each with a bivariate normal distribution of intercepts and slopes. The various latent class approaches – which also rely on normality assumptions for the measures in this analysis – also revealed more than one class. Suppose, however, we had assumed that the distribution of intercepts and slopes was bivariate log-normal. In that case, we may have found that a single class provided better fit to the data than a three (or more) class model. For example, Bauer and Curran (2003) found via simulation that if the data arose from one class but were t-distributed rather than normal, there was a tendency to find more than one class under the common normality assumption. Thus, it is very important to realize that results of trajectory modeling may be quite sensitive to one's fundamental distributional assumptions.

A key consideration, then, should be a theoretical one: does theory suggest that the population is composed of subpopulations that fundamentally differ, or is the population homogeneous but with extensive and perhaps asymmetric variation around a common centerpoint (Nagin & Tremblay, 2005)? This decision determines whether a GMM or LCGA should be performed versus a single-class growth model; but there are two additional considerations that should inform this decision: (1) whether trajectory shapes are *qualitatively distinct* between segments of the population, and (2) how covariates influence the outcome – directly, via class assignment, or both.

Regarding whether trajectory shapes are qualitatively distinct, it is common to see researchers either mistakenly or imprecisely claim that growth modeling and latent class approaches are distinguished by whether one assumes that a single trajectory underlies the population. Indeed, latent class analysis is commonly discussed as if it is a more general methodology because it may capture more than one trajectory, whereas growth modeling is limited to one. Another claim is that it may capture trajectories with different shapes, whereas growth modeling is limited to a common shape for all individuals in the population (indeed, we made a similar initial claim in introducing latent class methods earlier). These claims are either poorly worded or simply false. Furthermore, to the extent that latent class analysis is able to capture nonparametric trajectories, it comes with a price: the within-class homogeneity assumption. Thus, it trades capturing greater variance within a single class for capturing more classes.

To be clear, the growth model assumes one *average* trajectory, as indicated by Eq. (2.4). However, the most important part of the model is not this average, just as the most important part of a linear regression model is not the intercept. What is important in the growth model is the variation around the average and the way covariates explain it. In a quadratic growth model with substantial variance in the random effects, individual trajectories may be flat, linearly increasing, linearly decreasing, accelerating upward, accelerating downward, tapering upward, tapering downward, increasing and then decreasing, or decreasing and then increasing. Furthermore, they may begin high or low. In short, there is no limit to the number and shape of trajectories assumed to exist in the population. Indeed, the growth model in fact assumes that there are *more* trajectories in the population than the LCGA assumes. The *only* advantage to the latent class approach in this respect is that it partitions the population into possibly meaningful subgroups whose trajectories differ in shape or starting location; yet, this partition comes at the price of the aforementioned assumption: homogeneity of variance within classes. And, again, the number of distinct subpopulations that are to be found will depend on the assumed distribution of the outcome, possibly but not necessarily after conditioning on covariates.

Finally, while the latent class model (but not the LCGA) may be able to capture nonparametric trajectories, it does this also with an assumption: that deviations from a parametric shape are meaningful and not simply noise. This assumption cannot generally be tested. Instead, researchers often rely on *ad hoc* methods such as considering classes consisting of less than 5% of the population not to be meaningful. We address this assumption below.

Extraction of Classes and Inclusion of Covariates

Beyond distributional assumptions, other factors are relevant in determining the number of latent classes that may exist in the population. Given a distributional assumption, how does one determine how many classes exist? Models with more classes will generally fit the data better than models with fewer. However, there is a tradeoff between fitting the data and parsimony. Thus, information criteria like the BIC discussed earlier, which penalize model fit for the number of parameters, are commonly used to compare fit across models (Geiser, 2013). The model with the lowest value of the statistic is generally considered best. However, agreement in the literature ends at that point. Models with higher BICs are sometimes preferred by researchers, as we have shown in our examples, because the BIC-preferred model may contain unreasonable results, like negative error variances or classes that are represented by only a handful of sample members. Some argue having such problems may indicate the over/under extraction of classes. In other words, the production of unreasonable results is symptomatic of model misspecification (Muthén, 2004; Tofighi & Enders, 2008). However, some argue such results do not offer a valid contribution to determining the number of classes (Pastor & Gagné, 2013). In short, there is ongoing debate in the literature regarding this topic.

Such debate is closely tied to another, perhaps broader, ongoing debate regarding the influence of covariates, specifically in latent class and, especially, GMM models. It is crucial to think carefully about whether covariates influence outcome measures directly or indirectly via latent class membership. This decision can easily influence the number of latent classes that can be extracted, but it also speaks to a more theoretical issue: what exactly is a latent class? In our latent class examples (excluding the GMM example), we used a two-stage approach in estimating our latent class models. First, we estimated models determining the number of latent classes and then assigned individuals deterministically to classes based on their highest posterior probability of class membership. Second, we estimated a logistic regression model predicting latent class membership with covariates. In contrast, in the GMM example, we estimated a single model that simultaneously extracted classes and estimated the influence of covariates on class assignment (and initially, even on the outcome measures themselves). These are the two extremes for how covariates can be incorporated in trajectory models. A variety of intermediate approaches exist, including non-deterministic assignment to classes in a two-stage (or three-stage) model that compensates for uncertainty in class assignment but keeps class extraction and covariate influences as separate processes, and single-stage GMM modeling in which covariates are allowed to influence outcome measures either indirectly via class assignment or directly, but not both. Which approach is correct, and does it matter which approach one chooses?

In part, the decision depends on how one defines a latent class. Suppose our outcome was simply body weight unadjusted for height, and sex was a covariate. If we extracted the latent classes first, we might find that there are two "latent" classes, with membership predicted well by sex: men tend to be heavier than women. If, instead, we controlled on sex first and estimated the latent classes based on the residuals, we may find only one latent class. The former, two-stage approach may fit well with a theory that there are two classes of people with women more likely to be in one type of class versus another. However, we would reach a very different conclusion if we found one latent class after controlling on sex. We would conclude simply that men are heavier for their heights compared to women, but they otherwise come from the same, single population (i.e., human). To some extent, extraction of latent classes before controlling on covariates can lead to finding classes that are only latent because known covariates that influence the outcome have been excluded! Thus, the classes are not truly "latent"; they are an artifact of important omitted variables. This matter is complicated by the fact that, if there are at least two latent classes, covariates may also differ in their effect on the outcome (y) across classes and not just in how they assign individuals to classes, as mentioned above in our discussion of the GMM. In short, the correct approach to the extraction of classes and to including covariates is not at all straightforward, and is as theoretical and substantive as it is statistical. Thus, trajectory methods should perhaps not be used as exploratory methods; instead, they should be employed under the guidance of strong theory for confirmatory testing.

CONCLUSION

In this chapter we have provided an overview of several types of trajectory methods that are commonly used in contemporary aging research and have illustrated them using six examples in order to highlight consistencies and differences between methods. We began with one of the simplest methods – the basic growth model – and we ultimately extended this model to the most general model, the GMM. Each of the methods we discussed can be seen as a special case of the GMM. That is, the other models can be obtained by imposing constraints on (1) the number of latent classes, (2) the parameterization of trajectories within latent classes, and (3) the sources that contribute to the variance in observed measures that form the basis of the trajectories.

These models assume that data have two potential sources of variance: individual time-specific variance in outcomes around the individual's trajectory (level-1 variance) and variance in the individual's trajectory around his/her class's average trajectory (level-2 variance). If we (1) restrict the number of latent classes in the GMM to one, (2) assume that individual trajectories are parametric in shape, and (3) allow variance at both level 1 and level 2, we have the basic growth model. If we (1) allow the number of classes to vary, (2) assume no parametric shape to trajectories but assume that means of the time-specific measures vary by class, and (3) assume that level-1 variance of the time-specific measures exists but that there is no level-2 variance, we have a nonparametric latent class model. If we (1) allow the number of classes to vary, (2) assume a parametric shape for trajectories, and (3) assume that level-1 variance exists but level-2 variance does not, we obtain an LCGA.

Although the GMM is the most general of these models, it is frequently difficult to get model estimation routines to converge to reasonable values, precisely because of its generality. That is, the GMM generally contains a considerable number of parameters, and they are often only weakly identified. For example, without imposing substantial constraints, the model must be able to partition the variance in the outcome measures into measurement error variance, level-2 random intercept and slope variance, and between class variation, while simultaneously distinguishing both how covariates influence assignment to classes and how they influence random effects within classes. This is a lot to expect from the model, especially when the number of observed time periods and cases is relatively small in many panels. Thus, considerable exploration of more restrictive models – that is, models with more stringent assumptions – is often required that may or may not conform to any *a priori* theory. As one changes the model assumptions, the substantive conclusions that are obtained will often also change, as we saw between the final and penultimate examples, and so one is left with a decision to make regarding which model is best. Unfortunately, this is an ongoing area of investigation, and there are no clear, definitive answers (Liu & Hancock, 2014).

At the same time that identifiability is a serious issue, researchers continue to push model capabilities in new directions. We have covered several popular strategies, but recent methodological advances have incorporated the estimation of mortality or other distal outcomes into these models (Taylor & Lynch, 2011; Zimmer,

Martin, Nagin, & Jones, 2012) and the ability to allow individuals to make transitions between latent classes over time (see Collins & Lanza, 2010, for discussion of latent transition analysis), to name only two advances. We expect that the ability to extend these methods will continue to outpace the ability to estimate the models, and, unfortunately, most likely substantive theory as well. Nonetheless, trajectory methods are increasingly popular and important tools for modeling life course processes with panel data.

References

Barban, N., & Billari, F. C. (2012). Classifying life course trajectories: A comparison of latent class and sequence analysis. *Journal of the Royal Statistical Association: Applied Statistics*, 61, 765–784.

Bauer, D. J., & Curran, P. J. (2003). Distributional assumptions of growth mixture models: Implications for overextraction of latent trajectory classes. *Psychological Methods*, 8, 338–363.

Bollen, K. A., & Curran, P. J. (2006). *Latent curve models: A structural equation perspective*. Hoboken, NJ: Wiley.

Clogg, C. C. (1995). Latent class models. In G. Arminger, C. C. Clogg, & M. E. Sobel (Eds.), *Handbook of statistical modeling for the social and behavioral sciences* (pp. 311–359). New York, NY: Plenum.

Collins, L. M., & Lanza, S. T. (2010). *Latent class and latent transition analysis: With applications in the social, behavioral, and health Sciences*. Hoboken, NJ: Wiley.

Elder, G. H., Jr. (Ed.). (1985). Perspectives on the life course. In *Life course dynamics: Trajectories and transitions, 1968–1980* (pp. 23–49). Ithaca, NY: Cornell University Press.

Geiser, C. (2013). *Data analysis with Mplus*. New York, NY: Guilford Press.

George, L. K. (2009). Conceptualizing and measuring trajectories. In G. H. Elder Jr. & J. Z. Giele (Eds.), *The craft of life course research* (pp. 163–186). New York, NY: The Guilford Press.

Goodman, L. A. (1974). Exploratory latent structure analysis using both identifiable and unidentifiable models. *Biometrika*, 61, 215–231.

Jung, T., & Wickrama, K. A. S. (2008). An introduction to latent class growth analysis and growth mixture modeling. *Social and Personality Psychology Compass*, 2, 302–317.

Kreuter, F., & Muthén, B. (2007). Longitudinal modeling of population heterogeneity: Methodological challenges to the analysis of empirically derived criminal trajectory profiles. In G. R. Hancock & K. M. Samuelsen (Eds.), *Advances in latent variable mixture models*. Charlotte, NC: Information Age Publishing.

Land, K. C. (2001). Introduction to special issue on finite mixture models. *Sociological Methods and Research*, 29(3), 282–318.

Lazarsfeld, P. F., & Henry, N. W. (1968). *Latent structure analysis*. Boston, MA: Houghton Mifflin.

Little, R. J. A., & Rubin, D. B. (2002). *Statistical analysis with missing data* (2nd ed.). Hoboken, NJ: Wiley.

Liu, M., & Hancock, G. R. (2014). Unrestricted mixture models for class identification in growth mixture modeling. *Educational and Psychological Measurement*, 74, 557–584. Available from: <http://dx.doi.org/10.1177/0013164413519798>.

Lynch, S. M. (2007). *Introduction to applied Bayesian statistics and estimation for social scientists*. New York, NY: Springer.

Magidson, J., & Vermunt, J. K. (2002). Latent class models for clustering: A comparison with K-means. *Canadian Journal of Marketing Research*, 20, 37–44.

Mehta, P. D., & West, S. G. (2000). Putting the individual back into individual growth curves. *Psychological Methods*, 5(1), 23–43.

Meredith, W., & Tisak, J. (1990). Latent curve analysis. *Psychometrica*, 55(1), 107–122.

Mirowsky, J., & Kim, J. (2007). Graphing age trajectories: Vector graphs, synthetic and virtual cohort projections, and cross-sectional profiles of depression. *Sociological Methods and Research*, 35(4), 497–541.

Muthén, B., & Asparouhov, T. (2006). Growth mixture analysis: Models with non-Gaussian random effects. In G. Fitzmaurice, M. Davidian, G. Verbeke, & G. Molenberghs (Eds.), *Advances in longitudinal data analysis* (pp. 143–165). Boca Raton, FL: Chapman & Hall/CRC Press.

Muthén, B., & Muthén, L. K. (2000). Integrating person-centered and variable-centered analyses: Growth mixture modeling with latent trajectory classes. *Alcoholism: Clinical and Experimental Research*, 24, 882–891.

Muthén, B. O. (2004). Latent variable analysis: Growth mixture modeling and related techniques for longitudinal data. In D. Kaplan (Ed.), *Handbook of quantitative methodology for the social sciences* (pp. 345–368). Newbury Park, CA: Sage Publications.

Muthén, L. K., & Muthén, B. O. (1998–2012). *Mplus user's guide* (7th ed.). Los Angeles, CA: Muthén & Muthén.

Nagin, D. S., & Odgers, C. L. (2010). Group-based trajectory modeling (nearly) two decades later. *Journal of Quantitative Criminology*, 26, 445–453.

Nagin, D. S., & Tremblay, R. (2005). Developmental trajectory groups: Fact or a useful statistical fiction? *Criminology*, 43, 873–904.

REFERENCES

National Institutes of Health (1998). Clinical guidelines on the identification, evaluation, and treatment of overweight and obesity in adults – The evidence report. *Obesity Research, 6*(S2), 51S–209S.

Nylund, K. L., Asparouhov, T., & Muthén, B. O. (2007). Deciding on the number of classes in latent class analysis and growth mixture modeling: A Monte Carlo simulation study. *Structural Equation Modeling, 14*(4), 535–569.

Pastor, D. A., & Gagné, P. (2013). Mean and covariance structure mixture models. In G. R. Hancock & R. O. Mueller (Eds.), *Structural equation modeling: A second course* (pp. 197–224, 2nd ed.). Charlotte, NC: Information Age.

RAND HRS Data, Version M. Produced by the RAND Center for the Study of Aging, with funding from the National Institute on Aging and the Social Security Administration. Santa Monica, CA, September 2013.

Raudenbush, S. W. (2001). Comparing personal trajectories and drawing causal inference from longitudinal data. *Annual Review of Psychology, 52*, 501–525.

Raudenbush, S. W., & Bryk, A. S. (2002). *Hierarchical linear models: Applications and data analysis methods* (2nd ed.). Thousand Oaks, CA: Sage.

Taylor, M. G., & Lynch, S. M. (2004). Trajectories of impairment, social support, and depressive symptoms in later life. *Journals of Gerontology: Social Sciences, 59B*, S238–S246.

Taylor, M. G., & Lynch, S. M. (2011). Cohort differences and chronic disease profiles of differential disability trajectories. *Journals of Gerontology: Social Sciences, 66B*, 729–738.

Tofighi, D., & Enders, C. K. (2008). Identifying the correct number of classes in a growth mixture model. In G. R. Hancock & K. M. Samuelsen (Eds.), *Advances in latent variable mixture models* (pp. 317–341). Charlotte, NC: Information Age.

Vermunt, J. K., & Magidson, J. (2002). Latent class cluster analysis. In J. A. Hagenaars & A. L. McCutcheon (Eds.), *Applied latent class analysis* (pp. 89–106). Cambridge, UK: Cambridge University Press.

Yang, Y., & Land, K. C. (2013). *Age-Period-Cohort analysis: New models, methods, and empirical applications*. Boca Raton, FL: CRC Press.

Zimmer, Z., Martin, L. G., Nagin, D. S., & Jones, B. L. (2012). Modeling disability trajectories and mortality of the oldest-old in China. *Demography, 49*(1), 291–314.

PART II

SOCIAL STRUCTURES AND PROCESSES

3 Biodemography: Adding Biological Insight into Social, Economic, and Psychological Models of Population and Individual Health Change with Age 55
4 Late-Life Disability Trends and Trajectories 77
5 Early Life Origins of Adult Health and Aging 101
6 Racial and Ethnic Inequalities in Health 123
7 Immigration, Aging, and the Life Course 143
8 Gender, Time Use, and Aging 163
9 Social Networks in Later Life 181

CHAPTER 3

Biodemography: Adding Biological Insight into Social, Economic, and Psychological Models of Population and Individual Health Change with Age

Eileen M. Crimmins[1] and Sarinnapha M. Vasunilashorn[2]

[1]Davis School of Gerontology, University of Southern California, Los Angeles, CA, USA [2]Beth Israel Deaconess Medical Center, Harvard Medical School, Brookline, MA, USA

OUTLINE

Introduction	56
Expansion of the Demographic Approach: The Process of Health Change	56
The Expanded Biodemographic Model of Health	57
Measuring Biomarkers in Population Studies	59
Use of Biomarkers in Assessing Population Health and Health Care Use and Its Effectiveness	60
Summary Indices of Biological Risk	63
Genetic Markers as a New Frontier	68
Summary	69
Appendix: Information on Biomarkers Often Used in Social Science Research	69
References	72

INTRODUCTION

In recent decades demographers interested in the health of aging populations have developed new approaches to understanding trends, differences, and changes in health based on biologically informed models. The multidisciplinary nature of demographic research on health and aging continues to grow in scope and importance as it has been enlightened by insights from sociology, economics, psychology, and now, biology. Biodemographic work in aging incorporates biological theory and measurement with traditional demographic approaches to better understand variation in health and mortality across and within populations as well as changes among aging individuals in those populations. Incorporating biology begins to uncover mechanisms through which traditional demographic variables (e.g., age, sex, race/ethnicity) are linked to subsequent health outcomes. The rapid aging of populations and extension of life expectancy in recent decades, coupled with the persistence of health differentials within populations and between countries such as the United States and other developed countries, has resulted in questions such as: Are people living longer healthy lives? How does one explain the variability in the aging process across individuals, groups, and contexts? Why do Americans appear to have relatively poor health? In other words, the focus of work in demographic analyses of mortality and health has changed from description to understanding how and why health change and differences occur and the identifying factors that contribute to the plasticity of health relative to environmental, social, behavioral, and technological changes (Vasunilashorn & Crimmins, 2008).

EXPANSION OF THE DEMOGRAPHIC APPROACH: THE PROCESS OF HEALTH CHANGE

The biodemographic focus arises, in part, from societal changes that have occurred over the past century. These changes have required a rethinking of how demographers conceptualize health and think about mortality. With the marked reduction in mortality from infectious conditions, chronic conditions dominate as causes of morbidity and mortality. Chronic conditions are concentrated among the older members of the population and they develop over the entire lifespan even though they may not be manifested until late in life. This has led to an emphasis on the process of health change over the lifespan.

In recent decades, mortality decline at older ages has been a major contributor to increasing life expectancy; and the addition of years of life to living into older ages has increased interest in: (1) determining how long we can live healthily, and (2) understanding the risks for mortality at the oldest ages. In addition, we have recently come to understand that some of the major chronic conditions of old age are disabling but not lethal conditions, leading to a shift in focus from mortality as an outcome to disability and functioning loss. Decrements in mental and physical conditions are now recognized as a major cause of disability and functioning loss, leading to an emphasis on understanding the etiology, prevention, and treatment of these conditions. All of these changes set the stage for new approaches to studying changes in late-life health and an improved understanding of the processes leading to health change from chronic disease to "aging" itself.

Until the advent of sample surveys about 60 years ago, demographic analysis of health was generally limited to the outcome of mortality. Data from surveys provided the opportunity to study additional health indicators including self-reports of functional difficulties, diseases, and overall health. The wealth of variables available to measure "health" in these surveys has led to multiple attempts to conceptualize the process of health change. Work produced by groups working under the auspices of the World Health Organization and the Institute of Medicine in the United States

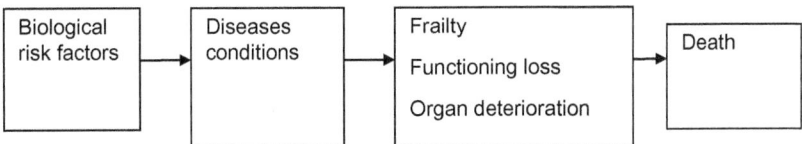

FIGURE 3.1 The morbidity process. *Source: Adapted from Crimmins et al. (2010).*

(World Health Organization, 1999), as well as by individual researchers (Verbrugge & Jette, 1994), has provided important foundations for understanding the meaning and linkages between various dimensions of health. Building on this work, we have outlined a model of the "morbidity process" which describes population-level changes in health (Crimmins, Kim, & Vasunilashorn, 2010). In populations, this process begins with physiological changes characterized by the onset of risk factors indicating physiological dysregulation. This then results in the onset of disease and functioning loss, followed by frailty (a state characterized by a loss of organ reserve and severe physiological deterioration), and the ultimate outcome: death (Figure 3.1). This model can be used to classify the physiological or biological changes that occur with aging to clarify that it is a process that involves multiple health dimensions, for populations dimensions change in a staged fashion, and that intervention to retard the process can occur at any stage of the process.

The dimensions of the morbidity process contrast with the disablement process in that the morbidity process focuses on health change intrinsic to a person while the disability process is, by definition, influenced by environmental circumstances (Verbrugge & Jette, 1994). Certainly the likelihood of having any health problem is affected by interaction with the environment, such as taking drugs or behaving in certain ways; but for the dimensions of morbidity, environmental conditions are not intrinsic to the definition as they are with disability, but rather the health change occurs "within the skin" of the individual. While much research on the health of older people has focused on the dimension of disability, and trends in it (Cai & Lubitz, 2007), Beltrán-Sánchez, Razak, and Subramanian (2014) recently argued for more focus on disease and biological risk in assessing trends in health. They make the case that both significant health change and significant health expenditures have generally occurred before disability. Our focus on the morbidity process dovetails with this suggestion.

With the integration of biological risk into demographic analyses, the study of health outcomes has expanded to include risk factors that occur prior to clinically diagnosed disease and disability, or the beginning of the process of change. Differential and change in the beginning of the process of health change is important to monitor since it indicates a potential time for intervention to delay or halt the process. Biomarkers have also been identified for processes later in the morbidity process (e.g., functioning loss, loss of organ function, or frailty).

THE EXPANDED BIODEMOGRAPHIC MODEL OF HEALTH

Multiple and complex life circumstances drive patterns of population morbidity and mortality. Demographic researchers have utilized increasingly broader models of population health though the integration of multidisciplinary perspectives. For decades, demographic analyses of mortality focused on population subgroup differences including age, sex, race, and socioeconomic status (SES) without considering the social, economic, psychological, and physiological pathways through which traditional "demographic" variables impact health and mortality. Recently a significant body of literature has introduced biological measurement as a way to improve or clarify how both the traditional demographic, and the social, economic, psychological, and environmental factors get "under the skin" to affect health (e.g., Seeman & Crimmins, 2001; Seeman, Epel, Gruenewald, Karlamangla, & McEwen, 2010).

The heuristic model presented in Figure 3.2 provides a general overview of the multiple and interacting lifetime circumstances through which these factors affect biology and downstream health outcomes (Crimmins & Seeman, 2004; Seeman & Crimmins, 2001; Vasunilashorn & Crimmins, 2008). The figure is heuristic in that there are many unrepresented arrows and concepts; there are influences in both directions, and there are some overarching variables that are likely related to all of the boxes and set the context in which other relationships occur: national, neighborhood, and policy contexts. Genetic endowment is also an overarching variable; not only determining sex and race but also predisposing to elevated risks for disease, loss of functioning, and mortality. Although race/ethnicity and SES are included in the model below as variables exogenous to the process of health change, these variables also provide information on overarching contextual factors. That is, the links between boxes in our model may be quite different for individuals in varied demographic and social subgroups.

The box labeled "Biological Risk" (Figure 3.2) depicts some of the physiological pathways through which social, economic, psychological, behavioral, and healthcare factors can affect subsequent health outcomes. Many additional biological mechanisms could be listed, as the ones illustrated serve merely as examples. Most large population studies with

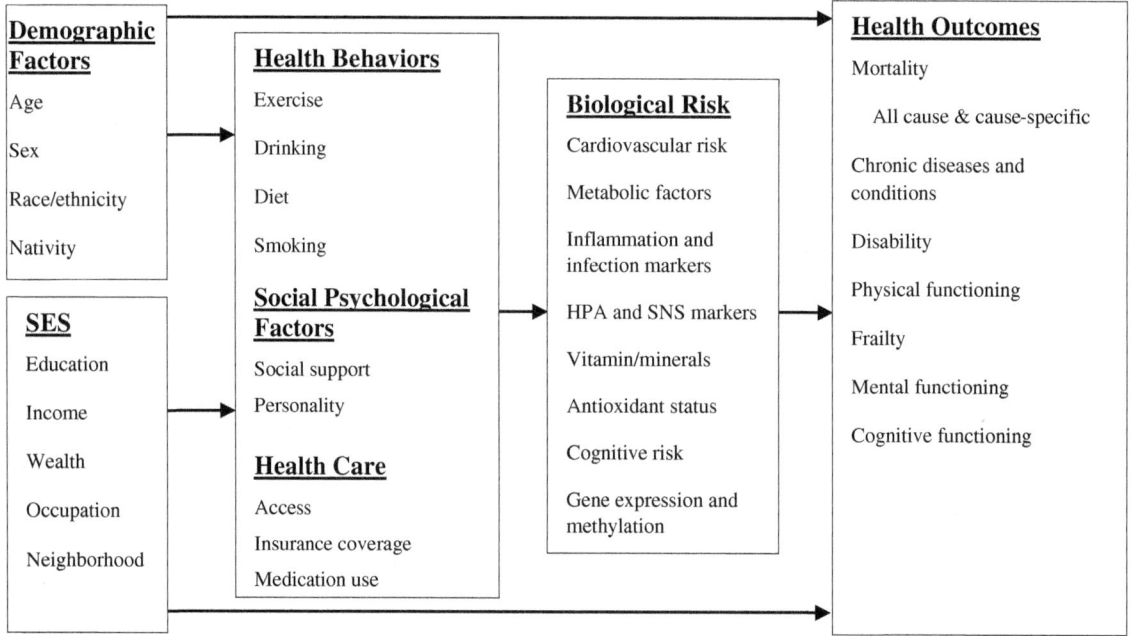

FIGURE 3.2 Demographic, socioeconomic, behavioral, and biological influences on health outcomes. SES, socioeconomic status; HPA, hypothalmic pituitary axis; SNS, sympathetic nervous system. *Source: Vasunilashorn and Crimmins (2008).*

indicators of biological risk include cardiovascular, metabolic, and inflammatory indicators (see the Appendix for a list of biomarkers often included in large studies, their sources, related conditions, and some of the studies in which they have been included).

The heuristic model also indicates how the outcomes investigated have grown in recent years. Demographers generally no longer feel that mortality provides a complete picture of health in an aging population. Declines in cognitive function, disability, functioning loss, and frailty are all important in understanding the costs and consequences of health change in old age. The recognition that a large proportion of disability is caused by nonfatal conditions including cognitive and physical functioning problems is one of the reasons that longer life is not necessarily accompanied by better health (Crimmins & Beltran-Sanchez, 2011).

MEASURING BIOMARKERS IN POPULATION STUDIES

A key characteristic of most biodemographic work is that it examines health in representative population samples. This allows generalizations to be drawn about population characteristics and processes, and it has resulted in incorporating biological measurement into the large sample surveys that are the major data source of current biodemographic work. Significant biological, medical, and epidemiological knowledge identifying major risk factors for the diseases and conditions important in older populations has been developed using large-scale community and population studies, including the Framingham Heart Study and the National Health and Nutrition Examination Survey, and this has provided the basis for including measurement of biology in large surveys. Changes in technology have also enabled collection of biomarkers using less invasive methods of collection, better preservation of samples in the field, and advancing technological developments in laboratory methods (McDade, Williams, & Snodgrass, 2007). These developments include: (1) the collection of blood samples using dried blood spots which can be accomplished by non-trained personnel, (2) the collection of DNA from saliva for genetic analysis using special containers housing a preservative that does not require special handling or refrigeration, (3) the collection of RNA using special collection and methods of preservation, and (4) new sources of samples such as hair.

Many of the health outcomes important in older age have broadly overlapping biological risk factors, for example, cardiovascular disease (CVD), disability, cognitive loss, and death, because physiological changes across systems in the body are often interrelated and because decline in one system may be related to declines in others. This has encouraged the inclusion of a small number of core biomarkers in many studies with an increasing proliferation of additional markers as techniques for collection and assay from populations improve (Weinstein, Vaupel, & Wachter, 2008).

Many large US studies of older populations, for example, the Health and Retirement Study (HRS) and the National Social Life, Health and Aging Project (NSHAP), now include biomarkers. These are studies of nationally representative samples whose data are collected by an interviewer and not trained medical personnel. Some smaller, less representative samples with extensive biomarker data collected in hospital or clinical settings also exist, for example, Midlife in the United States (MIDUS). Many of the HRS family of international studies, including the English Longitudinal Study of Ageing (ELSA), the Survey of Health, Ageing and Retirement in Europe (SHARE), the China Health and Retirement Longitudinal Study (CHARLS), the Mexican Health and Aging Study (MHAS), and the Longitudinal Study of Aging in India (LASI), have or are in the process of collecting biomarkers. Other studies that also collect biologic information include the Social Environment and Biomarkers of Aging Study (SEBAS) and the Study on Global Ageing and adult health (SAGE), and this list expands each year (see the Appendix for a list of studies).

The biomarkers included in surveys generally include cardiovascular and metabolic risk factors and increasingly include markers of inflammation. Markers of organ functioning (e.g., lung, kidney, and liver function) can also be collected at the same time as the other measures. Some are based on exam measures (e.g., blood pressure, height, weight, waist circumference), some on a blood sample (e.g., cholesterol, glycosylated hemoglobin (HbA1c), C-reactive protein (CRP), cystatin C (to indicate kidney function)), and some are based on performance (e.g., peak flow for lung function). These markers include indicators of frailty, organ functioning, and disability as well as risks for conditions associated with aging. Performance tests of balance, walking speed, grip strength, and ability to stand from a seated position are also indicators of both frailty and disability which are increasingly being included in the set of studies with other biomarkers. For a more detailed discussion of biomarkers of aging and the biomarkers included in studies, see Crimmins, Vasunilashorn, Kim, and Alley (2008) and Harris, Gruenewald, and Seeman (2007).

While biomarkers may more or less reflect one type of physiological activity, they interact within an organism. This interactive and cumulative aspect of physiological change has led to a view that small changes across a number of systems may be as, or more, important to the health and functioning of a person as a large change in a single system. This is the idea behind the concept of allostatic load, which has been an important framework in guiding the development of many biodemography studies (Crimmins & Seeman, 2004; McEwen, 2004; Seeman, McEwen, Rowe, & Singer, 2001). The literature on allostatic load initially focused on stress as a disruption in regulation of several physiological systems, including the sympathetic nervous system (SNS), immune system, hypothalamic–pituitary–adrenal (HPA) axis, inflammatory responses, and cardiovascular system. Chronically high levels of stress can result from continued exposure to adverse environmental conditions (e.g., occupations that expose individuals to harmful ultraviolet rays) or psychological stressors (e.g., job stress, family stress, financial constraints, or being a caregiver for a family member). The measurement of stress in large populations has lagged behind measurement of some of the other relevant processes. But the advent of assays from hair that reflect stress over the period when the hair was growing make it likely that this type of measurement will be accomplished in large populations in the near future.

USE OF BIOMARKERS IN ASSESSING POPULATION HEALTH AND HEALTH CARE USE AND ITS EFFECTIVENESS

Data on some biomarkers allows researchers to assess population health and effectiveness of health care at the same time. Blood pressure is one of the most commonly measured biomarkers, and there are clear clinical guidelines above which blood pressure is considered a risk for CVD. Such guidelines for clinical practice dictate that taking medication is appropriate and incorporating both blood pressure measurement and self-reported information can provide estimates of both population health and effectiveness of treatment at the population level. Information on measured biomarkers, respondent reports of diagnosis by a medical clinician, and reports of prescription drug usage for blood pressure can be used to divide the population into four groups: (1) those who do not have high blood pressure, (2) those with high

TABLE 3.1 Hypertensive States Based on Measurement of Systolic Blood Pressure (≥140), Report of Doctor Diagnosis, and Use of Medication

Panel A. Samples of the Health and Retirement Study and the Nihon University Japanese Longitudinal Study of Aging: 68 years of age and older

	United States (2006) (%)	Japan (2006) (%)
Healthy	24.0	17.8
Controlled	36.5	8.0
Undiagnosed	11.9	44.3
Uncontrolled	27.6	29.9

Panel B. Samples of the China Health and Retirement Longitudinal Study of Aging and the Indonesia Family Life Survey: 68 years of age and older

	China (2011) (%)	Indonesia (2007/2008) (%)
Healthy	41.6	29.8
Controlled	12.6	3.6
Undiagnosed	24.0	44.2
Uncontrolled	21.9	22.4

blood pressure who are unaware due to no prior diagnosis, (3) those previously diagnosed with high blood pressure who are currently taking medication and currently do not have elevated blood pressure or those with controlled blood pressure, and (4) those who are currently taking medication and still have elevated blood pressure. These groups can be collapsed to highlight two subpopulations: (1) those who have ever had or who now have high blood pressure provide an indication of the health of the population, and (2) those who were not diagnosed or not controlled with medication indicate problems in the medical care system.

Using HRS data for the US population age 68+, we see that about a quarter of Americans in this age range currently have no problem with high blood pressure (24%) (Table 3.1, Panel A). Only a small number of Americans who have high blood pressure are not diagnosed (11.9%). The largest group of Americans is those who are taking antihypertensives and who have controlled blood pressure (36.5%). It is also true that a large group of Americans who are on medication still have high blood pressure (27.6%). One can draw the conclusion from this that three-fourths of Americans within this age group have a problem with high blood pressure and that about 40% are not diagnosed or not being treated effectively.

This approach can be used to compare the United States to other countries. Blood pressure was recently measured in a national sample of older Japanese who can be compared to Americans of the same age. Japan is a particularly appropriate comparison as it is well-known that Japan is a country with exceptionally long survival and high life expectancy while, relative to other developed countries, the United States has relatively low life expectancy (Crimmins, Preston, & Cohen, 2011). Surprisingly, when categories for blood pressure and treatment management for Japan are determined, Japan looks neither healthy nor effective in their medical system. Relative to Americans, an even smaller percentage of the Japanese have no problem with high blood pressure (only 17.8%). On the other hand, almost half of the Japanese have undiagnosed high blood pressure (44.3%) and control of blood pressure is relatively rare among those who take medications (8.0% out of 37.9%). This would indicate that the relative mortality of Americans and Japanese occurs in spite of the fact that Americans have less high blood pressure and are more effective at treating it.

Similar data from two studies in developing countries that have recently collected biomarkers in national samples further indicate the importance of adding these measured data to respondent reports in these countries where self-reports do not indicate the extent of the blood pressure problem. The China Health and Retirement Longitudinal Study and the

Indonesia Family Life Study both include measured blood pressure in their studies, along with reports of medication use and knowledge of the presence of hypertension (Table 3.1, panel B). The proportion without high blood pressure is relatively high in China (41.6%) and higher in Indonesia than the United States or Japan (29.8%). However, the majority of those who have high blood pressure in Indonesia do not know they are hypertensive (44.2% out of 70.2%). In contrast, this is true for about 41% (24.0% out of 58.5%) of adults in China with hypertension. It becomes clear that it is difficult to understand population blood pressure without measuring it directly and asking questions about prior diagnosis and drug usage.

Similar analyses can be conducted for cholesterol and glycosylated hemoglobin (HbA1c), which is an indicator for diabetes and/or the control of glucose among diabetics. As an example of an investigation for cholesterol, Merkin et al. (2009) reported that educational differentials in the prevalence of high cholesterol in the United States arose from cholesterol screening rather than awareness among those screened, medication use, or effectiveness of medication use (Table 3.2). People who had less than a high school education had screening rates for high cholesterol that were about two-thirds of people with more than a high school education (53.0% versus 78.6%). Once screening was accounted for, the education groups had similar levels of awareness, medication use, and control. This highlights where in the process attention is needed to equalize risk from high cholesterol. It also indicates that once Americans have gotten into the medical care system, all education groups appear to do the same at having cholesterol control.

Similar indicators can be used to clarify trends in risk and treatment of risk. There has been a dramatic drop in recent years in measured high blood pressure in the older American population. From 1999–2000 to 2003–2006, measured high systolic blood pressure (SBP)

TABLE 3.2 Levels of High Cholesterol Screening, Awareness, Treatment, and Control in the United States. Population by Education. NHANES 19990 2002, Ages 20–65, Age-Adjusted

	Screened %	Awareness % (among the screened with high cholesterol)	Medication use % (among screened, prevalent, aware)	Control % (of those screened, aware, prevalent, treated)
Total	70.5	52.4	77.2	60.9
<High school education	53.0	56.4	80.7	53.1
High school	66.2	57.5	75.6	55.9
>High school education	78.6	48.4	80.1	71.5

Source: Merkin et al. (2009).

dropped from being observed in more than half the population (52.3%) to just over a third (36%) (Crimmins et al., 2010). This change in measured hypertension is largely explained by the increasing use of and improving efficacy of drugs rather than a decrease in the likelihood of being diagnosed with either of these risk factors. Figure 3.3 indicates that the proportion of the older male and female populations diagnosed with hypertension stayed virtually constant from 1999–2000 through 2005–2006 while the prevalence of measured hypertension fell. At the same time medication use increased among both men and women and users of medication were more likely to have controlled blood pressure at the later date (Crimmins et al., 2010). The story for cholesterol in this period is similar: the use of drugs increased markedly, and the measured levels of cholesterol have decreased.

As data from longitudinal studies with biomarkers at multiple time points become available, change in biomarkers within individuals over time is starting to be examined. Using

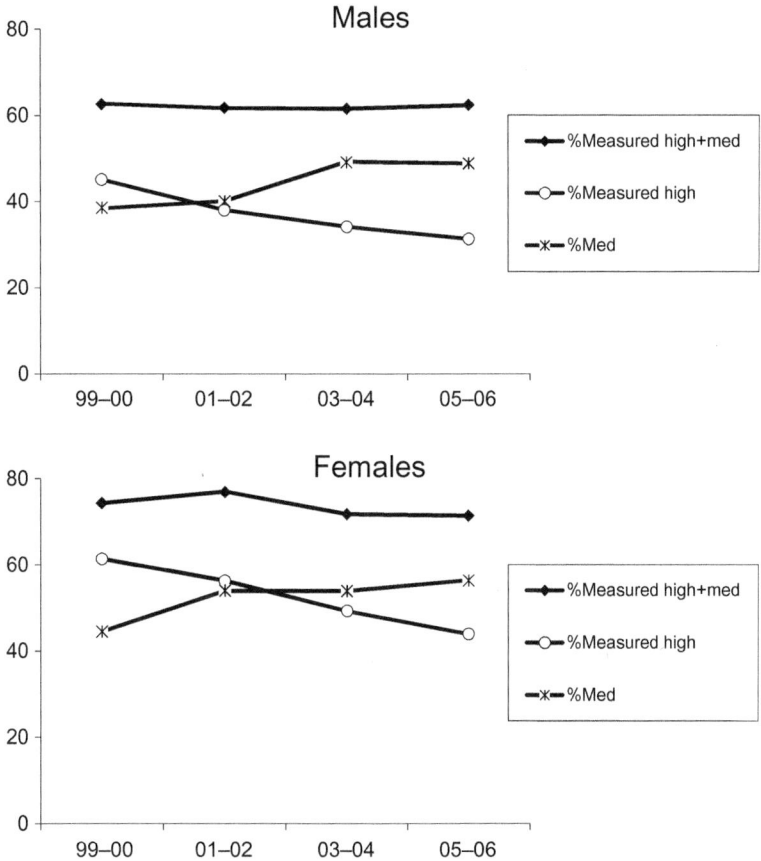

FIGURE 3.3 US time trend in high blood pressure: measured high, use of medication, and measured high + medication use (NHANES 1999–2006). *Source: Crimmins et al. (2010).*

data on body mass index (BMI) over 15 years from the nationally representative American's Changing Lives study, Ailshire and House (2011) examined social disparities in trajectories of BMI among individuals during a time of rapid weight gain in the United States. They find increasing social disparities in weight resulting from complex interactive effects of gender, race, SES, and age. Among people aged from 25–39 to 45–54, black women with low-education and income experienced the greatest increase in BMI, while white men with high education and income experienced the least increase in BMI. It is important to understand the life cycle patterning in change in risk in order to understand the timing of downstream outcomes.

SUMMARY INDICES OF BIOLOGICAL RISK

While individual indicators of biological risk have been predictive of poor subsequent health outcomes such as morbidity, disability, and mortality among older persons, researchers have attempted to develop summary indicators of risk across multiple systems to better predict downstream outcomes and assess population health

(Crimmins & Seeman, 2004; Goldman, Turra, Glei, Lin, & Weinstein, 2006; Karlamangla, Singer, McEwen, Rowe, & Seeman, 2002; Karlamangla, Singer, Seeman, 2006; Seeman et al., 2001; Wang et al., 2007). The rationale behind these measures is that small physiological change in multiple systems may be very important in preventing older persons from maintaining health and functioning and that the accumulation of dysregulation across multiple systems provides the best indicator of an individual's overall health.

One of the earliest attempts to summarize multiple biological indicators into an index predicting risk of mortality and cardiovascular events was the Framingham Risk Score (Wilson et al., 1998). This score has been widely used both in research and in clinical practice. It combines biological indictors with a behavioral indicator of smoking to predict the 10-year likelihood of a cardiovascular event or mortality. Metabolic syndrome is another summary measure that is in use in both clinical practice and research (Karlamangla, Merkin, Crimmins, & Seeman, 2010).

Allostatic load has become a highly used summary index for research that combines some of the indicators from the Framingham Index and metabolic syndrome with others. This index of physiological dysregulation across many systems was developed in the MacArthur Study of Successful Aging and has been related to the major outcomes of aging including mortality, disease onset, loss of physical and cognitive functioning, and mortality (Gruenewald, Seeman, Karlamangla & Sarkisian, 2009; Karlamangla et al., 2006; Seeman et al., 2001). The initial allostatic load score included 10 biological markers representing activity in the metabolic, cardiovascular, inflammation, HPA, and SNS (Seeman, Singer, Rowe, Horowitz, & McEwen, 1997). Subsequent adoption of the index in studies with fewer measured biomarkers has led to operationalization in multiple ways but with intent to include many physiological systems. In addition, while allostatic load was initially the equally weighted sum of the total number of elevated-risk biomarkers, alternative formulations have been developed based on canonical correlation analysis and recursive partitioning in order to address the issue that some biomarkers may differentially predict health outcomes (Gruenewald, Seeman, Ryff, Karlamangla, & Singer, 2006; Gruenewald et al., 2006; Karlamangla et al., 2002, 2006; Turra et al., 2005; Vasunilashorn, Best, Kim, & Crimmins, 2014; Zhang & Singer, 1999).

In a somewhat different approach, Levine (2013) has combined information on multiple biological systems to quantify "Biological Age" or the age of an individual indicated by his or her physiological dysregulation relative to those in the population. It can be compared to chronological age to indicate whether someone is aging faster or slower than the average person of that age. Ten indicators covering multiple physiological systems are included in Biological Age: glycocylated hemoglobin (HbA1c) total cholesterol, systolic blood pressure, forced expiratory volume, serum creatinine, serum urea nitrogen, serum alkaline phosphatase, serum albumin, C-reactive protein (CRP), and cytomegalovirus (CMV). A recent comparison of mortality prediction using biological age, the Framingham Risk Score, and a version of allostatic load found that Biological Age had the strongest association with all-cause and cancer mortality, while the Framingham Risk Score had the strongest association with mortality from CVD (Levine & Crimmins, 2014b).

Another approach to classifying biological risk uses factor analysis to group biomarkers into system-specific factors (Kubzansky, Kawachi, & Sparrow, 1999; Nakamura & Miyao, 2003; Sakkinen, Wahl, Cushman, Lewis, & Tracy, 2000). Following on this, Vasunilashorn et al. (2014) used latent class analysis to identify older individuals who share similar physiological indicators and functioning profiles. They include both biological and functioning indicators since both are predictive of future health problems and mortality in the older population. Indicators included are diastolic and systolic blood pressure, glycated hemoglobin, BMI, albumin, CRP, fibrinogen, glycated hemoglobin, total cholesterol, high-density lipoprotein cholesterol, timed walk, timed test of balance, tests of cognitive functioning, and indicators of kidney and lung function. People with profiles that include high-risk levels of inflammation, blood pressure, and frailty were more likely to die within the next 5 years than individuals classified as not high-risk. This methodological approach to using multiple indicators obtained at one time point seems an appropriate means of classifying individuals based on multiple biological and functioning indicators that represent functioning across several physiological systems.

Another approach to computing a biomarker summary score includes recursive partitioning of individuals into low, intermediate, and high allostatic load categories. This identifies a set of predictor biomarker variables and defines a well-established outcome, such as mortality. The algorithm searches among the predictor biomarker variables and their cutpoints to determine the best single predictor variable and its corresponding cutpoint, with individuals partitioned to two groups based on the cutpoint and outcome categories (Zhang & Singer, 1999). Gruenewald et al. (2006) illustrated the utility of regressive partitioning techniques to identify biomarker classifications predictive of 12-year mortality in the MacArthur Studies of Successful Aging.

All of the summary measures have been used to provide evidence of socioeconomic and racial differences in biological risk. A review of the use of allostatic load in evaluating differentials by SES concludes that allostatic load is higher among those of lower SES and lower among those of higher status (Szanton, Gill, & Allen, 2005). A recent paper has shown that perceptions of status, as well as absolute status, also affect the level of allostatic load (Seeman, Merkin, Karlamangla, Koretz, & Seeman, 2014). Several analyses have provided evidence of the life cycle linkages between SES and allostatic load (Gruenewald et al., 2012; Merkin, Karlamangla, Diez Roux, Shrager, & Seeman, 2014).

Biological risk measures comprised of a subset of the markers generally included in allostatic load have demonstrated differences by education, income, and race/ethnicity (Geronimus, Hicken, Keene, & Bound, 2006; Seeman et al., 2008) and clarified that the Hispanic paradox does not extend to biological risk (Crimmins, Kim, Alley, Karlamangla, & Seeman, 2007). This approach has also been used to demonstrate changes over time in the risk experienced by men and women. The finding that women appear to have experienced relative deterioration in biological risk helps to explain why they were also experiencing deterioration in relative mortality (Kim, Alley, Seeman, Karlamangla, & Crimmins, 2006). An examination of differences in a summary indicator of biological risk by age and poverty status found higher biological risk among the poor before old age; however, differences became nonsignificant at older ages. Linking biological risk and mortality showed that socioeconomic differences in risk disappeared at older ages because of mortality differences at earlier ages related to higher biological risk. Life expectancy is associated with both biological risk and poverty status. In the US population aged 20 and older, those with high biological risk had a life expectancy 6 years lower than those with low risk who had the same poverty status and gender (Crimmins, Kim, & Seeman, 2009).

Measurement of Biological Age in a national sample of American whites and blacks showed that at a given chronological age, blacks are approximately 3 years older biologically than whites (Levine & Crimmins, 2014a) (Figure 3.4). The size of the differences varies by age, increasing from age 30–39 to ages 60–69 and then declining, presumably due to mortality selection. In addition, racial differences in biological age were found to completely account

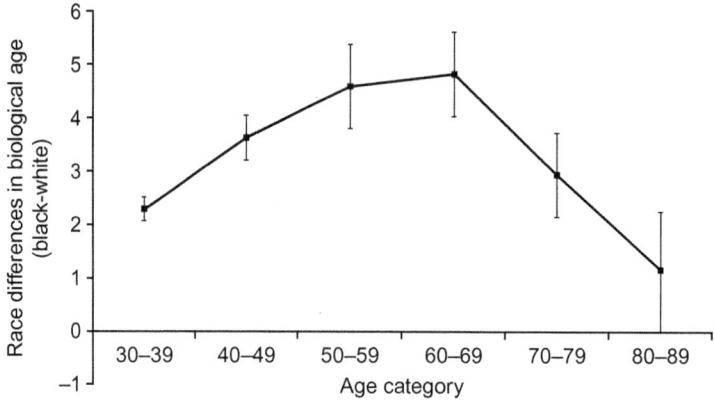

FIGURE 3.4 Race differences in adjusted mean biological age by 10-year chronological age groups. *Source: Levine and Crimmins (2014a). Models adjusted for age, sex, education, BMI, and smoking. Bars represent standard errors of adjusted means.*

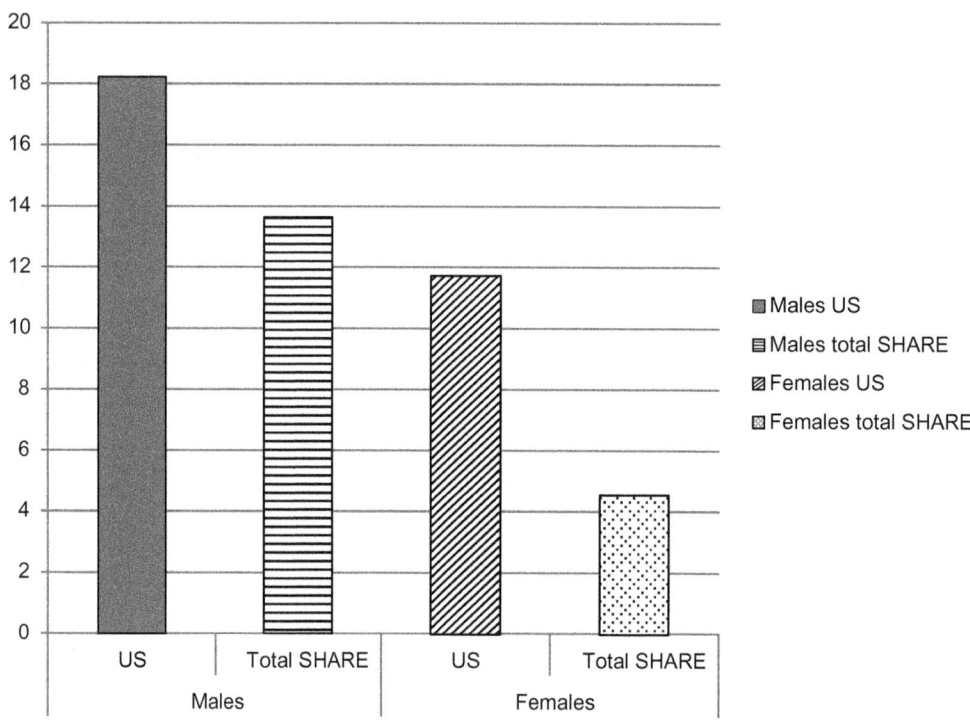

FIGURE 3.5 Percent with high or very high cardiovascular risk, United States and Europe (SHARE): 50–54 year olds. *Source: Crimmins, Solé-Auró, et al. (2011).*

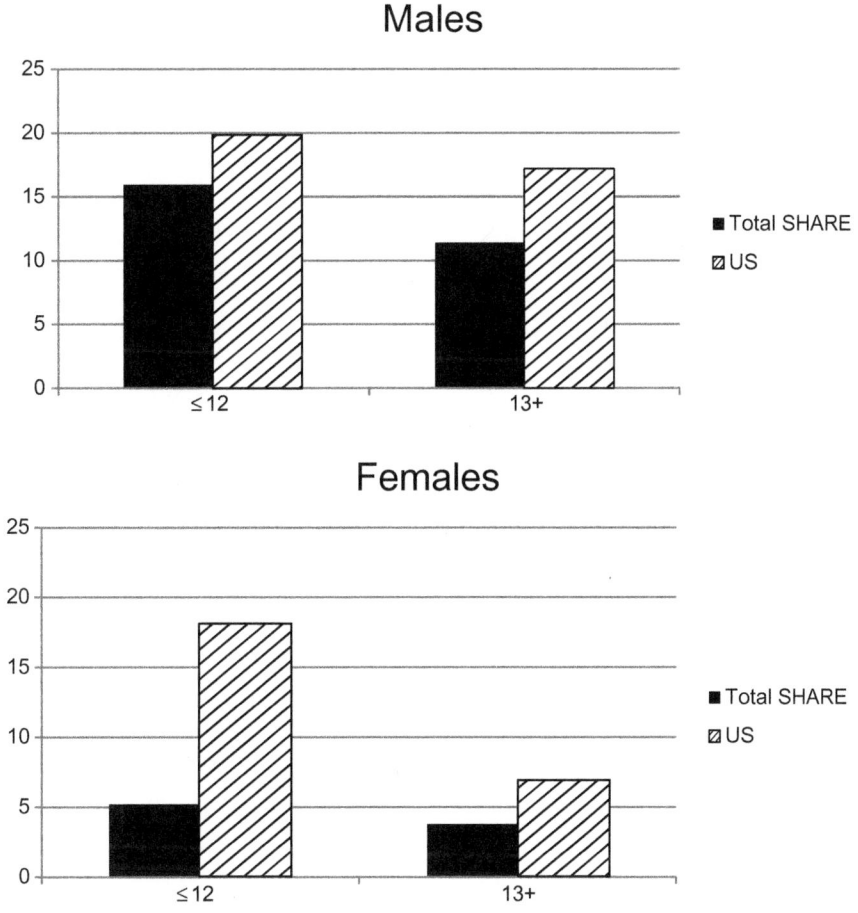

FIGURE 3.6 Percent with high or very high cardiovascular risk by education, United States and Europe. *Source: Crimmins, Solé-Auró, et al. (2011).*

for higher levels of all-cause, cardiovascular, and cancer mortality among blacks.

The Framingham risk score has also been used to provide evidence of strong socioeconomic and race/ethnic differences in cardiovascular risk in the United States (Karlamangla et al., 2010). An adaptation of the Framingham risk score has been used to address differences in cardiovascular risk between the United States and Europe (Crimmins, Solé-Auró, & Kim, 2011). This analysis of cardiovascular risk for persons aged 50–54 supported the idea that Americans enter older age less healthy that Europeans; and that women of low education in the United States are particularly disadvantaged. The percentage of US men at high risk for a cardiovascular event exceeds the European percentage by about 30%; for women the US excess is 150% (Figure 3.5).

The educational differences are particularly large in the United States. Among American women with low education, three to four times as many are in the high cardiovascular risk group compared to European women with the same education (Figure 3.6). The difference between the two categories of education is particularly large for American women (Figure

3.6). This disadvantage in risk factors and disease by middle age is undoubtedly at the root of the differences in life expectancy between Europe and the United States, the strikingly poor performance of the trends in mortality for American women, and the larger SES differences in mortality in the United States relative to Europe (Crimmins, Preston, et al., 2011).

GENETIC MARKERS AS A NEW FRONTIER

Genetic markers are another group of indicators that have recently been included in a number of population studies. The inclusion of genetic markers in analysis is likely to increase markedly over the next few years. Most markers to date are derived from DNA and thus represent inherent health risks to the individual. Until recently only a few genetic indicators of single nucleotide polymorphisms (SNPs) or other markers had been included in population studies. The most commonly examined candidate genetic indicator, and the marker with the greatest evidence of association with multiple age-related health outcomes is apolipoprotein E (APOE). Studies have found elevated risk for late-onset Alzheimer's disease among people with the APOE-ε4 allele (Corder et al., 1993; Poirier et al., 1993), as well as increased risk for CVDs (Schilling et al. 2013). While the effects of the APOE gene are relatively strong, in general the results of research examining the association of candidate genes on health outcomes has not indicated clear-cut relationships between health and longevity and specified candidate markers from DNA (Christensen, Johnson, & Vaupel, 2006).

In the last 2 years the availability of genetic information on large samples has increased rapidly and has changed the approach of many studies. The US Health and Retirement Study currently presents the largest sample with the most genetic information in adult populations. Genotyping of samples from HRS was performed by the NIH Center for Inherited Disease Research using the Illumina Human Omni-2.5 Quad beadchip, with coverage of approximately 2.5 million SNPs. This now allows genome-wide association studies (GWAS) to explore links between genetic markers and traits, behaviors, biological indicators, or health outcomes. In this capacity, most social scientists are not interested in discovering genes but in determining the relevance of genetics within social science research (Freese, 2008; Freese & Shostak, 2009). The ultimate aim of many is to uncover gene–environment interactions that can help explain why some people are at greater risk for certain outcomes, and this knowledge can be used to counsel people at high risk for specific behaviors or lifestyles (Boardman, Blalock, & Pampel, 2010; Boardman et al., 2011).

Up to now, researchers searching for links between social science outcomes, for example, obesity, depression, have not found many genetic markers that meet the significance level expected by geneticists (0.05×10^{-8}). This has been true even with very large samples. For this reason, it is becoming increasingly common for social scientists to combine the effect of many genes with somewhat lower levels of significance into a risk score with the view that genetic influences on the complex conditions of old age result from the small effects of many genes. This approach based on GWAS results combines the effects of multiple genetic markers into a Polygenic Risk Score (PRS) which represents the "genetic burden" associated with a phenotype (Belsky & Israel, 2014; Belsky, Moffit, & Caspi, 2013; Wray, Goddard, & Visscher, 2008). PRS constructed by weighting the SNP-specific coefficients from the GWAS (Dudbridge, 2013) have been used to estimate genetic links to obesity (Domingue et al., 2014), obesity patterns over many years (Belsky et al., 2012), depressive symptoms across multiple waves in an aging population (Levine, Crimmins, Prescott, Arpawong, & Lee, 2014), and the course of childhood asthma (Belsky & Sears, 2014).

Additional genetic measures that are changed with life circumstances are increasingly becoming available in large population studies. Telomere length is seen as a generalized marker of aging that changes with the stress of life circumstances and the individual rate of aging. Telomeres shorten with replication, suggesting that shorter telomeres are an indication of more rapid aging. Shorter telomeres have been related to morbidity (Demissie et al., 2006), mortality (Cawthon, Smith, O'Brien, Sivatchenko, & Kerber, 2003), and stressful circumstances (Epel et al., 2004).

Among the US National Health and Nutrition Examination Study (NHANES) sample of adults, telomeres have been found to be shorter among those with lower education, those who smoke, and the obese (Needham et al., 2013). Recent research in England has suggested that shorter telomeres are associated with allostatic load and reduced psychosocial resources (Zalli et al., 2014).

The next frontier for social researchers is analysis of gene expression (Cole, 2013). Expression analysis, based on RNA, indicates that the human genome responds to life circumstances with different gene expression programs. Negative conditions such as stress (Creswell et al., 2012), loneliness (Cole et al., 2007), and caring for a cancer patient (Rohleder, Marin, Ma, & Miller, 2009) have been associated with changes in gene expression that increase risk for a variety of poor health outcomes. Positive conditions such as psychological well-being can change gene expression in a way that should promote health (Fredrickson et al., 2013). Low SES and adverse childhood circumstances have also been related to changes in gene profiling among children (Chen, Miller, Kobor, & Cole, 2010; Miller & Chen, 2006) and these effects have been shown to persist until adulthood (Chen et al., 2010; Miller et al., 2009).

SUMMARY

The breadth and depth of biodemographic research is undergoing rapid advances in methodologic developments, access to population-level data, and formulation of theoretical frameworks. As our knowledge of the complexities of the biological, psychological, social, and environmental factors influencing aging expands, the need to address the more nuanced questions pertaining to population-level health becomes increasingly vital to improving health both within and across populations. The availability of several on-going, longitudinal population-based surveys with biological information across multiple countries will open new doors for intellectual inquiry into the general aging process, differences in health and aging across populations, trends in health over time, and mechanistic pathways that exacerbate differences in population health. This represents a marked outgrowth from the origins of demography, and the next few years bring great promise to addressing the next frontiers of biodemographic research.

APPENDIX: INFORMATION ON BIOMARKERS OFTEN USED IN SOCIAL SCIENCE RESEARCH

There are several publications that provide general information on this topic. Volumes produced under the National Academy of Sciences serve as a good introduction: Between Zeus and the Salmon (Wachter, Finch, & National Research Council Committee on Population, 1997); Cells and Surveys: Should Biological Measures Be Included in Social Science Research? (Kinsella, Vaupel, & Finch, 2001); Biosocial Surveys (Weinstein, Vaupel, & Wachter, 2007).

Extensive information on individual biomarkers is included in several articles including Crimmins et al. (2008). Below a list of biomarkers regularly included in social studies along with their health-relatedness, source, and inclusion in a specified multisystem index is provided (Table A1). Table A2 provides a listing of some

TABLE A1 Selected Biomarkers, Related Conditions, Sample Source, and Inclusion in Composite Indices

Category	Biomarker	Related condition/assessment	Sample source	Multisystem index
Blood pressure	Systolic BP	Cardiovascular functioning	Exam	AL, BA, FRS
	Diastolic BP			AL
	Pulse			AL
Lipids	Total cholesterol	Cardiovascular/metabolic risk	Blood	AL, BA, FRS
	HDL cholesterol			AL, FRS
	LDL cholesterol			
	Triglycerides			
Obesity, metabolic, insulin–related	HbA1c	Metabolic risk, diabetes	Blood	AL, BA
	Plasma glucose		Blood	
	BMI and waist–hip ratio	Metabolic risk	Exam	AL
Inflammation	Albumin	Inflammation, nutrition	Blood	AL, BA
	C-reactive protein	Inflammation, nutrition	Blood	AL, BA
	IL-6	Inflammation		AL
Infection	Epstein–Barr virus	Immune response, stress	Blood	
	Cytomegalovirus	Immune response	Blood	BA
Endocrine/neuroendocrine	Cortisol	Stress	Blood, urine, saliva	AL
	Epinephrine		Urine	Al
	Norepinephrine		Urine	AL
Kidney	Cystatin C	Kidney functioning	Blood	
	Creatinine		Blood/urine	AL, BA
	Urea nitrogen BUN		Blood	BA
Liver	Alkaline phosphatase	Liver functioning	Blood	BA
Lung	Forced expiratory volume	Lung functioning	Performance	BA
Genetic markers	DNA	Aging, disease, mortality	Blood, saliva	
	RNA			
	Telomere length			
	Methylation		Blood	

BP = blood pressure; HDL = high density lipoprotein; LDL = low density lipoprotein; HbA1c = glycosylated hemoglobin; BMI = body mass index; IL = interleukin; BUN = blood urea nitrogen; DNA = deoxyribonucleic acid; RNA = ribonucleic acid; AL = allostatic load; BA = biological age; FRS = Framingham risk score.

TABLE A2 National Study, Type of Blood Samples Collected, Blood-Based Assays Done

Study	Samples collected	Assays done
HRS FAMILY OF STUDIES		
Health and Retirement Study – HRS	Blood (DBS), saliva	Total cholesterol, HDL cholesterol, HbA1c, CRP, cystatin C, DNA
English Longitudinal Study of Ageing – ELSA	Blood (venous)	Total cholesterol, HDL cholesterol, LDL cholesterol, triglycerides, HbA1c, glucose, CRP, fibrinogen, white cell count, hemoglobin, ferritin, IGF-1, DHEA-S, vitamin D, DNA
Chinese Health and Retirement Longitudinal Study – CHARLS	Blood (venous)	Total cholesterol, HDL cholesterol, LDL cholesterol, triglycerides HbA1c, CRP, cystatin C, complete blood count
The Irish Longitudinal Study of Ageing – TILDA	Blood (venous)	Total cholesterol, HDL cholesterol, LDL cholesterol, triglycerides, HbA1c, DNA, RNA, PBMCs
Mexican Health and Aging Study – MHAS	Blood (DBS, point of care meter)	Total cholesterol, HDL cholesterol, glucose, HbA1c, TSH, Hb, CRP, vitamin D
Indonesia Family Life Study – IFLS	Blood (DBS, point of care meter)	CRP, hemoglobin
OTHER US STUDIES		
National Social Life, Health and Aging – NSHAP	Blood (DBS, microtainers)	HbA1c, CRP, EBV, Hb, cytokines
Midlife in the United States – MIDUS	Blood (venous)	Total cholesterol, HDL cholesterol, LDL cholesterol, triglycerides, HbA1c, CRP, ICAM, IL-6, s-IL-6r, fibrinogen, E-selectin, PINP, BSAP, NTx, vitamin D, DHEA, DHEA-S, creatinine
OTHER NON-US STUDIES		
Costa Rican Longevity and Healthy Aging Study – CRELES (Costa Rica)	Blood (venous)	Total cholesterol, HDL cholesterol, triglycerides, CRP, glucose, HbA1c, DHEA-S, creatinine
Social Environment and Biomarkers of Aging Study – SEBAS (Taiwan)	Blood (venous)	Total cholesterol, HDL cholesterol, triglycerides, HbA1c, glucose IGF-1, IL-6, IL-6sr, ICAM-1, E-selectin, CRP, fibrinogen, leucocytes, lymphocytes, platelet counts, DHEA-S, Hb, CBC, creatinine, homocysteine, folate

DBS = dried blood spot; HDL = high density lipoprotein; LDL = low density lipoprotein; HbA1c = glycosylated hemoglobin; Hb = hemoglobin; CRP = C-reactive protein; DNA = deoxyribonucleic acid; RNA = ribonucleic acid; IGF = insulin-like growth factor; DHEA-S = dehydroepiandrosterone sulfate; PBMCs = peripheral blood mononuclear cells; TSH = thyroid stimulating hormone; EVB = Epstein-Barr virus; ICAM = intracellular cell adhesion molecule; IL = interleukin; PINP = procollagen type I N-terminal propeptide; BSAP = bone-specific alkaline phosphatase; NTx = N-terminal peptide; CBC = complete blood count

of the studies with biomarker data along with information on samples collected and assays produced from blood in each study. We do not include biomarkers from exam (e.g., blood pressure) or performance (e.g., lung function) but focus on bodily fluids such as blood, saliva, and other sources. This table generally includes national studies with some existing data that are now available; however, in the near future, many of these studies will have data from more assays and many more studies will be added. In other cases the type of assays and source is expected to change in the future (e.g., HRS which will collect whole blood in 2016). A more extensive list of studies with included biomarkers is contained in Love, Seeman, Weinstein, and Ryff (2010). We only include blood-based markers here. A number of studies include urine-based markers (e.g., MIDUS, NSHAP, CRELES) and some include saliva-based markers (e.g., MIDUS, NSHAP). Extensive information on biomarkers in individual studies along with collection procedures and details on the studies is provided on the NIA Biomarker Network web site:

http://gero.usc.edu/CBPH/network/resources/studies.

References

Ailshire, J. A., & House, J. S. (2011). The unequal burden of weight gain: An intersectional approach to understanding social disparities in BMI trajectories from 1986 to 2001/2002. *Social Forces, 90*(2), 397–423.

Belsky, D. W., & Israel, S. (2014). Integrating genetics and social science: Genetic risk scores. *Biodemography and Social Biology, 60*(2), 137–155.

Belsky, D. W., Moffitt, T. E., & Caspi, A. (2013). Genetics in population health science: Strategies and opportunities. *American Journal of Public Health, 103*(S1), S73–S83.

Belsky, D. W., Moffitt, T. E., Houts, R., Bennett, G. G., Biddle, A. K., Blumenthal, J. A., et al. (2012). Polygenic risk, rapid childhood growth, and the development of obesity: Evidence from a 4-decade longitudinal study. *Archives of Pediatrics & Adolescent Medicine, 166*(6), 515–521.

Belsky, D. W., & Sears, M. R. (2014). The potential to predict the course of childhood asthma. *Expert Review of Respiratory Medicine, 8*(2), 137–141.

Beltrán-Sánchez, H., Razak, F., & Subramanian, S. V. (2014). Going beyond the disability-based morbidity definition in the compression of morbidity framework. *Glob Health Action, 7*, 24766.

Boardman, J. D., Blalock, C. L., & Pampel, F. C. (2010). Trends in the genetic influences on smoking. *Journal of Health and Social Behavior, 51*(1), 108–123.

Boardman, J. D., Blalock, C. L., Pampel, F. C., Hatemi, P. K., Heath, A. C., & Eaves, L. J. (2011). Population composition, public policy, and the genetics of smoking. *Demography, 48*(4), 1517–1533.

Cai, L., & Lubitz, J. (2007). Was there compression of disability for older Americans from 1992 to 2003? *Demography, 44*(3), 479–495.

Cawthon, R. M., Smith, K. R., O'Brien, E., Sivatchenko, A., & Kerber, R. A. (2003). Association between telomere length in blood and mortality in people aged 60 years or older. *The Lancet, 361*(9355), 393–395.

Chen, E., Miller, G. E., Kobor, M. S., & Cole, S. W. (2010). Maternal warmth buffers the effects of low early-life socioeconomic status on pro-inflammatory signaling in adulthood. *Molecular Psychiatry, 16*(7), 729–737.

Christensen, K., Johnson, T. E., & Vaupel, J. W. (2006). The quest for genetic determinants of human longevity: Challenges and insights. *Nature Reviews Genetics, 7*, 436–448.

Cole, S. W. (2013). Social regulation of human gene expression: Mechanisms and implications for public health. *American Journal of Public Health, 103*(S1), S84–S92.

Cole, S. W., Hawkley, L. C., Arevalo, J. M., Sung, C. Y., Rose, R. M., & Cacioppo, J. T. (2007). Social regulation of gene expression in human leukocytes. *Genome Biology, 8*(9), R189.

Corder, E. H., Saunders, A. M., Strittmatter, W. J., Schmechel, D. E., Gaskell, P. C., Small, G. W., et al. (1993). Gene dose of apolipoprotein E type 4 allele and the risk of Alzheimer's disease in late onset families. *Science, 261*(5123), 921–923.

Creswell, J. D., Irwin, M. R., Burklund, L. J., Lieberman, M. D., Arevalo, J. M., Ma, J., et al. (2012). Mindfulness-based stress reduction training reduces loneliness and pro-inflammatory gene expression in older adults: A small randomized controlled trial. *Brain, Behavior, and Immunity, 26*(7), 1095–1101.

Crimmins, E., & Beltran-Sanchez, H. (2011). Trends in mortality and morbidity: Is there a compression of morbidity? *Journal of Gerontology: Social Sciences, 66*, 75–86.

Crimmins, E. M., Kim, J. K., Alley, D. E., Karlamangla, A., & Seeman, T. (2007). Hispanic paradox in biological risk profiles. *American Journal of Public Health, 97*(7), 1305–1310.

Crimmins, E. M., Kim, J. K., & Seeman, T. E. (2009). Poverty and biological risk: The earlier "aging" of the poor. *The Journals of Gerontology. Series A, Biological Sciences and Medical Sciences, 64*(2), 286–292.

Crimmins, E. M., Kim, J. K., & Vasunilashorn, S. (2010). Biodemography: New approaches to understanding trends and differences in population health and mortality. *Demography, 47*(1), S41–S64.

Crimmins, E. M., Preston, S. H., & Cohen, B. (2011). *Explaining divergent levels of longevity in high-income countries*. Washington, DC: National Academies Press.

Crimmins, E. M., & Seeman, T. E. (2004). Integrating biology into the study of health disparities. *Population and Development Review, 30*, 89–107.

Crimmins, E. M., Solé-Auró, A., & Kim, J. K. (2011). Cardiovascular risk among those near age 50: The US relative to Europe. Unpublished paper prepared for the NAS/IOM Panel on Understanding Health Differences among High Income Countries.

Crimmins, E. M., Vasunilashorn, S., Kim, J. K., & Alley, D. (2008). Biomarkers related to aging in human populations. *Advances in Clinical Chemistry, 46*, 161–216.

Demissie, S., Levy, D., Benjamin, E., Cupples, L., Gardner, J., Herbert, A., et al. (2006). Insulin resistance, oxidative stress, hypertension, and leukocyte telomere length in men from the Framingham Heart Study. *Aging Cell, 5*(4), 325–330.

Domingue, B. W., Belsky, D. W., Harris, K. M., Smolen, A., McQueen, M. B., & Boardman, J. D. (2014). Polygenic risk predicts obesity in both white and black young adults. *PloS One, 9*(7), e101596.

Dudbridge, F. (2013). Power and predictive accuracy of polygenic risk scores. *PLoS Genetics, 9*(3), e1003348.

Epel, E. S., Blackburn, E. H., Lin, J., Dhabhar, F. S., Adler, N. E., Morrow, J. D., et al. (2004). Accelerated telomere shortening in response to life stress. *Proceedings of the National Academy of Sciences of the United States of America, 101*(49), 17312–17315.

Fredrickson, B. L., Grewen, K. M., Coffey, K. A., Algoe, S. B., Firestine, A. M., Arevalo, J. M., et al. (2013). A functional genomic perspective on human well-being. *Proceedings of the National Academy of Sciences of the United States of America, 110*(33), 13684–13689.

Freese, J. (2008). Genetics and the social science explanation of individual outcomes. *American Journal of Sociology, 114*(S1), S1–S35.

Freese, J., & Shostak, S. (2009). Genetics and social inquiry. *Annual Review of Sociology, 35*, 107–128.

Geronimus, A. T., Hicken, M., Keene, D., & Bound, J. (2006). "Weathering" and age patterns of allostatic load scores among Blacks and Whites in the United States. *American Journal of Public Health, 96*(5), 826–833.

Goldman, N., Turra, C. M., Glei, D. A., Lin, Y. H., & Weinstein, M. (2006). Physiological dysregulation and changes in health in an older population. *Experimental Gerontology, 41*, 862–870.

Gruenewald, T. L., Karlamangla, A. S., Hu, P., Stein-Merkin, S., Crandall, C., Koretz, B., et al. (2012). History of socioeconomic disadvantage and allostatic load in later life. *Social Science & Medicine, 74*(1), 75–83.

Gruenewald, T. L., Seeman, T. E., Karlamangla, A. S., & Sarkisian, C. A. (2009). Allostatic load and frailty in older adults. *Journal of the American Geriatrics Society, 57*(9), 1525–1531.

Gruenewald, T. L., Seeman, T. E., Ryff, C. D., Karlamangla, A. S., & Singer, B. H. (2006). Combinations of biomarkers predictive of later life mortality. *Proceedings of the National Academy of Sciences of the United States of America, 103*(38), 14158–14163.

Harris, J., Gruenewald, T. L., & Seeman, T. E. (2007). An overview of biomarker research from community and population-based studies of aging. In M. Weinstein, J. W. Vaupel, & K. W. Wachter (Eds.), *Biosocial surveys* (pp. 96–135). Washington, DC: National Academies Press.

Karlamangla, A. S., Merkin, S. S., Crimmins, E. M., & Seeman, T. E. (2010). Socioeconomic and ethnic disparities in cardiovascular risk in the United States, 2001–2006. *Annals of Epidemiology, 20*(8), 617–628.

Karlamangla, A. S., Singer, B. H., McEwen, B. S., Rowe, J. W., & Seeman, T. E. (2002). Allostatic load as a predictor of functional decline: MacArthur Studies of Successful Aging. *Journal of Clinical Epidemiology, 55*(7), 696–710.

Karlamangla, A. S., Singer, B. H., & Seeman, T. E. (2006). Reduction in allostatic load in older adults is associated with lower all-cause mortality risk: MacArthur Studies of Successful Aging. *Psychosomatic Medicine, 68*(3), 500–507.

Kim, J. K., Alley, D., Seeman, T., Karlamangla, A., & Crimmins, E. (2006). Recent changes in cardiovascular risk factors among women and men. *Journal of Women's Health, 15*(6), 734–746.

Kinsella, K., Vaupel, J. W., & Finch, C. E. (Eds.), (2001). *Cells and surveys: Should biological measures be included in social science research?*. Washington, DC: National Academies Press.

Kubzansky, L. D., Kawachi, I., & Sparrow, D. (1999). Socioeconomic status, hostility, and risk factor clustering in the normative aging study: Any help from the concept of allostatic load? *Annals of Behavioral Medicine, 21*(4), 330–338.

Levine, M. E. (2013). Modeling the rate of senescence: Can estimated biological age predict mortality more accurately than chronological age? *The Journals of Gerontology. Series A, Biological Sciences and Medical Sciences, 68*(6), 667–674.

Levine, M. E., & Crimmins, E. M. (2014a). Evidence of accelerated aging among African Americans and its implications for mortality. *Social Science & Medicine, 118*, 27–32.

Levine, M. E., & Crimmins, E. M. (2014b). A comparison of methods for assessing mortality risk. *American Journal of Human Biology, 26*(6), 768–776.

Levine, M., Crimmins, E. M., Prescott, C., Arpawong, T., & Lee, J. (2014). A polygenic risk score associated with measures of depressive symptoms among older adults. *Biodemography and Social Biology, 60*(2), 199–211.

Love, G. D., Seeman, T. E., Weinstein, M., & Ryff, C. D. (2010). Bioindicators in the MIDUS national study: Protocol, measures, sample, and comparative context. *Journal of Aging and Health, 22*(8), 1059–1080.

McDade, T. W., Williams, S., & Snodgrass, J. J. (2007). What a drop can do: Dried blood spots as a minimally invasive method for integrating biomarkers into population-based research. *Demography*, 44(4), 899–925.

McEwen, B. S. (2004). Protective and damaging effects of mediators of stress and adaptation: Allostasis and allostatic load. In J. Schulkin (Ed.), *Allostasis, homeostasis and the costs of physiological adaptation* (pp. 65–98). Cambridge: Cambridge University Press.

Merkin, S. S., Karlamangla, A., Crimmins, E., Charette, S. L., Hayward, M., Kim, J. K., et al. (2009). Education differentials by race and ethnicity in the diagnosis and management of hypercholesterolemia: A national sample of US adults (NHANES 1999–2002). *International Journal of Public Health*, 54(3), 166–174.

Merkin, S. S., Karlamangla, A., Diez Roux, A. V., Shrager, S., & Seeman, T. E. (2014). Life course socioeconomic status and longitudinal accumulation of allostatic load in adulthood: Multi-ethnic study of atherosclerosis. *American Journal of Public Health*, 104(4), e48–e55.

Miller, G. E., & Chen, E. (2006). Life stress and diminished expression of genes encoding glucocorticoid receptor and beta2-adrenergic receptor in children with asthma. *Proceedings of the National Academy of Sciences*, 103(14), 5496–5501.

Miller, G. E., Chen, E., Fok, A. K., Walker, H., Lim, A., Nicholls, E. F., et al. (2009). Low early-life social class leaves a biological residue manifested by decreased glucocorticoid and increased proinflammatory signaling. *Proceedings of the National Academy of Sciences of the United States of America*, 106(34), 14716–14721.

Nakamura, E., & Miyao, K. (2003). Further evaluation of the basic nature of the human biological aging process based on a factor analysis of age-related physiological variables. *The Journals of Gerontology Series A: Biological Sciences and Medical Sciences*, 58(3), B196–B204.

Needham, B. L., Adler, N., Gregorich, S., Rehkopf, D., Lin, J., Blackburn, E. H., et al. (2013). Socioeconomic status, health behavior, and leukocyte telomere length in the national health and nutrition examination survey, 1999–2002. *Social Science & Medicine*, 85, 1–8.

Poirier, J., Davignon, J., Bouthillier, D., Kogan, S., Bertrand, P., & Gauthier, S. (1993). Apolipoprotein E polymorphism and Alzheimer's disease. *Lancet*, 342, 697–699.

Rohleder, N., Marin, T. J., Ma, R., & Miller, G. E. (2009). Biologic cost of caring for a cancer patient; Dysregulation of pro and anti-inflammatory signaling pathways. *Journal of Clinical Oncology*, 18, 2909–2915.

Sakkinen, P. A., Wahl, P., Cushman, M., Lewis, M. R., & Tracy, R. P. (2000). Clustering of procoagulation, inflammation, and fibrinolysis variables with metabolic factors in insulin resistance syndrome. *American Journal of Epidemiology*, 152(10), 897–907.

Schilling, S., DeStefano, A. L., Sachdev, P. S., Choi, S. H., Mather, K. A., DeCarli, C. D., et al. (2013). APOE genotype and MRI markers of cerebrovascular disease. *Neurology*, 81(3), 292–300.

Seeman, M., Merkin, S., Karlamangla, A., Koretz, B., & Seeman, T. (2014). Social status and biological dysregulation: The "status syndrome" and allostatic load. *Social Science & Medicine*, 118, 143–151.

Seeman, T., & Crimmins, E. (2001). Social environment effects on health and aging: Integrating epidemiologic and demographic approaches and perspectives. *Annals of the New York Academy of Sciences*, 954(1), 88–117.

Seeman, T., Epel, E., Gruenewald, T., Karlamangla, A., & McEwen, B. (2010). Socio-economic differentials in peripheral biology: Cumulative allostatic load. *Annals of the New York Academy of Sciences*, 1186, 223–239.

Seeman, T. E., McEwen, B. S., Rowe, J. W., & Singer, B. H. (2001). Allostatic load as a marker of cumulative biological risk: MacArthur Studies of Successful Aging. *Proceedings of the National Academy of Sciences of the United States of America*, 98(8), 4770–4775.

Seeman, T. E., Merkin, S. S., Crimmins, E., Koretz, B., Charette, S., & Karlamangla, A. (2008). Education, income and ethnic differences in cumulative biological risk profiles in a national sample of US adults: NHANES III (1988–1994). *Social Science & Medicine*, 66(1), 72–87.

Seeman, T. E., Singer, B. H., Rowe, J. W., Horwitz, R. I., & McEwen, B. S. (1997). Price of adaptation – allostatic load and its health consequences: MacArthur Studies of Successful Aging. *Archives of Internal Medicine*, 157(19), 2259–2268.

Szanton, S. L., Gill, J. M., & Allen, J. K. (2005). Allostatic load: A mechanism of socioeconomic health disparities? *Biological Research for Nursing*, 7(1), 7–15.

Turra, C. M., Goldman, N., Seplaki, C. L., Glei, D. A., Lin, Y., & Weinstein, M. (2005). Determinants of mortality at older ages: The role of biological markers of chronic disease. *Population and Development Review*, 31(4), 675–698.

Vasunilashorn, S., Best, L. E., Kim, J. K., & Crimmins, E. M. (2014). Predicting mortality from profiles of biological risk and performance measures of functioning. In J. Anson & M. Luy (Eds.), *Mortality in an international perspective* (pp. 119–135). Cham, Switzerland: Springer International Publishing.

Vasunilashorn, S., & Crimmins, E. M. (2008). Biodemography: Integrating disciplines to explain aging. In V. L. Bengtson, D. Gans, N. M. Putney, & M. Silverstein (Eds.), *Handbook of theories of aging* (pp. 63–85). New York, NY: Springer.

Verbrugge, L. M., & Jette, A. M. (1994). The disablement process. *Social Science & Medicine*, 38(1), 1–14.

Wachter, K. W., Finch, C. E., & National Research Council Committee on Population, (1997). *Between Zeus and the salmon: The biodemography of longevity*. Washington, DC: National Academies Press.

Wang, T. J., Gona, P., Larson, M. G., Levy, D., Benjamin, E. J., Tofler, G. H., et al. (2007). Multiple biomarkers and the risk of incident hypertension. *Hypertension, 49*(3), 432–438.

Weinstein, M., Vaupel, J. W., & Wachter, K. W. (Eds.), (2007). *Biosocial surveys*. Washington, DC: National Academies Press.

Weinstein, M., Vaupel, J. W., & Wachter, K. W. (Eds.), (2008). *Biosocial surveys*. Washington, DC: National Academies Press.

Wilson, P. W., D'Agostino, R. B., Levy, D., Belanger, A. M., Silbershatz, H., & Kannel, W. B. (1998). Prediction of coronary heart disease using risk factor categories. *Circulation, 97*(18), 1837–1847.

World Health Organization. (1999). *International classification of functioning and disability*, Beta-2, Geneva. Available at <http://www.who.int/icidh/index.htm>.

Wray, N. R., Goddard, M. E., & Visscher, P. M. (2008). Prediction of individual genetic risk of complex disease. *Current Opinion in Genetics & Development, 18*(3), 257–263.

Zalli, A., Carvalho, L. A., Lin, J., Hamer, M., Erusalimsky, J. D., Blackburn, E. H., et al. (2014). Shorter telomeres with high telomerase activity are associated with raised allostatic load and impoverished psychosocial resources. *Proceedings of the National Academy of Sciences of the United States of America, 111*(12), 4519–4524.

Zhang, H., & Singer, B. (1999). *Recursive partitioning in the health sciences*. New York, NY: Springer-Verlag.

CHAPTER 4

Late-Life Disability Trends and Trajectories

Douglas Wolf

Aging Studies Institute, Syracuse University, Syracuse, NY, USA

OUTLINE

Introduction	77	*Potential Explanations for "True" Change*	88
The Conceptualization of Disability	79	Individual-Level Disability Trajectories	89
Measuring Disability	81	*Time to Death as an Additional "Time" Variable*	93
Evidence on Trends in Disability Prevalence	84	*Accounting for Sample Losses due to Death*	94
Overall Trends	84	Conclusion	94
Between-Group Differences, Compositional Change, and "True" Change	86	References	95

INTRODUCTION

Late-life disability, broadly defined as difficulties in carrying out everyday tasks or participating in social activities due to a lack of "… fit between individuals and their environments" (Iezzoni & Freedman, 2008, p. 334), is a topic of intense and sustained interest. Under this broad definition disability is complex and multidimensional, and can be associated with any of numerous physiological, cognitive, or health conditions ranging from arthritis to advanced dementia. Disability is a key indicator of population health, and the population of people with disabilities is an important target group for monitoring and intervention. For example, the interagency *Healthy People 2020* initiative includes several goals aimed at those with disabilities (US Department of Health and Human Services, 2014). These goals include enhancing opportunities for such people to be included in community and health promotion activities; to receive well-timed interventions and services; to interact with their environment without barriers; and to participate in everyday life.

Disability is also costly, imposing financial burdens on individuals with disabilities, their families, and society as a whole. One important component of the costs of disability is the time costs associated with the provision of "informal care" to those whose disabling conditions require it. It is well known that family members, and particularly children, continue to provide the majority of help received by older people with personal care needs (Wolff & Kasper, 2006). Concerns have been raised about the prospective passage of large numbers of individuals in the Baby Boom generation through the ages of greatest potential need for personal assistance, at a time when the relative number of potential caregivers – the children of the Baby Boomers – is at a historically low level (Redfoot, Feinberg, & Houser, 2013).

Documenting disability trends can also be useful as a basis for forecasting, helping to identify turning points and the causes of such developments. Moreover, trends in the characteristics of the younger population have implications for the characteristics of the older adult population of the future. For example, several studies have pointed out that recent increases in the prevalence of obesity among younger age groups may portend rising rates of disability among the older population of the future (Martin & Schoeni, 2014).

The recent and rapid growth in the literature on late-life disability has been encouraged, in part, by a growing availability of public-use survey data with which to study the topic, along with new methodological tools and statistical software that facilitate appropriate analyses. This chapter focuses on two main areas of empirical research. The first area relates to trends in the *prevalence* of disability among the older population. A prevalence measure is a point-in-time "snapshot" of the population's situation, and the study of disability *trends* therefore requires a sequence of two or preferably more such snapshots, in order to establish whether prevalence has changed over time. The large literature on this topic dates mainly from the mid-1980s, with scattered earlier contributions. Interest in population-level disability-prevalence trends can be expected to be of enduring interest, because trends are by their very nature continually evolving, and are subject to redirection and reversal at any point. Disability prevalence is, by its very nature, a characteristic of a population or subpopulation, that is, an *aggregate* phenomenon.

The second focus of the chapter is on individual-level disability *"trajectories"* – that is, "… individual patterns of stability and change" (George, 2009) in disability status or severity that are revealed with the passage of time. Growth in the disability-trajectory literature has been rapid and diffuse, reflecting the several methodological approaches that can be used to study them, along with a proliferation of new panel data sets containing the type of information needed to study them. In contrast to the inherently "macro" focus of disability prevalence studies, there is a natural "micro" focus in disability trajectory analyses. Yet there is a clear connection between these two objects of study: at any point in time, population-level disability prevalence is a summary of the current location along their dynamic disability trajectories of all the individuals that comprise the population. Both are emphatically empirical phenomena, so measurement and survey methodology issues figure prominently in the literature.

Given the size and breadth of the literature on disability trends and trajectories, it is necessary to limit the scope of this chapter in order to make the task manageable. Accordingly, I have limited my review mainly to empirical research based on population samples drawn from the United States. This means that many excellent and important studies based on regional or small-area samples, or group- or disease-specific samples (such as the Established

Populations for the Epidemiologic Study of Aging samples, the Women's Health and Aging Study, the Cardiovascular Health Study, or the Precipitating Events Project, among many others) as well as the growing literature based on data from other countries, are largely excluded. Another substantial body of research omitted from this review is that relating to active (or disability-free) life expectancy or ALE (Crimmins, Hayward, Hagedorn, Saito, & Brouard, 2009). ALE is a summary measure that reflects, at least in principle, both the "prevalence" and the "trajectory" aspects of late-life disability, yet it is rarely directly observable; instead, it entails a substantial computational burden as well as the imposition of many additional assumptions. ALE, like point-in-time prevalence, is regarded as a useful indicator of population health; for an extensive review of the approach, see Molla, Madans, Wagener, and Crimmins (2003).

THE CONCEPTUALIZATION OF DISABILITY

The ability to conduct empirical research on disability trends or trajectories rests on the availability of empirical measures of disability, and those measures should reflect an underlying conceptualization of the meaning and observable manifestations of disability. The task of developing such a conceptualization is complicated by the great variety of situations to which it may apply – to domains as diverse as childhood development, the full range of educational settings, the workplace, self-care in a domestic setting, and social interactions – across the life course, and across time, cultural settings, and space. The criteria by which an individual is classified as having, or not having, a disability, or with which the severity of disability is characterized, may also differ according to the purpose to which such classifications and characterizations are to be put, including determination of the need for, and the eligibility for, services; or for legal protections; for medical treatment; or for scientific research. It is therefore unsurprising that current conceptualizations of disability are both complex and multidimensional.

The most recent, and in many respects most fully elaborated, conceptual framework for disability is found in the World Health Organization's *International Classification of Functioning, Disability, and Health* (or ICF), released in 2001 (WHO, 2001). The main components of the ICF and the interrelationships among them are depicted in Figure 4.1.

Of the six components shown, two – health conditions, and body functions and structure – are exclusively "internal" to an individual. *Health conditions* include diseases, disorders, injuries, trauma, and genetic or congenital anomalies; they may, therefore, include unobservable or unrecognized as well as readily detectable or diagnosable conditions. Health conditions may give rise to problematic or abnormal features of *body structures* – anatomical features such as organs and limbs – and to *body functions* – the "operating characteristics" of those structures. Importantly, these first three elements can, in principle, be assessed without respect to a person's environment, social setting, or his or her preferences regarding desired or valued activities. Yet none are sufficient for classifying someone as "having a disability." Instead, the ICF, like several conceptual frameworks that preceded it, recognizes that disability depends on the interactions between an individual and his or her environment, which may be configured so as to help or hinder the performance of important or necessary *activities*, or one's *participation* in various life situations. Personal factors, the final component of the conceptual framework, are not classified in the ICF "… because of the large cultural variance associated with them" (WHO, 2001, p. 8). The latter decision seems to create a role for individual preferences, personality traits,

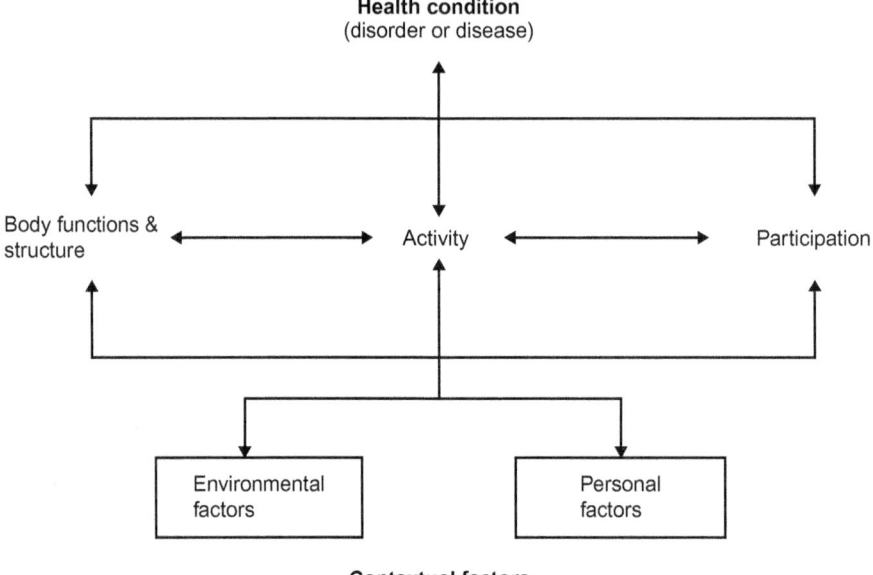

FIGURE 4.1 Components of the International Classification of Functioning, Disability and Health (ICF) and their interrelationships. *Source: World Health Organization (2001, p. 18).*

aspirations, ability to tolerate pain or discomfort, and many other possibly idiosyncratic factors that might play a role in disablement.

The ICF is vast in scope, containing eight chapters each on "body functions" and "body structures," nine chapters on "activities and participation," and five on "environmental factors." As an indication of its level of detail, the fourth of the nine chapters on activities and participation deals with mobility and includes 52 separate "indicators." One such indicator, "walking" (d450) encompasses subindicators such as "walking on different surfaces" (d4502), for which five levels of ability – no difficulty; mild difficulty; moderate difficulty; severe difficulty; and complete inability – are distinguished. In its entirety, the ICF provides hundreds of potential dimensions along which to characterize a person's degree of disability.

The ICF framework has been criticized for a number of reasons. For example, the ICF is unclear on how to distinguish between "activities" and "participation" (Guralnik & Ferrucci, 2009). Moreover, as can be seen in Figure 4.1, all the arrows representing "interactions" are two-headed, which hinders any attempt to characterize a developmental pathway for disability. More generally, the ICF pays scant attention to the dynamic nature of disability, that is, the "disablement process" (Verbrugge & Jette, 1994). Nevertheless, an Institute of Medicine report (Field & Jette, 2007) urged the US research community to adopt it, while recognizing many of its limitations and simultaneously calling for further improvements in it (see also Jette, 2009). Finally, it bears mentioning that while the ICF is characterized as a "bio-psycho-social model," one that encompasses both a medical model – one focused on individuals – and a social model – one focused on the environment, its conceptualization of disability remains contested (WHO, 2011).

Three key conclusions can be drawn from consideration of this conceptualization of disability: first, there is an inevitable, and potentially large, element of subjectivity in disability classification, whether it results from self- or clinical assessments; second, we can expect there to be a great deal of heterogeneity within any group of individuals classified as having a disability; and third, trends in disability prevalence could be due to many factors – such as environmental and technological features, the organization of work, or even bureaucratic behavior – that have little to do with disease, chronic health conditions, or physiological capacity.

MEASURING DISABILITY

The literature on trends and trajectories in late-life disability is based almost exclusively on self- or proxy-reported responses to survey questions. Given the costs of generating high-quality data in a population-level survey sample, it is not surprising that available data sources contain relatively few data elements in comparison to the numerous potential items suggested by a framework such as the ICF. Moreover, there are several variations on the wording that can be used to elicit a respondent's assessment of their "impairments" or "limitations" or whether they face "restrictions" (the quoted terms reflecting ICF language). The fact that minor variations in question wording can alter the estimated prevalence of disability is well established (Freedman, 2000; Freedman, Aykan, & Kleban, 2003; Jette, 1994). The consistency of survey question items over time is particularly relevant to research on trends in disability prevalence, which requires combining data, either from different surveys in different years or from different waves of a repeated cross-sectional survey, and especially for research on individual-level disability trajectories, in which the disability status of the same individuals at different times must be assessed. Accordingly, close attention to the various ways in which disability-related survey data are collected is necessary. Three such dimensions are emphasized here: the specific domains for which limitations or deficits are elicited; the specific verbal descriptor used to elicit limitations; and the time period referenced in the questions.

The great majority of relevant survey data has addressed two of the elements of the ICF, body *functions* and routine *activities*. Physiological functions often measured include: lifting or carrying heavy objects, sitting, walking, use of hands and fingers, climbing stairs, and stooping, bending, crouching or kneeling, items that trace back to early work by Nagi (1969, 1976). Additional domains of functioning that may be assessed in a survey include sensory (e.g., vision and hearing), cognitive (e.g., memory, language, executive function), psychological, and mental-health functioning. For these additional domains, however, a potential problem is posed by the fact that the presence of a deficit may itself undermine or hinder the attempt to measure it. All such domains have the advantage of being "intrinsic" to the person – they represent underlying capacity, independent of the setting in which activities are carried out – but this is, as well, a disadvantage inasmuch as it disregards the interaction between person and environment widely understood to characterize disability, as presently conceptualized. Furthermore, there are numerous potential domains of body functioning, and none are necessary or – arguably – sufficient, by themselves, to produce disability. For this reason measures of disability appear to be more often based on *activity* limitations than on *functional* limitations.

Two sets of activities have come to be widely used to assess disability at older ages: the first consists of Activities of Daily Living (ADL) – most often, bathing, dressing, toileting, transfer from bed or chair, walking indoors, and eating – initially developed by Katz et al. (1959); the second is a set of Instrumental Activities of Daily Living (IADL), which may include use of a telephone, shopping, meal preparation, housework, laundry, travel, medication management, and money management (Lawton & Brody, 1969). Problems performing ADL tasks would generally be viewed as more serious than would IADL problems; moreover, ADL tasks are truly "daily" in nature, and entail hands-on assistance if help is provided by others, whereas IADL tasks may be less intense, require less frequent attention, and in some cases can be done by others without face-to-face contact between the provider and recipient of assistance.

It is important to notice that among the many population-level surveys that have included questions on disability, the word "disability" is rarely used. Instead, the usual approach is to ask one or more questions that address specific functions or activities, leaving to the analyst the task of coding the survey responses so as to indicate those that have a disability, or to quantify the severity of disability. Indeed, some research has shown a lack of correspondence between individual characteristics that might generally be interpreted as indicators of having a disability, and the perception that one actually has a disability. For example, the 1994–1995 disability supplement to the National Health Interview Survey (NHIS), one of few studies that asked directly about disability as a label, included the questions "Do you consider yourself to have a disability?" and "Would other people consider you to have a disability?" Using those data, Iezzoni, McCarthy, Davis, and Siebens (2000) showed that among those with "major mobility problems" only about 71% viewed themselves as having a disability, and only 67% of that same group reported that others would perceive them in that way. Even among those with major mobility problems who used an electric wheelchair, only 87% viewed themselves as having a disability; about 83% of that group reported that others would perceive them in that way.

The wording of survey questions aimed at identifying problems with functioning or with carrying out activities can take several forms. Some surveys ask whether respondents have "any *difficulty*" – preferably, with an explicit qualification that the difficulty must be associated with or result from a health or physiological problem – performing a task; the degree of difficulty (e.g., "a little," "some," or "a lot") may also be elicited. A second approach uses questions about whether the respondent "requires assistance" or "*needs help*" from another person in order to carry out the task. Finally, a survey may ask whether the respondent *receives help* from another person, or whether he or she uses special equipment such as canes, walkers, grab bars, and so on, when performing the functions or activities. Some surveys combine two or all three of these approaches. Each leads logically to an attribution of disability, albeit using differing criteria. The differences in question wording are expected to produce differences in the estimated prevalence of disability; for example, it is likely that a larger proportion of a sample will report "difficulty" than will report that they "need help" to perform a particular activity (Freedman et al., 2004). Depending on the disability construct being measured, different question-wording approaches might produce under- or overestimated prevalence levels. For example, "false positive" attributions of disability may be produced if someone reports that they receive help from another person with a daily activity but could, nevertheless, perform that same task without help with little or no "difficulty" (Bootsma-van der Weil et al., 2001; Wolf, Hunt, & Knickman, 2005).

A third source of variability across disability-related questions relates to the time period for which difficulties, or the need for assistance, or the receipt of assistance, is to be reported. A question may omit any such temporal reference (e.g., "do you have any difficulty ...?"), implicitly referring to the present, but leaving to the respondent considerable freedom to interpret the reference period (e.g., it could be interpreted as "usually" or "recently" rather than "now"). More typically, a specific reference period is mentioned, although different surveys use very different such periods. For example, the National Long Term Care Survey (NLTCS) and the Health and Retirement Study (HRS), both of which have been extensively used to study disability trends and trajectories, adopted survey questions intended to limit reports of activity limitations to those that have lasted, or are expected to last, at least 3 months. Gill and Gahbauer (2005) have argued that measures based on limitations that are expected to last 3 months are substantially biased upward, because of respondents' inability to accurately foresee the duration of their limitations. In contrast, the more recent National Health and Aging Trends Study (NHATS) asks whether "in the last month" the respondent has used equipment or has received help from others in carrying out ADL activities.

In addition to the wording of questions, other aspects of survey methodology could affect the resulting measures of disability. One such methodological issue is the *mode* of survey administration, particularly the distinction between in-person and telephone surveys. One possible concern is that people with disabilities may be underrepresented in telephone-survey samples in comparison to face-to-face survey samples, although Kinne and Topolski (2005) found the opposite. A second concern is that survey respondents may respond differently to questions about disability if asked in person rather than by telephone. Both Wolf et al. (2005) and Kelley-Moore (2006) found differences in the prevalence of reported disability by mode – with in-person reports nearly always exceeding telephone reports – but neither study reported tests of significance for those differences. However, in what seems to be the only randomized-assignment test of survey mode effects, Herzog and Wallace (1999) found no significant differences by mode in cognitive functioning or in either ADL or IADL difficulties.

The NHATS survey, which collected its first wave of annual survey data in 2011, adopts a new measurement protocol that permits an expanded representation of the environmental aspects of disability (Freedman et al., 2011). Analysis of the 2011 data produced a five-category hierarchy: those "fully able" to carry out self-care and mobility activities (31% of the 65-and-older population); a "successful accommodation" group able to carry out those activities using assistive devices (24.5%); a small group characterized by "activity reduction," indicating a decline in their quality of life (5.6%); a "difficulty despite accommodation" group that experiences difficulties even when using devices but gets no help from others (18.4%); and a group which uses "assistance from others" for these tasks (20.5%; Freedman, Kasper et al., 2014). The second, third, and fourth of these groups contain people who report not being fully "able", yet are managing daily life without assistance from others; using previous measurement approaches a substantial fraction of these groups would not be viewed as having a disability despite their evident behavioral adaptations to diminished capacity. A second paper based on the 2011 NHATS data has also appeared, presenting a somewhat different measurement of disability that incorporates the receipt of help with household tasks along with the mobility and self-care measures considered above (Freedman & Spillman, 2014). Both papers show that nearly half of the 65-and-older population reported receiving help with, or having difficulty despite any supports used, this set of activities. Both studies are cross-sectional; however, current plans are to produce a series of eight annual waves of data using the NHATS disability protocol, providing data that should prove valuable in both trends and trajectories research.

It is evident that researchers face a wide range of possibilities in coding survey data for use in studies of disability trends or trajectories. Undoubtedly the simplest such coding scheme is a binary indicator of having a disability, for example, 1 = "yes" and 0 = "no." However, some studies limit the "having a disability" category to those with ADL limitations only, while others allow for ADL and IADL limitations, separately or in combination. Still others may include those with no reported ADL or IADL limitations, if selected functional limitations are nevertheless reported. Alternatively, a count of activity limitations, or of functional status problems, may be used to provide a more refined measure. Even when a count (e.g., 0 through 6 ADLs) is used, readers may be tempted to interpret results as if the population could be divided into those that do, and those that do not, have a disability. Continuous measures of disability severity, possibly based on weighted counts or factor loading scores, would have obvious utility in applied research, but few such disability scales appear to exist. When disability is an outcome variable, as is the case throughout the literature reviewed in this chapter, the choice of coding scheme will be driven by the purpose of the analysis (within the limits imposed by the original survey data). For example, for purposes of policy analysis, a carefully chosen binary variable that closely corresponds to the distinction between "eligible" and "not eligible" for program benefits may be useful. In contrast, for purposes of evaluating the efficacy of an intervention – for example, expanded subsidies for home-modification or other environmental enhancements – a more sensitive measure that captures the spectrum of difficulty with everyday tasks may be more appropriate. In the next two sections of this chapter, we consider first the trends-in-prevalence, and then the trajectories, literatures in which the various disability measures are used.

EVIDENCE ON TRENDS IN DISABILITY PREVALENCE

Overall Trends

Some of the earliest studies to call attention to an apparent decline in the prevalence of late-life disability (specifically, among the 65-and-older population) in the United States used data from the NLTCS; Manton, Corder, and Stallard (1993) reported a downward trend in disability prevalence – based on an indicator that incorporated both ADL and IADL limitations – for the 1982–1989 period. Manton, Corder, and Stallard (1997) and Manton and Gu (2001) showed that this downward trend continued through 1994 and 1999, respectively. These findings were enthusiastically received (see, e.g., Cutler, 2001), heralded as a leading indicator of increased quality of life and a moderating factor in the future growth of healthcare costs.

However, other research appearing around 2000, using a number of different data sources, did not consistently find such declines. If IADL-related disability were considered separately, then evidence from multiple sources showed, with remarkable consistency, a downward trend in prevalence during the 1990s (Freedman, Martin, & Schoeni, 2002). With respect to ADL disability, however, the evidence on direction of change over time was inconsistent: for example, in contrast to the NLTCS, analyses of the NHIS (Crimmins, Saito, & Reynolds, 1997; Schoeni, Freedman, & Wallace, 2001) and the Medicare Current Beneficiary Survey (Waidmann & Liu, 2000) failed to find significant changes in the prevalence of ADL disability over time, while the Supplements on Aging to the NHIS (Crimmins & Saito, 2000; Liao, McGee, Cao, & Cooper, 2001) reported significant increases in ADL disability. For additional details about these and other national-level sources of disability data see Freedman et al. (2004) or Livermore, Whalen, Prenovitz, Aggarwal, and Barbos (2011).

In an attempt to reconcile these contradictory findings, a working group was assembled in 2002 to carry out coordinated analyses. The group reanalyzed data from five of the major surveys on disability, investigating a number of potential sources of discrepancies such as question wording and the treatment of the institutionalized population, response rates, loss to follow-up, proxy response rates, and use of sampling weights. Considerable effort was expended in an attempt to develop strongly comparable measures and methods across data sources. With respect to the group's main goal, reconciling the inconsistent findings regarding the trend in ADL disability prevalence, the group's report concluded that while

> ... it is difficult to pinpoint when the declines began, we consistently found the following for the mid- and late 1990s:
> - declines on the order of 1–2.5% per year, on average, in the proportion of the community-based older population who reported *difficulty* with daily activities. We observed these declines in two different data sets even when we omitted the institutionalized population and did not adjust for shifts in the age composition of the older population.
> - declines on the order of 1–2.5% per year, on average, in the proportion of the community-based older population who reported the *use of help* with daily activities. We observed these declines in four different data sets. Adding the institutionalized population and adjusting for shifts in the age composition of the older population strengthened these findings.
> - increases in the proportion of the community-based older population who *used equipment but not personal care* to bathe (a trend that continued from the 1980s) and declines in the proportion who got *help* with bathing on the order of 2–4% per year, on average. These consistencies were observed across four different data sets. Limitations in other highly prevalent ADLs, particularly walking, did not display a clear and consistent pattern. (Freedman et al., 2004, p. 435)

Thus, considerable support exists for the conclusion that severe (ADL) disabilities *declined* among the older population during the mid- and late 1990s, a finding that holds for multiple measures of disability and over several data sources. Together with the previously established consensus that IADL disability declines were observed across multiple data sources, this provided a favorable picture of trends in disability prevalence at the end of the twentieth century.

However, later research suggested that those downward trends might have slowed or even flattened. The 2002 working group reconvened to consider early-2000s trends, again conducting coordinated analyses of five different data sources, covering the period 2000–2008. The report of the second reconciliation study concluded that

> ... the percentage of the older population with one or more activity limitations has been flat since 2000. Yet, for the oldest-old who have the highest rates of activity limitations, we found evidence of continuing declines in both IADL and ADL limitations and of contractions in the size of the institutional population. At the same time, adults poised to enter late life over the next decade appear to have rates of activity limitations that, albeit low relative to the older age groups, are about 1 percentage point higher than the same age group born 10 years earlier. (Freedman et al., 2013, p. 669)

Other recent research contributes to this less-rosy picture of post-2000 trends. Kaye (2013), who analyzed data from the Survey of Income and Program Participation (a source not included in either of the reconciliation projects described above), concluded that the "overall prevalence of disability for [the elderly] has stabilized since 2000" (p. 127). Freedman, Wolf, and Spillman (2014) compared disability prevalence using NLTCS "screener" items – administered to the full NLTCS sample only in 1982 and 2004 – and 2011 data from NHATS, which, in an effort to obtain comparability with NLTCS estimates, reproduced its screener disability questions exactly. The Freedman et al. (2014) study uses an "any disability" indicator reflecting an underlying list of 15 ADL and IADL activities; for the ADL activities, respondents were asked whether they had "any problem" performing the task without help from another person, and for the IADL tasks whether they were unable to perform the task due to a "health or disability problem." For men, the 2011 prevalence of disability was not significantly different from the 2004 prevalence, in keeping with Kaye's finding of "stable" prevalence rates. For women, however, the 2011 prevalence was significantly higher than the

2004 prevalence, a result which emerged both with and without age-standardization of the period rates. Even more troublingly, the 2001 prevalence rates (measured using NHATS data) were not significantly lower than the 1982 rates found in NLTCS data, suggesting that for women, at least, the gains of the 1990s have been completely eliminated.

Taken together, the literature reviewed here reveals a pattern of downward trends in late-life disability – a favorable result, from the perspectives of population health, individuals facing old age, and their families – followed by a period of lesser or no further improvement in the prevalence of disability, and even the prospects of a reversal of the previous favorable trends. Even without the apparent cessation and possible reversal of the trends exhibited in the 1990s, the search for explanatory factors would have presented challenges; given the changing temporal patterns, the problem of explanation becomes even more difficult.

Between-Group Differences, Compositional Change, and "True" Change

Many studies have investigated between-group differences, or disparities, in the level of or trends in disability prevalence, attempting to isolate factors that have contributed to overall trends in disability prevalence. If the criteria defining the overall population of interest are held constant – for example, and most broadly, all persons 65 and older – then the members of the population must necessarily change over time. Changes over time in the population prevalence of disability must therefore occur either because those joining the population differ from those leaving it, or because of within-person changes that arise with the passage of time – possibly, but not necessarily, changes associated with aging – or both. In contrast, if a defined cohort is followed over time, then changes in the prevalence of disability due to aging cannot be unambiguously disentangled from other types of temporal change.

At some risk of oversimplification, a trend in disability prevalence within a fixed-definition population can be produced exclusively by within-population *compositional* change, or exclusively by "*true*" (or "*rate*") change, or (more realistically) by a mixture of these two. If the prevalence of disability remains constant for each of several nonoverlapping groups over time, but the relative shares of each group in the population change over time (i.e., if population composition changes) then there can appear to be a trend in overall prevalence despite the absence of such trends in every constituent group. In contrast, changed behavior (e.g., lifestyle patterns) or external forces (e.g., environmental, treatment, or technological changes) could produce upward or downward trends in disability prevalence in one or more (or all) groups, producing "true" overall trends in the absence of compositional change. Both sorts of intertemporal patterns of change can co-occur. Obviously, an important feature of research intended to reveal the factors contributing to trends in disability prevalence is the definition of the groups across which between-group differences are investigated.

Many sources of point-in-time differences in disability prevalence within the older population have been examined. Such cross-sectional differences are a natural place to begin searching for explanations for disability trends, and particularly for identifying potential compositional effects. Differences by *age* (with older groups more likely to have a disability than younger groups), *gender* (with women more likely to have a disability than men), *race* (with blacks more likely to have a disability than whites, although the evidence on these differences is mixed), and *education* (with the more-educated less likely to have a disability than the less-educated) have been found in several studies; for example, see West, Cole, Goodkind, and He (2014), Newman and Brach (2001), or Liao, McGee, Cao, and

Cooper (1999). Marital status has also been shown to be associated with disability, with married people less likely to report having a disability than those in other marital-status groups (He & Larsen, 2014). Differences by parental status have also been documented: Henretta (2007) shows that high-parity women (those with four or more children) are significantly more likely than lower-parity women to report selected disabling conditions such as stroke and diabetes, while Spence (2008) finds that women's early childbearing (defined as a first birth before age 21) increases their odds of having ADL disabilities. Other less-often-studied differences in disability prevalence include:

- Area or country of Hispanic origin among immigrants (Coustasse, Bae, Arvidson, Singh, & Trevino, 2009);
- Area or country of birth among Asian immigrants (Mutchler, Prakash, & Burr, 2007);
- Age at immigration, among Asian immigrants (Lee, 2011; Ro & Gee, 2012); and
- The interaction between level of education and nativity (Lee, 2011).

The latter set of studies has found differences for very detailed representations of explanatory variables; for example, significant differences in outdoor mobility (usually considered an IADL task) between immigrants born in India, China, and Korea (Mutchler, Prakash, & Burr, 2007). Findings such as these indicate the potential information loss associated with use of more aggregated categories such as "Asian immigrant." The ability to detect such differences, however, rests on the use of very large samples such as those offered by the decennial US census; this data requirement, in turn, rules out an opportunity to explore the role of these fine-grained distinctions as sources of *trends* in disability prevalence.

One way to distinguish between the compositional and the true change components of trends is to use the demographic tool of decomposition analysis (Chevan & Sutherland, 2009; Das Gupta, 1978). An example of decomposition analysis is reported in Freedman, Schoeni, Martin, and Cornman (2007), which addresses the roles of demographic risk factors (age, gender, education, race and ethnicity, and marital status) and of selected chronic diseases on the change from 1997 to 2004 in overall disability prevalence; they find evidence of compositional effects (both positive and negative) as well as of "rate" effects (again, both positive and negative) among both sets of factors. For example, growth in the share of the older population found in the oldest age groups (80 and older) contributed to an increase in disability prevalence, while a reduction in the share with the lowest educational attainment (0–8 years of school) helped bring down the trend in disability prevalence; these are examples of compositional effects. However, the compositional effect of low educational attainment was more than offset by a corresponding increase in the rate effect for this group.

A second, and more widely used, way to isolate the role of potential sources of trends in disability prevalence is to use statistical models, for example, regression analysis. A major problem with both approaches is the problem that arises in virtually all multivariable studies: have all relevant factors been included in the model? And, assuming that the answer to the previous question is no, are any omitted relevant factors correlated with any of the included factors, producing the well-known (but not always acknowledged) phenomenon of omitted variable bias?

The potential for compositional change to explain overall disability prevalence is greatest for those attributes exhibiting the most compositional shifting over time. Schoeni et al. (2001) used data for the 15-year period 1982–1996, and considered five such dimensions: educational attainment, race, age group, gender, and marital status. Only in the case of education were substantial compositional shifts revealed: the percentage with 8 or fewer years of schooling fell from 46% to 22%, while those with exactly 12, and 13 or more, years of schooling rose dramatically. Both Schoeni et al. (2001) and Waidmann and Liu (2000), who studied a shorter period (1992–1996) using Medicare Current Beneficiary Study data, concluded that changes in the educational composition of the population contributed to declining disability prevalence over these years. Interestingly, education has also been shown to exhibit "rate" changes in addition to its compositional effects: in a later study Schoeni, Martin, Andreski, and Freedman (2005) found significant differences in the average annual rate of decline in disability prevalence *between* educational-attainment groupings, with the most highly educated groups (and especially college graduates) experiencing the largest declines.

Many of the demographic and socioeconomic attributes that have been studied represent fixed characteristics, or traits that are established well before reaching old age. In addition to the characteristics already mentioned, research has found other factors for which either compositional change or true prevalence trends have occurred, including family income (Schoeni et al., 2005) and early-life factors such as mother's education (Freedman, Martin, Schoeni, & Cornman, 2008). Another class of factors contributing to disability trends that has received considerable attention is chronic diseases and conditions. During the same period that disability prevalence was declining, the reported prevalence of disabling diseases and conditions such as arthritis, heart disease, cancer, diabetes, stroke, and osteoporosis increased (Martin, Schoeni, & Andreski, 2010). Those upward trends may, to an unknown extent, reflect increased detection and awareness of those conditions. Moreover, some of the most disabling conditions, in particular arthritis, vision impairment, and heart disease, have become less disabling during this same time period (Martin et al., 2010). Thus chronic diseases, like demographic and socioeconomic attributes, appear to have had both compositional and true effects on disability prevalence, at least during the period of downward trends in prevalence.

As informative as these studies are, they are individually and collectively limited with respect to their conclusiveness about the sources of trends in disability prevalence. Due to limitations on available data – either the failure of any given survey-data source to include relevant measures that are comparable over time, or a sample size inadequate to support fine-grained analyses, or both – it is not possible to simultaneously control for, and therefore to isolate differences in partial effects of, all the possibly relevant factors that might contribute to our understanding of the trends. Therefore, however finely we may partition the overall population into groups in our search for *between*-group differences in disability trends, it will remain possible that an even finer partition of the population would reveal still more *within*-group differences in trends.

Potential Explanations for "True" Change

Although the problem of isolating the groups for which disability trends have occurred – that is, the question of *what* has happened – presents analytic challenges, the problem of attributing those changes to underlying causes – that is, the question of *why* it happened – is even harder. Given the evident importance of education, many researchers have offered suggestions about its role in health and disability. Having more education implies command of, or easier access to, a broader knowledge base, and can also improve cognition; more-educated people also have greater material resources and higher levels of social integration. All four of these mechanisms have been shown to play substantial roles in explaining health behaviors (Cutler & Lleras-Muney, 2010), although their specific causal roles remain uncertain.

Medical treatments and medication use are also thought to have reduced the prevalence of disability late in life. For example, the disabling effects of the leading two chronic conditions implicated in disability – arthritis and cardiovascular disease (Field & Jette, 2007, p. 95) – have been greatly reduced through anti-inflammatory drugs and joint replacement surgery, and by antihypertension medication, respectively (Cutler, 2001). Innovations in, and growing access to, nonmedical services and technologies are also relevant. Spillman (2004) discusses the growing availability of electronic banking and shopping services, in particular the Social Security Administration's 1987 adoption of direct deposit as the default method of making benefit payments, as factors that could explain declining rates of IADL disability. Growing use of assistive devices, technologies, and home modifications such as canes, walkers, grab bars, and ramps has also helped reduce the prevalence of disability when measured by getting help from others (Freedman, Agree, Martin, & Cornman, 2006).

These various categories of underlying factors are, however, interrelated: more-educated people, with their greater command over material resources, are better able to acquire treatment and assistive devices, and their enhanced cognitive skills should also improve their ability to understand and adhere to the treatments, and to make appropriate and effective use of the devices and technologies. Therefore, the relative contribution of each potential explanatory factor is likely to remain unknown, while the specific causal mechanisms underlying trends in disability prevalence will remain somewhat speculative. And, we are left with the fact that the trends widely agreed to have occurred during the 1990s have subsequently slowed or stopped, and may even have reversed. Thus it is almost certainly the case that no single set of allegedly causal factors can explain the full spectrum of late-life disability trends.

INDIVIDUAL-LEVEL DISABILITY TRAJECTORIES

Empirical research on late-life disability trajectories is a relatively recent development, and it is accumulating at a rapid rate. This growth can be attributed to a combination of increasing availability of panel- or longitudinal-data sources that support such research, along with the development of new statistical methods and computer programs that implement them. The area encompasses a number of distinct approaches and issues. As George (2009) points out, trajectories can be either *transition*-based or *level*-based. The two are of course related: the occurrence of a disability transition (or "event") generally entails a move from one level (of disability severity) to another; it is assumed that the level of disability remains constant between event times. Transition indicators typically denote changes within a small set of discrete "states" – for example, the levels of disability among which transitions might occur could include "no disability," "low level of disability," "high level of disability, and (for completeness) "dead."

In order to study a sequence of disability events, the researcher must have data in which all such events are recorded. However, the several panel-study data sets that have been used in disability trajectory research rarely if ever record events; instead, they obtain measures of respondents' disability status at each interview time (and, in a few cases, limited further information on the elapsed duration of time spent with that status). Accordingly, researchers who wish to analyze transitions are forced to impose rather strong assumptions: if someone's disability status at two successive interviews is the same, then it is typically assumed that no transitions have occurred between the interviews, while if the disability status at two successive interviews is different, then one and only one event – a direct transition from the status held at the first interview to the status held at the second interview – must have occurred. When the two interviews are widely spaced in time, these assumptions are likely to be quite inappropriate, and to produce large biases in estimated rates of transition (Wolf & Gill, 2009). Moreover, it is not even apparent that everyone's disablement process includes any "events": for an unquestionably disabling disease such as dementia, underlying change can be quite irregular and gradual, involving small and hard-to-detect changes in underlying capacity. This process may give rise to considerable ambiguity in disability status over both short- and medium-term time intervals. Some panel surveys include questions that elicit information on the types of events that bring about a disability transition – for example, stroke – but they rarely include details about the timing of those events.

Another important distinction between transition-based and levels-based trajectory analysis relates to model complexity. A model of the *level* of some disability outcome generally requires only a single statistical relationship, that is, a single multivariable relationship per outcome. In contrast, with even the simplest three-state model (with states "no disability," "has a disability," and "dead") there are three possible transitions or "*events*," suggesting a need for at least three distinct statistical relationships connecting explanatory variables to events. For these reasons my review of the trajectory literature focuses exclusively on trajectories of disability *levels*.

However logical or intuitively appealing it may be in general, the application of the term "trajectory" to patterns of temporal change in disability is in some ways unfortunate. In physics and engineering, a "trajectory" is the path followed by a moving object as its position changes with time (von Oppen and Melchert, 2007). The path traced by a moving object is intrinsically continuous, and it unfolds in a continuous time domain. In contrast, disability, however measured, is an attribute rather than an object, and it need not trace a continuous path over time. As noted previously, even with a "levels" perspective there can be catastrophic health events that produce rapid and large discrete shifts in the level of disability. Moreover, virtually any instrument devised for the assessment or coding of disability status in a human population will, at best, be periodically applied at a sequence of discrete times. In practice, population-level surveys collect data with which to measure disability at widely space intervals, most often intervals of 1 or 2 years. Because disability can change rapidly and substantially, and could, in principle, change frequently, in general researchers wishing to analyze individual-level disability trajectories will have only a small number of measures, taken at time intervals that are large in relation to the underlying dynamics of change, with which to classify trajectories or to fit models of their direction and shape.

When considering disability trajectories, our focus broadens from an interest in between-person (or between-group) differences in disability status at specified times, to encompass an interest in within-person change over time. True individual-level disability trajectories cannot be observed except with continuous monitoring, which might be possible in a limited case study. Instead, the usual situation presents us with a "repeated measures" problem – a set of disconnected points along a presumed underlying continuum. There are relatively few patterns that a sequence of individual-level disability indicators taken at discrete times can exhibit: the indicator may remain unchanged, at either a high or a low level (with additional intermediate possibilities, depending on the measurement properties of the indicator); it may rise or fall; it may rise or fall and then reverse direction; and it may follow an erratic path. For sequences that exhibit change, the change may occur rapidly or gradually. And, in all the available data sources that include such sequences, for most if not all individuals included, the observed sequence covers only a small part of the person's lifetime – it is, by necessity, incomplete.

The usual starting point for analyzing disability trajectories is to develop a summary of the patterns of change over time in the level of disability. This summary may consist of a single population mean pathway – that is, a representation of the change over time in the expected level of the disability outcome. Alternatively, the summary may consist of the enumeration of a set of two or more distinctive patterns of change over time. The categories in this set of patterns are sometimes treated as observable, with individuals placed into categories using predetermined criteria; for example, Doblhammer and Hoffmann (2009) assigned individuals to one of seven trajectory classes according to the initial value and subsequent direction of change in the health outcome. If the latter approach is adopted, a logical second step is to characterize the relative size of each distinctive time pattern of change in the population. Finally, researchers will nearly always want to investigate how additional factors besides the passage of time influence the level and profile of disability over time.

The great majority of studies of late-life disability trajectories uses statistical modeling, of which three main forms exist; the three differ mainly with respect to their representation of between-person heterogeneity. The first consists of latent growth curve models (LGCM), which postulate the existence of a single population average pathway of the expected value of the disability outcome. The unobserved components of between-person variability in LGCMs take the form of time-invariant person-level components of regression intercepts and, possibly, slopes – that is, individual-level random effects. Many examples of this approach to modeling disability trajectories exist, including Haas (2008), Bowen and González (2008), Yang and Lee (2009), Warner and Brown (2011), and Wickrama, Mancini, Kwag, and Kwon (2013). A second modeling approach, latent class trajectory models (LCTM), uses group-level instead of individual-level random effects. The number of groups is usually not specified in advance, but is, instead, determined through exploratory analysis. Within (unobserved) groups, individuals share a common pathway of the expected value of the disability outcome. Several examples of LCTMs of disability trajectories exist in the literature, including Aneshensel, Botticello, and Yamamoto-Mitani (2004), Liang, Xu, Bennett, Ye, and Quiñones (2009), Taylor and Lynch (2011), and Tseng, Shyu, and Liang (2012). The third, growth mixture models (GMMs), includes both forms of unobserved heterogeneity: a set of discrete classes, each with its own distinctive disability trajectory, and person-level random effects within discrete classes (e.g., see Han et al., 2013; Hill, Burdette, Taylor, & Angel, in press; or Lewis, Heckman, & Himawan, 2011). Technical details of these three modeling approaches, accompanied by empirical examples, can be found in Chapter 2.

In addition to their choice of modeling approach, disability trajectory studies have addressed a wide variety of substantive questions, in particular investigating the influence of additional explanatory variables – some of which are fixed prior to old age, and some of which can themselves vary during late life – on the disability outcome or (in the case of LCTM or GMM) on the chances that a person will fall into one or another of the distinctive trajectory classes. This diversity of research questions effectively rules out any possibility of presenting a comprehensive review or summary of the literature. For example, studies have investigated the presence of gender, race, and ethnicity differences in disability trajectories (Bowen & González, 2008; Castora-Binkley, Peronto, Edwards, & Small, 2013; Liang et al., 2009; Rohlfsen & Kronenfeld, 2014; Warner & Brown, 2011; Yang & Lee, 2009), or have emphasized differences by education (Alley, Suthers, &

Crimmins, 2007), childhood health status (Haas, 2008) or birth cohort in disability trajectories (Taylor & Lynch, 2011; Yang & Lee, 2009, 2010), or have studied the effect of diabetes (Chiu & Wray, 2011) on trajectories. Remarkably many of these studies use data from the same source – the Health and Retirement Study – yet they use a variety of disability outcomes, include different survey years and age groups, and include a wide range of explanatory variables.

One thing that all the disability trajectory studies have in common, however, is that they depict the passage of time. However, here as well there are at least three distinct approaches to the representation of time dynamics: the passage of time may appear in the form of chronological age, or as time from baseline (or, "survey time"), or, less often, as time since a key event such as becoming widowed (Aneshensel et al., 2004; Sasson & Umberson, 2014), or admission to a nursing home (Banaszak-Holl et al., 2011). Age as a measure of time has the advantages of being continuous, of being commonly measured across observations, even if their ages do not overlap during the study period, and of being comparable across data sources. Time from baseline is often measured arbitrarily (e.g., wave 1, wave 2, … , and so on) and in discrete intervals. Moreover, if time is measured as survey wave, it will generally be necessary to control for age (at baseline) as well. Korn, Graubard, and Midthune (1997) recommend that age rather than time since baseline be used as the time variable when analyzing longitudinal data. Incorporating time since a key event generally entails the use of two time dimensions, both of which proceed at exactly the same pace; this is not necessarily problematic, given that life trajectories are generally governed by two or more – possibly several more – "clocks," but it may create problems of collinearity or small sample-size problems.

The use of these different ways to represent time makes it difficult to compare findings across studies. Among LGCMs using age as the time variable, all the cited studies found monotonic age effects: as age increases, the level of disability, however measured, also increases. This is an unsurprising finding; more refined comparisons across studies, however, are difficult in view of the fact that each includes a very different set of (possibly age-related) covariates. Comparisons of trajectory pathways across LCTMs and GMMs are complicated by the fact that different models end up with different numbers of latent classes. The papers cited previously found 3–5 trajectory classes, and in most cases all classes displayed either stable or monotonic patterns of change with the passage of time; in some cases, however, a trajectory class that allows for recovery of function – possibly temporarily – emerges from the analysis. A final possible point of comparison – the shares of the population in each of the trajectory classes – is also complicated by the fact that different studies find different numbers of classes.

As noted in the earlier discussion of trends in disability prevalence, chronic diseases and conditions contribute importantly to both point-in-time prevalence and to trends in prevalence of having a disability. Trajectory models offer a means for a deeper understanding of these population-level phenomena, by representing the *change* in one's trajectory associated with the onset and presence of chronic conditions. However, to date very few disability-trajectory studies have investigated such effects. Chiu and Wray (2011) show that among the 50-and-older population, having diabetes increases both the intercept and the age-slope of one's disability trajectory; diabetics are, in other words, more likely than nondiabetics to have a disability at every age, and this difference becomes larger with age. Gaugler et al. (2013) found analogous results with respect to physician-reported memory loss: during the years following such a diagnosis, people with memory loss lost ADL and IADL functioning at a faster rate than prior to the diagnosis. Available data present many opportunities for further research of this type, considering a broader set of health conditions.

Time to Death as an Additional "Time" Variable

All of the disability trajectory research reviewed to this point considered only the passage of time along either an age or calendar axis, which is the usual conception of time as running "forwards." In contrast, some researchers have conceptualized change with respect to *remaining lifetime* – that is, time to death (TTD) – thus invoking a time dimension that runs "backwards." For example, Lunney, Lynn, Foley, Lipson, and Guralnik (2003) classified decedents from a panel study into one of four predetermined trajectory groups ("sudden death," "terminal illness," "organ failure," and "frailty") based on disease diagnosis, cause of death, or nursing home residence (along with a fifth, "other," group of unassigned cases). Each displays a distinctive pattern of functional change as one approaches the moment of death. Disability trajectory models formulated on a TTD axis have often used samples of decedents. Teno, Weitzen, Fennell, and Mor (2001), for example, used a mortality follow-back survey administered to kin of a 1993 sample of decedents. Examples of prospective designs include Lunney et al.'s (2003) analysis of decedents from the Established Populations for Epidemiologic Studies of the Elderly study; Gill, Gahbauer, Han, and Allore (2010), who used decedents from the Precipitating Events Project; and Schoeni, Freedman, and Wallace (2002), who used NHIS respondents (1986–1994) linked to National Death Index (NDI) records (1986–1997). Other health and functioning outcomes have also been shown to covary with TTD. In the area of cognitive aging, researchers have called attention to the phenomenon of "terminal decline" or "terminal drop" (Palmore & Cleveland, 1976). Some argue that TTD is more important than age in explaining cognitive change (Thorvaldsson, Hofer, Hassing, & Johansson, 2008) and others have attempted to identify a "change point" on the age axis, at which the influence of age on cognition ceases, or is outweighed by the influence of TTD (Sliwinski et al., 2006). Other phenomena for which distinctive end-of-life patterns have been observed but for which the relative importance of aging versus dying have not yet been explored include oral health (Chen, Clark, Preisser, Naorungroj, & Shuman, 2013) visual acuity (Gerstorf, Ram, Lindenberger, & Smith, 2013), and weight change (Alley et al., 2010).

Analysis of TTD effects on disability trajectories based on decedent samples may suffer from selection bias. The main source of potential selection bias arises from the fact that those who die in a given year (or within the period of a panel study) are likely to be in worse health, and also more likely to have a disability, than those who survive that year (or the study's observation period), creating systematic differences between the decedents and the survivors. To deal with this issue, Wolf, Freedman, Ondrich, Seplaki, and Spillman (2015) propose a model of disability trajectories that represent change along both the age and the TTD dimension. They use panel data from the HRS, in which TTD is recorded for about 80% of the sample, but unobserved (due to right censoring) for 20% of the observations. They use a predictive equation for remaining lifetime to fill in missing values for TTD among the right-censored cases, and use multiple-imputation techniques to account for uncertainty regarding the true value of TTD. Their LCTM finds three latent classes, in all three of which disability monotonically worsens with both age and as one gets nearer to death. The three classes are associated with disability trajectories that differ with respect to the steepness of decline (from fully "able" to complete disability). A salient feature of the findings in Wolf et al. (2015) is that TTD is more powerfully predictive of disability than age is. A similar finding emerged in an earlier study (Diehr, Williamson, Burke, & Psaty, 2002).

Accounting for Sample Losses due to Death

A final point regarding the empirical literature on disability trajectories is the fact that most studies published through 2014 have done little to account for mortality among their sample members. Losses due to death inevitably occur within panel survey samples, and the disability trajectories of decedents are likely to differ systematically from those of survivors: if members of a given cohort are followed over time, for example, a pattern of disability decline might be attenuated or even reversed if the experience of those that die during the period of follow-up is ignored. Losses due to death produce abbreviated disability trajectories compared to those observed for surviving members of the study sample. If the abbreviated trajectories of deceased individuals are included in the analysis without any adjustments for selectivity, biased parameter estimates will likely be obtained. If, instead, the decedent subsample is simply omitted from the analysis, sample selectivity will again lead to biased estimates. Both such approaches have been used in many published works, but neither can be viewed as satisfactory.

A variety of methods have been used in the handful of studies that have tried to account for mortality selection in a disability trajectory model. A few have included in their trajectory model a dummy variable indicating those cases that die during the follow-up period (e.g., Warner & Brown, 2011). The latter approach is a form of control for TTD, albeit with considerable loss of information as well as substantial measurement error. Others have estimated joint models of disability trajectories and mortality, allowing for a statistical dependency between the two outcomes (Taylor & Lynch, 2011; Xue et al., 2012; Zimmer, Martin, Nagin, & Jones, 2012). The introduction of TTD as a regressor, as in Wolf et al. (2015), represents yet another way of approaching this problem. Accounting for mortality selection requires advanced statistical techniques, and the relative merits of the various approaches in the context of disability trajectory modeling have not yet been investigated. The biostatistics literature includes many methodological papers devoted to the problem of jointly determined repeat measures of a clinical outcome (e.g., "having a disability") and time to a censoring or terminal event (e.g., "death") – reviews can be found in Hogan, Roy, and Korkontzelou (2004) or Tsiatis and Davidian (2004) – but these methods have yet to be adopted in the gerontological literature.

CONCLUSION

The literatures on disability trends and trajectories address many common issues, particularly with respect to the conceptualization and measurement of disability, and have often used the same data sources, although in rather different ways. Nevertheless, the two literatures have evolved somewhat independently; moreover, both are vast and show no sign of slowed growth. Accordingly, a chapter-length review necessarily compresses much of its discussion while overlooking much of the relevant literature.

With respect to the literature on disability trends, there is considerable agreement that in the older population as a whole, the trend in disability prevalence was downward in the 1990s, but also an emerging consensus that those trends have not continued through or beyond the first decade of the twenty-first century. Aggregate-level trends are, to be sure, easier to establish than are explanations for those trends, because explanatory models require careful attention to the estimation of partial effects, while controlling for relevant covariates. However, different studies use different sets of control variables, leaving a residual of uncertainty about which relevant variables are omitted and the possible biases that these omissions produce. New and improved sources of data continue to develop, and this will enhance our ability to monitor and refine our knowledge about future disability trends. We cannot, however, create new and improved measures of disability prevalence in the past, so there are inherent and unavoidable limits on our ability to understand long-term trends in disability prevalence. Still, the continuing importance of late-life disability seems to ensure that research on this topic will remain a fixture of the scientific enterprise.

The growth in the literature on disability trajectories has been extremely rapid, and very much concentrated in the last few years. This body of work employs such a wide range of statistical methods, and addresses such a diverse set of substantive issues, that few generalizations about what has been learned from it are, as yet, possible. The most consistent finding seems to be that on average – whether averaging over the whole population or over most or all subgroups within it (where group membership is generally unobserved) – people become more likely to have a disability as they age. But this is neither a revelation nor a result that requires a complex model in order to be demonstrated. Yet there are multiple "clocks" ticking along with the one that records chronological age, including elapsed time since major life events and health shocks; there is, as well, a backwards-ticking time-to-death clock, and all of these time dimensions seem worth pursuing as we attempt to deepen our understanding of processes of late-life disability change.

References

Alley, D. E., Metter, E. J., Griswold, M. E., Harris, T. B., Simonsick, E. M., Longo, D. L., et al. (2010). Changes in weight at the end of life: Characterizing weight loss by time to death in a cohort study of older men. *American Journal of Epidemiology, 172,* 558–565.

Alley, D., Suthers, K., & Crimmins, E. (2007). Education and cognitive decline in older Americans. *Research on Aging, 29,* 73–94.

Aneshensel, C. S., Botticello, A. J., & Yamamoto-Mitani, N. (2004). When caregiving ends: The course of depressive symptoms after bereavement. *Journal of Health and Social Behavior, 45,* 422–440.

Banaszak-Holl, J., Liang, J., Quiones, A., Cigolle, C., Lee, I., & Verbrugge, L. M. (2011). Trajectories of functional change among long stayers in nursing homes: Does baseline impairment matter? *Journal of Aging and Health, 23,* 862–882.

Bootsma-van der Weil, A., Gusselkoo, J., de Craen, A. J. M., van Exel, E., Knook, D. L., Lagaay, A. M., et al. (2001). Disability in the oldest old: "Can do" or "do do?" *Journal of the American Geriatrics Society, 49,* 909–914.

Bowen, M. E., & González, H. M. (2008). Racial/ethnic differences in the relationship between the use of health care services and functional disability: The Health and Retirement Study (1992–2004). *The Gerontologist, 48,* 659–667.

Castora-Binkley, M., Peronto, C. L., Edwards, J. D., & Small, B. J. (2013). A longitudinal analysis of the influence of race on cognitive performance. *The Journals of Gerontology, Series B: Psychological Science and Social Science.*

Chen, X., Clark, J. L., Preisser, J. S., Naorungroj, S., & Shuman, S. K. (2013). Dental caries in older adults in the last year of life. *Journal of the American Geriatrics Society, 61,* 1345–1350.

Chevan, A., & Sutherland, M. (2009). Revisiting Das Gupta: Refinement and extension of standardization and decomposition. *Demography, 46,* 429–449.

Chiu, C.-J., & Wray, L. A. (2011). Physical disability trajectories in older Americans with and without diabetes: The role of age, gender, race/ethnicity and education. *The Gerontologist, 51,* 51–63.

Coustasse, A., Bae, S., Arvidson, C., Singh, K. P., & Trevino, F. (2009). Disparities in ADL and IADL disabilities among elders of Hispanic subgroups in the United States: Results from the National Health Interview Survey 2001–2003. *Hospital Topics: Research and Perspectives on Healthcare, 87,* 15–23.

Crimmins, E. M., Hayward, M. D., Hagedorn, A., Saito, Y., & Brouard, N. (2009). Change in disability-free life expectancy for Americans 70 years old and older. *Demography, 46,* 627–646.

Crimmins, E. M., & Saito, Y. (2000). Change in the prevalence of diseases among older Americans: 1984–1994. *Demographic Research, 3(9),* 1–20.

Crimmins, E. M., Saito, Y., & Reynolds, S. L. (1997). Further evidence on recent trends in the prevalence and incidence of disability among older Americans from two sources: The LSOA and the NHIS. *Journals of Gerontology, 52B,* S59–S71.

Cutler, D. M. (2001). Declining disability among the elderly. *Health Affairs, 20*, 11–27.

Cutler, D. M., & Lleras-Muney, A. (2010). Understanding differences in health behaviors by education. *Journal of Health Economics, 29*, 1–28.

Das Gupta, P. (1978). A general method of decomposing a difference between two rates into several components. *Demography, 15*, 99–111.

Diehr, P., Williamson, J., Burke, G. L., & Psaty, B. M. (2002). The aging and dying process and the health of older adults. *Journal of Clinical Epidemiology, 55*, 269–278.

Doblhammer, G., & Hoffmann, R. (2009). Gender differences in trajectories of health limitations and subsequent mortality. A study based on the German Socioeconomic Panel 1995–2001 with a mortality follow-up 2002–2005. *Journal of Gerontology: Social Sciences, 65B*, 482–491.

Field, M. J., & Jette, A. M. (2007). *The future of disability in America*. Washington, DC: National Academies Press.

Freedman, V. A. (2000). Implications of asking "ambiguous" difficulty questions: An analysis of the second wave of the Asset and Health Dynamics of the Oldest Old Study. *Journal of Gerontology: Social Sciences, 55B*, S288–S297.

Freedman, V. A., Agree, E., Martin, L. G., & Cornman, J. (2006). Trends in the use of assistive technology and personal care for late-life disability, 1992–2001. *The Gerontologist, 46*, 124–127.

Freedman, V. A., Aykan, H., & Kleban, M. H. (2003). Asking neutral versus leading questions: Implications for functional limitation measurement. *Journal of Aging and Health, 15*, 661–687.

Freedman, V. A., Crimmins, E., Schoeni, R. F., Spillman, B. C., Aykan, H., Kramarow, E., et al. (2004). Resolving inconsistencies in trends in old-age disability: Report from a technical working group. *Demography, 41*, 417–441.

Freedman, V. A., Kasper, J., Cornman, J., Agree, E., Bandeen-Roche, K., Mor, V., et al. (2011). Validation of new measures of disability and functioning in the National Health and Aging Trends Study. *Journals of Gerontology, Series B: Psychological Sciences and Social Sciences, 66*, S1013–S1021.

Freedman, V. A., Kasper, J. D., Spillman, B. C., Agree, E. M., Mor, V., Wallace, R. B., et al. (2014). Behavioral adaptation and late-life disability: A new spectrum for assessing public health impacts. *American Journal of Public Health*, e1–e7.

Freedman, V. A., Martin, L. G., & Schoeni, R. F. (2002). Recent trends in disability and functioning among older Americans: A critical review of the evidence. *Journal of the American Medical Association, 288*, 3137–3146.

Freedman, V. A., Martin, L. G., Schoeni, R. F., & Cornman, J. C. (2008). Declines in late–life disability: The role of early- and mid-life factors. *Social Science & Medicine, 66*, 1588–1602.

Freedman, V. A., Schoeni, R. F., Martin, L. G., & Cornman, J. C. (2007). Chronic conditions and the decline in late-life disability. *Demography, 44*, 459–477.

Freedman, V. A., & Spillman, B. C. (2014). Disability and care needs among older Americans. *The Milbank Quarterly, 92*, 509–541.

Freedman, V. A., Spillman, B. C., Andreski, P. M., Cornman, J. C., Crimmins, E. M., Kramarow, E., et al. (2013). Trends in late-life activity limitations in the United States: An update from five national surveys. *Demography, 50*, 661–671.

Freedman, V. A., Wolf, D. A., & Spillman, B. (2014). Examining late-life disability prevalence trends for men and women the US over three decades. Presented at the 2014 meeting of the Population Association of America, Boston.

Gaugler, J. E., Hovater, M., Roth, D. L., Johnston, J. A., Kane, R. L., & Sarsour, K. (2013). Analysis of cognitive, functional, health service use, and cost trajectories prior to and following memory loss. *Journals of Gerontology Series B: Psychological Sciences and Social Sciences, 68*, 562–567.

George, L. K. (2009). Conceptualizing and measuring trajectories. In G. H. Elder Jr. & J. Z. Giele (Eds.), *The craft of life course research* (pp. 163–186). New York, NY: The Guilford Press.

Gerstorf, D., Ram, N., Lindenberger, U., & Smith, J. (2013). Age and time-to-death trajectories of change in indicators of cognitive, sensory, physical, health, social, and self-related functions. *Developmental Psychology, 49*, 1805–1821.

Gill, T. M., & Gahbauer, E. A. (2005). Overestimation of chronic disability among elderly persons. *Archives of Internal Medicine, 165*, 2625–2630.

Gill, T. M., Gahbauer, E. A., Han, L., & Allore, H. G. (2010). Trajectories of disability in the last year of life. *New England Journal of Medicine, 362*, 1173–1180.

Guralnik, J. M., & Ferrucci, L. (2009). The challenge of understanding the disablement process in older persons. *Journal of Gerontology: Medical Sciences, 64A*, 1169–1171.

Haas, S. (2008). Trajectories of functional health: The "long arm" of childhood health and socioeconomic factors. *Social Science & Medicine, 66*, 849–861.

Han, L., Allore, H., Murphy, T., Gill, T., Peduzzi, P., & Lin, H. (2013). Dynamics of functional aging based on latent-class trajectories of activities of daily living. *Annals of Epidemiology, 23*, 87–92.

He, W., & Larsen, L. J. (2014). *Older Americans with a disability: 2008–2012*. Washington, DC: US Census Bureau. American Community Survey Reports, ACS-29.

Henretta, J. C. (2007). Early childbearing, marital status, and women's health and mortality after age 50. *Journal of Health and Social Behavior, 48*, 254–266.

Herzog, A. R., & Wallace, W. L. (1999). Cognitive performance measures in survey research on older adults. In N. Schwarz, D. Park, B. Knauper, & S. Sudman (Eds.), *Aging, cognition, and self-reports* (pp. 327–340). Philadelphia, PA: Psychology Press.

Hill, T. D., Burdette, A. M., Taylor, J., & Angel, J. L. (in press). Religious attendance and the mobility trajectories of older Mexican Americans: An application of the growth mixture model. *Journal of Health and Social Behavior*.

Hogan, J. W., Roy, J., & Korkontzelou, C. (2004). Handling drop-out in longitudinal studies. *Statistics in Medicine, 23*, 1455–1497.

Iezzoni, L. I., & Freedman, V. A. (2008). Turning the disability tide: The importance of definitions. *Journal of the American Medical Association, 299*, 332–334.

Iezzoni, L. I., McCarthy, E. P., Davis, R. B., & Siebens, H. (2000). Mobility problems and perceptions of disability by self-respondents and proxy respondents. *Medical Care, 38*, 1051–1057.

Jette, A. M. (1994). How measurement techniques influence estimates of disability in older populations. *Social Science and Medicine, 7*, 937–942.

Jette, A. M. (2009). Towards a common language of disablement. *Journal of Gerontology: Medical Sciences, 64A*, 1165–1168.

Katz, S., & Staff of the Benjamin Rose Hospital (1959). Multidisciplinary studies of illness in aging persons II. A new classification of functional status in Activities of Daily Living. *Journal of Chronic Disease, 9*, 55–62.

Kaye, H. S. (2013). Disability rates for working-age adults and for the elderly have stabilized, but trends for each mean different results for costs. *Health Affairs, 32*, 127–134.

Kelley-Moore, J. A. (2006). Assessing racial health inequality in older adulthood: Comparisons from mixed-mode panel interviews. *Journal of Gerontology: Social Sciences, 61B*, S212–S220.

Kinne, S., & Topolski, T. D. (2005). Inclusion of people with disabilities in telephone health surveillance surveys. *American Journal of Public Health, 95*, 512–517.

Korn, E. L., Graubard, B. I., & Midthune, D. (1997). Time-to-event analysis of longitudinal follow-up survey: Choice of the time-scale. *American Journal of Epidemiology, 145*, 72–80.

Lawton, M. P., & Brody, E. M. (1969). Assessment of older people: Self-maintaining and instrumental activities of daily living. *The Gerontologist, 9*, 179–186.

Lee, M. A. (2011). Disparity in disability between native-born non-Hispanic White and foreign-born Asian older adults in the United States: Effects of educational attainment and age at immigration. *Social Science & Medicine, 72*, 1249–1257.

Lewis, K. N., Heckman, B. D., & Himawan, L. (2011). Multinomial logistic regression analysis for differentiating 3 treatment outcome trajectory groups for headache-associated disability. *International Association for the Study of Pain, 152*, 1718–1726.

Liang, J., Xu, X., Bennett, J. M., Ye, W., & Quiñones, A. R. (2009). Ethnicity and changing functional health in middle and late life: A person-centered approach. *Journal of Gerontology: Social Sciences, 65B*, 470–481.

Liao, Y., McGee, D. L., Cao, G., & Cooper, R. S. (1999). Black–White differences in disability and morbidity in the last years of life. *American Journal of Epidemiology, 149*, 1097–1103.

Liao, Y., McGee, D. L., Cao, G., & Cooper, R. S. (2001). Recent changes in the health status of the older US population: Findings from the 1984 and 1994 Supplement on Aging. *Journal of the American Geriatrics Society, 49*, 443–449.

Livermore, G., Whalen, D., Prenovitz, S., Aggarwal, R., & Barbos, M. (2011). *Disability data in national surveys*. Mathematica Policy Research: Report prepared for the Office of Disability, Aging and Long-Term Care Policy, US Department of Health and Human Services.

Lunney, J. R., Lynn, J., Foley, D. J., Lipson, S., & Guralnik, J. M. (2003). Patterns of functional decline at the end of life. *Journal of the American Medical Association, 289*, 2387–2392.

Manton, K. G., Corder, L., & Stallard, E. (1993). Estimates of change in chronic disability and institutional incidence and prevalence rates in the US elderly population from the 1982, 1984, and 1989 National Long Term Care Survey. *Journal of Gerontology: Social Sciences, 48*, S153–S166.

Manton, K. G., Corder, L., & Stallard, E. (1997). Chronic disability trends in elderly United States populations: 1982–1994. *Proceedings of the National Academy of Sciences, 94*, 2593–2598.

Manton, K. G., & Gu, X. (2001). Changes in the prevalence of chronic disability in the United States black and nonblack population above age 65 from 1982 to 1999. *Proceedings of the National Academy of Sciences, 98*, 6354–6359.

Martin, L. G., & Schoeni, R. F. (2014). Trends in disability and related chronic conditions among the forty-and-over population: 1997–2010. *Disability and Health Journal, 7*(Suppl.), S4–S14.

Martin, L. G., Schoeni, R. F., & Andreski, P. M. (2010). Trends in health of older adults in the United States: Past, present, future. *Demography, 47*(Suppl.), S17–S40.

Molla, M. T., Madans, J. H., Wagener, D. K., & Crimmins, E. M. (2003). *Summary measures of population health: Report of findings on methodologic and data issues*. Hyattsville, MD: National Center for Health Statistics.

Mutchler, J. E., Prakash, A., & Burr, J. A. (2007). The demography of disability and the effects of immigrant history: Older Asians in the United States. *Demography, 44*, 251–263.

Nagi, S. Z. (1969). *Disability and rehabilitation: Legal, clinical, and self-concepts and measurement*. Columbus, OH: Ohio State University Press.

Nagi, S. Z. (1976). An epidemiology of disability among adults in the United States. *Milbank Memorial Fund Quarterly, 54*, 439–467.

Newman, A. B., & Brach, J. S. (2001). Gender gap in longevity and disability in older persons. *Epidemiologic Reviews, 23*, 343–350.

Palmore, E., & Cleveland, W. (1976). Aging, terminal decline, and terminal drop. *Journal of Gerontology, 31*, 76–81.

Redfoot, D., Feinberg, L., & Houser, A. (2013). *The aging of the baby boom and the growing care gap: A look at future declines in the availability of family caregivers*. Washington, DC: AARP Public Policy Institute.

Ro, A., & Gee, G. C. (2012). Disability status differentials among Asian immigrants in the United States: The added dimensions of duration and age. *Race and Social Problems, 4*, 83–92.

Rohlfsen, L. S., & Kronenfeld, J. J. (2014). Gender differences in functional health: Latent curve analysis assessing differential exposure. *Journals of Gerontology, Series B: Psychological Sciences and Social Sciences, 69*, 590–602.

Sasson, I., & Umberson, D. J. (2014). Widowhood and depression: New light on gender differences, selection, and psychological adjustment. *Journals of Gerontology, Series B: Psychological Sciences and Social Sciences, 69*, 135–145.

Schoeni, R. F., Freedman, V. A., & Wallace, R. B. (2001). Persistent, consistent, widespread, and robust? Another look at recent trends in old-age disability. *Journal of Gerontology: Social Sciences, 56B*, S206–S218.

Schoeni, R. F., Freedman, V. A., & Wallace, R. B. (2002). Late-life disability trajectories and socioeconomic status. In S. Crystal & D. Shea (Eds.), *Annual Review of Gerontology and Geriatrics* (Vol. 22, pp. 184–206). New York, NY: Springer.

Schoeni, R. F., Martin, L. G., Andreski, P. M., & Freedman, V. A. (2005). Persistent and growing socioeconomic disparities in disability among the elderly: 1982–2002. *American Journal of Public Health, 95*, 2065–2070.

Sliwinski, M. J., Stawski, R. S., Hall, C. B., Katz, M., Verghese, J., & Lipton, R. (2006). Distinguishing preterminal and terminal cognitive decline. *European Psychologist, 11*, 172–181.

Spence, N. J. (2008). The long-term consequences of childbearing: Psychical and psychological well-being of mothers in later life. *Research on Aging, 30*, 722–751.

Spillman, B. C. (2004). Changes in elderly disability rates and the implications for health care utilization and cost. *The Milbank Quarterly, 82*, 157–194.

Taylor, M. G., & Lynch, S. M. (2011). Cohort differences and chronic disease profiles of differential disability trajectories. *Journal of Gerontology: Social Sciences, 66*, 729–738.

Teno, J. M., Weitzen, M. H. A., Fennell, M. L., & Mor, V. (2001). Dying trajectory in the last year of life: Does cancer trajectory fit other diseases? *Journal of Palliative Medicine, 4*, 457–464.

Thorvaldsson, V., Hofer, S. M., Hassing, L. B., & Johansson, B. (2008). Cognitive change as conditional on age heterogeneity in onset of mortaliy-related processes and repeated testing effects. In S. M. Hofer & D. F. Alwin (Eds.), *Handbook of cognitive aging: Interdisciplinary perspectives* (pp. 284–297). Los Angeles, CA: Sage Publications.

Tseng, M. Y., Shyu, Y. I. L., & Liang, J. (2012). Functional recovery of older hip-fracture patients after interdisciplinary intervention follows three distinct trajectories. *The Gerontologist, 52*, 833–842.

Tsiatis, A. A., & Davidian, M. (2004). Joint modeling of longitudinal and time-to-event data: An overview. *Statistica Sinica, 14*, 809–834.

US Department of Health and Human Services. (2014). *Healthy People 2020*. Retrieved from: <http://www.healthypeople.gov/2020/topics-objectives/topic/disability-and-health>.

Verbrugge, L. M., & Jette, A. M. (1994). The disablement process. *Social Science and Medicine, 38*, 1–14.

von Oppen, G., & Melchert, F. (2007). *Physics for engineers and scientists*. Hingham, MA: Infinity Science Press.

Waidmann, T. A., & Liu, K. (2000). Disability trends among elderly persons and implications for the future. *Journal of Gerontology: Social Sciences, 55B*, S298–S307.

Warner, D. F., & Brown, T. H. (2011). Understanding how race/ethnicity and gender define age-trajectories of disability: An intersectionality approach. *Social Science & Medicine, 72*, 1236–1248.

West, L. A., Cole, S., Goodkind, D., & He, W. (2014). *65+ in the United States: 2010*. Washington, DC: US Census Bureau. P23-212.

Wickrama, K. A. S., Mancini, J. A., Kwag, K., & Kwon, J. (2013). Heterogeneity in multidimensional health trajectories of late old years and socioeconomic stratification: A latent trajectory class analysis. *Journals of Gerontology, Series B: Psychological Sciences and Social Sciences, 68*, 290–297.

Wolf, D. A., Freedman, V. A., Ondrich, J. I., Seplaki, C. L., & Spillman, B. C. (2015). Disability trajectories at the end of life: A "countdown" model. *Journal of Gerontology, Series B: Psychological and Social Sciences*.

Wolf, D. A., & Gill, T. M. (2009). Modeling transition rates using panel current-status data: How serious is the bias? *Demography, 46*, 371–386.

Wolf, D. A., Hunt, K., & Knickman, J. (2005). Perspectives on the recent decline in disability at older ages. *The Milbank Quarterly, 83*, 365–395.

Wolff, J. L., & Kasper, J. D. (2006). Caregivers of frail elders: Updating a national profile. *The Gerontologist, 46,* 344–356.

World Health Organization. (2001). *International classification of functioning, disability and health ICF.* Geneva, Switzerland: WHO.

World Health Organization. (2011). *World report on disability.* Geneva, Switzerland: WHO.

Xue, Q.-L., Bandeen-Roche, K., Mielenz, T. J., Seplaki, C. L., Szanton, S. L., Thorpe, R. J., et al. (2012). Patterns of 12-year change in physical activity levels in community-dwelling older women: Can modest levels of physical activity help older women live longer? *American Journal of Epidemiology, 176,* 534–543.

Yang, Y., & Lee, L. C. (2009). Sex and race disparities in health: Cohort variations in life course patterns. *Social Forces, 87,* 2093–2124.

Yang, Y., & Lee, L. C. (2010). Dynamics and heterogeneity in the process of human frailty and aging: Evidence from the US older population. *Journal of Gerontology: Social Sciences, 65B,* 246–255.

Zimmer, Z., Martin, L. G., Nagin, D. S., & Jones, B. L. (2012). Modeling disability trajectories and mortality of the oldest-old in China. *Demography, 49,* 291–314.

CHAPTER 5

Early Life Origins of Adult Health and Aging

Diana Kuh[1] and Yoav Ben-Shlomo[2]

[1]MRC Unit for Lifelong Health and Ageing and MRC National Survey of Health and Development, University College London, London, UK [2]School of Social and Community Medicine, University of Bristol, Bristol, UK

OUTLINE

Introduction	101
Historical Overview	103
Early Origins of Adult Disease: From FOAD Through DOHaD to a Life Course Perspective	104
The Intrauterine Environment	*104*
The Postnatal Environment	*105*
A Genetic and Evolutionary Perspective	*107*
Early Life Origins of Functional Aging in a Life Course Perspective	109
An Integrated Life Course Model of Aging	109
Structural Reserve and Compensatory Mechanisms	*112*
Endocrine System	*112*
Vascular Function	*114*
Physical and Cognitive Capability	*115*
Conclusions and Future Directions	116
References	118

INTRODUCTION

The idea that factors early in life, whether *in utero*, childhood or adolescence, can have long-term effects on later health has a long and interdisciplinary history. The term *life course epidemiology* was not coined until 1997 (Kuh & Ben-Shlomo, 1997), even though research applying a life course perspective to population health had been growing rapidly since the 1970s. Life course epidemiology has a particular interest in early social and biological factors that affect adult health, aging, and disease risk, and how these early effects are mediated or modified by later life risk. Growing empirical evidence that early life matters for adult health and disease, set within conceptual life course models and evolutionary frameworks, has strengthened the life course perspective as a general paradigm for the study of development and aging, health, and disease. This evidence has depended on the increasing wealth and richness of maturing cohort studies (Power, Kuh, & Morton, 2013) that follow population samples from pregnancy or birth into adult life, or even across generations; descriptions of the cohort studies that have contributed most to the development of life course epidemiology can be found elsewhere (see Kuh & Ben-Shlomo, 1997; Power et al., 2013; and www.halycon.ac.uk).

After a brief historical overview to set the scene, our first aim is to summarize some of the key findings and discuss the major advances in scientific knowledge of the early origins of adult chronic disease, particularly cardiovascular (CV), metabolic, and respiratory diseases, which revived the epidemiological life course perspective in the last quarter of the twentieth century. However, by the early twenty-first century, life course epidemiology had widened to include many other aspects of health and aging, and had become more integrated within a broader interdisciplinary life course perspective. Health during development and aging (postmaturity) is increasingly being assessed using tests of function at the individual, body system, and molecular levels, which allows the early life determinants of later function and functional change (often the precursors to chronic disease) to be studied. So our second aim is to set some of the key findings on the early life origins of adult function within an integrated life course model of aging, which builds on our earlier life course models (Kuh, Ben Shlomo, Lynch, Hallqvist, & Power, 2003), and which is still evolving (Ben-Shlomo, Mishra, & Kuh, 2014; Kuh, Ben-Shlomo, Tilling, & Hardy, 2015; Kuh, Richards, Cooper, Hardy, & Ben-Shlomo, 2014).

Our earlier life course models distinguished between critical or sensitive period models (with or without later life modifiers) and risk accumulation models, attempting to identify both the timing and type of exposures and when in the life course these exposures "get under the skin" and leave biological imprints that later may manifest as adult chronic conditions. Critical or sensitive period models hypothesize that specific exposures (such as maternal undernutrition) during times of rapid growth and development affect chronic disease by programming the structure or function of organs, tissues, and body systems, and that these biological imprints may or may not be modified by later risk factors. Risk accumulation models hypothesize that correlated or uncorrelated adverse exposures either cause damage to biological systems at any life stage which eventually manifests as chronic disease, or there is a sequence of linked exposures or environmental continuity and only the final exposure in the chain or the most recent environment causes damage to health. Key tasks of study designs and life course analysis are to distinguish between these models, separate potential exposures from confounders, and identify mediators along the causal pathway. Other key methodological tasks are: to model repeat exposures and outcomes; deal with missing data in longitudinal studies; harmonize and link data from across a number of cohort studies; and strengthen causal inference from observational studies. These tasks are discussed in more detail elsewhere (Kuh, Cooper, Hardy, Richards, & Ben Shlomo, 2014; Lawlor & Mishra, 2009). Relevant to epidemiology more generally has been the development of systematic reviews and meta-analyses, which are commonly used to synthesize the growing evidence using standardized guidelines. These methods increase the power to test robustness and generalizability across cohorts as well as to examine for interactions. We will cite recent systematic reviews where available and the pioneering studies that set the research agenda, and highlight areas where there remains most debate about exposures, confounders, and mediators.

HISTORICAL OVERVIEW

As part of a debate on national efficiency and fitness, population health and vitality in the first 40 years of the twentieth century in a number of newly industrialized countries, there was concern about the falling birth rate, the poor condition of children, and the fitness of army recruits, and a number of epidemiological studies or monitoring of child health and development were initiated. The idea that adult health and longevity were dependent on child health and environment gained ground, but the empirical evidence was limited, relying on early research on cohort analysis, human constitution and disease susceptibility, and behavioral and animal studies into developmental critical periods (Kuh & Ben-Shlomo, 2004). In population health, the focus of the chapter, this idea went out of vogue in the immediate postwar period as research on adult lifestyles and chronic disease took center stage, but empirical evidence, scattered across a number of disciplinary fields, continued to accumulate, and several birth cohort studies were established in the United Kingdom and in Europe.

From the 1970s, several strands of health research were emerging that were initially separate but would come together in the 1990s to build a life course perspective on adult health, disease, and aging, driven in part by the perceived social and economic costs of population aging (see section "Early Life Origins of Functional Ageing in a Life Course Perspective"). One strand, led by the United States, was the growing interest in assessing CV risk in cohort studies of children (e.g., the Bogalusa study set up in 1972, Berenson et al., 1980), given the difficulty of instigating behavior change in adults. The second strand, led by the United Kingdom, was the development of a *chronic disease model* based on social causation that competed with the *adult life style model* (Marmot, Rose, Shipley, & Hamilton, 1978), but the focus of attention remained on adult socioeconomic circumstances. In North America, the research interest in health disparities came to the fore later, accompanied by calls for interdisciplinary research into biopsychosocial models of health and aging to understand mechanisms underlying socioeconomic gradients in health and disease (National Research Council, 2001). The third strand was the growing evidence from maturing birth cohort studies that adverse early conditions and developmental health could affect the risk of adult chronic disease (Colley, Douglas, & Reid, 1973; Wadsworth, Cripps, Midwinter, & Colley, 1985) and might partly explain the social gradients in adult health and disease (e.g., Hertzman, 1995).

However, the fourth strand of epidemiological research did most to fire up interest in the early origins of adult chronic disease: the research undertaken by David Barker and his team in Southampton, UK. We describe this body of work, first called the Fetal Origins of Adult Disease (FOAD) and then the Developmental Origins of Health and Disease (DOHaD) in more detail (see sections "The Intrauterine Environment" and "A Genetic and Evolutionary Perspective") because of the impact of these findings on research within and beyond epidemiology into the early origins of adult health and disease. We show how this work acted as the catalyst for the revival of life course epidemiology which studies how physical or social exposures during gestation, childhood, adolescence, young adulthood and later adult life, and across generations, independently, cumulatively, and interactively impact on later health and disease risk (see sections "The Postnatal Environment" and "Early Life Origins of Functional Ageing in a Life Course Perspective").

These developments set the stage for an explosion of interdisciplinary life course research on health, disease, and aging which is still continuing in the second decade of the twenty-first century. Many cohorts have been established since the 1990s in Europe and North America, whether beginning in pregnancy or birth, childhood or later; and they have generally taken a more interdisciplinary biomedical, psychological, and social approach from the start.

EARLY ORIGINS OF ADULT DISEASE: FROM FOAD THROUGH DOHaD TO A LIFE COURSE PERSPECTIVE

Research into the fetal and the wider developmental origins of adult disease has paid particular attention to CV, metabolic, and respiratory diseases and their risk factors. We shall do likewise in our summary of the findings while acknowledging that the early origins of other diseases, such as osteoporosis, dementia, and breast cancer have also been investigated.

The Intrauterine Environment

Like others, Barker's team first undertook ecological studies, showing links between infant or maternal mortality and deaths from ischemic heart disease (IHD) and stroke 70 years later (Barker & Osmond, 1986, 1987). Then, from medical records kept by health visitors or hospitals in the 1920s and 1930s, they linked data on body size at birth (birthweight, birthweight to placental weight ratio, and ponderal index) and in infancy to individual mortality records. These showed that lower weight at birth or in infancy and the other birth measures were associated with higher adult CV mortality (Barker, Winter, Osmond, Margetts, & Simmonds, 1989). They went on to assess surviving cohort members and showed that these associations also held for CV disease, type II diabetes or insulin resistance, and hypertension (Barker, Bull, Osmond, & Simmonds, 1990; Fall, Vijayakumar, Barker, Osmond, & Duggleby, 1995; Hales et al., 1991; Phillips, Barker, Hales, Hirst, & Osmond, 1994).

Barker argued that these initial research findings pointed to an important role for the intrauterine environment in the development of CV and respiratory diseases, although others argued that continuity of adverse conditions across life could explain these findings (Ben-Shlomo & Davey Smith, 1991). Barker published his FOAD hypothesis in the early 1990s hypothesizing that poor maternal nutrition could program (i.e., permanently alter) the structure or function of organs, systems, tissues, or cells during the intrauterine period (Barker, 1995). He established a worldwide research organization to stimulate further research with the involvement of basic scientists to investigate the underlying processes. This group also went on to show that these associations held in non-Western populations, initially India, where babies, although smaller, were born with a higher proportion of fat to lean mass and increased risk was also seen with low maternal bodyweight (Stein et al., 1996). Fittingly, the first FOAD World Congress took place in Mumbai in 2001.

A growing number of researchers inside and outside the United Kingdom investigated these associations in other populations. They found that these associations were generally independent of gestational age or socioeconomic circumstances, and were not necessarily mediated by the classic adult risk factors. The early evidence was reviewed in the first book on life course epidemiology (Kuh & Ben-Shlomo, 1997) where contributors also discussed the role of other pre-adult factors associated with adult chronic diseases (such as poor early social conditions, postnatal growth, and childhood infections). They highlighted a number of possible interpretations of these associations based on the notion of risk accumulation as well as critical or sensitive periods.

Over time, as more findings were published, systematic reviews and meta-analyses of all the available evidence were undertaken. For example, a systematic review and meta-analysis of birthweight in relation to all-cause and cause-specific mortality covering 22 studies and over 36 000 deaths showed an inverse association between birthweight and all-cause mortality and a stronger inverse association with CV mortality (Risnes et al., 2011). Among men, a positive association was observed between birthweight and cancer mortality, raising the notion as others had done before that there are health trade-offs (see section "An integrated life course model of aging"). A systematic review and meta-analysis in 2007 found that all 16 published and two unpublished studies showed an inverse relationship between birthweight to IHD with no significant heterogeneity across studies (Huxley et al., 2007). A systematic review of 31 studies covering over 152 000 participants and 6000 cases of diabetes showed that 23 studies found an inverse association between birthweight and diabetes, whereas eight found a positive association (Whincup et al., 2008). The inverse and graded association extended to birthweights of at least 3 kg (6.6 lbs) and there was little evidence that that the birthweight effects interacted with, or were dependent upon, current size. Excluding participants with birthweights greater than 4 kg and/or maternal diabetes strengthened the inverse association, and in later-born cohorts there was a possible increased risk of diabetes at higher birthweights; these findings confirmed increased risks at both ends of the birthweight distribution. For both IHD and diabetes, adjustment for socioeconomic position (SEP) had no appreciable impact on the findings.

Systematic reviews and meta-analyses were also undertaken for birthweight in relation to the conventional adult risk factors for CV disease and diabetes. There is evidence of inverse associations with insulin resistance (Newsome et al., 2003) and blood pressure (Huxley, Neil, & Collins, 2002; Law & Shiell, 1996) where the effects are generally modest, and only weak evidence of an association with cholesterol levels (Huxley et al., 2004). The reviews by Huxley and colleagues suggested that the evidence may be affected by publication bias. The positive associations between birthweight and adult obesity (Rogers & the Euro-BLCS Study Group, 2003) appear inconsistent with the fetal origins hypothesis, though may partially explain the J-shaped associations seen with diabetes.

Birthweight has the benefit of being ubiquitously recorded but is generally agreed to be a poor indicator of fetal growth and development, and cannot disentangle the underlying mechanisms (see section "A Genetic and Evolutionary Perspective"). Over time, there has been a growing literature on other fetal exposures that may be indicative of intrauterine programming, such as maternal diabetes, obesity and smoking in pregnancy, placental size and shape (Burton, Barker, Moffett, & Thornburg, 2010), and preterm birth. For example, a recent systematic review showed that infants born preterm have modestly increased blood pressure in later life and an increased risk of hypertension (de Jong, Monuteaux, van Elburg, Gillman, & Belfort, 2012).

The Postnatal Environment

Life course epidemiologists developed conceptual frameworks that distinguished life course models based on critical or sensitive periods (prenatally or postnatally, and with or without later effect modification) from those based on risk accumulation in order to test the fetal and developmental theories of adult chronic disease (Ben-Shlomo & Kuh, 2002; Kuh et al., 2003). They were interested in the interplay between social and biological risk factors, and in potentially sensitive periods of development in infancy, childhood, and adolescence when exposures, such as poor nutrition, infections, and other stressors may have long-term effects on adult health and disease. They prompted those engaged in FOAD research to expand their interest into the early origins of adult disease into the postnatal environment and to acknowledge alternative social and behavioral pathways. By the Third World Congress of 2005, FOAD had metamorphosed into DOHaD in recognition of the long-term effects of postnatal as well as prenatal experience and a focus on health as well as disease; however, the wider life course field was not yet fully embraced.

Direct measures of postnatal nutrition, infection, and stress are often not available and indicators of poor growth and development and of adverse environmental factors were and still are commonly used as risk markers. An exception is the systematic reviews that have shown that infants breastfed as opposed to formula fed had lower CV risk in adult life in terms of lower total cholesterol levels, lower risk of type II diabetes, and marginally lower levels of adiposity and blood pressure (Owen, Whincup, & Cook, 2011).

Physical Growth and Development

Shorter adult height and its components, especially leg length, are indicators of adverse early life risk exposure and have been regularly associated with adult all-cause and cardiovascular disease (CVD) mortality. The most powerful study to date has examined adult height in relation to cause-specific mortality and vascular morbidity (The Emerging Risk Factors Collaboration, 2012). It showed broadly consistent results across 121 prospective cohorts in 24 countries and confirmed many of the earlier smaller studies: adult height was inversely related to mortality from CVD and chronic obstructive pulmonary disease but was positively associated with mortality for several cancer sites.

Studies of physical growth are increasingly separated into infancy, childhood, and adolescence in order to identify the most sensitive time windows in terms of their associations with adult health and disease. As far as infant growth is concerned, a review of relatively few studies, all conducted by Barker and his colleagues, suggested that: smaller infant size or slower infant growth were associated with IHD, particularly in men; the evidence for type II diabetes was weaker; and overall in relation to the burden of adult chronic disease there was no single optimal pattern of infant growth (Fisher et al., 2006).

Data on heights and weights through childhood and adolescence have been used to study growth patterns in relation to CVD and its risk factors (Andersen et al., 2010; Bann et al., 2014; Barker, Osmond, Forsen, Kajantie, & Eriksson, 2005; Bhargava et al., 2004; Hardy, Kuh, Langenberg, & Wadsworth, 2004; Raghupathy et al., 2010) and also in relation to other adult health outcomes like osteoarthritis, bone density, and fractures (Javaid et al., 2011; Kuh, Wills, et al., 2014; Tandon et al., 2012; Wills et al., 2012). Systematic reviews are not common because there are few studies on each outcome, and the analysis and interpretation of conditional models of change are more complex. However, they generally show that linear growth in the first few years of life is beneficial for several health outcomes, in cohorts based in high-income countries and increasingly in low–middle-income countries (Adair et al., 2013). The role of body mass index (BMI) or weight gain varies by health outcome and by its timing, but from around mid-childhood onwards there appear to be adverse associations with most adult health outcomes. Some studies show that these adverse outcomes are more likely if preceded by lower birthweight.

Early Adverse Environments and the Stress Response

Parallel to these research findings on physical growth and adult disease has been a growing epidemiological and demographic literature linking adverse early socioeconomic circumstances with higher risks of adult CV, metabolic, and respiratory disease, as well as other adult health outcomes. Much of this literature is dependent on aggregate measures of childhood SEP, such as parental social class, education, and income. For example, systematic reviews led by Galobardes, Smith, and Lynch (2006) revealed strong evidence of associations between childhood SEP and all-cause and CV mortality and CVD. For example, 31 out of 40 studies showed an inverse association between childhood SEP and CVD, although social patterning differed by CV sub-type (Galobardes et al., 2006).

Although there is a long psychological tradition of demonstrating the impact of early psychosocial adversity on adult mental health (Fryers & Brugha, 2013; Koenen, Rudenstine, Susser, & Galea, 2013; Rutter, Kim-Cohen, & Maughan, 2006), its impact on adult chronic disease and aging (our focus here) has been less studied. Different features of the adverse environment have been emphasized at different times, the focus shifting from parental separation or divorce, to the quality of the early family environment, identifying characteristics of "risky families" such as overt family conflict, disorganized and unpredictable family settings, and physical and emotional mistreatment (Repetti, Taylor, & Seeman, 2002). Being able to separate out the effects of different adversities is challenging because they cluster together and are often imprecisely measured or based on retrospective recall in older cohorts. It is also challenging to test whether different adversities operate through continuity in the social environment, behavioral pathways, or through processes of biological embedding (i.e., how they "get under the skin"), and to what extent they explain the adult social gradient in health. Possible mediators of biological embedding, such as epigenetics, neural structure and function, the hypothalamo–pituitary–adrenal (HPA) axis, and inflammatory processes (see Rutter, 2012 and accompanying articles), are increasingly being investigated in younger cohorts with more precisely assessed exposures and stress responses.

A related research area has linked lower childhood characteristics such as cognitive ability with a range of health outcomes, from CV and respiratory disease to physical functioning and reproductive aging. A systematic review based on 16 studies covering over one million participants followed for 17–69 years and 22 000 deaths, showed a 24% lower risk of death for a one standard deviation increase in IQ measured in childhood or youth (Calvin et al., 2011). Whether these associations are due to childhood cognitive ability being a mediator of early adversity, or a determinant of subsequent educational attainment, SEP and other life chances are an ongoing and policy-relevant debate. Childhood antisocial behavior has also been linked with all-cause and CV mortality, possibly through its effects on a range of life chances (Maughan, Stafford, Shah, & Kuh, 2014).

Characterizing early stress responses and to what extent they are common to later physical and mental health and provide explanations for health inequalities is a major research area (Cohen, Janicki-Deverts, Chen, & Matthews, 2010; Hertzman & Boyce, 2010; Shonkoff et al., 2012), given greater recognition in the second edition of our life course book (Kuh & Ben-Shlomo, 2004).

A Genetic and Evolutionary Perspective

Evolutionary biology has come to provide a broad interpretative framework for the developmental origins of adult disease (Gluckman & Hanson, 2004). Developmental biology (or "life course biology," Godfrey, Inskip, & Hanson, 2011) is furthering mechanistic understanding through increased knowledge of developmental plasticity and its long-term effects, and the emerging field of epigenetics suggests that the debate about genetics versus environmental factors is based on a false dichotomy.

In the early research on the FOAD, the critical stimulus was seen as being poor maternal nutrition operating through its effects on fetal nutrition (Barker et al., 1993). Much research effort in animals and humans (in particular studying famine exposure) has attempted to shed light on the role of maternal nutrition but suffice to say here that results remain inconsistent. Other fetal exposures such as defective placentation, maternal glucose intolerance, or hypertension are now given more consideration. One of the early debates was whether the epidemiological associations reflected genetic rather than environmental explanations (the "thrifty phenotype hypothesis") (Hattersley & Tooke, 1999). Genetic epidemiology has shown that genetic polymorphisms that are associated with type II diabetes are also associated with reduced size at birth (Freathy et al., 2009). One of the first publications from UK Biobank has confirmed, in 257000 participants, that birthweight was inversely and linearly associated with type II diabetes in adult life. It also showed that paternal diabetes, consistent with a genetic interpretation, was associated with lower birthweight whereas maternal diabetes was associated with higher birthweight, and that birthweight was a partial mediator of the effects of parental diabetes on risk of offspring diabetes (Tyrrell, Yaghootkar, Freathy, Hattersley, & Frayling, 2013). The relatively new field of epigenetics provides a further mechanistic bridge that could explain how early exposures, such as diet, could modify genetic predisposition, by silencing or activating protein products and hence modifying long-term risks (Ng et al., 2012).

Both DOHaD and life course epidemiology have benefited from the broader application of evolutionary theories to development, reproduction, aging, and disease, providing a unifying research framework in which to make sense of the new research findings. Such an evolutionary framework calls for a dynamic not static concept of health – where health and disease reflect the ability of an organism to respond to environmental challenges. Others previously (Dubos, 1965) and more recently (Huber et al., 2011) have also called for such a dynamic health concept. Gluckman proposed a model that captures the responses to environmental challenges over different timescales: *homeostasis* (the "delicate equilibrium") being the short-term response of the organism to challenge; *developmental plasticity* being a key response to current and predicted challenges (predictive adaptive response – PAR) across the life course that encapsulates the adaptation of the organism in early life to maximize the chance of survival and reproduction but that may have costs in later life; and natural selection which occurs over many generations (Gluckman et al., 2009). Using the perspective of evolutionary biology, Gluckman and Hanson have argued that where there is a match between the PAR and the actual mature adult environment, then the PAR aids survival. Conversely, an inappropriate PAR may result in greater disease burden if the later environment is maladaptive; hence a PAR assuming a nutrient-poor environment will be counter-productive if the organism finds itself in an obesogenic environment. This hypothesis is similar, though distinct, to the "antagonistic pleiotropy" model of aging. Pleiotropic genes with good early effects would be favored by selection even if these genes had bad effects at later ages on the assumption that by the time these effects became functionally important, the organism had successfully passed on these genes to future generations (Kirkwood & Austad, 2000). The "live fast, die young" strategy is successful when there are strong external influences on mortality so that few survive much beyond reproductive age but, with increased life expectancy, will result in variations in morbidity. The PAR model does not assume a genetic mechanism and is dependent on an environmental mismatch unlike the antagonistic pleiotropy concept. Both models assume that survival into later life has no benefits on the survival advantage of first- or even second-generation offspring, which may not be true for human societies where kinship networks can play an important part in the health and well-being of children and grandchildren.

EARLY LIFE ORIGINS OF FUNCTIONAL AGING IN A LIFE COURSE PERSPECTIVE

The research we have described so far has focused almost exclusively on adult disease rather than change in function. Life course epidemiology has a fundamental interest in human systems and functional change across life, during development and during aging and their lifetime determinants. Biological aging is the progressive deterioration in function that occurs across various body systems postmaturity, and is generally thought to be caused by accumulating cellular and molecular damage (Kirkwood & Austad, 2000). Life course studies ideally should capture function at the individual, body systems, and cellular levels, repeated at regular intervals so that functional change can be detected. These measures of function also capture the earliest and intermediate changes that eventually may lead to chronic disease. Investigating functional aging and its lifetime determinants offers opportunities to intervene earlier to increase the chance of healthy aging by improving survival, delaying the onset of chronic diseases, and maintaining optimum functioning for the maximal period of time.

For most functional capacities, we do not yet have a clear description of their lifetime trajectory, the periods of life when specific changes are observed, or their variation by gender and ethnicity. Rather we make the assumption that many aspects of function (e.g., muscle or lung function) have a period of rapid growth and development reaching a peak or plateau at maturity, and then gradually decline during senescence. We recognize that there is a danger in focusing on mean levels of change because of the vast heterogeneity that is seen in aging. But further research is needed to reveal if we can distinguish systematic patterns for population subgroups (that will aid intervention strategies) from random chance (Davey Smith, 2011). Early life factors can potentially affect the peak or structural reserve or the rate of decline. There is growing knowledge about the early life factors that are associated with level of function at the peak or plateau or at various ages in later life (briefly described below). But we know little about the early life determinants that drive functional change, either independently, cumulatively, or interactively with later life factors.

The main reason for the research gap is that birth cohort studies have generally not had sufficient repeat measures taken during the period of functional decline, whereas aging studies may have sufficient repeat measures to characterize the decline but do not generally have pre-adult risk markers. The research that does exist is usually based on single studies and most of it, as we shall see in the section "Endocrine System", is in research on vascular and metabolic function. However, longitudinal functional data from several cohort studies are now being combined to better understand the shape of the lifetime trajectory (Wills et al., 2011).

An Integrated Life Course Model of Aging

Here we further develop our integrated life course model of aging recently published, referencing only key or additional publications (Kuh, Cooper, et al., 2014). Figure 5.1 is a diagrammatic representation of this model within which to place current knowledge on the early origins of adult health; identify research gaps; develop our original life course models; design and test hypotheses; and make research findings relevant for policy and practice.

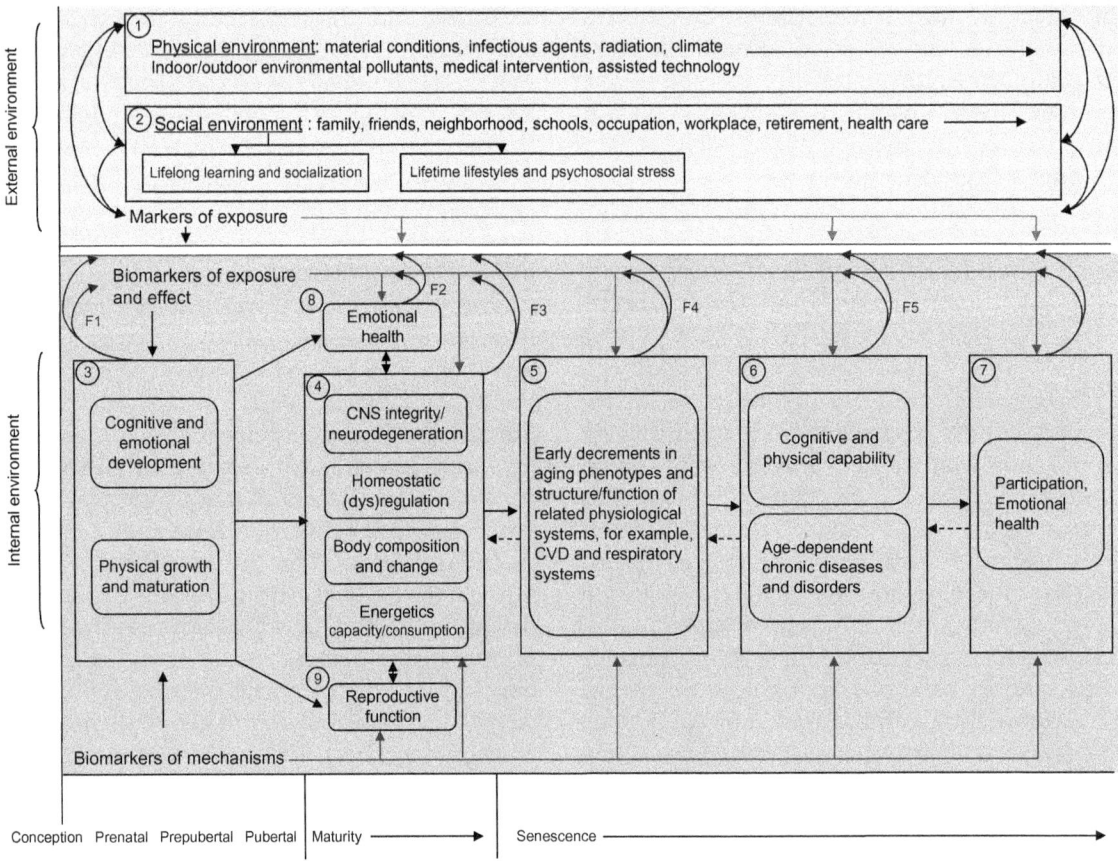

FIGURE 5.1 An integrated life course approach to aging.

The numbers in the boxes are there to guide the reader through our description of this model.

The model highlights the fundamental importance of the external physical environment (box 1) and the social environment (box 2) across the entire life course. We need to capture environmental characteristics before the start of functional decline and to characterize the unfolding interaction between individuals and their environments.

The early social and physical environment is shown to affect physical growth and maturation, and cognitive and emotional development (box 3) which in turn affect four common aging phenotypes: central nervous system degeneration, homeostatic dysregulation in endocrine and immunological systems, changes in body composition (muscle, fat, and bone), and the changing balance between energy production and consumption (box 4 and section "Endocrine System") (Ferrucci & Studenski, 2012).

These, in turn, lead to changes in related physiological systems (box 5), such as the structure and function of the CV system (see "Vascular Function"), which underlie the subsequent changes in physical and cognitive capabilities and the manifestation of age-dependent diseases (box 6, see section

"Physical and Cognitive Capability"). The reverse hashed arrows from box 6 back to box 4 allow for bidirectionality of these processes.

Our model of aging also incorporates the emotional health of older people and their continued participation in society (box 7). Emotional health encompasses both positive and negative emotions and the capacity to organize one's life. The reversed hash arrow indicates that adult emotional health is a determinant as well as a consequence of physiological aging (and also probably for the capacity for adaptation); and a recent review of the evidence confirmed modest bidirectional associations between well-being and capability in later life: while well-being is adversely affected by declines in physical and cognitive capability, it is also protective of functional decline (Gale, Deary, & Stafford, 2014). Emotional health earlier in adult life is shown separately (box 8) with direct pathways from developmental experiences, as the aging phenotypes (box 4) that precede functional decline and chronic disease risk may be less important mediators. As noted in section "Early Adverse Environments and the Stress Response," a wide variety of childhood adversities and responses influence adult mental disorders, and evidence indicates that mental disorders generally precede and contribute to the development of chronic disease risk and reduced capability, either through direct pathophysiological effects on disease risk or through harmful health-related behaviors (Kubzansky & Winning, 2013).

A life course model of aging that sits within an evolutionary perspective must pay attention to reproductive function (box 9), given that the evolutionary purpose of developmental plasticity is to ensure reproductive success. Rich-Edwards was one of the first to integrate this into a life course perspective (Rich-Edwards, 2002); since then further evidence has shown that reproductive characteristics are both shaped by early experience and then, in turn, act as sentinels for chronic disease and functional aging postmaturity.

Reproductive characteristics, such as hypertension in pregnancy, offspring low birthweight, and early menopause, have all been linked to the risk of later disease and accelerated aging (Rich-Edwards, McElrath, Karumanchi, & Seely, 2010). For example, one large systematic review showed that pre-eclampsia was associated with subsequent CV disease (but not cancer) (Bellamy, Casas, Hingorani, & Williams, 2007); another showed that offspring birthweight was associated with maternal CV mortality (Davey Smith, Hypponen, Power, & Lawlor, 2007). Interestingly, the overall associations were stronger with maternal than paternal mortality, consistent with specific maternal factors affecting intrauterine growth and supporting the fetal origins hypothesis. Shared environmental or genetic effects may also account for some of the associations in both parents. There is also growing evidence of the developmental origins of reproductive aging, reviewed elsewhere (Hardy, Potischman, & Kuh, 2012). For example, breastfeeding, better early socioeconomic circumstances, infant weight gain, and childhood cognitive ability all delayed age at menopause in the Medical Research Council National Survey of Health and Development (NSHD), even taking account of known adult circumstances, not smoking, and parity which are also associated with delayed menopause.

The underlying cellular and molecular "hallmarks" of aging-genomic instability, telomere attrition, epigenetic alterations, loss of proteostasis, deregulated nutrient sensing, mitochondrial dysfunction, cellular senescence, stem cell exhaustion, and altered intercellular communication (Lopez-Otin, Blasco, Partridge, Serrano, & Kroemer, 2013) are not shown on this figure but are captured by the label "biomarkers of mechanisms." Rather our model focuses on incorporating aging phenotypes at the biological systems and individual levels; prompting investigations of the dynamic inter-relationships across these levels, and their shared markers of early risk.

Structural Reserve and Compensatory Mechanisms

One specific aspect of studying diseases or pathophysiology associated with aging is that, by definition, research is undertaken only on survivors who live long enough to get old. So what determines the changing resilience to environmental challenges that accompany aging? It is helpful to distinguish two different aspects of intrinsic compensation. These are "structural reserve," such as having two kidneys, that may be programmed through embryological development and "compensatory reserve," the ability of body systems to compensate physiologically with varying degrees of success when faced with acute or chronic low-level challenges in order to maintain function or limit its decline. There may or may not be a positive association between structural and compensatory reserve, so individuals who experience a stroke but had better pre-morbid synaptic connectivity may be more able to reconnect adjacent circuits and hence recover function; in this hypothetical example, better structural reserve itself influences a better compensatory response.

The concept of reserve can also be applied to psychological traits (Matthews, Gallo, & Taylor, 2010), and in the second edition of our life course book, we used the concept of personal capital or reserve to capture the individual's capacity to mobilize available resources, exploit opportunities, and be resilient to adversity. This capacity reflects the accumulation of social and cognitive skills, self-esteem, coping strategies, attitudes and values, and its foundations are laid down in childhood. It is helpful to consider how these types of reserve may influence extrinsic responses when faced with age-related functional declines; "adaptations" of behavior or the environment may modify the effect of decline on activities of daily living and social participation. These adaptations could also slow the rate of decline by changing the level of exposures. The net effect of both intrinsic compensation and extrinsic adaptations to environmental challenges (through feedback loops labeled F1–5 in the figure) may be summarized by the over-arching term "resilience."

Endocrine System

Of the aging mechanisms captured in box 4 of the figure, we focus here on neuroendocrine mechanisms which are excellent candidates for mediating some of the phenotypic changes seen with aging, such as loss of physical performance and cognitive decline. There are a variety of different endocrine systems and these play a fundamental role in maintaining homeostasis as well as being tightly regulated by feedback loops. They are also clearly critical in terms of growth and development, reproduction, and responding to stressors and therefore may be mediators of the PAR. We will focus on the HPA axis, but it is very likely that these multiple systems intimately modulate each other; therefore consideration of any one system alone is overly simplistic. For example two studies have already demonstrated the additive effects of being in the lowest quartiles for insulin-like growth factor-I and testosterone (Maggio et al. also included dehydroepiandrosterone-sulfate) on increased risk of all-cause mortality (Friedrich et al., 2012; Maggio et al., 2005).

Overactivity of adrenal glucocorticoid secretion, due to Cushing's syndrome, can cause sarcopenia, glucose intolerance, osteoporosis, CV disease, cognitive decline, and mood disturbance, features that are all seen in older populations. There is now a moderate literature suggesting that chronic stress through activation of the HPA axis can increase the "wear and tear" or allostatic load on the body and may mediate effects of chronic stress on the medical disorders associated with aging, such as CV disease (Davey Smith et al., 2005; McEwen & Wingfield, 2003).

The HPA axis is complex as it is tightly regulated by a peripheral hourly pituitary adrenal oscillatory mechanism and a central circadian oscillator located in the brain. Older individuals have a reduced circadian gradient of cortisol so that the normal diurnal decline (high in morning and low at night) can be blunted (Kumari et al., 2010). In addition there is a well-recognized cortisol awakening response (CAR) that is seen as an elevation of cortisol between waking and 30 min later. We have previously postulated that this "normative" profile may change both acutely and chronically under differing stressor conditions (Ben-Shlomo, Gardner, & Lightman, 2014). So, for example, in the short term one may observe a heightened CAR, which may or may not be physiologically appropriate, but with preserved normal diurnal variability. With more chronic exposure, there is a subsequent reduction in diurnal variability such that nighttime levels are now elevated compared to the normative pattern. Finally there may be a stage of reduced variability so that both the CAR and diurnal variability are reduced, leading to greater cortisol exposure over a 24-h period. Age-related dysregulation would also impact these responses. Unfortunately, most studies have single cross-sectional data only and so cannot track any longitudinal differences in responsiveness. Similarly, some studies only collect data on acute psychological challenges and do not capture diurnal variability.

We have recently reported an individual participant data meta-analysis from the Healthy Ageing across the Life Course (HALCyon) collaboration where we have examined if various dimensions of HPA activity predict worse physical or cognitive performance. In simple age- and sex-adjusted models we found that slower walking speed was associated with lower morning cortisol, higher nighttime cortisol, smaller diurnal drop, and a smaller CAR. Slower chair rise speed was also associated with a lower morning cortisol, higher nighttime cortisol, and smaller diurnal drop. Poor balance was only predicted by higher nighttime levels and grip strength was not associated with any of the cortisol measures (Gardner et al., 2013). These were cross-sectional associations but in the Caerphilly Prospective Cohort study we found that higher fasting clinic morning cortisol at phase one was associated with faster walking speed at phase five (20 years later). In this case, we believe the ability to mount a good cortisol response to the stressor of fasting indicated a healthier HPA axis which predicted better performance measures (Gardner et al., 2011).

The HALCyon collaboration has also examined HPA activity and cognitive performance. In this case a greater diurnal decline was associated with better fluid capability but no associations were found between morning cortisol, nighttime cortisol or diurnal decline and crystallized capability. A bigger CAR was weakly associated with better fluid and crystallized capability.

Life Course Physical and Psychological Influences on the HPA Axis

Studies that have examined life course influences on the HPA axis can be divided into those that have examined physical risk factors, such as sub-optimal development, and those that have examined psychological exposures. Some studies have shown that lower birthweight is associated with higher morning cortisol levels though this is not consistent, and follow-up of the Dutch famine survivors failed to find an association with an acute stress task. A 25-year follow-up of babies with good-quality measures of childhood infections found that those exposed to more respiratory infections had lower nighttime levels, suggesting that a greater burden of infection may have resulted in a more reactive or healthier HPA (Vedhara et al., 2007). A systematic review of early psychological adversity (Essex et al., 2011) found inconsistent results for prenatal substance use (alcohol, tobacco, illicit drugs). Similarly, a systematic review found that maternal factors such as anxiety and depression during pregnancy, life time history of depression, and poor mother–child attachment were associated with both a heightened and decreased response (Hunter, Minnis, & Wilson, 2011). There is also some evidence for cumulative effects of adversity rather than just a sensitive period model. The Midlife in the United States (MIDUS) and National Study of Daily Experiences (NSDE) sub-sample found that those subjects reporting social strain with family and friends across two data collection waves (10 years apart) had a lower peak level and flatter diurnal variation cross-sectionally (Friedman, Karlamangla, Almeida, & Seeman, 2012). However, subjects reporting social strain only at wave two did not show an effect whereas those with strain at wave one or persistent strain at both time points had a more pathological pattern of response.

This evidence supports the role of a range of exposures acting across the life course in different ways and which may modulate the HPA reactivity. Their relative importance in terms of disease causation or functioning is still not fully elucidated, and it is possible that this may be modest or that our current methods of measuring the HPA axis are limited, thereby underestimating the true effects. Embedding HPA measures as secondary outcomes into intervention studies will provide more robust evidence of causal effects.

Vascular Function

The CV system or vascular aging (specifically blood or pulse pressure) may be a helpful paradigm to consider aging in the broader sense. In a simplistic engineering framework, the heart can be considered as a mechanical pump and the arteries are elastic tubes that have to cope with the repetitive expansion and contraction of the lumen with obvious "wear and tear" consequences over time. Thus life course influences on blood pressure or arterial stiffness, as measured by pulse pressure or pulse wave velocity, an independent predictor of CV morbidity even after adjusting for blood pressure (Ben-Shlomo, Spears et al., 2014), are of particular interest (see Lawlor and Hardy, 2014). There is much variability in how blood pressure increases with age suggesting that high blood pressure is not an inevitable consequence of ageing per se with some populations apparently not demonstrating much age-related change (Gurven, Blackwell, Rodriguez, Stieglitz, & Kaplan, 2012).

Lifetime Influences on Vascular Function

Meta-analyses have shown consistent associations between birthweight and systolic blood pressure (Huxley et al., 2002), though not usually diastolic pressure and some, but not all studies, that have examined postnatal growth suggest that accelerated growth in the early postnatal period may also be important (Ben-Shlomo et al., 2008). If these effects are due to structural alterations such as an abnormal ratio of elastin to collagen in the arterial wall, then one should be able to detect differences from birth albeit with difficulty as the physiological effects at this age may be trivial.

Cross-sectional associations suggested an "amplification" of birthweight blood pressure effects so that the magnitude of differences increases with age, though longitudinal studies in general do not support the amplification phenomenon (see Lawlor and Hardy, 2014). Even if we accept the presence of amplification, this in itself may indicate either an early life influence or the interaction between early life and later life modifiable factors and the latter may be of far greater public health relevance. For example, pulse wave velocity is strongly related to the cumulative exposure of pulse pressure and heart rate (McEniery et al., 2010), both of which can be reduced by either lifestyle or pharmacological modifications. It is possible that birthweight as a proxy marker of arterial development modifies this association because it alters either the structural or compensatory reserve. It is very likely that different factors influence the natural history of blood pressure trajectory seen across the life course.

Adolescence is associated with a more rapid increase in systolic blood pressure than other phases of life and this is more marked for men than women (see Lawlor and Hardy, 2014; Wills et al., 2011). Given the marked structural change seen with the pubertal growth spurt, it may be that such patterns are physiologically adaptive and explain birth cohort effects (e.g., the increase in height and blood pressure; McCarron, Okasha, McEwen, & Smith, 2001) as well as possible differences between high-income and low- and middle-income countries. The second period of blood pressure acceleration seen in middle age, however, may reflect other influences such as increases in adiposity and dietary factors that may be unrelated to early life but more reflective of adult socioeconomic conditions.

The final phase of life has been associated with a decline in blood pressure and excluding treatment effects, this probably reflects co-morbidity, for example, heart failure or autonomic dysregulation or other changes such as weight loss. At this age conventional risk factors can show paradoxical effects, increased BMI is associated with increased rather than decreased survival among post-operative patients (Hunt et al., 2013) and it may be that this is also true for blood pressure. Though trials of blood pressure lowering still show benefits (Muller, Smulders, de Leeuw, & Stehouwer, 2014), recent data from The National Health and Nutrition Examination Survey (NHANES) highlights increased mortality with lower blood pressure among frail elderly subjects (Odden, Peralta, Haan, & Covinsky, 2012). Similar arguments can be applied to other organ systems such as lung function due to common mechanisms. In at least one study, lung function measured in earlier life was a better predictor of arterial stiffness than that measured in later life, suggesting a common developmental pathway with vasculature (Bolton et al., 2009).

Physical and Cognitive Capability

An area for further research suggested by the figure is to examine how common aging mechanisms (box 4) and early decrements in CV, metabolic, and respiratory function (boxes 5) impact adult physical and cognitive capability and which of these pathways account for the early life origins of capability. The initial US longitudinal studies of aging based on measures of functional status started no earlier than midlife and the role of early life factors was initially neglected (see "Historical Overview"). In contrast, UK researchers were slower in extending their gaze to functional aging, but have brought with them a life course perspective, as exemplified by the HALCyon network (www.halycon.ac.uk). HALCyon researchers used the term "capability" to refer to functioning at the individual level, as opposed to the functioning of body systems that may underpin capability, and to draw attention to the benefits of studying those who maintain their capacity for the physical and mental tasks of daily living as they age, as well as those who are frail or have disabilities.

Lifetime Influences on Physical and Cognitive Capability

Systematic reviews and meta-analyses from HALCyon and elsewhere have contributed to the growing evidence that early life factors affect adult capability and its change. For example, there is strong evidence from a systematic review that birthweight is positively related to subsequent grip strength in children, young adults, and those at older ages (Dodds et al., 2012); research on the NSHD suggests this is most likely through its influence on the maximum level of function achieved at maturity rather than on the rate of decline (Cooper, Muniz-Terrera, & Kuh, 2013). In our overviews, we also describe evidence from single cohort studies showing that prepubertal growth is associated with better adult chair rise and standing balance performance, while pubertal growth is associated with better grip strength, independent of socioeconomic circumstances. There is also strong evidence of associations between disadvantaged childhood SEP and poorer walking speed and chair rise time which remained after adjustment for adult body size and SEP (Birnie et al., 2011) and growing evidence from single studies that area-based, as well as individual-level, socioeconomic characteristics influenced adult capability. Neurodevelopmental indicators, such as childhood cognitive ability and motor milestones, as well as indicators of childhood growth and early home environment are likely mediators of these early socioeconomic effects. Recent work on NSHD indicates that these neurodevelopmental indicators may be associated with decline in adult physical capability as well as the initial level (Cooper et al., 2013).

Recent findings from the 1946 cohort show striking similarities in childhood relative social inequalities in physical and cognitive capability in the early sixties with a range of pathways implicated (Hurst et al., 2013). Across the British 1946, 1958 and the US Wisconsin cohort studies, midlife fluid cognition was associated with childhood cognition and level of educational qualifications independent of social class (Clouston et al., 2012); the additional benefit of attaining degree-level qualifications was shown to vary across these cohorts, being smallest in the 1946 cohort and largest for the US cohort. Cross cohort research on cognitive aging in adult cohorts has embraced the lifespan perspective (Hofer & Piccinin, 2010); and the emergence of a growing network of life course and aging studies is likely to accelerate research into the early origins of cognitive aging and neurodegenerative disorders in this area over the next few years (www.ialsa.ac.uk). Hopefully this will also build on the limited research base, recently reviewed (Clouston et al., 2013), on whether changes in physical and cognitive capability covary, and if so, the role of shared early life experience, and the underlying mechanisms. As discussed in the section "Endocrine System", the HPA axis is one of the likely mechanisms.

CONCLUSIONS AND FUTURE DIRECTIONS

There has been a rapid rise in research investigating the early life origins of adult health, disease, and aging over the last 25 years, and we have tried to identify the key strands, show how they inter-relate, and capture the excitement of this research field. We have provided a context and a conceptual model in which these findings can be set and which prompts further life course hypotheses to test. The central questions that must be addressed are whether certain early life experiences or exposures make individuals more vulnerable to later life adverse exposures, or whether the long-term effects of early adversity can be ameliorated by later environmental or behavioral improvements. Looking for "effect modification" has been a longstanding interest of life course epidemiology in order to reveal any systematic patterns underlying heterogeneity in aging and disease risk so that subgroups, which might benefit most from interventions, can be identified early.

This research area is by its very nature interdisciplinary, but there is still too little overlap between social and biomedical scientists, and between researchers studying different specific diseases. For example, within the DOHaD paradigm, understanding the social mechanisms has not been given a high priority. Social scientists, while contributing to the epidemiological studies, have been rather conspicuous by their absence on the explanatory side except by testing for socioeconomic confounding. However, the growing number of studies of lifetime influences on functional aging attracts researchers from many disciplines and offers the challenge of linking the individual, body system, and cellular functional levels over a lifetime – and in relation to the changing external environment.

Life course researchers across disciplines also need to join forces in the joint study of common risk and protective factors across life that drive changes in physical and cognitive capability and in emotional health, and that affect subsequent disease development. This will encourage the design, implementation, and evaluation of interventions that may improve multiple outcomes and promote an integrated approach to health and social care.

The field is benefiting from new techniques for environmental and behavior exposure measurement and functional assessment, particularly dynamic assessment; these are emerging rapidly. In turn, life course epidemiology promotes the development of longitudinal and life course methods to make the best use of these new data.

A methodological challenge in life course research is the need to combine longitudinal data from several cohorts as no cohort study has all the data required to investigate functional or behavioral trajectories across the whole of life, and to assess the determinants and consequences. Lifetime trajectories of blood pressure (Wills et al., 2011), body size (Johnson, Li, Kuh, & Hardy, 2015), and alcohol consumption (Britton, Ben-Shlomo, Benzeval, Kuh, & Bell, 2015) have recently been derived by combining data in this way and are excellent examples of the way forward in this area. Ideally there should be overlapping data so that potential cohort effects can be identified.

This research field will further develop as more cohort studies with repeat assessments of the childhood environment and developmental characteristics mature enough to have collected repeated functional measures in adult life. We see two related and growing challenges. One is the balance between these cohort studies with information reaching back to early life, and the much larger cohort studies or very large biobanks that generally start in midlife and are developing their "omics" capabilities in the (possibly elusive) hunt for biomarkers of specific diseases and the development of personalized medicine. In the ideal world both types of study are required and they are complementary, but in the real world they compete for limited resources; and recent funding cuts to some long-established US cohort studies are cause for concern. The second challenge, given this climate, is to ensure we continue to attract and train primary cohort investigators who will invest time in these maturing cohort studies to capture richer and multiple measures in ways that are not too burdensome for study participants so that response rates and data of high quality are maintained; and who will apply the relevant scientific innovations and new techniques that are rapidly emerging.

The ultimate challenge, however, is to translate the growing body of evidence on the early origins of adult health, disease, and aging into practice or policy-relevant guidelines and intervention studies to improve the population health. This is increasingly recognized by policymakers. For example, the chief medical officer for England in the foreword to her annual report in 2012 stated "There is a growing knowledge of the complex interplay between psychosocial events and biological factors, and we now understand that events that occur as a fetus and in early life play a fundamental part in later life, and indeed in the lives of future generations" (Chief Medical Officer, 2013). We dedicate this chapter to Professors David Barker and Clyde Hertzman. Both scientists were responsible for imaginative and ground-breaking research that led to fresh insights into the early life origins of adult health; and they both shared a passion, drive, and commitment to use their findings to make the world a better place.

References

Adair, L. S., Fall, C. H., Osmond, C., Stein, A. D., Martorell, R., Ramirez-Zea, M., et al. (2013). Associations of linear growth and relative weight gain during early life with adult health and human capital in countries of low and middle income: Findings from five birth cohort studies. *The Lancet, 382*, 525–534.

Andersen, L. G., Angquist, L., Eriksson, J. G., Forsen, T., Gamborg, M., Osmond, C., et al. (2010). Birth weight, childhood body mass index and risk of coronary heart disease in adults: Combined historical cohort studies. *PLoS ONE, 5*, e14126.

Bann, D., Wills, A., Cooper, R., Hardy, R., Aihie, S. A., Adams, J., et al. (2014). Birth weight and growth from infancy to late adolescence in relation to fat and lean mass in early old age: Findings from the MRC National Survey of Health and Development. *International Journal of Obesity (Lond), 38*, 69–75.

Barker, D. J. P. (1995). Fetal origins of coronary heart disease. *British Medical Journal, 311*, 171–174.

Barker, D. J. P., Bull, A. R., Osmond, C., & Simmonds, S. J. (1990). Fetal and placental size and risk of hypertension in adult life. *British Medical Journal, 301*, 259–262.

Barker, D. J. P., Gluckman, P. D., Godfrey, K. M., Harding, J. E., Owens, J. A., & Robinson, J. S. (1993). Fetal nutrition and cardiovascular disease in adult life. *The Lancet, 341*, 938–941.

Barker, D. J. P., & Osmond, C. (1986). Infant mortality, childhood nutrition, and ischaemic heart disease in England and Wales. *The Lancet, 327*(8489), 1077–1081.

Barker, D. J. P., & Osmond, C. (1987). Death rates from stroke in England and Wales predicted from past maternal mortality. *British Medical Journal, 295*, 83–86.

Barker, D. J. P., Osmond, C., Forsen, T. J., Kajantie, E., & Eriksson, J. G. (2005). Trajectories of growth among children who have coronary events as adults. *New England Journal of Medicine, 353*, 1802–1809.

Barker, D. J. P., Winter, P. D., Osmond, C., Margetts, B., & Simmonds, S. J. (1989). Weight in infancy and death from ischaemic heart disease. *The Lancet, 334*(8663), 577–580.

Bellamy, L., Casas, J. P., Hingorani, A. D., & Williams, D. J. (2007). Pre-eclampsia and risk of cardiovascular disease and cancer in later life: Systematic review and meta-analysis. *British Medical Journal, 335*(7627), 974.

Ben-Shlomo, Y., & Davey Smith, G. (1991). Deprivation in infancy or adult life: Which is more important for mortality risk? *The Lancet, 337*, 530–534.

Ben-Shlomo, Y., Gardner, M., & Lightman, S. (2014). A life course approach to neuroendocrine systems: The example of the HPA axis. In D. Kuh, R. Cooper, R. Hardy, M. Richards, & Y. Ben-Shlomo (Eds.), *A life course approach to healthy ageing* (1st ed., pp. 133–145). Oxford: Oxford University Press.

Ben-Shlomo, Y., & Kuh, D. (2002). A life course approach to chronic disease epidemiology: Conceptual models, empirical challenges and interdisciplinary perspectives. *International Journal of Epidemiology, 31*, 285–293.

Ben-Shlomo, Y., McCarthy, A., Hughes, R., Tilling, K., Davies, D., & Smith, G. D. (2008). Immediate postnatal growth is associated with blood pressure in young adulthood: The Barry Caerphilly Growth Study. *Hypertension, 52*, 638–644.

Ben-Shlomo, Y., Mishra, G., & Kuh, D. (2014). Life course epidemiology. In W. Ahrens & I. Pigeot (Eds.), *Handbook of epidemiology* (2nd ed., pp. 1521–1549). New York, NY: Springer.

Ben-Shlomo, Y., Spears, M., Boustred, C., May, M., Anderson, S. G., Benjamin, E. J., et al. (2014). Aortic pulse wave velocity improves cardiovascular event prediction: An individual participant meta-analysis of prospective observational data from 17,635 subjects. *Journal of the American College of Cardiology, 63*, 636–646.

Berenson, G. S., MacMahan, C. A., Voors, A. W., Webber, L. S., Frank, G. S., Foster, T. A., et al. (1980). *Cardiovascular risk factors in children: The early natural history of atherosclerosis and essential hypertension*. New York, NY: Oxford University Press.

Bhargava, S. K., Sachdev, H. S., Fall, C. H., Osmond, C., Lakshmy, R., Barker, D. J., et al. (2004). Relation of serial changes in childhood body-mass index to impaired glucose tolerance in young adulthood. *New England Journal of Medicine, 350*, 865–875.

Birnie, K., Cooper, R., Martin, R. M., Kuh, D., Sayer, A. A., Alvarado, B. E., et al. (2011). Childhood socioeconomic position and objectively measured physical capability levels in adulthood: A systematic review and meta-analysis. *PLoS ONE, 6*, e15564.

Bolton, C. E., Cockcroft, J. R., Sabit, R., Munnery, M., McEniery, C. M., Wilkinson, I. B., et al. (2009). Lung function in mid-life compared with later life is a stronger predictor of arterial stiffness in men: The Caerphilly Prospective Study. *International Journal of Epidemiology, 38*, 867–876.

Britton, A., Ben-Shlomo, Y., Benzeval, M., Kuh, D., & Bell, S. (2015). Life course trajectories of alcohol consumption in the United Kingdom using longitudinal data from nine cohort studies. *BMC Medicine, 13*, 47.

Burton, G. J., Barker, D. J. P., Moffett, A., & Thornburg, K. (2010). *The placenta and human developmental programming*. Cambridge: Cambridge University Press.

Calvin, C. M., Deary, I. J., Fenton, C., Roberts, B. A., Der, G., Leckenby, N., et al. (2011). Intelligence in youth and all-cause-mortality: Systematic review with meta-analysis. *International Journal of Epidemiology, 40*, 626–644.

Chief Medical Officer. (2013). *Chief Medical Officer's annual report 2012: Our children deserve better: Prevention pays*. Department of Health. Retrieved from: <https://www.gov.uk/government/publications/chief-medical-officers-annual-report-2012-our-children-deserve-better-prevention-pays>.

Clouston, S. A. P., Brewster, P., Kuh, D., Richards, M., Cooper, R., Hardy, R., et al. (2013). The dynamic relationship between physical function and cognition in longitudinal aging cohorts. *Epidemiologic Reviews, 35*, 33–50.

Clouston, S. A. P., Kuh, D., Herd, P., Elliott, J., Richards, M., & Hofer, S. M. (2012). Benefits of educational attainment on adult fluid cognition: International evidence from three birth cohorts. *International Journal of Epidemiology, 41*, 1729–1736.

Cohen, S., Janicki-Deverts, D., Chen, E., & Matthews, K. A. (2010). Childhood socioeconomic status and adult health. *Annals of New York Academy of Sciences, 1186*, 37–55.

Colley, J. R. T., Douglas, J. W. B., & Reid, D. D. (1973). Respiratory disease in young adults; Influence of early childhood lower respiratory tract illness, social class, air pollution, and smoking. *British Medical Journal, 2*, 195–198.

Cooper, R., Muniz-Terrera, G., & Kuh, D. (2013). Neurodevelopmental pathways are associated with changes in physical capability in early old age. *The Gerontologist, 53*(S1), 630.

Davey Smith, G. (2011). Epidemiology, epigenetics and the "Gloomy Prospect": Embracing randomness in population health research and practice. *International Journal of Epidemiology, 40*, 537–562.

Davey Smith, G., Ben-Shlomo, Y., Beswick, A., Yarnell, J., Lightman, S., & Elwood, P. (2005). Cortisol, testosterone, and coronary heart disease: Prospective evidence from the Caerphilly study. *Circulation, 112*, 332–340.

Davey Smith, G., Hypponen, E., Power, C., & Lawlor, D. A. (2007). Offspring birth weight and parental mortality: Prospective observational study and meta-analysis. *American Journal of Epidemiology, 166*, 160–169.

de Jong, F., Monuteaux, M. C., van Elburg, R. M., Gillman, M. W., & Belfort, M. B. (2012). Systematic review and meta-analysis of preterm birth and later systolic blood pressure. *Hypertension, 59*, 226–234.

Dodds, R., Denison, H. J., Ntani, G., Cooper, R., Cooper, C., Sayer, A. A., et al. (2012). Birth weight and muscle strength: A systematic review and meta-analysis. *The Journal of Nutrition, Health and Ageing, 16*, 609–615.

Dubos, R. (1965). *Man adapting*. New Haven, CT: Yale University Press.

Essex, M. J., Shirtcliff, E. A., Burk, L. R., Ruttle, P. L., Klein, M. H., Slattery, M. J., et al. (2011). Influence of early life stress on later hypothalamic-pituitary-adrenal axis functioning and its covariation with mental health symptoms: A study of the allostatic process from childhood into adolescence. *Development and Psychopathology, 23*, 1039–1058.

Fall, C. H. D., Vijayakumar, M., Barker, D. J. P., Osmond, C., & Duggleby, S. (1995). Weight in infancy and prevalence of coronary heart disease in adult life. *British Medical Journal, 310*, 17–19.

Ferrucci, L., & Studenski, S. (2012). Clinical problems in aging: *Harrison's principles of internal medicine* (18th ed., pp. 570–585). New York, NY: McGraw-Hill.

Fisher, D., Baird, J., Payne, L., Lucas, P., Kleijnen, J., Roberts, H., et al. (2006). Are infant size and growth related to burden of disease in adulthood? A systematic review of literature. *International Journal of Epidemiology, 35*, 1196–1210.

Freathy, R. M., Bennett, A. J., Ring, S. M., Shields, B., Groves, C. J., Timpson, N. J., et al. (2009). Type 2 diabetes risk alleles are associated with reduced size at birth. *Diabetes, 58*, 1428–1433.

Friedman, E. M., Karlamangla, A. S., Almeida, D. M., & Seeman, T. E. (2012). Social strain and cortisol regulation in midlife in the US. *Social Science & Medicine, 74*, 607–615.

Friedrich, N., Schneider, H. J., Haring, R., Nauck, M., Volzke, H., Kroemer, H. K., et al. (2012). Improved prediction of all-cause mortality by a combination of serum total testosterone and insulin-like growth factor I in adult men. *Steroids, 77*, 52–58.

Fryers, T., & Brugha, T. (2013). Childhood determinants of adult psychiatric disorder. *Clinical Practice and Epidemiology in Mental Health, 9*, 1–50.

Gale, C. R., Deary, I. J., & Stafford, M. (2014). A life course approach to psychological and social wellbeing. In D. Kuh, R. Cooper, R. Hardy, M. Richards, & Y. Ben-Shlomo (Eds.), *A life course approach to healthy ageing* (1st ed., pp. 46–61). Oxford: Oxford University Press.

Galobardes, B., Smith, G. D., & Lynch, J. W. (2006). Systematic review of the influence of childhood socioeconomic circumstances on risk for cardiovascular disease in adulthood. *Annals of Epidemiology, 16*, 91–104.

Gardner, M. P., Lightman, S., Sayer, A. A., Cooper, C., Cooper, R., Deeg, D., et al. (2013). Dysregulation of the hypothalamic pituitary adrenal (HPA) axis and physical performance at older ages: An individual participant meta-analysis. *Psychoneuroendocrinology, 38*, 40–49.

Gardner, M. P., Lightman, S. L., Gallacher, J., Hardy, R., Kuh, D., Ebrahim, S., et al. (2011). Diurnal cortisol patterns are associated with physical performance in the Caerphilly Prospective Study. *International Journal of Epidemiology, 40*, 1693–1702.

Gluckman, P. D., & Hanson, M. A. (2004). Living with the past: Evolution, development, and patterns of disease. *Science, 305*, 1733–1736.

Gluckman, P. D., Hanson, M. A., Bateson, P., Beedle, A. S., Law, C. M., Bhutta, Z. A., et al. (2009). Towards a new developmental synthesis: Adaptive developmental plasticity and human disease. *The Lancet, 373*, 1654–1657.

Godfrey, K. M., Inskip, H. M., & Hanson, M. A. (2011). The long-term effects of prenatal development on growth and metabolism. *Seminars in Reproductive Medicine, 29*, 257–265.

Gurven, M., Blackwell, A. D., Rodriguez, D. E., Stieglitz, J., & Kaplan, H. (2012). Does blood pressure inevitably rise with age?: Longitudinal evidence among forager-horticulturalists. *Hypertension, 60*, 25–33.

Hales, C. N., Barker, D. J. P., Clark, P. M. S., Cox, L. J., Fall, C., Osmond, C., et al. (1991). Fetal and infant growth and impaired glucose tolerance at age 64. *British Medical Journal, 303*, 1019–1022.

Hardy, R., Kuh, D., Langenberg, C., & Wadsworth, M. E. J. (2004). Birth weight, childhood growth and blood pressure at 43 years in a British birth cohort. *International Journal of Epidemiology, 33*, 121–129.

Hardy, R., Potischman, N., & Kuh, D. (2012). Life course approach to research in women's health. In L. Goldman, R. Troisi, & K. M. Rexrode (Eds.), *Women and health* (2nd ed., pp. 119–129). London: Elsevier.

Hattersley, A. T., & Tooke, J. E. (1999). The fetal insulin hypothesis: An alternative explanation of the association of low birthweight with diabetes and vascular disease. *The Lancet, 353*, 1789–1792.

Hertzman, C. (1995). The biological embedding of early experience and its effects on health in adulthood. *Annals of the New York Academy of Sciences, 896*, 85–95.

Hertzman, C., & Boyce, T. (2010). How experience gets under the skin to create gradients in developmental health. *Annual Review of Public Health, 31*, 329–347.

Hofer, S. M., & Piccinin, A. M. (2010). Toward an integrative science of life-span development and aging. *The Journals of Gerontology Series B: Psychological Sciences and Social Sciences, 65B*, 269–278.

Huber, M., Knottnerus, J. A., Green, L., van der Horst, H., Jadad, A. R., Kromhout, D., et al. (2011). How should we define health? *British Medical Journal, 343*, d4163.

Hunt, L. P., Ben-Shlomo, Y., Clark, E. M., Dieppe, P., Judge, A., MacGregor, A. J., et al. (2013). 90-day mortality after 409,096 total hip replacements for osteoarthritis, from the National Joint Registry for England and Wales: A retrospective analysis. *The Lancet, 382*, 1097–1104.

Hunter, A. L., Minnis, H., & Wilson, P. (2011). Altered stress responses in children exposed to early adversity: A systematic review of salivary cortisol studies. *Stress, 14*, 614–626.

Hurst, L., Stafford, M., Cooper, R., Hardy, R., Richards, M., & Kuh, D. (2013). Lifetime socioeconomic inequalities in physical and cognitive aging. *American Journal of Public Health, 103*, 1641–1648.

Huxley, R., Neil, A., & Collins, R. (2002). Unravelling the fetal origins hypothesis: Is there really an inverse association between birthweight and subsequent blood pressure? *The Lancet, 360*, 659–665.

Huxley, R., Owen, C. G., Whincup, P. H., Cook, D. G., Colman, S., & Collins, R. (2004). Birth weight and subsequent cholesterol levels: Exploration of the "fetal origins" hypothesis. *The Journal of the American Medical Association, 292*, 2755–2764.

Huxley, R., Owen, C. G., Whincup, P. H., Cook, D. G., Rich-Edwards, J., Smith, G. D., et al. (2007). Is birth weight a risk factor for ischemic heart disease in later life? *American Journal of Clinical Nutrition, 85*, 1244–1250.

Javaid, M. K., Eriksson, J. G., Kajantie, E., Forsen, T., Osmond, C., Barker, D. J., et al. (2011). Growth in childhood predicts hip fracture risk in later life. *Osteoporosis International, 22*, 69–73.

Johnson, W., Li, L., Kuh, D., & Hardy, R. (2015). How has the age-related process of overweight or obesity development changed over time? Co-ordinated analyses of individual participant data from five United Kingdom birth cohorts. *PLoS Medicine*.

Kirkwood, T. B., & Austad, S. N. (2000). Why do we age? *Nature, 408*, 233–238.

Koenen, K., Rudenstine, S., Susser, E., & Galea, S. (2013). *Life course approach to mental disorders* (1st ed.). Oxford: Oxford University Press.

Kubzansky, L. D., & Winning, A. (2013). Mental disorders and the emergence of physical disorders. In K. Koenen, S. Rudenstine, E. Susser, & S. Galea (Eds.), *A life course approach to mental disorder* (pp. 291–305). Oxford: Oxford University Press.

Kuh, D., & Ben-Shlomo, Y. (1997). *A life course approach to chronic disease epidemiology: Tracing the origins of ill-health from early to adult life* (1st ed.). Oxford: Oxford University Press.

Kuh, D., & Ben-Shlomo, Y. (2004). *A life course approach to chronic disease epidemiology* (2nd ed.). Oxford: Oxford University Press.

Kuh, D., Ben Shlomo, Y., Lynch, J., Hallqvist, J., & Power, C. (2003). Life course epidemiology. *Journal of Epidemiology and Community Health, 57*, 778–783.

Kuh, D., Ben-Shlomo, Y., Tilling, K., & Hardy, R. (2015). Life course epidemiology and analysis. In R. Detels, M. Gulliford, Q. Abdool Karim, & C. Tan (Eds.), *The Oxford Textbook of Public Health* (6th ed.). Oxford: Oxford University Press. pp. 679–692.

Kuh, D., Cooper, R., Hardy, R., Richards, M., & Ben Shlomo, Y. (2014). *A life course approach to healthy ageing*. Oxford: Oxford University Press.

Kuh, D., Richards, M., Cooper, R., Hardy, R., & Ben-Shlomo, Y. (2014). Life course epidemiology, ageing research, and maturing cohort studies: A dynamic combination for understanding healthy ageing. In D. Kuh, R.

Cooper, R. Hardy, M. Richards, & Y. Ben-Shlomo (Eds.), *A life course approach to healthy ageing* (1st ed., pp. 3–15). Oxford: Oxford University Press.

Kuh, D., Wills, A. K., Shah, I., Prentice, A., Hardy, R., Adams, J. E., et al. (2014). Growth from birth to adulthood and bone phenotype in early old age: A British birth cohort study. *Journal of Bone and Mineral Research, 29,* 123–133.

Kumari, M., Badrick, E., Sacker, A., Kirschbaum, C., Marmot, M., & Chandola, T. (2010). Identifying patterns in cortisol secretion in an older population. Findings from the Whitehall II study. *Psychoneuroendocrinology, 35,* 1091–1099.

Law, C. M., & Shiell, A. W. (1996). Is blood pressure inversely related to birth weight? The strength of evidence from a systematic review of the literature. *Journal of Hypertension, 14,* 935–941.

Lawlor, D. A., & Hardy, R. (2014). Vascular and metabolic function across the life course. In D. Kuh, R. Cooper, R. Hardy, M. Richards, & Y. Ben-Shlomo (Eds.), *A life course approach to healthy ageing* (pp. 146–161). Oxford: Oxford University Press.

Lawlor, D., & Mishra, G. (2009). *Family matters: Designing, analysing and understanding family based studies in life course epidemiology* (1st ed.). Oxford: Oxford University Press.

Lopez-Otin, C., Blasco, M. A., Partridge, L., Serrano, M., & Kroemer, G. (2013). The hallmarks of aging. *Cell, 153,* 1194–1217.

Maggio, M., Cappola, A. R., Ceda, G. P., Basaria, S., Chia, C. W., Valenti, G., et al. (2005). The hormonal pathway to frailty in older men. *Journal of Endocrinology Investigation, 28,* 15–19.

Marmot, M. G., Rose, G. A., Shipley, M. J., & Hamilton, P. J. S. (1978). Employment grade and coronary heart disease in British civil servants. *Journal of Epidemiology and Community Health, 32,* 244–249.

Matthews, K. A., Gallo, L. C., & Taylor, S. E. (2010). Are psychosocial factors mediators of socioeconomic status and health connections? A progress report and blueprint for the future. *Annals of New York Academy of Sciences, 1186,* 146–173.

Maughan, B., Stafford, M., Shah, I., & Kuh, D. (2014). Adolescent conduct problems and premature mortality: Follow-up to age 65 years in a national birth cohort. *Psychological Medicine, 44*(5), 1077–1086.

McCarron, P., Okasha, M., McEwen, J., & Smith, G. D. (2001). Changes in blood pressure among students attending Glasgow University between 1948 and 1968: Analyses of cross sectional surveys. *British Medical Journal, 322,* 885–889.

McEniery, C. M., Spratt, M., Munnery, M., Yarnell, J., Lowe, G. D., Rumley, A., et al. (2010). An analysis of prospective risk factors for aortic stiffness in men: 20-year follow-up from the Caerphilly Prospective Study. *Hypertension, 56,* 36–43.

McEwen, B. S., & Wingfield, J. C. (2003). The concept of allostasis in biology and biomedicine. *Hormones and Behavior, 43,* 2–15.

Muller, M., Smulders, Y. M., de Leeuw, P. W., & Stehouwer, C. D. (2014). Treatment of hypertension in the oldest old: A critical role for frailty? *Hypertension, 63,* 433–441.

National Research Council, (2001). *New horizons in health: An integrative approach.* Washington, DC: National Academy Press.

Newsome, C. A., Shiell, A. W., Fall, C. H. D., Phillips, D. I. W., Shier, R., & Law, C. M. (2003). Is birth weight related to later glucose and insulin metabolism? – A systematic review. *Diabetic Medicine, 20,* 339–348.

Ng, J. W., Barrett, L. M., Wong, A., Kuh, D., Smith, G. D., & Relton, C. L. (2012). The role of longitudinal cohort studies in epigenetic epidemiology: Challenges and opportunities. *Genome Biology, 13,* 246.

Odden, M. C., Peralta, C. A., Haan, M. N., & Covinsky, K. E. (2012). Rethinking the association of high blood pressure with mortality in elderly adults: The impact of frailty. *Archives of Internal Medicine, 172,* 1162–1168.

Owen, C. G., Whincup, P. H., & Cook, D. G. (2011). Breastfeeding and cardiovascular risk factors and outcomes in later life: Evidence from epidemiological studies. *Proceedings of the Nutrition Society, 70,* 478–484.

Phillips, D. I. W., Barker, D. J. P., Hales, C. N., Hirst, S., & Osmond, C. (1994). Thinness at birth and insulin-resistance in adult life. *Diabetologia, 37,* 150–154.

Power, C., Kuh, D., & Morton, S. (2013). From developmental origins of adult disease to life course research on adult disease and aging: Insights from birth cohort studies. *Annual Review of Public Health, 34,* 7–28.

Raghupathy, P., Antonisamy, B., Geethanjali, F. S., Saperia, J., Leary, S. D., Priya, G., et al. (2010). Glucose tolerance, insulin resistance and insulin secretion in young south Indian adults: Relationships to parental size, neonatal size and childhood body mass index. *Diabetes Research and Clinical Practice, 87,* 283–292.

Repetti, R. L., Taylor, S. E., & Seeman, T. E. (2002). Risky families: Family social environments and the mental and physical health of offspring. *Psychological Bulletin, 128,* 330–366.

Rich-Edwards, J. W., McElrath, T. F., Karumanchi, S. A., & Seely, E. W. (2010). Breathing life into the lifecourse approach: Pregnancy history and cardiovascular disease in women. *Hypertension, 56,* 331–334.

Rich-Edwards, J. (2002). A life course approach to women's reproductive health. In D. Kuh & R. Hardy (Eds.), *A life course approach to women's health* (pp. 23–43). Oxford: Oxford University Press.

Risnes, K. R., Vatten, L. J., Baker, J. L., Jameson, K., Sovio, U., Kajantie, E., et al. (2011). Birthweight and mortality in adulthood: A systematic review and meta-analysis. *International Journal of Epidemiology, 40,* 647–661.

Rogers, I., & the Euro-BLCS Study Group, (2003). The influence of birthweight and intrauterine environment on adiposity and fat distribution in later life. *International Journal of Obesity, 25*, 755–777.

Rutter, M. (2012). Achievements and challenges in the biology of environmental effects. *Proceedings of the National Academy of Sciences, USA, 109*(Suppl. 2), 17149–17153.

Rutter, M., Kim-Cohen, J., & Maughan, B. (2006). Continuities and discontinuities in psychopathology between childhood and adult life. *Journal of Child Psychology and Psychiatry, 47*, 276–295.

Shonkoff, J. P., Garner, A. S., The Committee on Psychosocial Aspects of Child and Family Health., Committee on Early Childhood, Adoption and Dependent Care and Section on Developmental and Behavioral Pediatrics, Siegel, B. S., Dobbins, M. I., & Wood, D. L. (2012). Life course approach to mental disorders. *Pediatrics, 129*, e232–e246.

Stein, C. E., Fall, C. H., Kumaran, K., Osmond, C., Cox, V., & Barker, D. J. (1996). Fetal growth and coronary heart disease in South India. *The Lancet, 348*, 1269–1273.

Tandon, N., Fall, C. H., Osmond, C., Sachdev, H. P., Prabhakaran, D., Ramakrishnan, L., et al. (2012). Growth from birth to adulthood and peak bone mass and density data from the New Delhi Birth Cohort. *Osteoporosis International, 23*, 2447–2459.

The Emerging Risk Factors Collaboration. (2012). Adult height and the risk of cause-specific death and vascular morbidity in 1 million people: Individual participant meta-analysis. *International Journal of Epidemiology, 41*, 1419–1433.

Tyrrell, J. S., Yaghootkar, H., Freathy, R. M., Hattersley, A. T., & Frayling, T. M. (2013). Parental diabetes and birthweight in 236 030 individuals in the UK Biobank Study. *International Journal of Epidemiology, 42*, 1714–1723.

Vedhara, K., Miles, J., Crown, A., McCarthy, A., Shanks, N., Davies, D., et al. (2007). Relationship of early childhood illness with adult cortisol in the Barry Caerphilly Growth (BCG) cohort. *Psychoneuroendocrinology, 32*, 865–873.

Wadsworth, M. E. J., Cripps, H. A., Midwinter, R. A., & Colley, J. R. T. (1985). Blood pressure at age 36 years and social and familial factors, cigarette smoking and body mass in a national birth cohort. *British Medical Journal, 291*, 1534–1538.

Whincup, P. H., Kaye, S. J., Owen, C. G., Huxley, R., Cook, D. G., Anazawa, S., et al. (2008). Birth weight and risk of type 2 diabetes: A systematic review. *The Journal of the American Medical Association, 300*, 2886–2897.

Wills, A. K., Black, S., Cooper, R., Coppack, R. J., Hardy, R., Martin, K. R., et al. (2012). Life course body mass index and risk of knee osteoarthritis at the age of 53 years: Evidence from the 1946 British birth cohort study. *Annals of the Rheumatic Diseases, 71*, 655–660.

Wills, A. K., Lawlor, D. A., Matthews, F. E., Sayer, A. A., Bakra, E., Ben-Shlomo, Y., et al. (2011). Life course trajectories of systolic blood pressure using longitudinal data from eight UK cohorts. *PLoS Medicine, 8*, e1000440.

CHAPTER 6

Racial and Ethnic Inequalities in Health

Jacqueline L. Angel[1], Stipica Mudrazija[2], and Rebecca Benson[1]

[1]The University of Texas at Austin, Austin, TX, USA [2]University of Southern California, Los Angeles, CA, USA

OUTLINE

Racial and Ethnic Inequalities in Health	123	Research Across Minority Groups	130
Theoretical Perspectives	125	Summary and Conclusion	133
Life-Course Perspectives on Health	*127*	References	137
Theories of Life-Course Racial and Ethnic Health Disparities	*129*		

RACIAL AND ETHNIC INEQUALITIES IN HEALTH

The population is aging in both the developed and developing world, and the United States is no exception. It is well established that Americans are living longer. US life expectancy at birth increased from 70.8 in 1970 to 78.7 in 2010 (Kochanek, Arias, & Anderson, 2013). For those aged 65, remaining life expectancy rose from 15.2 years to 19.1 years over the same time period (National Center for Health Statistics, 2012). Perhaps most importantly, and the focus of this chapter, as the nation grows older it is rapidly becoming more racially and ethnically diverse. Currently, there is less racial/ethnic diversity in the population 65 and older than among younger Americans. However, the older population is expected to be 42% minority by 2050 (US Census Bureau, 2011). The dual trends of increased life expectancy and the growing diversity of the aging population raise important theoretical and practical questions about race-based inequalities in health and their consequences for health care spending in light of projected deficits in Medicare.

In this chapter, we present a historical overview of the study of minority health and aging in order to examine the implications of extended morbidity for minority groups. We trace the evolution of the field, delineating contemporary theoretical perspectives used to understand the underlying social processes associated with health inequality across the adult life course for the largest American minority groups. We end with a summary of what is already known and a discussion of core research questions that merit investigation for future work.

Even though the United States is becoming increasingly racially and ethnically diverse, until relatively recently the focus of the research on health and aging across minority groups was predominantly on the Black–White population differences, and only over the past two decades has the study of the Hispanic population come to the forefront of the research (Mutchler & Burr, 2011). (We refer herein to non-Hispanic Black and non-Hispanic White populations as Black and White populations, respectively.) Studies of aging in Asian–American and Native American groups are comparatively few. Some issues are common among multiple minority groups – Hispanic and Asian elders with limited English proficiency experience similar barriers to accessing health services (Kim et al., 2011) – but there remains a danger of homogenizing Asian experiences of aging when in fact there is considerable diversity between subgroups (Kim et al., 2010). Just as the unique issues faced by aging Asian Americans, as well as their projected numerical increase, make research on aging in this group challenging, so the historical disadvantage experienced by Native Americans makes the lack of research on the aging experiences of this group even more problematic.

We address several specific issues related to Latinos and individuals of Mexican origin, primarily because they represent the largest segment of the fast-growing population of Hispanic elders (Angel, Torres-Gil, & Markides, 2012). This emphasis is motivated by the fact that this population on average has extremely low levels of education, health insurance, income, and wealth, and it suffers from high rates of diabetes, hypertension, and disability (Angel & Angel, 2009; Hummer, Benjamins, & Rogers, 2004; Palloni, 2007; Wu et al., 2003). Yet compared to Black and even White populations, its mortality experience is remarkably favorable – a so-called Hispanic paradox (Markides & Eschbach, 2005; Palloni & Arias, 2004). In 2006, life expectancy at birth for Hispanic individuals, the majority of whom are of Mexican origin (64.5%), was 80.6 years compared to 78.1 for White individuals (Arias, 2010). By contrast life expectancy at birth for Black persons was only 72.9 years (Arias, 2010). By age 65, life expectancy for Hispanic men and women was 84 and 86.7 years, respectively, while for White men and women it was 82.1 and 84.7 and for Black men and women it was 80 and 83.4. In light of these demographic trends, examining racial and ethnic differences in health is becoming ever more relevant. As we discuss later, differences in socioeconomic position contribute to race-based differences in health and may persist in later life (Ferraro, 2011).

Although mortality has declined across all racial and ethnic groups (Crimmins et al., 2009), increasing life expectancy may not add "life to years" as well as "years to life" (Fries, Bruce, & Chakravarty, 2011). Functional disability remains very high among older adults 65 and over and is becoming prevalent among persons in late middle age in large part due to the rise in obesity rates. The possibility of an increased number of years characterized by protracted periods of impaired health and functioning and increasing dependency poses potentially serious economic, social, and political problems related to the care of a large dependent elderly population (Cutler & Landrum, 2011; Fried, Ferrucci, Darer, Williamson, & Anderson, 2004; Manton & Stallard, 1991). Yet, the evidence suggests that such a conclusion would be premature, especially for certain subgroups of the older population (Gorin & Lewis, 2004). Significant health and disability differentials exist among racial and ethnic groups (Hayward, Miles, Crimmins, & Yang, 2000; Kelley-Moore & Ferraro, 2004).

Epidemiologic data on elderly minority groups, although limited, show that Black and Hispanic individuals 65 and over suffer more limitations in activities of daily living (ADL) and instrumental activities of daily living (IADL) on average than White individuals (Song et al., 2007). Black (11.8%) and Hispanic persons (11.2%) aged 65 and over were more likely than White persons (5.5%) to report ADL disability (Markides, Eschbach, Ray, & Peek, 2007; US Census Bureau, 2003). A higher percentage of Hispanic (17%) and Black individuals (18.7%) also reported difficulties with performing IADLs than White individuals (11.6%) (US Census Bureau, 2003). In the 1998–2004 Health and Retirement Study (HRS) higher rates of ADL disability due to arthritis were found in the Black (28%), Spanish-speaking Hispanic (28.5%), and English-speaking Hispanic populations (19.1%) than in the White population (16.2%) (Song et al., 2007). Markides et al. (2007) disaggregated disability rates for older adults by race/Hispanic ethnicity and gender. Among the Hispanic population in the 1990 Census, Mexican-American individuals over 65 reported a higher need for ADL assistance than White individuals: 12.9% and 17.1% for Mexican-origin men and women, respectively, versus 9.5% and 12.6% for White men and women, respectively. Similar rates of ADL disability were found in the Hispanic Established Populations for the Epidemiologic Study of the Elderly (H-EPESE) (Angel & Angel, 1997).

Co-morbid physical conditions for the Mexican-origin population in particular are lower and more protracted than for White Americans in the 1995–2006 HRS (Quiñones, Liang, Bennett, Xu, & Ye, 2011). The data suggest a more protracted accumulation of physical diseases among the Black and Hispanic populations compared to the White population that strongly relate to educational differences. Investigations have also used the HRS to estimate growth curve models of physical mobility and ADL and IADL disability trajectories. The results reveal higher levels of protracted infirmity for the Hispanic and Black than for the White population (Chiu & Wray, 2011).

Other studies explore within-group heterogeneity among the Hispanic population. The data reveal important gender and immigrant differences in morbidity trajectories. Although women and men are similarly advantaged at baseline in terms of cognitive status, immigrant men tend to maintain their cognitive advantage for a longer period of time with a slower rate of decline in cognitive impairment (Hill, Angel, & Balistreri, 2012; Zeki et al., 2012). Some of these associations were larger in first- and second-generation immigrant families (Zeki et al., 2012). These findings point to the heterogeneity of cognitive aging among diverse race/ethnic groups that may be influenced by intergenerational changes in socioeconomic status (SES), cultural norms and values, and health behaviors related to changes in the social and physical environment. Even so, the extent to which these race/ethnic differences reflect *genetic* or social factors are nuanced and remain unclear (Kelley-Moore & Ferraro, 2004).

THEORETICAL PERSPECTIVES

While recent scientific developments are making it possible to explore the unique contributions of genetic characteristics to variation in health outcomes by racial and ethnic groups (Frank, 2007), this area of research is still in its early stage. The research on racial and ethnic disparities in health is still primarily focused on different aspects of SES as the major explanatory factor of the observed variation in health. It suggests that the disproportionate burden of ill-health borne by minorities can be attributed to the fact that they are also disproportionately likely to have low incomes and low education and to live in neighborhoods characterized by low income and low levels of education.

The study of socioeconomic position and health in old age has established that, in addition to changes in health behavior, socioeconomic position indeed shapes successful aging (House, 2002). There is a strong association between socioeconomic disparities and the expansion of morbidity and disability rates across the adult life course. In the 1986 American's Changing Lives study, House and colleagues observed over a 15-year study period that the more-educated fared far better than the less-educated in terms of reporting better physical and mental health (House, Lantz, & Herd, 2005). Education works in complex ways, however: while lower education can amplify early-life adversities, higher education may contribute to attenuating them (O'Rand & Hamil-Luker, 2005). Education is a key component to one's life chances because it is associated with higher paying jobs and increased job security. Many jobs held by low-income people subject them to unhealthful conditions. Education also brings a greater propensity to engage in healthy life styles across the life course. Higher education fosters a greater sense of self-efficacy, leading to a higher locus of health control (Leigh & Fries, 1994; Thorpe & Angel, 2013). American minority groups represent a disproportionate share of the disadvantaged. For ethnic minorities, who are disproportionately represented in the lower strata of SES, low levels of education, low incomes, and little wealth may hinder access to treatment and prevention of chronic illness. In general, longer life contributes substantially to diminished health (Thorpe & Angel, 2013).

Despite its importance for the study of racial and ethnic variation in health, SES cannot account for the full extent of the observed differences. In fact, some differences in health outcomes across different racial and ethnic groups (e.g., the Hispanic health paradox) seem to exist largely independently from the socioeconomic position of those groups. Therefore, the debate in the literature continues as to whether there is an effect of race/ethnicity independent of socioeconomic characteristics (Do, Frank, & Finch, 2012; Farmer & Ferraro, 2005; Geruso, 2012; Hayward et al., 2000; Kawachi, Daniels, & Robinson, 2005; LaVeist, 2005).

Moreover, other characteristics, most notably gender, interact with socioeconomic resources in ways that may lead to different health outcomes for older minority and White individuals of comparable SES, but the intersection of race/ethnicity, gender, and SES has received relatively little scholarly attention (Krekula, 2007; Venn, Davidson, & Arber, 2011). Recent empirical research using a nationally representative sample of older White, Black, and Mexican-origin Americans emphasizes complexities of this relationship that warrant a careful consideration of their combined effects rather than adopting an implicit assumption of separate, additive effects of gender and race/ethnicity (Warner & Brown, 2011). Warner and Brown's findings suggest that women of a minority group suffer from more functional limitations relative to minority men than do White women relative to White men, and Black women follow a unique trajectory of accelerated disability with age. System-level factors like scarcity of health care providers and facilities in geographic areas with higher concentrations of minority populations are also important contributors to the observed racial and ethnic disparities in health outcomes (Smedley, Stith, & Nelson, 2009). Although it is beyond the scope of this chapter to examine these and other modifying factors in depth, their importance has to be recognized and taken into consideration.

Life-Course Perspectives on Health

Although the research focus on socioeconomic origins of racial and ethnic inequalities in health has been instrumental in establishing the existence of strong correlation between socioeconomic disparities and late-life health outcomes, it has had only limited success in describing the causal mechanism linking the two. As the evidence of the effects of education on health in old age suggests, early-life conditions and events may have decisive impacts on adult and late-life outcomes. The observed within-cohort disparities can be explained by structural and institutional constraints on individual decisions over the life course that determine costs and benefits of different choices for different individuals. The life-course perspective, then, provides the necessary framework for studying differences within and across cohorts by examining the nexus of structural and institutional arrangements with individual characteristics (O'Rand, 1996).

Figure 6.1 depicts the organizing framework for understanding racial and ethnic-based differences in health over the life course. It demonstrates that the link between early-life, mostly ascribed characteristics, and late-life health outcomes is mediated by mid-life, mostly achieved characteristics, and affected by system-level characteristics (e.g., the availability and quality of health care) throughout the life course. It is by necessity an overly simplified representation of an otherwise immensely complex life-course process where the timing and sequencing of different life-course events (e.g., education, marriage, or childbearing) can produce numerous iterations of this life-course process with different health outcomes for individuals with similar initial endowment and other ascribed characteristics. Nonetheless, the general framework is valuable in as much as it highlights the overarching importance of incorporating a life-course perspective in any attempt to explain the causal mechanism responsible for substantial variation in health across racial and ethnic groups.

While theoretically appealing, the life-course perspective can be incorporated in rigorous empirical research only to the extent the appropriate longitudinal data are available. Moreover, for the purposes of examining minority health and racial/ethnic disparities in health at old age, it is critical that sufficient numbers of minorities are included in surveys used to study health and aging. Recruiting

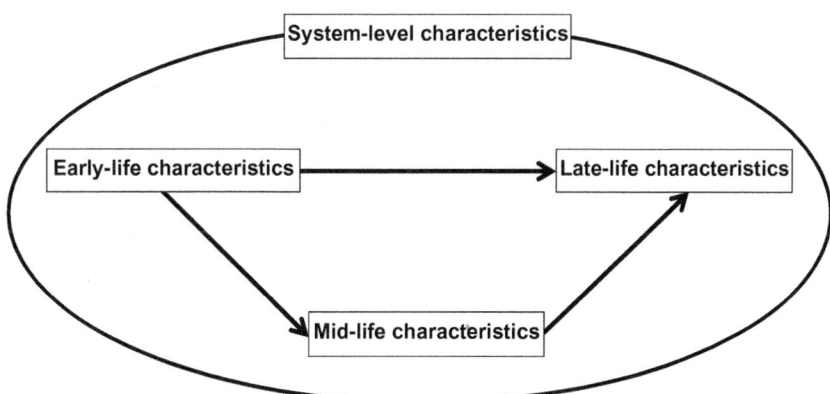

FIGURE 6.1. Organizing framework for explaining racial and ethnic-based disparities in health.

sufficient numbers of minority participants to longitudinal data collection projects is difficult for two reasons. The first is that minorities, by definition, comprise a smaller proportion of the population and are thus less likely to be selected in any random process compared to the majority. A second reason for unsuccessful recruitment relates to the lack of culturally and racially sensitive approaches for outreach into ethnically diverse communities (Levkoff, Prochaska, Weitzman, & Ory, 2000).

Given these barriers, existing data sets containing large numbers of minority participants are even more valuable. The H-EPESE is one such example, following community-dwelling Mexican-origin Americans aged 65 and over in Texas, Arizona, New Mexico, California, and Colorado since 1993–1994. More than 300 published studies have used these data to provide insight into the health of older Mexican-origin Americans, a group who will become one of the largest population groups in the United States in coming years. The HRS has also been a valuable source of data for research on minority health and aging. The HRS is not limited to minorities, but oversamples Black and Hispanic individuals in order to enable the use of inferential statistics on racial/ethnic subgroups. This approach is becoming standard and should ensure that future research can confidently study minority health.

Empirical studies based on the life-course approach have provided important insights into the mechanisms linking early- and midlife conditions with mid- and late-life health outcomes. Maternal exposure to discrimination, stress, and poverty predicts low birth weight (Geronimus, 1996), and childhood SES predicts adult mortality (Hayward & Gorman, 2004). Early childhood conditions affect health significantly in late life (Crimmins, Hayward, & Seeman, 2004). Although health disparities appear modest in early adulthood, they increase through middle age and early old age. In later old age, these disparities lessen again.

Healthful aging may partly reflect differential exposure to all known psychosocial factors, as well as a wide range of biobehavioral and environmental factors, for disease and illness (Crimmins, Kim, Alley, Karlamangla, & Seeman, 2007). Vigorous exercise and weight control, even if begun in middle age, can delay the onset of disabling conditions by more than a decade (Hubert, Bloch, Oehlert, & Fries, 2002).

Another important insight of this literature is that the amount of social support available provides a health advantage at different life stages. Supportive intimate relationships provide emotional support that mitigates the exposure to social isolation and loneliness (Umberson & Montez, 2010). They also allow individuals to develop a strong self-identity that gives life meaning and purpose (Umberson & Montez, 2010). Married people of both sexes enjoy longer, fuller, happier lives than do single people, and they also have more satisfied and successful children than those who divorce (Waite & Gallagher, 2001). Marriage benefits both men and women at older ages because they have a larger network of friends and family to turn to in times of need (Waite & Das, 2010). The benefits persist even during stressful life events, at least for women, because they tend to rebuild supportive networks after the death of a spouse (Umberson, Wortman, & Kessler, 1992). Thus, besides the clear practical benefits of social integration, the availability of relatives and friends with whom to interact directly influences mental health and morale as one ages (Fowler & Christakis, 2008).

Notwithstanding the simplifying nature of Figure 6.1, this schemata clearly suggests the complexity of studying racial and ethnic differences in late-life health as the number of micro- and macro-level characteristics related to different health outcomes is large and their possible permutations seemingly endless. The goal of this chapter, therefore, is not to describe each of these characteristics and document their role in the causal mechanism explaining racial/ethnic late-life health differences, but rather to provide an overview of the theoretical perspectives most salient for the study of minority late-life health, to highlight some of the key contributions of research on health disparities across minorities, and to identify the areas of particular interest of future research adopting the life-course perspective as the underlying theoretical guideline.

Theories of Life-Course Racial and Ethnic Health Disparities

If we consider older people's health from a life-course perspective and recognize the importance of early experiences on later outcomes, then the next question is how such experiences aggregate over the life course. One framework to explain this is cumulative (dis)advantage. Cumulative (dis)advantage theory (also described by Quesnel-Vallée, Willson, and Reiter-Campeau in Chapter 23 of this book) provides explanations not only for the connections between social forces and health, but also for the connections between early-life experiences and later-life health outcomes (Kelley-Moore & Ferraro, 2004; Walsemann, Geronimus, & Gee, 2008). A key feature of the cumulative (dis)advantage paradigm is that exposures to advantages (or disadvantages) have an additive effect on health. Repeated or prolonged exposure to, for example, socioeconomic disadvantage or racial discrimination will have greater health effects than fewer or shorter exposures. Thus early-life experiences are important in considering health and aging because they contribute to the total "exposure" experienced across the life course. Further, social and environmental exposures – advantageous or disadvantageous – create a degree of path dependency so that (dis)advantage at one point in time predicts not only future health but also future (dis)advantage, in turn contributing to future health. A modification of cumulative (dis)advantage theory, cumulative inequality, pays more attention to the role of individual agency while still recognizing the cumulative nature of exposure to disadvantageous social forces (Ferraro & Shippee, 2009).

Using cumulative (dis)advantage or cumulative inequality as a framework, we can think of exposures such as discrimination, poverty, and low SES as having a dose–response relationship with health (Dannefer, 2003). Minority groups in the United States face discrimination and a high probability of having low socioeconomic position over their lifetimes. Not only are they more likely to experience disadvantage, but they are also likely to experience more of it than the White population. These experiences, according to cumulative (dis)advantage or cumulative inequality theories, are likely to lead to the accumulation of negative health effects over the life course.

One possible mechanism linking life-course socio-environmental exposures to health outcomes is thought to be through physiological reactions to stress. These include increased heart rate, changes in hormone levels, and other reactions to enable the "fight or flight" response. However, with prolonged exposure to stress, for example in an ongoing situation of poverty, observed physiological changes are also prolonged and potentially damaging. The mechanism, which is the basis for the "weathering hypothesis" (Geronimus, Bound, & Waidmann, 1999), posits that African Americans experience earlier disease onset due to greater exposure to stressors such as economic adversity, causing "physiological deterioration." Similarly, the "toxic stress" pathway emphasizes the detrimental health effects of prolonged experience of heightened stress, especially in childhood (Shonkoff, Boyce, & McEwen, 2009; Shonkoff et al., 2012). Proximal measures of the physiological effects of the heightened stress experienced by disadvantaged groups, including minorities, are often labeled "allostatic load" (Beckie, 2012; Gruenewald et al., 2012; Seeman, Epel,

Gruenewald, Karlamangla, & McEwen, 2010). The markers of this load include such measures as high blood pressure, high blood glucose, high BMI, and other markers that are precursors of and risk factors for disease.

Having established that it is beneficial to adopt a life-course perspective in the study of late-life health outcomes, describing the pattern of inequalities over the life course is a research task of particular interest and practical importance. Several competing hypotheses have thus far been suggested. First, the double jeopardy hypothesis suggests old age will be the time of greatest disparities because of the interaction of two distinct health disadvantages: those experienced by racial/ethnic minorities at any age, and those experienced by older adults of any racial/ethnic group (Ferraro, 1987; Ferraro & Farmer, 1996a). Facing both these conditions simultaneously magnifies the disadvantage, resulting in greater disparities in old age compared with other stages in the life course. Conversely, the aging-as-leveler hypothesis proposes that health disparities in later life weaken relative to earlier stages in the life course (Dowd & Bengtson, 1978). Because the processes involved in aging are, to a degree, inevitable, health declines are experienced universally. These declines are greater in magnitude than the health effects of disadvantage or minority status, so that in old age the latter are overwhelmed by the former, making disparities smaller at this stage of the life course than earlier (Hoffmann, 2011). The third possible pattern of inequality over the life course is a fairly stable one, with later life no better and no worse in this regard. This view is known as the persistent inequality hypothesis (Ferraro & Farmer, 1996b).

Tests of these hypotheses have failed to produce a clear favorite, but have highlighted the importance of methodological and data considerations in studies of health and aging. For example, while studies sampling only older adults tend to find either persistent inequality or even a crossover effect, those which sample the whole adult population observe widening disparities in older age (Ferraro, 2011). In studies of self-rated health and chronic illness, unadjusted results pointed to aging-as-leveler but adjusting for differential attrition produced results more consistent with persistent inequality (Ferraro & Farmer, 1996b; Willson, Shuey, & Elder, 2007).

RESEARCH ACROSS MINORITY GROUPS

With the growing racial and ethnic diversity of the United States, understanding the unique aspects of health and aging across minority groups becomes ever more important. Thus far, however, the research has mostly focused on differences between Black and White populations and, more recently, between these groups and the Hispanic population. While limited, the lessons of this research are still invaluable as they form the foundation for the future study of differences in health for other racial and ethnic groups, and point to the limits of generalized research on any single minority group. It is therefore important to highlight the common themes in the research focusing on Black and Hispanic populations as well as the themes that are exclusively or predominantly the subject of study for only one of these two populations.

Research on racial and ethnic differences in health has an inherent life-course perspective given the importance of health outcomes in childhood and mid-life for the health in late life across all racial and ethnic groups (Ferraro, 2011). Consequently, the major theories intended for explaining the nexus of age with race and ethnicity are not group-specific but rather universal in nature. Cumulative advantage/disadvantage (Dannefer, 1987; Ferraro & Kelley-Moore, 2003; O'Rand, 1996;

Walsemann et al., 2008), double jeopardy (Dowd & Bengtson, 1978; Ferraro & Farmer, 1996a), persistent inequality (Ferraro & Farmer, 1996b; Henretta & Campbell, 1976; Kelley-Moore & Ferraro, 2004), and aging-as-leveler (Dowd & Bengtson, 1978) hypotheses, while offering largely competing explanations of the effects of early-life inequalities in SES, health, and well-being on health outcomes in late life, are inherently framed as theories aimed at explaining the general pattern of race and ethnic differences in health over the life course and especially in old age. As such, they should be tested across different racial and ethnic groups. However, as noted earlier, the literature review reveals that the overwhelming majority of empirical work has thus far focused on the study of Black–White population differences in health (e.g., Clark & Maddox, 1992; Ferraro & Farmer, 1996a; Kelley-Moore & Ferraro, 2004; Kim & Miech, 2009). This is most likely due to the lack of sufficiently rich data on other racial and ethnic groups. Only in recent years has the research expanded to include persons of Hispanic origin (e.g., Haas & Rohlfsen, 2010) as well as, more specifically, the Mexican-origin population (e.g., Brown, O'Rand, & Adkins, 2012). While the studies focusing on the Black–White differences in aging and health revealed the importance of careful study design and longitudinal data to understand the differences in the trajectories of White people and minorities over time and across cohorts, the inclusion of Hispanic and Mexican-origin populations has revealed that the interaction of health and age may operate differently across different racial and ethnic minorities.

An example of how such differences may arise is the issue of residential segregation. While the research focusing on Black residential segregation established a clear link with larger health disparities and higher mortality rates (e.g., Collins & Williams, 1999; Fang, Madhavan, Bosworth, & Alderman, 1998; Jackson, Anderson, Johnson, & Sorlie, 2000; Williams & Collins, 2001), the research on the Hispanic population and other racial and ethnic groups suggests that the effects of segregation may not be as detrimental. In fact, some studies of Mexican-origin populations (e.g., Eschbach, Ostir, Patel, Markides, & Goodwin, 2004) suggest possible benefits of living in ethnically homogenized areas on various health outcomes such as stroke, cancer, or hip fracture. Furthermore, immigrant enclaves have been related to healthier diet (e.g., less high-fat foods) (Osypuk, Diez Roux, Hadley, & Kandula, 2009). Therefore, the same phenomenon, racial segregation, can contribute to different health outcomes for different racial and ethnic groups, which suggest that findings for any single minority group cannot always be considered relevant for all racial and ethnic minorities.

Another example of the importance of studying different racial and ethnic minorities is the patterns of institutionalization versus home- and community-based long-term care. The research suggests that both Black and Hispanic adults may have stronger family ties than White adults in terms of filial responsibility toward older persons (Burr & Mutchler, 1999), and both Hispanic (Fennell, Feng, Clark, & Mor, 2010) and Black older adults (Howard et al., 2002; Smith, Feng, Fennell, Zinn, & Mor, 2008) are at high risk of placement in low-quality nursing homes, yet the likelihood of nursing-home use substantially differs between the two groups. While the White–Black gap in nursing-home placement has been narrowing (Akamigbo & Wolinsky, 2007) and may have even closed (Smith et al., 2008), nursing-home utilization among Hispanic persons remains significantly lower (Angel & Angel, 1997). Strong family-oriented culture with the traditional role of daughters (and daughters-in-law) as primary caregivers to older individuals (Angel & Hogan, 2004; Fennell et al., 2010) and language barriers encountered by Hispanic persons (Flores, 2006; Timmins, 2002) are among the explanations advanced to account for this

Black–Hispanic difference in the patterns of nursing-home placement and home- and community-based care use. As in the example of residential segregation, therefore, the unique characteristics of various racial and ethnic minorities interact with the challenges common to all of them and result in substantially different outcomes across different minorities.

The research on Black–White differences in health over the life course has made a number of unique contributions to the literature. One such contribution is identifying the phenomenon of selective survival, that is, the fact that by the time they reach old age, Black adults have been exposed to higher mortality rates than White adults throughout the life course, which leaves only a highly selected group of comparatively healthy Black people in advanced old age that may have lower risk of dying than White people of the same age – known as the "crossover effect" (Eberstein, Nam, & Heyman, 2008; Manton & Stallard, 1991). Furthermore, despite being disadvantaged across many dimensions that affect health and quality of life, Black adults have been found to better cope with stress and have better mental health (e.g., Keyes, 2009; Roff et al., 2004). Research has also identified cultural and psychosocial resources that foster resilience. For example, religious involvement can enhance health in the face of adversity and racial discrimination, dampening the negative effects of interpersonal discrimination on health (Bierman, 2006). On the other hand, lower average birth weight among Black newborns has been linked to conditions such as hypertension and diabetes in later life (e.g., Cruickshank et al., 2005; Lopes & Port, 1995). While these issues are not unique to the Black population (see Hummer et al., 2004, for crossover effect and Acevedo-Garcia, Soobader, & Berkman, 2005, 2007, for analysis of low birth weight across different racial and ethnic groups), the lack of the appropriate data has likely contributed to the comparative lack of studies focusing on other racial and ethnic groups.

Another important contribution of the literature focusing on Black–White health differences is the study of the links between race-based discrimination and health. This research has thoroughly documented how experiences of racism and discrimination are associated with poorer mental and physical health (e.g., Mays, Cochran, & Barnes, 2007; Williams & Mohammed, 2009). Research shows that the subjective experience of poverty and racism by African Americans appears to have adverse health and health care consequences (Williams & Sternthal, 2010). For several decades sociologists have comprehensively characterized social stressors associated with racial disparities in health. In particular, they have examined how *discrimination* and *racism* affect disease, illness, and health care use (Williams & Mohammed, 2009). Data from the Commonwealth Fund's Minority Health Survey show that Black respondents are far more likely to perceive age discrimination in the US health care system than White respondents (LaVeist, Rolley, & Diala, 2003). The harmful health effects of internalized racism, in which minority groups accept the dominant society's ideology of their lower status, correlates strongly with, for example, high blood pressure and cardiovascular disease (Williams & Leavell, 2012). Such risk factors help to explain the size and persistence of disparities in late-life infirmity and disabilities (Whitfield & McClearn, 2005). Yet, the researchers have not been able to fully describe the mechanism that links exposure to race-based discrimination with health deterioration, in part due to the lack of appropriate measures of discrimination and study design limitations, but also because of the lack of consideration of the life-course evolution of the discrimination–health relationship (Williams, Neighbors, & Jackson, 2003).

The study of aging and health for the Hispanic population and, somewhat less often, the Mexican-origin population has emerged as an important and fruitful subfield of research within the broader field of research on racial and ethnic disparities in health. Given the US demographic trends, it will continue to grow in importance. As discussed earlier, a major focus of this research has been on the "Hispanic health paradox," a phenomenon initially described by Markides and Coreil (1986), who noted more favorable mortality profiles of Hispanic persons than White or Black persons of similar SES. This observation spurred a significant increase in the study of health and aging of Hispanic persons, which in turn resulted in a major advancement of the understanding of the interconnectedness of race/ethnicity, immigration, and age in shaping the health outcomes of the Hispanic population. These are lessons that may be of some relevance for other (minority) immigrant populations that are experiencing fast growth in the United States.

Explanations of the apparent Hispanic epidemiological paradox focus on the issue of selective immigration (i.e., "healthy immigrant" effect). This includes selective return migration of less-healthy older immigrants back to their countries of origin (i.e., "salmon bias" effect), social network support that provides a health-protective environment, and a traditional culture that encourages health-promoting practices related, for example, to nutrition. Immigration selectivity, whereby it is assumed that on average healthier individuals are more likely to move internationally as good health may increase the likelihood of success in their new country of residence, is a particularly potent explanation that finds substantial support in the literature (e.g., Eschbach, Stimpson, Kuo, & Goodwin, 2007; Singh & Hiatt, 2006). However, it is not by itself sufficient to account for the full extent of the Hispanic mortality advantage (Turra & Elo, 2008). Salmon bias effect had also been found to be an important part of the mechanism that explains Hispanic health advantage (e.g., Palloni & Arias, 2004).

Furthermore, unique culture and strong family/social bonds are found to be associated with positive health outcomes among Hispanic population, especially the non-US-born (Lara, Gamboa, Kahramanian, Morales, & Hayes-Bautista, 2005; Magana & Clark, 1995). Yet, over time this Hispanic immigrant health advantage decreases or completely disappears, and the Hispanic population converges to the health trends of the native-born population (Antecol & Bedard, 2006; Cho, Frisbie, Hummer, & Rogers, 2004). This suggests that immigrants adopt relatively quickly some of the unhealthy behaviors characteristic of the general population.

While many of the lessons learned studying Black and Hispanic adults can be applied to the study of Asian adults, a rapidly growing segment of the US population, or other racial/ethnic minorities, one would be ill-advised to assume that the same models of behavior can be uncritically assumed to be equally valid across all racial and ethnic minorities. In fact, one of the major lessons of this overview of research on Black and Hispanic populations is that each population faces somewhat different challenges and addresses them in ways that are considered appropriate in the context of that population's culture, but may not be appropriate or even viable for other populations. The task of researchers is to clearly discern between the common and the unique elements of various populations' response to challenges and to offer a convincing explanation of their behavior with sufficient predictive power.

SUMMARY AND CONCLUSION

In this chapter we have identified the major theories and empirical research to investigate racial and ethnic disparities in health. Explanations for racial/ethnic health differences have largely been theoretical accounts of social processes and, more recently, methodological innovations that allow the effects of timing to be considered. The chief emphasis of these theories has been the distribution of a single health measure – or occasionally, a group of closely related measures – between racial/ethnic groups, with the most common result being that racial/ethnic minorities have higher prevalence and/or earlier onset of an outcome. However, racial/ethnic differences in health have been observed not just in the prevalence or severity of a given outcome, but in the complex relationships between outcomes. The Hispanic epidemiologic paradox is just one example of how racial/ethnic differences in anticipated life expectancy do not correspond closely to actual differences in life expectancy (Bulanda & Zhang, 2009).

As we discussed, most of the literature on racial disparities in health is built on the architecture of cumulative disadvantage theory or a close relative of it. Although this is widely used for understanding the nature and extent of racial-based inequalities in health, the perspective is limited in terms of its explanatory power (Dannefer, 2003). As such, new perspectives deserve consideration. Theories such as cumulative inequality will enable systematic examination of the intricacies of health trajectories, how they are influenced by the life course and social inequalities, and how they are altered by resources, social networks, psychosomatic processes, human agency, and incentives (Ferraro & Shippee, 2009; Kelley-Moore & Ferraro, 2004). It is becoming increasingly clear that health at a given point in life is determined not only by current or recent exposures, but by exposures at earlier stages in the life course. Work in molecular biology has suggested intergenerational epigenetic effects whereby an individual's pattern of gene expression can be shaped by the environmental exposures of their parents even before conception (Daxinger & Whitelaw, 2012; Harper, 2005). Although this latter finding has yet to be applied in the context of human health, it provides an example of a mechanism by which the social environment can affect much later health outcomes, in this case across generations. Paradigm shifts of this sort, by necessity, extend existing theory and organizing frameworks to account not only for the risk of health and illness over the life course, but also for what appear to be fundamental causes of different health profiles (Phelan, Link, & Tehranifar, 2010).

Any framing of the problem, then, must acknowledge age and sex differences in physiological status and dysregulation and investigate how socioeconomic factors, behaviors, cultural norms, and changes in the social and physical environment associate with variability in biological indicators in minority group health. A contemporary perspective on biogerontology must focus on the core questions that reflect on some of the most critical gaps in our understanding of organismal senescence and human longevity (Waters, 2007). Examining how morbidity can be compressed, culminating in the extension of a healthy life span, is essential in the next generation of research. Racial differences in the link between morbidity and mortality point to the importance of investigating how chronic diseases, life style, and modifiable risk factors are related to active life expectancy for subgroups of the minority aging population (Hayward & Heron, 1999). Even so, researchers have only recently begun to investigate the extent to which the quality of life can be enhanced by delaying the onset of chronic and disabling diseases and conditions until the last years, or perhaps even months, of life for elderly minorities (Haas, Krueger, & Rohlfsen, 2012). Research is needed to disentangle the health consequences of age-period-cohort effects (Yang & Land, 2013).

Building on recent longitudinal panel studies to determine whether the risk due to new morbidities, such as childhood obesity, for example, is persistent or modifiable can draw on cumulative disadvantage theory and biogerontology. Research along these lines must also pay closer attention to compensatory mechanisms in the development of cumulative disadvantage theory. They must identify what specific behavioral interventions work to minimize or eliminate the effects of risk exposure due to aging and in diverse populations. Gerontological health research should focus on the interventions and mechanisms that lead to longer and healthier lives by identifying the social factors in epidemiology and basic health services and their implications for health policy. In addition, there is a clear need to determine the role of the community in protecting and improving the health of older people of Mexican origin, Cuban origin, and Puerto Rican ancestry in light of the fact that the Hispanic population has become the largest minority group in the US population and will continue to grow in number. Examining the implications of a healthy and active life style and extended life expectancy among Hispanic workers who need to feed into the social security system merits attention.

Our knowledge of racial and ethnic differences in health has grown significantly in the past 25 years. But gaps remain. In charting the territory of future research, new models of the complex interconnectedness of cultural and structural factors deserve scientific development. Specifically, multi-level models are needed to compare social structural factors and social institutions that place certain minority groups at high risk of illness, and that impede their access to the highest-quality care. Studies should examine institutionalized disadvantages that manifest themselves most obviously as occupational, income, and asset disadvantages across the life course. Such disadvantages translate directly into impaired health care access and poorer health among minority Americans and especially underserved and understudied populations, such as Native Americans. Caution is warranted, though, in generalizing findings to the broader population due to small sample sizes. One approach is to combine multiple years of data to produce reliable estimates for small groups.

There are other methodological challenges and opportunities in addressing these issues.

The effort requires the imaginative use of existing data sets and an enhancement of their samples to include larger oversamples of Hispanic Americans and their subgroups (Angel & Angel, 2006). Fortuitously, a variety of publicly available data sources can be used to investigate the implications of longevity for older minorities, partly as the result of National Institute on Aging (NIA) funding of major Hispanic studies carried out in the early 1990s. Focused data collection efforts, such as the Border Epidemiologic Study of Aging (BESA) and the H-EPESE, an almost 20-year *longitudinal panel* study of older Mexican Americans, provide important in-depth health and function information for a single Hispanic group. The HRS and the Study of Assets and Health Dynamics among the Oldest Old (AHEAD) make it possible to investigate the impact of individuals' economic situations in the years before retirement on their health in their post-retirement years. The Mexican Health and Aging Study (MHAS) allows for comparative analyses of the health of older Mexicans in the United States and Mexico and can be harmonized with the HRS. The sample design of the 2011 National Health and Aging Trends Study provides critical data on changing patterns in late-life disability in a large cohort of African American and White Medicare beneficiaries ages 65 and 90.

Other federal government initiatives include the National Health and Nutrition Examination Survey (NHANES), a periodic survey of the health and nutritional status of minority groups across the life span. A special feature of the national survey is the combination of different types of heath data, such as interviews and clinical tests as well as physical examinations for each respondent in the sample. The seventh wave (NHANES III) oversamples older persons 60 and over, African Americans, and Mexican Americans and is a particularly useful source of information on racial and ethnic differences in the incidence and prevalence of type

II diabetes and other chronic diseases among the young-old, people aged 65–74. Lastly, the Hispanic Community Health Study (HCHS) provides opportunities to examine the role of cultural adaptation and the ecology of poverty in the development of chronic disease for about 16 000 Hispanic persons 18–74 years old in four American communities. This study is unique in that it includes physical examinations and interviews to help identify the prevalence of and risk factors for a wide variety of diseases, disorders, and disabling conditions in urban areas with high concentrations of Latinos: the Bronx, Chicago, Miami, and San Diego.

As important as these data sets are, there is a need for those interested in research on aging to make use of data from other stages of life. This effort will entail the collection of new and specialized data sets dedicated to specific vulnerabilities among minority groups living and working in specific ecological and social niches. To inform the development of the research agenda on healthful minority aging, it is becoming increasingly clear that triangulation using a variety of data sources is necessary to understand the implications of longevity for the minority population. To improve the quality of minority health and aging studies, data collection and research must supplement and cross-validate self-reported measures of health and the anchoring vignette approach with objective measures. This will improve comparability of self-reported health and disability measures through measured performance tests. Biomarkers improve reliability of self-reported data on morbidity, risk factors, and assessments of interventions.

Future data collections must focus on the way in which social aspects of racial and ethnic diversity are affected by the life course and social policy. Despite a growing body of literature documenting that SES may be a core factor in explaining racial and ethnic group differences in health, there is also reason to suspect that cultural groups differ in the availability of social support that may modify or mitigate adverse mental and physical health sequelae (Angel, Angel, McClellan, & Markides, 1996; Waite & Das, 2010). Yet, we have relatively little good comparative data on how cultural differences in levels and types of social support and social integration affect the health of older minority group members, and especially among aging Hispanics (Dilworth-Anderson, Williams, & Gibson, 2002). The condition is particularly acute for late-life Mexican-origin immigrants who lack English language proficiency and suffer high rates of depression (Woodward et al., 2012). There are clear practical benefits of social integration; the availability of relatives and friends with whom to interact directly influences mental health and morale (Falcón, Todorova, & Tucker, 2009; Umberson & Montez, 2010). But more social ties may not necessarily mean better health if the relationships are of poor quality. In that case, having more contacts creates interpersonal conflicts and role strain (Christakis & Fowler, 2009). Further research is needed to work through this puzzle.

Future research must also sample large numbers of minorities, including Asians and Native Americans, who have been neglected in the health and aging literature compared to Hispanic and Black persons, and it must cover the whole life course, with recruitment beginning as early in life as is practicable. Information on early-life events and markers is important to health disparities research not only in terms of potential exposures, but also because age at disease incidence varies by race/ethnicity (Taylor, 2008; Willson et al., 2007). Participation in and attrition from research studies is correlated with both health and race/ethnicity. Studies that lack comprehensive life-course data may fail to capture the burden of disease among older minority groups.

As we progress into the twenty-first century, new and important medical innovations will further increase life spans and hopefully improve the quality of those additional years of life. Much of that progress will no doubt result from a better understanding of the genetic contribution to disease. Although such progress and understanding of biology is clearly admirable, analysis of health disparities in older minority populations must be informed by high correlations among cultural and structural factors. The over-emphasis on cultural factors runs the risk of cultural essentialism, in which non-structural factors are seen as the major causes of differential levels of health and illness. Finding a balance among approaches will require cooperation among all of those who conduct research in this increasingly important area of gerontology.

References

Acevedo-Garcia, D., Soobader, M.-J., & Berkman, L. F. (2005). The differential effect of foreign-born status on low birth weight by race/ethnicity and education. *Pediatrics*, 115(1), 20–30.

Acevedo-Garcia, D., Soobader, M.-J., & Berkman, L. F. (2007). Low birthweight among US Hispanic/Latino subgroups: The effect of maternal foreign-born status and education. *Social Science & Medicine*, 65(12), 2503–2516.

Akamigbo, A. B. M., & Wolinsky, F. D. (2007). New evidence of racial differences in access and their effects on the use of nursing homes among older adults. *Medical Care*, 45(7), 672–679.

Angel, J. L., & Angel, R. J. (2006). Minority group status and healthful aging: Social structure still matters. *American Journal of Public Health*, 96(7), 1152–1159.

Angel, J. L., Angel, R. J., McClellan, J. L., & Markides, K. S. (1996). Nativity, declining health, and preferences in living arrangements among elderly Mexican Americans: Implications for long term care. *The Gerontologist*, 36(4), 464–473.

Angel, J. L., & Hogan, D. (2004). Population aging and diversity in a new era. In K. Whitfield (Ed.), *Closing the gap: Improving the health of minority elders in the new Millennium* (pp. 1–12). Washington, DC: Gerontological Society of America.

Angel, J. L., Torres-Gil, F., & Markides, K. (2012). *Aging, health and longevity in the Mexican-origin population*. New York, NY: Springer Sciences.

Angel, R. J., & Angel, J. L. (1997). *Who will care for us? Aging and long-term care in multicultural America*. New York, NY: New York University Press.

Angel, R. J., & Angel, J. L. (2009). *Hispanic families at risk: The new economy, work, and the welfare state*. New York, NY: Springer Sciences.

Antecol, H., & Bedard, K. (2006). Unhealthy assimilation: Why do immigrants converge to American health status levels? *Demography*, 43(2), 337–360.

Arias, E. (2010). *United States Life Tables by Hispanic Origin* (p. Table G). Hyattsville, MD: National Center for Health Statistics. Retrieved from: <http://www.cdc.gov/nchs/data/series/sr_02/sr02_152.pdf>.

Beckie, T. M. (2012). A systematic review of allostatic load, health, and health disparities. *Biological Research for Nursing*, 14(4), 311–346.

Bierman, A. (2006). Does religion buffer the effects of discrimination on mental health? Differing effects by race. *Journal for the Scientific Study of Religion*, 45(4), 551–565.

Brown, T. H., O'Rand, A. M., & Adkins, D. E. (2012). Race-ethnicity and health trajectories tests of three hypotheses across multiple groups and health outcomes. *Journal of Health and Social Behavior*, 53(3), 359–377.

Bulanda, J. R., & Zhang, Z. (2009). Racial-ethnic differences in subjective survival expectations for the retirement years. *Research on Aging*, 31(6), 688–709.

Burr, J. A., & Mutchler, J. E. (1999). Race and ethnic variation in norms of filial responsibility among older persons. *Journal of Marriage and the Family*, 61(3), 674–687.

Chiu, C.-J., & Wray, L. A. (2011). Physical disability trajectories in older Americans with and without diabetes: The role of age, gender, race or ethnicity and education. *The Gerontologist*, 51(1), 51–63.

Cho, Y., Frisbie, W. P., Hummer, R. A., & Rogers, R. G. (2004). Nativity, duration of residence, and the health of Hispanic adults in the United States. *International Migration Review*, 38(1), 184–211.

Christakis, N. A., & Fowler, J. H. (2009). *Connected: The surprising power of our social networks and how they shape our lives*. New York, NY: Little, Brown and Company.

Clark, D. O., & Maddox, G. L. (1992). Racial and social correlates of age-related changes in functioning. *Journal of Gerontology*, 47(5), S222–S232.

Collins, C. A., & Williams, D. R. (1999). Segregation and mortality: The deadly effects of racism? *Sociological Forum*, 14(3), 495–523.

Crimmins, E. M., Hayward, M. D., Hagedorn, A., Saito, Y., Brouard, N., & Leen, M. D. (2009). Change in disability-free life expectancy for Americans 70 years old and older. *Demography*, 46(3), 627–646.

Crimmins, E. M., Hayward, M. D., & Seeman, T. E. (2004). Race/ethnicity, socioeconomic status, and health. In N. B. Anderson, R. A. Bulatao, & B. Cohen (Eds.), *Critical perspectives on racial and ethnic differences in health in late life* (pp. 310–352). Washington, DC: National Academies Press.

Crimmins, E. M., Kim, J. K., Alley, D. E., Karlamangla, A., & Seeman, T. (2007). Hispanic paradox in biological risk profiles. *American Journal of Public Health, 97*(7), 1305–1310.

Cruickshank, J. K., Mzayek, F., Liu, L., Kieltyka, L., Sherwin, R., Webber, L. S., et al. (2005). Origins of the "'Black/White'" difference in blood pressure: Roles of birth weight, postnatal growth, early blood pressure, and adolescent body size: The Bogalusa Heart Study. *Circulation, 111*(15), 1932–1937.

Cutler, D. M., & Landrum, M. B. (2011). *Dimensions of health in the eldelry population*. Cambridge: National Bureau of Economic Research. Retrieved from: <www.nber.org/papers/w17148>.

Dannefer, D. (1987). Aging as intracohort differentiation: Accentuation, the Matthew effect, and the life course. *Sociological Forum, 2*(2), 211–236.

Dannefer, D. (2003). Cumulative advantage/disadvantage and the life course: Cross-fertilizing age and the social science theory. *Journal of Gerontology Series B: Psychological and Social Sciences, 58B*(6), S327–S337.

Daxinger, L., & Whitelaw, E. (2012). Understanding transgenerational epigenetic inheritance via the gametes in mammals. *Nature Reviews Genetics, 13*(3), 153–162.

Dilworth-Anderson, P., Williams, I. C., & Gibson, B. E. (2002). Issues of race, ethnicity, and culture in caregiving research: A twenty-year review (1980–2000). *The Gerontologist, 42*(2), 237–272.

Do, D. P., Frank, R., & Finch, B. K. (2012). Does SES explain more of the black/white health gap than we thought? Revisiting our approach toward understanding racial disparities in health. *Social Science & Medicine, 74*(9), 1385–1393.

Dowd, J. J., & Bengtson, V. L. (1978). Aging in minority populations: An examination of the double jeopardy hypothesis. *Journal of Gerontology, 33*(3), 427–436.

Eberstein, I. W., Nam, C. B., & Heyman, K. M. (2008). Causes of death and mortality crossovers by race. *Biodemography and Social Biology, 54*(2), 214–228.

Eschbach, K., Ostir, G. V., Patel, K. V., Markides, K. S., & Goodwin, J. S. (2004). Neighborhood context and mortality among older Mexican Americans: Is there a barrio advantage? *American Journal of Public Health, 94*(10), 1807–1812.

Eschbach, K., Stimpson, J. P., Kuo, Y.-F., & Goodwin, J. S. (2007). Mortality of foreign-born and US-born Hispanic adults at younger ages: A reexamination of recent patterns. *American Journal of Public Health, 97*(5), 1297–1304.

Falcón, L. M., Todorova, I., & Tucker, K. L. (2009). Social support, life events, and psychological distress among the Puerto Rican population in the Boston area of the United States. *Aging and Mental Health, 13*(6), 863–873.

Fang, J., Madhavan, S., Bosworth, W., & Alderman, M. H. (1998). Residential segregation and mortality in New York City. *Social Science & Medicine, 47*(4), 469–476.

Farmer, M. M., & Ferraro, K. F. (2005). Are racial disparities in health conditional on socioeconomic status? *Social Science & Medicine, 60*(1), 191–204.

Fennell, M. L., Feng, Z., Clark, M. A., & Mor, V. (2010). Elderly Hispanics more likely to reside in poor-quality nursing homes. *Health Affairs, 29*(1), 65–73.

Ferraro, K. F. (1987). Double jeopardy to health for black older adults? *Journal of Gerontology, 42*(5), 528–533.

Ferraro, K. F. (2011). Health and aging: Early origins, persistent inequalities?. In R. A. Settersten & J. L. Angel (Eds.), *Handbook of sociology of aging* (pp. 465–475). New York, NY: Springer.

Ferraro, K. F., & Farmer, M. M. (1996a). Double jeopardy to health hypothesis for African Americans: Analysis and critique. *Journal of Health and Social Behavior, 37*(1), 27–43.

Ferraro, K. F., & Farmer, M. M. (1996b). Double jeopardy, aging as leveler, or persistent health inequality? A longitudinal analysis of white and black Americans. *The Journals of Gerontology Series B: Psychological Sciences and Social Sciences, 51B*(6), S319–S328.

Ferraro, K. F., & Kelley-Moore, J. A. (2003). Cumulative disadvantage and health: Long-term consequences of obesity? *American Sociological Review, 68*(5), 707–729.

Ferraro, K. F., & Shippee, T. P. (2009). Aging and cumulative inequality: How does inequality get under the skin? *The Gerontologist, 49*(3), 333–343.

Flores, G. (2006). Language barriers to health care in the United States. *New England Journal of Medicine, 355*(3), 229–231.

Fowler, J. H., & Christakis, N. A. (2008). Estimating peer effects on health in social networks: A response to Cohen-Cole and Fletcher, and Trogdon, Nonnemaker, and Pais. *Journal of Health Economics, 27*(5), 1400–1405.

Frank, R. (2007). What to make of it? The (re)emergence of a biological conceptualization of race in health disparities research. *Social Science & Medicine, 64*(10), 1977–1983.

Fried, L. P., Ferrucci, L., Darer, J., Williamson, J. D., & Anderson, G. (2004). Untangling the concepts of disability, frailty, and comorbidity: Implications for improved targeting and care. *Journals of Gerontology Series A: Biological Sciences and Medical Sciences, 59*(3), 255–263.

Fries, J. F., Bruce, B., & Chakravarty, E. (2011). Compression of morbidity: 1980–2011: A focused review of paradigms and progress. *Journal of Aging Research, 2011*, 1–10.

Geronimus, A. T. (1996). Black/white differences in the relationship of maternal age to birthweight: A population-based test of the weathering hypothesis. *Social Science & Medicine, 42*(4), 589–597.

Geronimus, A. T., Bound, J., & Waidmann, T. A. (1999). Poverty, time, and place: Variation in excess mortality across selected US populations, 1980–1990. *Journal of Epidemiology and Community Health, 53*(6), 325–334.

Geruso, M. (2012). Black-white disparities in life expectancy: How much can the standard SES variables explain? *Demography, 49*(2), 553–574.

Gorin, S. H., & Lewis, B. (2004). The compression of morbidity: Implications for social work. *Health & Social Work, 29*(3), 249–254.

Gruenewald, T. L., Karlamangla, A. S., Hu, P., Stein-Merkin, S., Crandall, C., Koretz, B., et al. (2012). History of socioeconomic disadvantage and allostatic load in later life. *Social Science & Medicine, 74*(1), 75–83.

Haas, S. A., Krueger, P. M., & Rohlfsen, L. (2012). Race/ethnic and nativity disparities in later life physical performance: The role of health and socioeconomic status over the life course. *Journal of Gerontology Series B: Psychological and Social Sciences, 66B*(1), i102–i110.

Haas, S., & Rohlfsen, L. (2010). Life course determinants of racial and ethnic disparities in functional health trajectories. *Social Science & Medicine, 70*(2), 240–250.

Harper, L. V. (2005). Epigenetic inheritance and the intergenerational transfer of experience. *Psychological Bulletin, 131*(3), 340–360.

Hayward, M. D., & Gorman, B. K. (2004). The long arm of childhood: The influence of early-life social conditions on men's mortality. *Demography, 41*(1), 87–107.

Hayward, M. D., & Heron, M. (1999). Racial inequality in active life among adult Americans. *Demography, 36*(1), 77–91.

Hayward, M. D., Miles, T. P., Crimmins, E. M., & Yang, Y. (2000). The significance of socioeconomic status in explaining the racial gap in chronic health conditions. *American Sociological Review, 65*(6), 910–930.

Henretta, J. C., & Campbell, R. T. (1976). Status attainment and status maintenance: A study of stratification in old age. *American Sociological Review, 41*(6), 981–992.

Hill, T., Angel, J., & Balistreri, K. (2012). Does the "healthy immigrant effect" extend to cognitive aging? In J. L. Angel, F. Torres-Gil, & K. Markides (Eds.), *Aging, health, and the Mexican-origin population* (pp. 19–33). New York, NY: Springer Sciences.

Hoffmann, R. (2011). Illness, not age, is the leveler of social mortality differences in old age. *The Journals of Gerontology Series B: Psychological Sciences and Social Sciences, 66B*(3), 374–379.

House, J. S. (2002). Understanding social factors and inequalities in health: 20th century progress and 21st century prospects. *Journal of Health and Social Behavior, 43*(2), 125–142.

House, J. S., Lantz, P. M., & Herd, P. (2005). Continuity and change in the social stratification of aging and health over the life course: Evidence from a nationally representative longitudinal study from 1986 to 2001/2002 (The Americans' Changing Lives Study). *Journals of Gerontology Series B: Psychological and Social Sciences, 60B*(Special issue 2), 15–26.

Howard, D. L., Sloane, P. D., Zimmerman, S., Eckert, J. K., Walsh, J. F., Buie, V., et al. (2002). Distribution of African Americans in residential care/assisted living and nursing homes: More evidence of racial disparity? *American Journal of Public Health, 92*(8), 1272–1277.

Hubert, H. B., Bloch, D. A., Oehlert, J. W., & Fries, J. F. (2002). Lifestyle habits and compression of morbidity. *Journal of Gerontology: Medical Sciences, 57*(6), M347–M351.

Hummer, R. M., Benjamins, M., & Rogers, R. (2004). *Race/ethnic disparities in health and mortality among the elderly: A documentation and examination of social factors.* Washington, DC: National Academies Press.

Jackson, S. A., Anderson, R. T., Johnson, N. J., & Sorlie, P. D. (2000). The relation of residential segregation to all-cause mortality: A study in black and white. *American Journal of Public Health, 90*(4), 615–617.

Kawachi, I., Daniels, N., & Robinson, D. E. (2005). Health disparities by race and class: Why both matter. *Health Affairs, 24*(2), 343–352.

Kelley-Moore, J. A., & Ferraro, K. F. (2004). The black/white disability gap: Persistent inequality in later life? *The Journals of Gerontology Series B: Psychological Sciences and Social Sciences, 59B*(1), S34–S43.

Keyes, C. L. M. (2009). The black–white paradox in health: Flourishing in the face of social inequality and discrimination. *Journal of Personality, 77*(6), 1677–1706.

Kim, G., Chiriboga, D., Jang, Y., Lee, S., Huang, C., & Parmelee, P. (2010). Health status of older Asian Americans in California. *Journal of the American Geriatrics Society, 58*(10), 2003–2008.

Kim, G., Worley, C., Allan, R., Vinson, L., Crowther, M., Parmelee, P., et al. (2011). Vulnerability of older Latino and Asian immigrants with limited English proficiency. *Journal of the American Geriatrics Society, 59*(7), 1246–1252.

Kim, J., & Miech, R. (2009). The black–white difference in age trajectories of functional health over the life course. *Social Science & Medicine, 68*(4), 717–725.

Kochanek, K. D., Arias, E., & Anderson, R. N. (2013). How did cause of death contribute to racial differences in life expectancy in the United States in 2010? *National Center for Health Statistics Data Brief, 125*, 1–8. Hyattsville, MD: National Center for Health Statistics.

Krekula, C. (2007). The intersection of age and gender: Reworking gender theory and social gerontology. *Current Sociology, 55*(2), 155–171.

Lara, M., Gamboa, C., Kahramanian, M. I., Morales, L. S., & Hayes-Bautista, D. E. (2005). Acculturation and Latino health in the United States: A review of the literature and its sociopolitical context. *Annual Review of Public Health*, 26(1), 367–397.

LaVeist, T. (2005). Disentangling race and socioeconomic status: A key to understanding health inequalities. *Journal of Urban Health*, 82(3), iii26–iii34.

LaVeist, T. A., Rolley, N. C., & Diala, C. (2003). Prevalence and patterns of discrimination among US health care consumers. *International Journal of Health Services*, 33(2), 331–344.

Leigh, J. P., & Fries, J. F. (1994). Education, gender, and the compression of morbidity. *International Journal of Aging and Human Development*, 39(3), 233–246.

Levkoff, S. E., Prochaska, T. R., Weitzman, P. F., & Ory, M. G. (2000). Recruitment and retention in minority populations: Lessons learned in conducting research on health promotion and minority aging. *Journal of Mental Health and Aging*, 6(1), 5–8.

Lopes, A. A. S., & Port, F. K. (1995). The low birth weight hypothesis as a plausible explanation for the black/white differences in hypertension, non-insulin-dependent diabetes, and end-stage renal disease. *American Journal of Kidney Diseases*, 25(2), 350–356.

Magana, A., & Clark, N. M. (1995). Examining a paradox: Does religiosity contribute to positive birth outcomes in Mexican American populations? *Health Education Quarterly*, 22(1), 96–109.

Manton, K. G., & Stallard, E. (1991). Cross-sectional estimates of active life expectancy for the US elderly and oldest-old populations. *The Journal of Gerontology*, 46(3), S170–S182.

Markides, K. S., & Coreil, J. (1986). The health of Hispanics in the southwestern United States: An epidemiologic paradox. *Public Health Reports*, 101(3), 253–265.

Markides, K. S., & Eschbach, K. (2005). Aging, migration, and mortality: Current status of research on the Hispanic paradox. *The Journals of Gerontology Series B: Psychological Sciences and Social Sciences*, 60(Special issue 2), S68–S75.

Markides, K. S., Eschbach, K., Ray, L. A., & Peek, M. K. (2007). Census disability rates among older people by race/ethnicity and type of Hispanic origin. In J. L. Angel & K. E. Whitfield (Eds.), *The health of aging Hispanics: The Mexican-origin population* (pp. 26–39). New York, NY: Springer.

Mays, V. M., Cochran, S. D., & Barnes, N. W. (2007). Race, race-based discrimination, and health outcomes among African Americans. *Annual Review of Psychology*, 58, 201–225.

Mutchler, J. E., & Burr, J. A. (2011). Race, ethnicity, and aging. In R. A. Settersten & J. L. Angel (Eds.), *Handbook of sociology of aging* (pp. 83–101). New York, NY: Springer.

National Center for Health Statistics. (2012). *Life expectancy at birth, at 65 years of age, and at 75 years of age, by sex, race, and Hispanic origin: United States, selected years 1900–2010*. Retrieved from: <http://www.cdc.gov/nchs/data/hus/hus12.pdf#018>.

O'Rand, A. M. (1996). The precious and the precocious: Understanding cumulative disadvantage and cumulative advantage over the life course. *The Gerontologist*, 36(2), 230–238.

O'Rand, A. M., & Hamil-Luker, J. (2005). Processes of cumulative adversity: Childhood disadvantage and increased risk of heart attack across the life course. *The Journals of Gerontology Series B: Psychological Sciences and Social Sciences*, 60(Special issue 2), S117–S124.

Osypuk, T. L., Diez Roux, A. V., Hadley, C., & Kandula, N. R. (2009). Are immigrant enclaves healthy places to live? The multi-ethnic study of atherosclerosis. *Social Science & Medicine*, 69(1), 110–120.

Palloni, A. (2007). Health status of elderly Hispanics in the United States. In J. L. Angel & K. E. Whitfield (Eds.), *The health of aging Hispanics: The Mexican-origin population* (pp. 17–25). New York, NY: Springer.

Palloni, A., & Arias, E. (2004). Paradox lost: Explaining the Hispanic adult mortality advantage. *Demography*, 41(3), 385–415.

Phelan, J. C., Link, B. G., & Tehranifar, P. (2010). Social conditions as fundamental causes of health inequalities: Theory, evidence, and policy implications. *Journal of Health and Social Behavior*, 51(Suppl. 1), S28–S40.

Quiñones, A. R., Liang, J., Bennett, J. M., Xu, X., & Ye, W. (2011). How does the trajectory of multimorbidity vary across black, white, and Mexican Americans in middle and old age? *The Journals of Gerontology Series B: Psychological Sciences and Social Sciences*, 66B(6), 739–749.

Roff, L. L., Burgio, L. D., Gitlin, L., Nichols, L., Chaplin, W., & Hardin, J. M. (2004). Positive aspects of Alzheimer's caregiving: The role of race. *Journal of Gerontology: Psychological Sciences*, 59(4), 185–190.

Seeman, T., Epel, E., Gruenewald, T., Karlamangla, A., & McEwen, B. S. (2010). Socio-economic differentials in peripheral biology: Cumulative allostatic load. *Annals of the New York Academy of Sciences*, 1186(1), 223–239.

Shonkoff, J. P., Boyce, W., & McEwen, B. S. (2009). Neuroscience, molecular biology, and the childhood roots of health disparities: Building a new framework for health promotion and disease prevention. *JAMA*, 301(21), 2252–2259.

Shonkoff, J. P., Garner, A. S., Siegel, B. S., Dobbins, M. I., Earls, M. F., Garner, A. S., et al. (2012). The lifelong effects of early childhood adversity and toxic stress. *Pediatrics*, 129(1), e232–e246.

Singh, G. K., & Hiatt, R. A. (2006). Trends and disparities in socioeconomic and behavioural characteristics, life expectancy, and cause-specific mortality of native-born and foreign-born populations in the United States, 1979–2003. *International Journal of Epidemiology*, 35(4), 903–919.

Smedley, B. D., Stith, A. Y., & Nelson, A. R. (Eds.), (2009). *Unequal treatment: Confronting racial and ethnic disparities in health care*. Washington, DC: National Academies Press.

Smith, D. B., Feng, Z., Fennell, M. L., Zinn, J., & Mor, V. (2008). Racial disparities in access to long-term care: The illusive pursuit of equity. *Journal of Health Politics, Policy and Law*, 33(5), 861–881.

Song, J., Chang, J. J., Tirodkar, M., Chang, R. W., Manheim, L. M., & Dunlop, D. D. (2007). Racial/ethnic differences in activities of daily living disability in older adults with arthritis: A longitudinal study. *Arthritis Care and Research*, 57(6), 1058–1066.

Taylor, M. G. (2008). Timing, accumulation, and the black/white disability gap in later life: A test of weathering. *Research on Aging*, 30(2), 226–250.

Thorpe, R. J., & Angel, J. L. (2013). Introduction: Sociology of minority aging. In K. E. Whitfield & T. Baker (Eds.), *Handbook of minority aging* (pp. 381–385). New York, NY: Springer Publishing.

Timmins, C. L. (2002). The impact of language barriers on the health care of Latinos in the United States: A review of the literature and guidelines for practice. *The Journal of Midwifery & Women's Health*, 47(2), 80–96.

Turra, C. M., & Elo, I. T. (2008). The impact of salmon bias on the Hispanic mortality advantage: New evidence from Social Security data. *Population Research and Policy Review*, 27(5), 515–530.

Umberson, D., & Montez, J. K. (2010). Social relationships and health: A flashpoint for public policy. *Journal of Health and Social Behavior*, 51(1), S54–S66.

Umberson, D. J., Wortman, C. B., & Kessler, R. C. (1992). Widowhood and depression: Explaining gender differences in vulnerability. *Journal of Health and Social Behavior*, 33(1), 10–24.

US Census Bureau. (2003). *No. 196. Persons 65 Years Old and Over With Limitation of Activity Caused by Chronic Conditions: 1999 to 2001*. Retrieved from: <http://www.census.gov/prod/2004pubs/03statab/health.pdf>.

US Census Bureau. (2011). *California: Quick Facts*. Washington, DC: US Census Bureau.

Venn, S., Davidson, K., & Arber, S. (2011). Gender and aging. In R. A. Settersten & J. L. Angel (Eds.), *Handbook of sociology of aging* (pp. 71–81). New York, NY: Springer.

Waite, L. J., & Das, D. A. (2010). Families, social life and well-being at older ages. *Demography*, 47(Suppl.), S87–S109.

Waite, L. J., & Gallagher, M. (2001). *The case for marriage: Why married people are happier, healthier, and better off financially*. New York, NY: Random House.

Walsemann, K. M., Geronimus, A. T., & Gee, G. C. (2008). Accumulating disadvantage over the life course: Evidence from a longitudinal study investigating the relationship between educational advantage in youth and health in middle age. *Research on Aging*, 30(2), 169–199.

Warner, D. F., & Brown, T. H. (2011). Understanding how race/ethnicity and gender define age-trajectories of disability: An intersectionality approach. *Social Science & Medicine*, 72(8), 1236–1248.

Waters, D. J. (2007). Cellular and organismal aspects of senescence and longevity. In J. M. Wilmoth & K. F. Ferraro (Eds.), *Gerontology: Perspectives and issues* (pp. 59–88) (3rd ed.). New York, NY: Springer Publishing.

Whitfield, K. E., & McClearn, G. (2005). Genes, environment, and race: Quantitative genetic approaches. *American Psychologist*, 60(1), 104–114.

Williams, D. R., & Collins, C. (2001). Racial residential segregation: A fundamental cause of racial disparities in health. *Public Health Reports*, 116(5), 404–416.

Williams, D. R., & Leavell, J. (2012). The social context of cardiovascular disease: Challenges and opportunities for the Jackson Heart Study. *Ethnicity and Disease*, 22(3 Suppl. 1), S1–15–21.

Williams, D. R., & Mohammed, S. A. (2009). Discrimination and racial disparities in health: Evidence and needed research. *Journal of Behavioral Medicine*, 32(1), 20–47.

Williams, D. R., Neighbors, H. W., & Jackson, J. S. (2003). Racial/ethnic discrimination and health: Findings from community studies. *American Journal of Public Health*, 93(2), 200–208.

Williams, D. R., & Sternthal, M. (2010). Understanding racial-ethnic disparities in health: Sociological contributions. *Journal of Health and Social Behavior*, 51(Suppl. 1), S15–S27.

Willson, A. E., Shuey, K. M., & Elder, G. H., Jr. (2007). Cumulative advantage processes as mechanisms of inequality in Life Course Health. *American Journal of Sociology*, 112(6), 1886–1924.

Woodward, A. T., Taylor, R. J., Bullard, K. M., Aranda, M. P., Lincoln, K. D., & Chatters, L. M. (2012). Prevalence of lifetime DSM-IV affective disorders among older African Americans, black Caribbeans, Latinos, Asians and non-Hispanic white people. *International Journal of Geriatric Psychiatry*, 27(8), 816–827.

Wu, J. H., Haan, M. N., Liang, J., Ghosh, D., Gonzales, H. M., & Herman, W. H. (2003). Diabetes as a predictor of change in functional status among older Mexican Americans. *Diabetes Care*, 26(2), 314–319.

Yang, Y., & Land, K. C. (2013). *Age-period-cohort analysis: New models, methods, and empirical applications*. CRC Press.

Zeki, A. A.-H., Odden, M., Mayeda, E. R., Aiello, A. E., Neuhaus, J. M., & Haan, M. N. (2012). Lifetime socioeconomic position and functional decline in older Mexican Americans: Results from the Sacramento Area Latino Study on Aging. In J. L. Angel, F. Torres-Gil, & K. S. Markides (Eds.), *Aging, health and longevity in the Mexican-origin population* (pp. 35–49). New York, NY: Springer Sciences.

CHAPTER 7

Immigration, Aging, and the Life Course

Judith Treas[1] and Zoya Gubernskaya[2]

[1]Department of Sociology, Center for Demographic & Social Analysis, University of California, Irvine, Irvine, CA, USA [2]Department of Sociology, University at Albany, State University of New York, Albany, NY, USA

OUTLINE

Introduction	143	Immigrants and Families	150
Immigration as a Life-Course Experience	144	Socioeconomic Outcomes of Older Immigrants	152
The Principle of Life-Span Development	146	The Health of Older Immigrants	154
The Principle of Agency	147	Conclusion	155
The Principle of Time and Place	147	References	157
The Principle of Timing	148		
The Principle of Linked Lives	149		

INTRODUCTION

Today, immigrants make up one-eighth of the US population. Although this share has not yet reached the high point seen a century ago during the heyday of US immigration, the rising numbers of foreign-born are remaking American society in numerous ways and raising concerns about how well the US incorporates newcomers. Globalization means that a similar story is playing out around the world. This chapter examines the critical issues of immigration and immigrant incorporation through the lens of aging and the life course. We address international migration within a general framework of historically situated adaptation and change occurring over the lives of individuals. We do not attempt a comprehensive survey of the rich literature on immigrants and immigration, because useful introductions to research on immigration and aging are already available (Jasso, 2004; Markides & Eschbach, 2005; Population Reference Bureau, 2013; Torres-Gil & Treas, 2008–2009; Treas & Batalova, 2009; Warnes, 2009). We focus largely, but not exclusively, on immigration to the United States, historically "a nation of immigrants" and the country leading the world in the number of immigrants today.

The chapter begins by framing immigration as a distinctive life-course event. We then review the analytic principles associated with the life-course approach and demonstrate ways in which they apply in the context of international migration. The strength of the life-course perspective is its integrated set of concepts and principles for making sense of the changes over time that are associated with growing up and growing older. Here, we draw on these principles to embed the immigrant experience in a perspective that recognizes the powerful temporal dimension of both individual development and historical change. Our objective is to demonstrate a systematic way to think about immigration and immigrants, and we rely on the research literature on international migration and immigrant incorporation to illustrate this point. In particular, we consider how life-course insights inform the understanding of immigrant families, health, and socioeconomic attainments.

IMMIGRATION AS A LIFE-COURSE EXPERIENCE

From birth to death, the life course is punctuated by milestones and turning points. Like marriage or retirement, immigration is a major life event catapulting the individual into a new status, namely, immigrant. This status has wide-ranging implications, because it embodies a different social position, an altered standing, and a changed situation. There are other important events experienced by immigrants that are also unique to them. Some immigrants change their legal immigration status (going from, say, refugees to permanent residents). Some become naturalized citizens, acquiring rights and opportunities not otherwise available. Some voluntarily emigrate, perhaps returning to their country of origin in a life-course countertransition. Others are traumatically deported, compelled by the government to leave. As these examples point out, the immigrant's life course is a genuinely state-managed one (Mayer & Schoepflin, 1989), but there are many other kinds of immigrant actions, transitions, and accomplishments, such as bankrolling a relative's immigration or learning enough of a new language to talk to the neighbors.

Compared to birth or marriage, the timing of the immigration event is ambiguous, because immigrants often come and go or make temporary visits before deciding to settle permanently (Redstone & Massey, 2004). Furthermore, international migrants may make a succession of moves, immigrating more than once over their lives. Based on our tabulations from the New Immigrant Survey (Jasso, Massey, Rosenzweig, & Smith, 2004), Figure 7.1 shows the migration routes for the cohort of immigrants, ages 50 and older, who became US lawful permanent residents in 2003. Most of these immigrants had come to the United States directly from the country where they were born, but nearly 17% had experienced multiple moves.

Over 7% moved from their birthplace to the United States, only to move later to another country before eventually returning to the United States. For nearly 6%, another country had been an intermediate destination between their birthplace and the United States. Some immigrants had lived in three (or more) countries before arriving in the United States. In short, the latest migration event sometimes fails to capture a trajectory of complex experience. The immigrant life course may reflect the multiple moves of some privileged transnational elite, the circular migration of seasonal agricultural laborers, or the harrowing expulsions and

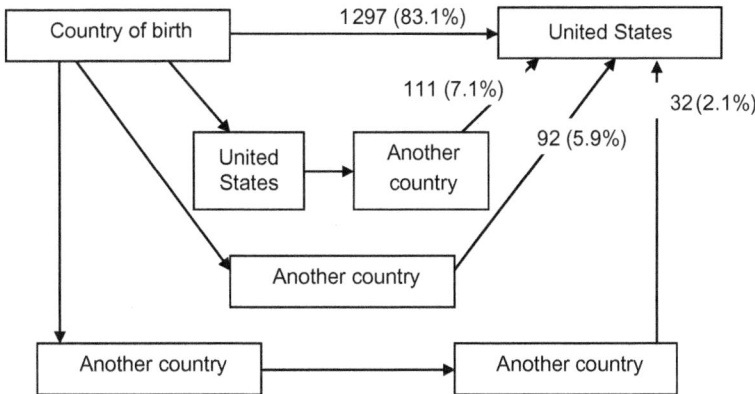

FIGURE 7.1 Most common migration routes to the United States: US lawful permanent residents age 50 and over.

resettlements of forced refugees. Thus, a richer and more consequential legacy of experience may shape the circumstances of immigrants than is implied by their nativity alone.

While the immigrant life course is marked by such events, it is not only transitions that matter, but also the length of spells in various states. The significance and meaning of immigrant status often depend on how long one has been an immigrant. Lawful permanent residents in the United States, for example, must clock 5 years living in the country before being eligible to apply for American citizenship. Nor is this the only temporal dimension relevant to naturalization, because immigrants must be at least 18 years of age to apply on their own. Consistent with the life-course principle that experiences are shaped by intimate ties, child immigrants become citizens automatically when their parents do. At least in the lives of adults, becoming a naturalized citizen is not just a gateway to rights, responsibilities, and benefits, but often a personally meaningful milestone in the life course. Taking pride in passing their citizenship exam, older immigrants have been observed to retrieve naturalization documents and show them off to visitors (Treas, 2008).

Naturalization is taken as only one indicator of the immigrant's broader incorporation into the host society. Since Thomas and Znaniecki's classic study of *The Polish Peasant in Europe and America* (1918), the central question for researchers studying immigrants has been how they adapt to the experience of living in a new country. This adaptation has been described primarily in terms of "acculturation" (adopting the dominant group's culture) and "assimilation" (gaining equal standing in the primary groups and social institutions of the receiving society) (Alba & Nee, 1999). Although immigration scholars today resist the idea of immigrants' inevitable march toward the end of ethnic differences (Portes & Zhou, 1993; Rumbaut, 2008), the broad concept of incorporation is founded on a profoundly temporal dynamic that parallels aging. Although events like naturalization serve as markers of incorporation, the duration of the immigrant's spell in the receiving country is a strong determinant. As a time-dependent process, immigrant adaptation is assumed to hinge on the length of exposure to the host society (Treas, 2015). Mirroring the classic "contact hypothesis" that Gordon Allport (1954) articulated for race relations, this argument predicts reduced social distance the more that majority and minority groups interact. Because the immigrant's exposure to the host society, that is, to its members and institutions, accumulates with the passage of time, long-time immigrants are more thoroughly incorporated than are newcomers.

Whether the focus is the event, the duration in a status, or a temporal process, the immigrant experience displays a familiar life-course dynamic that invites further attention from researchers on aging. The life-course framework offers conceptual tools for analysis. Integrating developmental trajectories, socially structured careers and pathways, and historical change, this perspective provides a holistic approach to understanding immigrant adaptation. To illustrate the embeddedness of immigrant experience in the life course, we point to five principles of life-course theory, as articulated by Elder, Johnson, and Crosnoe (2003).

THE PRINCIPLE OF LIFE-SPAN DEVELOPMENT

Life-course theorizing demands attention to the whole of human lives. Development and adaptation, including immigrant incorporation, transpire over the long term. Early experience sets the stage for later outcomes. Immigration is a defining experience with implications for the remainder of the life course. For example, although advances in transportation and communication technology make it easier for immigrants to stay in touch with their homelands, immigration still imposes painful separations on kin (Baldassar, Kilkey, Merla, & Wilding, 2014). As demonstrated by parents who immigrate to better their children's prospects (Dreby, 2010), decisions made today are influenced not merely by current circumstances, but also by an anticipation of the future.

Few would deny the value of a long view of lives, but practical constraints often discouraged immigrant studies from considering the whole life span. Although more than one in eight Americans is an immigrant, their numbers in national surveys are often too small for reliable analyses, particularly when a subgroup (e.g., high-skilled workers, Central American immigrants, recent arrivals) is of interest. Even the ambitious New Immigrant Survey, which follows arriving or visa-adjusting cohorts of legal immigrants and their children, misses undocumented immigrants who were unable to legalize. Although there have been ambitious efforts to study immigrants in their sending and receiving societies (e.g., the Mexican Migration Project; Massey & Alarcon, 1987), US surveys capture the American experience, reconstructing the earlier life course with retrospective data that are less accurate than contemporary reports.

Although the birth-to-death articulation of experience offers the fullest account of human lives, the expedient approach is often to focus on the particular developmental stage or socially organized age-grade most salient for the problem at hand. Schools are the main window on early socialization of immigrant and nonimmigrant children alike. Thus, children in immigrant families are studied in terms of their cognitive preparedness, academic achievement, and behavioral outcomes (Rumbaut, 2008). Given a long-standing preoccupation with how well immigrants and their children accommodate to the host society, research on immigrants has singled out adolescence and young adulthood as a sensitive period for identity formation, lifestyle choice, work force entry, relationship building, and family formation (Gonzales, 2011; Kasinitz, Mollenkopf, Waters, & Holdaway, 2008; Rumbaut & Portes, 2001). Similarly, the prime working years come in for attention when the focus is on the economic incorporation and wage growth of immigrants (Batalova, 2006; Donato & Sisk, 2012; Massey & Gelatt, 2010). Older foreign-born adults are considered in the context of health (Carrasquillo, Carrasquillo, & Shea, 2000; Gubernskaya, Bean, & Van Hook, 2013; Markides & Eschbach, 2005), family (Angel, Angel, Lee, & Markides, 1999; Treas & Mazumdar, 2004), and program participation (Burr, Gerst, Kwan, & Mutchler, 2008–2009; Yoo, 2001).

THE PRINCIPLE OF AGENCY

Individuals shape the course of their lives through the choices that they make. The life-course principle of agency recognizes that decision-making is subject to the limits and opportunities posed by the environment, but it also acknowledges that individuals have some discretion in the paths they make and take. Immigration, for example, is facilitated by immigrant social networks that help newcomers with housing and jobs (Massey, 1990; Winters, De Janvry, & Sadoulet, 2001), but people still have the choice of staying or going. Stricter border enforcement "locks in" unauthorized Mexican immigrants who fear being unable to re-enter if they return home (Massey & Pren, 2012), but some choose to take that risk. Even given structural constraints, the action and initiative of individuals help us to understand why two people who start with similar circumstances can diverge markedly by the end of their lives (Dannefer, 1987).

The decision to immigrate or not epitomizes human agency, and this immigration decision sets the stage for subsequent choices. Admittedly, imperfect information and unexpected contingencies can derail plans. What is initially seen as a short-term stint of labor migration may eventually lead to a decision to settle permanently in the receiving society (Chavez, 1994). Or, parent's plans to bring children to the United States may be pushed further and further into the future by financial reversals or growing youngsters' developmental needs (Dreby, 2010).

Contemporary immigrant incorporation theories have recognized human agency by abandoning the mechanistic, straight-line model that assumed greater assimilation necessarily came from greater time in the host country. Instead, in a process referred to as "selective acculturation," immigrants are acknowledged to pick and choose which aspects of the mainstream culture to accept and which to reject (Portes & Rumbaut, 2001; Portes & Zhou, 1993). Another revision of classic assimilation theory points to "reactive ethnicity," whereby immigrants' children and grandchildren, alienated by racism, choose to embrace ethnicity as a political statement (Rumbaut, 2008). Conversely, return migrants who find themselves marginalized and discriminated against in their homelands sometimes respond with diminished attachment to their roots (Tsuda, 2009). In short, incorporation, perhaps the central tenet of immigration research, has increasingly been cast as personal choice, mediated by family and community, rather than as an inevitable process (Waters, Tran, Kasinitz, & Mollenkopf, 2010).

THE PRINCIPLE OF TIME AND PLACE

In acknowledging the constraints on human agency, the life-course perspective emphasizes that the broader context influences how lives play out across the life course. Historical eras differ in their social structure and in the cultural scaffolding that directs both individual pursuits and understandings of experience. The key concepts of cohort (Ryder, 1965) and generation (Mannheim, 1997(1952)) are founded on historical distinctions. Those arriving without legal authorization before 1982 were able to legalize under the Immigration Reform and Control Act of 1986, but later cohorts of unauthorized immigrants lack a pathway to legal status. Their marginal status will not only diminish their own prospects, but also those of their US citizen children (Massey & Pren, 2012).

Lives also occur in contexts, and the diversity in contexts means that immigrant experiences differ from place to place. Even grand historical events are typically fairly local in their influence and, thus, in their implications for personal trajectories. German Reunification, for example, was a political and economic transformation that had the greatest consequences for the occupational prospects of East Germans. Akin to immigration, even immobile East Germans confronted an abrupt change in structure and expectations, which Diewald, Goedicke, and Mayer (2006) liken to immigration, that is, as the transition between "the society of departure and the society of destination."

The principle of time and place defines the life trajectories of international migrants. Consider one aspect of immigrant incorporation – becoming a naturalized citizen. Bloemraad (2006) explains why naturalization rates in the United States are considerably lower than in neighboring Canada. While American public policy takes a *laissez faire* approach to immigrant incorporation, newcomers to Canada are consciously welcomed with material support, language learning programs, job search assistance, and a multicultural ethos that encourages political engagement. To take another example, the security and economic opportunities for an unauthorized immigrant in the United States may hinge on whether she lives in a state where she can get a driver's license or in a community where police report passengers in routine traffic stops to the immigration authorities. Given the life-course emphasis on place, where immigrants come from deserves as much attention as where they go. As demonstrated by the ethnic Koreans moving back home from the United States and China, return migrants coming from wealthy countries receive warmer receptions in South Korea than their counterparts from other countries (Kim, 2009; Song, 2009).

Immigrants' experiences depend on where they settle upon arrival. Residing in an ethnic enclave may ease immigrants' adaptation by providing social support, access to familiar food and services, information about public services, employment opportunities and entertainment in their native language (Hao & Kawano, 2001; Logan, Zhang, & Alba, 2002). However, residence in a community with a high concentration of co-ethnics may also stall economic incorporation and acculturation (Chiswick & Miller, 2005; Xie & Gough, 2011).

The context of immigration itself depends on the historical era. Early in the twentieth century, US immigrants were overwhelmingly European, if only because of exclusionary provisions aimed at Asians. By the start of the twenty-first century, the United States attracted few workers from the developed economies of Europe. Instead, immigrants haled primarily from Asia and Latin America. Many Asians began to migrate after the Immigration and Nationality Act of 1965 replaced race-based quotas with labor and family reunification criteria for immigration. Because of continuing labor demand, the end of the popular temporary worker (Bracero) program as well as newly limited visas for Western Hemisphere migrants created the phenomenon of unauthorized immigration from South of the border (Massey & Pren, 2012).

THE PRINCIPLE OF TIMING

The implications of an event or experience differ depending on its timing in the life course. Workers in late middle-age at the time of German Reunification, for example, suffered the greatest financial reversals, being too young to fall back on a public pension but too old to adapt to the new job market with additional education or retraining (Diewald et al., 2006). Immigrants reaffirm the importance of the life-course principle of timing. In 2012, the Obama Administration announced the Deferred Action for Childhood Arrivals Initiative, which offered a reprieve from deportation for unauthorized immigrants who, among other things, were under the age of 31 and entered the United States before age 16. While some young people brought to the United States as children stand to benefit, their slightly older siblings will not, because of inauspicious timing – being too old at immigration or at the initiation of the program (Batalova, Hooker, Capps, Bachmeier, & Cox, 2013).

Sometimes, timing is consequential because of the legal reliance on chronological age to establish rights of residency and citizenship. In other cases, timing of immigration matters, because it defines immigrants' participation in age-graded social structures, their developmental capacities, or beliefs about their abilities. Immigrants who arrive as children assimilate and acculturate more readily than do their parents or older siblings. Rumbaut (2004) coined the term "1.5 generation" to describe first-generation child migrants, because they grow up to be much like the US-born second generation in terms of their incorporation into the host society. By contrast, immigrants who arrive late in life are unlikely to catch up to other first-generation immigrants. As Treas (2015) puts it, they are the ".5 generation." Older adults' slower incorporation implicates "normal" aging processes, such as poorer memory which newcomers say makes it hard to learn a new language (Treas & Mazumdar, 2002). Age-graded social institutions, however, amplify the importance of timing in the life course. Late-life immigrants to the United States are too old to benefit from the Americanizing influences of school and workplace.

Bureaucratic wait times to legal immigration introduce inequality and uncertainty into life plans by virtue of their impact on the life-course timing of immigration (Bergeron, 2013). Although the immigration of spouses, minor children, and parents of US citizens is relatively straight-forward, US law subjects other family immigrants to annual numerical limits depending on the specific immigration category. More people want to immigrate than can be accommodated by these limits. Official, albeit incomplete, estimates of those waiting to immigrate stood at 4.4 million in 2012. Among the siblings of Filipino immigrants or the adult unmarried children of Mexicans, the wait even to apply for a visa is two decades. People in other countries or other immigration categories face comparatively short waits. A longer wait translates to a longer period "at risk" of life-course disruptions. By delaying the timing of immigration, the long wait likely works against incorporation by shortening the number of years living in the United States and pushing immigration to older ages when it is harder to start over. Although temporary visits and even unauthorized migration address some immediate needs, immigration backlogs affect the course of lives. During waiting time, a normal life-course transition – a marriage or a career change – may undermine immigration prospects. Upon marrying, those wanting to immigrate cease to qualify as the unmarried adult child of a legal permanent resident. They face the choice of starting over on a new pathway, say, an employment-based channel, or giving up on immigrating altogether. Although turning points in the life course are inevitable, the life course is hostage to the timing uncertainties associated with the delay in translating agency into an immigration event.

THE PRINCIPLE OF LINKED LIVES

Because of the interdependence created by human relationships, the course of an individual's life is shaped by others. Indeed, the influence of the broader socio-historical context is often mediated by the experience of those who are more directly affected. This fact was illuminated by Glen Elder (1974), whose classic book, *Children of the Great Depression*, showed that these children felt the legacy of their father's unemployment throughout their lives.

Immigration exemplifies the importance of linked lives. Social networks, for example, are known to facilitate migration (Massey, 1990; Winters et al., 2001). Family members may urge kin to immigrate, loan money to pay for the trip, or look after children in their parents' absence. In communities where many people have immigrated, they influence others indirectly by creating an "immigration culture," which may make an extended stay in another country a normative part of the life course (Kandel & Massey, 2002). A foreign posting can become a routine career step in a multinational corporation. Meanwhile, young people from poor countries can come to look forward to a stint of unskilled labor in a wealthy society as an adventurous *rite de passage*. Even those who stay behind have their lives touched by immigrants who send money to support kin, bankroll schools, or provide the capital for local businesses (Eckstein & Najam, 2013).

Becoming an immigrant has secondary or incidental effects, often quite profound, on the life course of intimates. Some children grow up with the benefits of citizenship because their mother relocated to the United States before they were born. On the other hand, young children brought to the United States by parents may not learn that they are unauthorized until late adolescence when their undocumented status blocks them from making life-course transitions to college and employment with their peers (Gonzales, 2011). Children left behind when parents immigrate may benefit from remittances, even as they miss out on close parental relationships (Dreby, 2010; Parreñas, 2001). Of course, children are not the only ones affected. Parents' immigration can leave grandparents with full-time responsibility for grandchildren left behind (Dreby, 2010).

The immigration of older adults increased as a result of US historical developments that have given greater weight to linked lives as a basis for immigration. As noted, the 1965 changes in immigration law prioritized family reunification as a criterion for immigrating. US citizens, 21 years and older, are permitted to sponsor the immigration of their parents. And, unlike the visas for adult children or siblings, parental visas are not subject to numerical limits. While the desire to be close to loved ones undoubtedly motivates immigration decisions, older adults are also sometimes invited by grown children in order to provide care for the younger generation in the United States (Treas & Mazumdar, 2004).

The principles distilled from the study of the life course inform the understanding of immigration and immigrant experience. They provide an overarching conceptual framework for the systematic analysis of a phenomenon characterized by complexity and contingency. Approaching established questions in research on immigrant incorporation from the perspective of life-course theorizing establishes fresh connections and reveals new insights, as we demonstrate by interrogating studies in three active areas of research – immigrant families, health, and socioeconomic attainment of older immigrants.

IMMIGRANTS AND FAMILIES

Although there is a large research literature on immigrant families, greater understanding has emerged from the articulation with life-course principles in studies of immigrant family life. Work in two fields demonstrates the intersection of family and life course. One literature concerns the influence of immigration on the formation of families. The second considers the households in which immigrants are embedded.

Because labor migration is concentrated at the ages when young adults typically start their families, the migration event might be assumed to disrupt the life course, because immigration demands adaptations likely to slow the usual progression to marriage or childbearing (Landale, 1994). The evidence does not confirm that immigration delays family formation in the United States. A study comparing Puerto Rican women in Puerto Rico and the United States found that migrants were more likely than nonmigrants to enter into an early union (Landale, 1994). This reflected the selective nature of migration: Disadvantages earlier in the life course (e.g., limited schooling, a mother who had children at a young age) predisposed women not only to migrate, but also to partner at a young age. Even though timing does not seem to be affected, migration does disrupt marriage patterns by altering the availability and desirability of potential mates. Mexican immigrants to the United States tend to marry up, because their own limited numbers lead them to marry nonmigrants, who tend to be generally better educated than they are (Choi & Mare, 2012). The timing of immigration in the life course is also consequential because of its implications for structural and cultural assimilation. Particularly for Mexicans and Chinese immigrants, those who arrive as children or preteens are more likely to marry non-Hispanic Whites than those who immigrate later as adolescents (Qian, Glick, & Batson, 2012).

As for fertility, there is a tendency for Hispanic women to have a birth shortly after US immigration (Parrado, 2011), a pattern also observed for immigrants in Sweden (Andersson, 2004). Mexican women migrating to the United States do not delay a first birth, perhaps because they schedule the two events together, want a US-born citizen child, or plan to have children quickly so they will be free for employment sooner (Lindstrom & Giorgiuli Saucedo, 2011). Research in West Germany finds no evidence of migration disrupting the timing of fertility either (Milewski, 2007), but Italian investigators report that migrating for work per se delays childbearing (Mussino & Strozza, 2012). The implication is that migration and family-building are interrelated, not independent, decisions (Andersson, 2004) in the life course.

That migration has different effects on family formation in sending than receiving societies serves to underscore the importance that the life-course approach attributes to place. Mexican municipalities with high rates of immigration to the United States have a lower proportion of women married, likely reflecting their lower ratio of men to women (White & Potter, 2013). When separated from husbands working in the United States, women in Mexico have lower fertility than those living with their partner (Lindstrom & Giorgiuli Saucedo, 2011). Communities with high female migration rates have lower fertility as well, perhaps because low fertility norms of the receiving society diffuse to women who remain behind.

Immigrant households are larger, more complex, and more often multigenerational than those of native-born Americans (Glick & Van Hook, 2002). Often debated is whether these household differences represent cultural differences and/or greater economic need (Glick, 2010), influences which can be expected to diminish with greater time and incorporation into the host society. Immigrant living arrangements, however, also relate to other aspects of the life course – above and beyond the obvious connection of linked lives.

The extended household has particular advantages at the time of arrival when living with others is dictated by limited income and by unique needs for help (e.g., language translation, job referrals) from trusted co-ethnics familiar with American society. With greater incorporation and more time in the United States, immigrants are in a better position to strike out on their own. As suggested by the principle of time and place, context also matters. Older Korean immigrants are more likely to live independently of kin in communities where they can count on subsidized housing and a higher concentration of Korean business establishments (Kim & Lauderdale, 2002).

As they age, most family householders experience a residential life cycle; their households eventually become smaller, because children grow up and move out to establish their own homes. In immigrant families, young adults linger longer in the parental household (Treas & Batalova, 2011). Instead of embracing residential independence as the marker of adulthood, they see continued co-residence with parents as evidence of their mature outlook:

shared housing allows them to save money and help out their families (Kasinitz et al., 2008). Whatever the economic and cultural constraints, continued co-residence demonstrates agency, consistent with immigrants' selective acculturation, that is, picking-and-choosing which practices of the host country to adopt (Portes & Zhou, 1993; Waters et al., 2010).

In some immigrant families, continued immigration from a homeland means that households never empty out; a large, complex household is maintained, because newcomers take the place of those who move out on their own (Myers & Lee, 1996). The supply of foreign-born, extended kin available to share households depends on the historical context. Compared to earlier cohorts, recent immigrants are sponsoring the immigration of comparatively more family members, including parents over age 50 (Carr & Tienda, 2013). Such older newcomers are particularly likely to reside in the homes of grown children (Angel, Angel, & Markides, 2000). Arriving at an advanced age, these latecomers have difficulties meeting the citizenship or work requirements for Social Security and Medicare – federal benefits that reduce older adults' economic dependency on kin. Of course, multigenerational immigrant families are not only shaped by the needs of older adults. Demonstrating the interdependence linking generations, younger adults depend on older household members, especially for housekeeping and help caring for children (Treas & Mazumdar, 2004).

SOCIOECONOMIC OUTCOMES OF OLDER IMMIGRANTS

Previous research on older immigrants' socioeconomic status aimed at comparing them to native-born older adults (Batalova, 2012; Leach, 2008; Treas & Batalova, 2009). On most indicators, the foreign-born do not fare as well. Overall, older immigrants have less education than the native-born although the share of older foreign-born with at least a bachelor's degree is similar. Even though foreign-born men are more likely to be in the labor force, immigrants' incomes are about 20% lower than US natives' (Batalova, 2012). Older immigrants are less likely to rely on economic resources earned during their working years (social security, income from assets, private pensions) and more likely to depend on public assistance, such as Supplementary Security Income (SSI) and the Supplemental Nutritional Assistance Program (SNAP or "food stamps") (Burr et al., 2008–2009; Gerst & Burr, 2011; Lee & Angel, 2002; Nam & Jung, 2008). About three-quarters of older immigrants are naturalized citizens (Batalova, 2012). This life-course transition has lasting implications, as citizenship status has been linked to higher income and to higher rates of welfare receipt in old age (Lee & Angel, 2002; Nam & Jung, 2008; Nam & Kim, 2012; Van Hook, 2000). Many older immigrants are not well integrated into the broader society. Over half report having limited English language proficiency, and about one-third live in linguistically isolated households where no adult speaks English very well (Batalova, 2012).

Although the demographic profile of the older foreign-born population is informative, the diversity within this group warrants against generalizations. Country of origin, type of migration, as well as past and present legal status have implications for immigrants' acculturation and incorporation over the life course, and consequently, for their socioeconomic outcomes in later life. Similar distinctions are seen for the diversity in life-course factors, including historical period of migration, age at migration, and length of stay in the United States. As the processes of acculturation and socioeconomic incorporation are often complex, dynamic, interdependent and mutually reinforcing, life-course principles offer a useful framework for understanding the socioeconomic disparities among the older foreign-born.

The principle of timing in lives points to the importance of age at migration. Older adults who migrated as children or young adults have had more time and opportunities to learn English and participate in mainstream social institutions through school, community, and workplace involvements. Those who migrated in midlife or old age, on the other hand, have had less time, but also fewer opportunities, to incorporate into the host society. Older newcomers face multiple barriers to gainful employment due to their age, a mismatch of skills, unfamiliarity with American society, and little or no English language proficiency (Rhee, Chi, & Yi, 2013). Not surprisingly, high rates of program participation are found primarily among the older immigrants who arrived in advanced age and are related to their low income (O'Neil & Tienda, 2015) and poor English language ability (Burr et al., 2008–2009; Nam & Jung, 2008). For similar reasons, older immigrant newcomers are less likely to be homeowners and much more likely to live in crowded households than their native or long-term immigrant counterparts (Burr, Mutchler, & Gerst, 2010, 2011). The usual correlates of socioeconomic well-being, such as education or citizenship, may have less predictive power for older adults who have recently immigrated. Instead, age and limited English are strong barriers to incorporation, even for well-educated older newcomers. Naturalization is generally viewed as a sign of incorporation. But for late-life immigrants, naturalizing (especially after the 1996 Welfare Reform, which restricted welfare participation for noncitizens) may signal their lack of incorporation, namely, their need to qualify for public assistance as a result of poor health and limited income (Gubernskaya et al., 2013; Nam & Kim, 2012; Van Hook, 2000).

As the principle of life-span development suggests, socioeconomic outcomes in later life reflect the legacy of education, occupation, and income earlier in the life course. For instance, young, poorly educated immigrants confined to low-skilled jobs and subsistence wages in informal labor markets struggle to save for retirement. Failure to qualify for retirement benefits not only forces them to stay in a labor force past normal retirement age, but also explains why they have little or no income from social security or savings in old age (Borjas, 2011; Burr et al., 2008–2009). Similarly, past legal status has implications for socioeconomic status in later life. Lawful permanent residents and citizens turn to unemployment benefits and temporary assistance during periods of unemployment. Being undocumented forces some foreign-born to take any job they can find or to stick with jobs having long hours, low wages, no benefits, and slim chances of promotion (Gleeson & Gonzales, 2012; Mehta, Theodore, Mora, & Wade, 2002; Orrenius & Zavodny, 2009). Thus, the accumulation of advantage and disadvantage across the entire life course leads to very disparate late-life outcomes even within the immigrant population.

As the principle of time and place underscored, the period of migration defines existing immigration policies that are critical for socioeconomic trajectories. Those legal permanent residents who immigrated before the 1996 Welfare Reform (PRWORA) were eligible for SSI and food stamps on arrival while those who migrated after this date could face a wait of at least 5 years (Borjas, 2002; Nam & Jung, 2008; Van Hook, 2003). Among the poorly educated foreign-born, especially those from Mexico and Latin America, economic changes since the 1980s are thought to have led to stagnating wages and diminishing economic returns to skills and assets, including English language proficiency, citizenship and general familiarity with American society (Donato & Sisk, 2012; Massey & Gelatt, 2010; Massey & Pren, 2012). It is likely that socioeconomic disparities among the older foreign-born will be even larger in the future.

THE HEALTH OF OLDER IMMIGRANTS

Most research on immigrants' health is framed around the persistent finding of an "immigrant health paradox." Despite their lower socioeconomic status, the foreign-born have lower mortality rates and comparable or better health than the native-born. Research on the general adult immigrant population finds that longer residence in the United States is associated with a diminishing health benefit, suggesting that the immigrants' better health may not carry over into old age. Studies that focus on older adults find an immigrant health advantage in mortality, but they produce mixed results regarding health status. There are also significant differences in health *among* the foreign-born. Detailed reviews of the immigrant health paradox literature can be found elsewhere (Acevedo-Garcia & Bates, 2008; Cunningham, Ruben, & Venkat Narayan, 2008; Riosmena & Dennis, 2012). Therefore, we focus on how life-course principles improve our understanding of immigrants' health in later life. Reviewing the most recent research points to the challenges of studying the health of older immigrants, for whom it is a cumulative product of life-course circumstances and experiences.

Older foreign-born persons enjoy a mortality advantage over native-born non-Hispanic Whites, especially if socioeconomic differences are taken into account (Angel, Angel, Diaz Venegas, & Bonazzo, 2010; Borrell & Lancet, 2012; Elo, Turra, Kestenbaum, & Ferguson, 2004; Palloni & Arias, 2004). Prevalence of chronic conditions is also lower among foreign-born Hispanics (Cantu, Hayward, Hummer, & Chiu, 2013; Choi, 2011; Swallen, 1997). At the same time, older immigrants often report worse self-rated health (Angel, Buckley, & Sakamoto, 2001; Wakabayashi, 2010) and have higher rates of disability (Hayward, Hummer, Chiu, González-González, & Wong, 2014; Markides, Eschbach, Ray, & Peek, 2007; Mendes de Leon, Eschbach, & Markides, 2011; Mutchler, Prakash, & Burr, 2007). On such measures as depression, chronic stress, self-rated emotional health and overall subjective well-being, older immigrants' health is no better (and sometimes worse) than the health of the US-born (Cuellar, Bastida, & Braccio, 2004; González, Haan, & Ladson, 2001; Ladin & Reinhold, 2013; Lum & Vanderaa, 2010). Older foreign-born persons have healthier behavioral profiles with lower rates of obesity, smoking and drinking, but their health may be compromised by their lack of access to health care (Nam, 2008; Reyes & Hardy, 2013).

Differences in data and methodology limit the comparability of results across different studies, and methodological challenges work against sorting out explanations for the immigrant health advantage. The "healthy immigrant effect" points to positive health selection. Because healthy people are more likely to migrate, immigrants are healthier than the average person in both the home and host countries, at least when they arrive (Crimmins, Kim, Alley, Karlamangla, & Seeman, 2007; Jasso, 2004; Palloni & Ewbank, 2004). According to an equally hard-to-test argument, immigrant culture — healthier diets, social support from family and kin, and risk-avoidant behavior — is protective of health. Still another explanation is related to selective return migration or the so-called "salmon bias." Failing to account for the out-migration of less healthy immigrants results in overly optimistic estimates of the health status of the foreign-born in the United States. Past research found some evidence of a "salmon bias," but concluded that it is unlikely to fully explain the immigrant health advantage (Aguila, Escarce, Leng, & Morales, 2013; Turra & Elo, 2008). Recall that the foreign-born enjoy longer lives, but experience more disability in old age (Hayward et al., 2014; Hayward, Warner, & Crimmins, 2007). One life-span explanation points to the lifetime accumulation of negative experiences, including adverse childhood conditions as well as health disadvantages due to low socioeconomic status, physically demanding work (leading to "weathering"), stress, and discrimination.

Of course, the older foreign-born constitute a diverse population. They differ by gender, race/ethnicity, and socioeconomic status – variation that also exists among the native-born. Differences unique to immigrants include age at migration, length of residence in the United States, country of origin, type of migration, legal status in the United States, and ability to speak English. These factors are likely to be related to health in old age, but small sample sizes and data limitations (e.g., on legal status or type of migration) often prevent researchers from exploring these differences. For example, research focuses on the health of the larger populations of older Hispanics, Mexicans, or Asians. Studies with a sufficient number of respondents, however, find significant differences by country of origin on most indicators of health within Hispanic and Asian categories (Borrell & Lancet, 2012; Lauderdale & Kestenbaum, 2002; Markides et al., 2007). Focusing exclusively on health disparities by nativity and neglecting immigrant diversity presents an incomplete picture of older immigrants' health. While some foreign-born are able to preserve good health in old age, others will come to experience a health disadvantage.

Life-course principles can be a useful tool for analyzing the health of older immigrants. The principle of life-span development emphasizes the importance of exploring health trajectories rather than focusing on health status or health disparities at a particular age. As the "cumulative disadvantage" and the "weathering" hypotheses suggest, health disparities can be expected to increase as immigrants get older. Not only current socioeconomic position, but also the socioeconomic status and environmental conditions earlier in the life course help explain the health disparities in later life. Although structural circumstances influence immigrant health, immigrants can be seen as active agents in their health trajectories. The lifestyles of the foreign-born (e.g., smoking, exercise, diet), their decisions regarding health care and medical treatment, their use of complementary and alternative medicine, their strategies of coping with illness and disability, and their decisions to stay or return in later life are micro-level factors calling for additional research.

Following the principle of timing, immigrants' health in later life is likely to depend on when in the life-course immigration occurs (Angel et al., 2010; Angel et al., 1999; Gubernskaya, 2015; Gubernskaya et al., 2013; Wakabayashi, 2010). For example, the degree of health selectivity of the foreign-born is apt to vary depending on age of migration (e.g., child, young adult, old age) or type of migration (e.g., labor, family, refugees). Presumably, poor health will be a bigger impediment for working age, labor migrants than for young or old family dependents. Ignoring these differences may bias estimates of health disparities, especially from cross-sectional data. Moreover, the effect of factors commonly associated with better health, such as race or education, is smaller not only for the foreign-born compared to US-born population, but also for the foreign-born who arrive at older ages compared to younger migrants (Gubernskaya, 2015; Gubernskaya et al., 2013; Riosmena & Dennis, 2012).

CONCLUSION

The rise in immigration surely ranks as one of the momentous developments of our times. Children arriving shortly after the 1965 legal reforms that ushered in a new wave of US immigration have grown old, but newcomers arrive every day. How these immigrants navigate the life course – growing up and growing old, building careers and families, caring for themselves and others, fitting into the host society while keeping faith with their homeland — are matters of significance to us all. For researchers who study aging and the life course, important issues posed by immigration and immigrants demand our attention and invite our contributions. They also offer a new strategic site for refining life-course theories and approaches. In this chapter, we have emphasized the potential of studies that take an explicit life-course approach to the phenomenon of immigration. In illustrating these possibilities, we have focused on only a few abstract ideas, the key principles that have guided life-course thinking (to the admitted neglect of many other relevant insights from the broad field of aging and the life course).

Five ideas have been central to our discussion. In the spirit of the *life-span development* principle that underpins the life-course perspective, we argue for research that takes a long view of immigrant adaptation as a multifaceted process, predating the actual migration event and extending through the end of life. Many of the outcomes of immigration cannot be known for decades or even generations. We situate immigration and the incorporation of immigrants in *historical time* and in *socio-cultural space*. This life-course precept requires that we acknowledge the consequential and on-going role not only of the host society, but also of the sending one. It also acknowledges the very different contexts across sending societies or receiving communities. Inevitably, this attention to time and place points to the changing social structure and the ways in which various social institutions channel the immigrant's progress over the life course.

Whatever the constraints and opportunities posed by particular institutional arrangements in a particular time and a particular place, the life-course principle of *agency* celebrates immigrants' initiative as well as the creativity they bring to bear on decisions shaping their own destinies. That most migration is voluntary migration underscores the importance of this life-course premise. The principle emphasizing the significance of the *timing* of experiences in the life course is arguably the distinguishing contribution of the life-course approach to understanding immigration and incorporation. Immigration may be a pivotal turning point in lives, but its implications depend on whether this transition comes early or late in the life course. Lastly, the reach of immigration extends well beyond the individual immigrant. Consistent with the tenet of *linked lives*, immigration has profound implications for the intimates of migrants. Its impact, however, extends beyond micro-level social networks to influence whole communities that send and receive immigrants.

Research opportunities come with challenges. The life-course approach to immigration will continue to demand entrepreneurial imagination and methodological innovation. The life-span development imperative calls for multifaceted, longitudinal data that can follow individuals over the long course of their lives. Of course, attrition bedevils all panel studies, but the difficulty of following informants though their lives is compounded for a population that moves back and forth, sometimes returning to live permanently in a homeland. Both the local and the global are the contemporary reality for today's immigrants, and this fact calls into question traditional research strategies bounded by the nation state. The most productive studies must not only drill down to portray local communities in which immigrants are embedded, but also jump borders to capture the reciprocal influences of origins and destinations in the lives of international migrants. The particularities of the migration streams from different parts of the world make it clear that genuinely comparative studies must be an aspiration in the quest to achieve a full understanding of the immigrant life course. Even within countries, research is frustrated by sample sizes that are too small for reliable estimates of immigrants who differ by origin (or on many other dimensions). As the linked lives principle suggests, there is also much to be gained from studies that extend to family members and to broader social networks. Perhaps the biggest threat to the realization of this research agenda comes from the marginalization of the foreign born. This problem ranges from the difficulty in interviewing non-English speakers to unauthorized immigrants' calculated efforts to avoid notice in administrative records.

The study of immigration and immigrants speaks to one of the great historical forces in individual lives. The orienting concepts and research approaches from the study of aging and the life-course complement the substantive demands of this field of study to inform our understanding of international migration. Whether the question is cohort differences in immigrant experience, the pathways to incorporation over a lifetime, or the implications of immigration for loved ones, the study of immigration and the study of the life-course dovetail and can only enrich one another.

References

Acevedo-Garcia, D., & Bates, L. (2008). Latino health paradoxes: Empirical evidence, explanations, future research, and implications. In H. Rodriguez, R. Saenz, & C. Menjivar (Eds.), *Latinas/os in the United States: Changing the face of America* (pp. 101–113). New York, NY: Springer Science + Business Media.

Aguila, E., Escarce, J., Leng, M., & Morales, L. (2013). Health status and behavioral risk factors in older adult Mexicans and Mexican immigrants to the United States. *Journal of Aging and Health, 2*, 136–158.

Alba, R., & Nee, V. (1999). Rethinking immigration theory for a new era of immigration. In C. Hirschman, P. Kasinitz, & J. D. Wind (Eds.), *The handbook of international migration: The American experience* (pp. 137–160). New York, NY: Russell Sage.

Allport, G. W. (1954). *The nature of prejudice*. Reading, NA: Addison-Wesley.

Andersson, G. (2004). Childbearing after migration: Fertility patterns of foreign-born women in Sweden. *International Migration Review, 38*, 747–775.

Angel, J. L., Angel, R. J., & Markides, K. S. (2000). Late-life immigration: Changes in living arrangements, and headship status among older mexican-origin individuals. *Social Science Quarterly, 81*, 389–403.

Angel, J. L., Buckley, C. J., & Sakamoto, A. (2001). Duration or disadvantage? Exploring nativity, ethnicity, and health in midlife. *The Journals of Gerontology Series B: Psychological Sciences and Social Sciences, 56*, S275–S284.

Angel, R. J., Angel, J. L., Diaz Venegas, C., & Bonazzo, C. (2010). Shorter stay, longer life: Age at migration and mortality among the older Mexican-origin population. *Journal of Aging and Health, 22*, 914–931.

Angel, R. J., Angel, J. L., Lee, G.-Y., & Markides, K. S. (1999). Age at migration and family dependency among older Mexican immigrants: Recent evidence from the Mexican American EPESE. *The Gerontologist, 39*, 59–65.

Baldassar, L., Kilkey, M., Merla, L., & Wilding, R. (2014). Transnational families. In J. Treas, J. Scott, & M. Richards (Eds.), *The wiley blackwell companion to the sociology of families* (pp. 155–175). Oxford: Wiley-Blackwell.

Batalova, J. (2006). *Skilled Immigrant and native workers in the United States: The economic competition debate and beyond.* New York, NY: LFB Scholarly Publishing.

Batalova, J. (2012). *Senior immigrants in the United States*. Migration Policy Institute. <http://www.migrationinformation.org/USfocus/display.cfm?ID=894>.

Batalova, J., Hooker, S., Capps, R., Bachmeier, J. D., & Cox, E. (2013). *Deferred action for childhood arrivals at the one-year mark: A profile of currently eligible youth and applicants. Issue brief.* Washington, DC: Migration Policy Institute.

Bergeron, C. (2013). *Going to the back of the line: A primer on lines, visa categories and wait times. Policy brief No. 1.* Washington, DC: Migration Policy Institute. <http://www.migrationpolicy.org/pubs/CIRbrief-BackofLine.pdf>.

Bloemraad, I. (2006). *Becoming a citizen: Incorporating immigrants and refugees in the United State and Canada.* Berkeley, CA: University of California Press.

Borjas, G. J. (2002). Welfare reform and immigrant participation in welfare programs. *International Migration Review, 36*, 1093–1123.

Borjas, G. J. (2011). Social security eligibility and the labor supply of older immigrants. *Industrial and Labor Relations Review, 64*, 485–501.

Borrell, L. N., & Lancet, E. A. (2012). Race/ethnicity and all-cause mortality in US adults: Revisiting the Hispanic paradox. *American Journal of Public Health, 102*, 836–843.

Burr, J., Gerst, K., Kwan, N., & Mutchler, J. (2008–2009). Economic well-being and welfare program participation among older immigrants in the United States. *Generations, 32*, 53–60.

Burr, J., Mutchler, J., & Gerst, K. (2010). Patterns of residential crowding among Hispanics in later life: Immigration, assimilation, and housing market factors. *The Journals* of *Gerontology Series B: Psychological Sciences and Social Sciences, 65B*, 772–782.

Burr, J., Mutchler, J., & Gerst, K. (2011). Homeownership among Mexican Americans in later life. *Research on Aging, 33*, 379–402.

Cantu, P. A., Hayward, M. D., Hummer, R. A., & Chiu, C.-T. (2013). New estimates of racial/ethnic differences in life expectancy with chronic morbidity and functional loss: Evidence from the National Health Interview Survey. *Journal of Cross-Cultural Gerontology, 28*, 283–297.

Carr, S., & Tienda, M. (2013). Family sponsorship and late-age immigration in aging America: Revised and expanded estimates of chained migration. *Population Research and Policy Review, 32*, 825–849.

Carrasquillo, O., Carrasquillo, A. I., & Shea, S. (2000). Health insurance coverage of immigrants living in the United States: Differences by citizenship status and country of origin. *American Journal of Public Health, 90*, 917.

Chavez, L. R. (1994). The power of the imagined community: The settlement of undocumented Mexicans and Central Americans in the United States. *American Anthropologist, 96*, 52–73.

Chiswick, B. R., & Miller, P. W. (2005). Do enclaves matter in immigrant adjustment? *City & Community, 4*, 5–35.

Choi, K. H., & Mare, R. D. (2012). International migration and educational assortative mating in Mexico and the United States. *Demography, 49*, 449–476.

Choi, S. H. (2011). Testing healthy immigrant effects among late life immigrants in the United States: Using multiple indicators. *Journal of Aging and Health, 24*, 475–506.

Crimmins, E. M., Kim, J. K., Alley, D. E., Karlamangla, A., & Seeman, T. (2007). Hispanic paradox in biological risk profiles. *American Journal of Public Health, 97*, 1305–1310.

Cuellar, I., Bastida, E., & Braccio, S. M. (2004). Residency in the United States, subjective well-being, and depression in an older Mexican-origin sample. *Journal of Aging and Health, 16*, 447–466.

Cunningham, A. S., Ruben, J. D., & Venkat Narayan, K. M. (2008). Health of foreign-born people in the United States: A review. *Health & Place, 14*, 623–635.

Dannefer, D. (1987). Aging as intracohort differentiation: Accentuation, the Matthew effect and the life course. *Sociological Forum, 2*, 211–236.

Diewald, M., Goedicke, A., & Mayer, K. U. (2006). *After the fall of the wall: Life courses in the transformation of East Germany*. Stanford, CA: Stanford University Press.

Donato, K. M., & Sisk, B. (2012). Shifts in the employment outcomes among Mexican migrants to the United States, 1976–2009. *Research in Social Stratification and Mobility, 30*, 63–77.

Dreby, J. (2010). *Divided by borders: Mexican migrants and their children*. Berkeley, CA: University of California Press.

Eckstein, S. E., & Najam, A. (Eds.). (2013). *How immigrants impact their homelands*. Durham, NC: Duke University Press.

Elder, G. H., Jr. (1974). *Children of the great depression: Social change in life experience*. Chicago, IL: University of Chicago Press.

Elder, G. H., Johnson, M. K., & Crosnoe, R. (2003). The emergence and development of life course theory. In J. T. Mortimer & M. J. Shanahan (Eds.), *Handbook of the life course* (pp. 3–19). Dordrecht: Springer.

Elo, I. T., Turra, C. M., Kestenbaum, B., & Ferguson, B. R. (2004). Mortality among elderly Hispanics in the United States: Past evidence and new results. *Demography, 41*, 109–128.

Gerst, K., & Burr, J. (2011). Welfare use among older Hispanic immigrants: The effect of state and federal policy. *Population Research and Policy Review, 30*, 129–150.

Gleeson, S., & Gonzales, R. G. (2012). When do papers matter? An institutional analysis of undocumented life in the United States. *International Migration, 50*, 1–19.

Glick, J. E. (2010). Connecting complex processes: A decade of research on immigrant families. *Journal of Marriage & Family, 72*, 498–515.

Glick, J. E., & Van Hook, J. (2002). Parents' coresidence with adult children: Can immigration explain racial and ethnic variation? *Journal of Marriage & Family, 64*, 240–253.

Gonzales, R. G. (2011). Learning to be illegal: Undocumented youth and shifting legal contexts in the transition to adulthood. *American Sociological Review, 76*, 602–619.

González, H. M., Haan, M. N., & Ladson, H. (2001). Acculturation and the prevalence of depression in older Mexican Americans: Baseline results of the Sacramento area Latino study on aging. *Journal of the American Geriatrics Society, 49*, 948–953.

Gubernskaya, Z. (2015). Age at migration and self-rated health trajectories after age 50: Understanding the older immigrant health paradox. *The Journals of Gerontology Series B: Psychological Sciences and Social Sciences, 70*, 279–290.

Gubernskaya, Z., Bean, F. D., & Van Hook, J. (2013). (Un) Healthy immigrant citizens naturalization and activity limitations in older age. *Journal of Health and Social Behavior, 54*, 427–443.

Hao, L., & Kawano, Y. (2001). Immigrants' welfare use and opportunity for contact with co-ethnics. *Demography, 38*, 375–389.

Hayward, M. D., Hummer, R. A., Chiu, C.-T., González-González, C., & Wong, R. (2014). Does the Hispanic paradox in US adult mortality extend to disability? *Population Research and Policy Review, 33*, 81–96.

Hayward, M. D., Warner, D. F., & Crimmins, E. M. (2007). Does longer life mean better health? Not for native-born Mexican Americans in the Health and Retirement Survey. In J. L. Angel & K. E. Whitfield (Eds.), *The health of aging Hispanics* (pp. 85–95). New York, NY: Springer.

Jasso, G. (2004). Migration, human development, and the life course. In J. T. Mortimer & M. J. Shanahan (Eds.),

Handbook of the life course (pp. 331–364). New York, NY: Springer.

Jasso, G., Massey, D. S., Rosenzweig, M. R., & Smith, J. P. (2004). Immigrant health: Selectivity and acculturation. In N. B. Anderson, R. A. Bulatao, & B. Cohen (Eds.), *Critical perspectives on racial and ethnic differences in health in late life* (pp. 227–266). Washington, DC: The National Academic Press.

Kandel, W., & Massey, D. S. (2002). The culture of Mexican migration: A theoretical and empirical analysis. *Social Forces, 80*, 981–1004.

Kasinitz, P., Mollenkopf, J., Waters, M. C., & Holdaway, J. (2008). *Inheriting the city: The children of immigrants come of age*. New York, NY: Russell Sage.

Kim, J., & Lauderdale, D. S. (2002). The role of community context in immigrant elderly living arrangements: Korean American elderly. *Research on Aging, 24*, 630–653.

Kim, N. Y. (2009). Finding our way home: Korean Americans, "homeland" trips, and cultural foreignness. In T. Tsuda (Ed.), *Diasporic homecomings: Ethnic return migration in comparative perspective* (pp. 305–324). Stanford, CA: Stanford University Press.

Ladin, K., & Reinhold, S. (2013). Mental health of aging immigrants and native-born men across 11 European countries. *The Journals of Gerontology Series B: Psychological Sciences and Social Sciences, 68*, 298–309.

Landale, N. S. (1994). Migration and the Latino family: The union formation behavior of Puerto Rican women. *Demography, 31*, 133–157.

Lauderdale, D. S., & Kestenbaum, B. (2002). Mortality rates of elderly Asian American populations based on medicare and social security data. *Demography, 39*, 529–540.

Leach, M. (2008). America's older immigrants: A profile. *Generations, 32*, 34–39.

Lee, G.-Y., & Angel, R. J. (2002). Living arrangements and Supplemental Security Income use among elderly Asians and Hispanics in the United States: The role of nativity and citizenship. *Journal of Ethnic and Migration Studies, 28*, 553–563.

Lindstrom, D., & Giorguili Saucedo, S. (2011). The interrelationship between fertility, family maintenance and Mexico-US migration. *Demographic Research, 17*, 821–858.

Logan, J. R., Zhang, W., & Alba, R. D. (2002). Immigrant enclaves and ethnic communities in New York and Los Angeles. *American Sociological Review, 67*, 299–322.

Lum, T. Y., & Vanderaa, J. P. (2010). Health disparities among immigrant and non-immigrant elders: The association of acculturation and education. *Journal of Immigrant and Minority Health, 12*, 743–753.

Mannheim, K. (1997(1952)). The problem of generations. In M. A. Hardy (Ed.), *Studying aging and social change* (pp. 22–65). London: Sage.

Markides, K. S., & Eschbach, K. (2005). Aging, migration, and mortality: Current status of research on the Hispanic paradox. *Journal of Gerontology B: Social Sciences, 60*(Special Issue 2), 68–75.

Markides, K. S., Eschbach, K., Ray, L., & Peek, M. (2007). Census Disability rates among older people by race/ethnicity and type of Hispanic origin. In J. L. Angel & K. E. Whitfield (Eds.), *The health of aging hispanics* (pp. 26–39). New York, NY: Springer.

Massey, D. S. (1990). Social structure, household strategies, and the cumulative causation of migration. *Population Index, 56*, 3–26.

Massey, D. S., & Alarcon, R. (1987). *Return to Aztlan: The social process of international migration from Western Mexico*. Berkeley, CA: University of California Press.

Massey, D. S., & Gelatt, J. (2010). What happened to the wages of Mexican immigrants? Trends and interpretations. *Latino Studies, 8*, 328–354.

Massey, D. S., & Pren, K. A. (2012). Origins of the new Latino underclass. *Race and Social Problems, 4*(1), 5–17.

Mayer, K. U., & Schoepflin, U. (1989). The state and the life course. *Annual Review of Sociology, 15*, 187–209.

Mehta, C., Theodore, N., Mora, I., & Wade, J. (2002). *Chicago's undocumented immigrants: an analysis of wages, working conditions, and economic contributions*. Chicago, IL: University of Illinois, Center for Urban Economic Development.

Mendes de Leon, C. F., Eschbach, K., & Markides, K. S. (2011). Population trends and late-life disability in Hispanics from the Midwest. *Journal of Aging and Health, 23*, 1166–1188.

Milewski, N. (2007). First child of immigrant workers and their descendants in West Germany: Interrelation of events, disruption, or adaptation? *Demographic Research, 17*, 859–896.

Mussino, E., & Strozza, S. (2012). The fertility of immigrants after arrival: The Italian case. *Demographic Research, 26*, 99–130.

Mutchler, J. E., Prakash, A., & Burr, J. A. (2007). The demography of disability and the effects of immigrant history: Older Asians in the United States. *Demography, 44*, 251–263.

Myers, D., & Lee, G.-Y. (1996). Immigration cohorts and residential overcrowding in Southern California. *Demography, 33*, 51–65.

Nam, Y. (2008). Welfare reform and older immigrants' health insurance coverage. *The American Journal of Public Health, 98*, 2029–2034.

Nam, Y., & Jung, H. J. (2008). Welfare reform and older immigrants: Food stamp program participation and food insecurity. *The Gerontologist, 48*, 42–50.

Nam, Y., & Kim, W. (2012). Welfare reform and elderly immigrants' naturalization: Access to public benefits as an incentive for naturalization in the United States. *International Migration Review, 46*, 656–679.

O'Neil, K., & Tienda, M. (2015). Age at immigration and the incomes of older immigrants, 1994–2010. *The Journals* of *Gerontology Series B: Psychological Sciences and Social Sciences, 70,* 291–302.

Orrenius, P. M., & Zavodny, M. (2009). Do immigrants work in riskier jobs? *Demography, 46,* 535–551.

Palloni, A., & Arias, E. (2004). Paradox lost: Explaining the Hispanic adult mortality advantage. *Demography, 41,* 385–415.

Palloni, A., & Ewbank, D. C. (2004). Selection processes in the study of racial and ethnic differentials in adult health and mortality. In N. B. Anderson, R. A. Bulatao, & B. Cohen (Eds.), *Critical perspectives on racial and ethnic differences in health in late life* (pp. 171–226). Washington, DC: The National Academic Press.

Parrado, E. A. (2011). How high is Hispanic/Mexican fertility in the United States? Immigration and tempo considerations. *Demography, 48,* 1059–1080.

Parreñas, R. S. (2001). *Servants of globalization: Women, migration and domestic work.* Stanford, CA: Stanford University Press.

Population Reference Bureau. (2013). *Elderly immigrants in the US* Today's Research on Aging, issue 29, Retrieved from: <http://www.prb.org/Publications/Reports/2013/us-elderly-immigrants.aspx>.

Portes, A., & Rumbaut, R. G. (2001). *Legacies: The story of the immigrant second generation.* Berkeley, CA: University of California Press.

Portes, A., & Zhou, M. (1993). The new second generation: Segmented assimilation and its variants among post-1965 immigrant youth. *Annals of the American Academy of Political and Social Science, 530,* 74–96.

Qian, Z., Glick, J. E., & Batson, C. D. (2012). Crossing boundaries: Nativity, ethnicity, and mate selection. *Demography, 49,* 651–675.

Redstone, I., & Massey, D. S. (2004). Coming to stay: An analysis of the US Census question on year of arrival. *Demography, 41,* 721–738.

Reyes, A. M., & Hardy, M. (2013). Another health insurance gap: Gaining and losing coverage among natives and immigrants at older ages. *Social Science Research, 43,* 145–156.

Rhee, M.-K., Chi, I., & Yi, J. (2013). Understanding employment barriers among older Korean immigrants. *The Gerontologist.*

Riosmena, F., & Dennis, J. A. (2012). A tale of three paradoxes: The weak socioeconomic gradients in health among Hispanic immigrants and their relation to the Hispanic health paradox and negative acculturation. In J. L. Angel, F. Torres-Gil, & K. Markides (Eds.), *Aging, health, and longevity in the Mexican-origin population* (pp. 95–110). New York, NY: Springer Science + Business Media.

Rumbaut, R. (2004). Ages, Life stages, and generational cohorts: Decomposing the immigrant first and second generations in the United States. *International Migration Review, 38,* 1160–1205.

Rumbaut, R. (2008). Reaping what you sew: Immigration, youth, and reactive ethnicity. *Applied Developmental Science, 22,* 108–111.

Rumbaut, R., & Portes, A. (2001). *Ethnicities: Children of Immigrants in America.* New York, NY: Russell Sage.

Ryder, N. B. (1965). The cohort as a concept in the study of social change. *American Sociological Review, 30,* 843–861.

Song, C. (2009). Brothers only in name: The alienation and identity transformation of Korean Chinese return migrants in South Korea. In T. Tsuda (Ed.), *Diasporic homecomings: Ethnic return migration in comparative perspective* (pp. 281–304). Stanford, CA: Stanford University Press.

Swallen, K. C. (1997). Do health selection effects last? A comparison of morbidity rates for elderly adult immigrants and US-born elderly persons. *Journal of Cross-Cultural Gerontology, 12,* 317–339.

Thomas, W., & Znaniecki, F. (1918). *The polish peasant in Europe and America.* Boston, MA: Richard G. Badger.

Torres-Gil, F., & Treas, J. (Eds.). (2008–2009). *Generations: Journal of the American Society on Aging.* 32.

Treas, J. (2008). Transnational older adults and their families. *Family Relations, 57,* 468–478.

Treas, J. (2015). Incorporating immigrants: Integrating theoretical frameworks of adaptation. *The Journals of Gerontology Series B: Psychological Sciences and Social Sciences, 70,* 269–278.

Treas, J., & Batalova, J. (2009). Immigrants and aging. In P. Uhlenberg (Ed.), *International handbook of population aging* (pp. 365–394). New York, NY: Springer Verlag.

Treas, J., & Batalova, J. (2011). Residential independence: Race and ethnicity on the road to adulthood in two US immigrant gateways. *Advances in Life Course Research, 16,* 13–24.

Treas, J., & Mazumdar, S. (2002). Older people in America's immigrant families: Dilemmas of dependence, integration, and isolation. *Journal of Aging Studies, 16,* 243–258.

Treas, J., & Mazumdar, S. (2004). Caregiving and kinkeeping: Contributions of older people to America's immigrant families. *Journal of Comparative Family Studies, 35,* 105–122.

Tsuda, T. (Ed.). (2009). *Diasporic homecomings: Ethnic return migration in comparative perspective.* Stanford, CA: Stanford University Press.

Turra, C., & Elo, I. (2008). The impact of salmon bias on the Hispanic mortality advantage: New evidence from Social Security data. *Population Research and Policy Review, 27,* 515–530.

Van Hook, J. (2000). SSI eligibility and participation among elderly naturalized citizens and noncitizens. *Social Science Research, 29,* 51–69.

Van Hook, J. (2003). Welfare reform's chilling effects on noncitizens: Changes in noncitizen welfare recipiency or shifts in citizenship status? *Social Science Quarterly, 84,* 613–631.

Wakabayashi, C. (2010). Effects of immigration and age on health of older people in the United States. *Journal of Applied Gerontology, 29,* 697–719.

Warnes, T. A. (2009). International retirement migration. In P. Uhlenberg (Ed.), *International handbook of population aging* (pp. 341–363). New York, NY: Springer Verlag.

Waters, M. C., Tran, V. C., Kasinitz, P., & Mollenkopf, J. H. (2010). Segmented assimilation revisited: Types of acculturation and socioeconomic mobility in young adulthood. *Ethnic and Racial Studies, 33,* 1168–1193.

White, K., & Potter, J. E. (2013). The impact of outmigration of men on fertility and marriage in the migrant-sending states of Mexico, 1995–2000. *Population Studies, 67,* 83–95.

Winters, P., De Janvry, A., & Sadoulet, E. (2001). Family and community networks in Mexico-US migration. *The Journal of Human Resources, 36,* 159–184.

Xie, Y., & Gough, M. (2011). Ethnic enclaves and the earnings of immigrants. *Demography, 48,* 1293–1315.

Yoo, G. J. (2001). Constructing deservingness: Federal welfare reform, Supplemental Security Income, and elderly immigrants. *Journal of Aging & Social Policy, 13,* 17–34.

CHAPTER

8

Gender, Time Use, and Aging

Liana C. Sayer[1], Vicki A. Freedman[2], and Suzanne M. Bianchi[3]

[1]Sociology Department and Maryland Population Research Center, University of Maryland, College Park, MD, USA [2]Institute for Social Research, University of Michigan, Ann Arbor, MI, USA [3]Department of Sociology and California Center for Population Research, University of California-Los Angeles, Los Angeles, CA, USA

OUTLINE

Introduction	163
Measuring Time Allocation in Later Life	165
"A Day in the Life" of Older Adults	167
Age and Gendered Time Use	168
Employment and Family Role Influences on Time Use	169
Work and Family Roles, Gender, and Leisure Activities	171
The Social versus Solitary Dimension of Time	173
Caregiving, Time Use, and Well-Being	174
Measuring Caregiving Time	174
Future Directions	176
Acknowledgments	178
References	178

INTRODUCTION

Do gendered time use patterns among older adults mirror those observed in earlier life stages? Or does time use allocation in later life become less gendered as life stages are reoriented away from paid work and raising children toward new pursuits? Later life disability and increased likelihood of living alone in young and older adulthood may alter the activities in which individuals engage, the amount of time spent on various activities, and time socializing and interacting with others. "Productive" uses of time may be redefined with implications for later life well-being. Both objective and subjective aspects of time likely change at later life stages.

Time is fundamentally a gendered resource. How women and men allocate their time expresses the extent to which different activities are valued and how time constraints are linked with power relations, cultural beliefs, and individual behavior. Examining gender differences in time use among older adults provides vital insight into continuity and change in gender equality, and implications of gendered work and family roles for well-being. Women's and men's differential allocation of time across the life course is interwoven with well-being because of the ways time use affects economic, social, and health capital and resources.

Women's greater responsibility for care work and household work adversely affects labor market outcomes by reducing labor force participation and rewards. Most salient for older women is the negative effect of household and care work on lifetime earnings, and retirement income (savings, pension, social security) (O'Rand & Shuey, 2007). Consequently, economic insecurity in old age is particularly acute among single, divorced, and widowed older women (Venn, Davidson, & Arber, 2011). Gendered work and family roles also reduce women's access to total leisure as well as leisure with health or social implications (e.g., time exercising, playing sports, or socializing) (Bianchi, Robinson, & Milkie, 2006; Sayer & Gornick, 2009).

Social and behavioral theories posit that decisions about how time is allocated reflect individual habits, preferences, social roles, and time-budget constraints. Sociological and psychological models of time use and leisure emphasize that individuals spend time in ways that affirm values and achieve goals, while also recognizing that these choices are restricted by socio-structural constraints (Bourdieu, 1984; Gershuny, 2000). In this perspective, gender differences emerge because time is activated in ways that reinforce gendered identities. Women, for example, will devote more time to housework, care work, and family-oriented leisure, whereas men will devote more time to paid work, household maintenance, and self-oriented leisure in part because these activities express essential aspects of gender.

Irrespective of theoretical perspective, gender differences in time use at older ages exist today against a backdrop of temporal shifts in economic, demographic, and cultural factors. Over the last four decades of the twentieth century, earlier retirement, increased education, health, and life expectancy have altered the life course ratio of paid work to leisure. The surge of women into the labor force, combined with lower fertility rates, delayed entry into marriage, and increases in non-marital childbearing, cohabitation, and divorce rates, altered the life course ratio of paid work to caregiving, as well as community and civic involvement. Specifically, the delayed entry until the late twenties and early thirties into marriage and childbearing among college-educated young adults combined with an average life expectancy of 79 for women and 76 for men mean adults have on average about 25 years outside peak work/family time demands. Despite these substantial changes, paid work and parenting continue to differentiate women's and men's life course choices and trajectories. Changes in family structure have increased the number of women raising children alone, and decreased the amount of time men spend living with residential children (Goldscheider & Hogan, 2001). Changes in employment opportunities have ushered in a reversal of early retirement ages, as the proportion of workers 55 and older has been increasing since 2002 (Toosi, 2013). Additionally, the recession has caused some older employed adults to postpone retirement and others to return to paid work (US Bureau of Labor Statistics, 2013a). Women retire at older ages than men and may be more likely to postpone retirement and/or return to employment because of their lower wages and probability of having sufficient retirement nest eggs. Thus, women's life pathways show greater complexity, variability, and "disorderliness" compared with those of men, and these differences have profound influences on gender and time use among older adults (Hagestad, 1990).

This chapter provides an overview of gender differences in time use at older ages. We begin with a brief introduction to time use measurement. Next, we compare gender differences in daily time allocated to paid work, care work, and leisure between adults 55 and older and those 25–54, emphasizing how gender differences change with age. We then examine social aspects of time use in later life, including how older men and women differ in their time spent alone. Last, we examine caregiving and how its influence on time use and well-being varies by gender. When data are available, we present more detailed comparison of how time use differs among adults ages 55–64, 65–74, and 75 and older. We conclude with a discussion of needed directions in research on gender, aging, and time use.

MEASURING TIME ALLOCATION IN LATER LIFE

Three main approaches for measuring time are currently in use (National Research Council, 2000): experiential sampling methods (Csikszentmihalyi & Larson, 1992), retrospective 24-hour time diary methods (Juster & Stafford, 1991), and stylized reporting (Pencavel, 1986). Particularly germane for studying older adults, each approach has a fundamentally distinct reference period, which in turn has implications for cognitive demands and hence measurement error. The methods also differ in how much detail is obtained about the activities (how fine-grained the activity codes ultimately are and how much of the day is captured), whether descriptors (e.g., location, other actors, emotions) can be collected, how well they capture uncommon activities, and whether multitasking – believed to be common for instance among caregivers – can be readily captured. Sources of concerns about measurement bias also vary by approach.

The approach with the shortest reference period, experiential sampling, involves contacting study participants at random times of day. Typically, participants are asked questions about what they were doing in a brief window just before the contact. Contact may be made by any programmable device (beeper, programmable wristwatch, smartphone, or other type of technology), which then triggers a series of questions. Respondents may be asked not only what they have been doing, but other descriptors such as who they were with, where they were, and how they felt. Studies vary with respect to number of contacts per day and number of collection days.

Of the three approaches, questions following experiential sampling pose the lowest cognitive demands since the reference window is very recent and often relatively short. Experience sampling also offers the benefit of being able to collect open-ended text that allows fine-grained coding of activities. Descriptors are easily captured and typically focus on affective states, although the method has been successfully combined with physiological measures as well (Stone, Shiffman, & DeVries, 1999). In addition, multiple activities (multitasking) are well captured for a given time period (Offer & Schneider, 2011). However, the approach does not provide a picture of the full day, is more likely to capture activities with longer durations, and, although it does not exclude them, is not designed to capture uncommon activities. Non-random non-response, which may be linked to the participant's activity level, is also a potential concern.

The retrospective 24-hour diary asks individuals to recall all activities on the prior day. The American Time Use Study, http://www.bls.gov/tus/home.htm, conducted by the US Bureau of Labor Statistics (ATUS 2010, see Phipps & Vernon, 2009), and the Disability and Use of Time (DUST) supplement to the Panel Study of Income Dynamics (PSID; http://www.psidonline.isr.umich.edu/), conducted by the University of Michigan (Freedman, Stafford, Schwarz, & Conrad, 2013), for example, ask respondents what they were doing starting at 4 a.m. the previous day, for how long they did that, where they were and who else was there. They then ask the respondent what they did next, and so on, until a 24-hour diary is completed. Supplemental questions can be used to target specific types of activities on the prior day, such as care for an older adult (US Bureau of Labor Statistics, 2013b). The 24-hour diary also forms the basis of the Day Reconstruction Method (DRM), which is designed to measure well-being as it is experienced through the day (Kahneman, Krueger, Schkade, Schwarz, & Stone, 2004). The DRM methodology also can be used to measure physical symptoms such as pain (Krueger & Stone, 2008). ATUS and PSID's DUST both have incorporated well-being measures into their diary collections.

The diary interview takes longer to administer than other methods (e.g., 15–25 min total, depending on the number of descriptors), but offers a complete picture of the day and allows descriptors to be collected for each episode. Typically open text (or some combination of pre-codes and open text) is captured and systematically coded so that researchers have maximum flexibility in aggregating various activities. In addition, multitasking may be addressed. Large samples, like those in the ATUS, allow investigation of less common activities. Because time use varies substantially from day to day, diaries are useful for studying average differences between groups and aggregate population trends, but not for studying within-person trajectories over time unless multiple diaries per person are collected at each time period.

Although the retrospective diary is more cognitively demanding than experiential sampling, the few studies that have explored its validity among older adults suggest that older adults are able to complete the diary with an acceptable level of accuracy. In the ATUS, for instance, the incidence of activity codes with "don't know" or "can't remember" responses tends to increase with respondent age (US Bureau of Labor Statistics, 2013a), but remains low on average even at very old ages (less than 1% among 65 and older; authors' tabulations). Moreover, in a sample of 83 older volunteers, Klumb and Baltes (1999) found considerable agreement between yesterday interviews and experience sampling measures. Where there were differences, cognitive functioning did not appear to account for discrepancies. Similarly, Freedman, Stafford, Schwarz, and Conrad (2013) found that in a national sample of same-day diaries collected in separate interviews from older husbands and wives (mean age nearly 70), 96% of all diary sequences successfully elicited a codeable answer. In a related analysis (Freedman, Stafford, Conrad, Schwartz, & Cornman, 2012), the authors found that 76% of activities described as joint by at least one respondent had a matching record in the spouse's diary.

Perhaps the most common approach to measuring time use, stylized questions ask respondents to estimate how much time was spent on various activities over a longer period of time. Participants may be asked to report typical amounts of time or actual amounts in a given reference period, typically a week, a month, or sometimes longer if the activity is rare. For example, the Health and Retirement Study's Consumption and Activities Mail Survey asks, "How many hours did you actually spend LAST WEEK …" and "Now think about the LAST MONTH. How many hours did you spend last month….".

Such stylized questions may pose greater cognitive demands than the experience sampling and time diary methods because respondents need to review a longer time period and may need to compute answers, for example, by multiplying, adding, averaging, or otherwise estimating to provide an answer (Kan & Pudney, 2008). Moreover, for some respondents and activities, responses may be prone to systematic bias – for example, normative behaviors such as attending church may be overestimated by those who identify the activity as important (Brenner, 2012). Because open text is not captured, this method relies on a broadly shared understanding of the meaning of activity categories. Descriptors are not easily incorporated into this approach, nor is multitasking generally explicit, since time is often being summed across distinct activity episodes. Nor is it straightforward to make comparisons over time or across samples, as time does not sum to 24 h or to another standard unit. However, this approach is better able than the other methods to capture activities that are carried out less frequently. Using a recent reference period (e.g., a week) and focusing on commonly occurring activities may help mitigate errors. Stylized questions also may be especially well suited for studying time spent in particular activities (such as work, see Juster, Ono, & Stafford, 2003), and, because they are less variable than 24-h measures, for investigating changes within individuals over time.

Finally, briefer hybrid approaches that bring elements of the stylized approach to the last 24 h have been used in population surveys (Juster et al., 2003). For instance, in the National Study of Daily Experiences, a companion study to the Midlife Development in the United States study (see http://www.midus.wisc.edu/midus2/project2/) participants were asked at the end of eight consecutive days how much time they spent on various activities during the past day (Hahn, Cichy, Almeida, & Haley, 2011) as well as stressful experiences on the prior day. Similarly, a short day reconstruction measure was incorporated into the Health and Retirement Study's 2012 Psychosocial Self-Administered Questionnaire and the English Longitudinal Study on Aging's Wave 6 (2012/2013) Self Completion Questionnaire. After establishing wake and sleep times, participants are asked about time spent in categories of common activities on the previous day, with follow-up questions about emotions, stress, and pain experienced during those activities. This approach is less cognitively demanding than the stylized approach with its longer reference period and descriptors could be incorporated by asking respondents to report on each instance yesterday if more than one occurred. However, like the stylized method, time does not sum to 24 h and benchmarking exercises with the full 24-h diary approach suggest systematic upward bias in limited cases (e.g., physical activity; see Smith, Ryan, Queen, Becker, & Gonzalez, 2014). Nevertheless, on balance, the approach holds promise as a brief measure of activity-linked well-being on the prior day.

"A DAY IN THE LIFE" OF OLDER ADULTS

This descriptive section compares and contrasts the daily time allocation of women and men aged 55 and older. We consider how aging affects gendered time use and how associations are conditioned by employment and caregiving roles. Most studies of gendered time use and aging rely on cross-sectional data – either stylized or time diary estimates – limiting our ability to directly examine how role transitions affect older adults' time use patterns. Nevertheless, in general, the time use literature suggests that some roles – such as being employed (versus retired) or having a spouse who requires care – have a stronger association with time use than does age per se. Indeed, at all ages, individual decisions about time allocations to paid work, care work, and leisure are influenced by micro-level characteristics like employment, education, and family structure, along with public policies that influence choices about work (including policies about taxes, pensions, retirement, care provision, to name a few). Although employment/leisure time tradeoffs operate differently for older adults than they do for younger adults, both older and younger Americans devote the majority of daily hours to paid work, leisure, and self-care activities. Men of all ages report higher work and leisure time than women; and women of all ages report higher care and household work time than men. Gender differences persist among adults aged 55 and older, despite transitions out of employment and parental roles, although they attenuate at much older ages.

Age and Gendered Time Use

Gender differences in time use take root early in the life course. Time use earlier in adulthood depends on access to market-based and family-based resources and rewards, as well as access to leisure. These earlier gendered life stages have cumulative and contemporary effects on economic and social resources in later life. Occupational gender segregation and the gender wage gap reduce women's lifetime earnings and thus depress savings, pension, and social security income older women have available in retirement (Venn et al., 2011). Despite transitions out of time-intensive mothering, older women continue to have more responsibility for caregiving, even when they remain employed. Women's longer life expectancy and higher likelihood of marrying a man older than themselves means they are more likely than men to be the primary caregiver for a frail or ill spouse/partner (Barusch & Spaid, 1989; Meyer & Parker, 2011). Older men who are responsible for the care of an ill or frail spouse report higher levels of care work and household work than other men, but still less than comparable older women (Calasanti, 2001). Although women's larger social networks should facilitate socializing and interacting with others, their caregiving responsibilities instead are associated with lower-quantity and -quality leisure (Michelson & Tepperman, 2003).

In general, older Americans allocate less time to paid work and more time to household work, leisure, and sleep than younger adults (Robinson & Godbey, 1999; Gauthier & Smeeding, 2010; Krantz-Kent & Stewart, 2007). Comparisons of pooled 2008–2012 data from the American Time Use Survey (ATUS) show that adults aged 55–64 report 3.4h per day in paid work, 2.2h more than those aged 65–74, and 3.1h more than adults aged 75 and older (http://www.bls.gov/TUS/CHARTS/OLDER.HTM). Time "freed" from employment is devoted to leisure – 7.5h per day among adults 75 and older compared with 5.3 among those aged 55–64 – and sleep – 9.2h a day among adults 75 and older compared with 8.4h a day among those aged 55–64 – but not household activities, at about 2h a day for each age group.

Older women and men allocate different amounts of time to these activities. Figure 8.1 shows women's and men's daily hours in paid work, household work (care and housework), volunteer activities, leisure and sports (socializing, television, sports, exercise, reading, and relaxing), sleep, and other activities (religious, grooming, eating, travel, and other not elsewhere classified or uncoded) across four age groups: 55–59, 60–64, 65–69, and 70 and older (author calculations of 2003–2004 ATUS data from Krantz-Kent and Stewart, 2007, Table 1, pp. 12–13). On average, men 55 and older do about an hour more per day of paid work than women; women do about 1.3h more daily household work than men. Men's lower household work means they enjoy 0.7h a day more leisure than women. Older women (like younger women) also report lower levels of active leisure compared with men, but television accounts for about one-half of the additional leisure hours for men and women ages 55 and older (not shown). Gender does not differentiate volunteering (0.2h for both) or sleep time (8.5h for both).

How do gender differences in time use change as women and men grow older? Across age groups shown in Figure 8.1, gender differences in paid and household work diminish with age, but older men continue to do more paid work and less household work compared with same-age women. For example, men aged 55–59 report 1.3h more paid work and 1.2h less household work (including many activities that constitute care work) compared with women aged 55–59, whereas men aged 70 and older report only 0.4 hours more paid work and 1h less household work. For both women and men, age differences in volunteering are modest, perhaps because of the low levels of time volunteering captured by daily time diary data: 0.1h among women and men aged 55–59 and 0.2 among those 60 and older. Time sleeping also increases with age similarly for women and men: rising from 8.1h among those aged 55–59 to 9.0h per day among those aged 70 and older. However, because decreases in men's paid work time are not offset by comparable increases of household work time, across age groups men continue to enjoy a leisure bonus of about 0.6–0.7h per day. In sum, gender differences attenuate with age due to life-stage transitions out of full-time employment and parenting and, among some, into care for a partner.

Employment and Family Role Influences on Time Use

Figure 8.1 presents only a partial answer to the question of how women's and men's time use changes with age. Considering how employment and family roles affect women's and men's time use as they age is paramount to presenting a more accurate picture. Employment and family roles affect time use more than age, per se, because of their enduring and contemporary influences on economic resources, time availability, and opportunities for social engagement and productive activities. A comparison of indices of dissimilarity in time use across four age groups of older adults by employment status indicates that life course transitions out of employment change time use, whereas the comparison across age groups within employment status shows marked similarity in time use (Krantz-Kent & Stewart, 2007). Similar results are reported in other studies examining aging and time use (Gauthier & Smeeding, 2010; McNamara, 2008; Sayer & Gornick, 2009).

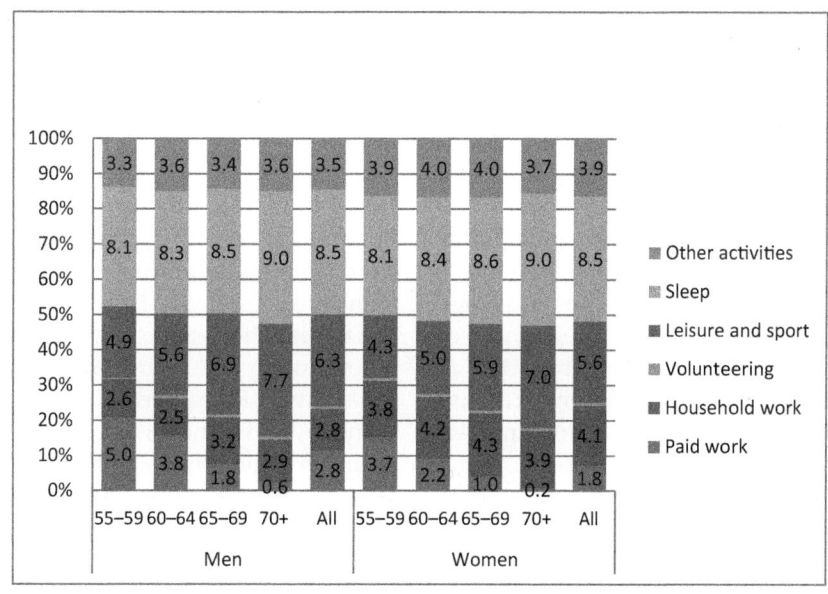

FIGURE 8.1 Older women's and men's daily time use across four age groups.

In general, older women and men who transition out of employment or reduce hours have more time available for household and care work. However, with increasing age, the demand for household work likely declines because of home downsizing, smaller household sizes, and at much older ages moves to residential care settings. Older adults may also experience disabilities that make physically or cognitively demanding chores (as well as other activities) more difficult. The demand for care work, however, may increase with age, because of the presence of a frail or ill spouse or partner. Offsetting supply and demand factors are suggested by the non-linear pattern of age and household work shown in Figure 8.1: household work time increases from 2.6h among men ages 55–59 to 3.2h per day among men aged 65–69, but then declines to 2.9h per day among men aged 70 and older. Comparable estimates are 3.8 for women aged 55–59, 4.3 for those aged 65–69, and 3.9 for those aged 70 and older. Detailed comparisons by employment status indicate household work time is relatively similar across age groups among women who are employed full- and part-time, and among men employed full-time. However, household work time declines with age for men employed part-time, and among men and women who are non-employed (Krantz-Kent & Stewart, 2007). This pattern points to health-related shifts that reduce women's and men's ability to do housework, more than supporting the idea that shifts of time out of employment are reallocated to household work. The amount of time in household work is also influenced by preferences or standards: women and men who report high participation and time in household work before retirement also spend more time in these activities post-retirement. In contrast, those who report limited pre-retirement household work do not increase their relatively modest allocations post-retirement (McNamara, 2008).

Gender differences in household and care work persist with age. For example, men's household and care work does not vary by the presence of spouse or partner, but among non-employed women ages 65 and older, living with a spouse increases household and care work by about 1 hour a day, with most of this time spent cooking and cleaning up (Krantz-Kent & Stewart, 2007). Further, spouses' time appears to complement, rather than substitute for, that of their spouses. Studies using couple-level time use data report that when husbands or male partners report high levels of household or care work, wives also report high levels, both among working-age couples (Bianchi et al., 2006; Craig & Mullan, 2011) and among couples 55 and older (Niemi, 2009).

Comparisons of older women and men in similar roles show diminished differences, especially in leisure and sleep time rather than in paid work and household work. For example, women who are employed full-time report slightly lower paid work hours (5.7 for women and 6.4 for men aged 55–59; 5.4 for women and 6.0 for men aged 65–69) and slightly higher household and care work hours (3.0 and 2.1 for women and men aged 55–59; 3.0 and 2.2 for women and men aged 65–69, see Table 1 in Krantz-Kent & Stewart, 2007). Similar to younger women, however, older women's combined work time (paid work and household and care work) is higher than comparable men, resulting in a gender leisure gap. For example, full-time employed women aged 55–59 report 3.6h per day of leisure compared with men's 4.2h per day; among those aged 70 and over, women report 4h of daily leisure, whereas men report 4.1h a day.

Despite the leisure gap, gender differences among older adults in paid and unpaid work time have diminished. One reason for movement toward gender time convergence among adults 55 and older is the life-stage progression of cohorts of women who were ages 25–44 in the 1980s. These women transitioned into adulthood during widespread expansion of women's educational and employment opportunities. Thus women aged 55 and older today experienced employment roles more similar to men's throughout their prime working years. As a result, their role transitions during later life are more similar to men's than they are to earlier cohorts of women.

The opportunity to segue out of paid work, and the experience of retirement, however, are distinctly different for women and men. Earlier studies adopted a "male stream" perspective and overlooked the implications of women's household work for their retirement and post-retirement status (Calasanti, 2001; Venn et al., 2011). Women's ability to retire and their post-retirement resources are depressed by gender employment inequalities and these affect mothers more than childless women. A national 2009 survey conducted by the Pew Research Center Social and Demographic Trends Projects (Kochar, 2009) reports that 54% of workers ages 65 and older remain employed because of social-psychological benefits (they feel more useful, have a sense of purpose and structure, and enjoy enhanced opportunities to interact with other adults), whereas younger workers are more likely to cite economic reasons for being employed. However, the breakdown by gender shows that 25% of older women report they need to work for the money, 43% want to work, and 32% indicate both need and want; whereas among older men, comparable estimates are 12% economic need, 63% want to work, and 22% both. Krantz-Kent and Stewart's (2007) comparison of time use patterns by employment status across four age groups (55–59, 60–64, 65–69, and 70 and older) indicates that men use part-time employment as a "bridge" between full-time work and retirement, whereas women's patterns are much more heterogeneous.

Basic patterns observed in the United States are also apparent in other industrialized countries, but there are differences. Decreases in paid work and household work time and increases in leisure and sleep time with age are observed across Western industrialized countries, but the level of time devoted to paid work and active leisure varies across countries. American women and men aged 55 and older report higher paid work hours than Western and Northern Europeans. The leisure activity mix also varies, with older Americans devoting a higher proportion of their leisure time to television and less to active leisure (Gauthier & Smeeding, 2010). Additionally, comparisons of 1970s and 1990s time diary and stylized survey data on participation in leisure activities indicates active leisure time declined among American women and men aged 55 and older whereas it increased among European older adults, particularly time devoted to social engagement with friends and public leisure, like frequenting pubs and movie theaters (Agahi & Parker, 2005; Gauthier & Smeeding, 2010; Gershuny, 2000).

Work and Family Roles, Gender, and Leisure Activities

Work and family roles also affect gendered experiences of leisure in old age. Employment and family responsibilities affect time available for leisure and differentiate types of leisure, as do the higher educational levels of today's older adults. Employment among older adults provides opportunities for social interaction and may increase funds available for leisure. In contrast, non-employed adults may experience greater constraints on activities because of poor health, more limited social networks, and income constraints (Sayer & Gornick, 2009). Gender disparities in caregiving time contribute to lower levels of leisure for older (and younger) women, and less time spent exercising and in social activities (Nomaguchi & Bianchi, 2004; Pinquart & Sorensen, 2007). Women's leisure is also more intertwined and fragmented with household work, and thus more "homebound," and less relaxing and refreshing (Venn et al., 2011).

Types of leisure have distinct implications for well-being because they affect the

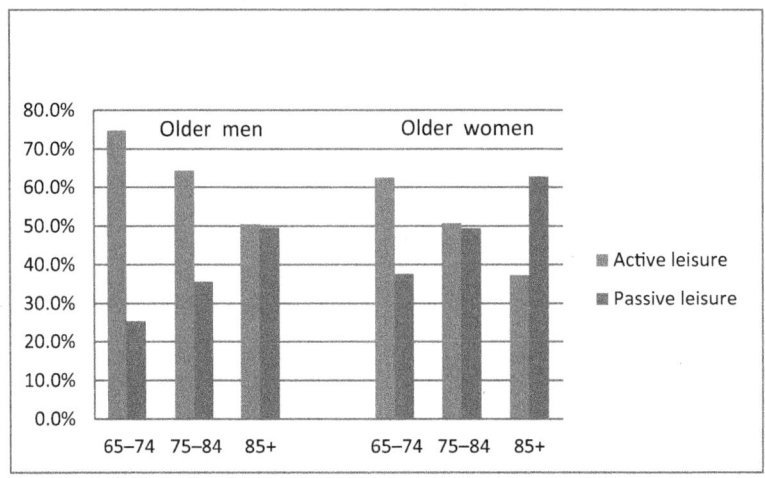

FIGURE 8.2 Older adults' favorite activity by age and sex: active versus passive leisure. *Source: Authors' tabulations of 2011 NHATS.*

accumulation of health, cultural, and social capital and other benefits related to participating in the social infrastructure of the world (Cutler & Hendricks, 1990). Civic, active, and social leisure develop and maintain social and physical capabilities and promote health, cultural, and social capital (Freysinger, 1995; Putnam, 2000). In contrast, sedentary leisure, such as watching television is less beneficial because it is more often socially isolated and, because of the zero-sum nature of time, crowds out time available for other activities. Recent data from the National Health and Aging Trends Study suggests gendered preferences in types of leisure activities in later life: when asked to name a favorite activity, older men are more likely than women to name an active leisure activity across all older age groups (see Figure 8.2).

Patterns of leisure engagement are established in adulthood and the small body of longitudinal research suggests gender differences persist over the life course (McNamara, 2008). Leisure in younger adulthood reinforces routines, capabilities, and interpersonal bonds with social network partners. McNamara (2008) reports that women and men who devote more time to leisure prior to retirement also devote more time to leisure post-retirement. Poor health reduces women's and men's time in active and social leisure (Agahi & Parker, 2005; Cutler & Hendricks, 1990; Jake, Davey, & Kleiber, 2005). Disability status in particular has pervasive negative influences on participation and time in discretionary activities like leisure, as well as time in paid work, care work, and volunteering. Research has not yet examined, however, how disability and time use patterns differ between women and men (Freedman, Stafford, Schwarz, Conrad, & Cornman, 2012; Verbrugge & Liu, 2014). Women experience higher morbidity in older adulthood, however, and this may be associated with gendered employment and caregiving patterns that reduce women's access to economic resources and increase obligatory time demands (Bird & Rieker, 2008). Older women may also be more reluctant than men to spend time alone in public social and/or active leisure.

THE SOCIAL VERSUS SOLITARY DIMENSION OF TIME

One of the issues for older adults is their connections to others, their ability to leave home to interact with others, and the frequency with which they live with others or are visited by others. The increased number of older adults living alone and class- and race-linked patterns of relationship formation and employment after age 55 suggest that inequalities in time with others may be one mechanism through which time use affects well-being.

Social versus solitary dimensions of time use are of growing interest to scholars studying gendered time use. Spending time with other adults enhances older adults' capabilities, life satisfaction, and general well-being. Women and men who have access to less time in shared activities face increased risks of negative health outcomes, social isolation, and reduced accumulation of social capital (Bird & Fremont, 1991; Bittman, 2002; Bourdieu, 1984; Miller & Brown, 2005).

The two factors most strongly associated with social time are employment and living with a spouse/partner. Both increase opportunities for contact with others and engagement in socially oriented or public leisure. Gendered aspects of social time are linked with earlier life-stage roles (Arber, Andersson, & Hoff, 2007). At older ages, women have stronger relationships with family, in particular adult children, in part because of mothers' earlier life-stage engagement in daily care of children. Older women are advantaged in terms of social relations vis-à-vis older men because they have more friends, better relationships not just with friends but also with family and neighbors, and larger social networks (Calasanti, 2001). Further, cultural beliefs that self-disclosure and expressions of emotion are not masculine limit men's friendship and social networks (Venn et al., 2011).

Although leisure time increases with age, time in social leisure is similar across age groups, clocking in at about 45 min a day. This means that as a proportion of all leisure time, social leisure declines with age (Gauthier & Smeeding, 2010; Krantz-Kent & Stewart, 2007). Net of employment and relationship status, however, few gendered age differences in social leisure are significant.

Comparing within employment groups, absolute and relative measures of time spent alone increase significantly for women but not for men with age. Men aged 55 and older spent about 50% of "available" time alone; in contrast, women of the same age group spent about 46.2% of available time alone (note that "available" time includes leisure and household activities that can be done with other people; time sleeping, grooming, and paid work is not considered because of the different function or context of these activities). Time alone increases as women age, by about 12 percentage points comparing women aged 55–59 to those aged 70 and older (Krantz-Kent & Stewart, 2007, Table 8, p. 23). The higher proportion of time spent alone among older women, relative to older men, is due to women's higher likelihood of outliving their spouses. Women and men who do not live with a spouse or partner spend 75% of available time alone, or about 10.3 h per day, twice the amount of alone time reported by older adults living with a partner. Time with friends and other family members is higher among older women and men not living with a spouse or partner, but does not fully compensate for the absence of a spouse or partner. Further, gender differences in time with family and time with friends are not significant, nor do they vary with age. However, older women spend more time with children than older men – among women aged 55–59, about 10.4% of available time, compared with 7.2% for men aged 55–59 (Krantz-Kent & Stewart, 2007).

CAREGIVING, TIME USE, AND WELL-BEING

Older adults are often the recipients of time in the form of caregiving. According to the 2011–2012 ATUS, nearly 40 million adults ages 15 and older report that they provide care to an adult age 65 or older who needs help because of a condition related to aging (US Bureau of Labor Statistics, 2013b). The ATUS defines care broadly to include hands-on care (such as assisting with grooming), assistance with household activities (e.g., preparing meals), transportation, companionship, and stand-by care. On a given day, one in four caregivers actually provides care for an average of 3.2h.

Most caregivers to older adults are middle-aged (45–64 years old) women; however, many care providers to older adults are of retirement age themselves (see Table 8.1). The intensity of care provided by older caregivers differs from that of their younger counterparts in several ways. Care providers ages 65 and over spent the most time providing care (4.1h per day versus 3.2h per day overall) and a higher percentage of caregivers in this age group helped the previous day (34.7% versus 23.0% overall).

Although the majority of older caregivers are women (3.8 out of 6.6 million), the percentage of older women and men who provide care is similar: 16.8% of older women versus 15.7% of older men provide care (not shown). This apparent similarity may mask differences in the types of caregiving tasks that men and women carry out; for instance older women are more likely than older men to cook, clean, and do laundry for a spouse with a disability whereas older men are more likely to do home repairs (Freedman, Cornman, & Carr, 2014).

Measuring Caregiving Time

Measuring time spent providing care poses unique challenges. Stylized time use questions that ask individuals to enumerate hours of care over some recent time period (e.g., last week or month) may provide an incomplete assessment of hours if care providers do not perceive their activities to be "care," per se (Bittman, Fast, Fisher, & Thomson, 2004). Retrospective diary-based measures, which obtain data on the specific activities performed over a 24-h period, can also be problematic if they do not identify *why* or *for whom* household activities are carried

TABLE 8.1 Caregiving to Older Adults in the United States: by Caregiver Age

Age of caregiver	Men (000s)	Women (000s)	Total (000s)	% of the population	% of caregivers who gave care yesterday	Average number of hours yesterday
15 years and older	17 500	22 064	39 564	16.1	23.0	3.2
15–24 years	2 569	2 761	5 330	12.5	15.4	1.3
25–34 years	2 035	2 015	4 050	9.8	12.9	3.0
35–44 years	2 301	2 758	5 060	12.8	20.5	2.5
45–54 years	4 267	5 839	10 106	23.1	22.3	3.3
55–64 years	3 517	4 849	8 366	22.2	26.0	3.5
65 years and older	2 810	3 842	6 652	16.3	34.7	4.1

Source: US Bureau of Labor Statistics (2013a).

out. For instance, a wife who does the laundry for her fully able husband may be doing housework, but if she does the laundry because he is unable due to health or functioning, she may be providing care.

Despite these measurement challenges, a number of studies have attempted to understand how caregiving influences other forms of time use. The most consistent focus has been on the tradeoff between care and labor force participation. A review of the literature by Lilly and colleagues (2007) concludes that caregivers are as likely to work as non-caregivers; that caregivers work fewer hours than non-caregivers, especially when care needs are substantial; and only those involved in substantial amounts of care are more likely than non-caregivers to withdraw from work altogether. Van Houtven and colleagues (2013) point out that given the largely cross-sectional evidence, whether a causal relationship between work and care exists and if so the specific underlying mechanisms remain unclear. Focusing on adult children who are care providers to parents or parents-in-law in the Health and Retirement Study, they find that female care providers who remain working decrease their work by 3–10h per week relative to non-caregivers. They find little effect of caregiving on men's working hours.

A secondary focus has been on the tradeoff between care and leisure time. Descriptive studies have noted that co-resident caregivers spent less time on leisure activities than non-caregivers (Bittman et al., 2004; Michelson & Tepperman, 2003). However, modeling efforts that explicitly recognize the tradeoffs between different types of time use question this tenet. For example, Arora and Wolf (2014) jointly modeled time spent caring, working, and engaged in physical activity for middle-aged adults with at least one living parent in the Health and Retirement Study. They found that parent characteristics associated with increased care were not inversely associated with the frequency of caregivers' physical activity. Further, unobserved factors influencing time transfers to parents and frequency of physical activity were positively correlated across the care and physical activity equations among men, suggesting that these two types of time-allocation decisions appear to be complementary rather than substitute for each other.

A third, and largely separate, literature has focused on the influence of caregiving on well-being. Findings are equivocal, with some studies suggesting positive influences and others negative effects on well-being. In their comprehensive review, Pinquart and Sorensen (2003) find that studies consistently report higher levels of depressive symptoms and stress and lower levels of self-efficacy and subjective well-being among caregivers than non-caregivers, although these differences are typically small and appear to be substantial only in non-representative samples. Nevertheless, female caregivers and older caregivers appear to be more disadvantaged than male caregivers and younger caregivers with regard to most outcomes. One reason women caregivers experience more negative influences of caregiving may be the repetitive nature of the caregiving tasks.

Why such gender differences exist has been a long-standing interest. Early studies drew upon non-representative samples and stylized questions about time spent caring, and concluded that husbands may enjoy the caregiver role more than wives, that they may be more effective in coping with interpersonal problems with their spouses, and that they are more willing to take on the caregiving role in later life (Barusch & Spaid, 1989; Fitting, Rabins, Lucas, & Eastham, 1986; Pruchno & Resch, 1989). Amirkhanyan and Wolf (2003) pointed out that these studies are problematic because they do not distinguish between the influence of having a relative with a disability and the effect of providing care to that relative. Freedman, Cornman, & Carr (2014) raise the additional complexity that such studies also confound *for whom* the activity is done with *what* is done, typically a combination of household chores and personal care tasks.

Moreover, from a time use perspective, studies have relied primarily on evaluative measures of life satisfaction or decontextualized affect measures (e.g., how happy are you?) and how they vary with caregiving intensity, typically measured with stylized questions about care given over the last week or month. Despite the fact that individuals may rely on general beliefs about experiences rather than true moment-to-moment experiences in such stylized assessments (Schwarz, Kahneman, & Xu, 2009), with few exceptions, experienced well-being of older caregivers has not been explored (Bittman et al., 2004; Freedman, Cornman, & Carr 2014; Poulin et al. 2010).

A recent study (Freedman, Cornman, & Carr, 2014) attempts to fill this gap by using time diary data from a national sample. Using the DUST Supplement to the 2009 PSID, the authors explore whether spousal caregiving is associated with reduced experienced well-being (how happy and how frustrated during particular activities) for older husbands and wives. They estimate three distinct effects: having a spouse with a disability; doing household or personal care tasks ("chores") for someone other than a spouse with a disability; and doing such tasks for a spouse with a disability ("care"). The authors find no significant effect on experienced well-being for husbands who have a wife with a disability, do chores, or care for a wife with a disability. For wives, they find no significant effect of having a husband with a disability. Yet, for wives, carrying out chores was associated with *lower* experienced well-being (happiness in particular) compared to other activities and care to one's husband was associated with *greater* experienced well-being than carrying out the same activities as chores. They conclude that caregiving per se does not erode the well-being of older spousal caregivers – but for women, the chores that often constitute daily care, may do so.

FUTURE DIRECTIONS

This chapter has identified several key themes in the literature related to gender, time use, and aging. First, the familiar gendered patterns observed among working-age adults appear to hold in later life: older men devote more time to paid work and enjoy more leisure; older women devote more time to care work. Time disparities in paid work attenuate with transitions from full-time employment to phased or full retirement. However, disparities in care time remain substantial, as older women transition from intensive mothering to providing care for an elderly spouse.

Second, although older women's disproportionate care time is associated with economic disadvantages, it may also be associated with social advantages in later life via more companionship and enhanced well-being. Nonetheless, women's higher life expectancy, and lower probability of re-partnering after widowhood, also means that "oldest old" women spend a larger proportion of time in old age in solitary pursuits. Time use researchers and social gerontologists offer mixed perspectives on how solitary and alone time influence well-being among older adults. One salient element appears to be whether older adults experience time with others as emotionally satisfying and socially integrative or instead as emotionally burdensome or intrusive (Hooyman & Kiyak, 2008; Larson, 1990). Time use studies that collect simultaneous data on respondents' affective states during activities have the potential to provide a deeper understanding of the emotional impact of solitary time.

Third, the consequences of providing care to older adults differ by gender. Adult daughters who work and provide care tend to reduce their work hours whereas adult sons are unaffected. Adult children's time spent in physical activity also seems largely unaffected, irrespective of gender. For wives but not husbands who provide care to a spouse, the act of caring for a spouse with a disability appears to enhance experienced well-being, whereas the tasks and activities that commonly constitute care appear to erode it.

Despite these findings, there continue to be important gaps in the literature on gender, time use, and aging that recent high-quality longitudinal data from the HRS, the PSID, and European panel studies may help to resolve. The cross-sectional design of research using time diary data precludes examination of transitions out of or into employment and family roles influence time use in later life. Nor are researchers able to disentangle selection factors from causal effects of role statuses and transitions. Indeed, the life course perspective has been notably absent from the literature on gender, time use, and well-being.

One important topic for future research is the implications of the Baby Boom generation's distinctive marriage, fertility, and household formation patterns for their time use in older adulthood. Although family and household patterns through the life course have been linked to women's and men's work and family roles in midlife, their indirect influence on time use and well-being at older ages has not been explored. Particularly important will be studies that move beyond cross-sectional descriptive patterns to parse out the contribution of selection into various family and work statuses versus causal influences.

Similarly, the influence of growing family inequality on time use and well-being in later life has not been fully explored. Transitions into marriage, parenthood, and stable employment are differentiated by education more so in the twenty-first century compared with the twentieth century. Specifically, college-educated women and men have higher marriage rates and are more likely to raise children within a stable marriage, and college-educated women and men are more likely to enjoy stable jobs with good pay and benefits. In contrast, women and men without a college degree are more likely to have and raise children outside of marriage, have more intermittent employment over the life course in jobs with low pay and few benefits (Kalleberg, 2011; McLanahan, 2004). In terms of influencing time use patterns in later life, at least two main pathways could be at play. First, economic opportunities at earlier ages lead to a lifetime of economic advantage, which in turn is likely to influence decisions about retirement that in turn influence time use and well-being in later life. Second, economic opportunities at earlier ages may influence how men and women structure their time in midlife, which may set time use patterns in motion that in turn enhance or impede their ability to "age well" and "age productively." Moreover, the interplay among economic opportunity, the unfolding of the disablement process, and time use in later life, and how these processes differ for men and women, also warrants additional attention.

The research on time with others is small in volume, particularly studies that focus on older adults. Future research should examine if, net of employment and marital status, gender differences in social time dissipate or grow as newer cohorts reach older age. Older men who experienced divorce or fathered children outside of legally recognized relationships may have less social support, more isolation, and also more identity issues because of the loss of "normative roles and relationships status, which usually emphasize masculine autonomy and power" (Venn et al., 2011). The disparate early-life experiences of women and men associated with increases in single parent families, non-marital and multi-partner births, and greater life expectancy for older women vis-à-vis older men point to the possibility of more substantial differences in women's and men's social time among future older adults.

Additional research is also needed on how family caregiving – across diverse family arrangements – intersects with the changing work status of older adults. According to the US Bureau of Labor Statistics (Toosi, 2013), 55% of men and 72% of women ages 50–61 say they may delay retirement because of the recession that began in 2008. Projections indicate that by 2022, 32% of workers ages 65–74 will be in the labor force, compared with 20% in 2002 and 27% in 2012. The implication of macroeconomic shifts for time use and well-being at older ages, including care patterns, will be an important topic of future research.

Finally, we note that individuals change more rapidly than social institutions, leading to cultural/structural lags that produce "dilemmas" (Riley, Johnson, & Foner, 1972; Riley & Riley, 1994). For older women and men, these include laws and employment policies that make it more difficult for those who prefer to stay employed to keep jobs and wage and benefit disparities that make it difficult for older workers who would prefer to retire to do so, particularly women and older adults with less than a college education. Another key dilemma is the structure of family, market, and state-subsidized care work that disproportionately burdens wives and adult daughters with care obligations and disproportionately excludes unmarried and divorced men from family care networks. Addressing the "structural lag" in research on role transitions, sequencing, and synchronicity is a first step in developing a blueprint for solving gendered work and family dilemmas of older adults.

Acknowledgments

This chapter was funded in part by the National Institute on Aging grant P01-AG-029409. The views expressed are those of the authors alone and do not represent their employers or funding agency. The authors thank Liz Kofman for research assistance.

References

Agahi, N. A., & Parker, M. G. (2005). Are today's older people more active than their predecessors? Participation in leisure-time activities in Sweden in 1992 and 2002. *Ageing & Society, 25*, 925–941.

Amirkhanyan, A., & Wolf, D. A. (2003). Caregiver stress and noncaregiver stress: Exploring the pathways of psychiatric morbidity. *The Gerontologist, 43*, 817–827.

Arber, S., Andersson, L., & Hoff, A. (2007). Changing approaches to gender and ageing: Introduction. *Current Sociology, 55*, 147–153.

Arora, K., & Wolf, D. A. (2014). Is there a trade-off between parent care and self-care? *Demography, 51*, 1251–1270.

Barusch, A. S., & Spaid, W. M. (1989). Gender differences in caregiving: Why do wives report greater burden? *The Gerontologist, 29*, 667–676.

Bianchi, S. M., Robinson, J. P., & Milkie, M. A. (2006). *Changing rhythms of American family life*. New York, NY: Russell Sage Foundation.

Bird, C. E., & Fremont, A. M. (1991). Gender, time use and health. *Journal of Health and Social Behavior, 32*, 114–129.

Bird, C. E., & Rieker, P. P. (2008). *Gender and health: The effects of constrained choices and social policies*. New York, NY: Cambridge University Press.

Bittman, M. (2002). Social participation and family welfare: The money and time costs of leisure in Australia. *Social Policy and Administration, 36*, 408–425.

Bittman, M., Fast, J. E., Fisher, K., & Thomson, C. (2004). Making the invisible visible: The life and time(s) of informal caregivers. In N. Folbre & M. Bittman (Eds.), *Family time: The social organization of care* (pp. 69–90). London: Routledge Press.

Bourdieu, P. (1984). *Distinction: A social critique of the judgement of taste*. Cambridge, MA: Harvard University Press.

Brenner, P. S. (2012). Investigating the effect of bias in survey measures of church attendance. *Sociology of Religion, 73*, 361–383.

Calasanti, T. M. (2001). *Gender, social inequalities, and aging*. Walnut Creek, CA: AltaMira Press.

Craig, L., & Mullan, K. (2011). How mothers and fathers share childcare. *American Sociological Review, 76*, 834–861.

Csikszentmihalyi, M., & Larson, R. W. (1992). Validity and reliability of the Experience Sampling Method. In M. deVries (Ed.), *The experience of psychopathology* (pp. 43–57). Cambridge: Cambridge University Press.

Cutler, S. J., & Hendricks, J. (1990). Leisure and time use across the life course. In R. H. Binstock & L. K. George (Eds.), *Handbook of Aging and the Social Sciences* (pp. 169–185, 3rd ed.). San Diego, CA: Academic Press, Inc.

Fitting, M., Rabins, P., Lucas, M. J., & Eastham, J. (1986). Caregivers for dementia patients: A comparison of husbands and wives. *Gerontologist, 26*, 248–252.

Freedman, V. A., Cornman, J. C., & Carr, D. S. (2014). Is spousal caregiving associated with enhanced well-being? New evidence from the Panel Study of Income Dynamics. *Journals of Gerontology Series B: Psychological Sciences and Social Sciences, 69*, 861–869.

Freedman, V. A., Stafford, F., Schwarz, N., Conrad, F., & Cornman, J. C. (2012). Disability, participation, and subjective well-being among older couples. *Social Science & Medicine, 74*, 588–596.

Freedman, V. A., Stafford, F. P., Conrad, F., Schwartz, N., & Cornman, J. (2012). Time together: An assessment of diary quality for older couples. *Annals of Economics and Statistics*, 271–289. *January–June 2012*.

Freedman, V. A., Stafford, F. P., Schwarz, N., & Conrad, F. (2013). Measuring time use of older couples: Lessons from the Panel Study of Income Dynamics. *Field Methods*, 25, 405–422.

Freysinger, V. J. (1995). The dialectics of leisure and development for women and men in mid-life: An interpretive study. *Journal of Leisure Research*, 27, 61–84.

Gauthier, A. H., & Smeeding, T. M. (2010). Historical trends in the patterns of time use of older adults. In S. Tuljapurkar, S. Ogawa, & A. H. Gauthier (Eds.), *Ageing in advanced industrial states: Riding the age waves* (vol. 3, pp. 289–310). Heidelberg: Springer Science + Business Media B.V.

Gershuny, J. (2000). *Changing times: Work and leisure in postindustrial society*. Oxford: Oxford University Press.

Goldscheider, F. K., & Hogan, D. P. (2001). Men's flight from children in the US? A historical perspective. *Advances in life Course Research*, 6, 173–191.

Hagestad, G. O. (1990). Social perspectives on the life course. In R. H. Binstock & L. K. George (Eds.), *Handbook of Aging and the Social Sciences* (pp. 151–168) (3rd ed.). San Diego, CA: Academic Press, Inc.

Hahn, E. A., Cichy, K. E., Almeida, D. M., & Haley, W. E. (2011). Time use and well-being in older widows: Adaptation and resiliance. *Journal of Women and Aging*, 23, 149–159.

Hooyman, N. R., & Kiyak, H. A. (2008). *Social gerontology: A multidisciplinary perspective*. New York, NY: Pearson Education.

Jake, M., Davey, A., & Kleiber, D. (2005). Modeling change in older adult's leisure activities. *Leisure Sciences*, 28, 285–303.

Juster, F. T., Ono, H., & Stafford, F. P. (2003). An assessment of alternative measures of time use. *Sociological Methodology*, 33, 19–54.

Juster, F. T., & Stafford, F. P. (1991). The allocation of time: Empirical findings, behavioral models, and problems of measurement. *The Journal of Economic Literature*, 29, 471–522.

Kahneman, D., Krueger, A. B., Schkade, D. A., Schwarz, N., & Stone, A. A. (2004). A survey method for characterizing daily life experience: The day reconstruction method. *Science*, 306, 1776–1780.

Kalleberg, A. L. (2011). *Good jobs, bad jobs: The rise of polarized and precarious employment systems in the United States, 1970s–2000s*. Russell Sage Foundation.

Kan, M. Y., & Pudney, S. (2008). Measurement error in stylized and diary data on time use. *Sociological Methodology*, 38, 101–132.

Klumb, P. L., & Baltes, M. M. (1999). Validity of retrospective time-use reports in old age. *Applied Cognitive Psychology*, 13, 527–539.

Kochar, R. (2009). Recession Turns a Graying Office Grayer. <http://www.pewsocialtrends.org/2009/09/03/recession-turns-a-graying-office-grayer/>.

Krantz-Kent, R., & Stewart, J. (2007). How do older Americans spend their time? *Monthly Labor Review, 2007*, 8–26.

Krueger, A. B., & Stone, A. A. (2008). Assessment of pain: A community-based diary survey in the USA. *The Lancet*, 371, 1519–1525.

Larson, R. W. (1990). The solitary side of life: An examination of the time people spend alone from childhood to old age. *Developmental Review*, 10, 155–183.

Lilly, M. B., LaPort, A., & Coyte, P. C. (2007). Labor market work and home care's unpaid caregivers: A systematic review of labor force participation rates, predictors of labor market withdrawal, and hours of work. *Milbank Quarterly*, 85, 641–690.

McLanahan, S. (2004). Diverging destinies: How children are faring under the second demographic transition. *Demography*, 41, 607–627.

McNamara, T. K. (2008). *Time use across the life course (Rep. No. Issue Brief 18)*. Boston, MA: Boston College.

Meyer, M. H., & Parker, W. M. (2011). Gender, aging, and social policy. In R. H. Binstock & L. K. George (Eds.), *Handbook of aging and the social sciences* (pp. 323–336, 7th ed.). San Diego, CA: Academic Press.

Michelson, W., & Tepperman, L. (2003). Focus on home: What time-use data can tell about caregiving to adults. *Journal of Social Issues*, 59, 591–610.

Miller, Y. D., & Brown, W. J. (2005). Determinants of active leisure for women with young children: An "ethic of care" prevails. *Leisure Sciences*, 27, 405–420.

National Research Council, (2000). *Time use measurement and research: Report of a workshop*. Washington, DC: National Academy Press.

Niemi, I. (2009). Sharing of tasks and lifestyle among aged couples. Electronic International. *Journal of Time Use Research*, 6, 286–305.

Nomaguchi, K. M., & Bianchi, S. M. (2004). Exercise time: Gender differences in the effects of marriage, parenthood, and employment. *Journal of Marriage and Family*, 66, 413–430.

O'Rand, A. M., & Shuey, K. M. (2007). Gender and the devolution of pension risks in the US. *Current Sociology*, 55, 287–304.

Offer, S., & Schneider, B. (2011). Revisiting the gender gap in time-use patterns. *American Sociological Review*, 76, 809–833.

Pencavel, J. (1986). Labor supply of men. In O. Ashenfelter & R. Layards (Eds.), *Handbook of labor economics* (pp. 3–102). Amsterdam: North Holland.

Phipps, P. A., & Vernon, M. K. (2009). Twenty-four hours: An overview of the recall diary method and data quality in the American Time Use Survey. In R. F. Belli, F. P. Stafford, & D. F. Alwin (Eds.), *Calendar and time diary methods in life course research* (pp. 109–128). Thousand Oaks, CA: Sage.

Pinquart, M., & Sorensen, S. (2003). Differences between caregivers and noncaregivers in psychological health and physical health: A meta analysis. *Psychology and Aging, 18*, 250–267.

Pinquart, M., & Sorensen, S. (2007). Correlates of physical health of informal caregivers: A meta-analysis. *Journal of Gerontology: Psychological Sciences, 62B*, P126–P137.

Poulin, M. J., Brown, S. L., Ubel, P. A., Smith, D. M., Jankovic, A., & Langa, K. M. (2010). Does a helping hand mean a heavy heart? Helping behavior and well-being among spouse caregivers. *Psychology and Aging, 25*, 108–117.

Pruchno, R. A., & Resch, N. L. (1989). Husbands and wives as caregivers: Antecedents of depression and burden. *The Gerontologist, 29*, 159–165.

Putnam, R. D. (2000). *Bowling alone: The collapse and revival of American community*. New York, NY: Simon & Schuster.

Riley, M. W., Johnson, M., & Foner, A. (1972). *Aging and society: A sociology of age stratification* (Vol. 3). New York, NY: Russell Sage Foundation.

Riley, M. W., & Riley, J. W. (1994). Age integration and the lives of older people. *The Gerontologist, 34*, 110–115.

Robinson, J. P., & Godbey, G. (1999). *Time for life: The surprising ways Americans use their time* (2nd ed.). University Park, PA: Pennsylvania State University Press.

Sayer, L. C., & Gornick, J. C. (2009). Older adults: International differences in housework and leisure. *Social Indicators Research, 93*, 215–218.

Schwarz, N., Kahneman, D., & Xu, J. (2009). Global and episodic reports of hedonic experience. In R. F. Belli, F. P. Stafford, & D. F. Alwin (Eds.), *Calendar and time diary methods in life course research* (pp. 157–174). Los Angeles, CA: Sage.

Smith, J., Ryan, L. H., Queen, T. L., Becker, S., & Gonzalez, R. (2014). Snapshots of mixtures of affective experiences in a day: Findings from the health and retirement study. *Journal of Population Ageing, 7*, 55–79.

Stone, A. A., Shiffman, S. S., & DeVries, M. W. (1999). Ecological momentary assessment. In D. Kahneman, E. Diener, & N. Schwarz (Eds.), *Well-being: The foundations of hedonic psychology* (pp. 26–39). New York, NY: Russell Sage Foundation.

Toosi, M. (2013). Labor force projections to 2022: The labor force participation rate continues to fall. *Monthly Labor Review*, 1–27.

US Bureau of Labor Statistics. (2013a). American Time Use Survey user's guide: Understanding ATUS 2003 to 2011. ATUS [Electronic version]. Available: <http://www.bls.gov/tus/atususersguide.pdf>.

US Bureau of Labor Statistics. (2013b). Unpaid eldercare in the United States 2011–2012: Data from the American Time Use Survey. ATUS [Electronic version]. Available: <http://www.bls.gov/news.release/pdf/elcare.pdf>.

Van Houtven, C., Coe, N. B., & Skira, M. M. (2013). The effect of informal care on work and wages. *Journal of Health Economics, 32*, 240–252.

Venn, S., Davidson, K., & Arber, S. (2011). Gender and aging. In R. A. Settersten & J. L. Angel (Eds.), *Handbook of sociology of aging* (pp. 71–82). New York, NY: Springer Verlag.

Verbrugge, L. M., & Liu, X. (2014). Midlife trends in activities and disability. *Journal of Aging and Health, 26*, 178–206.

CHAPTER 9

Social Networks in Later Life

Benjamin Cornwell[1] and Markus H. Schafer[2]

[1]Department of Sociology, Cornell University, Ithaca, NY, USA [2]Department of Sociology, University of Toronto, Toronto, ON, Canada

OUTLINE

Introduction	181
Network Concepts and Definitions	182
Basic Social Network Data Concepts	182
Elements of Network Structure and Composition	184
Composite Network Measures	185
Why and How Social Networks Matter	185
Access to Social Resources	185
Health and Well-Being	186
Aging and Social Network Change	189
Theories of Network Change in Later Life	189
Consequences of Network Change for Older Adults	190
Social Networks and Stratification	191
Race/Ethnicity	191
Socioeconomic Status	192
Gender	193
Emerging Topics in Network-Gerontology	193
Electronic Networks	193
Whole Networks	194
Network Diffusion Processes	195
Negative Network Ties	196
Conclusions	196
References	197

INTRODUCTION

For close to two decades now, research on older adults' social connectedness has been expanding on conceptualizations of social integration that focus on roles and activities in order to further examine the nature of older adults' "social networks" (Cornwell, Laumann, & Schumm, 2008). A *social network* refers to a defined set of social actors – which may include individuals, organizations, or other entities – and the social relationships that connect them to each other in a larger structure (Wasserman & Faust, 1994). There are several reasons for the shift toward this approach, including the explosive growth of social network analysis as a paradigm throughout the social sciences. Beyond this, researchers continue to uncover evidence of powerful consequences of structural features of social networks for health and well-being, especially in later life (e.g., Ashida & Heaney, 2008; Berkman & Syme, 1979; Cheng, Lee, Chan, Leung, & Lee, 2009; Fiori, Antonucci, & Cortina, 2006; Holtzman et al., 2004; Litwin & Shiovitz-Ezra, 2011; Seeman & Syme, 1987; York Cornwell & Waite, 2009).

An increasing number of scholars have become particularly interested in the structural properties of social networks and the role of aging in shaping them. Over the past few years, several large, publicly available datasets have been developed to facilitate the measurement and analysis of key structural and compositional features of older adults' social networks. These include, among others, the National Social Life, Health, and Aging Project (NSHAP) and the Survey of Health, Ageing and Retirement in Europe (SHARE). These studies are yielding detailed social network data that expand on the more general network and relationship data that have long been available in studies such as the Americans' Changing Lives study and the Health and Retirement Study (HRS).

This explosion of network-focused gerontological research calls for a survey of what we know about social networks in later life. Our overarching goal in this chapter is to provide an overview of insights that have emerged in recent years along these lines. Social gerontologists are increasingly interested in not only the structural features and functions of older adults' networks, but also their consequences for important outcomes such as well-being. This has sparked renewed interest in the factors that shape social networks in later life – processes such as retirement, bereavement, and health decline – which may help to explain why older adults' networks look different from those of younger and middle-aged adults. Furthermore, it is increasingly apparent that network connectedness and processes of network change are experienced differently by different segments of the older adult population. A growing body of research suggests that race, gender, and socioeconomic status (SES) all affect the nature of network connectedness in later life. We review some of these differences in this chapter. Finally, we cover emerging topics in social-gerontological network research.

NETWORK CONCEPTS AND DEFINITIONS

It is useful to begin by defining and illustrating key network terms – many of which have only recently gained visibility in social gerontology, and most of which are easier to understand with visual aids. To that end, Figure 9.1 presents an example of an actual social network. This network depicts the social structure of an independent-living retirement community in the Midwestern United States in 2009, defined in terms of residents' discussion relations (see Schafer, 2011). We will come back to this figure throughout the chapter.

Basic Social Network Data Concepts

A distinguishing feature of social network research is the formal conceptualization of individuals or other actors as elements, or *nodes*, in a larger social structure. The network shown in Figure 9.1 includes 123 residents between 74 and 96 years of age. These constitute the nodes, which are represented here as circles (the lighter circles represent women). Networks are composed of both nodes and the ties, or *links*, that connect them together. These links are the social relationships that exist between the actors in the social setting being examined. They may reflect friendship, social support, social exchange, conflict, formal relations, kinship, and many others. To identify the network

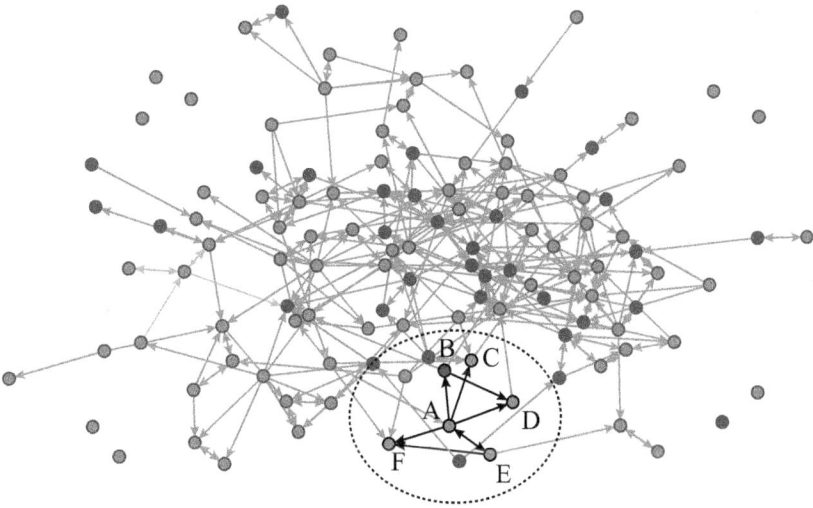

FIGURE 9.1 Social network from an independent-living retirement community ($N=123$), with a sample egocentric network circled and bolded. *Note*: Darker circles represent men, lighter circles represent women. This diagram layout was created using the Kamada-Kawai spring-embedder algorithm in Pajek (Batagelj & Mrvar, 1998), which arranges nodes such that nodes that are connected to each other are placed close together, whereas nodes that are not connected to each other are placed far apart from each other.

shown in Figure 9.1, for example, Schafer (2011) recorded links wherever a given resident reported that s/he talked to another resident about "important issues." For example, resident A indicated that she talked to residents B, C, D, E, and F about such issues.

There are different ways to analyze a social network such as the one depicted in Figure 9.1. One approach is to describe and examine the network as a single, or *whole*, network structure. For example, one might undertake to compare the network shown in Figure 9.1 with a network from a different retirement community. This may provide clues about the role of population composition, neighborhood features, geographic region, or other contextual factors in shaping the networks in which older adults are embedded. Looking at an entire network can also shed light on how diverse phenomena, such as information or disease, diffuse through a group of people. The more common approach in social-gerontological research (partly due to data availability) is to study networks from an *egocentric* perspective. Because social gerontologists are often concerned with individual-level outcomes – such as loneliness, health and well-being, and related issues – they tend to study networks from the perspective of individuals. This involves identifying a given individual – referred to as *ego* – within a given population and the set of other people – referred to as *alters* – to whom ego is connected.

To illustrate, the egocentric social network of resident A (a woman) is circled and shown with the links and node borders in bold. Note that the egocentric network only includes those residents to whom A is *directly* (not indirectly) connected, and it also includes any links that exist among those residents. It does not include anyone whom ego did not name, even if those people named ego. This egocentric information is sufficient to assess the size, composition, and density of ego's personal network, as well as the strengths of her network ties – concepts to which we now turn.

Elements of Network Structure and Composition

Several features of social networks are consequential for the actors in a network. Below we will discuss why these features matter and how they relate to aging. One important concept is *network size*, which refers to the number of other actors in the network to which a given actor is connected. Another key issue involves the strengths of the network ties within a network (Granovetter, 1973). *Tie strength* may refer to the emotional intensity of a tie, duration, frequency of contact, the degree of *reciprocity* in the relationship, and potentially other factors. Accordingly, tie strength may be measured in numerous ways.

Social network *composition* refers to the types of contacts maintained by network members. Depending on the context of the study and the characteristics that interest the researcher, this can refer to the extent to which an actor's network is composed of kin ties, as well as other features such as racial or gender composition. For example, Figure 9.1 includes 88 women (shown in light gray) and 35 men (dark gray). The tendency for actors to form and maintain ties to others who are similar in some way (e.g., same race) is referred to as *homophily*. Figure 9.1, for example, reveals a tendency for same-sex clustering within the network.

One thing that distinguishes social network research from research on related concepts such as social support and loneliness is that it is concerned with the internal wiring of networks – that is, how actors' network members are in turn connected to each other. An important concept here is that of the *triad*, which refers to a set of three nodes and the links that may or may not exist between them. Consider the case of the triad consisting of residents A, C, and E in Figure 9.1. Residents A and C are linked, as are A and E. C and E, however, are not connected. We refer to this as an *open* or empty triad, because not all of the connections in this triad are realized. If C and E had a social relationship with each other, A–C–E would constitute a *closed* triad. The extent to which a given person's network members are connected to each other has major implications for how that network functions and the opportunities it presents to individuals. A network that has a large number of closed triads is referred to as a dense network. The *density* of a network simply refers to the proportion of possible links in the set that are actually realized. In the network in Figure 9.1, for example, there are 310 links among 123 nodes. In total, $123 \times 122 = 15\ 006$ links were possible. Thus, the density of this network is $310/15\ 006 = 0.021$. That is, only 2.1% of the links that could have existed in this community actually existed.

Social network data provide valuable insight into individuals' positions within a larger social structure. One concept that is important here is *centrality* (Freeman, 1979). This refers to the extent to which actors are able to access a larger number of other actors in the network (through relatively few intermediaries). In the network in Figure 9.1, for instance, resident C occupies a more central position in the overall retirement community network than A. Resident C was named as a discussion partner by four community residents – a measure of her *degree centrality* – whereas A was named by only two. Resident C also has a relatively high *closeness centrality*, meaning that she can contact other residents in the community through relatively few intermediaries. Most other actors – such as residents E and F – are less central in the retirement community network. As such, they are located toward the periphery of the network.

While a given actor may not be central, s/he may still occupy other important positions within a network. Note that there are numerous pairs of alters in A's network who are not connected to each other, including B–C, B–E, B–F, C–D, C–E, C–F, D–E, and D–F. As discussed below, while some research implies that this situation may reduce B's access to valuable social support functions (e.g., coordinative capacity among network members), it also presents opportunities. It puts A in what is known as a *bridging* position in the network structure, as A spans the gap that exists between these other pairs of otherwise unconnected actors (Burt, 1992). In these cases, resident A may be able to facilitate or otherwise control the flow of information or resources between those four actors. Likewise, this affords greater independence from either of those parties, which is an important option for some older adults (Cornwell, 2011).

Composite Network Measures

Not all social gerontologists decompose networks into their individual features, however. Some of the first and most influential studies to document the link between social networks and mortality, for example, utilized what are known as *social network indices* (e.g., Berkman & Syme, 1979). In this approach, researchers record the extent to which respondents have various types of network contacts (e.g., a spouse, children, friends, neighbors, fellow group members) and how often they interact with those contacts. The social network index is then generated by combining these different components into a composite measure that amalgamates the network's size, range, and frequency (see also Lubben, 1988).

Some recent studies of older adults employ a related *network typology* approach. This approach looks beyond the simple presence/absence of one or more network ties (e.g., spouse) and instead forms a network profile on the basis of multiple measures. Much research in this vein uses a range of proxy network indicators such as marital status, number of children and other relatives, number of friends, and frequency of informal and formal social activity (Litwin, 2001; Litwin & Shiovitz-Ezra, 2011; Park, Smith, & Dunkle, 2014; Shiovitz-Ezra & Litwin, 2012). A set of categorical indicators reflecting the overall type of network (e.g., family-based, friend-focused, restricted, wide-ranging) is then created. Several recent studies have developed typologies that also combine the general types of questions that are used to generate network indices with the egocentric network data that are drawn from name-generator methods. The name-generator approach means identifying a type of relationship – for example, confidants, social support providers, sexual partners – and eliciting the names of specific people who meet that criterion. Scholars who adopt the social "convoy" model (Kahn & Antonucci, 1980) have also combined more conventional information about social network structure with respondents' assessments of support availability and satisfaction with one's social contacts to generate more refined typologies that reflect both subjective and objective dimensions of networks (e.g., see Fiori, Smith, & Antonucci, 2007).

WHY AND HOW SOCIAL NETWORKS MATTER

A guiding assumption behind the network perspective is that the structure of social relationships is consequential for individuals. This section addresses two of the most salient arguments along these lines in social gerontology – the implications of older adults' networks both for their access to valuable social resources (especially social support) and for their health and well-being.

Access to Social Resources

Social networks are conduits for flows of resources, and the internal wiring of networks affects the channeling, coordination, and control of those flows (Wellman, 1983). This insight provides an important starting point in justifying why networks matter and are worth studying, especially in the context of later life. *Social support* – which is distinct from, but flows through, social network ties – is a major focal point of gerontological research. It can be defined as resources that are "accessible to an individual through social ties to other individuals, groups, and the larger community" (Lin, Ensel, Simeone, & Kuo, 1979, p. 109). Scholars distinguish among different types of support, including instrumental (e.g., material aid), emotional (e.g., listening to problems), companionship, and informational (e.g., problem-solving) – as well as between support actually received versus perceived availability of support.

As will be addressed in greater detail later, social networks do not automatically yield social resources, but they condition the delivery of support resources (House, Umberson, & Landis, 1988). Network composition is one important feature of networks in this regard. Different types of network members (e.g., spouse, friend) are associated with different forms of support (Wellman & Wortley, 1990). In general, kinship ties – especially spouses and offspring – provide more reliable and wider-ranging support when needed (Fischer, 1982; Wellman & Wortley, 1990). At the same time, friendship ties provide emotional support and companionship (Lapierre & Keating, 2013), and are often especially emotionally rewarding for older adults (Litwin, 2001). Less research has been done to determine the extent to which maintaining weak ties such as organizational memberships (Granovetter, 1973), in addition to strong ones, affects older adults' access to support. Access to resources through networks also depends on the SESs of network members. This is suggested by research on the link between social networks and *social capital* (Lin, 2001). Research in this vein asks respondents which of their social network members fit a given occupational profile (e.g., whether they are a lawyer, doctor, or other professional) (Lin & Dumin, 1986), and/or whether they have network members who can provide them with specific resources (e.g., a loan) (Van Der Gaag & Snijders, 2005).

Structural features of networks shape access to valuable social resources as well. All else being equal, larger networks make a wider range of support resources available. Denser networks may also help to facilitate support delivery. Referring again to Figure 9.1, resident A is connected to B, C, D, E, and F, but note that few of them are connected to each other. Residents B and D are connected, as are residents E and F, but those factions are not connected to each other, or to C. This may present problems in terms of those individuals' capacities to provide high-quality social support to A. Denser networks are more effective at delivering social support due to the greater coordinative capacity that exists when members are connected to each other (Ashida & Heaney, 2008). Dense networks make it possible for members to share and compare information, to coordinate their activity, and to maintain ongoing monitoring and communication efforts. Thus, the coordinated support that is common among family members who take care of older adults is in part owing to their pre-existing social connections to *each other* (Cornwell, 2012).

Health and Well-Being

Attention to the health-protective effects of social networks began with a set of path-breaking articles published in the late 1970s and 1980s (e.g., Berkman & Syme, 1979; House, Robbins, & Metzner, 1982). These studies reached conclusions that foreshadowed findings from decades of subsequent research: "Socially isolated individuals are less able than others to buffer the impact of health stressors and consequently are at greater risk for negative health outcomes such as illness and death" (Smith & Christakis, 2008, p. 408). A recent meta-analysis of 148 studies (average age = 63) concluded that structural and functional aspects of people's social networks have at least as strong a link to mortality as conventional factors such as physical activity or excess alcohol consumption (Holt-Lunstad, Smith, & Layton, 2010).

Support Pathways. Social networks may affect health through a number of psychological and physiological pathways, many of which involve social support (House et al., 1988; Thoits, 2011). The isolation that comes with a lack of close network members typically involves a lack of emotional support and leads to greater loneliness, which is detrimental to mental health (York Cornwell & Waite, 2009). Having or perceiving access to support helps people cope with stress, and self-esteem is boosted by supportive social contacts (Krause, 1987). The sense of mattering to other people that comes with being connected to a network also helps protect against mental health problems (Thoits, 2011). Loneliness, depression, low self-esteem, and associated conditions also suppress immune function, which has downstream consequences for physical health.

Support networks also work through direct cardiovascular, neuroendocrine, and immunological pathways (Uchino, Bowen, Carlisle, & Birmingham, 2012). For example, receiving emotional support in stressful situations can reduce cardiovascular reactivity (e.g., blood pressure), as close social ties can produce a "calming" effect on the cardiovascular system (Gerin, Milner, Chawla, & Pickering, 1995; Holt-Lunstad, Uchino, Smith, Olson-Cerny, & Nealey-Moore, 2003). Observational research suggests a link between social support and stress-related biomarkers, including norepinephrine, epinephrine, salivary cortisol, and C-reactive protein, though this association may differ by gender (Ford, Loucks, & Berkman, 2006; Loucks, Berkman, Gruenewald, & Seeman, 2006). The lack of close network ties may constitute a chronic stressor that produces hormonal wear-and-tear and ultimately hastens biological aging (Berkman, 1988).

Social networks also shape health through pathways that involve *informal social control* and *monitoring* functions. For one, without regular social contact, subtle health changes can go undetected, particularly among those who are hesitant to seek medical care (see Henry & Rosenthal, 2013, for an interesting example related to nighttime breathing). In this sense, daily health appraisal is a valuable function of network ties. Social network ties are also more beneficial to the extent that they yield *instrumental* support, which involves such things as financial aid, physical assistance, or even rides to the hospital. Instrumentally supportive ties also aid during the recovery process from an illness or a medical procedure (e.g., Glass, Matchar, Belyea, & Feussner, 1993). It is worth highlighting several studies, however, which suggest that the availability of instrumental assistance from social network members may reduce the need for independent physical activity and thus reduce functional health (Litwin & Stoeckel, 2013; Mendes de Leon, Gold, Glass, Kaplan, & George, 2001).

Bear in mind that access to close contacts does not automatically yield beneficial social support. For one, network ties may produce what is sometimes called "negative support," which can involve criticism, excessive demands, or oppressive social control (e.g., Krause, 2007). At the same time, some network members may be unable to provide support, depending on their own circumstances. When they face the same stressors as ego, for example, close contacts may have the same need for support as ego, making it necessary for both of them to turn to others for support (e.g., see Gottlieb & Wagner, 1991). In these circumstances, weaker ties may constitute better sources of support. Some recent work suggests that people tend to be disproportionately connected to others with similar health statuses (e.g., Daw, Margolis, & Verdery, 2015; Monden, 2007; see also Smith & Christakis, 2008). This form of network homophily (McPherson, Smith-Lovin, & Cook, 2001) may therefore shape the types of support that are available to older adults who experience health problems. Research also shows that some social groups are more likely to receive support from their network members than others. For example, much research has examined whether women receive less support from their spouses than men do (e.g., Antonucci & Akiyama, 1987). These issues are still being addressed in ongoing research.

Network-Structural Pathways. Other research focuses less on the role of social support and more on specific compositional or structural features of networks that have distal associations with health. For one, network composition shapes the array of social obligations that are embedded in one's pre-existing network ties, and thus shape the availability of social support. Close contacts (e.g., kin) have more role expectations and thus provide a wider range of support (Wellman & Wortley, 1990). Close ties also provide more direct health-protective support in the form of social control (Umberson, 1987). This may manifest as the deterrence of harmful health behavior, such as heavy drinking, illicit drug use, and overeating. It may also involve encouraging an individual to see a physician or receive a treatment or screening (Neimann & Schmitz, 2010) or enforcing a diet or exercise regimen to deal with chronic illness (Stephens, Rook, Franks, Khan, & Iida, 2010). But much of this is contingent on social network density – as having interconnected network members facilitates their monitoring and coordination efforts (Ashida & Heaney, 2008; Coleman, 1988; DiMatteo, 2004; Haines, Hurlbert, & Beggs, 1996).

Research on social network typologies has found that older adults who have restricted network types are at greater risk of early mortality (Litwin & Shiovitz-Ezra, 2006), depression (Fiori & Jager, 2012; Park et al., 2014) and low self-rated health (Fiori & Jager, 2012). Greater *network diversity* – having different types of contacts – provides unique health benefits for older adults. One potential benefit is exposure to new information, including knowledge of uncommon preventative and health-promoting treatment options (Shiovitz-Ezra & Litwin, 2012). Maintaining a diverse social network also contributes to psychological well-being (Fiori et al., 2006). Networks that are composed of a wide range of people provide access to different social roles, which in turn can increase self-esteem, self-efficacy, and other psychosocial resources that protect mental health (Avlund et al., 2004). Several observers emphasize the special importance of (nonkin) *friends* and other weak ties, noting that "the absence of family in the context of some community support (e.g., friends) is less detrimental than the absence of friends in the context of familial support" (Fiori et al., 2006, p. 31). Nonfamilial social contact is typically based on shared interests and rewarding activity, while interaction with family is often more mundane and obligatory (e.g., Larson, Mannell, & Zuzanek, 1986). This may explain why the nonkin aspect of network diversity is so important for general measures of health in samples of older adults (Fiori et al., 2006; Shiovitz-Ezra & Litwin, 2012). It also helps to explain why many older adults value having social network members who are largely external to their family networks, and why lacking such ties can lead to health problems (Cornwell, 2011; Cornwell & Laumann, 2011). However, it is worth noting that the health implications of having family-centered networks may be different in non-Western contexts (see Cheng et al., 2009; Fiori, Antonucci, & Akiyama, 2008).

Researchers are also paying greater attention to network content. Rather than assuming that greater network frequency and size are automatically beneficial, scholars are increasingly recognizing that social ties can have variable effects depending on the extent to which health is a matter of shared interest and concern among alters (see Perry & Pescosolido, 2010; Schafer, 2013a; York Cornwell & Waite, 2012). These considerations are especially relevant for detecting and managing chronic disease. In a study assessing older adults' risk of undiagnosed or uncontrolled hypertension, for example, York Cornwell and Waite (2012) show that the protective effects of social network size are only evident for those who discuss health matters with their social network members. In fact, the authors argue that "when lines of communication about health are closed, relationships may present more costs than benefit" (p. 227).

Social networks are also crucial in shaping individuals' exposure to health-related behavioral norms as well as to heath conditions. One important mechanism linking social networks to health is person-to-person contact (Berkman, Glass, Brissette, & Seeman, 2000; Smith & Christakis, 2008). Epidemiologists have long studied the spread of health conditions such as air-borne illnesses and sexually transmitted diseases through populations, and social network analyses have made it possible to model how this transmission operates through person-to-person contact and is shaped by factors such as social proximity, demographic similarity (homophily), and joint participation in activities (Christakis & Fowler, 2010; Morris, Zavisca, & Dean, 1995). This approach has recently been generalized to understand the spread of conditions that are noninfectious yet subject to social influence, such as obesity and smoking (Christakis & Fowler, 2007; Smith & Christakis, 2008; Valente, 2010). One's network ties constitute role models and sources of information and norms that can gradually but indelibly influence health-related behavior over long periods of time. This research shows that only by studying older adults' positions within a broader configuration of network members – who may vary in terms of behavior and health status – is it possible to understand the full range of social forces that operate to shape older adults' health and well-being.

AGING AND SOCIAL NETWORK CHANGE

The available evidence suggests that, in general, older adults have smaller and denser networks, lower rates of social contact, and greater loneliness – yet, older adults are also more involved in community networks involving local groups and associations (Ajrouch, Blandon, & Antonucci, 2005; Cornwell et al., 2008; Schnittker, 2007; Shaw, Krause, Liang, & Bennett, 2007). A key question is why older adults' networks differ in these respects. Scholars are increasingly cognizant of the fact that change is endemic in social networks, and thus there is growing interest in how social networks evolve (see Snijders & Doreian, 2010). This is something that social gerontologists have understood for a long time, partly because the issue of network change seems to be particularly salient in later life. Many scholars are interested in how networks change during this period of the life course because it is a time when health declines and the need for social support increases, when network ties are a primary source of sense of belonging, and when life transitions affect network structure (Fiori et al., 2008; Litwin & Stoeckel, 2013). Much research on this is motivated by concern over the social implications of transitions like retirement, bereavement, and health decline.

Theories of Network Change in Later Life

There are conflicting views regarding what happens to individuals' social networks in later life. Echoing the early concerns of the oft-castigated theory of social disengagement (Cumming & Henry, 1961), some scholars emphasize that the older population is vulnerable to the kinds of life-course experiences (e.g., bereavement, health decline) that can seriously hamper network connectedness and affect tendencies to maintain certain types of network ties in the first place (Cornwell, 2009; Ikkink & van Tilburg, 1999; Lyyra, Lyyra, Lumme-Sandt, Tiikkainen, & Heikkinen, 2010; Schafer, 2011, 2013b; Shaw et al., 2007; Stevens & van Tilburg, 2011). Socioemotional selectivity theory holds that older adults intentionally shift their priorities to stronger, emotionally rewarding ties (Charles & Carstensen, 2010). There is also evidence that aging partners restrict their social networks to shared contacts through a process of dyadic withdrawal (Kalmijn, 2003), which results in a narrowing of the range of aging partners' networks. In this sense, it is not surprising that older adults' social networks tend to be smaller and more kin-centered than the networks of young and middle-aged adults. In much of this work, there is a strong emphasis on mechanisms through which older adults' networks shrink and collapse.

At the same time, social gerontologists are increasingly aware of the fact that many older adults want to remain independent and socially active, so that the very aspects of aging that affect social relationships only strengthen their resolve to establish social connections. Some research has found, for example, that retirement can both reduce network range and access to weak ties and increase connectedness to both kin and to the broader community, often by increasing involvement in voluntary associations (e.g., Cornwell et al., 2008; McDonald & Mair, 2010; van Groenou & van Tilburg, 2012). There is also mixed evidence regarding the aftermath of bereavement, as some scholars have highlighted increased loneliness while others have documented increased social activity (Donnelly & Hinterlong, 2010; Zettel & Rook, 2004).

The notion that older adults' networks can persist and even grow is present in continuity and activity theories, which point out that people grow accustomed to certain social roles and activities during their lives and attempt to maintain them in the wake of later-life transitions (Atchley, 1989; Donnelly & Hinterlong, 2010). The loss of social connections sometimes sparks efforts to adapt to and/or compensate by developing new social ties (e.g., Lamme, Dykstra, & van Groenou, 1996). Indeed, an analysis of longitudinal egocentric network data from the NSHAP study reported that most older adults (81.8%) added at least one new network member during the study period, and most (59.4%) cultivated multiple new confidant relationships (Cornwell & Laumann, 2015). In fact, more respondents (37.9%) reported a net expansion of their networks than a net decrease (26.6%).

Consequences of Network Change for Older Adults

How exactly people's social networks change in later life is important to understand for a variety of reasons. In particular, a growing body of research demonstrates that changes in social networks have health effects for individuals above and beyond baseline levels (Cornwell & Laumann, 2015; Holtzman et al., 2004; Seeman et al., 2011; Zhang, Yeung, Fung, & Lang, 2011). Seeman et al. (2011), for example, find that decreases in social engagement lead to poorer executive function and memory performance. Using longitudinal network data, Cornwell and Laumann (2015) find that the loss of confidants is significantly associated with declines in self-rated health as well as increasing functional impairment over a 5-year period, net of baseline network connectedness, health, and a variety of other risk factors.

What consequences social network change has for health is an especially important issue in the context of aging, given the particular challenges, upheaval, and adaptation that can occur in later life. Several lines of research provide clues about the link between changes in networks and health during this period of life. Life-course research sheds light on the effects of common later-life transitions that involve separation from key social roles (especially retirement) and the loss of contacts (especially widowhood). This work suggests the presence of unique dynamic mechanisms that link social networks to health. Social network change – especially if it involves losing ties – may be not only a psychologically destabilizing experience, but also one that disrupts the internal support functions of a network that are so important to older adults, including established routines of coordination and communication among one's network members (Cornwell & Laumann, 2015; Gerstorf, Röcke, & Lachman, 2010; Stroebe, Schut, & Stroebe, 2007). This may lead to a variety of problems, including difficulty coordinating support and care among network members, conflict, and a sense of instability or normlessness. Unfortunately, few scholars have examined this issue using empirical data.

While much of the available research on network change in later life emphasizes loss and isolation, other research suggests that these later-life transitions may actually spur network *growth*, as cultivating new social connections is one way that many older adults adapt to later-life challenges and compensate for loss (Atchley, 1989; Donnelly & Hinterlong, 2010; Zettel & Rook, 2004). People who adjust to later-life transitions by remaining socially active tend to be happier and healthier, both because they maintain their access to social resources and because of the physical activity and mental stimulation that comes with social adaptation (Kahana, Kelley-Moore, & Kahana, 2012). Analysis of the longitudinal egocentric network data from the NSHAP study shows that cultivating new confidant relationships leads to significant improvements in functional, self-rated, and psychological health, net of any concurrent network losses (Cornwell & Laumann, 2015). Cultivating even weak ties in the wake of bereavement may be more beneficial to older adults than turning inward to one's closest contacts, as older adults value the weaker ties that they maintain with acquaintances, neighbors, and group members because these ties help them maintain their independence and nonfamilial connectedness (Cornwell, 2011). Just as the finding that network changes have unique individual-level consequences calls to mind a variety of implications for research and policy, the notion that social networks are just as likely to grow as they are to collapse in later life suggests new avenues for research in social gerontology that move beyond unfounded assumptions about the social experiences of later life.

SOCIAL NETWORKS AND STRATIFICATION

One of the most important topics in social gerontology is variation in different older adults' connectedness to social networks. We have already addressed several factors that shape networks in later life, including life-course changes associated with marital status, labor force status, and health. Several sociodemographic characteristics also appear to give rise to inequality with respect to social resources in later life – including race/ethnicity, SES, and gender. Motivated in part by theoretical perspectives within social gerontology that underscore heightened inequality at older ages – such as the *double jeopardy hypothesis* (Carreon & Noymer, 2011) and the *cumulative dis/advantage framework* (Dannefer, 2003) – some researchers have begun to look more closely at the role of social disadvantage in deepening inequalities with respect to some older adults' access to valuable network resources.

Race/Ethnicity

A 1992 Detroit-based study found that black older adults have smaller networks than Whites, a racial disparity that held steady across age (Ajrouch, Antonucci, & Janevic, 2001). The same study found that black adults had a greater proportion of kin in their networks than did Whites, though the difference shrank at older ages. Another study reveals similar findings regarding Black–White differences in network size, also concluding that the disparity shrinks with age (Barnes, Mendes de Leon, Bienias, & Evans, 2004). Specifically, black seniors see fewer children, relatives, and friends on a monthly basis than do white adults, but this gap is more pronounced for 75-year olds than it is for 85-year olds. Recent findings using the nationally representative NSHAP data also indicate that black older adults have smaller networks than Whites (Cornwell et al., 2008). Latinos name fewer close discussion partners than do others. At the same time, black and Latino seniors see their network alters more frequently than do white older adults.

It is likely that different kinds of social network ties are more available and/or useful in different social contexts for different race/ethnic groups. For example, due to higher rates of mortality, residential mobility, incarceration, and other forms of environmental instability among Blacks, older black adults may be more likely to have to rely on what might otherwise be thought of as weaker (e.g., nonspousal) sources of support, though evidence on this point is mixed (see Krause, 2002; Peek & O'Neill, 2001). Because people disproportionately form ties with others who share similar socio-demographic characteristics (e.g., race/ethnicity, age, income), racial minorities are more likely to be in a situation in which their network members face the same disadvantages as they do. This, in turn, may further complicate minorities' abilities to deal with later-life challenges.

Recent work also suggests that racial minorities experience social network change differently and to a greater extent than others (Shaw et al., 2007). In addition (and related) to aspects of social disorganization (e.g., higher rates of residential turnover, incarceration), racial minorities suffer from more health problems, contributing to higher mortality rates (see Adler & Rehkopf, 2008). By implication, racial minorities experience bereavement more frequently due to the health-related incapacitation, institutionalization, and death of their network members (Hawkins & Abrams, 2007). Longitudinal data from the NSHAP study show that older black adults experienced a significantly greater rate of loss from their confidant networks than older Whites over a 5-year time period (Cornwell, 2015). Likewise, black respondents reported higher rates of death among their confidants during this period. Higher rates of network loss among black older adults may also help explain greater rates of cultivating new network ties among older black adults. These patterns are not present, however, in older Latino adults' personal networks. In fact, they experience only about half the rate of network member deaths as non-Latinos, and there is no connection between Hispanic ethnicity and the addition of new network members (Cornwell, 2015).

Socioeconomic Status

Research suggests that individuals of low SES also experience persistent network disadvantages. While low-SES individuals have more kin-centered networks, they also have less access to forms of social capital that can affect health and access to other resources, such as ties to medical experts (Lin, 2000; McDonald, Lin, & Ao, 2009). This situation continues into later life. Older adults who have less formal education, for example, have smaller, less diverse networks, although they also experience higher rates of contact with and closer physical proximity to their network ties (Ajrouch et al., 2005; Kohli, Hank, & Künemund, 2009; Krause & Borawski-Clark, 1995). It is unclear to what extent these differences reflect differences in material resources (e.g., owning a car), cognitive skills (Broese van Groenou & van Tilburg, 2003), or cultural differences. Regardless, this is an important issue, as recent work suggests that SES-related differences in both social network connectedness and social support increase in later life (Fischer & Beresford, 2015). Given that people tend to be disproportionately connected to other people who have similar educational backgrounds and social class status (McPherson et al., 2001), low-SES older adults also face more challenges with respect to the procurement of instrumental social support.

As with racial minorities, low-SES older adults also experience less stable network environments. Recent analyses of the second NSHAP wave (2010/2011) reveal that older adults who did not go to college lost network members at a 14.1% greater rate than those who had college degrees (Cornwell, 2015). Moreover, the lowest-SES group experienced network member deaths at a 37.5% greater rate than those who had college degrees. Longitudinal analyses of cross-national data from the SHARE study also suggest that higher-SES older adults are more likely to start (and less likely to stop) involvement in formal groups than others (Kohli et al., 2009). Finally, older adults who have less formal education experience enduring disadvantages with respect to levels of social contact with friends, as opposed to increasing or decreasing disadvantage over time (Shaw et al., 2007).

Gender

For the most part, research demonstrates the greater network connectedness of older women compared to older men. Older women have larger social networks in general (McLaughlin, Vagenas, Pachana, Begum, & Dobson, 2010), and while older men often maintain more ties to co-workers, older women maintain more connections to family members, friends, and neighbors (Shaw et al., 2007). Older women have more connections to formal groups and are more involved in the community (Cornwell et al., 2008; Kohli et al., 2009). There are also gender differences in network structure. Despite their closer connections to kin, older women are also more likely to maintain contacts with people who do not know each other (or their spouse), providing them with both more bridging opportunities and more social independence (e.g., from their spouses) (Cornwell, 2011).

These gender differences may decrease over time (e.g., Ajrouch et al., 2005), which may reflect the structural effects of the fact that older couples' networks tend to converge over time (Kalmijn, 2003). Recent analyses show that older women experience more net increases in confidant network size than do older men (Cornwell, 2015). Analyses of cross-national data also reveal that older women are significantly more likely to start formal group connections than older men (Kohli et al., 2009). In addition, older women are less likely to begin to disengage from existing family relations (de Jong Gierveld & Perlman, 2006). These connections may provide older women with more opportunities to adapt to network losses.

EMERGING TOPICS IN NETWORK-GERONTOLOGY

As the social network paradigm grows increasingly dominant within the field, social gerontologists have begun to use it as a framework both for providing fresh insight into problems facing older adults and for understanding how recent social changes are affecting older adults.

Electronic Networks

Electronic social networks play an increasingly important role in linking people to social resources, though the extent to which they integrate older adults is still unclear. A 2012 Pew Internet report reveals that about one-third of American adults aged 65+ report using social networking sites (18% said they had used a site such as Facebook or LinkedIn in the past day) – much lower than rates for younger adults (Zickuhr & Madden, 2012; see also Brenner,

2012). At the same time, older adults have been the fastest-growing age group for online social network adoption over the past 3 years. As recently as 2009, only 13% of adults aged 65+ reported using online social networks (Zickuhr & Madden, 2012). Despite this uptick in online social network usage, a sizeable proportion of older adults continue to express concerns over privacy issues online and many do not see the point of online social networks (Norvall, 2012; Xie, Watkins, Goldbeck, & Huang, 2012).

Older adults' integration into online networks is important to study for several reasons. Scholars emphasize the practical and clinical virtues of social network usage for older adults, and some anticipate that sites such as Facebook may reduce loneliness and maintain contact with family members or close friends (Erickson, 2011; Farkas, 2010). This may be especially important when health problems limit seniors' travel and when close contacts are geographically dispersed. Online social network platforms may also be especially helpful for preserving and strengthening intergenerational ties (Nef, Ganea, Müri, & Mosimann, 2013). For instance, photo-sharing is one of the most popular activities on Facebook (Joinson, 2008), so instant access to pictures of grandchildren and children (e.g., while they are on their family vacations, or during graduations) can produce a sense of intergenerational involvement and participation in meaningful events. Australian scholars are currently working on identifying older adults' social networks from online content. The Social Networks and Ageing Project (SNAP) uses trace data from Australian senior social networking web sites (e.g., Greypath) to identify online network ties among older adults (Ackland, 2010). With supplemental survey data regarding participants' health and well-being, the ultimate goal is to understand the association between online social networking and healthy aging in general (see http://adsri.anu.edu.au/groups/ageing/snap). Related work is exploring the possibility of harnessing electronic networks to help older adults "age in place" by using new technologies within older adults' homes to alert network members when something is amiss in the household, and to help alleviate caregiver burden when nothing is wrong (see Kang et al., 2010). A promising direction for future research is to use controlled experiments and other methods to determine whether older adults' participation in online networks and their positions within them are related to relationship quality, sense of belonging, and other aspects of well-being.

Whole Networks

Earlier we discussed structural features of networks – including cluster, bridging potential, homophily, and closeness centrality – many of which can only be examined using whole network data (e.g., Schafer, 2011, 2013b). A key task for network-oriented social gerontologists is to track and examine these networks, and older adults' positions within them. Older adults form and maintain their social network ties within different contexts than others – for example, in retirement communities – and must do so under different circumstances (e.g., in the wake of health decline). As such, it is not only likely that the whole networks that are composed solely of older adults differ structurally from those that are commonly examined elsewhere in the social sciences, such as adolescent friendship networks, sexual contact and risk networks, and electronic networks. Because analyses of these other kinds of networks provide the basis for much of researchers' understanding of social networks themselves (e.g., see Wasserman & Faust, 1994), analysis of the unique networks formed by older adults may provide new insights into the structural properties of social networks. Recent statistical advances – including exponential random graph models for the analysis of network structure, and Simulation Investigation for Empirical Network Analysis (SIENA) for the study of structural change – have made this type of work far more feasible than in the past. These tools overcome many of the problems that social network analysis introduces for traditional statistical methods (e.g., the non-independence of network actors) and enable a host of novel hypothesis tests about triads, homophily, reciprocity, and other quintessential network phenomena. See Robins, Pattison, Kalish, and Lusher (2007) and Snijders, van de Bunt, and Steglich (2010) for nontechnical introductions.

More importantly, however, analyses of whole networks will provide fresh insight into the roles and positions of older adults within broader society. For reasons already discussed, older adults likely occupy different positions within the networks of their broader community and which include young, middle-aged, and older adults alike. This, in turn, has important implications for the social influence of and opportunities available to older adults. These are issues that can only be explored with the aid of whole-network data.

Analyses of whole networks may also yield information about more peripheral members of older adults' networks. Social-gerontological research is heavily oriented toward social support and strong ties. Rarer is explicit discussion about the full range of diverse resources located within and between seniors' weaker social ties. Some recent work has sought to tap into the very weakest, transient social ties that exist in older people's social worlds – including the individuals encountered in cafes, at bus stops, or in the grocery store that acknowledge their presence or participate in small talk (Gardner, 2011). Individually, each of these "consequential strangers" (Fingerman, 2009) may contribute very few resources or directly offer much social support, but, in aggregate, they can contribute to people's sense of belonging in their community (Gardner, 2011) as well as their sense of independence from dense family-only networks (Cornwell, 2011).

Network Diffusion Processes

One analytic opportunity that only whole social network data can provide is the ability to study diffusion processes across older adult populations. With data on the social connections that exist throughout an entire community, it is often possible to trace flows of information, exchanges, attitudes and beliefs, social norms, and other phenomena (Valente, 2010), and thus to predict rates and locations of adoption, infection, and other important outcomes. This could be particularly valuable for gaining insight into understanding how social norms (e.g., regarding health-related behavior) emerge within these communities, and especially for understanding the (risk of) spread of illnesses (e.g., influenza).

Natural contexts for such research in gerontology are senior congregate housing and healthcare facilities. Nursing homes, assisted living facilities, and active-lifestyle communities place older adults in a relatively confined space, so diffusion processes – biological and otherwise – may be apparent in these settings. Surprisingly few studies have investigated person-to-person contact in explicitly older populations, despite the insight this would provide into the risk of epidemic spread of infection. For example, risky sexual behavior within congregate housing has been a growing concern amidst rising rates of HIV/AIDS among older adults (see Orel, Spence, & Steele, 2005), so there is a need for even descriptive studies of the sexual networks that emerge within assisted living facilities and retirement communities. There are other promising opportunities for research in this area as well. Electronic sensors, for instance, can be used in geriatric hospital units to investigate patterns of interpersonal contact between patients and healthcare workers and to identify the actors who are most likely to spread infection (Vanhems et al., 2013). The practical and ethical implications of this type of research need to be explored.

Negative Network Ties

Above we discussed the link between social network connections and desirable social resources, such as social support and social capital. Not all network ties are unequivocally supportive or advantageous, and a growing number of studies investigate how more negatively valenced interactions occur between social network members (Akiyama, Antonucci, Takahashi, & Langfahl, 2003; Birditt, Jackey, & Antonucci, 2009; Krause, 2007; Rook, 1997). Social network members can be over-controlling, make excessive demands, be overly critical, or simply get on each other's nerves (Antonucci, Akiyama, & Lansford, 1998). Interestingly, however, research suggests that the prevalence of such negative experiences is somewhat lower within older people's networks than in the networks of younger or middle-aged adults (Akiyama et al., 2003; Birditt et al., 2009). This may be due to underreporting biases, as older adults who are dependent upon family or friends could be hesitant to report actions that could be construed as elder mistreatment (see Chapter 16). A lower prevalence of negative social interactions may also reflect the effects of successful network selection processes in later life (e.g., Charles & Carstensen, 2010), but we are not aware of any research that has examined this possibility. To date, this line of work has focused on the "milder" side of negativity, and few studies have examined the prevalence and consequences of outright belittlement, hostility, and even abuse within older adults' networks. Furthermore, more research is needed on the potential role of social network composition and structure in shaping negative network phenomena. Social gerontologists may benefit by considering the possibility that network-structural features such as size or density may decrease or increase the risk that physical abuse goes undetected (see Schafer & Koltai, in press), or the possibility that older adults' greater integration into electronic networks may reduce such risk.

CONCLUSIONS

The introduction of an explicit network-analytic paradigm (Wasserman & Faust, 1994) has yielded valuable insights in social gerontology, and is moving the field's approach to the study of social relationships in exciting new directions. Network-oriented gerontological research has come a long way in the past few years. We have witnessed a rapid increase in scholarly interest in the internal wiring of older adults' networks, the structure of whole community networks, electronic networks, and renewed interest in the causes and consequences of network change in later life. This chapter has therefore sought to increase social gerontologists' familiarity with the particular concepts and methods of social network analysis, the unique analytic questions it asks, and the opportunities it provides to answer important questions about the challenges of aging.

The implementation of formal analyses of important features of older adults' social networks is increasingly feasible thanks to developments in data collection and technology. The move toward a network paradigm is apparent in the development of national and international population-based studies that focus specifically on older adults' social networks. Recent datasets that make it possible to analyze basic structural and compositional features of older adults' networks include the NSHAP study – which includes unique longitudinal egocentric network data – and the SHARE study – which will have longitudinal egocentric network data in the near future. Fortunately, the widespread availability of these data makes research on the structure and importance of older adults' egocentric social networks more feasible. Some researchers have also begun to collect whole community-level social network data (e.g., Schafer, 2011), including ongoing efforts to study the structure of older adults' electronic networks like the SNAP.

What is most badly needed is a study of whole networks of older adults across multiple communities, much like the National Longitudinal Study of Adolescent Health (Add Health) did with adolescent friendship networks.

Network-oriented social gerontology has yet to achieve its full potential for identifying consequential aspects of older adults' social lives. In this respect, we believe that the analysis of whole social network data is the most important goal. Otherwise, it will be impossible to assess older adults' positions within larger communities and thus their access to and control over important resources, and their exposure to others' behaviors and attitudes. Available research suggests that analyses of larger networks will provide valuable insight into not only the structural positions of older adults relative to younger and middle-aged adults, but also the unique social constraints and opportunities that face subgroups within the older adult population. There is much need for additional research on the nature and implications of racial/ethnic, socioeconomic, and gender-based differences in older adults' social network resources. For it is possible that differences in access to certain kinds of network ties, as well as the experience of network change, may help to explain parallel differences in older adults' health and mortality. Disproportionate turnover within their social networks in particular may create a cycle that amplifies pre-existing instability within disadvantaged groups' broader social environments. Social network analysis thus presents gerontologists with a unique opportunity to examine how a consequential yet under-examined set of factors shapes the experiences of particularly vulnerable older adults.

References

Ackland, R. (2010). Using digital trace data to research social networks and ageing. *Presentation at ACSPRI Social Science Methodology Conference.* December 2, 2010, University of Sydney.

Adler, N. E., & Rehkopf, D. H. (2008). U.S. disparities in health: Descriptions, causes, and mechanisms. *Annual Review of Public Health, 29,* 235–252.

Ajrouch, K. J., Antonucci, T. C., & Janevic, M. R. (2001). Social networks among blacks and whites: The interaction between race and age. *The Journals of Gerontology Series B: Psychological Sciences and Social Sciences, 56,* S112–S118.

Ajrouch, K. J., Blandon, A. Y., & Antonucci, T. C. (2005). Social networks among men and women: The effects of age and socioeconomic status. *Journals of Gerontology Series B: Psychological Sciences and Social Sciences, 60,* S311–S317.

Akiyama, H., Antonucci, T. C., Takahashi, K., & Langfahl, E. S. (2003). Negative interactions in close relationships across the life span. *Journals of Gerontology Series B: Psychological Sciences and Social Sciences, 58,* P70–P79.

Antonucci, T. C., & Akiyama, H. (1987). An examination of sex differences in social support among older men and women. *Sex Roles, 17,* 737–749.

Antonucci, T. C., Akiyama, H., & Lansford, J. E. (1998). Negative effects of close social relations. *Family Relations, 47,* 379–384.

Ashida, S., & Heaney, C. A. (2008). Differential associations of social support and social connectedness with structural features of social networks and the health status of older adults. *Journal of Aging and Health, 20,* 872–893.

Atchley, R. C. (1989). The continuity theory of normal aging. *The Gerontologist, 29,* 183–190.

Avlund, K., Lund, R., Holstein, B. E., Due, P., Sakari-Rantala, R., & Heikkinen, R. L. (2004). The impact of structural and functional characteristics of social relations as determinants of functional decline. *Journals of Gerontology Series B: Psychological Sciences and Social Sciences, 59,* S44–S51.

Barnes, L. L., Mendes de Leon, C. F., Bienias, J. L., & Evans, D. A. (2004). A longitudinal study of black-white differences in social resources. *The Journals of Gerontology Series B: Psychological Sciences and Social Sciences, 59,* S146–S153.

Batagelj, V., & Mrvar, A. (1998). Pajek: Program for large network analysis. *Connections, 21,* 47–57.

Berkman, L. F. (1988). The changing and heterogeneous nature of aging and longevity: A social and biomedical perspective. *Annual Review of Gerontology and Geriatrics, 8,* 37–68.

Berkman, L. F., Glass, T., Brissette, I., & Seeman, T. E. (2000). From social integration to health: Durkheim in the new millennium. *Social Science & Medicine, 51,* 843–857.

Berkman, L. F., & Syme, S. L. (1979). Social networks, host resistance, and mortality: A nine-year follow-up study of Alameda county residents. *American Journal of Epidemiology, 109,* 186–204.

Birditt, K. S., Jackey, L. M. H., & Antonucci, T. C. (2009). Longitudinal patterns of negative relationship quality across adulthood. *The Journals of Gerontology Series B: Psychological Sciences and Social Sciences, 64B*, 55–64.

Brenner, J. (2012). *The Demographics of Social Media Users – 2012*. Pew Research Center's Internet & American Life Project. Available at: <http://www.pewinternet.org/files/old-media//Files/Reports/2013/PIP_Social MediaUsers.pdf> (Last retrieved December 9, 2013).

Broese van Groenou, M. I., & Van Tilburg, T. G. (2003). Network size and support in old age: Differentials by socio-economic status in childhood and adulthood. *Ageing & Society, 23*, 625–645.

Burt, R. S. (1992). *Structural holes: The social structure of competition*. Cambridge, MA: Harvard University Press.

Carreon, D., & Noymer, A. (2011). Health-related quality of life in older adults: Testing the double jeopardy hypothesis. *Journal of Aging Studies, 25*, 371–379.

Charles, S. T., & Carstensen, L. L. (2010). Social and emotional aging. *Annual Review of Psychology, 61*, 383–409.

Cheng, S.-T., Lee, C. K. L., Chan, A. C. M., Leung, E. M. F., & Lee, J.-J. (2009). Social network types and subjective well-being in Chinese older adults. *Journals of Gerontology Series B: Psychological Sciences and Social Sciences, 64B*, 713–722.

Christakis, N. A., & Fowler, J. H. (2007). The spread of obesity in a large social network over 32 years. *New England Journal of Medicine, 357*, 370–379.

Christakis, N. A., & Fowler, J. H. (2010). Social network sensors for early detection of contagious outbreaks. *PLoS ONE, 5*, e12948.

Coleman, J. S. (1988). Social capital in the creation of human capital. *American Journal of Sociology, 94*, S95–S120.

Cornwell, B. (2009). Good health and the bridging of structural holes. *Social Networks, 31*, 92–103.

Cornwell, B. (2011). Independence through social networks: Bridging potential among older women and men. *Journals of Gerontology Series B: Psychological Sciences and Social Sciences, 66*, 782–794.

Cornwell, B. (2012). Spousal network overlap as a basis for spousal support. *Journal of Marriage and Family, 74*, 229–238.

Cornwell, B. (2015). Social disadvantage and network turnover. *Journals of Gerontology Series B: Psychological Sciences and Social Sciences, 70*, 132–142.

Cornwell, B., & Laumann, E. O. (2011). Network position and sexual dysfunction: Implications of partner betweenness for men. *American Journal of Sociology, 117*, 172–208.

Cornwell, B., & Laumann, E. O. (2015). The health benefits of network growth: New evidence from a national survey of older adults. *Social Sciences & Medicine, 125*, 94–106.

Cornwell, B., Laumann, E. O., & Schumm, L. P. (2008). The social connectedness of older adults: A national profile. *American Sociological Review, 73*, 185–203.

Cumming, E., & Henry, W. E. (1961). *Growing old*. New York, NY: Arno Press.

Dannefer, D. (2003). Cumulative advantage/disadvantage and the life course: Cross-fertilizing age and social science theory. *Journal of Gerontology Series B: Psychological Sciences and Social Sciences, 58*, S328–S337.

Daw, J., Margolis, R., & Verdery, A. (2015). Siblings, friends, course-mates, club-mates: How adolescent health behavior homophily varies by race, class gender, and health status. *Social Science & Medicine, 125*, 32–39.

de Jong Gierveld, J., & Perlman, D. (2006). Long-standing nonkin relationships of older adults in the Netherlands and the United States. *Research on Aging, 28*, 730–748.

DiMatteo, M. R. (2004). Social support and patient adherence to medical treatment: A meta-analysis. *Health Psychology, 23*, 207–218.

Donnelly, E. A., & Hinterlong, J. E. (2010). Changes in volunteer activity among recently widowed older adults. *The Gerontologist, 50*, 158–169.

Erickson, L.B. (2011). Social media, social capital, and seniors: The impact of Facebook on bonding and bridging social capital of individuals over 65. In *AMCIS 2011 Proceedings*, Paper 85. Available at: <http://aisel.aisnet.org/amcis2011_submissions/85>

Farkas, P.A. (2010). Senior social platform – An application aimed to reduce the social and digital isolation of seniors. In *Proceedings of REAL CORP*, Vienna, Austria, May 18–20, 2010.

Fingerman, K. L. (2009). Consequential strangers and peripheral ties: The importance of unimportant relationships. *Journal of Family Theory & Review, 1*, 69–86.

Fiori, K. L., Antonucci, T. C., & Akiyama, H. (2008). Profiles of social relations among older adults: A cross-cultural approach. *Ageing & Society, 28*, 203–231.

Fiori, K. L., Antonucci, T. C., & Cortina, K. S. (2006). Social network typologies and mental health among older adults. *Journals of Gerontology Series B: Psychological Sciences and Social Sciences, 61*, P25–P32.

Fiori, K. L., & Jager, J. (2012). The impact of social support networks on mental and physical health in the transition to older adulthood: A longitudinal, pattern-centered approach. *International Journal of Behavioral Development, 36*, 117–129.

Fiori, K. L., Smith, J., & Antonucci, T. C. (2007). Social network types among older adults: A multidimensional approach. *The Journals of Gerontology Series B: Psychological Sciences and Social Sciences, 62*, P322–P330.

Fischer, C. S. (1982). The dispersion of kinship ties in modern society: Contemporary data and historical speculation. *Journal of Family History, 7*, 353–375.

Fischer, C. S., & Beresford, L. (2015). Changes in support networks in late middle age: The extension of gender and educational differences. *Journals of Gerontology Series B: Psychological Sciences and Social Sciences, 70*, 123–131.

Ford, E. S., Loucks, E. B., & Berkman, L. F. (2006). Social integration and concentrations of C-reactive protein among US adults. *Annals of Epidemiology, 16*, 78–84.

Freeman, L. C. (1979). Centrality in social networks: Conceptual clarification. *Social Networks, 1*, 215–239.

Gardner, P. J. (2011). Natural neighborhood networks – Important social networks in the lives of older adults aging in place. *Journal of Aging Studies, 25*(3), 263–271.

Gerin, W., Milner, D., Chawla, S., & Pickering, T. G. (1995). Social support as a moderator of cardiovascular reactivity in women: A test of the direct effects and buffering hypotheses. *Psychosomatic Medicine, 57*, 16–22.

Gerstorf, D., Röcke, C., & Lachman, M. E. (2010). Antecedent-consequent relations of perceived control to health and social support: Longitudinal evidence for between-domain associations across adulthood. *Journals of Gerontology: Psychological Sciences and Social Sciences, 66B*, 61–71.

Glass, T. A., Matchar, D. B., Belyea, M., & Feussner, J. R. (1993). Impact of social support on outcome in first stroke. *Stroke, 24*, 64–70.

Gottlieb, B. H., & Wagner, F. (1991). Stress and support processes in close relationships. In J. Eckenrode (Ed.), *The social context of coping* (pp. 165–188). New York, NY: Springer.

Granovetter, M. S. (1973). The strength of weak ties. *American Journal of Sociology, 78*, 1360–1380.

Haines, V. A., Hurlbert, J. S., & Beggs, J. J. (1996). Exploring the determinants of support provision: Provider characteristics, personal networks, community contexts, and support following life events. *Journal of Health and Social Behavior, 37*, 252–264.

Hawkins, R. L., & Abrams, C. (2007). Disappearing acts: The social networks of formerly homeless individuals with co-occurring disorders. *Social Science & Medicine, 65*, 2031–2042.

Henry, D., & Rosenthal, L. (2013). "Listening for his breath:" The significance of gender and partner reporting on the diagnosis, management, and treatment of obstructive sleep apnea. *Social Science & Medicine, 79*, 48–56.

Holt-Lunstad, J., Smith, T. B., & Layton, J. B. (2010). Social relationships and mortality risk: A meta-analytic review. *PLoS Med, 7*, e1000316.

Holt-Lunstad, J., Uchino, B. C., Smith, T. W., Olson-Cerny, C., & Nealey-Moore, J. B. (2003). Social relationships and ambulatory blood pressure: Structural and qualitative predictors of cardiovascular function during everyday social interactions. *Health Psychology, 22*, 388–397.

Holtzman, R. E., Rebok, G. W., Saczynski, J. S., Kouzis, A. C., Doyle, K. W., & Eaton, W. W. (2004). Social network characteristics and cognition in middle-aged and older adults. *Journals of Gerontology Series B: Psychological Sciences and Social Sciences, 59*, P278–P284.

House, J. S., Robbins, C., & Metzner, H. L. (1982). The association of social relationships and activities with mortality: Prospective evidence from the Tecumseh Community Health Study. *American Journal of Epidemiology, 116*, 123–140.

House, J. S., Umberson, D., & Landis, K. R. (1988). Structures and processes of social support. *Annual Review of Sociology, 14*, 293–318.

Ikkink, K. K., & van Tilburg, T. (1999). Broken ties: Reciprocity and other factors affecting the termination of older adults' relationships. *Social Networks, 21*, 131–146.

Joinson, A.N. (2008). Looking at, looking up or keeping up with people?: Motives and use of facebook. *Proceedings of the SIGCHI conference on Human Factors in Computing Systems* (pp. 1027–1036), Available at: <http://dl.acm.org/citation.cfm?id=1357213>.

Kahana, E., Kelley-Moore, J., & Kahana, B. (2012). Proactive aging: A longitudinal study of stress, resources, agency, and well-being in late life. *Aging & Mental Health, 16*, 438–451.

Kahn, R. L., & Antonucci, T. C. (1980). Convoys over the life course: Attachment, roles, and social support. In P. B. Baltes & O. B. Brim (Eds.), *Life-span development and behavior* (pp. 253–268). New York, NY: Academic Press.

Kalmijn, M. (2003). Shared friendship networks and the life course: An analysis of survey data on married and cohabiting couples. *Social Networks, 25*, 231–249.

Kang, H. G., Mahoney, D. F., Hoenig, H., Hirth, V. A., Bonato, P., Hajjar, I., et al. (2010). In situ monitoring of health in older adults: Technologies and issues. *Journal of the American Geriatrics Society, 58*, 1579–1586.

Kohli, M., Hank, K., & Künemund, H. (2009). The social connectedness of older Europeans: Patterns, dynamics and contexts. *Journal of European Social Policy, 19*, 327–340.

Krause, N. (1987). Life stress, social support, and self-esteem in an elderly population. *Psychology and Aging, 2*, 349–356.

Krause, N. (2002). Church-based social support and health in old age: Exploring variations by race. *The Journals of Gerontology Series B: Psychological Sciences and Social Sciences, 57*, S332–S347.

Krause, N. (2007). Longitudinal study of social support and meaning in life. *Psychology and Aging, 22*, 456–465.

Krause, N., & Borawski-Clark, E. (1995). Social class differences in social support among older adults. *The Gerontologist, 35*, 498–508.

Lamme, S., Dykstra, P. A., & van Groenou, M. I. B. (1996). Rebuilding the network: New relationships in widowhood. *Personal Relationships, 3*, 337–349.

Lapierre, T. A., & Keating, N. (2013). Characteristics and contributions of non-kin carers of older people: A closer look at friends and neighbours. *Ageing & Society, 33*, 1442–1468.

Larson, R., Mannell, R., & Zuzanek, J. (1986). Daily well-being of older adults with friends and family. *Psychology and Aging, 1*, 117–126.

Lin, N. (2000). Inequality in social capital. *Contemporary Sociology, 29*, 785–795.

Lin, N. (2001). *Social capital: A theory of social structure and action*. Cambridge, MA: Cambridge University Press.

Lin, N., & Dumin, M. (1986). Access to occupations through social ties. *Social Networks, 8*(4), 365–385.

Lin, N., Ensel, W. M., Simeone, R. S., & Kuo, W. (1979). Social support, stressful life events, and illness: A model and an empirical test. *Journal of Health and Social Behavior, 20*, 108–119.

Litwin, H. (2001). Social network type and morale in old age. *The Gerontologist, 41*, 516–524.

Litwin, H., & Shiovitz-Ezra, S. (2006). Network type and mortality risk in later life. *The Gerontologist, 46*, 735–743.

Litwin, H., & Shiovitz-Ezra, S. (2011). Social network type and subjective well-being in a national sample of older Americans. *The Gerontologist, 51*, 379–388.

Litwin, H., & Stoeckel, K. J. (2013). Social network and mobility improvement among older Europeans: The ambiguous role of family ties. *European Journal of Ageing, 10*, 159–169.

Loucks, E. B., Berkman, L. F., Gruenewald, T. L., & Seeman, T. E. (2006). Relation of social integration to inflammatory marker concentrations in men and women 70 to 79 years. *The American Journal of Cardiology, 97*, 1010–1016.

Lubben, J. E. (1988). Assessing social networks among elderly populations. *Family & Community Health, 11*, 42–52.

Lyyra, T. M., Lyyra, A. L., Lumme-Sandt, K., Tiikkainen, P., & Heikkinen, R. L. (2010). Social relations in older adults: Secular trends and longitudinal changes over a 16-year follow-up. *Archives of Gerontology and Geriatrics, 51*, e133–e138.

McDonald, S., Lin, N., & Ao, D. (2009). Networks of opportunity: Gender, race, and job leads. *Social Problems, 56*, 385–402.

McDonald, S., & Mair, C. A. (2010). Social capital across the life course: Age and gendered patterns of network resources. *Sociological Forum, 25*, 335–359.

McLaughlin, D., Vagenas, D., Pachana, N. A., Begum, N., & Dobson, A. (2010). Gender differences in social network size and satisfaction in adults in their 70s. *Journal of Health Psychology, 15*, 671–679.

McPherson, M., Smith-Lovin, L., & Cook, J. M. (2001). Birds of a feather: Homophily in social networks. *Annual Review of Sociology, 27*, 415–444.

Mendes de Leon, C. F., Gold, D. T., Glass, T. A., Kaplan, L., & George, L. K. (2001). Disability as a function of social networks and support in elderly African Americans and whites: The Duke EPESE 1986–1992. *Journals of Gerontology Series B: Psychological Sciences and Social Sciences, 56*, S179–S190.

Monden, C. (2007). Partners in health? Exploring resemblance in health between partners in married and cohabiting couples. *Sociology of Health & Illness, 29*, 391–411.

Morris, M., Zavisca, J., & Dean, L. (1995). Social and sexual networks: Their role in the spread of HIV/AIDS among young gay men. *AIDS Education and Prevention, 7*(Suppl. 5), 24–35.

Nef, T., Ganea, R. L., Müri, R. M., & Mosimann, U. P. (2013). Social networking sites and older users – A systematic review. *International Psychogeriatrics, 25*, 1041–1053.

Neimann, S., & Schmitz, H. (2010). Honey, why don't you see a doctor? Spousal impact on health behavior. *Presentation at the annual meeting of the Social Policy Association*, Kiel, Germany.

Norvall, C. (2012). Understanding the incentives of older adults' participation on social networking sites. *ACM SIGACCESS Accessibility and Computing, 102*, 25–29.

Orel, N. A., Spence, M., & Steele, J. (2005). Getting the message out to older adults: Effective HIV health education risk reduction publications. *Journal of Applied Gerontology, 24*, 490–508.

Park, S., Smith, J., & Dunkle, R. E. (2014). Social network types and well-being among South Korean older adults. *Aging & Mental Health, 18*, 72–80.

Peek, M. K., & O'Neill, G. S. (2001). Networks in later life: An examination of race differences in social support networks. *International Journal of Aging and Human Development, 52*, 207–230.

Perry, B. L., & Pescosolido, B. A. (2010). Functional specificity in discussion networks: The influence of general and problem-specific networks on health outcomes. *Social Networks, 32*, 345–357.

Robins, G., Pattison, P., Kalish, Y., & Lusher, D. (2007). An introduction to exponential random graph (p^*) for social networks. *Social Networks, 29*, 173–191.

Rook, K. S. (1997). Positive and negative social exchanges: Weighing their effects in later life. *Journals of Gerontology Series B: Psychological Sciences and Social Sciences, 52B*, S167–S169.

Schafer, M. H. (2011). Health and network centrality in a continuing care retirement community. *Journals of Gerontology B: Psychological and Social Sciences, 66B*, 795–803.

Schafer, M. H. (2013a). Discussion networks, physician visits, and non-conventional medicine: Probing the relational correlates of health care utilization. *Social Science & Medicine, 87*, 176–184.

Schafer, M. H. (2013b). Structural advantages of good health in old age: Investigating the health-begets-position hypothesis with a full social network. *Research on Aging, 35*, 348–370.

Schafer, M.H. & Koltai, J. (in press). Does embeddedness protect? Personal network density and vulnerability to mistreatment among older American adults. *Journals of Gerontology Series B: Psychological Sciences and Social Sciences.*

Schnittker, J. (2007). Look (closely) at all the lonely people: Age and the social psychology of social support. *Journal of Aging and Health, 19,* 659–682.

Seeman, T. E., Miller-Martinez, D. M., Merkin, S. S., Lachman, M. E., Tun, P. A., & Karlamangla, A. S. (2011). Histories of social engagement and adult cognition: Midlife in the U.S. Study. *Journals of Gerontology, B: Psychological Sciences and Social Sciences, 66B,* 141–152.

Seeman, T. E., & Syme, S. L. (1987). Social networks and coronary artery disease: A comparison of the structure and function of social relations as predictors of disease. *Psychosomatic Medicine, 49,* 341–354.

Shaw, B. A., Krause, N., Liang, J., & Bennett, J. (2007). Tracking changes in social relations throughout late life. *Journal of Gerontology: Psychological Sciences and Social Sciences, 62B,* S90–S99.

Shiovitz-Ezra, S., & Litwin, H. (2012). Social network type and health-related behaviors: Evidence from an American national survey. *Social Science & Medicine, 75,* 901–904.

Smith, K. P., & Christakis, N. A. (2008). Social networks and health. *Annual Review of Sociology, 34,* 405–429.

Snijders, T. A. B., & Doreian, P. (2010). Introduction to the special issue on network dynamics. *Social Networks, 32,* 1–3.

Snijders, T. A. B., van de Bunt, G. G., & Steglich, C. E. G. (2010). Introduction to stochastic actor-based models for network dynamics. *Social Networks, 32,* 44–60.

Stephens, M. A. P., Rook, K. S., Franks, M. M., Khan, C., & Iida, M. (2010). Spouses use of social control to improve diabetic patients' dietary adherence. *Families, Systems & Health: The Journal of Collaborative Family Healthcare, 28,* 199–208.

Stevens, N. L., & van Tilburg, T. G. (2011). Cohort differences in having and retaining friends in personal networks in later life. *Journal of Social and Personal Relationships, 28,* 24–43.

Stroebe, M., Schut, H., & Stroebe, W. (2007). Health outcomes of bereavement. *Lancet, 370,* 1960–1973.

Thoits, P. A. (2011). Mechanisms linking social ties and support to physical and mental health. *Journal of Health and Social Behavior, 52,* 145–161.

Uchino, B. N., Bowen, K., Carlisle, M., & Birmingham, W. (2012). Psychological pathways linking social support to health outcomes: A visit with the "ghosts" of research past, present, and future. *Social Science & Medicine, 74,* 949–957.

Umberson, D. (1987). Family status and health behaviors: Social control as a dimension of social integration. *Journal of Health and Social Behavior, 28,* 306–319.

Valente, T. W. (2010). *Social networks and health.* Oxford: Oxford University Press.

Van Der Gaag, M., & Snijders, T. A. B. (2005). The resource generator: Social capital quantification with concrete items. *Social Networks, 27,* 1–29.

van Groenou, M. B., & van Tilburg, T. (2012). Six-year follow-up on volunteering in later life: A cohort comparison in the Netherlands. *European Sociological Review, 28,* 1–11.

Vanhems, P., Barrat, A., Cattuto, C., Pinton, J. F., Khanafer, N., Régis, C., et al. (2013). Estimating potential infection transmission routes in hospital wards using wearable proximity sensors. *PloS ONE, 8*(9), e73970.

Wasserman, S., & Faust, K. (1994). *Social network analysis: Methods and applications.* New York, NY: Cambridge University Press.

Wellman, B. (1983). Network analysis: Some basic principles. *Sociological Theory, 1,* 155–200.

Wellman, B., & Wortley, S. (1990). Different strokes from different folks: Community ties and social support. *American Journal of Sociology, 96*(3), 558–588.

Xie, B., Watkins, I., Goldbeck, J., & Huang, M. (2012). Understanding and changing older adults' perceptions and learning of social media. *Educational Gerontology, 38*(4), 282–296.

York Cornwell, E., & Waite, L. J. (2009). Social disconnectedness, perceived isolation, and health among older adults. *Journal of Health and Social Behavior, 50,* 31–48.

York Cornwell, E., & Waite, L. J. (2012). Social network resources and management of hypertension. *Journal of Health and Social Behavior, 53,* 215–231.

Zettel, L. A., & Rook, K. S. (2004). Substitution and compensation in the social networks of older widowed women. *Psychology and Aging, 1,* 433–443.

Zhang, X., Yeung, D. Y., Fung, H. H., & Lang, F. R. (2011). Changes in peripheral social partners and loneliness over time: The moderating role of interdependence. *Psychology and Aging, 26,* 823–829.

Zickuhr, K., & Madden, M. (2012). *Older adults and internet use. For the first time, half of adults ages 65 and older are online.* Washington, DC: Pew Internet & American Life Project.

PART III

SOCIAL FACTORS AND SOCIAL INSTITUTIONS

10 *Stability, Change, and Complexity in Later-Life Families 205*
11 *The Influence of Military Service on Aging 227*
12 *Religion, Health, and Aging 251*
13 *Evolving Patterns of Work and Retirement 271*
14 *Productive Engagement in Later Life 293*
15 *Aging, Neighborhoods, and the Built Environment 315*
16 *Abusive Relationships in Late Life 337*
17 *The Impact of Disasters: Implications for the Well-Being of Older Adults 357*
18 *End-of-Life Planning and Health Care 375*

CHAPTER 10

Stability, Change, and Complexity in Later-Life Families

J. Jill Suitor[1], Megan Gilligan[2], and Karl Pillemer[3]

[1]Department of Sociology, Center on Aging and the Life Course, Purdue University, West Lafayette, IN, USA [2]Human Development and Family Studies, Iowa State University, Ames, IA, USA [3]Department of Human Development, Cornell University, Ithaca, NY, USA

OUTLINE

Introduction	206
Theoretical Roots and Conceptual Advances	207
Intergenerational Solidarity	207
The Life-Course Perspective in Later-Life Family Relationships	208
Within-Family Complexity	209
Substantive Advances	209
Supportive Exchanges Between Generations	209
Relationship Quality Between Older Parents and Adult Children	212
Social Structural Characteristics	212
Ascribed Characteristics	212
Achieved Structural Characteristics	213
Value Similarity	215
Exchange Processes	215
Sibling Relations	215
New Directions in the Study of Relationship Quality Between Siblings	216
Patterns of Support Between Siblings	216
Grandparent–Grandchild Relations	217
New Directions in the Study of Relationship Quality Between Grandparents and Grandchildren	218
Patterns of Support Between Grandparents and Grandchildren	218
Marriage in the Later Years	219
Conclusion	220
References	221

INTRODUCTION

Family relations in later life have been an issue of concern throughout recorded history. In fact, intergenerational roles, responsibilities, and conflict have permeated both legend and literature from Greek mythology to the Bible and through Shakespeare to the modern age. These deeply human and highly salient relationships have drawn significant scientific interest as well. The dramatic increase in life expectancy across the twentieth century, combined with more recent changes in patterns of marriage and divorce, childbearing, and women's employment, have created ever-increasing complexity in individuals' personal and family lives. In this chapter, we identify and discuss changing patterns of intergenerational relations emanating from these sociodemographic transitions and their consequences on family members.

We begin by highlighting the theoretical roots and recent conceptual developments in the study of later-life families. We then review new directions in research on four family relationships across the middle and later years of the life course. These are relationships between: (a) parents and adult children; (b) grandparents and grandchildren; (c) siblings; and (d) spouses.

Major sociodemographic changes leading up to and continuing through the first decade of the twenty-first century suggest that all four of these relationships are of increasing importance in individuals' lives. For example, as young adults (especially young women) postpone or eschew marriage, the parent–child tie may remain their most salient source of expressive and instrumental support for several decades. Further, the absence of a marital or romantic partner means that sibling ties may also retain greater importance in adulthood than would be the case for people with partners. The high rate of births to unmarried women also increases the likelihood that both parents and siblings remain central sources of support. Furthermore, when new mothers are single, the salience of the grandmother role is often far greater than when mothers are married.

Increasing longevity also means that individuals have the opportunity to enact all four of these roles for a much longer period than did previous generations. For example, the number of years that women are likely to live after becoming grandmothers may now constitute more than half of their adult lives, spanning 30 or more years. Further, married individuals are likely to spend twice as many years together after their children enter adulthood (Vespa, 2014), and they may even find that one or more of their offspring remain or return to the parental home at various points. Finally, siblings may serve as important members of one another's sources of support for eight decades or more.

In reviewing scholarship on the aging family over the past two decades, an overarching progression is discernable: a movement from simpler models of family relationships to approaches and orientations that take complexity into account (Pillemer et al., 2007; Suitor, Sechrist, Gilligan, & Pillemer, 2011). In their attempts to shed new light on how the family shapes its members' development in the second half of life and affects their well-being, researchers have strived to fill conceptual and methodological gaps that limit our understanding of the complex and sometimes contradictory world of the aging family.

In particular, contemporary researchers increasingly attend to the following themes. First, the problem of endogeneity is now widely recognized. Given the impossibility of experimental designs in many areas of interest, distinguishing correlation from causation is challenging (e.g., reported poor relationship quality with an adult child may be a cause of mothers' depression, but the reverse path is equally plausible). Increasing reliance on large-scale, long-duration panel studies has helped greatly in addressing this problem. Second, in this field of study it is important to distinguish between age and cohort effects; this problem also has been addressed increasingly through longitudinal designs. Third, more researchers are recognizing that parents' relationships with individual children within the same family differ. Study designs have increasingly accounted for this development by permitting an examination of multiple children in the same family. Fourth, theory and empirical investigations now more typically take into account the existence of both positive and negative sentiments in family relationships in later life. Finally, attention to the critically important role of gender in shaping relations among parents and adult children, older spouses, and siblings permeates research over the past 20 years. Throughout this chapter, these core themes will re-emerge, reflecting the growth of more sophisticated theoretical and methodological approaches.

THEORETICAL ROOTS AND CONCEPTUAL ADVANCES

Intergenerational Solidarity

Theoretical perspectives on intergenerational solidarity have strongly influenced the study of later-life family relations for five decades. Informed by classical theories of group dynamics (Durkheim, 1933 (1883); Heider, 1958; Homans, 1950), solidarity theory emerged in response to popular allegations of a "generation gap" and dire predictions about the dissolution of intergenerational family ties. A major contribution of solidarity theory has been inspiring a large body of research on the emotional dimensions of family life in the later years, shedding light on the interrelationships among support, contact, value similarity, and the affective quality of intergenerational relationships (Fingerman, Sechrist, & Birditt, 2013). This trend stands in contrast to earlier research, which placed greater emphasis on family structure than process (Sussman, 1976).

The most influential scholarship on this topic has been conducted by Bengtson and colleagues (cf. Bengtson, 2001; Bengtson, Olander, & Haddad, 1976; Bengtson & Roberts, 1991). Bengtson and his collaborators developed both a measurement model and a conceptual scheme to identify the links between family members in different generations (Silverstein, Conroy, & Gans, 2012). The core of the model incorporates six interrelated components of family solidarity: (a) affectional (emotional closeness), (b) associational (frequency of contact), (c) normative (norms of obligation), (d) consensus (agreement about values), (e) structural (geographical proximity), and (f) functional (exchange of support). Thus, the model is generally characterized as emphasizing positive and supportive ties among the generations. As Silverstein and colleagues note: "Indeed, the term 'intergenerational solidarity' has come to mean emotional cohesion between generations" (2012, p. 1249).

Research using this model developed in several directions. Initially, solidarity research largely focused on single dimensions, such as affect or support, emphasizing how they develop within the family and influence outcomes for parents and children (Giarrusso, Feng, Silverstein, & Bengtson, 2001; Silverstein, Chen, & Heller, 1996). In addition, considerable research effort was devoted to examining the reciprocal relationships between the six dimensions, demonstrating the interdependence among them (Bengtson & Roberts, 1991). Another notable finding was differential views of solidarity depending on generational position, with parents viewing relationships more positively due to their higher investment, or "intergenerational stake" in the relationship (Giarrusso, Feng, & Bengtson, 2005).

In recent years, research informed by the solidarity perspective has expanded in two areas. First, responding to critiques regarding the positive bias of solidarity theory, attempts have been made to incorporate conflict into the model (Bengtson, Giarrusso, Mabry, &

Silverstein, 2002). Second, scholars (particularly from Europe) have emphasized linking societal-level and family-level intergenerational solidarity, examining how social stratification, class, and national differences affect patterns of solidarity (Szydlik, 2012; Timonen, Conlon, Scharf, & Carney, 2013). Recent work goes a step further by integrating these two emerging themes in an international, comparative analysis of the intersection of solidarity and conflict (Silverstein, Gans, Lowenstein, Giarrusso, & Bengtson, 2010). For example, Silverstein and colleagues have shown that nations that share particular characteristics, such as their degree of industrialization, can be classified on the basis of the levels of solidarity and conflict within families (Silverstein et al., 2010). These developments suggest the enduring ability of the solidarity model to generate innovative research questions.

The Life-Course Perspective in Later-Life Family Relationships

Across the past three decades, the life-course perspective has been one of the most influential approaches in the social sciences. The life-course perspective draws from both sociological theories of social change and psychological theories of individual and family development. This perspective highlights the importance of historical and social contexts and individual time and development on family relationships (Settersten, 2003); further, it addresses individual change within the family context as well as how these changes are linked to other family members (Conger & Elder, 1994). An important distinction made by life-course scholars is between cohort and generation. For example, mothers and daughters are not only members of different generational groups, but also of distinct birth cohorts, resulting in variations in the ways in which their lives have been shaped by historical experiences unique to the periods in which they were born and grew to adulthood. This perspective is complementary to Bengtson's solidarity model in that they both emphasize the importance of time and generation in explaining the relationships between members of all dyads within the family at any point in the life course (George, 2004).

Empirical studies of intergenerational relations have typically drawn upon the life-course perspective to address two issues. First, studies have examined continuity in family relations across the life course, reporting that closer and more harmonious relationships between parents and children in early life are associated with higher relationship quality and exchange of support at later points (Rossi & Rossi, 1990). The second line of inquiry has addressed the notion of "linked lives," focusing on the ways in which life events experienced by any family member affect the lives and relationships of other family members. In some cases, transitions have uniformly positive or negative effects on other family members, as in the case of job loss; however, in many cases, the direction and extent of the effects are conditional, as in the case of children's divorce, which affects parent–child relations depending upon the extent to which offspring must become heavily dependent upon their parents (Gilligan, Suitor, & Pillemer, 2013).

Thus far, much of the research that has drawn upon the life-course framework has been cross-sectional, despite the obvious longitudinal character of its basic tenets, and the difficulty in ascertaining causal ordering with data collected at only one point in time. The benefits of studying multiple generations across time are particularly well illustrated by the vast body of knowledge generated from Bengtson's Longitudinal Study of Generations. The breadth and impact of the findings from this study point to the importance of considering both birth cohort and generational position when investigating family processes. This perspective is likely to become even stronger when more studies of intergenerational relations follow families across time.

Within-Family Complexity

A recent development regarding the complexity of later-life families is the emphasis on variations in intergenerational relations within families. Until the early 2000s, the majority of studies asked parents about their relationships with their adult children in the aggregate or asked parents to report on their relationship with only one target child. Scholarship from developmental psychology suggests that such an approach masks variations in parent–child relationships within the family, thus providing an incomplete and potentially inaccurate picture of relationships in later-life families. Consistent with this argument, studies of younger families in the 1980s and 1990s showed that parents of young and adolescent children differentiated between their children in terms of both affection and disapproval (Suitor, Sechrist, Plikuhn, Pardo, & Pillemer, 2008). A small number of studies conducted in the same period suggested that such variations in parent–child relations within the family continued into adulthood (cf. Aldous, Klaus, & Klein, 1985); however, it was not until the early 2000s that attention was devoted to understanding the prevalence, predictors, and consequences of within-family differences in the later years.

The study of within-family differences in adulthood is grounded in classic theories of social interaction in both sociology (Simmel, 1964) and psychology (Heider, 1958), which can be used to argue that the relationship between a parent and any one of his or her adult children is likely to be affected by the parent's relationships with other adult children in the family. Research using within-family designs has shed new light on later-life family relations. First, this line of research has shown that there is substantial variation in parents' relationships with adult children in the same family across a wide range of dimensions, including emotional closeness (Suitor, Gilligan, & Pillemer, 2013b; Suitor & Pillemer, 2006; Ward, Spitze, & Deane, 2009), contact (Ward, Deane, & Spitze, 2014), exchange of support (Fingerman et al., 2011; Spitze, Ward, Deane, & Zhuo, 2012; Suitor, Pillemer, & Sechrist, 2006), and preferences for support (Suitor, Gilligan, & Pillemer, 2013a; Suitor & Pillemer, 2006). Further, such patterns of variation occur in both large and small families (Suitor, Sechrist, & Pillemer, 2007).

Differentiation within the family has been found to be consequential for both parents and children. For example, adult children's perceptions that their parents make distinctions among them have been found to have detrimental effects on sibling relations (Boll, Ferring, & Filipp, 2003; Gilligan, Suitor, Kim, & Pillemer, 2013; Suitor et al., 2009) and psychological well-being (Pillemer, Suitor, Pardo, & Henderson, 2010). Further, within-family differentiation has been found to predict which offspring become caregivers when their parents experience serious health events (Leopold, Raab, & Engelhardt, 2014; Pillemer & Suitor, 2014) as well as the degree of interpersonal stress that adult children experience when serving as caregivers (Suitor, Gilligan, Johnson, & Pillemer, 2014b), and mothers' psychological well-being when being provided care (Suitor et al., 2013b). The broad consequences of within-family differentiation for both children and parents suggest that further research on such variations has the potential for expanding understanding of the conditions under which parents and children affect one another's relational and psychological well-being.

SUBSTANTIVE ADVANCES

Supportive Exchanges Between Generations

Supportive exchanges between parents and adult children are a central topic in the study of later-life families. Exchange can take a number of forms between parents and adult offspring, including material support (such as money or housing); practical help (such as assistance with chores or running errands); and expressive support and advice (Seltzer & Bianchi, 2013). An important form of exchange is provision of care, which in the later stages of life often shifts from assistance from parent to child to help from offspring to parents. In our discussion, we focus on expressive and instrumental support, with special attention to caregiving by adult children.

Consistent with contingent exchange theory (Davey & Norris, 1998), both parents' and children's needs are important predictors of patterns of intergenerational exchange. Recent studies have shown that both mothers and fathers provide instrumental and expressive support to offspring with greater needs (Fingerman et al., 2011; Suitor et al., 2007). In fact, need often trumps other factors that generally play a strong role in intergenerational exchanges, such as affection, perceptions of relational equity, and child's gender. For example, parents appear to be almost equally likely to provide support when the problems children face are involuntary, as in the case of health problems (Seltzer, Greenberg, Orsmond, Lounds, & Smith, 2008; Suitor et al., 2007), as when the problems result from children's poor choices, as in the case of deviant behaviors (Fingerman, Cheng, Birditt, & Zarit, 2012). Children's divorce, which may be either voluntary or involuntary, is an event that is especially likely to increase parents' support to their adult children, particularly in the form of housing (cf. Smits, van Gaalen, & Mulder, 2010).

Parents' needs also shape exchanges between the generations. Parents sometimes experience negative health events at a point where support would have been expected to flow to rather than from adult children, as in the case of a diagnosis of serious health conditions such as cancer, multiple sclerosis, or accidents. Further, parents who have been recently widowed receive increased expressive and instrumental support (Fuller-Thomson, 2000).

History of support plays an important role in exchange processes. Parents and children are more likely to provide support when they previously received support from one another (Schwarz, Trommsdorff, Albert, & Mayer, 2005; Suitor et al., 2007). Further, this pattern has been found in both black and white families (Fingerman et al., 2011; Suitor et al., 2007). The history of exchange affects not only current support but the support parents expect to receive in response to future need (Lin & Wu, 2014). Further, parents and adult children report higher levels of support exchange when they experience greater closeness and affection (Fingerman et al., 2011); in turn, equitable exchanges of support enhance positive affect between parents and adult children (Sechrist, Suitor, Howard, & Pillemer, 2014; Silverstein et al., 2013).

Recent studies have shown that history of exchange is also an important predictor of which children become caregivers (Leopold et al., 2014; Pope, Kolomer, & Glass, 2012). It has been argued that the history of supportive exchanges affects caregiving patterns for a combination of reasons, including that adult children feel greater obligation to provide care to parents who have been a source of support in the past and that a well-established history of support creates greater mutual positive affect that translates into children's desire to provide support when parents need care (Silverstein, Conroy, Wang, Giarrusso, & Bengtson, 2002). The effects of this factor, however, weaken somewhat when value similarity between adult children and parents is taken into consideration because similarity is a strong predictor of which offspring provide both current and previous support (Pillemer & Suitor, 2014; Suitor et al., 2013a).

Additionally, gender remains one of the strongest predictors of the flow of support between the generations. Consistent with classic theories of gender-role development (cf. Chodorow, 1978), mothers and daughters report the highest level of exchange of both expressive and instrumental support

(Davey, Eggebeen, & Salva, 2007; Suitor et al., 2011). Further, when both adult sons and daughters hold a strong sense of filial responsibility, these feelings are more likely to result in the actual provision of support from daughters to mothers than other gender combinations (Silverstein, Gans, & Yang, 2006).

As with exchange more generally, gender of both parents and children continues to play the greatest role in patterns of caregiving. Mothers receive more care from their adult children than do fathers (Leopold et al., 2014; Silverstein et al., 2002), and daughters are more likely than sons to be the source of that support (Leopold et al., 2014; Pillemer & Suitor, 2014). Fathers may be less likely than mothers to receive support from adult children because spouses typically rely primarily on one another for care when both are living. Women, on average, outlive men, and thus are likely to be available to perform these caregiving responsibilities (Carr & Moorman, 2011). Daughters are also parents' preferred caregivers, particularly among mothers (Suitor et al., 2013a; Suitor & Pillemer, 2013). Earlier studies questioned whether women Baby Boomers' high level of labor-force participation would affect their likelihood of continuing to serve as primary caregivers to their parents. However, recent studies have shown that women's employment has fewer effects than were feared (Pavalko, 2011), although there is evidence that daughters who provide care are less likely to be employed (Leopold et al., 2014) and may alter their employment responsibilities to accommodate their parents' care needs (Barnett, 2013).

Longitudinal studies of pathways to caregiving have revealed several factors that consistently explain patterns of adult children's caregiving to their parents. Proximity is a strong predictor of which adult children become caregivers. This finding is robust regardless of the measure of proximity (Leopold et al., 2014; Pillemer & Suitor, 2006, 2014), with coresident offspring most likely to serve as caregivers (Leopold et al., 2014). It is interesting to note, however, that residential proximity does not play a role in which children mothers prefer as their caregivers (Suitor et al., 2013a, 2013b), and that mothers, in fact, often report discrepancies between the offspring that they would *prefer* care for them and the offspring they expect or actually experience caring for them due to proximity (Suitor et al., 2013b).

These studies have also revealed three socioemotional factors that predict which children are likely to care for their parents: mothers' expectations for care, mothers' preferences for care, and perceived value similarity between mothers and their offspring. Studies by both Leopold and colleagues (2014) and Pillemer and Suitor (2014) found that parents' expectations that particular offspring would provide care at a later point were strong predictors of which children in the family assumed this role when parents required assistance. Finally, Pillemer and Suitor's (2014) study revealed the salience of value similarity in determining which children become caregivers to their parents. Adult children who later became caregivers were substantially more likely to have been perceived by their mothers as sharing their values at the first wave of data collection 7 years earlier.

In summary, exchanges between parents and adult children are shaped greatly by the combination of parents' and children's characteristics, particularly need, affection, gender, and the history of support between the generations. It is worth noting the high level of consistency between the patterns revealed by recent studies and those conducted across the past several decades (Suitor et al., 2011). Further, recent studies reveal patterns similar to earlier research regarding the roles of gender, history of support, and residential proximity in explaining pathways to caregiving, as well as shed new light on the ways in which parents' expectations for care and perceptions of value similarity lead particular adult children to assume caregiving whereas their siblings do not.

RELATIONSHIP QUALITY BETWEEN OLDER PARENTS AND ADULT CHILDREN

As the large body of scholarship inspired by the intergenerational solidarity perspective demonstrates, most parents and adult children report that their relationships are meaningful and supportive. Nevertheless, new research has shown that troubled relationships and negative interactions are not uncommon between the generations. Further, such positive and negative elements often coexist in relationships between parents and adult children. Thus, in this section, we consider the factors that shape a wide range of dimensions of parent–adult child relations. In particular, we include closeness, conflict, preferences for support, and ambivalence, which represents the simultaneous presence of both positive and negative feelings (Lüscher & Pillemer, 1998). We take this approach in recognition of the complex and sometimes contradictory emotions and assessments common to family relationships in later life.

Studies of later-life families have shown that the processes shaping the quality of ties between parents and their adult children greatly mirror those of other interpersonal relationships. Thus, the quality of these ties is affected by a combination of social structural and socioemotional characteristics of both parents and their adult children (Suitor et al., 2011). Further, like other interpersonal ties, the parent–child relationship is subject to change across time, particularly in response to life events and status transitions experienced by one of the role partners. However, because most research on the parent–adult child tie continues to be cross-sectional, we typically must assess the effects of both parties' characteristics at the time of data collection, missing the processes that occur during transitions. Thus, in our discussion we will need to draw primarily from studies that look at only one point in time in the parent–child relationship, but will pay particular attention to recent studies that have examined changes in this tie across time.

We organize our discussion of the predictors of the quality of parent–adult child relations by key dimensions of the life-course perspective: (a) social structural positions; (b) value similarity between parents and children; and (c) support processes.

Social Structural Characteristics

We divide our discussion of social structural characteristics by whether the characteristics are ascribed (typically those over which the individual has little choice) or achieved. We make this distinction because the presumed voluntary nature of achieved characteristics leads both parents and children to assign credit and culpability to their presence or absence.

Ascribed Characteristics

Gender. We begin with gender because this characteristic has been found to be the most consistent structural predictor of the quality of relations between parents and children. Beginning with the earliest studies of parent–child affect in the later years (cf. Adams, 1968), the preponderance of investigations have reported stronger affectional ties between adult children and their mothers than their fathers (Birditt, Tighe, Fingerman, & Zarit, 2012; Rossi & Rossi, 1990; Sarkisian & Gerstel, 2008; Ward et al., 2009) as well as greater closeness and confiding between mothers and daughters than any other gender combination (Rossi & Rossi, 1990; Suitor & Pillemer, 2006, 2013). Interestingly, although the mother–daughter tie is typically the most close, that does not mean that it is without tension. Some studies have shown that mothers and daughters also sometimes have higher levels of conflict (Birditt et al., 2012) and ambivalence (Pillemer, Munsch, Fuller-Rowell, Riffin, & Suitor, 2012) due to their greater investment in the relationship. Nevertheless, mothers' relationships are more likely to become closer over time with daughters than with sons (Suitor et al., 2013a), and relationships involving either mothers or daughters tend to become less ambivalent over time than do those involving either fathers or sons (Schenk & Dykstra, 2012).

These findings are consistent with classic arguments developed by Chodorow (1978) and Gilligan (1982) regarding socialization, which have often been used to explain both girls' and women's stronger emphasis on interpersonal relations across the life course relative to their male counterparts (Suitor et al., 2011). Although there is evidence that older cohorts hold more traditional gender-role attitudes than do younger cohorts (Powers et al., 2003), recent findings suggest greater stability in the pattern of stronger ties between mothers and daughters than other parent–child gender combinations in the family (Suitor et al., 2013a).

Age. Theories and research suggest that as children move across the adult life course, their relationships with their parents are less conflictual (Fingerman et al., 2011) as well as less likely to be considered ambivalent (Fingerman, Hay, & Birditt, 2004). In contrast, children's age is a less consistent predictor of *positive* dimensions of the parent–child relationship and when differences are found, they show greater closeness with younger rather than older children in both between-family (Merz, Schuengel, & Schulze, 2009) and within-family (Suitor et al., 2013a) studies. Parents' age has received less attention in the literature than has the age of their offspring. However, the few studies that have examined parents' age have found more positive relations with older parents (Ward et al., 2009).

Race and Ethnicity. Studies of variations in parent–adult child relationship quality by race and ethnicity have revealed consistent patterns regarding differences between black and white families, but not between Hispanic and white families. Recent studies have found greater closeness in black than in white parent–child ties (Sarkisian & Gerstel, 2008; Sechrist, Suitor, Henderson, Cline, & Steinhour, 2007) as well as lower levels of ambivalent feelings (Kiecolt, Blieszner, & Salva, 2011). In contrast, findings comparing Hispanic and white families have been mixed; whereas Sechrist and colleagues (2007) reported greater closeness between Hispanic than between white parent–child ties, Sarkisian and Gerstel (2008) reported no difference.

Achieved Structural Characteristics

A powerful social norm holds that children should attain adult statuses in a timely fashion, establish independent lives, and ultimately become potential sources of support for parents (Suitor et al., 2011). When children do not achieve such normative benchmarks, parents question the success of their socialization efforts, and feel obligated to assist their errant offspring (Greenfield & Marks, 2006). Thus, having achieved adult benchmarks such as marriage, parenthood, and employment would be expected to predict greater closeness and less conflict and ambivalent feelings between parents and their adult children. However, alternatively, theoretical and empirical work on the "greedy nuclear family" can be used to argue that marriage, parenthood, and employment have the potential to limit the time and attention focused on relations with parents (Coser & Coser, 1974; Sarkisian & Gerstel, 2008), resulting in less close and more negative intergenerational ties.

Further, children also hold expectations that their *parents* will maintain normative adult statuses, such as marriage and employment, until a point at which transitions from these statuses occur in an expected manner, through widowhood and retirement. Thus, both parents' and children's achievement and maintenance of normative adult status characteristics may play roles in explaining relationship quality. Because the effects of achieved social structural characteristics vary considerably, we give greater attention to some than others.

Marital Status. Marital status is the achieved structural characteristic found to be the best predictor of parent–adult child relationship quality, although the effects are clearly more consistent for parents' than adult children's marital statuses. Studies of both midlife and older families have shown that the parent–child relationship is affected by changes in *parents'* marital status, although the effects vary considerably by whether the marriage was terminated by death or divorce.

When parents become widowed, adult children report greater closeness to the surviving parent (Carr & Boerner, 2013; Fuller-Thomson, 2000), and, in the case of mothers, are also more responsive to mothers' wishes regarding caregiving preferences (Suitor, Gilligan, Johnson, & Pillemer, 2014a). In contrast, when parents become divorced, even in the later years, the parent–child relationship often becomes less close and more conflictual. These effects, however, differ by gender. For fathers, parents' divorce, both in childhood and adulthood, appears to result most often in less close parent–child relations when offspring are adults (Amato & Sobolewski, 2001; Shapiro & Cooney, 2007; Yu, Pettit, Lansford, Dodge, & Bates, 2010). It might be expected that the greatest toll would be for the father–son tie, but instead, the father–daughter tie seems to be the most vulnerable (Amato & Sobolewski, 2001). For mothers, the evidence regarding the effect of divorce on their relationships with their adult children is mixed, with some studies showing closer parent–child relations with divorced rather than married mothers (Yu et al., 2010), others showing a modest negative effect on closeness (Amato & Afifi, 2006), and yet others showing no effect (Amato & Sobolewski, 2001). Further, research has revealed that parental divorce is also associated with children's development of interpersonal behaviors that affect the stability of their own marriages, as well as other negative outcomes in adulthood (Amato, 1996; Fomby & Cherlin, 2007).

Findings regarding the effect of *adult children's* marital status on parent–adult child relations are more mixed. Parents feel less ambivalent toward adult children who are married (Kiecolt et al., 2011; Pillemer et al., 2012; Pillemer et al., 2007). In contrast, married children are less emotionally close to their mothers (Suitor & Pillemer, 2006). Further, panel data revealed that children who moved into first marriages or remarriages were substantially less likely to be reported as most emotionally close than those who never married or remained divorced (Suitor et al., 2013a). This pattern of findings can be explained by the argument noted above regarding the limits to time and attention on the parent–child tie that is introduced by children's entrances into marriage (Coser & Coser, 1974; Sarkisian & Gerstel, 2008).

Some studies have suggested that the effect of marriage on parent–child relations differs by child's gender. Sarkisian and Gerstel (2008) reported that for daughters, those who were divorced had lower relationship quality than did marrieds, whereas there were no differences between marrieds and never-marrieds. Never-married sons had higher relationship quality than did marrieds, whereas there was no difference between those who were married and divorced.

Other Achieved Social Structural Characteristics. In contrast to marital status, few patterns emerge regarding the effects of children's educational attainment, employment, or parental status on parent–child relationship quality. For all three of these structural characteristics, most studies find no effects on parent–adult child relations, and those that do report effects do not reveal consistent patterns. Thus, we conclude that these factors, which are typically important predictors of relationship quality in other interpersonal ties, such as those between marital partners and between friends, have surprisingly little role in shaping affective ties between parents and their adult children.

Value Similarity

Consistent with classic theories of homophily (Homans, 1950; Lazarsfeld & Merton, 1954), value similarity is an important predictor of relationship quality between parents and adult children. In fact, it has been found to be among the strongest and most consistent predictors of a wide range of dimensions of parent–adult child relations, including closeness, conflict, ambivalence, and preferences for confiding and caregiving (Pillemer et al., 2007, 2012; Rossi & Rossi, 1990; Suitor et al., 2013a; Suitor & Pillemer, 2006). Further, longitudinal research on changes in parental differentiation in the later years has demonstrated that value similarity is the best predictor of which offspring remain the most close to their mothers across time, as well as which offspring become most close over time (Suitor et al., 2013a). It is not surprising that homophily of values is of such importance to the parent–child tie – theories of child development propose that a primary goal of parental socialization is imbuing the values that mothers and fathers believe are necessary to be a "good adult" (Grusec, 2002). Because parents typically attempt to socialize their children to hold the values that they hold themselves, closeness would be expected to be greater among dyads in which there is high value consensus.

Exchange Processes

Based on our earlier discussion of intergenerational exchanges, it is not surprising that parents and adult children generally express better relationship quality when there is a flow of support between the generations. Both parents and adult children report higher relationship quality when there have been exchanges of either expressive or instrumental support (Merz et al., 2009; Schenk & Dykstra, 2012; Sechrist et al., 2014). However, the strongest effects have been consistent with theories of equity rather than exchange. Mothers were substantially more likely to report greater closeness and less tension and ambivalence toward children with whom they perceived they had more equitable relationships (Pillemer et al., 2007; Sechrist et al., 2014). Further, a direct comparison of the effects of perceptions of equity versus equal exchanges of instrumental and expressive support revealed the primacy of equity over exchange in predicting both closeness and conflict with adult children (Sechrist et al., 2014).

In summary, relationship quality between parents and adult children is shaped by the same factors that play important roles in other interpersonal relations – value similarity, social structural positions, and support exchange. Consistent with the findings of the broader literature on interpersonal relations, socioemotional factors, such as value similarity and perceptions of equity, are stronger and more consistent predictors of relationship quality than are structural characteristics, with the exception of gender (Suitor et al., 2011).

SIBLING RELATIONS

Until the late 1990s, scholarship on sibling relations was constrained to early stages of the life course, consistent with Parsons' (1943) essays depicting sibling relationships as salient only in childhood and adolescence. More recently, scholars have started to recognize that siblings play an important role in individuals' lives across the life course, as evidenced by the growing body of literature examining this relationship in middle and later years (Bedford & Avioli, 2012). In fact, for many individuals, the sibling tie is one of the most enduring kin relationships. Further, increased life expectancy and recent demographic shifts in family patterns regarding marriage and parenthood suggest that siblings may be becoming increasingly important sources of support. In this section, we review new directions in research on adult sibling relationships, paying particular attention to how demographic changes may affect this tie.

New Directions in the Study of Relationship Quality Between Siblings

Scholars have documented that although the intensity of sibling relationships is often the strongest in young adulthood, siblings continue to express feelings of closeness, conflict, and ambivalence toward one another throughout the life course (Antonucci, Akiyama, & Takahashi, 2004; Fingerman et al., 2004). Empirical research has demonstrated that closeness and conflict among siblings in childhood and adolescence continue to affect well-being in adulthood (Bedford, 1998; Waldinger, Vaillant, & Orav, 2007), and that the quality of current sibling ties predicts well-being in both midlife and the later years (Cicirelli, 1989; Paul, 1997).

Studies predicting the quality of adult sibling relations have focused primarily on the role of social structural characteristics, such as gender, marital status, and parental status in shaping these ties. Specifically, studies have demonstrated that sisters have higher levels of contact with their siblings than do brothers, and also report greater closeness in these relationships (Connidis & Campbell, 1995; Spitze & Trent, 2006; White & Riedmann, 1992). Siblings who are single or childless have particularly high levels of contact and confiding with their brothers and sisters, relative to those who are married with children (Connidis & Campbell, 1995).

Consistent with the life-course perspective outlined above, recent attention has been directed toward understanding how other family ties – especially those between parents and adult children – affect relationships between siblings. For example, research on within-family differences has demonstrated that adult children's perceptions of parental favoritism are associated with less closeness and greater conflict among siblings (Boll et al., 2003; Suitor et al., 2009). Further, perceptions of fathers' favoritism have been found to play an even greater role in shaping adult children's sibling relations than do perceptions of mothers' favoritism in families in which both parents are living (Gilligan et al., 2013).

It is well documented that the transition to caregiving for parents is an event that often increases tension between siblings (Connidis & Kemp, 2008; Lashewicz & Keating, 2009). Recent research indicates that the effects of caregiving on sibling tension are especially strong when perceptions of parental favoritism are present (Suitor, Gilligan, Johnson, & Pillemer, 2014b). The impact of parents on sibling relationships appears to remain even after parental death. Khodyakov and Carr (2009) found that parental death had a negative effect on sibling closeness, and that parents' use of advance directives did not reduce these negative consequences.

Patterns of Support Between Siblings

Studies reveal that brothers and sisters play an important role in the system of support on which individuals rely in adulthood. Most adults receive both instrumental and emotional support from their siblings on a fairly regular basis (Eriksen & Gerstel, 2002) in patterns that can be explained by a combination of life-course variations and contingent needs. Comparisons suggest that siblings' exchanges of support decrease in early adulthood but resurge in later life (White, 2001). Further, some siblings benefit more from this tie than do others. For example, single, widowed, and childless women appear to have especially involved sibling relationships (Campbell, Connidis, & Davies, 1999), and are more likely to provide support to siblings who have additional familial responsibilities (Voorpostel, van der Lippe, Dykstra, & Flap, 2007). We suggest that such patterns are consistent with theories of contingent need exchange (Davey & Norris, 1998) in that individuals with fewer competing responsibilities provide more support to their siblings who face higher levels of role strain. This perspective can also explain the finding that siblings provide one another with greater support when they are not receiving support from their parents (Voorpostel & Blieszner, 2008). Further, the finding that single adults provide more support to their siblings than do their married counterparts suggests that the concept of the "greedy nuclear family" (Coser & Coser, 1974; Sarkisian & Gerstel, 2008) can be applied to other kin relations as well as to the parent–child tie.

Recent attention has also been directed toward examining support between siblings when one member of the dyad experiences serious physical or psychological health problems. Based on the contingent need exchange perspective, it would be expected that the flow of support would be unilaterally from the healthier to the less healthy sibling; however, this does not appear to be the case. For example, Burbidge and Minnes (2014) found that adults with developmental disabilities both received support from and provided support to their siblings. Similarly, Voorpostel, van der Lippe, and Flap (2012) found that individuals with physical illnesses also engaged in supportive exchanges with their siblings. It may be that such reciprocal exchanges among siblings are easier to maintain when parents are living and sufficiently healthy to provide high levels of support to the sibling with physical or mental health problems. However, healthier siblings often anticipate taking on the primary responsibility of care for brothers and sisters with severe disabilities after parental death (Smith, Greenberg, & Seltzer, 2007), which would substantially change the balance of support.

In summary, recent research has shown that siblings play an important role in one another's lives in adulthood, serving as sources of affection and support. Further, these patterns of reciprocal support occur even when one of the siblings has chronic conditions. The sociodemographic changes occurring in life expectancy and marriage patterns discussed earlier suggest that the sibling tie may become more important in individuals' lives in coming decades. Further, the high prevalence of reconfigured families in the cohorts of Baby Boomers and their children, as a result of both divorce and increasing rates of births to unmarried parents, suggests that relations among both biological and step-siblings may become more complex. These changes in family structure call for new research that investigates the role of siblings in the lives of adults in the twenty-first century.

GRANDPARENT–GRANDCHILD RELATIONS

Perhaps more than any other family relationship, increases in life expectancy have altered the experience of the grandparent–grandchild tie. In particular, many children grow up with multiple living grandparents, and an increasing number of individuals can expect to continue the grandparent–grandchild relationship well into adulthood (Uhlenberg, 2004). These shifts in longevity are reflected in the large body of scholarship on grandparenting that has developed in recent years, beginning with Cherlin and Furstenberg's classic study of this tie (1986).

This line of research has shown that although many grandparent–grandchild ties can be characterized as emotionally close and highly involved (Mueller, Wilhelm, & Elder, 2002; Silverstein & Marenco, 2001), the relationship can be experienced as ambivalent and sometimes detached (Mason, May, & Clarke, 2007; Mueller et al., 2002). Further, consistent with the literature on parent–adult child relations, scholarship has also started to show that there is often variability in grandparent–grandchild relationships within the same family (Davey, Savla, Janke, & Anderson, 2009). In this section, we highlight some of the recent research examining the complexity of ties between grandparents and grandchildren.

New Directions in the Study of Relationship Quality Between Grandparents and Grandchildren

Studies have shown that it is important to consider the grandparent–grandchild tie in the context of the larger familial network. In particular, the relationship between grandparents and their own children appears to be highly consequential for the grandparent–grandchild relationship. Research examining the effects of the middle generation on grandparent–grandchild ties in childhood suggests that parents often facilitate opportunities for contact between grandparents and young children (Barnett, Scaramella, Neppl, Ontai, & Conger, 2010). Further, studies examining this association in late adolescence and early adulthood indicate that parents, particularly mothers, may continue to facilitate this relationship into adulthood (Monserud, 2008; Mueller & Elder, 2003). Some evidence suggests that the relationship between grandparents and their children-in-law may be even more influential on grandparent–grandchild ties than grandparents' relationships with their own children, further demonstrating the need to examine this relationship in the environment of the larger family (Fingerman, 2004).

Also consistent with the life-course concept of "linked lives," transitions in the lives of the middle generation shape relationships between grandparents and grandchildren. In particular, because the tie between grandparents and grandchildren relies heavily on the middle generation to facilitate opportunities for interaction, parental divorce can have deleterious effects on the grandparent–grandchild relationship (Shapiro & Cooney, 2007). Middle generation migration, especially international migration, also poses a threat to grandparent–grandchild relations (Uhlenberg, 2004). Recent research has explored the challenges that American grandparents encounter when their grandchildren live in Israel (Sigad & Eisikovits, 2013). The salience of the grandparent–grandchild relationship for the older generation is shown by the findings that such loss of contact, as a result of marital dissolution or relocation, negatively impacts grandparents' psychological well-being (Drew & Silverstein, 2007).

Patterns of Support Between Grandparents and Grandchildren

The most common form of support that grandparents provide to grandchildren in early life is caregiving (Silverstein & Marenco, 2001). Grandparents have traditionally participated in the informal care of grandchildren; however, in recent years there has been a sharp increase in the practice of grandparents taking primary responsibility for raising grandchildren (Hayslip & Kaminski, 2005). Grandparents who assume these responsibilities, particularly those who coreside with grandchildren, are more likely to be women, black, Mexican American or Native American, and have low incomes and educational attainment (Fuller-Thomson & Minkler, 2005, 2007; Minkler & Fuller-Thomson, 2005). A great deal of attention has been directed toward understanding the costs and benefits of this care (Hayslip & Kaminski, 2005); however, recent research indicates that many of the detrimental health consequences previously attributed to caregiving are actually the result of grandparents' preexisting conditions (Hughes, Waite, LaPierre, & Luo, 2007).

An emerging area of research is grandchildren caring for grandparents (Fruhauf, Jarrott, & Allen, 2006; Hamill, 2012). This research suggests that the middle generation is also influential in the process in that grandchildren provide more care when parents are caregivers (Hamill, 2012). As life expectancy increases, this is a topic that deserves additional attention in future research.

MARRIAGE IN THE LATER YEARS

The past 50 years have witnessed dramatic changes in marital patterns, including delayed and foregone marriage, alterations in normative family structure, and an increase in cohabitation and nonmarital childbearing (Cherlin, 2010). One outcome of these trends has been an impact on family structure in later life: a greater proportion of individuals reaching retirement age will be unmarried. Divorce rates escalated in the 1970s, and the number of persons never marrying has also increased. These trends are compounded by a decrease in rates of remarriage and an increase in the rate of failure of second marriages (Tamborini, 2007). Gender is a major factor in these patterns. Because of women's greater longevity and lower likelihood of remarriage after divorce or widowhood, 72% of men aged 65 and older are married, but this is the case for only 42% of women. There are also substantial race differences in marital status in later life, with 25% of African-American women compared to 42% of white women living with their spouses.

Although most research on later-life relationships has focused on marriage, in recent years other forms of partner relationships have become increasingly common (i.e., civil unions, cohabitation, committed dating relationships, and "living alone together"). This trend is likely to continue over the coming decades (Amato, Booth, Johnson, & Rogers, 2009; Powell, Bolzendahl, Geist, & Steelman, 2012). The proportion of unmarried adults aged 45–54 today is three times what is was in 1986 (Kreider & Ellis, 2011). At present, one in three Baby Boomers is unmarried. The decline in marriage rates is especially steep for economically disadvantaged groups and ethnic minorities, with African-Americans experiencing the largest drop in matrimony (Cherlin, 2010). In contrast, cohabitation has become increasingly common, nearly doubling since the mid-1990s (Cohn, Passel, Wang, & Livingston, 2011); however, such unions tend to provide fewer of the psychological benefits of marriage, especially for men (Brown, Bulanda, & Lee, 2005). These trends will be reflected in the nature and dynamics of later life over the coming two decades as unmarried older individuals are more vulnerable (compared to marrieds) to such problems as social isolation, loneliness, health problems, and economic disadvantage (Lin & Brown, 2012).

Divorce. Divorced older people are still in a relatively small minority, representing only 12% of all older persons in 2010. However, although the proportion is small, over the past two decades the actual *number* of divorced older individuals has increased dramatically. Among men, the percentage doubled from 5% to 10%, and among women the percentage tripled from 4% to 12%. This increase is striking when compared to widowhood, the prevalence of which has not changed among men since 1980 and has actually declined among women during that time. The numbers of divorced elders is certain to increase given that the divorce rate for adults aged 50 and over doubled between 1990 and 2010 (Brown & Lin, 2012).

Thus, as a life-course transition, divorce will figure much more prominently in the lives of older Americans over the coming decades. Most research suggests that increases in later-life divorce are likely to lead to a weakening of intergenerational ties. Although there is some evidence that both divorced mothers and fathers provide less support to their adult children, the evidence on this issue is mixed (Shapiro, 2012). Research has found lower contact from adult children to divorced parents, particularly fathers (Lin, 2008; Ward et al., 2014), although the effects on the quality of the parent–child tie vary by parents' and children's genders (as we discussed in detail in an earlier section). In general, accumulating research shows that people who experience multiple marriage transitions have lower levels of intergenerational solidarity than once-married individuals (Shapiro, 2012).

Cohabitation. Rates of cohabitation remain low among older Americans (approximately 4%), but the numbers are rapidly increasing with the aging of the Baby Boomers. Cohabitation can be an attractive option to older persons, some of whom express a desire for companionship but not for marriage. People may wish to keep their assets separate, and widowed individuals can be hesitant to enter into a marriage that may result in a second spell of caregiving. Unlike younger people, cohabiting unions of older people appear to be stable and are a long-term alternative to marriage (Brown, Bulanda, & Lee, 2012), although as noted earlier, it is not clear that they provide the level of psychological benefit found in marriage.

Marital Quality. Research consistently shows higher levels of marital satisfaction among older couples in comparison to younger and middle-aged couples (Korporaal, van Groenou, & van Tilburg, 2013). Further, rates of marital conflict are substantially lower in older couples than in younger ones (Smith et al., 2009). Cross-sectional studies of marital satisfaction have described a U-shaped curve, involving high satisfaction early in the marriage, a decline in the middle years, and a rebound in the later years. Longitudinal research, however, suggests that decline occurs following the early years of marriage and remains flat or rebounds only slightly in later life (Silverstein & Giarrusso, 2010), suggesting that some of the positivity in older marriages can be explained by survival bias or cohort effects.

A major factor negatively affecting the quality of later-life marriages is declining health of a partner, particularly for women with ill husbands (Iveniuk, Waite, Laumann, McClintock, & Tiedt, 2014; Korporaal et al., 2013). Spouse's health problems can disrupt prior patterns of shared activities and emotional closeness as well as create the need for personal assistance. Recent research has identified the experience of chronic pain as a risk factor for marital discord and dissatisfaction, a potentially serious problem given the high prevalence of chronic pain conditions among older people (Riffin, Suitor, Reid, & Pillemer, 2012). Further, the onset of dementia is a particularly potent stressor for marriages (Clare et al., 2012). Thus, the picture of marital quality in later life is complex, with a tendency toward positive relationships sometimes diminished by the negative effects of health stressors.

A rapidly growing body of research has examined the impact of predictable life-course transitions on the quality of marriages in later life (cf. Bookwala, 2012). Despite popular portrayals of a dysphonic "empty-nest syndrome," studies show a generally positive impact of launching children, with a reduction in marital conflict and an increase in marital happiness. Research on the impact of retirement shows a more complex picture, with better marital quality among couples where both are retired or the wife is retired and poorer marital quality in husband-retired/wife-employed couples (Moen, Kim, & Hofmeister, 2001). Another common transition is becoming a caregiver for a spouse, with the preponderance of studies indicating that marital quality decreases in caregiving situations (Bookwala, 2012). The increasing availability of large-scale longitudinal data sets will make it possible to shed light further on the effects of life-course transitions on marital quality among older people.

CONCLUSION

The nature, dynamics, and contexts for family relationships in later life are currently undergoing dramatic change. New demographic realities have generated increasing interest in this topic on the part of researchers, policy makers, and the general public. As the average lifespan increases, most people can anticipate continued relationships with parents well into late middle-age, relationships with siblings for seven or more decades, and relationships with grandchildren well into their adulthood. Marriages are less stable, and a decrease in marriage rates coupled with high divorce rates and increased cohabitation will lead to a growing population of unmarried older people. The implications of these patterns over the coming decades provide a rich opportunity for research that is greatly needed to inform practice and policy.

Perhaps the most pressing research need is studies that better capture the complexity of relationships within families. Expanding research on the ways in which individual relationships within the family differ shows that it is important to collect data from two (or three) generations and from multiple members of each generation to fully understand relations in later life. Given the importance of life-course transitions on family relations, researchers should also emphasize longitudinal analyses that track changes in parent–child relations across time.

Further, the move away from a unidimensional focus on either closeness *or* conflict in understanding the quality of relationships should be encouraged. Given mounting evidence about the importance of both positive and negative feelings, attitudes, and behaviors in parent–child relationships (Pillemer et al., 2007) and in marriage (Uchino, Smith, & Berg, 2014), future research must take such complexity into account. Although the inclusion of multiple family members and multiple generations in longitudinal studies may be time- and cost-intensive, the knowledge that has been derived from panels such as the Longitudinal Study of Generations (Bengtson, 2001; Bengtson et al., 2002), the Family Exchanges Study (Fingerman, Miller, Birditt, & Zarit, 2009), and the Within-Family Differences Study (Suitor et al., 2013a), demonstrates the need of such designs to more fully understand later-life families in the twenty-first century.

Finally, we would note that the family in later life is likely to be of increasing interest to both policy makers and to organizations that provide assistance to older people. The growth in the older population creates serious concerns as the Baby Boomers move from providing care to needing care themselves. The Baby Boom cohort will have more limited informal support than their parents given smaller family size and the greater likelihood of entering the later years unmarried. These factors, combined with the other demographic shifts in family reconfiguration discussed in this chapter, suggest that later-life relationships are becoming increasingly complex. Research is therefore greatly needed that will allow society to anticipate and plan for the mental, physical, and social well-being of older people, in which family relationships play a critical part (Fingerman, Pillemer, Silverstein, & Suitor, 2012).

References

Adams, B. N. (1968). *Kinship in an urban setting*. Chicago, IL: Markham Publishing.

Aldous, J., Klaus, E., & Klein, D. M. (1985). The understanding heart: Aging parents and their favorite children. *Child Development, 56*, 303–316.

Amato, P. R. (1996). Explaining the intergenerational transmission of divorce. *Journal of Marriage and Family, 58*, 628–640.

Amato, P. R., & Afifi, T. D. (2006). Feeling caught between parents: Adult children's relations with parents and subjective well-being. *Journal of Marriage and Family, 68*, 222–235.

Amato, P. R., Booth, A., Johnson, D. R., & Rogers, S. J. (2009). *Alone together: How marriage in America is changing*. Cambridge, MA: Harvard University Press.

Amato, P. R., & Sobolewski, J. M. (2001). The effects of divorce and marital discord on adult children's psychological well-being. *American Sociological Review, 66*, 900–921.

Antonucci, T., Akiyama, H., & Takahashi, K. (2004). Attachment and close relationships across the life span. *Attachment & Human Development, 6*, 353–370.

Barnett, A. E. (2013). Pathways of adult children providing care to older parents. *Journal of Marriage and Family, 75*, 178–190.

Barnett, M. A., Scaramella, L. V., Neppl, T. K., Ontai, L., & Conger, R. D. (2010). Intergenerational relationship quality, gender, and grandparent involvement. *Family Relations, 59*, 28–44.

Bedford, V. H. (1998). Sibling relationship troubles and well-being in middle and old age. *Family Relations, 47*, 369–376.

Bedford, V. H., & Avioli, P. S. (2012). Sibling relationships from midlife to old age. In R. Blieszner & V. H. Bedford (Eds.), *Handbook of families and aging* (pp. 125–151, 2nd ed.). Santa Barbara, CA: Praeger.

Bengtson, V. L. (2001). Beyond the nuclear family: The increasing importance of multigenerational bonds. *Journal of Marriage and Family, 63*, 1–16.

Bengtson, V. L., Giarrusso, R., Mabry, J. B., & Silverstein, M. (2002). Solidarity, conflict, and ambivalence: Complementary or competing perspectives on intergenerational relationships? *Journal of Marriage and the Family, 64*, 568–576.

Bengtson, V. L., Olander, E. B., & Haddad, A. A. (1976). The "generation gap" and aging family members: Toward a conceptual model. In J. F. Gubrium (Ed.), *Time, roles, and self in old age* (pp. 237–263). New York, NY: Human Sciences Press.

Bengtson, V. L., & Roberts, R. E. L. (1991). Intergenerational solidarity in aging families: An example of formal theory construction. *Journal of Marriage and Family, 53*, 856–870.

Birditt, K. S., Tighe, L. A., Fingerman, K. L., & Zarit, S. H. (2012). Intergenerational relationship quality across three generations. *The Journals of Gerontology, Series B: Psychological Sciences and Social Sciences, 67*, 627–638.

Boll, T., Ferring, D., & Filipp, S. H. (2003). Perceived parental differential treatment in middle adulthood: Curvilinear relations with individuals' experienced relationship quality to sibling and parents. *Journal of Family Psychology, 17*, 472–487.

Bookwala, J. (2012). Marriage and other partnered relationships in middle and late adulthood. In R. Blieszner & V. H. Bedford (Eds.), *Handbook of families and aging* (pp. 91–124, 2nd ed.). Santa Barbara, CA: Praeger.

Brown, S. L., Bulanda, J. R., & Lee, G. R. (2005). The significance of nonmarital cohabitation: Marital status and mental health benefits among middle-aged and older adults. *Journals of Gerontology Series B: Psychological Sciences and Social Sciences, 60*, 21–29.

Brown, S. L., Bulanda, J. R., & Lee, G. R. (2012). Transitions into and out of cohabitation in later life. *Journal of Marriage and Family, 74*, 774–793.

Brown, S. L., & Lin, I.-F. (2012). The gray divorce revolution: Rising divorce among middle-aged and older adults, 1990–2010. *Journal of Gerontology Series B: Psychological Sciences and Social Sciences, 67*, 731–741.

Burbidge, J., & Minnes, P. (2014). Relationship quality in adult siblings with and without developmental disabilities. *Family Relations, 63*, 148–162.

Campbell, L. D., Connidis, I. A., & Davies, L. (1999). Sibling ties in later life: A social network analysis. *Journal of Family Issues, 20*, 114–148.

Carr, D., & Boerner, K. (2013). The impact of spousal loss on parent-child relations in later life: Are effects contingent upon the quality of the late marriage? *Family Science, 4*, 37–49.

Carr, D., & Moorman, S. M. (2011). Social relations and aging. In R. A. Settersten & J. L. Angel (Eds.), *Handbook of sociology of aging* (pp. 145–160). New York, NY: Springer.

Cherlin, A. J. (2010). *The marriage-go-round: The state of marriage and the family in America today*. New York, NY: Vintage Books.

Cherlin, A. J., & Furstenberg, F. F. (1986). *The new American grandparent: A place in the family, a life apart*. New York, NY: Basic Books.

Chodorow, N. J. (1978). *The reproduction of mothering*. Berkeley, CA: University of California Press.

Cicirelli, V. G. (1989). Feelings of attachment to siblings and well-being in later life. *Psychology and Aging, 4*, 211–216.

Clare, L., Nelis, S. M., Whitaker, C. J., Martyr, A., Markova, I. S., Roth, I., et al. (2012). Marital relationship quality in early-stage dementia: Perspectives from people with dementia and their spouses. *Alzheimer Disease & Associated Disorders, 26*, 148–158.

Cohn, D., Passel, J.S., Wang, W., & Livingston, G. (2011, December 14). *Barely half of US adults are married – A record low*. Pew Research Social & Demographic Trends. Retrieved from: <http://www.pewsocialtrends.org/2011/12/14/barely-half-of-u-s-adults-are-married-a-record-low/>.

Conger, R. D., & Elder, G. H., Jr. (1994). *Families in troubled times: Adapting to change in rural America*. New York, NY: Aldine de Gruyter.

Connidis, I. A., & Campbell, L. D. (1995). Closeness, confiding, and contact among siblings in middle and late adulthood. *Journal of Family Issues, 16*, 722–745.

Connidis, I. A., & Kemp, C. L. (2008). Negotiating actual and anticipated parental support: Multiple sibling voices in three-generation families. *Journal of Aging Studies, 22*, 229–238.

Coser, L., & Coser, R. (1974). *Greedy institutions: Patterns of undivided commitment*. New York, NY: Free Press.

Davey, A., Eggebeen, D. J., & Salva, J. (2007). Parental marital transitions and instrumental assistance between generations: A within-family longitudinal analysis. In T. J. Owens & J. J. Suitor (Eds.), *Interpersonal relations across the life course: Advances in life course research* (Vol. 12, pp. 221–242). New York, NY: Elsevier.

Davey, A., & Norris, J. E. (1998). Social networks and exchange norms across the adult life-span. *Canadian Journal on Aging, 17*, 212–233.

Davey, A., Savla, J., Janke, M., & Anderson, S. (2009). Grandparent-grandchild relationships: From families in contexts to families as contexts. *International Journal of Aging and Human Development, 69*, 311–325.

Drew, L. M., & Silverstein, M. (2007). Grandparents' psychological well-being after loss of contact with their grandchildren. *Journal of Family Psychology, 21*, 372–379.

Durkheim, E. (1933(1883)). *The division of labor in society*. New York, NY: Free Press.

Eriksen, S., & Gerstel, N. (2002). A labor of love or labor itself care work among adult brothers and sisters. *Journal of Family Issues, 23*, 836–856.

Fingerman, K., Miller, L., Birditt, K., & Zarit, S. (2009). Giving to the good and the needy: Parental support of grown children. *Journal of Marriage and Family, 71*, 1220–1233.

Fingerman, K., Sechrist, J., & Birditt, K. (2013). Changing views on intergenerational ties. *Gerontology, 59*, 64–70.

Fingerman, K. L. (2004). The role of offspring and in-laws in grandparents' ties to their grandchildren. *Journal of Family Issues, 25*, 1026–1049.

Fingerman, K. L., Cheng, Y. P., Birditt, K., & Zarit, S. (2012). Only as happy as the least happy child: Multiple grown children's problems and successes and middle-aged parents' well-being. *The Journals of Gerontology Series B: Psychological Sciences and Social Sciences, 67B*, 184–193.

Fingerman, K. L., Hay, E. L., & Birditt, K. S. (2004). The best of ties, the worst of ties: Close, problematic, and ambivalent social relationships. *Journal of Marriage and Family, 66*, 792–808.

Fingerman, K. L., Pillemer, K. A., Silverstein, M., & Suitor, J. J. (2012). The baby boomers' intergenerational relationships. *The Gerontologist, 52*, 199–209.

Fingerman, K. L., Pitzer, L. M., Chan, W., Birditt, K. S., Franks, M. M., & Zarit, S. (2011). Who gets what and why: Help middle-aged adults provide to parents and grown children. *Journal of Gerontology: Social Sciences, 66B*, 87–98.

Fomby, P., & Cherlin, A. J. (2007). Family instability and child well-being. *American Sociological Review, 72*, 181–204.

Fruhauf, C. A., Jarrott, S. E., & Allen, K. R. (2006). Grandchildren's perceptions of caring for grandparents. *Journal of Family Issues, 27*, 887–911.

Fuller-Thomson, E. (2000). Loss of the kin-keeper: Sibling conflict following parental death. *OMEGA: The Journal of Death and Dying, 40*, 547–559.

Fuller-Thomson, E., & Minkler, M. (2005). American Indian/Alaskan Native grandparents raising grandchildren: Findings from the Census 2000 supplementary survey. *Social Work, 50*, 131–139.

Fuller-Thomson, E., & Minkler, M. (2007). Mexican American grandparents raising grandchildren: Findings from the census 2000 American community survey. *Families in Society, 88*(4), 567–574.

George, L. (2004). Life course research: Achievements and potential. In J. T. Mortimer & M. J. Shanahan (Eds.), *Handbook of the life course* (pp. 671–680). New York, NY: Springer.

Giarrusso, R., Feng, D., & Bengtson, V. L. (2005). The intergenerational-stake phenomenon over 20 years. In K. W. Schaie & M. Silverstein (Eds.), *Annual review of gerontology and geriatrics: Focus on intergenerational relations across time and place* (pp. 55–76). New York, NY: Springer.

Giarrusso, R., Feng, D., Silverstein, M., & Bengtson, V. L. (2001). Grandparent-adult grandchild affection and consensus. *Journal of Family Issues, 22*, 456–477.

Gilligan, C. (1982). *In a different voice: Psychological theory and women's development*. Cambridge, MA: Harvard University Press.

Gilligan, M., Suitor, J. J., Kim, S., & Pillemer, K. (2013). Differential effects of perceptions of mothers' and fathers' favoritism on sibling tension in adulthood. *The Journals of Gerontology Series B: Psychological Sciences and Social Sciences, 68*, 593–598.

Gilligan, M., Suitor, J. J., & Pillemer, K. (2013). Recent economic distress in midlife: Consequences for adult children's relationships with their mothers. *Contemporary Perspectives in Family Research, 7*, 159–184.

Greenfield, E. A., & Marks, N. F. (2006). Linked lives: Adult children's problems and their parents' psychological and relational well-being. *Journal of Marriage and Family, 68*, 442–454.

Grusec, J. E. (2002). Parental socialization and children's acquisition of values. In M. H. Bornstein (Ed.), *Handbook of parenting: Practical issues in parenting* (Vol. 5, pp. 143–167). Mahwah, NJ: Lawrence Erlbaum Associates.

Hamill, S. B. (2012). Caring for grandparents with Alzheimer's disease: Help from the "Forgotten" generation. *Journal of Family Issues, 33*, 1195–1217.

Hayslip, B., Jr., & Kaminski, P. L. (2005). Grandparents raising their grandchildren: A review of the literature and suggestions for practice. *The Gerontologist, 45*, 262–269.

Heider, F. (1958). *The psychology of interpersonal relations*. New York, NY: Wiley.

Homans, G. C. (1950). *Social behavior: Its elementary forms*. New York, NY: Harcourt Brace Jovanovich.

Hughes, M. E., Waite, L. J., LaPierre, T. A., & Luo, Y. (2007). All in the family: The impact of caring for grandchildren on grandparents' health. *Journals of Gerontology: Social Sciences, 62B*, S108–S119.

Iveniuk, J., Waite, L. J., Laumann, E., McClintock, M. K., & Tiedt, A. D. (2014). Marital conflict in older couples: Positivity, personality, and health. *Journal of Marriage and Family, 76*, 130–144.

Khodyakov, D., & Carr, D. (2009). The impact of late-life parental death on adult sibling relationships: Do parents' advance directives help or hurt? *Research on Aging, 31*, 495–519.

Kiecolt, J. K., Blieszner, R., & Salva, J. (2011). Long-term influences of intergenerational ambivalence on midlife parents' psychological well-being. *Journal of Marriage and Family, 73*, 369–382.

Korporaal, M., van Groenou, M. I. B., & van Tilburg, T. G. (2013). Health problems and marital satisfaction among older couples. *Journal of Aging and Health, 25*, 1279–1298.

Kreider, R.M., & Ellis, R. (2011). Number, timing, and duration of marriages and divorces: 2009. *Current population reports*. US Census Bureau, Washington, DC, pp. P70–P125.

Lashewicz, B., & Keating, N. (2009). Tensions among siblings in parent care. *European Journal of Ageing, 6*, 127–135.

Lazarsfeld, P. F., & Merton, R. K. (1954). Friendship as a social process: A substantive and methodological analysis. *Freedom and Control in Modern Society, 18*, 18–66.

Leopold, T., Raab, M., & Engelhardt, H. (2014). The transition to parent care: Costs, commitments, and caregiver selection among children. *Journal of Marriage and Family, 76*, 300–318.

Lin, I.-F. (2008). Consequences of parental divorce for adult children's support of their frail parents. *Journal of Marriage and Family, 70*, 113–128.

Lin, I.-F., & Brown, S. L. (2012). Unmarried boomers confront old age: A national portrait. *The Gerontologist, 52*, 153–165.

Lin, I.-F., & Wu, H. S. (2014). Intergenerational exchange and expected support among the young-old. *Journal of Marriage and Family, 76*, 261–271.

Lüscher, K., & Pillemer, K. (1998). Intergenerational ambivalence: A new approach to the study of parent-child relations in later life. *Journal of Marriage and Family, 60*, 413–425.

Mason, J., May, V., & Clarke, L. (2007). Ambivalence and the paradoxes of grandparenting. *The Sociological Review, 55*, 687–706.

Merz, E.-M., Schuengel, C., & Schulze, H.-J. (2009). Intergenerational relations across 4 years: Well-being is affected by quality, not by support exchange. *The Gerontologist, 49*, 536–548.

Minkler, M., & Fuller-Thomson, E. (2005). African American grandparents raising grandchildren: A national study using the Census 2000 American Community Survey. *Journal of Gerontology: Social Sciences, 60B*, S82–S92.

Moen, P., Kim, J. E., & Hofmeister, H. (2001). Couples' work/retirement transitions, gender, and marital quality. *Social Psychology Quarterly, 64*, 55–71.

Monserud, M. A. (2008). Intergenerational relationships and affectual solidarity between grandparents and young adults. *Journal of Marriage and Family, 70*, 182–195.

Mueller, M. M., & Elder, G. H., Jr. (2003). Family contingencies across the generations: Grandparent-grandchild relationships in holistic perspective. *Journal of Marriage and Family, 65*, 404–417.

Mueller, M. M., Wilhelm, B., & Elder, G. H., Jr. (2002). Variations in grandparenting. *Research on Aging, 24*, 360–388.

Parsons, T. (1943). The kinship system of the contemporary United States. *American Anthropologist, 45*, 22–38.

Paul, E. L. (1997). A longitudinal analysis of midlife interpersonal relationships and well-being. In M. E. Lachman & J. B. James (Eds.), *Multiple paths of midlife development* (pp. 171–206). Chicago, IL: University of Chicago Press.

Pavalko, E. K. (2011). Caregiving and the life course: Connecting the personal and the public. In R. A. Settersten & J. L. Angel (Eds.), *Handbook of sociology and aging* (pp. 603–616). New York, NY: Springer.

Pillemer, K., Munsch, C. L., Fuller-Rowell, T., Riffin, C., & Suitor, J. J. (2012). Ambivalence toward adult children: Differences between mothers and fathers. *Journal of Marriage and Family, 74*, 1101–1113.

Pillemer, K., & Suitor, J. J. (2006). Making choices: A within-family study of caregiver selection. *The Gerontologist, 46*, 439–448.

Pillemer, K., & Suitor, J. J. (2014). Who provides care? A prospective study of caregiving by adult children. *The Gerontologist, 54*, 589–598.

Pillemer, K., Suitor, J. J., Mock, S. E., Sabir, M., Pardo, T. B., & Sechrist, J. (2007). Capturing the complexity of intergenerational relations: Exploring ambivalence within later-life families. *Journal of Social Issues, 63*, 775–791.

Pillemer, K., Suitor, J. J., Pardo, S., & Henderson, C. (2010). Mothers' differentiation and depressive symptoms among adult children. *Journal of Marriage and Family, 72*, 333–345.

Pope, N. D., Kolomer, S., & Glass, A. P. (2012). How women in late midlife become caregivers for their aging parents. *Journal of Women & Aging, 24*, 242–261.

Powell, B., Bolzendahl, C., Geist, C., & Steelman, L. C. (2012). *Counted out: Same-sex relations and Americans' definitions of family*. New York, NY: Russell Sage Foundation.

Powers, R., Suitor, J. J., Guerra, S., Shackelford, M., Mecom, D., & Gusman, K. (2003). Regional differences in gender–role attitudes: Variations by gender and race. *Gender Issues, 21*(2), 40–54.

Riffin, C., Suitor, J. J., Reid, M. C., & Pillemer, K. (2012). Chronic pain and parent-child relations in later life: An important, but understudied issue. *Family Science, 3*, 75–85.

Rossi, A. S., & Rossi, P. H. (1990). *Of human bonding: Parent-child relations across the life course.* New York, NY: Aldine de Gruyter.

Sarkisian, N., & Gerstel, N. (2008). Till marriage do us part: Adult children's relationships with their parents. *Journal of Marriage and Family, 70*, 360–376.

Schenk, N., & Dykstra, P. A. (2012). Continuity and change in intergenerational family relationships: An examination of shifts in relationship type over a three-year period. *Advances in Life Course Research, 17*, 121–132.

Schwarz, B., Trommsdorff, G., Albert, I., & Mayer, B. (2005). Adult parent-child relationships: Relationship quality, support, and reciprocity. *Applied Psychology: An International Review, 54*, 396–417.

Sechrist, J., Suitor, J. J., Henderson, A. C., Cline, K. M., & Steinhour, M. (2007). Regional differences in mother-adult-child relations: A brief report. *The Journals of Gerontology Series B: Psychological Sciences and Social Sciences, 62*, 388–391.

Sechrist, J., Suitor, J. J., Howard, A., & Pillemer, K. (2014). Perceptions of equity, balance of support exchange, and mother-adult child relations. *Journal of Marriage and Family, 76*, 285–299.

Seltzer, J. A., & Bianchi, S. M. (2013). Demographic change and parent-child relationships in adulthood. *Annual Review of Sociology, 39*, 275–290.

Seltzer, M. M., Greenberg, J. S., Orsmond, F. I., Lounds, J., & Smith, M. J. (2008). Unanticipated lives: Inter- and intragenerational relationships in families with children with disabilities. In A. Booth, A. Crouter, S. Bianchi, & J. A. Seltzer (Eds.), *Intergenerational caregiving* (pp. 233–242). Washington, DC: Urban Institute Press.

Settersten, R. A. (2003). Propositions and controversies in life-course scholarship. In R. A. Settersten (Ed.), *Invitation to the life course: Toward new understandings of later life* (pp. 15–45). Amityville, NY: Baywood Publishing Company.

Shapiro, A. (2012). Rethinking marital status: Partnership history and intergenerational relationships in American families. *Advances in Life Course Research, 17*, 168–176.

Shapiro, A., & Cooney, T. (2007). Divorce and intergenerational relations across the life course. *Advances in Life Course Research, 12*(0), 191–219.

Sigad, L. I., & Eisikovits, R. A. (2013). Grandparenting across borders: American grandparents and their Israeli grandchildren in a transnational reality. *Journal of Aging Studies, 27*, 308–316.

Silverstein, M., Chen, X., & Heller, K. (1996). Too much of a good thing? Intergenerational social support and the psychological well-being of older parents. *Journal of Marriage and the Family, 58*, 970–982.

Silverstein, M., Conroy, S. J., Wang, H., Giarrusso, R., & Bengtson, V. L. (2002). Reciprocity in parent-child relations over the adult life course. *The Journals of Gerontology Series B: Psychological Sciences and Social Sciences, 57*, S3–S13.

Silverstein, M., Gans, D., Lowenstein, A., Giarrusso, R., & Bengtson, V. L. (2010). Older parent-child relationships in six developed nations: Comparisons at the intersection of affection and conflict. *Journal of Marriage and Family, 72*, 1006–1021.

Silverstein, M., Gans, D., & Yang, F. M. (2006). Intergenerational support to aging parents: The role of norms and needs. *Journal of Family Issues, 27*, 1068–1084.

Silverstein, M., & Giarrusso, R. (2010). Aging and family life: A decade review. *Journal of Marriage and Family, 72*, 1039–1058.

Silverstein, M., Lowenstein, A., Katz, R., Gans, D., Fan, Y. K., & Oyama, P. (2013). Intergenerational support and the emotional well-being of older Jews and Arabs in Israel. *Journal of Marriage and Family, 75*, 950–963.

Silverstein, M., & Marenco, A. (2001). How Americans enact the grandparent role across the family life course. *Journal of Family Issues, 22*, 493–522.

Silverstein, M. D., Conroy, S. J., & Gans, D. (2012). Beyond solidarity, reciprocity and altruism: Moral capital as a unifying concept in intergenerational support for older people. *Ageing and Society, 32*, 1246–1262.

Simmel, G. (1964). *The sociology of Georg Simmel.* New York, NY: Free Press.

Smith, M. J., Greenberg, J. S., & Seltzer, M. M. (2007). Siblings of adults with schizophrenia: Expectations about future caregiving roles. *American Journal of Orthopsychiatry, 77*, 29–37.

Smith, T. W., Berg, C. A., Florsheim, P., Uchino, B. N., Pearce, G., Hawkins, M., et al. (2009). Conflict and collaboration differences in agency and communion during marital interaction. *Psychology and Aging, 24*, 259–273.

Smits, A., van Gaalen, R. I., & Mulder, C. H. (2010). Parent-child coresidence: Who moves in with whom and for whose needs? *Journal of Marriage and Family, 72*, 1022–1033.

Spitze, G., & Trent, K. (2006). Gender differences in adult sibling relations in two-child families. *Journal of Marriage and Family, 68*, 977–992.

Spitze, G., Ward, R., Deane, G., & Zhuo, Y. (2012). Cross-sibling effects in parent-adult child exchanges of socioemotional support. *Research on Aging, 34*, 197–221.

Suitor, J. J., Gilligan, M., Johnson, K., & Pillemer, K. (2014a). How widowhood shapes adult children's responses to mothers' preferences for care. *Journals of Gerontology Series B: Psychological Sciences and Social Sciences, 69,* 95–102.

Suitor, J. J., Gilligan, M., Johnson, K., & Pillemer, K. (2014b). Caregiving, perceptions of maternal favoritism, and tension among siblings. *The Gerontologist, 54,* 580–588.

Suitor, J. J., Gilligan, M., & Pillemer, K. (2013a). Continuity and change in mothers' favoritism toward offspring in adulthood. *Journal of Marriage and Family, 75,* 1229–1247.

Suitor, J. J., Gilligan, M., & Pillemer, K. (2013b). The role of violated caregiver preferences in psychological well-being when older mothers need assistance. *The Gerontologist, 53,* 388–396.

Suitor, J. J., & Pillemer, K. (2006). Choosing daughters: Exploring why mothers favor adult daughters over sons. *Sociological Perspectives, 49,* 139–160.

Suitor, J. J., & Pillemer, K. (2013). Differences in mothers' and fathers' parental favoritism in later-life: A within-family analysis. In M. Silverstein & R. Giarrusso (Eds.), *From generation to generation: Continuity and discontinuity in aging families* (pp. 11–31). Baltimore, MD: Johns Hopkins University Press.

Suitor, J. J., Pillemer, K., & Sechrist, J. (2006). Within-family differences in mothers' support to adult children. *Journal of Gerontology: Social Sciences, 61B,* S10–S17.

Suitor, J. J., Sechrist, J., Gilligan, M., & Pillemer, K. (2011). Intergenerational relations in later-life families. In R. Settersten & J. Angel (Eds.), *Handbook of sociology of aging* (pp. 161–178). New York, NY: Springer.

Suitor, J. J., Sechrist, J., & Pillemer, K. (2007). Within-family differences in mothers' support to adult children in Black and White families. *Research on Aging, 29,* 410–435.

Suitor, J. J., Sechrist, J., Plikuhn, M., Pardo, S., Gilligan, M., & Pillemer, K. (2009). The role of perceived maternal favoritism in sibling relations in midlife. *Journal of Marriage and Family, 71,* 1026–1038.

Suitor, J. J., Sechrist, J., Plikuhn, M., Pardo, S., & Pillemer, K. (2008). Within-family differences in parent-child relations across the life course. *Current Directions in Psychological Science, 17,* 334–338.

Sussman, M. B. (1976). The family life of old people. In R. H. Binstock & E. Shanas (Eds.), *Handbook of aging and the social sciences* (pp. 218–243). New York, NY: Van Nostrand Reinhold.

Szydlik, M. (2012). Generations: Connections across the life course. *Advances in Life Course Research, 17,* 100–111.

Tamborini, C. R. (2007). The never-married in old age: Projections and concerns for the near future. *Social Security Bulletin, 67,* 25–40.

Timonen, V., Conlon, C., Scharf, T., & Carney, G. (2013). Family, state, class and solidarity: Reconceptualising intergenerational solidarity through the grounded theory approach. *European Journal of Ageing, 10,* 171–179.

Uchino, B. N., Smith, T. W., & Berg, C. A. (2014). Spousal relationship quality and cardiovascular risk dyadic perceptions of relationship ambivalence are associated with coronary-artery calcification. *Psychological Science, 25,* 1037–1042.

Uhlenberg, P. (2004). Historical forces shaping grandparent-grandchild relationships: Demography and beyond. In M. Silverstein & K. W. Schaie (Eds.), *Annual review of gerontology and geriatrics* (pp. 77–97). New York, NY: Springer.

Vespa, J. (2014, February 10). *Marrying older, but sooner?* [Web log post]. Retrieved from: <http://blogs.census.gov/2014/02/10/marrying-older-but-sooner/>

Voorpostel, M., & Blieszner, R. (2008). Intergenerational solidarity and support between adult siblings. *Journal of Marriage and Family, 70,* 157–167.

Voorpostel, M., van der Lippe, T., Dykstra, P. A., & Flap, H. (2007). Similar or different? The importance of similarities and differences for support between siblings. *Journal of Family Issues, 28,* 1026–1053.

Voorpostel, M., van der Lippe, T., & Flap, H. (2012). For better or worse: Negative life events and sibling relationships. *International Sociology, 27,* 330–348.

Waldinger, R. J., Vaillant, G. E., & Orav, E. J. (2007). Childhood sibling relationships as a predictor of major depression in adulthood: A 30-year prospective study. *American Journal of Psychiatry, 164,* 949–954.

Ward, R. A., Deane, G., & Spitze, G. (2014). Life-course changes and parent-adult child contact. *Research on Aging, 36,* 568–602.

Ward, R. A., Spitze, G., & Deane, G. (2009). The more the merrier? Multiple parent-adult child relations. *Journal of Marriage and Family, 71,* 161–173.

White, L. (2001). Sibling relationships over the life course: A panel analysis. *Journal of Marriage and Family, 63,* 555–568.

White, L. K., & Riedmann, A. (1992). Ties among adult siblings. *Social Forces, 71,* 85–102.

Yu, T., Pettit, G. S., Lansford, J. E., Dodge, K. A., & Bates, J. E. (2010). The interactive effects of marital conflict and divorce on parent-adult children's relationships. *Journal of Marriage and Family, 72,* 282–292.

CHAPTER 11

The Influence of Military Service on Aging

Janet M. Wilmoth and Andrew S. London

Department of Sociology, Aging Studies Institute, Center for Policy Research, and Institute for Veterans and Military Families, Syracuse University, Syracuse, NY, USA

OUTLINE

Introduction	227	An Overview of Military Service and Aging Among Specific War Cohorts	234
Cohort Flow, Periods of War, and the Composition of the US Older Adult Population	228	WWI	234
		WWII	236
		Korean War and Post-Korean War	237
Military Service as a "Hidden Variable" in Aging Research	230	Vietnam War	238
		All-Volunteer Force	240
Mechanisms Through Which Military Service Influences Aging	232	Studying Military Service and Aging	242
		References	246

INTRODUCTION

Much of what is known about aging processes is based on scientific evidence that has been generated since the middle of the twentieth century. Military service was prevalent among men in the cohorts upon which this evidence is based due to the succession of wars that occurred during the twentieth century. For example, over 50% of men born between 1915 and 1935 served in the armed forces, with rates of military service peaking at over 75% among men born between 1920 and 1926, who transitioned to adulthood during World War II (WWII) (Hogan, 1981). Despite its prevalence in lived experience, military service is generally a "hidden variable" in aging research because it is rarely examined explicitly or taken into account as a control variable (Settersten, 2006; Spiro, Schnurr, & Aldwin, 1997). The lack of scholarly attention to the influence of military service on aging is due, in part, to insufficient nationally representative data on military service experiences. But even the data that are available have been under-analyzed, which could potentially contribute to omitted variable bias in the majority of studies on aging that focus on men, as well as women, many of whom lived their lives linked to men with histories of military service. These considerations recently led Settersten and Patterson (2006) to argue that "wartime experiences may be important but largely invisible factors underneath contemporary knowledge about aging" (p. 5). Building on their argument, we have suggested elsewhere that the "institutional influence of military service on lives has more generally been under-acknowledged among life-course scholars, and more broadly within Sociology and other disciplines that study human lives" (Wilmoth & London, 2013, p. 2).

Similar to other social institutions that engage individuals during the transition to adulthood, like educational, economic, family, and penal systems, military service has enduring effects on the life course that can influence the aging process (Wilmoth & London, 2013). The primary aim of this chapter is to highlight the military as a social force that shapes later-life outcomes. We first examine the extent to which military service is a "hidden variable" in aging research by: considering how cohort flow in relation to periods of war has altered the composition of the older adult population in the United States and presenting the findings of a systematic search of publications in selected gerontology journals from 1980 through 2013. Then, we review the mechanisms through which military service influences the aging process, focusing particularly on selection into service, service-related experiences, and post-service access to benefits. This provides a foundation for considering what is known about military service and aging among those who served in particular wars. We consider various life-course outcomes, including educational attainment, employment, marriage, civic engagement, health, and mortality. The majority of the empirical evidence we review is based on samples of men because women's participation in the military was low prior to the military's transition to an All-Volunteer Force (AVF) in 1973. We conclude with some observations about studying the impact of military service on aging, including an integrated overview of available data, methodological challenges, and potential directions for future research.

COHORT FLOW, PERIODS OF WAR, AND THE COMPOSITION OF THE US OLDER ADULT POPULATION

Figure 11.1 presents a lexis diagram of twentieth century cohorts in relation to periods of war and major military conflict in which the United States was engaged. We refer to these as "wars" throughout the chapter, even though there was not an official declaration of war for some military conflicts. Age is arrayed along the vertical axis, with a light gray bar extending along the diagram to indicate the peak ages of military service from 18 to 29. Time period is arrayed along the horizontal axis, with dark gray bars indicating periods of war based on dates from the US Code of Federal Regulations (2014).

Seven cohorts are shown along the diagonal intersection of age and time period, each of which has a unique historical relationship to the wars and other historical events of the twentieth and early twenty-first centuries (Carlson, 2008). The cohorts include:

1. "New Worlders" (born from 1871 to 1889), who were young adults during World War I (WWI) and reached retirement age by the time of WWII and the Korean War.
2. "Hard Timers" (born from 1890 to 1908), the oldest of whom would have been of age to serve in the military during WWI, lived through the Great Depression during young adulthood, experienced WWII and

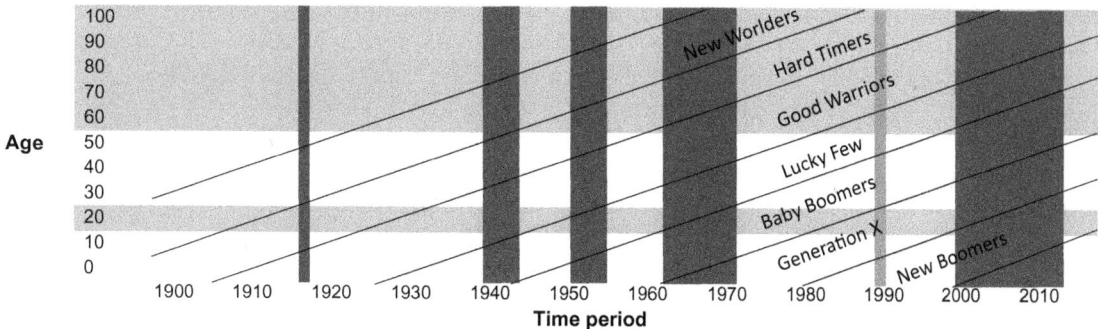

FIGURE 11.1 Lexis diagram of twentieth century cohorts in relation to periods of US wars and military conflicts. *Notes*: Carlson (2008) defines seven twentieth century "American Generations" (which are analogous to birth cohorts): New Worlders (born from 1871 to 1889); Hard Timers (born from 1890 to 1908); Good Warriors (born 1909 to 1928); Lucky Few (born 1929 to 1945); Baby Boomers (born 1946 to 1964); Generation X (born 1965 to 1982); and New Boomers (born 1983 to 2001). Periods of US wars and military conflicts are shown in dark gray. According to the US Code of Federal Regulations (2014) the periods of wars and military conflicts are as follows: WWI = 4/6/1917–11/11/1918; WWII = 12/7/1941–12/31/1946; Korean War = 6/27/1950–1/31/1955; Vietnam War = 8/5/1964–5/7/1975; Gulf War = 8/2/1990–4/6/1991; OEF/OIF/OND = 10/7/2001–End date to be determined. Peak ages of military service, which are from 18 to 29, and ages over 60 indicating later-life are indicated by light gray.

the Korean War during middle adulthood, and moved into retirement during the Cold War and Vietnam War years when social insurance programs were being expanded.

3. "Good Warriors" (born from 1909 to 1928), most of whom were children during the Great Depression, young adults during WWII and the Korean War, middle-aged during the Cold War and the Vietnam War, and retirement-aged during the golden age of defined benefit plans.

4. "Lucky Few" (born from 1929 to 1945) was a relatively small cohort due to low fertility rates throughout the Depression Era; it had a leading edge that was eligible to serve in the Korean War, but most of the cohort hit the peak years of military service during the Cold War. Members of this cohort were beneficiaries of the nation's longest period of economic expansion during their young adult years and reached retirement age during the economic boom of the 1990s.

5. "Baby Boomers" (born from 1946 to 1964) consists of the large number of people born during the post-WWII fertility boom, most of whom came of age during the Vietnam War, experienced the economic recessions of the 1970s and 1980s during their young adult years, and moved into retirement after the shift to defined contribution plans.

6. "Generation X" (born from 1965 to 1982) has a leading edge that was eligible to serve during the Gulf War. All members of this cohort were young adults on 9/11/2001 and were eligible to volunteer for service in the recent wars in Iraq and Afghanistan – Operation Enduring Freedom (OEF), Operation Iraqi Freedom (OIF), and Operation New Dawn (OND). They are reaching middle age as the country recovers from the Great Recession of 2008.

7. "New Boomers" (born from 1983 to 2001), the oldest of whom were eligible to serve in OEF/OIF/OND and were transitioning to adulthood during the 2008 recession.

The profiles Carlson and Andress (2009) provide of these cohorts' military service experiences, which we expand upon later in the chapter, are consistent with Hogan's (1981) estimates of the percentage serving on active duty in the armed forces for 6 months or longer among American males born 1907–1946. Based on data from the 1973 Current Population Survey (CPS), Hogan reported that: approximately 25% of the youngest members of the "Hard Timers" cohort served in the military; the percent of men with military service among the "Good Warriors" cohort ranged from a low of 27% for those born in 1909 to a high of nearly 80% for the those born in 1926; the percentage of the "Lucky Few" men with military service ranged from a high of 68% for those born in 1929 to a low of 40% for those born in 1944; and 53% of men born in 1946, the first year of the Baby "Boom" served in the military.

Taken together, the evidence provided by Carlson and Andress (2009) and Hogan (1981) indicates that military service was a normative part of the transition to adulthood during the twentieth century for the majority of American men. This has implications not only for those who were members of the military, but also those whose lives are linked to military personnel through marriage, family, and community ties. Therefore, given the military was a major institutional force that influenced the lives of Americans throughout the twentieth century, it is important to consider how cohort flow in relation to periods of war might be shaping our understanding of aging.

Focusing on ages 60 and over in the upper portion of Figure 11.1 reveals how the composition of the older adult population has shifted over time as successive cohorts have moved through later-life. During the middle of the twentieth century, when research on aging began to gain momentum, the older adult population was primarily comprised of the "New Worlders" and "Hard Timers" cohorts who experienced WWI. As the aging research enterprise expanded during the 1970s and 1980s, the "Hard Timers" cohort moved into the oldest-old ages and the "Good Warriors" cohort, who experienced WWII and the Korean War, aged into later life. Since the 1990s, the "Lucky Few" and "Baby Boomers," who served during the Cold War and the Vietnam War, have aged into later life, while the "Good Warrior" cohort has become the oldest-old. As the discussion in the next section indicates, gerontology scholars rarely consider these substantial cohort shifts in the composition of the older adult population in relation to the potential role of military service in shaping later-life outcomes.

MILITARY SERVICE AS A "HIDDEN VARIABLE" IN AGING RESEARCH

In order to gauge empirically the extent to which military service is a "hidden" variable in aging research, we conducted a systematic search of selected gerontology journals. We focused on the 34-year period from January 1980 through December 2013 and included the five Gerontological Society of America (GSA)-sponsored journals (*The Journals of Gerontology: Series A*, *The Journals of Gerontology: Series B*, *The Gerontologist*, *The Journal of Gerontology*, and *Public Policy & Aging Report*), *Research on Aging*, and *Journal of Aging and Health*. For the GSA-sponsored journals, we were able to search the title and abstract for all of the journals at the same time; however, we were not able to include multiple terms at the same time. Thus, we did separate searches using each of the following terms: veteran, military, armed forces, war, Army, Navy, Air Force, Marines, and Coast Guard. We compiled an unduplicated list of articles. For both *Research on Aging* and *Journal of Aging and Health*, we used the advanced search option and a string of search terms connected by "or" in order to identify all articles that included any one of the terms listed above. We searched separately in the title and abstract fields, and then compared what we identified.

Over the 34-year period, we identified 101 military-related articles that were published in the seven journals we considered (a list of the articles is available from the first author upon

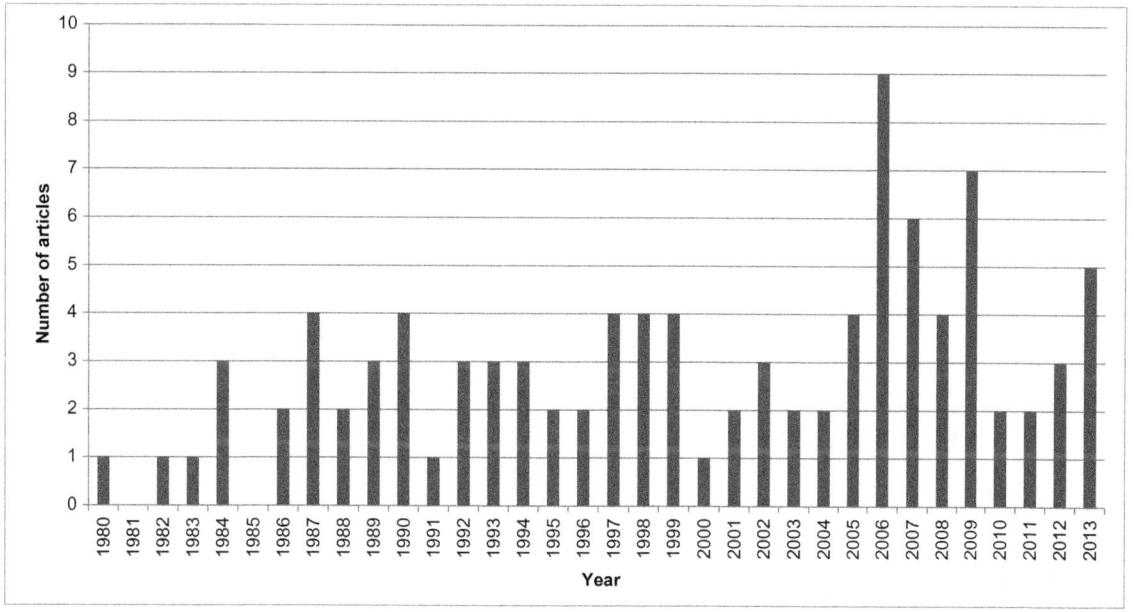

FIGURE 11.2 Number of military service-related articles published per year, selected aging journals, 1980–2013. *Note:* Journals include: *The Journals of Gerontology: Series A, The Journals of Gerontology: Series B, The Gerontologist, The Journal of Gerontology, Public Policy & Aging Report, Research on Aging,* and *Journal of Aging and Health*.

request). This translates into a rate of 3.0 articles per year and represents less than 1% of the nearly 12000 articles published in these mainstream aging-focused journals over this period. As seen in Figure 11.2, the number of military-related articles has increased in more-recent years. Over the 20-year period from 1980 to 1999, there are 5 years when four or more articles were published; over the 14-year period from 2000 to 2013, there are 6 years when four or more articles were published. The nine articles published in the peak year – 2006 – include the introduction and six research articles that were published in the seminal special issue of *Research on Aging* on "Military Service, the Life Course, and Aging" that was guest-edited by Richard A. Settersten, Jr.

Although a systematic content analysis of these articles is beyond the scope of this chapter, our analysis of the abstracts indicates that the vast majority of these studies are based on cross-sectional research designs and non-representative samples, and are focused exclusively on veterans or Department of Veteran Affairs (VA) health care facilities and programs. Although these studies address important, policy-relevant questions pertaining to older veterans, the degree to which prior service in the armed forces affects variation in aging-related outcomes remains virtually unexamined because almost none of these studies directly compare veterans to nonveterans in well-controlled models based on population-representative data [e.g., only four studies use data from the Health and Retirement Study (HRS)].

Given the dearth of studies that make direct comparisons between veterans and nonveterans, relatively little is known about how aging veterans fair relative to their nonveteran counterparts. We understand even less about how differences between veterans and nonveterans vary across cohorts. But evidence about the mechanisms through which military service might influence aging, as well as differences in aging among veterans who served in different wars, can be gleaned from the extant literature, much of which is based on a life-course perspective (Wilmoth & London, 2013; see also, Camacho & Atwood, 2007; Card, 1983; Carlson & Andress, 2009; Hogan, 1981; London & Wilmoth, in press; MacLean & Elder, 2007; Mettler, 2005; Modell & Haggerty, 1991). The earliest conceptual foundations in this area appear in the seminal work of Glen H. Elder, Jr. (Elder, 1986, 1987), whose ongoing work with a range of colleagues has explored various aspects of the potential influence of military service on the aging process (Clipp & Elder, 1996; Dechter & Elder, 2004; Elder, 1986, 1987; Elder & Bailey, 1988; Elder & Clipp, 1988a, 1988b, 1989; Elder, Clipp, Brown, Martin, & Friedman, 2009; Elder, Gimbel, & Ivie, 1991; Elder & Johnson, 2002; Elder, Johnson, & Crosnoe, 2003; Elder, Shanahan, & Clipp, 1994, 1997; Elder, Wang, Spence, Adkins, & Brown, 2010; Pavalko & Elder, 1990). This research, in conjunction with the research of others from a variety of disciplines, has yielded critical insights, but in some cases has produced conflicting findings given differences across samples in the composition of veterans studied and statistical techniques employed. Below, we broadly discuss the factors that this body of research identifies as potential mechanisms through which military service influences aging. Then, we summarize what is known about military service and aging for specific war cohorts.

MECHANISMS THROUGH WHICH MILITARY SERVICE INFLUENCES AGING

When considering the mechanisms through which military service influences aging processes, it is important to begin by acknowledging the selective nature of military service. The selection process starts with military recruiters, who screen prospects and only encourage applications from the strongest candidates who meet age and minimum education requirements. There is evidence that recruiters have become more selective during the AVF period (US Department of Defense, 2006). After applying for military service, individuals are required to take an aptitude test to ensure they have sufficient skills to complete training programs and perform required duties (Defense Manpower Data Center, 2014). All branches of the US military employ pre-induction physical exams that are designed to identify a range of physical attributes that preclude military service, such as specific health conditions, vision or hearing impairments, physical abnormalities, and illicit drug use (Sackett & Mayer, 2006). Rejection rates based on pre-induction physical exams vary by year depending on the personnel needs of the specific branches of service, but even during years with low rejection rates (which tend to coincide with periods of war), a large portion of individuals do not meet the physical requirements of service. For example, between 1950 and 1971, the pre-induction rejection rate ranged from a high of nearly 70% in 1950 at the beginning of the Korean War to a low of 32% in 1953 during the middle of the Korean War (Wolf, Wing, & Lopoo, 2013). Collectively, recruitment, physical, and aptitude screening processes may result in a veteran population that is healthier on average than the non-veteran population. Once in military service, individuals in the best physical condition are selected into combat, often referred to as the "healthy warrior" or "healthy deployer" effect (Armed Forces Health Surveillance Center, 2007). Selection processes also govern the duration of military service, with those best-suited to effective performance in the military and most receptive to financial inducements to reenlist being more likely to pursue an armed forces career. Each of these selection processes – into the military, into combat, and into career service – shape the characteristics of the active-duty force and the veteran population in ways that could produce variation in the outcomes of studies of military service and aging.

Second, the primary mechanism through which military service is expected to influence the aging process is the experiences of military personnel during service. Military experiences are shaped by many factors, including time period of service, branch of service, rank, military occupational specialty, and combat exposure. As a result, military service potentially has a range of positive and negative implications for the aging process. On the positive side, military personnel are required to maintain a relatively high level of physical fitness. Training programs and exercise regimens engaged in during military service may set the stage for life-long physical activity (MacLean, 2013). Military service also involves specialized training that potentially facilitates occupational attainment, which can lead to improved work environments, increased earnings, and access to high-quality health insurance (Kleykamp, 2013; Street & Hoffman, 2013). Military personnel and veterans also have access to various health care, housing, and educational benefits designed to compensate those who have served in the armed forces and ameliorate the negative effects of service-connected injuries. Veterans with 20 or more years of military service also receive retirement pay and additional health care benefits. Collectively, these veteran benefits work in tandem with federal programs (e.g., Social Security, Medicare, and Medicaid) to provide an additional safety net to military personnel, veterans, and their families (for a more detailed discussion of benefits, see Street & Hoffman, 2013; Wilmoth & London, 2011). On the negative side, military service is a risky endeavor that can compromise health through training accidents and over-use injuries, hazardous work assignments (e.g., involving radiation, agent orange, infectious diseases), and exposure to a culture of substance use (tobacco, alcohol, and other drugs) (Bedard & Deschênes, 2006; Clipp & Elder, 1996; Elder & Clipp, 1989; Elder et al., 2009; MacLean, 2013; Miech, London, Wilmoth, & Koester, 2013). In addition, wartime service often involves combat exposure, which increases the risk of short-term injury and long-term disability (Elder et al., 1997; MacLean, 2010, 2013), physical and mental health problems (Elder et al., 2009; Vogt, King, King, Savarese, & Suvak, 2004), inadequate sleep (London, Burgard, & Wilmoth, 2014), and mortality (Bedard & Deschênes, 2006; Dobkin & Shabani, 2009; Elder et al., 2009).

The extent to which military service produces positive or negative outcomes depends in part on the timing of service during young adulthood and pre-service family socioeconomic status. The *military-as-turning-point hypothesis* states that young age at entry into the military maximizes the chances for redirection of the life course and, assuming no service-connected injury or disability, minimizes disruption to established life-course trajectories. Elder (1986, 1987) argues that early entry into the military delays the transition to adulthood and allows for the maximum utilization of VA benefits. Furthermore, for youth from socioeconomically disadvantaged backgrounds, who tend to join the military at younger ages, military service may be a route out of difficult life circumstances because the transition to the military both "knifes off" certain social ties and patterns of behavior, and provides a "bridging environment" in which service members can obtain education, training, skills, and resources that put them on better life-course trajectories than they otherwise would have followed (Bennett & McDonald, 2013; Browning, Lopreato, & Poston, 1973; Laub & Sampson, 2003; Sampson & Laub, 1996). The *life-course-disruption hypothesis* focuses on how military service creates disorder in the lives of those who enter service after families and careers have been established (Elder et al., 1994). Later entrants are more likely to be from relatively advantaged backgrounds and therefore have less to gain from military training and GI Bill benefits. In addition, the disruption to established trajectories can generate discontinuity that leads to worse post-service outcomes (Dechter & Elder, 2004). It is noteworthy that the majority of evidence about the disruptive effects of late entry into military service is based on samples who were subject to the WWII draft, which made men ages 18–45 subject to military service (Selective Service, 2015a). Whether later entry is as disruptive to the lives of contemporary military personnel has not been examined.

A third mechanism by which experiences rooted in military service are expected to have enduring effects on the lives of military personnel and their families operates through their continued impact on post-service employment, family, and health trajectories. However, as Teachman (2013) notes: "the effects of military service may not be proportionate across the life course … military service may produce results that vary according to stage in the life course … What may appear to be a null effect of military service at one point in the life course may be very different at earlier or later points in time" (pp. 282–283). Evidence of delayed effects are demonstrated in Davison et al.'s (2006) study of late-onset stress symptoms among aging combat veterans. Given the potential variable effect of military service over the life course, it is essential to consider how the long-term effects of military service change with age. In addition to the long-term repercussions of the previously discussed positive and negative experiences during military service, veterans and their families have access to various VA benefits, including health care, service-connected disability compensation, pensions, education and training, home loan guaranty, and life insurance (US Department of Veterans Affairs, 2014a). These benefits meet the needs of aging veterans by providing an additional layer of support beyond public social insurance and assistance programs, while providing targeted programs designed to meet the unique needs of specific subgroups of aging veterans, such as veterans with service-connected disabilities, veterans from specific wars, and other veterans with unique service-related experiences (Wilmoth & London, 2011).

Taken together, selection into military service, military service experiences, and post-service access to benefits and life-course trajectories combine to shape the aging process in ways that lead to later-life differences between veterans and nonveterans. We are limited in our ability to characterize aging experiences of specific war-service cohorts because research that makes explicit comparisons between veterans and nonveterans is rare and studies often group together veterans who served in different historical time periods. However, we can, to a limited extent, glean some insight from what is known about the historical circumstances of their service and their characteristics after separation from service. In the section that follows, we do not provide an exhaustive review of the extensive literature on aging veterans, but instead focus on highlighting exemplary research that demonstrates the *long-term consequences* of military service by providing insight into the differences between veterans and nonveterans in later life. Collectively, this research demonstrates the mixed and countervailing impacts of military service on later-life outcomes, as well as variation across historical time and cohort.

AN OVERVIEW OF MILITARY SERVICE AND AGING AMONG SPECIFIC WAR COHORTS

WWI

The duration of US involvement in the war was a relatively short 19 months, but the fighting in the European theater, where ground, sea, and air operations were carried out during the war, was intense. Over 4.7 million Americans served in WWI, of which more than 204 000 were reported as wounded and more than 116 000 died during service (MacLean, 2013). Although the majority of those who served were white men of Northern and Western European descent, the armed forces during WWI were relatively diverse due in part to shifting immigrant patterns during the first part of the twentieth century. However, minority group members encountered extensive discrimination and segregation (Lutz, 2013). It is also noteworthy that the role of nurses expanded during WWI, increasing the participation of women in war zones and laying the foundation for women to play larger roles in subsequent wars.

Knowledge about the long-term impact of service in WWI on aging veterans is limited in large part because this group was part of the "Hard Timers" cohort, who reached middle and late life before the scientific study of aging and the life course was widespread. What is known about these veterans is based on historical documents and oral histories. As noted by Campbell (2013), the oral history interviews of men who served in WWI held at the Army War College "are too haphazard to be useful" (p. 66) and "there is very little scholarship about the impact of WWI on women" (p. 48). Some insight can be gleaned from studies of US military history. This literature underscores the importance of the militia model through WWI; the armed forces were characterized by a small standing army supplemented as needed by citizen soldiers recruited through local militia units, National Guard mobilization, and selective national conscription (Kelty & Segal, 2013). Consistent with this military model, the large WWI uniformed force was demobilized after the end of the war, which returned a large number of military personnel to civilian life without sufficient advanced planning for reintegration or compensation for service.

The paltry compensation WWI veterans received upon their return led to demonstrations in 1919 and lobbying on the part of the American Legion, which resulted in the 1924 passage of the World War Compensation Act (Keene, 2001). This act provided monetary compensation that was primarily in the form of certificates that matured over 20 years (American Red Cross, 1926). This form of compensation proved to be unpopular and led to subsequent demonstrations in 1932–1933, which prompted the passage of the Adjusted Compensation Payment Act of 1936 and replaced the certificates with US Treasury bonds that could be redeemed at any time (New York Times, 1936). The political debate over WWI veteran compensation is pertinent to this discussion of veteran aging because it demonstrates the lack of governmental support provided to WWI veterans, particularly during their young-adult years immediately following service. This may have compromised their financial well-being as they entered later-life. Even though most of these veterans would have had access to Social Security Benefits after 1936 and some lived long enough to benefit from the establishment of Medicare and Medicaid in 1965, the federal government's slow response to veterans' financial needs likely had long-term consequences for their well-being.

The federal government was more responsive to the health needs of WWI veterans, although it took many years to get the necessary infrastructure and services established. Congress consolidated all WWI veteran programs into the Veteran's Bureau in 1921, which subsequently became the Veterans Administration in 1930 (US Department of Veteran Affairs, 2013). During this time period, veterans hospitals were built; services were developed to meet the unique needs of WWI veterans – many of whom were suffering from the effects of mustard gas exposure, tuberculosis, and post-traumatic stress, which was known as "shell shock," "battle fatigue," or "irritable heart" prior to the Vietnam War (Dean, 1997; Shay, 2002); and women, National Guard, and militia veterans were allowed admission into National Homes that previously had been available only to male disabled veterans.

Overall, the experiences and advocacy of those who served during WWI challenged the federal government to develop well-coordinated and appropriate programs designed to meet the unique needs of veterans and their families. As adeptly noted by Keene (2001, p. 214):

> Starting with adjusted compensation and ending with the GI Bill, World War I veterans forced the government to accept responsibility for redistributing profits and opportunities from advantaged civilians to disadvantaged veterans in the aftermath of total war. The executive branch found itself transformed forever by its experience in conscripting a mass national army to fight the Great War. World War I soldiers shaped critical policies and practices of the army and modern state, and created the piece of social welfare legislation that played a key role in generating the unprecedented prosperity Americans enjoyed in the second half of the twentieth century.

Thus, WWI veterans' experiences of war and subsequent political efforts laid foundations for the expansion of veteran's programs after WWII and the establishment of the modern welfare state, including various old age policies that now provide an essential safety net for all Americans.

WWII

Although there were several geopolitical conflicts in the late 1930s, the start of WWII is generally agreed to be September 1, 1939, when Germany invaded Poland (Axelrod, 2007). The United States officially entered WWII on December 7, 1941, after the Japanese bombed Pearl Harbor. The extensive US war effort continued until December 31, 1946 (US Code of Federal Regulations, 2014). The scope and length of the war required a substantial mobilization of men, mostly from the "Good Warrior" cohort, who were primarily aged 18–39 and typically were assigned to long stretches of overseas duty. Consequently, the "Good Warrior" cohort had the highest intensity of lifetime military service (measured by person-years lived on active duty) of any twentieth century cohort (Carlson, 2008). Over 16 million Americans served in WWII, with over 670 000 reported as wounded and 405 000 dying during service (MacLean, 2013). WWII marked a turning point for many racial and ethnic minority groups, as military service provided them with strong collective claims to equal citizenship, which laid a foundation for the post-war desegregation of the military and the subsequent civil rights movement of the 1960s (Lutz, 2013). It was also a turning point for women, approximately 350 000 of whom served in the Women's Army Corps, Navy WAVES, Marines, Coast Guard SPARS, Army Nurses, and Navy Nurses (Campbell, 2013).

In many respects, we know the most about how military service affected the lives of WWII veterans, but there are substantial gaps in that knowledge. Members of the "Good Warrior" cohort were the young adult subjects of early social science surveys in the middle of the twentieth century – for example, see Stouffer, Suchman, DeVinney, Star, and Williams' (1949) extensive research on WWII soldiers – who became the mid- to late-life subjects of gerontological inquiry during the latter half of the twentieth century. Our understanding of this cohort benefited from the methodological, statistical, and technological advancements of the latter twentieth century that enabled social scientists to effectively model age-related changes with longitudinal data. In many respects, our current understanding of aging as a life-long process is based on the experiences of this cohort.

There is no doubt that WWII profoundly shaped the lives of the "Good Warrior" cohort in ways that influenced their later-life experiences. Due to the educational benefits of the GI Bill, which covered costs associated with job training and tuition for up to 48 months in addition to providing a monthly stipend (for details about changes in provisions over time, see Bennett & McDonald, 2013), WWII veterans had higher educational attainment than their peers (Bound & Turner, 2002;

Martindale & Poston, 1979; Stanley, 2003). WWII veterans, particularly those from disadvantaged backgrounds, were more likely to be employed and had higher earnings than nonveterans (Fredland & Little, 1985; Little & Fredland, 1979; Moskos & Butler, 1996), although some of the veteran advantage in earnings was due to selection (Angrist & Krueger, 1994; Teachman & Tedrow, 2004) and benefits did not accrue equally to all veterans. For example, those who enlisted at older ages or were from working-class backgrounds were less likely to use the educational benefits of the GI Bill than those who enlisted at younger ages or were from middle and upper classes (Stanley, 2003). Similarly, although the GI Bill provided low-interest loans, African American veterans experienced discrimination in the housing market that made it difficult for them to take full advantage of the benefit (Lutz, 2013).

Beyond educational and economic outcomes, WWII service appears to have had enduring effects on veteran's social relationships and health. The evidence regarding divorce is mixed. Pavalko and Elder (1990) found WWII service was associated with higher rates of divorce, but Ruger, Wilson, and Waddoups (2002) found that WWII veterans had similar rates of divorce as nonveterans married during the same time period and that WWII service lowered the risk of divorce. Rates of fertility were high among this cohort, who went on after WWII to be the parents of the "Baby Boom" generation. Studies have consistently found WWII veterans are more politically involved and engaged in their local communities, which has been attributed to their increased sense of civic duty due to receipt of the GI Bill (Mettler, 2005). In terms of health outcomes, WWII veterans who survived into late life are in better health around age 70 than veterans from subsequent wars, but experienced steeper age-related health declines (Wilmoth, London, & Parker, 2010). WWII service has also been linked to higher mortality across mid- to late-life, particularly for those who experienced overseas duty, service in the Pacific theater of operations, and/or combat exposure (Elder et al., 2009). This higher mortality is due in part to high rates of smoking among WWII veterans, which was promoted by a pro-tobacco military culture that included cigarette distribution to overseas troops and subsidized tobacco products on US military bases (Bedard & Deschênes, 2006).

In many respects, the experience of those who served in WWII was unique relative to those who served in WWI or subsequent wars. Given the extensive war mobilization effort, military service was a normative experience among an overwhelming majority of men in the "Good Warrior" cohort. The long-term outcomes associated with serving in the military during WWII were mostly positive, in large part due to the extensive benefits provided by the GI Bill. In addition, WWII veterans were well-positioned to take advantage of the postwar economic expansion. Given this, veterans from this era demonstrated considerable resilience over the life course and seem to have fared relatively well as they aged into later-life (Ardelt, Landes, & Vaillant, 2010).

Korean War and Post-Korean War

Service in the military between June 27, 1950, and January 31, 1955, is considered Korean War era by the US government, even though the armistice agreement was signed July 27, 1953 (US Code of Federal Regulations, 2014). Over 5.7 million Americans served in the Korean War, with over 103000 reported as wounded and nearly 37000 dying while in service (MacLean, 2013). Although racial equality in the military was established by President Truman via executive order in 1948, the armed forces remained largely segregated at the start of the Korean War. However, the services were almost fully integrated by its end (Lutz, 2013). Despite the Women's Armed Services Integration Act of

1948, which permitted regular and reserve status in the armed forces, women's military participation continued to be limited primarily to nurses, who often served in the army's mobile surgical hospitals (US Department of Defense, 1998).

Some of the older members of the "Good Warrior" cohort who had served in WWII went on to serve in the Korean War, often in more senior positions, while the younger members served in the enlisted ranks alongside the oldest members of the "Lucky Few" cohort. At the start of the Korean War, 41% of the men on active duty were from the latter cohort and that percentage rose to 72% by the armistice (Carlson, 2008). The "Lucky Few" also served during the Post-Korean War period (a.k.a. the "Cold War" period). Military service continued to be a normative part of the transition to adulthood during this time period, although the military changed its mobilization strategy during this time from collective mass mobilization to a rotation system in which a steady stream of troops shifted through tours of duty (Carlson, 2008). This practice effectively limited the widespread societal impact of the war effort on communities and families. Military personnel who served in the Korean War gained access to the GI Bill, the provisions of which were similar to those provided to those who served in WWII (Bennett & McDonald, 2013).

Although the literature that explicitly examines Korean War veterans is relatively small, the broader literature on veterans suggests Korean War veterans are similar in many respects to WWII veterans in terms of education, employment, earnings, and health outcomes (Angrist, 1993; Martindale & Poston, 1979; Stanley, 2003; Wilmoth et al., 2010). However, there is evidence Korean War veterans had higher divorce rates than nonveterans married during the same time period (Ruger et al., 2002). Veterans from the Cold War were not exposed to the risks of wartime service, but they were also not able to take full advantage of the GI Bill because it was not initially available to those who served after the Korean War through 1965. When the program was reinstated in 1966, Cold War veterans were retroactively covered (Bennett & McDonald, 2013). However, uptake of the benefits was low and this group consequently had lower levels of educational attainment than their non-veteran counterparts (MacLean, 2005). Despite this, older veterans aged 65–70 in 2008–2010 – who would have been eligible to first serve during 1956–1963 – are better off economically than nonveterans: they have higher median total income, Social Security income, and retirement income; higher home ownership rates, but lower median home values; higher rates of employer-provided pensions and Medigap coverage; and lower rates of Medicaid use (Street & Hoffman, 2013). Furthermore, among veterans, career veterans who served for at least 20 years did better on these indicators than non-career veterans, which was due in part to their access to additional Department of Defense benefits (Street & Hoffman, 2013).

Overall, the long-term impact of service during the Korean War and Cold War periods was positive. Although there were negative health consequences for some veterans who served in the Korean War, as a group, they were able to translate GI Bill benefits into upward social mobility during the post-WWII period of economic expansion. Cold War veterans were not able to take full advantage of these benefits, but did not experience the long-term negative health consequences of wartime service. Although there was socioeconomic and health heterogeneity within both groups, veterans of these two time periods generally arrived at later life in good health with sufficient economic resources and strong social ties.

Vietnam War

The United States had military advisors in Vietnam (known as French Indochina at the time) as early as 1950, but substantial troop involvement did not occur until the early

1960s. The US Code of Federal Regulations (2014) indicates that a start date of February 28, 1961, is applicable to veterans who served in the Republic of Vietnam, but in all other cases, August 5, 1964, is used to determine the start of war service. For both groups, the end date is May 7, 1975. Over 8.7 million Americans served in the Vietnam War, the majority of whom were from the "Baby Boom" cohort, with over 300 000 reported as wounded and approximately 58 000 dying in service (MacLean, 2013). It is noteworthy that due to improvements in military medicine the relative odds of being wounded versus dying were substantially higher during the Vietnam War (5.22) than prior wars (1.65 and 2.82 for WWII and the Korean War, respectively) (MacLean, 2013). The average age of military personnel was also younger during the Vietnam War (approximately 19 years) compared to earlier wars (e.g., 26 years during WWII) (D. Segal, personal communication, January 28, 2015). Although deferments from military service due to occupation, dependants, and other limited reasons were granted in earlier time periods (Yoder, 2014), attempts to avoid military service in Vietnam through educational deferments, exemptions, enlistment in the Guard or Reserves, and emigration to Canada were more common during the Vietnam War era (Selective Service, 2015b).

Reflecting the broad civil unrest in the United States during this time period, the military experienced widespread racial strife during the Vietnam War: there were widespread allegations that African Americans were disproportionately drafted, fighting in combat, being wounded and dying; racial discrimination on military bases and within units was rampant; and there were race riots on military installations (Lutz, 2013). Similar to prior wars, women's participation in the Vietnam War was numerically limited by law to 2% of each branch of service and functionally limited primarily to nursing roles (Bellafaire, 2006). Regardless of race or gender, service members experienced a hostile reception when they returned home from this contentious war, which made reintegration into civilian life challenging for Vietnam veterans (Card, 1983). Reintegration was particularly difficult for those who experienced combat. Vietnam veterans who served in theater had significantly more symptoms of post-traumatic stress disorder (PTSD), higher conviction rates for misdemeanors and felonies, and lower life satisfaction in general than their counterparts who served during the same time period outside of Vietnam and nonveterans (Card, 1983). However, differences in how these groups rated various aspects of their lives converged over time, which suggests substantial resilience among combat veterans.

The benefits of military service that were evident among veterans of prior wars did not materialize for Vietnam-era veterans, particularly those from advantaged backgrounds (Bookwala, Frieze, & Grote, 1994). Vietnam veterans had less educational attainment than comparable peers at the time of discharge, but the educational gap did close over time (Angrist & Chen, 2011; Card, 1983; Cohen, Segal, & Temme, 1986; Teachman, 2005; Teachman & Call, 1996). There were modest economic returns to Vietnam veterans' schooling given their midlife earnings were not significantly different than nonveterans (Angrist, 1990; Angrist & Chen, 2011). However, the positive effect of military service on earnings was evident among veterans who were black and less educated, while white men experienced an earnings penalty (Berger & Hirsch, 1983; Cohany, 1992; Martindale & Poston, 1979; Rosen & Taubman, 1982; Teachman, 2004). Vietnam-era veterans finished school, got married, and became fathers at significantly older ages than their non-veteran classmates (Card, 1983). They also had higher divorce rates overall than nonveterans married during the same time period, but service in Vietnam had little effect on the risk of divorce after controlling for confounding factors

(Ruger et al., 2002). Similar to veterans of previous wars, Vietnam veterans have higher levels of political activism and civic participation than nonveterans despite their alienation after returning to civilian life (Kelty & Segal, 2013).

There is a large literature, based primarily on samples of veterans who use VA services, that examines the physical and mental health of Vietnam veterans, and it includes topics such as exposure to Agent Orange and cancer risk, combat and PTSD, suicide, and substance abuse. Research that directly compares veterans to nonveterans consistently indicates that Vietnam veterans are in worse health than nonveterans, and that this health disparity increases with age (Dobkin & Shabani, 2009). In later life, Vietnam veterans have more conditions and poorer self-rated health around retirement age than WWII veterans (Wilmoth et al., 2010). Vietnam veterans also had worse psychological outcomes (Dohrenwend et al., 2006) and higher mortality rates, especially due to external causes (i.e., accidents, suicide, and homicide), immediately following separation from service (Boehmer, Flanders, McGeehin, Boyle, & Barrett, 2004; Boscarino, 2006).

The majority of Vietnam-era veterans have entered later-life relatively recently as the "Baby Boom" cohort ages into the prime retirement ages. Extant evidence suggests that Vietnam-era military service negatively impacted the aging process through a number of different mechanisms; however, some research has also suggested a substantial amount of resiliency among veterans from this era (Aldwin, Levenson, & Spiro, 1994). Of course, there is substantial heterogeneity among this group of veterans and negative life course outcomes are more likely among those who served overseas and experienced combat. Unlike the two prior wars, men from more advantaged backgrounds were less likely to experience positive returns to service, but military service continued to be a positive turning point in the life course for many men from disadvantaged backgrounds.

All-Volunteer Force

The context of military service changed substantially as the Vietnam War wound down and the draft ended in 1973. With the end of the draft, the US military shifted to an All-Volunteer Force (AVF). The AVF era is characterized by three broad periods: the peacetime period prior to 1990; the Gulf War period, which began with the war in Iraq in 1991–1992 and extended through the September 11, 2001 terrorists attacks on the United States; and the War on Terror, which encompasses OIF, OEF, and OND. Since 1991, the military has relied heavily on activating the Reserves and National Guard for active-duty service, and has increasingly deployed units multiple times. Of the more than 2.2 million personnel who served during the War on Terror (Institute of Medicine, 2013), over one-third have deployed more than once and 16% have deployed five or more times (Adams, 2013). This repeated deployment approach has minimized the impact of the war on the general public and concentrated the burden of service on a relatively small portion of individuals, families, and local communities.

Over the AVF era, military personnel have primarily been drawn from the "Baby Boom" and "Generation X" cohorts. The average age of active component military personnel has increased during the AVF period, from 25 years in 1974 to 27 years in 2011 for enlisted members, and from 32 years in 1974 to 34.5 years in 2011 for commissioned officers (US Department of Defense, 2011). The military has become more diverse in terms of race, ethnicity, and gender (Kelty & Segal, 2013; Lutz, 2013): in 2010, 16%, 10%, and 18% of military personnel were African American, Hispanic, and female, respectively (Hurt, Ryan, & Strayley, 2011). Additionally, the 2011 repeal of "Don't Ask, Don't Tell" has allowed lesbian and gay persons to serve openly in the armed forces for the first time (Brown, 2013). Even though overall rates of military service have declined (Wilmoth &

London, 2013), the military continues to be the single largest employer of young adults (Angrist, 1998). Military service is no longer a normative experience for young men, but it continues to be a salient pathway to adulthood that is particularly important for youth from disadvantaged backgrounds (Bennett & McDonald, 2013; Kelty & Segal, 2013).

There is some evidence that the benefits of service experienced by earlier cohorts have been diminished during the AVF period. As higher education became more accessible to the general population, military service no longer provides a primary pathway to obtaining a college degree. Consequently, AVF-era veterans have lower levels of education at discharge than their non-veteran peers, but those who were older at enlistment, had higher Armed Forces Qualification Test scores, were African American, or were officers generally fared better in terms of educational attainment (Teachman, 2007a). AVF-era veterans overall do not appear to experience positive returns to employment and earnings. There is consistent evidence that the labor market outcomes of AVF-era veterans are worse than those of nonveterans, although African American men and those with less education fare better (Angrist, 1998; Hirsch & Mehay, 2003; Phillips, Andrisani, Daymont, & Gilroy, 1992; Teachman, 2004; Teachman & Tedrow, 2007; Xie, 1992). Recently returning veterans were particularly hard-hit by the 2008 Great Recession (London & Wilmoth, 2012).

The military during the AVF period has developed more family-friendly policies in an effort to retain personnel who are eligible for reenlistment, causing some to argue that AVF-era military service has promoted family formation among young adults in the "Baby Boom" and "Generation X" cohorts (Burland & Lundquist, 2013). Marriage is more prevalent and occurs at younger ages among those who have served in the AVF period compared to their civilian counterparts (Lundquist, 2004; Teachman, 2007b). With the increase of women serving in the military during the AVF period, the rate of marriage involving two active-duty personnel has also increased. Such marriages represent approximately 10% of military marriages (Burland & Lundquist, 2013). AVF-era service is also associated with a higher likelihood of marrying among African American veterans in fragile families (Usdansky, London, & Wilmoth, 2009). During the AVF era, military personnel had lower divorce rates than their civilian counterparts, particularly among African American men, but after separation from service, the divorce rates of AVF-era veteran and nonveterans are similar (Lundquist, 2006; Teachman & Tedrow, 2008). AVF-era military personnel have children at younger ages (Lundquist & Smith, 2005) and larger completed family sizes than their civilian counterparts (Military Family Resource Center, 2000). Consequently, over half of service members are parents (Burland & Lundquist, 2009). The growing number of military dependants and spouses has important implications for the lives of children and spouses, who are subjected to the demands of military life, including frequent moves and successive separations from loved ones during deployments (Bailey, 2013; Burland & Lundquist, 2013). For example, military spouses often experience weaker labor force attachments than their civilian counterparts, including greater rates of unemployment, fewer hours worked, and lower earnings (Kleykamp, 2013), which are likely to have serious implications for their long-term financial well-being in later life.

Although the health-related risks of service in the early AVF period were low, veterans who served during this time period are not in better mental or physical health over the long-term. AVF-era veterans have poorer mid-life physical health than nonveterans (Teachman, 2010). They exhibit lower levels of depression after discharge, but their depression levels become similar to the non-veteran population during the 10 years after service (Whyman, Lemmon, & Teachman, 2010). Service since 1990 has been quite risky. Some military personnel serving in Iraq during the early 1990s have exhibited a range of highly debilitating health problems that are commonly referred to as "Gulf War Syndrome." Sexual assault, which is pervasive throughout the armed forces, disproportionately affects female military personnel and generates trauma that is associated with a variety of negative consequences (US Department of Veteran Affairs, 2015). Those serving in the War on Terror have been subject to nontraditional combat in urban settings, with combatants relying on difficult-to-detect and often deadly improvised explosive devices. Due to substantial improvements in military medicine, military personnel who have been injured during the War on Terror have been much more likely to survive than was the case in previous wars. The odds of being wounded relative to dying during this time period have been over 7–1 (Hurt et al., 2011). Rates of traumatic brain injury, amputations, and PTSD have been high among returning veterans. Consequently, the number (and percentage) of veterans with a service-connected disability rating has been growing steadily, from approximately 2 million (8%) in 1990 to over 3.5 million (15%) in 2012 (US Department of Veterans Affairs, 2014b). Moreover, the highest rate of growth is among those with a disability rating of 70–100%, which represents nearly 1 million veterans in 2012 (US Department of Veterans Affairs, 2014b). These service-connected disabilities will have life-long consequences for veterans and their families. For example, recent research suggests that households containing a disabled veteran are significantly more likely to experience poverty and material hardship (Heflin, London, & Wilmoth, 2012; London, Heflin, & Wilmoth, 2011; Wilmoth, Heflin, & London, 2015).

Overall, military service during the AVF era has increasingly been associated with negative short- and long-term outcomes. Although the veterans who served prior to 1990 are currently reaching mid- to late life, most Gulf War and War on Terror veterans are in young and middle adulthood. Their experiences thus far suggest that they and their families may face a more challenging set of circumstances during later life than pre-AVF veterans. The extent to which this is the case will depend in part on the effectiveness of VA programs and services. The experiences of WWI veterans nearly a century ago demonstrated the lasting effects of war service in the context of limited and poorly implemented government policies. The safety net developed for veterans in the middle of the twentieth century effectively mitigated many of the negative long-term consequences of war service thereby benefiting veterans, their families, and the broader society. The challenge moving forward will be to update programs and services in ways that are responsive to the needs of younger AVF-era veterans and facilitate subsequent later-life well-being, while continuing to meet the current needs of older pre-AVF veterans.

STUDYING MILITARY SERVICE AND AGING

To develop the most effective policies to meet the diverse needs of aging veterans, we must improve our understanding of the long-term impact of military service on lives. This means our studies should capture the experiences of the cohorts who are currently moving through later life, and consider younger cohorts who are currently in young adulthood and middle age. As shown in Figure 11.3, the composition of the overall veteran population will change rapidly over the next 30 years as the WWII, Korean War, and Vietnam War cohorts are replaced by cohorts who served during the AVF era. Now is the time to lay the foundations for understanding how AVF-era service will affect the aging of the Baby Boom and Generation X cohorts.

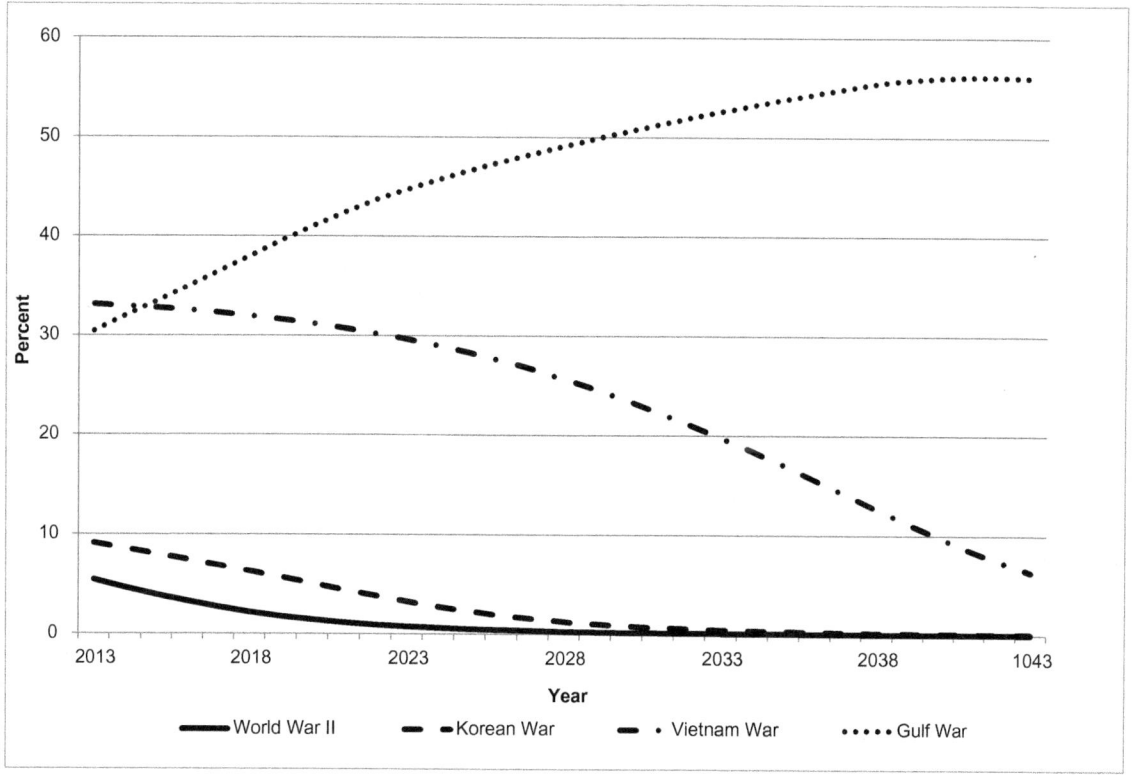

FIGURE 11.3 Projected distribution of period of service among all US veterans, 2013–2043. *Note*: Each line represents the percentage of veterans who ever served in a given war. Veterans who served in multiple wars are counted in each war in which they served. Veterans who never served during war are not included. *Source: National Center for Veterans Analysis and Statistics (2014).*

It is critical that we develop high-quality, nationally representative, longitudinal data that include sufficient measures of military service experiences and various selection processes. As previously mentioned, the majority of research on veterans utilizes specialized samples that are often drawn from individuals who use VA hospitals. While these studies have yielded important information about aging veterans, particularly related to specific health conditions, these samples are not representative because veterans who use VA facilities tend to be sicker and poorer than the general population. Three notable exceptions are the Normative Aging Study (NAS), the National Survey of Veterans (NVS), and the Millennium Cohort Study, all of which are unique in the depth and breadth of data that are provided about aging veterans. The NAS is a longitudinal study of 2280 men ages 21–81 living in the Boston area in the 1960s. The majority of these men served in WWII and the Korean War. Clinical health data have been collected at 3–5-year intervals, with supplemental surveys, interviews, and exams focusing on a range of targeted topics (Spiro & Vokonas, 2007). The NVS is a cross-sectional survey that has been collected six times since 1979. The most recent survey in 2010 was based on self-administered, mailed questionnaires that were completed by 8710 veterans and an additional

2262 spouses (US Department of Veterans Affairs, 2010). The Millennium Cohort Study is a longitudinal study that is based on a random sample of over 200000 military personnel and veterans who have served since 2001 (US Department of Defense, 2015). The primary goal of this study is to determine the relationship between military service, particularly deployment, and chronic illness. It is one of the few large-scale surveys that links participant self-reports of demographic characteristics, medical conditions and symptoms, and health behaviors to military records of deployment, occupation, health care utilization, and disability (Gray et al., 2002).

A number of large-scale, mostly nationally representative datasets that are commonly used by aging and life-course researchers contain measures of veteran status, although there is considerable variation across these datasets in the time periods of military service covered and level of detail provided about military service experiences. Repeated cross-sectional data that contain veteran status on an ongoing basis include the CPS, American Community Survey (ACS), and General Social Survey (GSS). Longitudinal studies, which facilitate the study of aging processes by tracking individual respondents over time, with veteran status measures include: the Wisconsin Longitudinal Study (WLS), Project Talent, the Longitudinal Study of Generations (LSOG), the Longitudinal Study on Aging (LSOA), the Panel Study of Income Dynamics (PSID), the National Longitudinal Study of Youth 1979 (NLSY79), the National Survey of Families and Households (NSFH), the Survey of Income and Program Participation (SIPP), the HRS, the National Health and Aging Trends Study (NHATS), and the National Longitudinal Study of Adolescent to Adult Health (ADD HEALTH).

Most of these secondary data sets have limited information about military service experiences. However, the PSID is noteworthy for its inclusion in 1994 of additional questions that inquire about branch of service, combat exposure, and rank. In addition, the HRS is unique in its breadth of cohorts, which span from WWII through the Vietnam War, as well as the supplemental veterans mail survey collected in 2013, which includes measures of service-connected disability ratings, use of VA services, experience in the military, and contact with friends from the military. An additional strength of the HRS is its ability to link public-access data with restricted-use administrative records (e.g., from the Social Security Administration and Medicare), which further enhances researchers' ability to understand later-life differences between veterans and nonveterans.

There are substantial methodological challenges to using survey data, including the datasets listed above, to examine the effect of military service on the aging process. A central methodological challenge is selection. As previously mentioned, military recruiters and entrance exams screen out individuals who are not fit for service, which results in a military population that is initially healthier, on average, than the civilian population. In addition, there is some evidence that youth from socioeconomically disadvantaged backgrounds are more likely to serve. If this is so, observed later-life differences between veterans and nonveterans could be due to the enduring effect of early-life factors that preceded military service. However, it is important to recognize that the forces that select young adults into military service have varied over time. During periods of war and military drafts, selection mechanisms are relaxed to meet the armed forces, high demand for personnel or to satisfy equity concerns about risky wartime service. Consequently, veterans from those eras should have early-life characteristics that are more similar to the nonveteran population than veterans who served during peacetime. Selection also plays a role in who serves in combat and who makes military service a career, which affects the composition of both active-duty and veteran populations.

In addition, the characteristics of baseline participants in studies that start in mid- to late life might be influenced by unobserved differential mortality. If veterans are subjected to higher mortality rates throughout adulthood, which is likely, those who survive to participate in later-life surveys might be more robust than nonveterans. Differential mortality can also operate between study waves, generating selective survival and attrition from longitudinal follow-ups. This process could further bias the results of studies that compare veterans and nonveterans if it is not sufficiently taken into account. Wolf et al. (2013) describe the challenges selection poses for research on military service in lives in more detail and offer several potential approaches for addressing the problem. All of these approaches are predicated on having adequate information about the participants, including a range of pre-service characteristics (such as family of origin socioeconomic status, personality characteristics, health status, and aptitude), as well as measures of historical circumstances and contextual factors (such as population characteristics, community circumstances, and policy changes).

Another methodological challenge is documenting the nature of military service experiences. Even when these questions are asked, it is often years after separation from service and therefore responses are subject to recall error. This is problematic because it makes it difficult to identify the conditions of military service that shape the aging process. In this case, veteran status becomes a black box of unmeasured effects that cannot be disentangled. Although the ideal source of data would be military personnel records, which would include health and performance measures, this information is very difficult to obtain because access is restricted by the Department of Defense. Therefore, researchers have to rely on self-reports. To the extent possible, future research must use subsample surveys to collect as much information about military service experiences as soon as possible after separation from service. Collected measures should include: start and stop dates of service, historical time period of service, branch of service, rank at entrance and exit, military occupational specialty, type of discharge (e.g., honorable, other than honorable), service outside of the United States, countries stationed, combat zone service, firing a weapon in combat, wounding or killing another person, being exposed to people who were dead, dying, or wounded, frequency of feeling one's life was in danger, having a service-connected disability rating and if so, the percent rating, use of VA benefits (including educational benefits, health care, disability payments), and receipt of DoD benefits (e.g., retirement income or TRICARE). Researchers also need to consider how to measure the timing of military events in relation to other major life events, such as the completion of education, marriage, and parenthood. The use of event history calendars might be useful in this regard.

A final challenge is examining how the lives of military personnel and veterans are linked to parents, spouses, children, and siblings. This requires collecting data from others in the respondent's social network and linking the data across respondents. The LSOG, PSID, and NLSY79 (all of which have data linked across generations), WLS and Project Talent (both of which include siblings), and the HRS (which includes spouses and select children) provide examples of how to approach data collection in ways that will facilitate research that uncovers how lives are linked across time. In conducting this type of research, it is important to develop comparable measures for veterans and nonveterans. For example, both groups can experience post-traumatic stress, but sometimes those measures are only collected from veteran populations. Careful consideration needs to be given to measuring outcomes in a way that will allow direct comparisons between veterans and nonveterans between and within families.

Future research that tackles these challenges will be able to identify the effects of military service on lives in ways that will yield insights into the aging process among current and future cohorts, and provide information that can produce more informed policies. In addition, as Teachman (2013) notes: "How veteran status plays out in the intermeshed lives of service members, and their families and communities, can provide a blueprint for the ways in which historical and social context influence individual choice and experience" (p. 290). Thus, research on military service and aging has the potential to make broader contributions to social science and the multidisciplinary science of aging by promoting a long view of lives that is informed by the life-course approach. Such an approach will enable scientists to separate what is unique about the aging experience of a given cohort from the common aspects of human aging across cohorts.

References

Adams, C. (2013). Millions went to war in Iraq, Afghanistan, leaving many with lifelong scars. *McClachy Newspapers*. Retrieved from: <http://www.mcclatchydc.com/2013/03/14/185880_millions-went-to-war-in-iraq-afghanistan.html?rh=1>.

Aldwin, C. M., Levenson, M. R., & Spiro, A., III. (1994). Vulnerability and resilience to combat exposure: Can stress have lifelong effects? *Psychology and Aging*, 9(1), 34–44.

American Red Cross. (1926). World War Adjusted Compensation Act. Retrieved from: <http://books.google.com/books?id=0WYXAAAAYAAJ&> (Accessed 19.07.26.)

Angrist, J. D. (1990). Lifetime earnings and the Vietnam era draft lottery: Evidence from social security administrative records. *American Economic Review*, 80(3), 313–336.

Angrist, J. D. (1993). The effect of veterans benefits on education and earnings. *Industrial & Labor Relations Review*, 46(4), 637–652.

Angrist, J. D. (1998). Estimating the labor market impact of voluntary military service using Social Security data on military applicants. *Econometrica*, 66(2), 249–288.

Angrist, J. D., & Chen, S. H. (2011). Schooling and the Vietnam-era GI Bill: Evidence from the draft lottery. *American Economic Journal: Applied Economics*, 3, 96–118.

Angrist, J. D., & Krueger, A. B. (1994). Why do World War II veterans earn more than nonveterans. *Journal of Labor Economics*, 12(1), 74–97.

Ardelt, M., Landes, S. D., & Vaillant, G. E. (2010). The long-term effects of World War II combat exposure on later life well-being moderated by generativity. *Research in Human Development*, 7(3), 202–220.

Armed Forces Health Surveillance Center, (2007). Healthy deployers: Nature and trends of health care utilization during the year prior to deployment to OEF/OIF, Active Components, US Armed Forces, January 2002 – December 2006. *Medical Surveillance Monthly Report*, 14(3), 2–5.

Axelrod, A. (2007). *Encyclopedia of World War II* (Vol. 1). New York. NY: Infobase Publishing. p. 659.

Bailey, A. K. (2013). Military employment and spatial mobility across the life course. In J. M. Wilmoth & A. S. London (Eds.), *Life-course perspectives on military service* (pp. 185–199). New York. NY: Routledge.

Bedard, K., & Deschênes, O. (2006). The long-term impact of military service on health: Evidence from World War II and Korean War veterans. *American Economic Review*, 96(1), 176–194.

Bellafaire, J. (2006). Volunteering for Vietnam: African-American servicewomen. *Women in military/service for America, Memorial Foundation, Inc., History and Collections*. <http://www.womensmemorial.org/H&C/History/afamvet.html>.

Bennett, P. R., & McDonald, K. B. (2013). Military service as a pathway to early socioeconomic achievement for disadvantaged groups. In J. M. Wilmoth & A. S. London (Eds.), *Life-course perspectives on military service* (pp. 119–143). New York, NY: Routledge.

Berger, M. C., & Hirsch, B. T. (1983). The civilian earnings experience of Vietnam-era veterans. *Journal of Human Resources*, 18(4), 455–479.

Boehmer, T. K. C., Flanders, D., McGeehin, M. A., Boyle, C., & Barrett, D. H. (2004). Postservice mortality in Vietnam veterans: 30-year follow-up. *Archives of Internal Medicine*, 164(17), 1908–1916.

Bookwala, J., Frieze, I., & Grote, N. (1994). The long-term effects of military service on quality of life: The Vietnam experience. *Journal of Applied Social Psychology*, 24(6), 529–545.

Boscarino, J. A. (2006). Posttraumatic stress disorder and mortality among US Army veterans 30 years after military service. *Annals of Epidemiology*, 16(4), 248–256.

Bound, J., & Turner, S. (2002). Going to war and going to college: Did World War II and the G.I. Bill increase educational attainment for returning veterans? *Journal of Labor Economics*, 20(4), 784–815.

Brown, M. T. (2013). Military service and lesbian, gay, bisexual, and transgender lives. In J. M. Wilmoth & A. S. London (Eds.), *Life-course perspectives on military service* (pp. 97–118). New York, NY: Routledge.

Browning, H. L., Lopreato, S. C., & Poston, D. L. (1973). Income and veteran status: Variations among Mexican

Americans, Blacks and Anglos. *American Sociological Review, 38*(1), 74–85.

Burland, D., & Lundquist, J. H. (2009). Family relationships and the military. In H. T. Reis & S. K. Sprecher (Eds.), *Encyclopedia of human relationships*. Newbury Park, CA: Sage Publications.

Burland, D., & Lundquist, J. H. (2013). The best years of our lives: Military service and family relationships – A life-course perspective. In J. M. Wilmoth & A. S. London (Eds.), *Life-course perspectives on military service* (pp. 165–184). New York, NY: Routledge.

Camacho, P. R., & Atwood, P. L. (2007). A review of the literature on veterans published in Armed Forces & Society, 1974–2006. *Armed Forces & Society, 33*(3), 351–381.

Campbell, D. (2013). Women's lives in wartime: The American Civil War and World War II. In J. M. Wilmoth & A. S. London (Eds.), *Life-course perspectives on military service* (pp. 48–67). New York, NY: Routledge.

Card, J. J. (1983). *Lives after Vietnam: The personal impact of military service*. Lexington, MA: Lexington Books.

Carlson, E. (2008). *The lucky few: Between the greatest generation and the baby boom*. New York, NY: Springer.

Carlson, E., & Andress, J. (2009). Military service by twentieth-century generations of American men. *Armed Forces & Society, 3*(2), 385–400.

Clipp, E. C., & Elder, G. H., Jr. (1996). The aging veteran of World War II: Psychiatric and life course insights. In P. E. Ruskin & J. A. Talbott (Eds.), *Aging and post-traumatic stress disorder* (pp. 19–51). Washington, DC: American Psychiatric Press.

Cohany, S. R. (1992). The Vietnam-era cohort: Employment and earnings. *Monthly Labor Review, 115*(6), 3–15.

Cohen, J., Segal, D. R., & Temme, L. V. (1986). The educational cost of military service in the 1960s. *Journal of Political & Military Sociology, 14*(2), 303–319.

Davison, E. H., Pless, A. P., Gugliucci, M. R., King, L. A., King, D. W., Salgado, D. M., et al. (2006). Late-life emergence of early-life trauma: The phenomenon of late-onset stress symptomatology among aging combat veterans. *Research on Aging, 28*(1), 84–114.

Dean, E. T. (1997). *Shook over hell: Post-traumatic stress, Vietnam, and the Civil War*. Cambridge, MA: Harvard University Press.

Dechter, A. R., & Elder, G. H., Jr. (2004). World War II mobilization in men's work lives: Continuity or disruption for the middle class? *American Journal of Sociology, 110*(3), 761–793.

Defense Manpower Data Center. (2014). *Armed services vocational aptitude battery: History of military testing*. Retrieved from: <http://official-asvab.com/history_res.htm>.

Dobkin, C., & Shabani, R. (2009). The health effects of military service: Evidence from the Vietnam draft. *Economic Inquiry, 47*(1), 69–80.

Dohrenwend, B. P., Turner, J. B., Turse, N. A., Adams, B. G., Koenen, K. C., & Marshall, R. (2006). The psychological risks of Vietnam for US Veterans: A revisit with new data and methods. *Science, 313*, 979–982.

Elder, G. H., Jr. (1986). Military times and turning points in men's lives. *Developmental Psychology, 22*(2), 233–245.

Elder, G. H., Jr. (1987). War mobilization and the life course: A cohort of World War II veterans. *Sociological Forum, 2*(3), 449–472.

Elder, G. H., Jr., & Bailey, S. L. (1988). The timing of military service in men's lives. In J. A. Aldous & D. M. Klein (Eds.), *Social stress and family development* (pp. 157–174). New York, NY: Guilford.

Elder, G. H., Jr., & Clipp, E. C. (1988a). Wartime losses and social bonding: Influences across 40 years in men's lives. *Psychiatry: Interpersonal and Biological Processes, 51*(2), 177–198.

Elder, G. H., Jr., & Clipp, E. C. (1988b). War experiences and social ties: Influences across 40 years in men's lives. In M. White Riley (Ed.), *Social change and the life course, Volume 1: Social structures and human lives* (pp. 306–327). Beverly Hills, CA: Sage Publications.

Elder, G. H., Jr., & Clipp, E. C. (1989). Combat experience and emotional health: Impairment and resilience in later life. *Journal of Personality, 57*(2), 311–341.

Elder, G. H., Jr., Clipp, E. C., Brown, J. S., Martin, L. R., & Friedman, H. W. (2009). The lifelong mortality risks of World War II experiences. *Research on Aging, 31*(4), 391–412.

Elder, G. H., Jr., Gimbel, C., & Ivie, R. L. (1991). Turning points in life: The case of military service and war. *Military Psychology, 3*(4), 215–231.

Elder, G. H., Jr., & Johnson, M. K. (2002). Perspectives on human development in context. In C. von Hofsten & L. Bergman (Eds.), *Psychology at the turn of the millennium, Vol. 2: Social, developmental, and clinical perspectives* (pp. 153–175). East Sussex, UK: Psychology Press Ltd.

Elder, G. H., Jr., Johnson, M. K., & Crosnoe, R. (2003). The emergence and development of life course theory. In J. T. Mortimer & M. J. Shanahan (Eds.), *Handbook of the life course* (pp. 3–19). New York, NY: Kluwer Academic/Plenum.

Elder, G. H., Jr., Shanahan, M. J., & Clipp, E. C. (1994). When war comes to men's lives: Life-course patterns in family, work, and health. *Psychology and Aging, 9*(1), 5–16.

Elder, G. H., Jr., Shanahan, M. J., & Clipp, E. C. (1997). Linking combat and physical health: The legacy of World War II in men's lives. *American Journal of Psychiatry, 154*(3), 330–336.

Elder, G. H., Jr., Wang, L., Spence, N. J., Adkins, D. E., & Brown, T. H. (2010). Pathways to the all-volunteer military. *Social Science Quarterly, 91*(2), 455–475.

Fredland, J. E., & Little, R. D. (1985). Socioeconomic status of World War II veterans by race: An empirical test of the bridging hypothesis. *Social Science Quarterly, 66*(3), 533–551.

Gray, G. C., Chesbrough, K. B., Ryan, M. A., Amoroso, P., Boyko, E. J., Gackstetter, G. D., et al. (2002). The Millennium Cohort Study: A 21-year prospective cohort study of 140,000 military personnel. *Military Medicine, 167*(6), 483–488.

Heflin, C., London, A. S., & Wilmoth, J. M. (2012). Veteran status and material hardship: The moderating influence of disability. *Social Service Review, 86*(1), 119–142.

Hirsch, B. T., & Mehay, S. L. (2003). Evaluating the labor market performance of veterans using a matched comparison group design. *Journal of Human Resources, 38*(3), 673–700.

Hogan, D. P. (1981). *Transitions and social change: The early lives of American man*. New York, NY: Academic.

Hurt, A., Ryan, E., & Strayley, J. (2011). By the numbers: Today's military. National Public Radio, Special Series: Those Who Serve. Retrieved from: <http://www.npr.org/2011/07/03/137536111/by-the-numbers-todays-military>.

Institute of Medicine. (2013). *Returning home from Iraq and Afghanistan: Readjustment needs of veterans, service members, and their families*. Retrieved from: <http://www.iom.edu/Reports/2013/Returning-Home-from-Iraq-and-Afghanistan.aspx>.

Keene, J. D. (2001). *Doughboys, the great war, and the remaking of America*. Baltimore, MD: Johns Hopkins Press.

Kelty, R., & Segal, D. R. (2013). The military as a transforming influence: Integration into or isolation from normal adult roles?. In J. M. Wilmoth & A. S. London (Eds.), *Life-course perspectives on military service* (pp. 19–47). New York, NY: Routledge.

Kleykamp, M. (2013). Labor market outcomes among veterans and military spouses. In J. M. Wilmoth & A. S. London (Eds.), *Life-course perspectives on military service* (pp. 144–164). New York, NY: Routledge.

Laub, J. H., & Sampson, R. J. (2003). *Shared beginnings, divergent lives: Delinquent boys to age 70*. Cambridge, MA: Harvard University Press.

Little, R. D., & Fredland, J. E. (1979). Veteran status, earnings, and race: Some long-term results. *Armed Forces & Society, 5*(2), 244–260.

London, A. S., Burgard, S. A., & Wilmoth, J. M. (2014). The influence of veteran status, psychiatric diagnosis, and traumatic brain injury on inadequate sleep. *Journal of Sociology & Social Welfare, 41*(4), 49–67.

London, A. S., Heflin, C., & Wilmoth, J. M. (2011). Work-related disability, veteran status, and poverty: Implications for family well-being. *Journal of Poverty, 15*, 1–20.

London, A.S., & Wilmoth, J.M. (2012). From soldier to civilian: Recently transitioning veterans experiences during the Great Recession. MacArthur Network on Transitions to Adulthood Working Paper. Retrieved from: <http://transitions.s410.sureserver.com/wp-content/uploads/2011/08/LondonWilmoth-From-Soldier-To-Civilian.pdf>.

London, A.S., & Wilmoth, J.M. (in press). Military service in lives: Where do we go from here? In M. Shanahan, J. Mortimer, & M. Johnson (Eds.), *Handbook of the life course*. New York, NY: Springer.

Lundquist, J. H. (2004). When race makes no difference: Marriage and the military. *Social Forces, 83*(2), 731–757.

Lundquist, J. H. (2006). The Black-White gap in marital dissolution among young adults: What can a counterfactual scenario tell us? *Social Problems, 53*, 421–441.

Lundquist, J. H., & Smith, H. L. (2005). Family formation among women in the US military: Evidence from the NLSY. *Journal of Marriage and the Family, 67*(1), 1–13.

Lutz, A. C. (2013). Race-ethnicity and immigration in the US military. In J. M. Wilmoth & A. S. London (Eds.), *Life-course perspectives on military service* (pp. 68–96). New York, NY: Routledge.

MacLean, A. (2005). Lessons from the Cold War: Military service and college. *Sociology of Education, 78*, 250–266.

MacLean, A. (2010). The things they carry: Combat, disability, and unemployment among US men. *American Sociological Review, 75*(4), 563–585.

MacLean, A. (2013). A matter of life and death: Military service and health. In J. M. Wilmoth & A. S. London (Eds.), *Life-course perspectives on military service* (pp. 200–220). New York, NY: Routledge.

MacLean, A., & Elder, G. H., Jr. (2007). Military service in the life course. *Annual Review of Sociology, 33*, 175–196.

Martindale, M., & Poston, D. L. (1979). Variations in veteran-nonveteran earnings patterns among World War II, Korea, and Vietnam War cohorts. *Armed Forces & Society, 5*(2), 219–243.

Mettler, S. (2005). *Soldiers to citizens: The G.I. Bill and the making of the greatest generation*. New York, NY: Oxford University Press.

Miech, R. A., London, A. S., Wilmoth, J. M., & Koester, S. (2013). The effects of the military's anti-drug policies over the life course: The case of past-year hallucinogen use. *Substance Use and Misuse, 48*, 837–853.

Military Family Resource Center, (2000 March 1). *Military families in the millennium*. Arlington, VA: MFRC. Retrieved from: <http://www.militaryhomefront.dod.mil/dav/lsn/LSN/BINARY_RESOURCE/BINARY_CONTENT/1725341.pdf>. Accessed 19.07.12.

Modell, J., & Haggerty, T. (1991). The social impact of war. *Annual Review of Sociology, 17*, 205–224.

Moskos, C. C., Jr., & Butler, J. S. (1996). *All that we can be: Black leadership and racial integration the Army way*. New York, NY: Basic Books.

National Center for Veterans Analysis and Statistics, Office of the Actuary. (2014). *Veteran Population Projection*

Model (VetPop2014), Table 2L. <http://www.va.gov/vetdata/Veteran_Population.asp>.

New York Times. (1936). *Bonus bill becomes law*. Retrieved from: <http://query.nytimes.com/gst/abstract.html?res=9A07E5DC1630E13BBC4051DFB766838D629EDE>.

Pavalko, E. K., & Elder, G. H., Jr. (1990). World War II and divorce: A life-course perspective. *American Journal of Sociology*, 95(5), 1213–1234.

Phillips, R. L., Andrisani, P. J., Daymont, T. N., & Gilroy, C. L. (1992). The economic returns to military service: Race-ethnic differences. *Social Science Quarterly*, 73(2), 340–359.

Rosen, S., & Taubman, P. (1982). Changes in life-cycle earnings: What do Social Security data show? *Journal of Human Resources*, 17(3), 321–338.

Ruger, W., Wilson, S., & Waddoups, S. (2002). Warfare and welfare: Military service, combat, and marital dissolution. *Armed Forces and Society*, 29, 85–107.

Sackett, P. R., & Mayer, A. S. (Eds.). (2006). *Assessing fitness for military enlistment: Physical, medical, and mental health standards*. Washington, DC: The National Academies Press.

Sampson, R. J., & Laub, J. H. (1996). Socioeconomic achievement in the life course of disadvantaged men: Military service as a turning point, circa 1940–1965. *American Sociological Review*, 61(3), 347–367.

Selective Service. (2015a). *Use of the draft*. Retrieved from: <http://www.selectiveservice.us/military-draft/7-use.shtml>.

Selective Service (2015b). *History*. Retrieved from: <http://www.selectiveservice.us>.

Settersten, R. A., Jr. (2006). When nations call: How wartime military service matters for the life course and aging. *Research on Aging*, 28(1), 12–36.

Settersten, R. A., Jr., & Patterson, R. S. (2006). Military service, the life course, and aging: An introduction. *Research on Aging*, 28(1), 5–11.

Shay, J. (2002). *Odysseus in America: Combat trauma and the trials of homecoming*. New York, NY: Scribner.

Spiro, A., Schnurr, P. P., & Aldwin, C. M. (1997). A life-span perspective on the effects of military service. *Journal of Geriatric Psychiatry*, 30(1), 91–128.

Spiro, A., & Vokonas, P. (2007). Normative aging study. In K. Markides (Ed.), *Encyclopedia of aging & health* (pp. 422–423). Newbury Park, CA: Sage Publications.

Stanley, M. (2003). College education and the midcentury GI Bills. *Quarterly Journal of Economics*, 118(2), 671–708.

Stouffer, S. A., Suchman, E. A., DeVinney, L. C., Star, A. S., & Williams, R. M., Jr. (1949). *Studies in social psychology in World War II, vol. I, The American soldier: Adjustment during army life*. Princeton, NJ: Princeton University Press.

Street, D., & Hoffman, J. (2013). Military service, social policy, and later-life financial and health security. In J. M. Wilmoth & A. S. London (Eds.), *Life-course perspectives on military service*. New York, NY: Routledge.

Teachman, J. D. (2004). Military service during the Vietnam era: Were there consequences for subsequent civilian earnings? *Social Forces*, 83(2), 709–730.

Teachman, J. D. (2005). Military service in the Vietnam era and educational attainment. *Sociology of Education*, 78(1), 50–68.

Teachman, J. D. (2007a). Military service and educational attainment in the all-volunteer era. *Sociology of Education*, 80(4), 359–374.

Teachman, J. D. (2007b). Race, military service, and marital timing: Evidence from the NLSY-79. *Demography*, 44(2), 389–404.

Teachman, J. D. (2010). Are veterans healthier? Military service and health at age 40 in the all-volunteer force era. *Social Science Review*, 40, 326–335.

Teachman, J. D. (2013). Setting an agenda for future research on military service and the life course. In J. M. Wilmoth & A. S. London (Eds.), *Life-course perspectives on military service* (pp. 254–274). New York, NY: Routledge.

Teachman, J. D., & Call, V. R. A. (1996). The effect of military service on educational, occupational, and income attainment. *Social Science Research*, 25(1), 1–31.

Teachman, J. D., & Tedrow, L. M. (2004). Wages, earnings, and occupational status: Did World War II veterans receive a premium? *Social Science Research*, 33(4), 581–605.

Teachman, J. D., & Tedrow, L. M. (2007). Joining up: Did military service in the early all volunteer era affect subsequent civilian income? *Social Science Research*, 36(4), 1447–1474.

Teachman, J. D., & Tedrow, L. M. (2008). Divorce, race, and military service: More than equal pay and equal opportunity. *Journal of Marriage and the Family*, 70(4), 1030–1044.

US Code of Federal Regulations. (2014). 38 §3.2, *Periods of war*. Available online at <http://www.ecfr.gov/cgi-bin/text-idx?rgn=div5&node=38:1.0.1.1.4#se38.1.3_12>.

US Department of Defense. (1998). *DoD marks 50th year of military women's integration*. <http://www.defense.gov/news/newsarticle.aspx?id=41380>.

US Department of Defense (2006). *Population representation in the military services, Fiscal Year 2004, Active component enlisted applicants and accessions*. Retrieved from: <http://prhome.defense.gov/portals/52/Documents/POPREP/poprep2004/enlisted_accessions/accessions.html>.

US Department of Defense. (2011). *Population representation in the military services*. Retrieved from: <http://prhome.defense.gov/RFM/MPP/AP/POPREP.aspx>.

US Department of Defense. (2015). *Millennium cohort study: About the study*. Retrieved from: <https://www.millenniumcohort.org/about>.

US Department of Veterans Affairs. (2010). *National survey of veterans, active duty service members, demobilized National Guard and Reserve members, family members, and surviving spouses.* Retrieved from: <http://www.va.gov/vetdata/docs/SurveysAndStudies/NVSSurveyFinalWeightedReport.pdf>.

US Department of Veteran Affairs. (2013). *History – Department of Veteran Affairs.* Retrieved from: <http://www.va.gov/about_va/vahistory.asp>.

US Department of Veterans Affairs. (2014a). *Federal benefits for veterans, dependents, and survivors 2011 edition.* Retrieved from: <http://www.va.gov/opa/publications/benefits_book.asp>.

US Department of Veterans Affairs. (2014b). *Trends in veterans with a service-connected disability: 1982–2012.* Retrieved from: <http://www.va.gov/vetdata/docs/QuickFacts/SCD_trends_FINAL_2012.pdf>.

US Department of Veteran Affairs. (2015). *Military sexual trauma.* Veterans Health Library, Heath Encyclopedia. Retrieved from: <http://www.veteranshealthlibrary.org/Encyclopedia/142,UG4317_VA>.

Usdansky, M. L., London, A. S., & Wilmoth, J. M. (2009). Veteran status, race-ethnicity, and marriage among fragile families. *Journal of Marriage and the Family, 71*(3), 768–786.

Vogt, D. S., King, D. W., King, L. A., Savarese, V. W., & Suvak, M. K. (2004). War-zone exposure and long-term general life adjustment among Vietnam veterans: Findings from two perspectives. *Journal of Applied Social Psychology, 34*(9), 1797–1824.

Wilmoth, J. M., Heflin, C. M., & London, A. S. (2015). Economic well-being among older households: Variation by veteran and disability status. *Journal of Gerontological Social Work, 58,* 399–410.

Wilmoth, J. M., & London, A. S. (2011). Aging veterans: Needs and provisions. In J. L. Angel & R. Settersten (Eds.), *Handbook of the sociology of aging* (pp. 445–461). New York, NY: Springer.

Wilmoth, J. M., & London, A. S. (2013). *Life course perspectives on military service.* New York, NY: Routledge.

Wilmoth, J. M., London, A. S., & Parker, P. (2010). Military service and men's health trajectories in later life. *Journal of Gerontology: Social Sciences, 56*(6), 744–755.

Wolf, D. A., Wing, C., & Lopoo, L. (2013). Methodological problems in determining the consequences of military service. In J. M. Wilmoth & A. S. London (Eds.), *Life-course perspectives on military service* (pp. 254–274). New York, NY: Routledge.

Whyman, M., Lemmon, M., & Teachman, J. D. (2010). Non-combat military service in the United States and its effects of depressive symptoms among men. *Social Science Research, 40,* 695–703.

Xie, Y. (1992). The socioeconomic-status of young male veterans, 1964–1984. *Social Science Quarterly, 73*(2), 379–396.

Yoder, A. (2014). *Military classifications for draftees.* Retrieved from: <http://www.swarthmore.edu/library/peace/conscientiousobjection/MilitaryClassifications.htm>.

CHAPTER 12

Religion, Health, and Aging

Neal Krause and R. David Hayward

School of Public Health, University of Michigan, Ann Arbor, MI, USA

OUTLINE

Introduction	251	Religion and a Sense of Meaning in Life	259
Religious Involvement over the Life Course	252	Religious Involvement and Forgiveness	260
		Prayer	261
Religion, Health, and Well-Being	254	Social Relationships in the Church	262
Religion and Physical Health	254		
Religion and Psychological Well-Being	255	Spiritual Struggles: Assessing the Dark Side of Religion	264
From Correlation to Explanation: Identifying the Health-Related Dimensions of Religion	256	Race/Ethnicity, Religion, and Health	265
		Conclusions	266
Religious Services Attendance	256	Acknowledgment	267
Religious Coping Responses	257		
God-Mediated Control Beliefs	258	References	267

INTRODUCTION

Philosophers, theologians, and researchers have discussed the relationship between religion and health for well over 100 years (e.g., Galton, 1872). However, it wasn't until more recently that empirical research on religion and health began to appear on a regular basis. A good deal of the early work focused on the relationship between attendance at worship services, health, and well-being. For example, Schmidt (1951) found that people who attend worship services more often tend to have better "personal adjustment" than individuals who go to church less frequently. Since that time, what was once a cottage industry has mushroomed into a vast enterprise that consists of thousands of studies (see Koenig, King, & Carson, 2012, for a review of this research). One of the more notable aspects of this literature is that a considerable number of studies focus specifically on religion and health in late life. As we will argue below, there are good reasons for studying religion and health among older people.

As the literature on religion and health evolved, researchers quickly realized that religion is a complex phenomenon that is comprised of a number of dimensions that can be measured in many ways (Fetzer Institute/National Institute on Aging Working Group, 1999). Although the development of detailed batteries to measure the complex nature of religion represents a significant advancement in the field, it also presented researchers with a challenge that is comprised of two issues. First, it is not clear which aspect(s) of religion are responsible for the presumed health benefits that researchers have observed. Second, the boundaries between measures of different dimensions of religion are not firm and as a result, there is significant overlap between them. This has led to further problems in isolating the aspects of religion that may promote better physical and mental health.

The purpose of the discussion that follows is to review research on religion and health that deals with five central issues in the field. First, we briefly examine studies on change in religiousness over the life course. Second, we summarize research on the relationship between religious involvement, physical health, and psychological well-being. Third, we provide a brief overview of eight dimensions of religion that, in our opinion, are likely to provide the most insight into how the potentially beneficial health-related effects of religion may arise. Fourth, we review research on race and ethnic differences in the relationship between religion and health. Fifth, we close by discussing a number of next steps that we think will further advance the field.

We would like to emphasize at the outset that a review of this scope is, of necessity, selective. Consequently, three steps were taken to place boundaries on the discussion that follows. First, we will focus on religion, but not spirituality. We exclude spirituality because, as Koenig (2008) convincingly argues, measures of spirituality are confounded with the outcomes they are designed to explain. For example, the widely cited Daily Spiritual Experiences Scale contains the following item: "I feel deep inner peace and harmony" (Fetzer Institute/National Institute on Aging Working Group, 1999). Obvious problems with measurement confounding arise when this type of measure is regressed on mental health and psychological well-being. Second, issues in the direction of causality have plagued research on religion and health from the outset. Recall that early studies indicate that more frequent attendance at worship services is associated with better health and well-being. But one could just as easily argue that people who enjoy better health are more likely to attend worship services in the first place. Although vexing issues in the direction of causality between religion and health are important, they will not be examined in the discussion that follows. Third, because we cannot provide a comprehensive review of all the research in the field, we will not cite long lists of studies. Instead, we briefly note a few studies that illustrate what has been learned and point to what we still need to know.

RELIGIOUS INVOLVEMENT OVER THE LIFE COURSE

As we discuss below, a number of researchers believe that people become more involved in religion as they grow older. It follows from this belief that if religion affects health and people become more involved in religion as they move through the life course, then the beneficial effects of religion should be most evident in samples that are comprised of older people.

Most studies on age differences in religious involvement employ a cross-sectional research design. This means that investigators only have a single observation of religiousness in samples that are comprised of people with a wide age-range (e.g., age 18 and older). Taken as a whole, this research reveals that people who are presently older are more deeply involved in religion than individuals who are currently younger (see Krause, 2008a, for a review of this research). Unfortunately, researchers who rely on cross-sectional studies cannot be sure if they are observing the effects of age per se or whether their findings are due to period or cohort effects.

Although there are clear limitations with cross-sectional studies on age differences in religion, one cross-sectional study stands out above the rest. This research, which was conducted by Deaton (2009), assesses age differences in religiousness among 300 000 people in 140 nations. The nations examined in this study range from the United States and the United Kingdom to Belize and Togo. Deaton (2009) reports a trend toward greater involvement in religion with advancing age across all the nations in his study. This study is noteworthy because the breadth of nations that were studied cast doubt on the notion that the findings may reflect cohort or period effects. It is difficult to imagine, for example, that people living in Togo have the same cohort experiences as individuals residing in the United States. But like many other studies that focus on age differences in religion, there are significant problems with the way Deaton (2009) assesses religion. He relies on only two measures. The first measure assesses whether religion is important in daily life while the second involves attendance at worship services.

In contrast to cross-sectional studies, research that is based on data that have been gathered at more than one point in time is potentially more informative. We briefly review three studies that are typical of the longitudinal research that has been done so far.

The first study was conducted by Wink and Dillon (2002). These investigators studied age differences in religiousness from mid-life through old age. They report that regardless of gender and cohort, religiousness increased from mid-life through late life.

The second longitudinal study on religiousness and aging was conducted by McCullough, Enders, Brion, and Jain (2005). These investigators analyzed data from the widely cited Terman Study, which consists of interviews with very bright individuals (i.e., an IQ above 135) from 1940 through 1991. A composite measure of religion was created that consisted primarily of interest in religion and satisfaction with religion. Using sophisticated growth mixture models, these investigators found three distinct trajectories of religious development. Some people experienced increases in religiousness in early adult life followed by a decline with advancing age while others exhibited low levels of religiousness in early adulthood and a subsequent age-related decline. And yet other study participants provided evidence of increasing religiousness with advancing age.

The third study on religion and aging was conducted by Argue, Johnson, and White (1999). Based on data over a 12-year period and sophisticated data analytic procedures, these investigators report that the importance of religion increases in a nonlinear fashion with age, with the steepest increase occurring in the middle adult years. This study is noteworthy because the authors controlled statistically for period and cohort effects.

Before making some concluding remarks on the literature that deals with aging and religious involvement, we would like to touch on an issue that has not received sufficient attention in the literature. This issue involves the need to focus on changes in religiousness within late life. Although researchers have yet to agree on a precise operational definition of old age, many would agree that it spans the period from age 65 to age 100 or so. It is hard to imagine that a person could live 35 years and not experience change in a number of life domains, including religion. Our recent research (Hayward & Krause, 2013a,b) points to the importance of examining this issue.

Based on four waves of data spanning a 7-year period, we found that feelings of God-mediated control tend to increase over time during late life (Hayward & Krause, 2013a). God-mediated control refers to the extent to which people believe that God works with them to manage the goals and challenges they face.

Using the same data, we also assessed change in church-based social relationships during late life (Hayward & Krause, 2013b). Based on a series of individual growth curve models, we observed increases in the amount of emotional support that older people receive and provide to fellow church members. But, in contrast, the amount of tangible help that was received from and provided to coreligionists declined over time.

Do people become more religious as they grow older? It is not possible to provide a definitive answer at this time. Problems with study samples (e.g., the Terman study), study measures (e.g., the importance of religion only), and research designs (i.e., cross-sectional studies) must first be overcome. Even so, based, in part, on the findings provided by McCullough et al. (2005), we suspect that there is more than one pattern of religious involvement over the life course. If for no other reason, this observation is consistent with simple observation. Perhaps, the best conclusion is that there appears to be some evidence that some people may become more involved in religion as they grow older. But until well-designed longitudinal studies are launched with the specific goal of assessing age differences in religion, the issue will remain unresolved.

RELIGION, HEALTH, AND WELL-BEING

In this section, we selectively review studies that focus on the relationship between various facets of religion, physical health status, and psychological well-being. Wherever possible, we restrict our review to studies with a longitudinal design.

Religion and Physical Health

Findings from a number of longitudinal studies indicate that greater involvement in religion is associated with better physical health. For example, Idler and Kasl (1997) found that more frequent church attendance is associated with lower levels of functional disability over time. The same conclusion was reached by Benjamins (2004) in her work with data from the Assets and Health Dynamics among the Oldest-Old Survey. Similarly, we found that more frequent church-based social support was associated with more favorable self-ratings of health over time (Krause, 2006a). Moreover, our research also reveals that stronger God-mediated control beliefs are associated with better self-rated health over time (Krause, 2010a).

Perhaps the most compelling findings from research on religion and health are found in studies that examine the relationship between religious involvement and mortality (e.g., Hill, Angel, Ellison, & Angel, 2005; Oman & Reed, 1998; Sullivan, 2010). Many of these studies focus specifically on the relationship between the frequency of church attendance and mortality. Taken as a whole, the findings from this research suggest that people who attend worship services more often are less likely to die over the course of the study follow-up period than individuals who do not go to church as often. Research on religion and mortality is compelling because issues involving the direction of causality are less of a concern.

If religion affects physical health, then researchers must ultimately link measures of religious involvement with biological markers of health. Fortunately, this type of research is beginning to appear with greater frequency. For example, Maselko, Kubzansky, Kawachi,

Seeman, and Berkman (2007) found that more frequent attendance at worship services is associated with a lower allostatic load. Allostatic load is a complex composite of physical dysregulation consisting of abnormalities in measures of things like systolic and diastolic blood pressure, waist/hip ratio, high-density lipoprotein, cortisol, and norepinephrine and epinephrine.

Although a growing number of studies suggest that greater involvement in religion is associated with better health, findings by other researchers do not support this conclusion. For example, Hunter and Merrill (2013) report that individuals with a pro-religious orientation had more trouble with physical functioning, more role limitations due to health problems, and greater fatigue than people who did not have a pro-religious orientation. Pro-religious people are those whose religious behaviors are motivated by both intrinsic and extrinsic religiousness. Atchley (1997) was unable to find significant relationships between religious affiliation, church attendance, and health in his longitudinal study that spanned 14 years. Self-rated health and functional disability served as the health outcomes in this research.

Taken as a whole, research suggests that greater involvement in religion may be associated with better physical health. The exceptions to this general trend can be explained by problems associated with the research design and the measures of religion that were utilized. More specifically, the study by Hunter and Merrill (2013) was based on cross-sectional data that were provided by a non-random sample of middle-aged and older people who were receiving a health screening. The restricted measures of religious involvement in the study by Atchley (1997) are a concern, as well. We believe that in order to take research on religion and physical health status to the next level, researchers must address two issues. First, research is needed on the interface between religion, a range of biomarkers, and health. Second, more sophisticated measures of religion should be employed in this research.

Religion and Psychological Well-Being

A good deal of the longitudinal research on religion and mental-health-related outcomes focuses specifically on depressive symptoms. The majority of these studies reveal that greater involvement in religion is associated with fewer symptoms of depression. For example, we found that stronger God-mediated control beliefs are associated with more gratitude, and older people who are more grateful tend to experience fewer symptoms of depression over time (Krause, 2009). In another study, we report that older people who feel they are valued by the members of their congregations tend to experience fewer symptoms of depression over time (Krause, 2012c). Using data that were gathered at three points in time, Law (2009) found that more frequent church attendance is associated with decreasing trajectories of depressive symptoms over time. Similarly, findings from a study by Sun et al. (2012) suggest that more frequent church attendance and higher levels of intrinsic religiousness were associated with a decline in the number of depressive symptoms over a 4-year period.

In contrast to studies of religiousness and depressive symptoms, fewer studies have assessed the relationship between religious involvement and clinical mental health problems in late life. One notable exception to this trend may be found in the research by Norton and her colleagues (2008). These investigators assessed the relationship between the frequency of church attendance and new episodes of major depression over a 3-year period. They report that older adults who attended worship services weekly or more often had a significantly lower risk of developing an episode of major depression over time.

Other investigators have focused on the relationships between religion and different mental health outcomes. For example, a study by Corsentino, Collins, Sachs-Ericsson, and Blazer (2009) reveals that more frequent church attendance slows the rate of cognitive decline over time. There is also a good deal of research on the relationship between religion and markers of positive mental health. For example, we examined the relationship between trust-based prayer expectancies and changes in life satisfaction over time (Krause & Hayward, 2013). Trust-based prayer expectancies refer to the belief that God knows the best way to answer a prayer and God knows the best time to do so. Our study reveals that stronger trust-based prayer beliefs are associated with greater life satisfaction among older people over time. Similarly, in a cross-sectional study, we report that stronger God-mediated control beliefs are associated with greater optimism, a greater sense of self-esteem, and lower levels of death anxiety (Krause, 2005).

Viewed in its entirety, the literature on religion and mental health appears to be more fully developed and the findings are more consistent than research on religion and physical health in late life. But this does not mean that there is no room for improvement. To move this literature forward, researchers should address two closely related issues. First, it is time to stop focusing solely on church attendance. Church attendance is a vast, global measure that encompasses many different facets of religion including group prayers, hymns, and interaction with fellow church members. Each of these facets of religion may have a beneficial relationship with mental-health-related outcomes. Second, researchers must begin to develop more complex conceptual models that provide greater insight into how church attendance might affect mental health. In an effort to encourage this type of work, we turn next to an assessment of the key dimensions of religion that may play a role in these conceptual schemas.

FROM CORRELATION TO EXPLANATION: IDENTIFYING THE HEALTH-RELATED DIMENSIONS OF RELIGION

During the past several decades, researchers have identified a number of religious factors that may have beneficial effects on health. We discuss eight dimensions of religion below which we believe have the greatest potential for explaining the link between religion and health: religious service attendance, religious coping responses, God-mediated control, a religious sense of meaning in life, forgiveness, prayer, social relationships in the church, and spiritual struggles. We will follow the same approach as we review each facet of religion. We will define each dimension (when necessary), we will briefly review research linking each facet of religion with health-related outcomes, and we will identify issues that can help improve the quality of the research that is being done.

Religious Services Attendance

Attendance at religious services is one of the most frequently used indicators in studies of religious involvement and health. Moreover, this research consistently reveals that more frequent attendance at religious services is associated with a range of health-related outcomes in late life including better self-rated health (Koenig et al., 1997), the adoption of beneficial health behaviors (Weaver & Gary, 1996), fewer symptoms of depression (Law & Sbarra, 2009), and greater life satisfaction (Morris, 1997).

Although these studies provide useful insight into the relationship between religion and health, there is an especially important way in which research on religious service attendance and health can be improved. As we argued earlier, attendance at worship services is a complex phenomenon that involves a number of different dimensions. Any (or all) aspects of church attendance may exert a beneficial effect on health and well-being. Research is needed to see which dimensions of religious involvement mediate the effects of church attendance on health. There have been some efforts to address this issue, but researchers typically assess whether a limited range of religious dimensions mediate the effects of church attendance on health-related outcomes. For example, our research suggests that church-based social support, trust in God, and feelings of forgiveness by God mediate the relationship between church attendance and death anxiety (Krause, in press a). But we have been unable to locate any studies that systematically sift through a full complement of dimensions of religion in order to see which dimensions are the most important.

Religious Coping Responses

Religious coping responses involve specific religious beliefs and religious practices that people engage in to either avoid or alter the course of stressors in their lives. Because the number of stressors that arise in life is quite diverse, scales that have been devised to assess religious coping responses can be quite long. For example, one of the most widely used religious coping scales was developed by Pargament, Koenig, and Perez (2000) (i.e., the RCOPE). This scale contains 105 different religious coping responses that involve things like asking for God's help, getting spiritual strength from others, and seeing a stressor as part of God's broader plan. This scale also captures negative aspects of religious coping, such as feeling abandoned by God.

Findings from a number of studies indicate that the use of religious coping responses tends to offset the deleterious effects of stress on physical and mental health. For example, Pargament, Smith, Koenig, and Perez (1998) report that medically ill older adults who rely on positive religious coping responses experience higher levels of stress-related growth while those who engage in negative religious coping have higher levels of depression and a lower quality of life. The findings from our own longitudinal work reveal that the deleterious effect of living in a rundown neighborhood on changes in self-rated health was offset completely for older adults who rely on positive religious coping strategies (Krause, 1998). And a study by Amy Ai and her colleagues suggests that middle-aged and older adults who used positive religious coping responses prior to cardiac surgery had better short-term postoperative functioning (Ai, Peterson, Bolling, & Rogers, 2006).

Although significant strides have been made in the study of religious coping responses, a considerable amount of research remains to be done. Closer attention should be given to two issues.

First, there may be an element of timing involved in the selection of religious coping responses. For example, a person might initially turn to God for guidance on how to handle a stressor. However, as time passes, the anticipated guidance may not be forthcoming. If this turns out to be the case, then perhaps these individuals subsequently turn to negative religious coping responses, such as feeling abandoned by God.

The second challenge facing researchers who use religious coping responses is complex. Of necessity, the scope of religious coping responses is quite broad. But this breadth comes at the expense of depth. Specific religious coping responses, such as seeking spiritual guidance from others, are typically evaluated with one or two items. Unfortunately, research reveals that the social support process is quite complex. As we have argued elsewhere, church-based support may offset the effects of stress by replenishing self-esteem, feelings of personal control, and the sense of meaning in life that have been eroded by the unwanted life event (Krause, 2008a). This is important because self-esteem, personal control, and meaning have, in turn, been associated with better health. None of these intervening variables is adequately captured by religious coping scales and, as a result, knowledge about the way in which religious coping operates is incomplete. Simply put, religious coping scales run the risk of being a mile wide and half an inch deep. Perhaps the following three-step process may provide a way out of this dilemma. First, a researcher might present the usual religious coping response scale items. Second, once this scale has been administered, the study participant could be asked to identify the particular coping response that he or she felt was most useful. Third, a battery of more detailed items might be administered to find out how the benefits of this particular coping response arise.

God-Mediated Control Beliefs

One function of religion is to help people feel they have greater control over their lives (Hood, Hill, & Spilka, 2009). The construct of God-mediated control was devised to assess this function. Recall that God-mediated control involves the extent to which people believe that God works with them to control the events they encounter in life (Krause, 2005). In addition to this conceptualization, Schieman, Pudrovska, and Milkie (2005) propose the closely related notion of divine control. The difference between the two rests on a straightforward issue: we focus on collaborative control whereby people work together with God whereas Schieman et al. (2005) are concerned with letting God take complete control of a person's life.

Regardless of how God-mediated and divine control are defined, the common ground they share may be especially important in late life. A number of studies indicate that feelings of personal control decline precipitously with advancing age (e.g., Mirowsky, 1995). The decline of personal control is disconcerting because a vast literature reveals that strong feelings of personal control are associated with a wide range of physical and mental health outcomes (see Fry & Debats, 2010, for a review of this research). Fortunately, as our research reveals, the decline in feelings of personal control may be offset by increases in God-mediated control beliefs (Hayward & Krause, 2013a).

Our research indicates that God-mediated control is associated with greater psychological well-being (Krause, 2005) and better self-rated health (Krause, 2010a). Similarly, Schieman and his colleagues report that divine control is associated with lower levels of distress (Schieman et al., 2005) and a stronger sense of self-worth (Schieman, Pudrovska, Pearlin, & Ellison, 2006).

Although research on God-mediated control may contribute to the literature, two issues have yet to be addressed. The first issue involves the way in which these religious notions of control are measured. Both God-mediated and divine control assess control over life as a whole. However, researchers have known for some time that feelings of control vary across different domains in life (Krause, 2003a). For example, an individual may feel that he or she can exercise more control over their family life and relatively less control over their job. Research is needed to see if feelings of God-mediated control vary across different domains in life.

Second, most of the research on religiously oriented feelings of control has focused solely on additive effects while fewer studies have been done on the potential stress-buffering effect of God-mediated control. This is curious because a fairly large body of research reveals that feelings of personal control tend to offset the noxious effects of stress on health and well-being (Skinner & Zimmer-Gembeck, 2011). We have been unable to find a stress-buffering effect of God-mediated control beliefs in our own unpublished research. The lack of a significant interaction effect with stress suggests that the benefits of God-mediated control may be found outside the stress process. For example, a strong sense of God-mediated control may be important for achieving personal plans and goals. Research is needed to see if this is so.

Religion and a Sense of Meaning in Life

Researchers in all the major social and behavioral science disciplines have argued that people have a compelling need to find a sense of meaning in life. For example, Victor Frankl (1959/1985) maintained that, "Man's search for meaning is the primary motivation in his life …" (p. 121). Similarly, Peter Berger (1967) argued that there is, "… a human craving for meaning that appears to have the force of instinct" (p. 22). Unfortunately, a sense of meaning in life has been notoriously difficult to define. We will not try to resolve this long-standing issue here. Instead, we rely on the definition devised by Reker (1997). He proposes that meaning in life involves having a sense of purpose, order, and direction, as well as the belief that there is a reason for our existence.

It is important to review research on religion and meaning in late life because a number of investigators maintain that a sense of meaning becomes increasingly important as people grow older. Evidence of this may be found, for example, in Erikson's (1959) theory of life course development. He divides the life course into eight stages. Each stage poses a unique developmental challenge or crisis. The final stage, which is typically encountered in late life, is characterized by the crisis of integrity versus despair. This is a time of deep introspection when older individuals review their lives and try to reconcile the inevitable gap between what they set out to do and what they were actually able to accomplish. Ultimately, the goal of this review is to imbue life with a deeper sense of meaning.

Researchers have taken one of two broad approaches to measuring meaning in life. Most studies so far assess meaning outside the context of religion. We will call these measures of secular meaning in life. For example, McDonald, Wong, and Gingras (2012) ask study participants to indicate whether the following item characterizes their own life: "I believe I can make a difference in the world." An effort is then made in this research to see if religion affects secular meaning and whether secular meaning, in turn, is associated with health. In contrast to this measurement strategy, other investigators explicitly weave religious themes into the question stems they have devised to assess meaning in life. For example, we have asked study participants how strongly they agree with the following item: "God put me in this life for a purpose" (Krause, 2007). We will call this type of measure religiously oriented measures of meaning in life. Researchers who focus on religiously oriented meaning simply relate this type of measure with health.

A small cluster of studies indicate that there is a significant association between religion, meaning in life, and health. For example, we report that older people with a deeper sense of religiously oriented meaning in life are less likely to experience a decline in physical functioning over time (Krause & Hayward, 2012a). Similarly, we found that older adults with a stronger sense of religiously focused meaning in life tend to have higher levels of life satisfaction, self-esteem, and optimism (Krause, 2003b). Using data from the US Congregational Life Survey, we recently reported that adults of all ages who get more social support at church develop a greater sense of secular meaning and a greater secular sense of meaning in life was, in turn, associated with more favorable self-rated health (Krause, Hayward, Bruce, & Woolever, 2013). And evidence from a study by Homan and Boyatzis (2010) suggests that a stronger sense of religiously oriented meaning in life is associated with better health behaviors (e.g., less smoking, more exercising) in late life.

Perhaps the most pressing challenge facing researchers who study religion, meaning, and health involves issues in the measurement of meaning. More specifically, researchers need to see if it is better to rely on secular or religiously oriented measures of meaning in life. Perhaps the best way to resolve this dilemma is to assess a model that contains both secular and religiously oriented measures of meaning in life and health. The effects of the two measures of meaning on health can then be compared and contrasted while controlling for the correlation between them.

Religious Involvement and Forgiveness

The literature on forgiveness is very broad (Toussaint, Worthington, & Williams, 2014). Moreover, a good deal of this work has been conducted outside the context of religion. Even so, forgiveness is an integral part of virtually every major faith tradition (Lundberg, 2010).

Viewed broadly, forgiveness is defined by Enright, Freedman, and Rique (1998) as "… a willingness to abandon one's right to resentment, negative judgment, and indifferent behavior toward one who unjustly injured us, while fostering the undeserved qualities of compassion, generosity, and even love toward him or her" (pp. 46–47). However, this definition focuses solely on forgiveness of others. As research in the field has progressed, researchers have also begun to study forgiveness by God, forgiveness of the self by others, and self-forgiveness.

A number of studies suggest that people may become more forgiving as they grow older (e.g., Steiner, Allemand, & McCullough, 2011). But there are some exceptions. For example, a recent meta-analysis uncovered only negligible effects of age on forgiveness of others (Fehr, Gelfand, & Nag, 2010). A few studies have examined the relationship between age and the other types of forgiveness. This research reveals that older adults are more likely than younger people to feel they have been forgiven by God (Toussaint, Williams, Musick, & Everson, 2001).

Further evidence that forgiveness may increase with age is provided in our study on change in forgiveness within late life (Hayward & Krause, 2013c). Based on a series of individual growth curve models, we found that forgiving others, forgiveness of self, feeling forgiven by God, and feeling forgiven by others all increase over the 7-year study period. Moreover, our work reveals that increases in forgiveness are more likely to be found among older adults who are more deeply committed to their faith.

There are relatively few studies that focus specifically on religion, forgiveness, and health-related outcomes in samples that are comprised of older people. Even so, the work that has been done so far indicates that religiously based forgiveness may be associated with better health and well-being. For example, Lawler-Row (2010) reports the results of two studies that were conducted with older adults. The first study reveals that forgiveness by God, self-forgiveness, and forgiveness of others mediate the relationships between church attendance and the frequency of prayer on five dimensions of successful aging. Findings from the second study suggest that forgiveness of others mediates the relationships between prayer and intrinsic religiousness on illness symptoms and the quality of life.

As North (1998) observed, forgiveness is a complex process that consists of a number of distinct stages. Unfortunately, most researchers focus solely on the direct effects of forgiveness on health and well-being without taking the finer nuances of the forgiveness process into account. One exception to this general tendency is found in our research on how people go about forgiving others (Krause & Ellison, 2003). Our work suggests that older adults who require transgressors to perform acts of contrition experience more psychological distress than older people who forgive unconditionally. Acts of contrition involve things like requiring a transgressor to make an apology, promise not to repeat the offense, and provide restitution, whenever possible. We also found that people who feel they have been forgiven by God are less likely to expect transgressors to perform acts of contrition.

Another of our studies indicates that performing acts of contrition in order to feel forgiven by God may not operate in the same manner (Krause & Hayward, 2015). More specifically, this study suggests that performing acts of contrition in order to feel forgiven by God may elevate feelings of self-worth among older Mexican Americans.

Given the complexity of the forgiveness process, it is not hard to see why a considerable amount of research remains to be done. We briefly touch on two issues that have yet to be addressed. First, we need to know more about how specific religious rituals affect the process of forgiveness. For example, the Catholic Church has instituted the practice of formal confession as well as a day of atonement. A day of atonement is also part of the Jewish faith. Research is needed to see if these religious rituals make it easier for older people to forgive themselves and others than other, more informal, avenues facilitating the process of forgiveness. Second, as we pointed out earlier, a good deal of the research on forgiveness has been conducted outside the context of religion. This suggests that studies are needed that directly compare and contrast the effects of religiously motivated forgiveness with forgiveness that is not associated with religion.

Prayer

William James (1902/1997) argued that prayer is, "…the very soul and essence of religion" (p. 486). Moreover, research consistently reveals that older adults pray more often than individuals who are younger (Barna, 2002). Like forgiveness, there is more than one type of prayer (Poloma & Gallup, 1991). Moreover, prayers can be directed in different ways: people may pray for themselves, they may pray for other people, and other people may pray for them.

A good deal of controversy has swirled around intercessory prayer, which involves prayers that are offered by strangers for individuals who are typically hospitalized for some type of cardiovascular surgery. Initially, the evidence seemed to suggest that intercessory prayer was effective in reducing treatment outcome problems (Byrd, 1988), but a number of methodological problems were found with this early research. More recently, a well-designed study by Benson and his colleagues (2006) put this issue to rest. Benson et al. (2006) report that intercessory prayer has no effect on whether coronary bypass surgery is free of complications. However, one aspect of this study is intriguing. The authors report that most study participants indicated their significant others would be praying for them and many were praying for themselves. So, in effect, Benson et al. (2006) were really studying whether intercessory prayer affects health over and above more naturally occurring prayer from significant others and the self. This could explain why they found that intercessory prayer did not work.

The fact that the study by Benson et al. (2006) could not rule out the influence of prayer that arises in informal social networks raises a host of questions. One is that intercessory prayer may be lacking some key element that is found in prayers that are offered by significant others. We will use this observation to draw out two aspects of prayer that have not received sufficient attention in the literature.

First, consider the following scenario. An older man is about to get bypass surgery. Shortly before the operation, his wife tells him that she will be praying for him. This situation may represent a unique type of religiously based emotional support. This is important because as we will discuss in the following section, a growing number of studies suggest that emotional support from fellow church members is associated with better physical and mental health.

The second issue builds on the scenario we just sketched out. If an older woman's husband is going to have bypass surgery, then this major surgical event may be a source of significant distress for her, as well. In the process of praying for her husband, she may find she has benefited as much as her husband. In fact, some of our research provides support for this issue. In one study, we found that the pernicious effects of financial strain on self-rated health are significantly offset for older people who pray for others (Krause, 2003c). Further support is found in a second study, which suggests that the deleterious effects of living in a rundown neighborhood on depressive symptoms are significantly reduced for older Mexican Americans who prayer for others often (Krause, in press b).

Like the other dimensions of religion that we have discussed so far, prayer is a complex phenomenon but researchers have yet to plumb the depths of this fundamental type of religious behavior. Recently, we have been assessing the relationship between trust-based prayer expectancies, health, and well-being. Trust-based prayer expectancies involve the extent to which people believe that God knows the best way to answer a prayer and whether they believe that God selects the best time to answer a prayer. In one study, we found that increases in trust-based prayer expectancies are associated with increases in life satisfaction over time (Krause & Hayward, 2013). In a second study, our findings reveal that among older Mexican Americans strong trust-based prayer expectancies are associated with more favorable self-rated health (Krause & Hayward, 2014).

A host of other issues involving prayer have yet to be examined. Two appear to be especially ripe for further research. First, people who are involved in religion frequently pray in groups. Many churches have formal prayer groups and congregational prayers are often part of worship services as well. Yet little is known about the relative effects of private prayer and group prayer on health and well-being in later life.

Second, Poloma and Gallup (1991) report that conversational prayer is the most common type of prayer. This type of prayer involves speaking to God in one's own words rather than relying on formal prayers that are found in religious texts (i.e., ritual prayer). We wonder if people use conversational prayer as a time to collect their thoughts, reflect on their lives, and in essence engage in religious meaning-making.

Social Relationships in the Church

Church-based social relationships form the cornerstone of our research programs. We define social relationships in the church as informal social ties that are developed and maintained within religious institutions.

We believe that social relationships that arise in the church may take on added importance as people grow older. This insight is based on two issues. First, church-based social ties may take on added meaning because they coalesce around shared religious rituals that signal key transitions in the life course including births, coming of age, marriages, and deaths. Social ties that are forged in the church may also take on added significance because they are sometimes forged in an intergenerational setting (i.e., people may worship where their parents and grandparents have worshiped). Second, socioemotional selectivity theory specifies that as people grow older they become increasingly aware that they have relatively little time left to live (Carstensen, 1992). Consequently, they reorder their social relationships by trimming social ties on the periphery of their networks so they can focus on a smaller core of more intimate relationships. Our unpublished analysis of the US Congregational Life Survey data suggests that people who are age 65 and older are likely to have worshiped longer in their current congregation than individuals who are less than 65 years of age ($r = 0.217$; $p < 0.001$). So if older people are more likely than younger adults to have worshiped for a longer period of time in the same congregation and if church-based social ties are built around core religious rituals that signal fundamental life course transitions, then perhaps fellow church members are more likely to become part of the smaller and more intimate social networks that are predicted by socioemotional selectivity theory (Carstensen, 1992).

In the discussion that follows we provide a brief overview of studies on church-based social support, we review research which shows that church-based social support is associated with the other dimensions of religion that have been discussed up to this point, and we identify several issues that should be addressed in future research.

Research on Church-Based Social Support. One of the most important types of social relationships that arise within the church involves the exchange of emotional help, tangible assistance, and spiritual encouragement. We call this church-based social support. Unfortunately, a number of researchers who study social support and religion rely on global measures that combine support from individuals inside and outside the church (e.g., Dulin, 2005). This strategy makes it difficult to isolate the potentially unique aspects of social ties in the church.

Fortunately, more researchers are beginning to use measures that assess social support that is exchanged specifically within religious institutions (e.g., Baruth, Wilcox, Hooker, Hussey, & Blair, 2013). The findings that have emerged so far indicate that church-based social support is associated with better physical and mental health in late life (see Krause, 2008a, for a review of this research). We have tried to push the envelope by addressing two issues that are essential for documenting the key role that church-based social support may play in late life. First, we provide evidence which suggests that church-based emotional support offsets the deleterious effects of financial strain on self-rated health while emotional support from people outside the church fails to have a similar stress-buffering effect (Krause, 2006a). Second, our findings reveal that providing emotional support to fellow church members reduces the effect of support providers' own financial problems on mortality while receiving emotional support from fellow church members does not perform a similar stress-buffering function (Krause, 2006b).

Organizing the Field around Church-Based Support. We have built our research programs around the notion that religious beliefs and behaviors are socially constructed and socially maintained (Berger, 1967). Consistent with this orientation, we have tried to show that church-based support is associated with many other aspects of religion, including seven of the eight dimensions of religion that are discussed above (i.e., all but church attendance). Based on findings from a series of longitudinal analyses, we provide evidence which suggests that more frequent church-based support is associated with greater use of positive religious coping responses (Krause, 2010b), stronger God-mediated control beliefs (Krause, 2007), a stronger sense of meaning in life (Krause, 2008b), greater forgiveness (Krause, 2010c), and increases in a wide range of prayer measures over time (Hayward & Krause, 2013d).

Unresolved Issues in the Study of Church-Based Support. Two issues should be addressed to move research on church-based social support forward. First, we need to know more about how social ties develop in the church. Focusing on the potentially important role that formal groups in the church (e.g., prayer groups and Bible study groups) may play in fostering the development of informal social relationships may provide a good point of departure. Second, research is needed to identify the composition of church-based social networks. Family members are a major source of social support. People often worship with members of their families. But since people typically have high levels of contact with family members outside of church, it is not clear if they view family members as fellow church members or whether they consider them to be part of their social networks outside of church. Resolving this issue is important for comparing and contrasting the health-related effects of church-based and secular social support.

SPIRITUAL STRUGGLES: ASSESSING THE DARK SIDE OF RELIGION

In their efforts to document the potential health-related benefits of religion, some researchers overlook the fact that religion may also have negative effects on health and well-being. Recently, researchers have begun to use the term "spiritual struggles" to refer to the detrimental aspects of religion (Ellison & Lee, 2010). As Ellison and Lee (2010) point out, spiritual struggles consist of three problems: (1) having a troubled relationship with God, (2) negative interaction with fellow church members, and (3) religious doubt. Unfortunately, most of the research on spiritual struggles has focused on the general population while fewer studies have been conducted with older adults. Moreover, with the exception of studies on negative religious coping responses, gerontologists have yet to assess the effects of having a troubled relationship with God.

A small cluster of studies focuses on the relationship between negative interaction in the church, health, and well-being in late life. Negative interaction in the church is defined as unpleasant social encounters with fellow church members that involve criticism, rejection, making unfair demands, and failing to provide help when it was needed. Our research suggests that older adults who encounter negative interaction with the clergy tend to have a lower sense of self-esteem (Krause, 2003d). We also found negative interaction with fellow church members is associated with more symptoms of depression in late life (Krause & Hayward, 2012b). And our work provides some evidence that more interpersonal conflict with a member of the clergy is associated with less favorable self-rated health and more acute and chronic health conditions.

Perhaps most of the research on spiritual struggles in late life has focused on religious doubt. Two intriguing patterns of findings have emerged from this research. First, we found that greater religious doubt is associated with more symptoms of depression and diminished feelings of psychological well-being (Krause, Ingersoll-Dayton, Ellison, & Wulff, 1999). Similarly, a study by Galek, Krause, Ellison, and Kudler (2007) suggests that greater doubts about religion are associated with a range of mental health outcomes including depression, general anxiety, paranoia, and obsessive-compulsive symptoms. However, both studies go on to suggest that the effects of religious doubt on mental health outcomes are lower for people who are currently older than individuals who are presently younger. These results highlight a potentially important aspect of religious doubt. Religious doubt is not inherently either good or bad. Instead, what matters is the way in which people choose to cope with it (Krause & Ellison, 2009). Perhaps people learn to cope more effectively with religious doubt as they grow older.

Second, there is some evidence that religious doubt is associated with other types of spiritual struggles. More specifically, data from a longitudinal study by Krause and Ellison (2009) reveal that older people who encounter more negative interaction with fellow church members tend to have more doubts about their faith over time.

Research on spiritual struggles in late life has just begun and, as a result, a number of important issues have not been examined. Recall that every major faith tradition in the world extols the virtue of forgiveness (Lundberg, 2010). We need to know if the noxious effects of negative interaction with a fellow church member are offset if the victim is willing to forgive the perpetrator for what he or she has done. Moreover, it would be interesting to see if religiously based forgiveness dispels the fallout from negative interaction more quickly than forgiveness that arises outside the context of religion. In addition, the research that is reviewed above indicates that negative interaction with rank-and-file church members and negative interaction with the clergy have undesirable consequences. We need to know if negative interaction with the clergy has a more noxious effect than interpersonal conflict with a fellow church member.

RACE/ETHNICITY, RELIGION, AND HEALTH

Research on race and ethnic differences in religion is important because it provides an opportunity to bring issues involving culture and history to the foreground. If greater involvement in religion is beneficial for health and if research reveals that members of disadvantaged racial and cultural groups are more involved in religion than Whites, then researchers may be able to use the insights from these studies to develop faith-based programs that benefit the health of the underserved (see Allicock, Resnicow, Hooten, & Campbell, 2013, for a discussion of faith-based initiatives). So far, the wide majority of studies in gerontology have focused on older Whites and older Blacks, but research on older Mexican Americans is beginning to appear with greater frequency.

Most studies so far have focused on levels of religious involvement across different racial and ethnic groups. Viewed broadly, this research reveals that there are pronounced differences in religious involvement between older Whites, older Blacks, and older Mexican Americans. Moreover, in every instance, older Whites appear to be less involved in religion than older Blacks or older Mexican Americans.

Research by Krause and Hayward (2012c) indicates older Blacks are the most likely to rely on religious coping responses, followed by older Mexican Americans and older Whites, respectively. These findings were partially replicated in a study by Taylor, Chatters, and Jackson (2007), who report that older African Americans and older Caribbean Blacks are more likely to rely on religious coping responses than older Whites.

Schieman et al. (2006) found that a sense of divine control is higher among older Blacks than older Whites. This finding was replicated by Krause (2005), who found that older Blacks have stronger God-mediated control beliefs than older Whites (Krause, 2005).

Research by Krause (2012a) also indicates that older Whites are less likely than older Blacks or older Mexican Americans to forgive themselves, to forgive others, and to believe they have been forgiven by God. In addition, this study reveals that older Whites are less likely to expect transgressors to perform acts of contrition and they are less likely to initiate the process of reconciliation. In contrast, fewer differences in forgiveness emerged between older Blacks and older Mexican Americans.

Krause (2012b) also evaluated differences between the three racial/ethnic groups on 12 different measures of prayer. In each instance, the prayer life of older Whites was less extensive than the prayer lives of older Blacks and older Mexican Americans. In contrast, relatively few differences emerged in the prayer lives of older Blacks and older Mexican Americans. Further support for these findings emerged in a study by Tait, Laditka, Nies, and Racine (2011). These investigators found that older Blacks and older Hispanics are more likely than older Whites to pray for their own health.

Krause and Bastida (2011) report that older Blacks have more extensive social relationships at church than older Whites or older Mexican Americans. In contrast, relatively few differences emerged between older Whites and older Mexican Americans.

A key issue that arises at this juncture involves explaining why older Whites are less involved in religion than older Blacks or older Mexican Americans. Viewed generally, these findings are consistent with the deprivation-compensation hypothesis, which suggests that socially disadvantaged groups tend to be more involved in religion and derive greater benefits from religion than more well-to-do people (Schieman et al., 2006).

In order to move research on race/ethnicity and religion forward, at last three issues should be addressed. First, research consistently reveals that racial discrimination has an adverse effect on health (Williams et al., 2012). We need to know if various aspects of religion help offset the deleterious effects of discrimination on health. Second, most of the studies to date have been concerned with race/ethnic differences in religious involvement or have focused on the relationships among religion, health, and well-being within a specific racial/ethnic group. For example, a study by Chatters et al. (2008) suggests that older Blacks who attend worship services more often have lower odds of having a lifetime mood disorder. What is needed now are more studies that assess the impact of religion on health and well-being across different racial and ethnic groups. Such studies exist (e.g., Krause, 2006a), but researchers have yet to take the more comprehensive approach that was followed in studies of race differences in levels of religious involvement. Third we need to know more about the culturally unique aspects of religious involvement across different racial/ethnic groups. For example, Krause and Bastida (2012) assessed the relationship between making mandas and health among older Mexican Americans. A manda is a religious quid pro quo whereby an older Mexican American promises to perform a religious act (e.g., a pilgrimage to a sacred shrine) if the Virgin Mary or one of the saints grants a request (e.g., improvement in the health of a loved one).

CONCLUSIONS

A good deal of research has been done on aging, religion, and health. So far, the literature suggests that greater involvement in religion is associated with better health and well-being among older people. But this is not true for everyone because there is some evidence that religion may be a source of distress for some older people. Although there are many ways to improve research in the field, we believe that attention should be given to five issues.

First, the overwhelming majority of studies that have been done so far have focused on older Christians. We need to know more about religion and health among older adults in other faith traditions. The work of Levin and Prince (2013) on Judaism and health provides a good example of the kind of work that needs to be done in this respect.

Second, researchers also need to know more about religion and health among other members of our burgeoning minority population. Some work is beginning to appear on religion and health among Asians of all ages (e.g., Ai, Huang, Bjorck, & Appel, 2013), but more research is needed on older people from the full range of cultural backgrounds.

Third, because the field is so broad, it is difficult get a firm grasp on what has been learned so far. Earlier, we proposed that focusing on the relationship between church-based social support and the other dimensions of religiousness helps weave the literature into a more coherent whole. But this perspective is not fully developed. We are in the process of rolling out a new research initiative that focuses on the interface between religion, a range of virtues, and health. In our view, another function of religion is to instill virtues such as humility, compassion, and gratitude. This makes it possible to specify a chain of "causal" influence that runs from religious involvement (e.g., church attendance) through virtues and church-based social relationships to health. Other researchers may prefer to specify different unifying models, but our basic point is that greater cohesiveness in the field cannot be attained without a specific plan.

Fourth, earlier we noted that an exciting new area of research involves focusing on the relationships among religion, biomarkers, and health. However, researchers are also beginning to discuss the interface between religion, genetics, and health (e.g., Sasaki, Heejung, & Xu, 2011). We believe that exploring the role that genes may play in the religion and health relationship is another promising area of inquiry.

Fifth, as we discussed in the section on religious meaning, many dimensions of religion have secular counterparts. This means, for example, that there are measures of explicitly religious coping responses as well as measures of coping that are not religiously based. The same is true with respect to God-mediated control and secular measures of control. A high priority for future research is to see whether dimensions of religion are more or less protective of health than their secular counterparts.

Any overview of the literature on religion, aging, and health is bound to fall short because the field is so broad. This is why we were unable to touch on many issues that have been raised by other investigators. A good deal has been left on the cutting room floor. Even so, we hope the discussion we have provided and the "next steps" we have woven throughout our review may be useful for those who are new to the field and old hands alike.

Acknowledgment

Work on this chapter was supported by two grants from the John Templeton Foundation.

References

Ai, A. L., Huang, B., Bjorck, J., & Appel, H. B. (2013). Religious attendance and major depression among Asian Americans from a national database: The mediation of social support. *Psychology of Religion and Spirituality, 5*, 78–89.

Ai, A. L., Peterson, C., Bolling, S. F., & Rogers, W. (2006). Depression, faith-based coping, and short-term postoperative global functioning in adult and older patients undergoing cardiac surgery. *Journal of Psychosomatic Research, 60*, 21–28.

Allicock, M., Resnicow, K., Hooten, E. G., & Campbell, M. K. (2013). Faith and health behavior: The role of the African American church in health promotion and disease prevention. In K. I. (2013). Pargament (Ed.), *APA handbook of psychology, religion, and spirituality* (Vol. 2, pp. 439–460). Washington, DC: American Psychological Association.

Argue, A., Johnson, D. R., & White, L. K. (1999). Age and religiosity: Evidence from a three-wave panel analysis. *Journal for the Scientific Study of Religion, 38*, 423–435.

Atchley, R. C. (1997). The subjective importance of being religious and its effect on health and morale 14 years later. *Journal of Aging Studies, 11*, 131–141.

Barna, G. (2002). *The state of the church 2002.* Ventura, CA: Issachar Resources.

Baruth, M., Wilcox, S., Hooker, S. P., Hussey, J. R., & Blair, S. N. (2013). Perceived environmental church support and physical activity among Black church members. *Health Education & Behavior, 40*, 712–720.

Benjamins, M. R. (2004). Religion and functional health among the elderly: Is there a relationship and is it constant? *Journal of Aging and Health, 16*, 355–374.

Benson, H., Dusek, J. A., Sherwood, J. B., Lam, P., Bethea, C. F., Carpenter, W., et al. (2006). Study of the therapeutic effects of intercessory prayer (STEP) in cardiac bypass patients. *American Heart Journal, 151*, 934–942.

Berger, P. L. (1967). *The sacred canopy: Elements of a sociological theory.* New York, NY: Doubleday.

Byrd, R. C. (1988). Positive therapeutic effects of intercessory prayer in a coronary care unit population. *Southern Medical Journal, 81*, 826–829.

Carstensen, L. L. (1992). Social and emotional patterns in adulthood: Support for socioemotional selectivity theory. *Psychology and Aging, 7*, 331–338.

Chatters, L. M., Bullard, K. M., Taylor, R. J., Woodward, A., Neighbors, H. W., & Jackson, J. S. (2008). Religious participation and DSM-IV disorders among older African Americans: Findings from the National Survey of American Life. *American Journal of Geriatric Psychiatry, 16*, 957–965.

Corsentino, E. A., Collins, N., Sachs-Ericsson, N., & Blazer, D. G. (2009). Religious attendance reduces cognitive decline among older women with high levels of depressive symptoms. *Journal of Gerontology, 64A*, 1283–1289.

Deaton, A. (2009). Aging, religion, and health. <http://www.nber.org/papers/w15271.pdf>.

Dulin, P. L. (2005). Social support as a moderator of the relationship between religious participation and psychological distress in a sample of community dwelling adults. *Mental Health, Religion & Culture, 8*, 81–86.

Ellison, C. G., & Lee, J. (2010). Spiritual struggles and psychological distress: Is there a dark side of religion? *Social Indicators Research, 98*, 501–517.

Enright, R. D., Freedman, S., & Rique, J. (1998). The psychology of interpersonal forgiveness. In R. D. Enright & J. North (Eds.), *Exploring forgiveness* (pp. 46–62). Madison, WI: University of Wisconsin Press.

Erikson, E. (1959). *Identity and the life cycle*. New York, NY: International University Press.

Fehr, R., Gelfand, M. J., & Nag, M. (2010). The road to forgiveness: A meta-analytic synthesis of its situational and dispositional correlates. *Psychological Bulletin, 136*, 894–914.

Fetzer Institute/National Institute on Aging Working Group, (1999). *Multidimensional measurement of religiousness/spirituality for use in health research*. Kalamazoo, MI: John E. Fetzer Institute.

Frankl, V. E. (1959/1985). *Man's search for meaning*. New York, NY: Washington Square Press.

Fry, P. S., & Debats, D. L. (2010). Sources of human strengths, resilience, and health. In P. S. Fry & C. L. Keyes (Eds.), *New frontiers in resilient aging: Life-strengths and well-being in late life* (pp. 15–59). New York, NY: Cambridge University Press.

Galek, K., Krause, N., Ellison, C. G., & Kudler, T. (2007). Religious doubt and mental health across the lifespan. *Journal of Adult Development, 14*, 16–25.

Galton, F. (1872). Statistical inquire into the efficacy of prayer. *Fortnightly Review, 12*, 124–135.

Hayward, R. D., & Krause, N. (2013a). Trajectories of late life change in God-mediated control: Evidence of compensation for declining personal control. *Journal of Gerontology: Psychological Sciences, 68B*, 49–58.

Hayward, R. D., & Krause, N. (2013b). Change in church-based social relationships during older adulthood. *Journal of Gerontology: Social Sciences, 68B*, 85–96.

Hayward, R. D., & Krause, N. (2013c). Trajectories of change in dimensions of forgiveness among older adults and their associations with religious commitment. *Mental Health, Religion & Culture, 16*, 643–659.

Hayward, R. D., & Krause, N. (2013d). Patterns of change in prayer activity, expectations, and content in older adulthood. *Journal for the Scientific Study of Religion, 52*, 17–34.

Hill, T. D., Angel, J. L., Ellison, C. G., & Angel, R. J. (2005). Religious attendance and mortality: An 8-year follow-up of older Mexican Americans. *Journal of Gerontology: Social Sciences, 60B*, S102–S109.

Homan, K. J., & Boyatzis, C. J. (2010). Religiosity, sense of meaning, and health behavior in older adults. *International Journal for the Psychology of Religion, 20*, 173–186.

Hood, R. W., Hill, P. C., & Spilka, B. (2009). *The psychology of religion: An empirical approach* (4th ed.). New York, NY: Guilford.

Hunter, B. D., & Merrill, R. M. (2013). Religious orientation and health among active older adults in the United States. *Journal of Religion and Health, 52*, 851–863.

Idler, E. L., & Kasl, S. V. (1997). Religion among disabled and nondisabled persons II: Attendance at religious services as a predictor of the course of disability. *Journal of Gerontology, 52B*, S306–S316.

James, W. (1902/1997). *Selected writings – William James*. New York, NY: Book-of-the-Month Club.

Koenig, H. G. (2008). Concerns about measuring 'spirituality' in research. *Journal of Nervous and Mental Disease, 196*, 349–355.

Koenig, H. G., King, D. E., & Carson, V. B. (2012). *Handbook of religion and health* (2nd ed.). New York, NY: Oxford University Press.

Koenig, H. G., Hays, J. C., George, L. K., Blazer, D. G., Larson, D. B., & Landerman, L. R. (1997). Modeling the cross-sectional relationships between religion, physical health, social support, and depressive symptoms. *American Journal of Geriatric Psychiatry, 5*, 131–144.

Krause, N. (1998). Neighborhood deterioration, religious coping, and changes in health during late life. *Gerontologist, 38*, 653–664.

Krause, N. (2003a). The social foundations of personal control in late life. In S. H. Zarit, L. I. Pearlin, & K. W. Schaie (Eds.), *Personal control in social and life course contexts* (pp. 45–70). New York, NY: Springer.

Krause, N. (2003b). Religious meaning and subjective well-being in late life. *Journal of Gerontology, 58B*, S160–S170.

Krause, N. (2003c). Praying for others, financial strain, and physical health status in late life. *Journal for the Scientific Study of Religion, 42*, 377–391.

Krause, N. (2003d). Exploring race differences in the relationship between social interaction with the clergy and feelings of self-worth in late life. *Sociology of Religion, 64*, 183–205.

Krause, N. (2005). God-mediated control and psychological well-being in late life. *Research on Aging, 27*, 136–164.

Krause, N. (2006a). Exploring the stress-buffering effects of church-based social support and secular social support on health in late life. *Journal of Gerontology: Social Sciences, 61B*, S35–S43.

Krause, N. (2006b). Church-based social support and mortality. *Journal of Gerontology: Social Sciences, 61B*, S140–S146.

Krause, N. (2007). Evaluating the stress-buffering function of meaning in life among older people. *Journal of Aging and Health, 19*, 792–812.

Krause, N. (2008a). *Aging in the church: How social relationships affect health*. West Conshohocken, PA: Templeton Foundation Press.

Krause, N. (2008b). The social foundations of religious meaning in life. *Research on Aging, 30*, 395–427.

Krause, N. (2009). Religious involvement, gratitude, and change in depressive symptoms over time. *International Journal for the Psychology of Religion, 19*, 155–172.

Krause, N. (2010a). God-mediated control beliefs and change in self-rated health. *International Journal for the Psychology of Religion, 20*, 267–287.

Krause, N. (2010b). The social milieu of the church and religious coping responses: A longitudinal investigation of older Whites and older Blacks. *International Journal for the Psychology of Religion, 20*, 109–129.

Krause, N. (2010c). Church-based social support and self-forgiveness in late life. *Review of Religious Research, 52*, 72–89.

Krause, N. (2012a). Studying forgiveness among older Whites, older Blacks, and older Mexican Americans. *Journal of Religion, Spirituality & Aging, 24*, 325–344.

Krause, N. (2012b). Assessing the prayer lives of older Whites, older Blacks, and older Mexican Americans. *International Journal for the Psychology of Religion, 22*, 60–78.

Krause, N. (2012c). Valuing the life experiences of older adults and change in depressive symptoms: Evaluating an overlooked benefit of involvement in religion. *Journal of Aging and Health, 24*, 227–249.

Krause, N. (in press a). Trust in God, forgiveness by God, and death anxiety. *Omega: Journal of Death and Dying*.

Krause, N. (in press b). Praying for others, living in run-down neighborhoods, and depressive symptoms among older Mexican Americans. *Review of Religious Research*.

Krause, N., & Bastida, E. (2011). Social relationships in the church during late life: Assessing differences between African Americans, Whites, and Mexican Americans. *Review of Religious Research, 53*, 41–63.

Krause, N., & Bastida, E. (2012). Religion and health among older Mexican Americans: Exploring the influence of making mandas. *Journal of Religion and Health, 51*, 812–824.

Krause, N., & Ellison, C. G. (2003). Forgiveness by God, forgiveness of others, and psychological well-being in late life. *Journal for the Scientific Study of Religion, 42*, 77–93.

Krause, N., & Ellison, C. G. (2009). The doubting process: A longitudinal study of the precipitants and consequences of religious doubt among older adults. *Journal for the Scientific Study of Religion, 48*, 293–312.

Krause, N., & Hayward, R. D. (2012a). Religion, meaning in life, and change in physical functioning during late adulthood. *Journal of Adult Development, 19*, 158–169.

Krause, N., & Hayward, R. D. (2012b). Negative interaction with fellow church members and depressive symptoms among older Mexican Americans. *Archive for the Psychology of Religion, 34*, 149–171.

Krause, N., & Hayward, R. D. (2012c). Social factors in the church and positive religious coping responses: Assessing differences between older Whites, older Blacks, and older Mexican Americans. *Review of Religious Research, 54*, 519–541.

Krause, N., & Hayward, R. D. (2013). Prayer beliefs and change in life satisfaction over time. *Journal of Religion and Health, 52*, 674–694.

Krause, N., & Hayward, R. D. (2014). Trust-based prayer expectancies and health among older Mexican Americans. *Journal of Religion and Health, 53*, 591–603.

Krause, N., & Hayward, R. D. (2015). Acts of contrition, forgiveness by God, and death anxiety among older Mexican Americans. *International Journal for the Psychology of Religion, 25*, 57–73.

Krause, N., Hayward, R. D., Bruce, D., & Woolever, C. (2013). Church involvement, spiritual growth, meaning in life, and health. *Archive for the Psychology of Religion, 35*, 169–191.

Krause, N., Ingersoll-Dayton, B., Ellison, C. G., & Wulff, K. M. (1999). Aging, religious doubt and psychological well-being. *The Gerontologist, 39*, 525–533.

Law, R. W. (2009). The effects of church attendance and marital status on the longitudinal trajectories of depressed mood among older adults. *Journal of Aging and Health, 21*, 803–823.

Law, R. W., & Sbarra, D. A. (2009). The effects of church attendance and marital status on the longitudinal trajectories of depressed mood among older adults. *Journal of Aging and Health, 21*, 803–823.

Lawler-Row, K. A. (2010). Forgiveness as a mediator of the religiosity-health relationship. *Psychology of Religion and Spirituality, 2*, 1–16.

Levin, J., & Prince, M. F. (2013). *Judaism and Health: A handbook of practical, professional, and scholarly resources*. Woodstock, VT: Jewish Lights Publishing.

Lundberg, C. D. (2010). *Unifying truths of the world's religions*. New Fairfield, CT: Heavenlight Press.

Maselko, J., Kubzansky, L., Kawachi, I., Seeman, T., & Berkman, L. (2007). Religious service attendance and allostatic load among high-functioning elderly. *Psychosomatic Medicine, 69*, 464–472.

McCullough, M. E., Enders, C. K., Brion, S. L., & Jain, A. R. (2005). The varieties of religious development in adulthood: A longitudinal investigation of religion and rational choice. *Journal of Personality and Social Psychology, 89*, 78–89.

McDonald, M. J., Wong, P. T., & Gingras, D. T. (2012). Meaning-in-life measures and the development of a brief version of the Personal Meaning Profile. In P. T. Wong (Ed.), *The quest for human meaning: Theories, research, and applications* (pp. 357–382, 2nd ed.). New York, NY: Routledge.

Mirowsky, J. (1995). Age and sense of control. *Social Psychology Quarterly, 58*, 31–43.

Morris, D. C. (1997). Health, finances, religious involvement, and life satisfaction of older adults. *Journal of Religious Gerontology, 10,* 3–17.

North, J. (1998). The ideal of forgiveness: A philosopher's exploration. In R. D. Enright & J. North (Eds.), *Exploring forgiveness* (pp. 15–34). Madison, WI: University of Wisconsin Press.

Norton, M. C., Singh, A., Skoog, I., Corcoran, C., Tschanz, J. T., Zandi, P. P., et al. (2008). Church attendance and new episodes of major depression in a community sample of older adults: The Cache County Study. *Journal of Gerontology, 63B,* P129–P137.

Oman, D., & Reed, D. (1998). Religion and mortality among the community-dwelling elderly. *American Journal of Public Health, 88,* 1469–1475.

Pargament, K. I., Koenig, H. G., & Perez, L. M. (2000). The many methods of religious coping: Development and initial validation of the RCOPE. *Journal of Clinical Psychology, 56,* 519–543.

Pargament, K. I., Smith, B. W., Koenig, H. G., & Perez, L. (1998). Patterns of positive and negative religious coping with major life stressors. *Journal for the Scientific Study of Religion, 37,* 710–724.

Poloma, M. M., & Gallup, G. H. (1991). *Varieties of prayer: A survey report.* Philadelphia, PA: Trinity Press International.

Reker, G. T. (1997). Personal meaning, optimism, and choice: Existential predictors of depression in community and institutional elderly. *The Gerontologist, 37,* 709–716.

Sasaki, J. Y., Heejung, S., & Xu, J. (2011). Religion and well-being: The moderating role of culture and the oxytocin (OXTR) gene. *Journal of Cross-Cultural Psychology, 42,* 1394–1405.

Schieman, S., Pudrovska, T., & Milkie, M. A. (2005). The sense of divine control and the self-concept. *Review of Religious Research, 27,* 165–196.

Schieman, S., Pudrovska, T., Pearlin, L. I., & Ellison, C. G. (2006). The sense of divine control and psychological distress: Variations across race and socioeconomic status. *Journal for the Scientific Study of Religion, 45,* 529–549.

Schmidt, J. F. (1951). Patterns of poor adjustment in old age. *American Journal of Sociology, 57,* 33–42.

Skinner, E. A., & Zimmer-Gembeck, M. J. (2011). Perceived control and the development of coping. In S. Folkman (Ed.), *The Oxford handbook of stress, health, and coping* (pp. 35–59). New York, NY: Oxford University Press.

Steiner, M., Allemand, M., & McCullough, M. E. (2011). Age differences in forgiveness: The role of transgression frequency and intensity. *Journal of Research in Personality, 45,* 670–678.

Sullivan, A. R. (2010). Mortality differentials and religion in the United States: Religious affiliation and attendance. *Journal for the Scientific Study of Religion, 49,* 740–753.

Sun, F., Oark, N. S., Roff, L. L., Klemmack, D. L., Parker, M., Koenig, H. G., et al. (2012). Predicting trajectories of depressive symptoms among southern community-dwelling older adults: The role of religiosity. *Aging and Mental Health, 16,* 189–198.

Tait, E. M., Laditka, S. B., Nies, M. A., & Racine, E. M. (2011). Praying for health by older adults in the United States: Differences by ethnicity, education, and income. *Journal of Religion, Spirituality & Aging, 23,* 338–362.

Taylor, R. J., Chatters, L. M., & Jackson, J. S. (2007). Religious and spiritual involvement among older African Americans, Caribbean Blacks, and Non-Hispanic Whites: Findings from the National Survey of American Life. *Journal of Gerontology, 52B,* S238–S250.

Toussaint, L. L., Williams, D. R., Musick, M. A., & Everson, S. A. (2001). Forgiveness and health: Age differences in a US probability sample. *Journal of Adult Development, 8,* 249–257.

Toussaint, L. L., Worthington, E., & Williams, D. R. (2014). *Forgiveness and health: Scientific evidence and theories relating forgiveness to better health.* New York, NY: Springer.

Weaver, G. D., & Gary, L. E. (1996). Correlates of health-related behaviors in older African American adults: Implications for health promotion. *Family & Community Health, 19,* 43–57.

Williams, D. R., John, D. A., Oyserman, D., Sonnega, J., Mohammed, S. A., & Jackson, J. S. (2012). Research on discrimination and health: An exploratory study of unresolved conceptual and measurement issues. *American Journal of Public Health, 102,* 975–978.

Wink, P., & Dillon, M. (2002). Spiritual development across the adult life course: Findings from a longitudinal study. *Journal of Adult Development, 9,* 79–94.

CHAPTER

13

Evolving Patterns of Work and Retirement

Kevin E. Cahill[1], Michael D. Giandrea[2], and Joseph F. Quinn[3]
[1]Sloan Center on Aging & Work at Boston College, Chestnut Hill, MA, USA [2]US Bureau of Labor Statistics, Office of Productivity and Technology, Washington, DC, USA [3]Department of Economics, Boston College, Chestnut Hill, MA, USA

OUTLINE

Introduction	271
The Beginning and End of Earlier and Earlier Retirement	273
A Closer Look at the Retirement Process in the Modern Era	275
Changes to the Traditional Pillars of Retirement Income and How They Relate to Labor Force Participation	277
The Increasing Importance of Macroeconomic Influences	282
The Potential Benefits of Continued Work Later in Life	284
Disclaimer and Acknowledgments	287
References	287

INTRODUCTION

The work and retirement patterns of older Americans have changed dramatically since the turn of the last century. From the early 1900s to the mid-1980s, the story of retirement among men in America was straightforward – steady declines in labor force participation at older ages (Burtless & Quinn, 2002; Purcell, 2009; Quinn, Burkhauser, & Myers, 1990). Growing prosperity permitted these earlier retirements, as Americans spent a portion of their newfound wealth on increased leisure later in life. Among older women, two trends – this early retirement trend and large increases in labor force participation after WWII – largely offset each other, leading to stable labor force participation rates between the 1960s and the mid-1980s (Purcell, 2009; Quinn, Cahill, & Giandrea, 2011). Since the mid-1980s, however, both older men and women have been working longer than prior trends would have predicted.

In addition to the timing of retirement, the pathways to labor force exit have evolved as well. The stereotypical view of retirement – a one-time transition from full-time work to complete labor force withdrawal – fails to capture the diversity of retirement patterns today (Cahill, Giandrea, & Quinn, 2006, 2012; Quinn, 2010; Shultz & Wang, 2011). Among those with career jobs later in life, the majority move to another job before leaving the labor force completely (Cahill, Giandrea, & Quinn, 2013c; Ruhm, 1990, 1991). In addition, a sizable minority reenter the labor force after an initial exit (Cahill, Giandrea, & Quinn, 2011a; Maestas, 2010). For most, retirement is not a one-time permanent event, but rather a process.

The factors behind these retirement decisions are many. Changes in economic incentives within the traditional pillars of retirement income – Social Security, private pensions, and savings – have altered the relative attractiveness of work and leisure later in life, almost always in favor of work (Burkhauser & Rovba, 2009; Gruber & Wise, 2004; Quinn et al., 2011). As a result, earnings have become a critical fourth leg of the retirement income stool (Munnell, 2007; Munnell & Sass, 2008). Recent evidence suggests that macroeconomic forces may also be playing a role among the Early Boomers, who experienced the recent Great Recession while on the cusp of retirement (Bosworth & Burtless, 2011; Butrica, Johnson, & Smith, 2011; Cahill et al., 2013c; Coile & Levine, 2011; Gorodnichenko, Song, & Stolyarov, 2013; Gustman, Steinmeier, & Tabatabai, 2011; McFall, 2011; Sass, Monk, & Haverstick, 2010).

All in all, continued work later in life among older Americans appears to be a good thing for many in light of the realities of our aging society. Individuals benefit from a more secure financial position, reducing the probability of reductions in living standards. Employers benefit from a larger pool of skilled workers with a lifetime of experience. The nation as a whole benefits as well, with more economically productive citizens. While continued work appears to be a good thing on balance, it is important to note that, for some, continued work may not be an option. Individuals with poor health and/or a history of intermittent, low-paying, and physically demanding jobs may simply not be able to continue working beyond traditional retirement ages. Their plight is of particular concern. For these older Americans at the lower end of the socioeconomic scale, few options may be available to counter earnings losses, resulting in lower living standards and the possibility of financial hardship at older ages.

As we highlight throughout this chapter, older Americans on the cusp of retirement today face a very different economic landscape than did prior cohorts and will need to adjust to these changing circumstances. Indeed, evidence suggests that the reversal of the early retirement trend has already been underway for nearly three decades, and that older workers remain productive (Burtless, 2013). Moreover, the aging of our society is only beginning (Board of Trustees of OASDI, 2014). When it comes to the retirement trends of older Americans, an old adage applies: the only constant is change.

The next section documents the beginning and decline of early retirement in America. The section "A Closer Look at the Retirement Process in the Modern Era" examines retirement in the modern era and older Americans' diverse patterns and determinants of labor force withdrawal. The next section documents changes to the traditional pillars of retirement income and how they relate to labor force participation. The section "The Increasing Importance of Macroeconomic Influences" describes recent research on the impact of macroeconomic influences on both the timing of retirement and retirement patterns. The final section looks ahead. With our aging population and increasing strains on our existing sources of retirement income, continued labor force participation later in life is the most likely and most promising outcome for many older Americans. Earnings are already a substantial income source for many and are likely to grow in importance.

THE BEGINNING AND END OF EARLIER AND EARLIER RETIREMENT

The retirement environment in the United States looks very different today than it did in the past. Older Americans today are responding to a new set of economic circumstances and incentives and are now working at rates not seen for nearly 30 years (Burkhauser & Rovba, 2009; Quinn et al., 2011). Just as the world of retirement today looks very different from what it was in the mid-1980s, the latter was very different from the world of retirement in the early part of the last century.

At the turn of the last century, the typical older American male worked as long as possible. The average retirement age among men in 1910 was 73 (Quinn et al., 2011), at a time when jobs were more physically demanding and life expectancy at birth was more than 20 years less than it is today (Arias, 2012). The reason most people worked later in life was fairly straightforward – to avoid poverty in a world without a financial safety net. As recently as 1959, the poverty rate among Americans aged 65 years and older was 35%, higher than any other age group (DeNavas-Walt, Proctor, & Smith, 2012).

These labor supply patterns late in life began to change in the early 1900s. Costa (1998, 1999) documented steady declines in labor force participation rates among older American men as far back as the late 1800s. Further, as private pensions became a reliable source of income for an increasing fraction of older American men in the mid-1900s, more people could leave the labor force without being thrust into poverty.

The Social Security Act of 1935 created the Old-Age and Survivors Insurance (OASI) program ("Social Security") (Perun & Dilley, 2011). Social Security provided insurance against poverty at older ages on a much larger scale than private pensions and enabled more older Americans to exit the labor force, even though their health might have permitted continued work. As America grew more prosperous, the OASDI program (which also provided disability insurance) expanded its coverage and employers began providing pension benefits more widely (Board of Trustees of OASDI, 2014; Congressional Budget Office, 2001; Sass, 1997). The availability of public and for many, private pensions, along with increases in other forms of wealth, such as housing and financial assets, allowed Americans to afford more leisure over their lifetimes, including earlier retirement (Burtless & Quinn, 2002).

In response to these changes, the labor force participation rates of older Americans declined precipitously through the mid-1980s (Purcell, 2009; Quinn, 1999). By 1990, the average retirement age of men – here, the youngest age at which one half of the population is out of the labor force – was just 63 years and reached a low of 62 in 1994 (Burtless & Quinn, 2002; Quinn et al., 2011). Many older Americans in the 1980s and 1990s could expect 20 or more years of retirement. Even as older Americans left the labor force earlier and earlier, their poverty rate declined dramatically, from 35% in 1959 to less than 15% in the mid-1980s, to 10% in 2005, and near 9% in 2012 (DeNavas-Walt et al., 2012).

The end of the era of progressively earlier retirement began with several important changes in the late 1970s and early 1980s. Mandatory retirement, which once covered about one half of the American workforce, was first delayed (from age 65 to 70 in 1978) and then eliminated for the vast majority of workers in 1986 (von Wachter, 2002). In 1983, Social Security amendments gradually raised the normal retirement age (NRA) to 66 and eventually to 67 for individuals born after 1959 (Congressional Budget Office, 2001). This increase in the NRA is equivalent to an across-the-board decline in lifetime Social Security benefits. In addition, Social Security's delayed retirement credit (DRC) had previously created a work disincentive (or retirement incentive) because increases in monthly benefits from postponing benefit receipt beyond the NRA were insufficient to compensate for the benefits foregone. The DRC was gradually increased from 3% to 8% for each year that benefits are postponed beyond the NRA, and now expected lifetime benefits remain more or less the same regardless of when benefits are first claimed between the ages of 62 and 70 (Munnell, 2013). This adjustment means that the work disincentive due to the DRC has been removed. Finally, the long-term financial outlook for Social Security may well require further reductions in benefits (perhaps through further increases in the NRA) and/or increases in the revenues that fund the program (Board of Trustees of OASDI, 2014; Congressional Budget Office, 2002, 2009; Lavery, 2009).

The 1980s also brought about a gradual shift in the nature of private pensions, and a move toward a "do-it-yourself" approach to retirement planning (Munnell, 2007). In the private sector, traditional defined-benefit (DB) plans, which provide lifetime annuities typically based on tenure with the firm and some measure of final salary, are being supplanted by defined-contribution (DC) plans, like 401(k)s (Copeland, 2009). While most traditional DB plans have age-specific work disincentives, usually at the earliest age of pension eligibility, DC plans are tax-deferred individual savings accounts with no such age-specific incentives (Munnell, 2006). In addition, individuals assume significant risks under DC plans, most notably investment risk and longevity risk, which are shouldered by employers under DB plans (Munnell & Sundén, 2004).

Private savings rates also began to decline in the 1980s and have since reached their lowest levels since the Great Depression, albeit with an uptick between 2008 and 2012, and then another decline (Federal Reserve Bank of St. Louis, 2014; US Department of Commerce, Bureau of Economic Analysis, 2013). In addition, fewer firms are providing post-retirement health benefits (Fronstin & Adams, 2012) and more employers are offering health plans with health savings accounts, which are tax-exempt accounts that can be used to pay health expenses (Fronstin, 2014). With these changes to Social Security, private pensions, and savings, along with potential cutbacks in Medicare and Medicaid generosity, many older Americans face a less attractive retirement environment than prior cohorts did, and many will face a choice between working longer or living poorer in retirement. For those unable to continue working, there may be little choice but to accept a lower standard of living at older ages.

The large influx of women into the labor force after WWII has had a substantial impact on retirement patterns. With labor force participation rates among older women now approaching those of older men (Quinn et al., 2011), and with more women's work histories including career employment, the labor force withdrawal patterns of older men and women are becoming more similar (Cahill, Giandrea, & Quinn, 2013a). Recently, however, the Great Recession and the subsequent sluggish recovery may have impacted older men and women differently, with the prevalence of bridge employment continuing to rise among women but declining slightly among men (Cahill et al., 2013a,c).

Retirement in the United States has changed over time, from a phenomenon enjoyed by the few at the beginning of the last century to the earlier and earlier retirements of most through the mid-1980s, and the reversal of that trend since then. The next section examines retirement in the modern era in more detail – when and how older Americans leave the labor force.

A CLOSER LOOK AT THE RETIREMENT PROCESS IN THE MODERN ERA

The stereotypical one-event view of retirement fails to capture the diversity of labor force withdrawal among older Americans today. First, even the term "retirement" is ambiguous and all the competing definitions have flaws. One straightforward and objective definition is complete labor force withdrawal (Cahill et al., 2006), but even this is inadequate when "retired" individuals later reenter the labor force.

Other researchers have based their definition of retirement on individuals' subjective assessments of their status. The flaw here is that two individuals with identical circumstances (e.g., working part time at age 68) could be classified differently, because one person views herself as retired (e.g., from a career job) while the other does not (still working). Other definitions have been based on the receipt of retirement benefits (Social Security or employer pension) or on declines in hours worked or earnings.

Given the numerous ways older Americans leave the labor force, assigning a point at which an individual is labeled "retired" is less useful than understanding the process of disengagement from the labor force. Researchers have identified three key categories of the retirement process (Kantarci & van Soest, 2008). One is phased retirement, in which an individual remains with the current employer but reduces the number of hours worked. Although more common in Europe, phased retirement is infrequent in the United States due to a combination of factors, including pension incentives (e.g., benefits based on the last several years of earnings), regulatory barriers, and resistance from employers (Hutchens & Chen, 2007; Hutchens & Grace-Martin, 2006; Johnson, 2011).

A second category is partial retirement which involves a change in employer. Bridge jobs – those that follow full-time career employment and precede complete labor force withdrawal – are very common in the United States (Giandrea, Cahill, & Quinn, 2009; Quinn, 1999; Ruhm, 1990, 1991; Shultz & Wang, 2011; Wang, Adams, Beehr, & Shultz, 2009). Approximately 60% of older Americans leave a career job for a bridge job prior to leaving the labor force completely, with approximately one half of these bridge jobs being part-time (Cahill et al., 2013c). A third though less common type of transition is labor market reentry or "unretirement" (Maestas, 2010). Recent estimates suggest that around 15% of older career workers who leave the labor force for at least 2 years later return to paid work (Cahill et al., 2011a).

The diversity of the retirement process for older Americans extends beyond these three categories. Many switch between wage-and-salary employment and self-employment, and the prevalence of self-employment rises steadily with age (Cahill, Giandrea, & Quinn, 2013b; Hipple, 2004; Karoly & Zissimopoulos, 2004; Zissimopoulos & Karoly, 2007). Among older Americans with career jobs in their fifties, approximately 20% of men and 10% of women are self-employed, but these numbers nearly double among those who remain working into their late sixties and seventies (Cahill et al., 2013b; Zissimopoulos & Karoly, 2007).

Self-employment rises with age for two reasons. First, the self-employed generally remain in the labor force longer than wage-and-salary workers do (Cahill et al., 2013b). One reason is hours flexibility. While many wage-and-salary workers are unable to reduce hours worked, sometimes promoting exits from their jobs and often from the labor force, self-employed individuals may have more control over their work hours (Bruce, Holtz-Eakin, & Quinn, 2000; Dunn & Holtz-Eakin, 2000; Evans & Jovanovic, 1989; Holtz-Eakin, Joulfaian, & Rosen, 1994a,b).

A second reason that self-employment rises with age is that many more career wage-and-salary individuals switch into self-employment later in life than vice versa. Approximately 10% of older Americans with wage-and-salary career jobs move to self-employment prior to leaving the labor force (Cahill et al., 2013b; Zissimopoulos & Karoly, 2009). In addition to choices based on hours flexibility, layoffs during economic downturns could push workers into self-employment. Another factor may be increased access to capital with age, both via accumulated personal savings and life experiences that improve the chances of success in self-employment (Holtz-Eakin et al., 1994a,b).

Transitions also take place in the other direction. Among older Americans who are self-employed on their career jobs, more than 25% switch into wage-and-salary bridge jobs. Although this percentage is higher than the reverse transition, there is a net increase in self-employment because of the much larger number of career wage-and-salary workers (Cahill et al., 2013b; Zissimopoulos & Karoly, 2007).

Transitions into self-employment have continued during the Great Recession and its aftermath, and recent evidence suggests that, among the Early Baby Boomers, transitions away from career self-employment to wage-and-salary bridge work later in life have declined (Cahill et al., 2013c). The reason might not be the appeal of self-employment per se, but rather the limited opportunities in wage-and-salary employment during the slow economic recovery.

Another important characteristic of retirement transitions pertains to occupational changes. For some, bridge employment late in life stems from financial necessity. For others, continued work is an opportunity to venture into a different line of work, referred to as "re-careering" (Johnson, Kawachi, & Lewis, 2009). One study found that among older American men in highly skilled white-collar career jobs, 14% moved into work that is not highly skilled (though still white collar) and another 17% transitioned into blue-collar occupations (Cahill, Giandrea, & Quinn, 2011b). Some older Americans in blue-collar jobs switch into white-collar work for some period of time before leaving the labor force. For example, among older Americans in highly skilled blue-collar, career jobs, Cahill et al. (2011b) found that 14% moved to white-collar bridge jobs. The fact that these transitions can occur later in life is a testament to the flexibility of older American workers and the labor markets they face.

Several decades ago, economists first made the case that, while health status is a significant determinant of retirement, the financial incentives that exist in Social Security and pension plans are also important (Quinn et al., 1990). After this insight, and considerable empirical evidence supporting it, health status received less attention. For a substantial portion of the workforce, health status was no longer a barrier to continued work later in life – at least not at traditional retirement ages. Trends in longevity help illustrate this point. In 1950 average life expectancy at age 65 was about 13 years for men and 15 years for women. By 2010, this life expectancy was nearly 18 years for men and over 20 years for women – an improvement of about one-third over only six decades (National Center for Health Statistics, 2013). As life expectancy increased, health status at traditional retirement ages improved as well (National Center for Health Statistics, 2013). Therefore, while poor health remains a statistically significant and important predictor of labor force exit (Conley & Thompson, 2013), a smaller proportion of older Americans fall into this category at any given age. In this way, health status has played a role in the retirement decisions of a smaller proportion of older Americans than it did in the early and mid-1900s.

Not only has the health status of older Americans improved, but work itself has evolved away from physically demanding jobs. The fraction of the US workforce in physically demanding occupations declined from 20% in 1950 to less than 8% in 1996 (Munnell, Cahill, Eschtruth, & Sass, 2004; Penner, Perun, & Steuerle, 2002; Steuerle, Spiro, & Johnson, 1999). In addition, the most recent (2010–2020) Bureau of Labor Statistics projections for the largest growth sectors of the workforce suggest that this trend will continue, with job growth strongest for the health care and social assistance sectors and professional and business services (Sommers & Franklin, 2012; Toossi, 2012). These two observations – that older Americans today are healthier than their counterparts decades ago and that physically demanding jobs make up a smaller fraction of all occupations – imply that continued work later in life could be a viable option for most older Americans.

The diversity of the retirement process, including transitions to bridge employment and reentry, and its value in promoting continued work later in life is predicated on a worker's ability to remain in the labor force. While a smaller fraction of the workforce in recent years compared with the past, those in poor health and in physically demanding occupations remain a vulnerable group, especially in light of changes to traditional retirement income sources.

Most research on retirement has been from the supply side, focusing on the preferences and labor market decisions of older workers. Until recently, the demand side of the market (i.e., the behavior of employers hiring older workers) received relatively little attention (Adler & Hilber, 2009; Blau & Shvydko, 2011). One reason is practical; while it is straightforward to gather information about labor supply decisions (and several excellent longitudinal surveys have done so), it is more difficult to gather information about labor demand decisions. In addition, labor demand was also less of an issue in the nearly two decades preceding 2008 because unemployment was comfortably low and job opportunities were available.

Labor demand became much more important when the unemployment rate spiked toward double digits during the Great Recession that began in December 2007 (Burtless & Looney, 2012; Elsby, Hobijn, Sahin, & Valletta, 2011). Unemployment for Americans aged 55 and older increased from 3% in November 2007 and reached a high of nearly 7% in August of 2010, and this excluded discouraged workers who were no longer searching for jobs. Unemployment among older Americans then remained near or above 5% through 2013 (US Bureau of Labor Statistics, 2014a). It was no longer the case that an older American who wanted to continue working could count on doing so. Further, for the older Americans who were officially unemployed, the average length of unemployment increased. The fraction of jobseekers aged 55 and older who were considered long-term unemployed – out of work for 27 weeks or more – increased from 23% in December 2007 to 49% in December 2012, and remained high at 44% as of January 2014 (Rix, 2013a,b, 2014). The average duration of unemployment among older workers in early 2014 – 44 weeks – far exceeded the average duration of 31 weeks among workers less than 55 years of age (Rix, 2014).

While the less physically intensive nature of jobs has meant more opportunities for older workers generally, weak labor markets have constrained some who want to continue working later in life. Moreover, some may be "locked" into career jobs, inhibiting the gradual retirements via bridge jobs that were seen among prior cohorts. The retirement patterns of those in their fifties and sixties today are evolving as older Americans adapt to different retirement incentives and changing macroeconomic circumstances.

CHANGES TO THE TRADITIONAL PILLARS OF RETIREMENT INCOME AND HOW THEY RELATE TO LABOR FORCE PARTICIPATION

While much of this chapter focuses on the many incentives encouraging work among older Americans today, it is important to understand that, prior to the mid-1980s, economic incentives within Social Security and private pensions, and public policies like mandatory retirement successfully encouraged earlier exits from the labor force (Burtless & Quinn, 2002; Gruber & Wise, 2004; Quinn, 1999). These incentives were widely considered to be positive, an expansion of the "golden years."

As noted above, at the start of the last century many worked as long as they could in order to avoid poverty (Costa, 1998, 1999; Sass, 1997). The Social Security Act of 1935 was designed to address this challenge, according to President Roosevelt, to protect older Americans against the "hazards and vicissitudes of old age" (Perun & Dilley, 2011). The linchpin of protection was a mandatory (for most) public DB plan that would allow older Americans to leave the labor force with some financial security.

The Social Security program was born out of a true social need to alleviate poverty in old age and the hardship of continued work that many older Americans endured (Perun & Dilley, 2011). Over the years Social Security became more inclusive and more generous and provided strong incentives to exit the labor force among those who might have otherwise chosen to continue working (Gruber & Wise, 2004).

A detailed overview of the Social Security program is beyond the scope of this chapter, so we instead focus on some key changes over time that pertain to work later in life. Social Security benefits were first available to individuals starting at age 65 using a progressive benefit formula based on a measure of average indexed monthly earnings over one's highest paid 35 years of work. Social Security benefits became available for spouses as a result of amendments passed in 1939, and because of age differences within the typical married couple, benefits were available at age 62 for wives of married men. In 1956, female workers were also able to claim actuarially reduced retirement benefits at age 62 and, beginning in 1961, men could to do so as well (Abbott, 1974). The percentage of workers covered by Social Security increased during the 1950s from 61% to 86%, and Congress passed benefit increases four times during the decade, more than offsetting inflation (Martin & Weaver, 2005). For a nearly 50-year period, from 1935 to 1983, the Social Security program expanded and increased benefits to older Americans and eliminated the once large gap between the poverty rates of older Americans and those in other age groups (Martin & Weaver, 2005).

The expansion of the Social Security program was costly and reforms were needed to bring revenues and expenditures into balance. The Social Security Amendments of 1983 increased payroll taxes and lowered benefits across-the-board by gradually raising the age at which full retirement benefits could be claimed – the NRA – from age 65 to 67. The increase in the NRA began with individuals born in 1938, and increased by 2 months per year until it reached age 66 for those born in 1943. The NRA for individuals born in 1955 is 66 and 2 months, and will increase 2 months per year until the NRA reaches 67 for those born after 1959 (Board of Trustees of OASDI, 2014; Martin & Weaver, 2005).

The second leg of the traditional retirement income stool is private pensions, whose incentives have also changed over time. Employer-provided pensions became widespread in the mid-1900s as a form of delayed compensation for workers with long-term commitments to their organizations (Sass, 1997; Seburn, 1991), with benefits typically based on years of service and some measure of the individual's final salary. These DB-type plans were the norm in both the public and private sectors – among the roughly 50% of workers with a pension – until the 1980s. Beginning in 1978, with a change in the Internal Revenue Taxation Code (subsection 401(k)), private pensions began a 30-year shift away from DB plans to DC plans (Copeland, 2009; Gale, Papke, & VanDerhei, 2005; Munnell & Sundén, 2004). DC plans are essentially tax-deferred savings accounts, often with an accompanying employer match. In 401(k) plans, individuals decide how much to contribute, how to invest their assets, and when and how quickly to spend them.

The speed of the shift away from DB to DC plans is staggering. According to the Federal Reserve's Survey of Consumer Finances, in 1983 more than 85% of workers with pension coverage had either a DB plan only (62%) or a DB plan plus some other plan (26%) (Munnell & Perun, 2006). Within 20 years, just 26% of covered workers had a DB plan only and 56% had a DC plan only (Copeland, 2009). Most recently, those with a DB plan only represented just 7% of private sector participants in an employment-based retirement plan and those with a DC plan only represented 69% (Employee Benefit Research Institute, 2014).

The shift from DB to DC plans is highly relevant to the retirement decisions of older Americans because incentives for continued work in old age are very different across plan types. In DB plans, individuals become eligible to receive benefits at a particular age, and can calculate how their stream of expected retirement benefits would change with each additional year of work. The cost of not claiming pension benefits in that year is the value of annual pension benefits foregone. The benefit is another year of service and higher annual benefits once they are claimed. Typically, the increase in future benefits is insufficient to compensate for the benefits foregone, resulting in a loss of expected lifetime benefits because of the additional year of work. This is equivalent to a pay cut and therefore is a work disincentive.

In DC plans, there are no such explicit incentives with respect to the timing of retirement, but other factors, like investment risk and longevity risk, do influence workers. Individuals must choose how to invest their assets, and assume the risk that comes with doing so. In contrast, employers assume investment risk under DB plans. Longevity risk presents the possibility of outliving one's assets. Unlike the lifetime annuities in DB plans, DC plans provide an asset that can be completely depleted. Both investment and longevity risks provide individuals with an incentive to work later in life in order to provide a buffer in case retirement assets fall due to market fluctuations and to insure against outliving their assets. Research has shown that, all else being equal, individuals with DC plans do indeed retire about 1–2 years later than those with DB coverage (Friedberg & Webb, 2005; Munnell, Cahill, & Jivan, 2003).

While private pensions are a key leg of the retirement income stool, it is important to keep in mind that, even in the world of DB plans that existed decades ago, only about one half of workers actually had a pension in their current job (Munnell & Perun, 2006). Retirees without pension coverage had to rely on Social Security and the third leg of the traditional retirement income stool – savings. Along with changes in Social Security and private pensions, there have been changes in the savings behavior of older Americans and, notably, not for the better.

The savings rate – savings as a percentage of personal disposable income – of Americans was centered around 11% for the three decades leading up to the mid-1980s (US Department of Commerce, Bureau of Economic Analysis, 2013). However, in the 1990s, savings rates began to decline steadily until reaching a low of 2.6% in 2005 – the lowest rate since the 1930s. Over the past decade, rates rose to about 6% in 2009, but then, following the recent recession and the subsequent sluggish recovery, they have fallen again and remain more than 5 percentage points below the average in the three decades preceding the 1990s.

Low savings rates translate into low levels of accumulated assets over time. A recent Retirement Confidence Survey found that among Americans age 45 and older, more than one half had less than $25 000 in financial assets – far below what is needed to finance many years of retirement (Helman, Adams, Copeland, & VanDerhei, 2014).

With such low levels of savings, what will sustain older Americans in retirement in the coming decades? One way to address this question is to examine the current income sources of today's older Americans. In 2012, Social Security provided 35% of the aggregate income of Americans aged 65 and over, with earnings another 34%, pensions 17%, and income from assets an additional 11% (US Social Security Administration, 2014a). But the relative importance of these sources differs dramatically by income level. In the top quintile, earnings provide 50% of aggregate income and Social Security only 16%. In the bottom quintile, Social Security is the overwhelmingly most important source, responsible for 83% of all their income, with only 3% coming from earnings. In fact, Social Security accounts for more than 80% of income for those in the second quintile as well, meaning that the lowest 40% of all aged units (married couples and non-married persons) rely on Social Security for 80% or more of their income. The Social Security program has been a critical component in keeping older Americans out of poverty, and all signs are that it will continue to serve this role in the years ahead, especially among those unable to remain working later in life.

As noted above, one way to help alleviate the strains of an aging society is through continued work, and the role of earnings later in life is apparent when one examines earnings as a percentage of total aggregate income. Among those aged 65–69 years, earnings account for more than half (55%) of aggregate income among married couples and 40% for non-married persons (US Social Security Administration, 2014a). Earnings make up more than 20% of aggregate income even among those aged 70–74 years. These percentages are much higher than those in the past, as the role of earnings has increased substantially over time. For example, the percentage of aggregate income of Americans aged 65–69 provided by earnings was just 32% in 1975 (McDonnell, 2010).

In addition to being a solid fourth leg of the retirement income stool for those who continue working, earnings can also serve as a safety net if planned retirement income falls short of expectations. Targets, such as income equal to 80% of pre-retirement income (i.e., an 80% replacement rate), can influence when one chooses to retire – or, equivalently, how long one needs to work. DB pensions and Social Security fit this view of planning, because they provide a predictable stream of benefits. DC plans can do the same, but only if they are converted into an annuity, something which few individuals do (Munnell & Sundén, 2004).

As we discuss in the next section, older Americans have become increasingly exposed to financial market risks over the past several decades. For many, work decisions may be based on an individual's best guess of what their retirement income will be in the years ahead. If retirement income falls short, a return to paid work can make up some or all of the difference.

This concept of earnings as a safety net has been identified as an important part of retirement transitions. In a study of gradual retirement, Maestas (2010) found that, even before leaving career employment, the majority of workers viewed a return to paid work as an option if actual retirement income did not meet expectations. This view of retirement once again differs from that of a one-time, permanent exit from the labor force. Instead, initial labor force withdrawal, for many, is a wait-and-see stage, based on realizations of retirement income sources and a host of non-pecuniary factors.

For some, this strategy is voluntary, and if successful is not a concern from a public policy perspective. But others who exit the workforce prematurely with insufficient resources may not have the option of re-employment, perhaps due to declines in health status, depreciated skills, or a weak labor market. Others may find work, but at wages substantially lower than what they once earned. Those who viewed earnings as a backup, but who cannot find adequate employment, can be left in a very vulnerable position at older ages.

Health insurance also provides important labor supply incentives later in life. Retiree health insurance – employer-provided plans that continue into retirement – allows retirees to leave a job without losing their coverage (Karoly & Rogowski, 1994). These policies, however, are becoming a rarity (Fronstin & Adams, 2012). The percentage of private sector workers with access to employer-provided retiree health insurance declined from 29% in 1997 to 18% in 2011 (Fronstin & Adams, 2012).

Employer-provided health insurance on the job is also an important determinant of a worker's decision whether to remain with a current employer (Robinson & Clark, 2010; Rogowski & Karoly, 2000). Some employees who would otherwise leave may stay with an employer to maintain health insurance (job-lock) while others without coverage may move to a new employer who does offer health insurance (job-push) (Madrian, 1994; Rogowski & Karoly, 2000).

The availability of health insurance encourages continued employment on career jobs. Blau and Gilleskie (2008) suggest that as many as 15% of older men stayed on the job longer due to lack of health insurance in retirement. The authors also observed that older men in poor health doubled their rate of exiting the labor force when post-retirement health insurance was available, while those in good health increased their rate by one-third. Pang, Warshawsky, and Weitzer (2008) found that older workers are about 8% more likely to retire if continued health insurance is provided by the employer or through the spouse's employer. Similarly, Boyle and Lahey (2010) found that VA medical coverage offered to veterans outside of employment decreased their probability of continuing work by 3%. In another study, French and Jones (2011) found that workers with retiree health benefits retire a half-year earlier than similar workers without such benefits.

Eligibility for Medicare benefits at age 65 permits workers to leave an employer (and the labor force) while maintaining health care coverage or to take a bridge job which does not provide health insurance. The net impact of the Affordable Care Act (ACA) on the retirement transitions of older Americans is uncertain. It will facilitate transitions from career jobs to bridge employment, and will also expedite labor force withdrawal. Many of the details about the ACA are still in flux and the impact of the ACA on retirement patterns will be a fruitful area of research for years to come.

Another topic that has received increased attention in recent years is the impact of disability programs on the retirement decisions of older Americans. The purpose of the Disability Insurance (DI) component of OASDI is to provide insurance against the inability to work due to injury or other qualifying reasons. Researchers have noted that DI might serve as an alternative route to early retirement (Autor, 2011; Burkhauser, Daly, McVicar, & Wilkins, 2014; Burkhauser & Rovba, 2009; Coile, in press; Maestas & Li, 2008; Milligan & Wise, 2011). Further, DI benefits are typically higher and are more permanent than unemployment insurance benefits, providing an economic incentive to switch from unemployment rolls to disability rolls for those who have been unemployed for extended periods of time. Research suggests that these incentives are important as the number of Americans claiming disability benefits has increased from 455 000 in 1960 to 3 million in 1990, and after the Great Recession, to nearly 9 million in 2012, a record high (Board of Trustees of OASDI, 2014).

The influence of the macroeconomy on labor supply later in life has recently received increased attention. New studies suggest that macroeconomic conditions have only a modest impact on the timing of retirement but may have a larger influence on retirement patterns, such as the prevalence of bridge jobs and labor market reentry (Cahill et al., 2013c).

THE INCREASING IMPORTANCE OF MACROECONOMIC INFLUENCES

The combination of expected reductions in Social Security replacement rates and the switch from DB to DC pension plans in the private sector means that older Americans today are in charge of the financial planning of retirement more than retirees in the past (Cahill, Giandrea, & Quinn, 2008). This has been labeled the "do-it-yourself" approach to retirement income planning (Munnell, Meme, Jivan, & Cahill, 2004). One consequence is that more older Americans are exposed to financial market fluctuations. If the market experiences a sudden downtown – as it did in the early 2000s and more dramatically in 2008 – those on the cusp of retirement could experience a sharp decline in the value of their retirement assets.

Asset values had minimal impact on most older Americans prior to the 1980s. Social Security retirement benefits were independent of the state of the financial market. For those with employer pensions, benefits were also independent of market forces, as employers assumed the investment risk, and benefits paid until death protected individuals from longevity risk. Individual savings were exposed to market fluctuations, but most retirees had little accumulation of financial assets. In summary, prior to the 1980s, older Americans' financial well-being in retirement was relatively predictable and largely independent of the state of asset markets.

As with many aspects of the retirement process, this landscape has evolved. One study suggests that long-term participation in a 401(k) plan, combined with Social Security retirement benefits, can provide sufficient income to replace 60% of pre-retirement income at age 64 (VanDerhei, 2014). For older workers who are not prescient about how to invest their 401(k) balances, however, a sharp reduction in retirement assets is possible, which can impact when and how individuals retire. As noted above, research has shown that the shift from DB to DC plans has led individuals to retire later than they otherwise would have (Friedberg & Webb, 2005; Munnell et al., 2003). Some studies that have looked at the direct impact of market declines on the timing of retirement show that the impact has been modest (Coile & Levine, 2006; Gustman et al., 2011). Other studies have demonstrated that market declines can alter the way individuals exit the labor force, for example, by truncating career employment and promoting job changes later in life (Cahill et al., 2013c).

Coile and Levine (2006) hypothesized that those with greater stock market holdings should have larger responses to the stock market boom – exiting the labor force at a higher rate than non-stockholders in the cohort – and larger responses to the stock market bust of 2000 and 2001 – delaying retirement or reentering the labor force at a higher rate than others in the cohort. Coile and Levine found that equity holdings were not very large among respondents in the Health and Retirement Study (HRS) – a large longitudinal nationally representative dataset of older Americans (Karp, 2007). In 2000, only about two-thirds owned stocks and of those who did, over 60% had holdings worth less than $50 000. The authors found no support for the hypothesis that those who owned equities were more likely to retire during the boom or less likely to retire (or more likely to reenter) during the bust.

The effect of the macroeconomy has been revisited since the Great Recession, with an emphasis on housing. Gustman et al. (2011) used the first nine biennial waves of the HRS (1992 through 2008) to examine differences across three cohorts of older Americans. Regarding labor force exit, the authors classified HRS respondents as not retired, partially retired, completely retired, not relevant, or not working but not retired. This last category included those officially unemployed and also discouraged workers – those not working or searching but willing to take employment if offered. The largest differences across cohorts were the changes in the unemployed and discouraged groups. The percentage of HRS respondents in these two groups increased during recessionary times and decreased during economic expansion. However, despite significant fluctuations in unemployment, the authors concluded that transitions into retirement were not accelerated as a result of macroeconomic and housing market declines.

Sass et al. (2010) noted that the stock market crash of 2008 and 2009 led to an average reduction of one-third in the values of 401(k) plans. To learn more about how this affected older Americans, the Center for Retirement Research at Boston College surveyed a nationally representative sample of over 1 300 workers between the ages of 45 and 59 during the summer of 2009. Only about 40% of workers reported that they expected to retire later than they had previously planned, because of the stock market crash. The authors suggested that the reason may be that most workers did not have enough financial assets to be significantly affected by the stock market downturn, which is consistent with the findings from Employee Benefit Research Institute's Retirement Confidence Survey (Helman et al., 2014).

McFall (2011) estimated the twin effects of decreased financial wealth and decreased housing values on expected retirement dates among a sample of over 300 older Americans interviewed in 2008 and 2009. McFall calculated the sustainable consumption level for each individual based on an annuity that could be purchased using the individual's expected stream of wages, Social Security, and DB pension income, and current financial and real estate wealth. The author estimated that average sustainable consumption between 2008 and 2009 fell about 6%, and this led to an increase in expected retirement age of only 2.5 months – a modest change.

Likewise, Bosworth and Burtless (2011) investigated the relationship between labor force participation rates and changes in home prices, state and national unemployment rates, and wealth. The authors also found only modest effects of changes in asset values on the labor force participation of older men. They estimated that as the unemployment rate during the Great Recession rose by 4.6 percentage points, the labor force participation rate for men aged 60–74 dropped by between 1.3 and 1.7 percentage points. They concluded that business cycles do influence, in the expected direction, both labor force participation rates and decisions to claim Social Security benefits, but that the effects are modest in size.

Other studies have focused on the impact of the economy on older workers' wages and the timing of Social Security benefit receipt. Butrica et al. (2011) used the Urban Institute's Dynamic Simulation of Income Model to estimate the impact of the Great Recession on the retirement incomes of workers in the labor force in 2008. The authors found that because of poor wage growth during and after the recession, annual incomes of workers when they reach age 70 will be reduced by 4.3% (about $2 300 per year). They found that the youngest workers were hardest hit because they were more likely to lose jobs, but older workers suffered as well. As the authors note, older workers who lose jobs have more difficulty becoming re-employed. This leads many to claim Social Security retirement benefits at an earlier age than they otherwise would have, resulting in a stream of permanently lower Social Security benefits.

Coile and Levine (2011) estimated the impact of a high unemployment rate on labor force participation and Social Security claims among older men, using 30 years of data from the March supplement of the Current Population Survey. They found that higher unemployment does increase the probability of men age 60–69 exiting the labor force, especially among men age 62 and older without any college education.

For example, for men age 62–64, a one percentage point increase in the unemployment rate led to a 1.1 percentage point increase in the number who exit the labor force and a 0.9 percentage point increase in the number who receive Social Security retirement benefits. For men 65–69, the analogous responses are 1.2 and 0.2 percentage points.

A couple of recent studies suggest that the state of the macroeconomy affects patterns of labor force withdrawal. Cahill et al. (2013c) found that the Great Recession has resulted in a higher frequency of involuntary transitions from career employment, more full-time bridge jobs, and an increase in bridge job activity among older career women. Using data from the Social Security Administration from 1960 to 2010 on white males ages 55–75, Gorodnichenko et al. (2013) found that a one percentage point increase in the unemployment rate was associated with a 7% increase in the number of men moving from full-time work to bridge jobs. The authors also found a positive relationship between inflation and partial retirement. So, while macroeconomic fluctuations might not greatly alter the timing of retirement among older Americans, they do appear to influence the ways in which older Americans leave the labor force.

Several factors suggest that macroeconomic volatility is likely to persist in the years ahead. While the economy has added jobs in recent years, new jobs have just kept pace with population growth. The fraction of the population that is working – the employment rate – has remained relatively unchanged, fluctuating between 58.2% and 58.9% between September 2009 and May 2014 (US Bureau of Labor Statistics, 2014b). In addition, the Federal Reserve has attempted to revive economic growth and reduce unemployment through a reduction in long-term interest rates. The Federal Reserve has done so through an unprecedented policy of direct purchases of long-term US government bonds and mortgage-backed securities. The manner in which the Federal Reserve sells these assets when the economy begins to recover will be a source of unpredictability in the markets, completely without precedent given the magnitude of the Federal Reserve's holdings (Bullard, 2013). This potential for increased macroeconomic volatility is yet a further sign that retirement patterns will continue to evolve in the years ahead.

THE POTENTIAL BENEFITS OF CONTINUED WORK LATER IN LIFE

The world of work and retirement changed considerably over the past 30 years. Mandatory retirement has been outlawed for the vast majority of American workers. The incentives within Social Security and employer pensions have changed in favor of continued work later in life. Social Security's NRA has increased from 65 to 66 and will increase further to 67 for those born after 1959. The earnings test – the amount that benefits are reduced for Social Security recipients with earnings – no longer applies for those working beyond the NRA and the amount that benefits are increased in return for delayed receipt (the DRC) is age neutral – meaning that, for a typical individual, one's expected lifetime benefits are about the same regardless of when they are initially claimed (Gruber & Orszag, 2003; Munnell, 2013). The decline of DB plans in the private sector means that fewer workers face incentives to claim benefits at a certain age. Within DC plans, workers have an incentive to remain working longer to insure against both longevity risk and investment risk, and to accumulate financial assets to make up for decades of low savings.

These changes provide a host of incentives that encourage work later in life. Not surprisingly, older Americans have been responding accordingly. The trend toward earlier and earlier retirements is over and has been for three decades. Labor force participation rates among older Americans are on the rise even in the face of the Great Recession and the ensuing sluggish recovery.

Is continued work later in life a change for the better? For society as a whole, the answer seems to be yes, in light of the challenges of our aging population. For individuals, the financial benefit of each additional year of continued work later in life is two-fold: an additional year to earn income and accumulate assets and one less year of retirement to be financed. Employers benefit from a larger pool of workers with a lifetime of experience. The nation benefits as fewer people rely on public programs and more goods and services are produced to be distributed across the population. As noted at various points in this chapter, continued work may not be an option for those in poor health at older ages or in physically demanding jobs. Moreover, those who are unable to continue working and who have had a lifetime of low-paid work – and, therefore, limited opportunities to accumulate substantial savings – will be particularly vulnerable in the face of reductions in traditional retirement income sources. Continued work later in life, while a valuable option for society generally, will do little to improve the financial outlook for some of the most vulnerable.

For younger workers, does the continued work of older Americans crowd out jobs that would have otherwise been available for them? The idea that older workers necessarily crowd out younger ones is based on a stagnant view of the economy, where the number of jobs is assumed to be fixed. In this view, each older worker who steps aside frees up a job, to be filled by the next-best experienced worker. This then frees up one for someone else, and the process continues until, finally, a job is available for the least experienced (younger) worker. This argument sounds logical, but is flawed. In reality, the number of jobs is not fixed. If they were, as Jonathan Gruber and his colleagues have pointed out, the large influx of women into the labor force post-WWII would have resulted in massive increases in unemployment among men (Gruber, Milligan, & Wise, 2009). This did not happen. Instead the US economy experienced an unprecedented boom with jobs available for almost anyone. While many questions exist with respect to how to expand the economy and reduce unemployment, one solution that should be rejected is to constrain the labor supply options of older workers.

Further, consider the alternative to the continued employment of older workers. A large number of older workers exiting the labor force in a short period of time would likely lead to skill shortages, in at least the near term, and result in a large-scale swing from net producers to net consumers. For some forced into retirement, their nest egg would be insufficient to finance their retirement years, resulting in a reduction in their standard of living. This means lower levels of consumption, slower economic growth, and fewer jobs. As these retirees age with limited financial resources, those fortunate to live long lives would have to rely on social programs for support – programs financed by the younger generation.

Looking beyond the near term, is continued work among older Americans likely to be a temporary phenomenon? Should we expect that, after the Baby Boomers, we will return to the world of retirement observed between WWII and the 1980s? All signs suggest that the world of work later in life is now fundamentally different from the past. A return to earlier patterns in which most Americans could expect 20 or more years of leisure later in life with little or no reduction in standard of living is unlikely.

The reason that recent retirement trends are unlikely to reverse is the permanent nature of the changes to key retirement determinants that have ushered in these changes. Mandatory retirement will not return, nor will DB plans in the private sector. Under current law, the Social Security NRA will increase to 67 for those born after 1959, creating a reduction in lifetime benefits. Policymakers have considered raising the NRA even further and perhaps benchmarking it to life expectancy (US Social Security Administration, 2014b).

With additional changes contemplated to Social Security's NRA, a logical question is what to do about the earliest eligibility age (EEA), currently age 62, when individuals can first claim benefits. When the NRA was 65, individuals who claimed at age 62 experienced a permanent annual reduction of 20% from full benefits. As the NRA increases to 67, with some changes in the benefit calculation rules, this permanent reduction will rise to 30% (Munnell, Meme, et al., 2004). For those who live into their nineties, as assets are depleted and as employee pension benefits that are not indexed to inflation decline in real value, this reduction could leave individuals in a precarious financial situation. If the NRA is increased further, for example to 69, a key policy question will be whether the EEA should change, and if not, how benefit reductions should be calculated. Without such changes, those who claim at the EEA with an NRA of 69 could have their monthly benefits reduced by 40% or more, which could greatly undermine the extent to which Social Security provides a safety net for the oldest old.

Since 2010, Social Security expenditures have exceeded its tax revenues (its primary source of income), and the most recent projections suggest that the OASDI trust fund will be exhausted in 203e (Board of Trustees of OASDI, 2014). Further benefit cuts are one way to stabilize the program's finances. Finally, with the Great Recession and the lackluster recovery, American households have been hit hard for several consecutive years and their financial situations will need time to heal. Unemployment was slow to decline and some researchers have concluded that a permanent decline in lifetime earnings is likely (Butrica et al., 2011; Rix, 2013a). If so, more years of work will be needed to accumulate the level of assets that would have otherwise been possible.

Moreover, our society will age rapidly over the next decade (Board of Trustees of OASDI, 2014) as the Baby Boom generation reaches traditional retirement ages. Throughout their lives, the Baby Boomers have challenged society – from large expansions in educational facilities and housing in the 1950s and 1960s to the huge influx of workers in the 1970s. These challenges have been overcome each step along the way. The retirement of the Baby Boomers is next and society is already adjusting to this.

If the Baby Boomers were to retire at the same ages as their parents did, many would experience a reduction in standard of living. As noted above, almost one half of workers aged 45 and over have total savings and investments of less than $25 000 (excluding home equity and DB plans) (Helman et al., 2014) and Social Security benefits for a worker with medium earnings are about $19 000 per year (Board of Trustees of OASDI, 2014). Without additional income, from employer pensions, social programs, or family members, the inevitable outcome is a reduction in living standards, especially for those who live into their nineties.

Of course, this reduction in consumption comes with an increase in leisure hours, so retirees might initially feel better off, as the additional leisure compensates for the reduction in consumption. Conversely, reductions in consumption do not just pertain to non-essential goods; a lower standard of living means fewer resources for necessities as well. Health care is one such necessity and in retirement, health care becomes increasingly important. In 2012, the average annual out-of-pocket health care expenditures of Medicare households was over $4 700 (Cubanski, Swoope, Damico, & Neuman, 2014), a sizable amount.

The poverty rate of older Americans is currently lower than that of individuals aged 18 years and younger and those aged 19–64 years, but this was not always the case. When the Social Security Act was signed in 1935, the estimated poverty rate among older Americans was nearly 80% (Smolensky, Danziger, & Gottschalk, 1988). Old age and poverty were closely linked. The risk of higher poverty rates among tomorrow's older Americans is real, given the dramatic reductions in DB pension plans and the financial risks in DC plans, possible Social Security and Medicare reductions, and further increases in longevity. Fortunately, the behavior of current and, we suspect, future retirees to work longer reduces the likelihood of significant declines in the living standards of future retirees. Workers are responding to the new world. As the retirement environment continues to evolve, society needs to adapt and continued work later in life, already underway, is a key component of this adaptation. Preparing for aging now will help lead to better outcomes for individuals, employers, and society as a whole in the years ahead.

Disclaimer and Acknowledgments

All views expressed in this chapter are those of the authors and do not necessarily reflect the views or policies of the US Bureau of Labor Statistics.

The authors would like to thank Sarah Jordan for excellent research assistance.

References

Abbott, J. (1974). Covered employment and the age men claim retirement benefits. *Social Security Bulletin, 37*, 3–16.

Adler, G., & Hilber, D. (2009). Industry hiring patterns of older workers. *Research on Aging, 31*(1), 69–88.

Arias, E. (2012). *United States life tables, 2008. National vital statistics reports, 61*(3). Hyattsville, MD: National Center for Health Statistics.

Autor, D. H. (2011). *The unsustainable rise of the disability rolls in the United States: Causes, consequences, and policy options*. Cambridge, MA: National Bureau of Economic Research. Working Paper No. 17697.

Blau, D. M., & Gilleskie, D. B. (2008). The role of retiree health insurance in the employment behavior of older men. *International Economic Review, 49*(2), 475–514.

Blau, D. M., & Shvydko, T. (2011). Labor market rigidities and the employment behavior of older workers. *Industrial and Labor Relations Review, 64*(3), 464–484.

Board of Trustees of OASDI. (2014). *The 2014 annual report of the board of trustees of the federal old-age and survivors insurance and federal disability insurance trust funds*. Washington, DC: US Government Printing Office.

Bosworth, B. P., & Burtless, G. (2011). *Recessions, wealth destruction, and the timing of retirement*. Chestnut Hill, MA: Center for Retirement Research at Boston College. Working Paper No. 2010–22.

Boyle, M. A., & Lahey, J. N. (2010). Health insurance and the labor supply decisions of older workers: Evidence from a US Department of Veterans Affairs expansion. *Journal of Public Economics, 94*(7-8), 467–478.

Bruce, D., Holtz-Eakin, D., & Quinn, J. F. (2000). *Self-employment and labor market transitions at older ages*. Chestnut Hill, MA: Center for Retirement Research at Boston College. Working Paper No. 2000–13.

Bullard, J. (2013). *The tapering debate: Data and tools*. St. Louis, MO: Federal Reserve Bank of St. Louis. Accessed at: <http://research.stlouisfed.org/econ/bullard/pdf/BullardStLRegChamberFinancialForum1November2013Final.pdf>.

Burkhauser, R. V., Daly, M. C., McVicar, D., & Wilkins, R. (2014). Disability benefit growth and disability reform in the United States: Lessons learned from other OECD nations. *IZA Journal of Labor Policy, 3*(4), 1–30.

Burkhauser, R. V., & Rovba, L. (2009). Institutional responses to structural lag: The changing patterns of work at older ages. In S. J. Czaja & J. Sharit (Eds.), *Aging and work* (pp. 9–34). Baltimore, MD: John Hopkins University Press.

Burtless, G. (2013). *The impact of population aging and delayed retirement on workforce productivity*. Chestnut Hill, MA: The Center for Retirement Research at Boston College. Working Paper No. 2013–11.

Burtless, G., & Looney, A. (2012). *The immediate jobs crisis and our long-run labor market problem*. Washington, DC: The Brookings Institution.

Burtless, G., & Quinn, J. F. (2002). *Is working longer the answer for an aging workforce?* Chestnut Hill, MA: The Center for Retirement Research at Boston College. Issue Brief No. 11.

Butrica, B. A., Johnson, R. W., & Smith, K. E. (2011). *The potential impact of the great recession on future retirement incomes*. Chestnut Hill, MA: Center for Retirement Research at Boston College. Working Paper No. 2011–9.

Cahill, K. E., Giandrea, M. D., & Quinn, J. F. (2006). Retirement patterns from career employment. *The Gerontologist, 46*(4), 514–523.

Cahill, K. E., Giandrea, M. D., & Quinn, J. F. (2008). *A micro-level analysis of recent increases in labor force participation among older workers*. Chestnut Hill, MA: The Center for Retirement Research at Boston College. Working Paper No. 8.

Cahill, K. E., Giandrea, M. D., & Quinn, J. F. (2011a). Reentering the labor force after retirement. *Monthly Labor Review, 134*(6), 34–42.

Cahill, K. E., Giandrea, M. D., & Quinn, J. F. (2011b). *How does occupational status impact bridge job prevalence?* Washington, DC: US Bureau of Labor Statistics. Working Paper No. 447.

Cahill, K. E., Giandrea, M. D., & Quinn, J. F. (2012). Older workers and short-term jobs: Employment patterns and determinants. *Monthly Labor Review, 135*(5), 19–32.

Cahill, K. E., Giandrea, M. D., & Quinn, J. F. (2013a). *Are gender differences emerging in the retirement patterns of the early boomers?* Washington, DC: US Bureau of Labor Statistics. Working Paper No. 468.

Cahill, K. E., Giandrea, M. D., & Quinn, J. F. (2013b). *New evidence on self-employment transitions among older Americans with career jobs.* Washington, DC: US Bureau of Labor Statistics. Working Paper No. 463.

Cahill, K. E., Giandrea, M. D., & Quinn, J. F. (2013c). Retirement patterns and the macroeconomy, 1992–2010: The prevalence and determinants of bridge jobs, phased retirement, and reentry among three recent cohorts of older Americans. *The Gerontologist.* doi: http://dx.doi.org/10.1093/geront/gnt146.

Coile, C. (in press). Disability insurance incentives and the retirement decision: Evidence from the US In D.A. Wise (Ed.), *Social security programs and retirement around the world: Disability insurance programs and retirement.* Chicago, IL: University of Chicago Press.

Coile, C., & Levine, P. B. (2006). Bulls, bears, and retirement behavior. *Industrial and Labor Relations Review, 59*(3), 408–429.

Coile, C., & Levine, P. B. (2011). Recessions, retirement, and social security. *American Economic Review: Papers & Proceedings, 101*(3), 23–28.

Congressional Budget Office. (2001). *Social security: A primer.* Washington, DC: US Government Printing Office.

Congressional Budget Office. (2002). *The looming budgetary impact of society's aging.* Washington, DC: US Government Printing Office.

Congressional Budget Office. (2009). *CBO's long-term projections for social security.* Washington, DC: US Government Printing Office.

Conley, D., & Thompson, J. (2013). The effects of health and wealth shocks on retirement decisions. *Federal Reserve Bank of St. Louis Review, 95*(5), 389–404.

Copeland, C. (2009). *Retirement plan participation and asset allocation, 2007.* Washington, DC: Employee Benefit Research Institute. EBRI Notes No. 30 (pp. 13–23).

Costa, D. (1998). *The evolution of retirement: An American economic history, 1880–1990.* Chicago, IL: University of Chicago Press.

Costa, D. (1999). *Has the trend toward early retirement reversed?* Paper presentation at the First Annual Joint Conference for the Retirement Research Consortium, Washington, DC.

Cubanski, J., Swoope, C., Damico, A., & Neuman, T. (2014). *Health care on a budget: The financial burden of health spending by Medicare households.* Menlo Park, CA: The Henry J. Kaiser Family Foundation.

DeNavas-Walt, C., Proctor, B. D., & Smith, J. C. (2012). *Income, poverty, and health insurance coverage in the United States: 2011.* Washington, DC: US Census Bureau.

Dunn, T., & Holtz-Eakin, D. (2000). Financial capital, human capital, and the transition to self-employment: Evidence from intergenerational links. *Journal of Labor Economics, 18*(2), 282–305.

Elsby, M. W., Hobijn, B., Sahin, A., & Valletta, R. G. (2011). The labor market in the Great Recession – an update to September 2011. *Brookings Papers on Economic Activity, 353–384.* Washington, DC: The Brookings Institution.

Employee Benefit Research Institute. (2014). *FAQs about benefits – retirement issues.* Washington, DC: Employee Benefit Research Institute. <http://www.ebri.org/publications/benfaq/index.cfm?fa=retfaqt14fig2>. Accessed 03.06.14.

Evans, D. S., & Jovanovic, B. (1989). An estimated model of entrepreneurial choice under liquidity constraints. *The Journal of Political Economy, 97*(4), 808–827.

Federal Reserve Bank of St. Louis. (2014). *Federal Reserve economic data.* St. Louis, MO: Federal Reserve Bank of St. Louis. <http://research.stlouisfed.org/fred2/>.

French, E., & Jones, J. B. (2011). The effects of health insurance and self-insurance on retirement behavior. *Econometrica, 79*(3), 693–732.

Friedberg, L., & Webb, A. (2005). Retirement and the evolution of pension structure. *Journal of Human Resources, 40*(2), 281–308.

Fronstin, P. (2014). *Health savings accounts and health reimbursement arrangements: Assets, account balances, and rollovers, 2006–2013.* Washington, DC: Employee Benefit Research Institute. Issue Brief No. 395.

Fronstin, P., & Adams, N. (2012). *Employment-based retiree health benefits: Trends in access and coverage, 1997–2010.* Washington, DC: Employee Benefit Research Institute. Issue Brief No. 377.

Gale, W. G., Papke, L. E., & VanDerhei, J. (2005). The shifting structure of private pensions. In W. G. Gale, J. B. Shoven, & M. J. Warshawsky (Eds.), *The evolving pension system: Trends, effects and proposals for reform* (pp. 51–76). Washington, DC: Brookings Institute Press.

Giandrea, M. D., Cahill, K. E., & Quinn, J. F. (2009). Bridge jobs: A comparison across cohorts. *Research on Aging, 31*(5), 549–576.

Gorodnichenko, Y., Song, J., & Stolyarov, D. (2013). *Macroeconomic determinants of retirement timing.* Cambridge, MA: National Bureau of Economic Research. Working Paper No. 19638.

Gruber, J., Milligan, K., & Wise, D. A. (2009). *Social Security programs and retirement around the world: The relationship* between youth employment, introduction and summary. Cambridge, MA: National Bureau of Economic Research. Working Paper No. 14647.

Gruber, J., & Orszag, P. (2003). Does the social security earnings test affect labor supply and benefit receipt? *National Tax Journal, 56*(4), 755–773.

Gruber, J., & Wise, D. A. (2004). *Social security programs and retirement around the world*. Chicago, IL: University of Chicago Press.

Gustman, A. L., Steinmeier, T. L., & Tabatabai, N. (2011). *How did the recession of 2007–2009 affect the wealth and retirement of the near retirement age population in the health and retirement study?* Cambridge, MA: National Bureau of Economic Research. Working Paper No. 17547.

Helman, R., Adams, J., Copeland, C., & VanDerhei, J. (2014). *The 2013 retirement confidence survey: Confidence rebounds for those with retirement plans*. Washington, DC: Employee Benefit Research Institute. Issue Brief No. 397.

Hipple, S. (2004). Self-employment in the United States: An update. *Monthly Labor Review, 127*(7), 13–23.

Holtz-Eakin, D., Joulfaian, D., & Rosen, H. S. (1994a). Sticking it out: Entrepreneurial survival and liquidity constraints. *The Journal of Political Economy, 102*(1), 53–75.

Holtz-Eakin, D., Joulfaian, D., & Rosen, H. S. (1994b). Entrepreneurial decisions and liquidity constraints. *The RAND Journal of Economics, 25*(2), 334–347.

Hutchens, R., & Grace-Martin, K. (2006). Employer willingness to permit phased retirement: Why are some more willing than others? *Industrial and Labor Relations Review, 59*(4), 525–546.

Hutchens, R. M., & Chen, J. (2007). The role of employers in phased retirement: Opportunities for phased retirement among white-collar workers. In T. Ghilarducci & J. Turner (Eds.), *Work options for older Americans* (pp. 95–118). Notre Dame, IN: University of Notre Dame Press.

Johnson, R. W. (2011). Phased retirement and workplace flexibility for older adults: Opportunities and challenges. *The ANNALS of the American Academy of Political and Social Science, 638*(1), 68–85.

Johnson, R. W., Kawachi, J., & Lewis, E. K. (2009). *Older workers on the move: Recareering in later life*. Washington, DC: AARP Public Policy Institute. Research Report No. 2009-08.

Kantarci, T., & Van Soest, A. (2008). Gradual retirement: Preferences and limitations. *De Economist, 156*, 113–144.

Karoly, L. A., & Rogowski, J. A. (1994). The effect of access to post-retirement health insurance on the decision to retire early. *Industrial and Labor Relations Review, 48*(1), 103–123.

Karoly, L. A., & Zissimopoulos, J. (2004). Self-employment among older US workers. *Monthly Labor Review, 127*(7), 24–47.

Karp, F. (2007). *Growing older in America: The Health and Retirement Study*. Washington, DC: US Department of Health and Human Services.

Lavery, J. (2009). *Social security finances: Findings of the 2009 Trustees Report*. Washington, DC: National Academy of Social Insurance. Social Security Brief No. 30.

Madrian, B. (1994). Employment-based health insurance and job mobility: Is there evidence of job-lock? *Quarterly Journal of Economics, 109* (February), 27–54.

Maestas, N. (2010). Back to work: Expectations and realizations of work after retirement. *Journal of Human Resources, 45*(3), 719–748.

Maestas, N., & Li, X. (2008). *Does the rise in the full retirement age encourage disability benefits applications? Evidence from the Health and Retirement Study*. Ann Arbor, MI: University of Michigan Retirement Research Center. Working Paper No. 2008-198.

Martin, P. P., & Weaver, D. A. (2005). Social security: A program and policy history. *Social Security Bulletin, 66*(1), 1–15.

McDonnell, K. (2010). *Income of the elderly population age 65 and over, 2008*. Washington, DC: Employee Benefit Research Institute. EBRI Notes No. 31(6).

McFall, B. H. (2011). Crash and wait? The impact of the great recession on the retirement plans of older Americans. *American Economic Review: Papers & Proceedings, 101*(3), 40–44.

Milligan, K. S., & Wise, D. A. (2011). *Social Security and retirement around the world: Mortality and health, employment, and disability insurance participation and reforms*. Cambridge, MA: National Bureau of Economic Research. Working Paper No. 16719.

Munnell, A. H. (2006). Employer sponsored plans: The shift from defined benefit to defined contribution. In G. L. Clark, A. H. Munnell, & M. Orszag (Eds.), *The Oxford handbook of pensions and retirement income* (pp. 359–380). Oxford: Oxford University Press.

Munnell, A. H. (2007). Working longer: A potential win-win proposition. In T. Ghilarducci & J. Turner (Eds.), *Work options for older Americans* (pp. 11–43). Notre Dame, IN: University of Notre Dame Press.

Munnell, A. H. (2013). *Social Security's real retirement age is 70*. Chestnut Hill, MA: The Center for Retirement Research at Boston College. Issue Brief No. 13-15.

Munnell, A. H., Cahill, K. E., Eschtruth, A., & Sass, S. A. (2004). *The graying of Massachusetts: Aging, the new rules of retirement, and the changing workforce*. Boston, MA: The Massachusetts Institute for a New Commonwealth.

Munnell, A. H., Cahill, K. E., & Jivan, N. A. (2003). *How has the shift to 401(k)s affected the retirement age?* Chestnut Hill, MA: The Center for Retirement Research at Boston College. Issue Brief No. 13.

Munnell, A. H., Meme, K. B., Jivan, N. A., & Cahill, K. E. (2004). *Should we raise Social Security's earliest eligibility age?* Chestnut Hill, MA: The Center for Retirement Research at Boston College. Issue Brief No. 18.

Munnell, A. H., & Perun, P. (2006). *An update on private pensions.* Chestnut Hill, MA: The Center for Retirement Research at Boston College. Issue Brief No. 50.

Munnell, A. H., & Sass, S. A. (2008). *Working longer: The solution to the retirement income challenge.* Washington, DC: Brookings Institution Press.

Munnell, A. H., & Sundén, A. (2004). *Coming up short: The challenge of 401(k) plans.* Washington, DC: Brookings Institution Press.

National Center for Health Statistics. (2013). *Health, United States, 2012: With special feature on emergency care.* Hyattsville, MD: US Department of Health and Human Services.

Pang, G., Warshawsky, M., & Weitzer, B. (2008). *The retirement decision: Current influences on the timing of retirement among older workers.* Arlington, VA: Watson Wyatt Worldwide. Working Paper.

Penner, R. G., Perun, P., & Steuerle, E. (2002). *Legal and institutional impediments to partial retirement and part-time work by older workers.* Washington, DC: The Urban Institute.

Perun, P. J., & Dilley, P. E. (2011). *Social security: The house that Roosevelt built.* Washington, DC: The Aspen Institute Initiative on Financial Security.

Purcell, P. (2009). *Older workers: Employment and retirement trends.* Washington, DC: Congressional Research Service.

Quinn, J. F. (1999). *Retirement patterns and bridge jobs in the 1990s.* Washington, DC: Employee Benefit Research Institute. Issue Brief No. 206.

Quinn, J. F. (2010). Work, retirement, and the encore career: Elders and the future of the American workforce. *Generations, 34,* 45–55.

Quinn, J. F., Burkhauser, R. V., & Myers, D. A. (1990). *Passing the torch: The influence of economic incentives on work and retirement.* Kalamazoo, MI: W.E. Upjohn Institute for Employment Research.

Quinn, J. F., Cahill, K. E., & Giandrea, M. D. (2011). *Early retirement: The dawn of a new era?* New York, NY: TIAA-CREF Institute. Policy Brief.

Rix, S. E. (2013a). *The employment situation, December 2012: Five years after the start of the Great Recession.* Washington, DC: AARP Public Policy Institute. Fact Sheet No. 276.

Rix, S. E. (2013b). *The employment situation, October 2013: Not much changed for older workers.* Washington, DC: AARP Public Policy Institute. Fact Sheet No. 294.

Rix, S. E. (2014). *The employment situation, January 2014, and a look back at 2013: Fewer older workers unemployed, more out of the labor force.* Washington, DC: AARP Public Policy Institute. Fact Sheet No. 302.

Robinson, C., & Clark, R. (2010). Retiree health insurance and disengagement from a career job. *Journal of Labor Research, 31*(3), 247–262.

Rogowski, J., & Karoly, L. (2000). Health insurance and retirement behavior: Evidence from the Health and Retirement Survey. *Journal of Health Economics, 19*(4), 529–539.

Ruhm, C. J. (1990). Bridge jobs and partial retirement. *Journal of Labor Economics, 8*(4), 482–501.

Ruhm, C. J. (1991). Career employment. *Industrial Relations, 30*(2), 193–208.

Sass, S. A. (1997). *The promise of private pensions: The first 100 years.* Cambridge, MA: Harvard University Press.

Sass, S. A., Monk, C., & Haverstick, K. (2010). *Workers' response to the market crash: Save more, work more?* Chestnut Hill, MA: Center for Retirement Research at Boston College. Issue Brief No. 10–3.

Seburn, P. A. (1991). Evolution of employer-provided defined benefit pensions. *Monthly Labor Review, 114*(12), 16–23.

Shultz, K. S., & Wang, M. (2011). Psychological perspectives on the changing nature of retirement. *American Psychologist, 66,* 170–179.

Smolensky, E., Danziger, S., & Gottschalk, P. (1988). The declining significance of age in the United States: Trends in the well-being of children and the elderly since 1939. In J. L. Palmer, T. Smeeding, & B. B. Torrey (Eds.), *The vulnerable.* Washington, DC: Urban Institute Press.

Sommers, D., & Franklin, J. C. (2012). Overview of projections to 2020. *Monthly Labor Review, 135*(1), 3–20.

Steuerle, E., Spiro, C., & Johnson, R. W. (1999). *Can Americans work longer?* Washington, DC: The Urban Institute.

Toossi, M. (2012). Labor force projections to 2020: A more slowly growing workforce. *Monthly Labor Review, 135*(1), 43–64.

US Bureau of Labor Statistics. (2014a). Unemployment of the population 55 and older, LNS14024230.

US Bureau of Labor Statistics. (2014b). Seasonally-adjusted employment-population ratio, LNS12300000.

US Department of Commerce, Bureau of Economic Analysis. (2013). National Income and Product Accounts, Table 5.1. <http://www.bea.gov/iTable/iTable.cfm?ReqID=9&step=1#reqid=9&step=1&isuri=1>.

US Social Security Administration. (2014a). *Income of the population aged 55 or older, 2012.* Washington, DC: US Social Security Administration.

US Social Security Administration. (2014b). Provisions affecting retirement age. Available at: <www.ssa.gov/OACT/solvency/provisions/retireage.html#C1>.

VanDerhei, J. (2014). *The role of Social Security, defined benefits, and private retirement accounts in the face of the retirement crisis*. Washington, DC: Employee Benefit Research Institute. EBRI Notes No. 35(1).

von Wachter, T. (2002). *The end of mandatory retirement in the US: Effects on retirement and implicit contracts*. Berkeley, CA: Center for Labor Economics, University of California Berkeley. Working Paper No. 49.

Wang, M., Adams, G. A., Beehr, T., & Shultz, K. S. (2009). Bridge employment and retirement: Issues and opportunities during the latter part of one's career. In S. Baugh & S. Sullivan (Eds.), *Maintaining focus, energy, and options over the career* (pp. 135–162). Charlotte, NC: Information Age Publishing, Inc.

Zissimopoulos, J. M., & Karoly, L. A. (2007). Transitions to self-employment at older ages: The role of wealth, health, health insurance and other factors. *Labour Economics, 14*, 269–295.

Zissimopoulos, J. M., & Karoly, L. A. (2009). Labor-force dynamics at older ages: Movements into self-employment for workers and nonworkers. *Research on Aging, 31*(1), 89–111.

CHAPTER 14

Productive Engagement in Later Life

Nancy Morrow-Howell[1] and Emily A. Greenfield[2]

[1]George Warren Brown School of Social Work, Washington University, St. Louis, MO, USA
[2]School of Social Work Affiliate of the Institute for Health, Health Care Policy, & Aging Research
Rutgers, The State University of New Jersey, New Brunswick, NJ, USA

OUTLINE

Introduction	293
Conceptual Issues	294
Defining the Term	294
Controversies in Defining the Term	295
Relevance of Productive Engagement in Later Life	296
Demographic Context	296
Prevalence of Productive Engagement in Later Life	297
Scholarship on the Antecedents and Outcomes of Productive Engagement	299
Conceptual Frameworks	299
Overview of the Current Evidence on Antecedents of Productive Engagement	301
Overview of the Literature on Outcomes of Productive Engagement	304
Challenges and Future Directions	306
References	309

INTRODUCTION

Discussions of the aging population have largely focused on the serious challenges of economic security, health care, and long-term care. In large part, this discourse takes the perspective of older adults as a social problem. Largely within the past two decades, the discussion has expanded to include consideration of another view on population aging – the growing human capital represented by the older population. As health, education, and economic security have generally increased with each cohort to date, and as environments become more inclusive of people with disabilities, so has the capacity of many individuals to initiate and continue valuable activities longer into the life course.

This growth in later life human capital has spawned the productive aging perspective, where the fundamental view is that older adults can and do serve as contributors. This perspective rests on the assumption that society cannot afford to dismiss the human capital of the older population and that the productive engagement of older adults as a population group is a necessity, not a luxury (Butler, 1997). In sum, the term "productive aging" reflects a paradigm shift in how we think of population aging – not as a social burden alone, but also as an opportunity to mobilize older adults as a resource for families, communities, and society at large.

This chapter will first provide an overview of the use of the term "productive aging" as well as specific operationalizations that have guided scholarship. We then discuss the prevalence of older adults' productive engagement. We further present a conceptual framework that organizes our review of theory and research on the antecedents and outcomes of productive engagement in later life. We conclude by discussing the challenges associated with this scholarship while also identifying directions for future research.

CONCEPTUAL ISSUES

Defining the Term

The term "productive aging" came into currency as one of many concepts of the "new gerontology," where more positive images of aging were put forth to confront the prevailing negative images (Johnson & Mutchler, 2014). Dr. Robert Butler, who is considered a founder of contemporary gerontology (Achenbaum, 2014), coined the phrase "productive aging" by arguing that in the face of the demographic revolution, we must transform retirement by extending work life and expanding volunteer roles for older adults (Butler, 1997). He asked how we can orient attention toward productivity rather than dependency (Butler & Gleason, 1985).

Just like other terms that emerged as part of positive gerontology – such as successful aging, vital aging, and conscious aging – productive aging has been used inconsistently. Morrow-Howell, Hingterlong, and Sherraden (2001) suggested that operationalizing the concept of *productive engagement in later life* rather than the concept of *productive aging* would better advance a research agenda given that "engagement" refers more to observable behaviors in comparison to "aging" as a broader theoretical construct. Further, the term "productive aging" implies there is "unproductive aging," with scholars warning that the term implies that productivity is the highest attainment of late life (Holstein & Minkler, 2007). The term "productive engagement" suggests that there are other types of engagements beyond those that most clearly indicate productivity – such as social, leisure, and spiritual activities – perhaps each with related-yet-distinct significance in later life.

Still, defining "productive" remains a challenge. The most commonly accepted definition involves monetary value: Productive activities are those that produce goods and services, whether paid or not (Morrow-Howell et al., 2001). In other words, older adults are productively engaged when they are involved in productive activities, that is, when they are performing paid and unpaid work that can be assigned a dollar value.

Typically, the value of unpaid productive activities is assessed by multiplying the time spent on these activities by an estimate of the dollar value of an hour of engagement (e.g., how much it would cost to purchase one hour of the service). These estimation techniques have led to statements, for example, that older adults provide $100 billion worth of care to parents, spouses, and grandchildren and $44.3 billion dollars of formal volunteer service a year (Johnson & Schaner, 2005). Folbre (2012) considers the complexity of valuing carework and describes several valuation methods, none of which fully captures the meaning of the work to the individual. She points out that carework for adults usually comes later in the life course than carework for children and therefore has fewer consequences for lifetime earnings for the care provider; however, the opportunity costs of caregiving in later life still come with substantial negative financial consequences for individuals.

Productive engagement is an umbrella term that encompasses discrete activities, and much of the scholarship on productive engagement has focused on three broad activities: working for pay (hereby referred to as "working"); volunteering (both formally for an organization and informally through one's private networks of neighbors and friends outside of the household); and caregiving (i.e., caring for an adult or child with a disability or health condition that limits their self-care). Recently, there has been growing attention to other forms of care provided by older adults, particularly within their families, such as grandparents' residential and nonresidential assistance with childcare.

There are multiple reasons that scholars have considered the discrete activities of working, volunteering, and caregiving as related and as a part of a research agenda on productive engagement. First, these activities all make direct economic contributions to society. For example, paid work contributes to the Gross Domestic Product, and volunteers and caregivers provide services that some party likely would otherwise need to finance. Second, these activities can be encouraged and facilitated through program and policy interventions, such as through the Senior Community Service Employment Program (Washko, Schack, Goff, & Pudlin, 2011) and caregiver support programs (Elmore & Talley, 2009).

Other scholars have taken a wider view of activities to include under the productive engagement umbrella. Bass and Caro (1996) included educational activities because they build the capacity to engage in productive activities. Others have argued that productive activities should include self-care and household chores, using the logic that if the individual cannot perform activities, such as managing their finances and getting in and out of bed, someone else would have to do this work for that individual (Butler & Gleason, 1985). Although public monies are involved in providing personal care to those that cannot do it themselves, productive activities are most often viewed as making a contribution outside the person doing the activity, including another person, a community, or an organization (Bass & Caro, 2001). At the broadest level, some scholars have included leisure and social activities as productive because these activities are valuable to the individual and to society to the extent that they promote well-being. However, these wider definitions are so inclusive that they threaten the development of a coherent and unified research agenda (Morrow-Howell & Wang, 2013). They also deviate from the emphasis on activities that can be readily monetized in terms of their economic value.

Controversies in Defining the Term

While the term "productive engagement" is valuable for advancing a broad area of research, policy, and program development to facilitate older adults' contributions, the term is not without its controversies. Below, we review arguments that have problematized this construct.

1. The concept has been criticized for suggesting that productivity might be considered the highest value of later life and that "unproductive" people could be devalued. Estes and Mahakian (2001) warned that "the aged may find themselves dogged by market judgments to the end of life, and in ways that become troubling

new sources of stigma and social control" (p. 211). Holstein and Minkler (2007) expressed concern that certain older adults will be marginalized, or continue to be marginalized, if certain expectations for productive engagement are not met.

2. Critics have emphasized that productive engagement is not unilaterally a "free choice." For example, individuals might need to continue to work in a job that they do not desire to meet their basic needs; or they might be overwhelmed by providing care to family and wish to be free of this activity. Moreover, social inequalities by race/ethnicity, gender, age, and health pattern the degree to which older adults have choice around engaging in productive activity, such as providing residential care to grandchildren (Luo, LaPierre, Huges, & Waite, 2012). In sum, some older adults have more and better choices to be productively engaged, or not, than others.

3. There has been concern regarding the degree to which certain productive activities are more celebrated than others. Historically, the scholarship on productive engagement in later life has emphasized work and formal volunteering more so than informal helping and caregiving – activities that are more invisible and are less rewarded at the public level. At the same time, it is these very types of activities that are more salient among less advantaged subgroups of older adults, including women and older adults from US ethnic minority groups (Martinson & Minkler, 2006).

4. Some have argued that the productive aging paradigm creates a detrimental dichotomy between negative and positive poles (Minkler & Fadem, 2002), viewing older adults either as dependants or as vital contributors. There is growing discourse that throughout life, individuals have varying and simultaneous states of dependence and independence and that greater scholarship on interdependence throughout the life course is needed (Settersten, 2005).

5. There remains controversy over how to assess the value – monetary or otherwise – of particular types of productive engagement. For example, volunteering for a particular cause might be deemed of great value by those who support that cause, yet viewed as detrimental to those who are against it (Greenfield, 2010). As another example, people might hold opposing views as to whether working for a particular company contributes to, or takes away from, economic value for individuals and society alike.

RELEVANCE OF PRODUCTIVE ENGAGEMENT IN LATER LIFE

Demographic Context

Interest in increasing the engagement of older adults in working, volunteering, and caregiving has grown in the last decade along with growing attention to demographic trends that will make our society in greater need of older adults' participation. Family caregivers will be in higher demand because of the rapid growth of the oldest old and the ongoing trend toward community living (Coughlin, 2010; National Alliance for Caregiving [NAC] & AARP, 2009). The average age of US caregivers continues to increase, especially as the Baby Boomer generation moves fully into caregiving for their parents and partners. Moreover, there is evidence that the number of grandparent-headed households with children has been rising in the US, with much of the growth occurring specifically from 2007 to 2008 at the start of the economic downturn (Livingston & Parker, 2010). Nonprofit and public agencies typically rely on volunteer labor; and in the face of shrinking public investments, it is assumed that more volunteers will be needed. For example, the National Association of Area Agencies on Aging (2014) launched an initiative to increase the number of volunteers to serve the aging services network in light of reduced public funding.

The demand for older workers has been more debated, especially in the context of the Great Recession with high unemployment rates. Some argue that lower birth rates will ultimately lead to labor shortages, which has been the case in other countries across the world (European Commission, 2011). The most common view is that labor shortages in specific industries in the United States will demand longer working lives as supplies of younger workers shrink and Baby Boomers leave the workplace (Lacey & Wright, 2009). Further, there is concern that Baby Boomers' retirement en masse will result in a major loss of experience and historical knowledge. Finally, our society might need older adults to work longer for the financial solvency of pensions and Social Security. Analysts have demonstrated that longer working lives reduce reliance on post-retirement income for the individual and reduce strain on public and private income support programs (Butrica, Smith, & Steuerle, 2006).

In sum, demographic shifts put pressure on job markets, on nonprofit and public service sectors, and on families; our society will likely need a greater number and percentage of older adults to be productively engaged as workers, volunteers, and caregivers. The social science agenda on productive engagement in later life is therefore highly salient at this time of demographic shift.

Prevalence of Productive Engagement in Later Life

There is a long research tradition of examining the extent to which older adults engage in activities that make contributions. This history is rooted, in part, in an advocacy agenda – an effort to dispel the long-standing images of the dependent, burdensome elderly (Butler & Gleason, 1985). In this section, we review the most recent research on rates of older adults' participation in various productive activities.

Paid Work. After decades of declining labor force participation among older adults, rates have risen since the mid-1990s. For persons aged 55 years and older, the labor force participation went from a low of 29.2% in 1993 to a high of 40.4% in 2009 (Sok, 2010). Among those over 65 years of age, 14% are in the labor force (United States Department of Labor: Bureau of Labor Statistics [US BLS], 2012). Trends among older workers include the growth of older women in the workforce, growth in full-time as opposed to part-time work, and growth in self-employment (Centers for Disease Control and Prevention, 2012; Rogoff, 2007).

Volunteering. The US Bureau of Labor Statistics [US BLS] (2013) estimates that the number of people 65 years and older who engage in formal volunteering ranges between 23.5% and 24.5%. This rate is lower than those among 35–64 years old. However, older adults who do volunteer put in more hours, averaging 90 hours a year, compared to 32 hours a year for young adults. Declining rates of any volunteering among older adults likely reflects, in part, disengagement from work and educational institutions, which are the major vehicles through which people are recruited to volunteer. Older adults are less likely to be asked to volunteer (Independent Sector, 2000). The most common site for volunteering by older adults is religious organizations (US BLS, 2013), where there is continuity of institutional connection and ongoing opportunity.

It is more difficult to determine rates of engagement in informal volunteering. This is partly due to ambiguity about when these activities are being performed (Taniguchi, 2012). That being said, Zedlewski and Schaner (2006) estimate that about half of people 55 years of age and older engage in informal volunteering. Time use studies indicate that midlife and older adults contribute, on average, 3.4 h per week of unpaid assistance and 2.5 h per week of emotional support to friends, neighbors, co-workers, and others outside of their families (Almeida & McDonald, 2005).

Caregiving and Other Carework. An estimated 65.7 million Americans, or 28.5%, provide unpaid care to an adult or child with functional impairment (NAC & AARP, 2009). Of those, about 35% are 50–64 years old, 9% are 65–74 years old, and 4% are over the age of 75. The average age of the caregiver has been rising and will continue to do so as more care will be needed for the growing 85-plus age group.

Research examining the rates of grandparents' care for their grandchildren historically has focused on custodial grandparenting, specifically where children live with grandparents with no parent residing in the household. Census data indicated that 1.6 million, or 2.2% of all children in the US, were being raised by a custodial grandparent, a number three times greater than the number of children in foster care (US Census, 2011). Scholars also have used US national survey data to indicate the salience of other types of grandparents' care. Luo and colleagues (2012) found that 7% of grandparents lived with grandchildren, with most of these households including three generations. Among non-coresiding grandparents, 28% of grandparents provided at least 50 h of care a year, 33% provided between 50 and 199 h of care, and 5% provided 500 or more hours of care a year.

Engagement in Multiple Types of Productive Activity. There is a growing body of research that documents the engagement of older adults in multiple productive activities. Employed older adults are more likely to volunteer than those not working, and highest rates of volunteering are among part-time older workers (Choi, 2003). Taniguchi (2012) documented that formal and informal volunteering are complementary and do not substitute for one another, surmising that similar forces, like altruism and social pressure, encourage both.

Researchers have produced conflicting evidence about the extent to which caregiving and volunteering compete. One study documented that caregivers volunteer just as much as non-caregivers (Burr, Choi, Mutchler, & Caro, 2005), while another found that female spousal caregivers are less likely to volunteer, either formally or informally (Choi, Burr, Mutchler, & Caro, 2007). Inconsistent results might be due to differences among studies in the extent to which they account for the demands and contexts of caregiving; however, Musick and Wilson (2008) conclude that there is no clear evidence that caregiving excludes volunteering. The majority of caregivers also work, and 17% of the American workforce report caregiving duties.

A few researchers have considered a wider range of productive activities simultaneously, often focusing on working, volunteering, informal helping, caregiving, and grandparenting (Baker & Silverstein, 2008; Hinterlong, 2008). For example, Burr, Mutchler, and Caro (2007) performed cluster analysis with formal volunteer work, informal help to others, unpaid domestic work, caregiving, and paid work. They identified four classes of older adults: helpers, home maintainers, workers/volunteers, and super-helpers. Almost half of the sample was classified as helpers, performing a moderate amount of formal volunteering, informal helping, and domestic work, with less time devoted to working and caregiving. A small group of super-helpers (less than 4%) were highly or moderately involved in all activities but working. Almost 15% were workers and volunteers. In sum, current research suggests that most older adults in the United States are involved in more than one productive activity and that these activities largely complement rather than substitute for each other.

SCHOLARSHIP ON THE ANTECEDENTS AND OUTCOMES OF PRODUCTIVE ENGAGEMENT

Over the last three decades, social scientists have sought to answer key questions about the productive engagement of older adults. Who engages in productive activities? What is the relative contribution of personal, social, and environmental conditions to productive engagement? What individual and societal outcomes are associated with productive engagement? Why are these outcomes produced? What conditions modify relationships between engagement and health outcomes for older adults? What programs and policies maximize engagement and promote the most positive outcomes? In this section, we provide an overview of theories that have guided this scholarship as well as current evidence regarding antecedents and outcomes of productive engagement.

Conceptual Frameworks

Bass and Caro (1996) presented one of the first explicit conceptual frameworks on productive engagement in later life, which focused on how social policy (e.g., government and employer policies), environmental (e.g., demographic changes), situational (e.g., socioeconomic status), and individual factors (e.g., motivation) contribute to levels of participation in productive activities. Sherraden, Morrow-Howell, Hinterlong, and Rozario (2001) built on this model by positing that sociodemographic factors (e.g., education) influence individual capacity for productive behaviors (e.g., health) and that public policy and programs influence institutional capacity for productive activity (e.g., information about opportunities for productive activity). This conceptual framework also included outcomes of productive engagement for individuals, families, communities, and society. More recently, Morrow-Howell and Wang (2013) extended this model through a cross-cultural perspective, noting how associations among sociodemographic factors; physical environments; economic environments; public policy and programs; productive behaviors; and outcomes are situated within broader sociocultural contexts.

We extend these developments by presenting a conceptual framework in Figure 14.1. We add specificity to antecedents and outcomes, as well as to the forms of the activity itself, which reflect more recent theoretical and empirical developments. For the sake of parsimony, we present each category as separate from each other, while recognizing that they are embedded and inter-related with each other. Also for the sake of parsimony, we formulate linkages among the constructs as largely unidirectional, with antecedents leading to productive engagement and outcomes following productive engagement. We recognize that many of the embedded associations are likely bidirectional, such as in instances in which older adults engage in formal volunteering, experience benefits, and therefore choose to continue.

This model is broadly guided by ecological systems theory – a perspective that has been developed and used mostly within human development and social work (Bronfenbrenner, 1993; Greene, 2008). The conceptual model draws on two primary insights from this framework. The first is that individuals constitute one level within a broader person–environment system, which comprises nested layers of environmental contexts that are organized "like a set of Russian dolls" (Bronfenbrenner, 1979, p. 3). Environments can be identified according to how proximal versus distal they are to individuals. More proximal environments are physically closer to individuals and afford more face-to-face interactions (Goodnow, 1995). Following from this insight, we suggest that antecedents to productive engagement can be organized at three basic levels, each of which influences the other: (a) individual, (b)

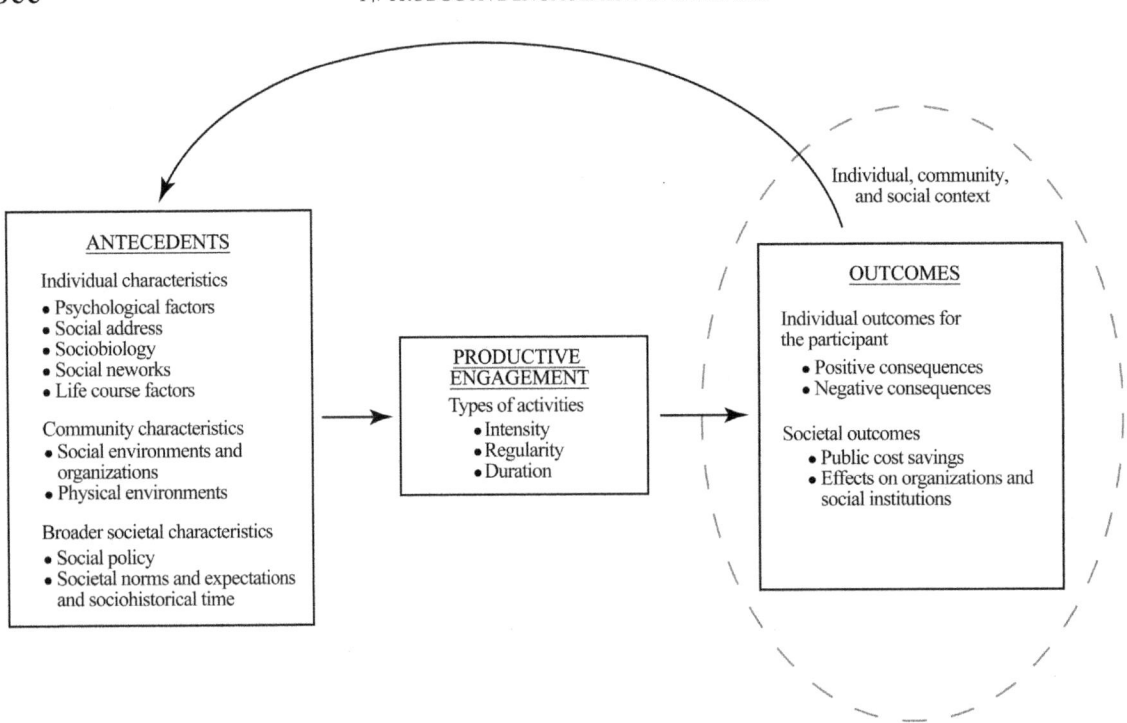

FIGURE 14.1 Conceptual framework guided by ecological systems theory on antecedents and consequences of productive engagement in later life.

community, and (c) societal. Individual characteristics refer to conditions that are specific to an individual or a group of individuals, such as their personal values about helping others and their socioeconomic resources. Community characteristics address circumstances that are unique to a geographic area, such as the availability of a local employer that makes an explicit effort to attract and retain older workers. Societal characteristics refer to circumstances that are shared across broad geographic areas that transcend any one subset of communities, including social policy, as well as widespread social norms regarding productive activity in later life.

The second insight from ecological systems theory that this conceptual model draws upon is that outcomes result from complex transactions among individuals and their environments, both of which are continuously changing over time (Bronfenbrenner & Morris, 2006). Core to the ecological perspective is the idea that individuals influence the very environments that influence them and that environments will have differential effects depending on people's biopsychosocial contexts. Accordingly, our conceptual model suggests that (a) the effects of productive activity on individuals and communities depend upon the contexts in which they are embedded, and (b) the effects of productive activity can, in turn, influence the very individual, community, and societal antecedents that influenced the activity in the first place.

It is our perspective that conceptual models such as this are especially important for making sense of research that addresses different types of productive activities and to identify the most important substantive gaps for social scientists to address in the future. Accordingly, we use this model to guide our review of the literature on productive activity in later life.

Overview of the Current Evidence on Antecedents of Productive Engagement

Theoretical Perspectives

Socioemotional Selectivity Theory. Socioemotional selectivity theory has been used to guide research on individual, intrapsychic characteristics that lead to productive behaviors (e.g., Hendricks & Cutler, 2004; Okun & Schultz, 2003). According to socioemotional selectivity theory, people in later life – who typically view time as more limited than endless – are oriented to goals, activities, and relationships that are most emotionally meaningful and relevant to their own identity (Carstensen, Isaacowitz, & Charles, 1999). Accordingly, this theory suggests that individuals who perceive productive activities as familiar, personally meaningful, and emotionally salient are most likely to engage in them. As an example of a study guided by this perspective, Hendricks and Cutler (2004) used US national data to test the hypothesis that older people will remain stable in their number of hours volunteering for their main organization, even though their number of volunteer hours to all organizations might decrease. Results generally supported this prediction, with older adults volunteering for fewer organizations, yet dedicating the same amount, if not more time, to the organizations for which they volunteered.

Social, Human, and Cultural Capital. Drawing on classic economic theory, the concept of capital has been developed within the social sciences to identify other forms of "value added" beyond financial markets. As a central theoretical construct, social capital addresses the yield when people, groups, and social institutions relate to each other in particular ways (Coleman, 1988). Social capital can be specific to individuals (e.g., an individual's private network of contacts to help them find a job) and also can exist at a community level (e.g., norms of trust within a neighborhood). Human capital addresses resources rooted within individuals, such as skills, knowledge, and functional ability (Coleman, 1988), whereas cultural capital indicates shared symbolic meanings, values, and ways of relating to others that advance one's position in status hierarchies (Lamont & Lareau, 1988). As an example of a study guided by this perspective, McNamara and Gonzales (2011) used US national survey data and found that human capital (e.g., greater education and assets), social capital (e.g., having a spouse who volunteers), and cultural capital (e.g., greater religiosity) were associated with volunteer engagement over time.

Life Course Perspective. The life course perspective focuses, in part, on patterns of continuity and change within major life domains, such as work, leisure, and family, that extend across long periods in people's lives; the perspective also addresses how these patterns are embedded within broader social contexts, including sociohistorical time and social positions (Elder, 1998). Researchers have applied concepts from this perspective to examine a variety of factors that potentially influence older adults' productive activity (e.g., Hirshorn & Settersten, 2013; Moen & Flood, 2013). Such concepts include: timing (e.g., how the timing of one's retirement influences volunteer activity), personal biography (e.g., how productive engagement at earlier phases of life influences engagement later in life), linked lives (e.g., how a younger family member's need for financial assistance leads to older adults continuing to work), and cohort (e.g., the expectation that Baby Boomers will have high rates of volunteerism in later life). As an example, Szinovacz and DeViney (2000) drew upon several principles from the life course perspective in their examination of how marital characteristics influence labor force withdrawal among adults in middle and later life. One such principle addressed how people's experiences in one life sphere, such as work, are likely influenced by those in others, such as family. Consistent with this perspective, the researchers found some evidence that marital partners adjust their work status according to each other's benefit eligibility, income, and health status, with patterns varying by gender.

Critical Perspectives. Critical perspectives focus on society as a whole, "concerned not merely with how things were but how they might be and should be" (Bronner, 2011, pp. 1–2). Critical perspectives are especially attuned to social inequalities in people's life choices and circumstances that result from larger political and economic contexts. Critical perspectives on productive activity suggest that social disparities among older adults – such as by gender, age, race, class, and disability – make some types of productive activity more celebrated and accessible to older adults than others (Martinson & Minkler, 2006). Critical gerontology also addresses ways in which major social institutions, such as the shrinking role of the federal government in providing benefits for older adults, reinforce social norms that expect older adults to engage in productive activity (Holstein, 1999). A study of retirement among women by Zimmerman, Mitchell, Wister, and Gutman (2000) demonstrates the application of critical perspectives in examining women's participation in the paid labor force. Oriented to gender as a social structure that provides unequal opportunities for role participation across the life course, the study found that family caregiving was associated with earlier retirement.

Empirical Findings

As indicated in Figure 14.1, antecedents of productive engagement are believed to be multi-level, although most empirical attention has been directed at individual characteristics of older adults. There has been greatest attention to sociodemographic characteristics and human capital, as well as social capital and other social network factors.

Findings regarding individual predictors of working and volunteering are similar to each other: Older adults with more personal and social resources are more likely to be productively engaged. Education and health are key factors in labor force participation across the life course, including later in life (Employment Benefit Research Institute, 2011; Schmitz, 2011). Similarly, compared to non-volunteers, older volunteers have higher levels of education, better health, more income, are married, have a spouse that volunteers, and work part-time (Butrica, Johnson, & Zedlewski, 2009; Rotolo &Wilson, 2006; Tang, 2006). In a study of Asian American elders, Mui and Shibusawa (2008) documented that those who volunteer have higher self-rated health, are more socially connected, and have higher levels of acculturation. Generally, the same antecedents apply to formal and informal volunteering (Finkelstein & Brannick, 2007), although Taniguchi (2012) found that more time spent with family and friends was associated with greater informal volunteering, while only time with friends was associated with more formal volunteering.

Research on predictors of caregiving in later life similarly has emphasized sociodemograhic factors. Older caregivers are more likely to be white and non-working (NAC & AARP, 2005). Compared to non-caregivers, caregivers are more likely to be female, although the number of male caregivers is growing (Coughlin, 2010). Research also has found that women are more likely to provide assistance to their social network members (Boerner & Reinhardt, 2003) and to provide nonresidential care to grandchildren (Luo et al., 2012).

Ethnicity also is associated with productive engagement. Older Hispanic and African Americans are less likely to be employed and formal volunteers (Tang, 2006; Urban Institute, 2011). In the Burr et al. (2007) study to identify categories of older adults' productive engagement, Whites were more likely to be involved in the groups performing more work and helping activities. Many factors associated with ethnicity explain these patterns, including disparities in health and education as well as historic segregation and structural discrimination (Sass-Lesser, Ghilarducci, & Richman, 2014).

When considering caregiving, non-white older caregivers provide more care to younger people who need assistance than Whites (NAC & AARP, 2009), and rates of custodial grandparenting are much higher among African Americans and Hispanics than Whites (Fuller-Thomson & Minkler, 2007; Whitley, Kelley, & Sipe, 2001). Neighboring and mutual aid are common among ethnic minority communities, where helping efforts often focus on family and community and are less often organized by nonprofit and public agencies (Carlton-LaNey, 2007).

Drawing attention to the other levels of predictors, as depicted in Figure 14.1, there has been less empirical study of community characteristics and broader societal characteristics. There is an emerging literature on how neighborhood characteristics are associated with behaviors that are related to productive activity, such as social participation (e.g., Richard et al., 2012). Krause (2011) found associations between neighborhood deterioration and older adults' provision of emotional support to others. Recent frameworks for efforts to make communities better places for aging include older adults' social inclusion, civic participation, and employment (e.g., World Health Organization, 2007). These perspectives highlight the importance of additional research that examines how social and physical dimensions of local areas influence older adults' productive engagement.

Broader societal characteristics that affect productive engagement include social policies and programs as well as societal norms and expectations. As exemplified by Social Security and Medicare, social policies affect individuals' decisions to work or retire. Raising the retirement age for full benefits from 65 to 67 years (legislation passed in 1983 to be fully implemented by 2022) demonstrates how public policy can influence working behaviors as well as societal expectations about productive engagement. In the face of the aging of the workforce and pending workforce shortages in some industries, new practices in the workplace are being demonstrated, with workplace flexibility, part-time arrangements, and bridge employment receiving the most attention (Cahill, Giandrea, & Quinn, 2013). Workforce development programs for older adults have existed since 1965 when Title V of the Older American's Act created a job training program for low-income older workers; and more recently, private foundations have supported community college programs to provide job counseling and training to older adults (Halvorsen & Emerman, 2013). The social context of the work, including age discrimination, job insecurity, and changing technology, contribute to older workers' experiences and retirement decisions (Roscigno, 2010). Decisions to work are affected by historical time, as demonstrated by the recession of 2008, after which real and expected retirement ages shifted upward (The Associated Press-NORC Center for Public Affairs and Research, 2011).

Social policies and programs also influence caregiving behavior, as exemplified by the Family Medical Leave Act and the advent of participant-directed care for Medicaid long-term care recipients. Federal policies for volunteering are more limited. For over 40 years, the federal government has supported stipended service programs for low-income older adults, such as the Foster Grandparent and Senior Companion Program. Moreover, recent legislation mandated the direction of AmeriCorps monies to older volunteers and allowed for the transfer of educational stipends to children and grandchildren (Public Law 111–13, 2009). Furthermore, the 2006 Reauthorization of the Older Americans Act explicitly called for greater support of older adults' engagement in volunteer and multigenerational activities (Public Law 109–365, 2006). There also is a rapid growth of volunteer opportunities aimed specifically at older adults through national non-profit organizations (e.g., Experience Corps, Environmental Alliance for Senior Involvement) as well as web sites to facilitate involvement (e.g., comingofage.org and volunteermatch.org).

Overview of the Literature on Outcomes of Productive Engagement

Theoretical Perspectives

Several of the theories used to guide research on antecedents also have been used within research on outcomes, such as the life course perspective (e.g., Marks, Lambert, & Choi, 2002). Two additional perspectives have been applied within the literature on the individual outcomes of productive activity: role theory and stress and coping. We summarize these perspectives below.

Role Theory. Role theory has developed within sociological paradigms that focus on how individuals' experiences are rooted within larger social institutions (Biddle, 1986). Following this perspective, social roles can be defined as a set of expectations for behavior associated with occupying a particular position, status, or category of persons within a larger social system (Thoits, 1983). Drawing on this concept, researchers have conceptualized productive engagement in terms of older adults occupying social roles, such as that of volunteer, worker, and caregiver. Researchers have drawn on insights regarding the social psychological processes through which roles influence individual outcomes to guide examinations of associations between productive activity and individual well-being (e.g., Greenfield & Marks, 2004; Lum & Lightfoot, 2005). Moreover, role theory's attention to variation within role experiences – such as how salient a role is to self-identity, how regularly an individual participates in the role, what other roles an individual has, and how much strain an individual experiences within the role – suggest why individual outcomes from productive activity might differ (e.g., Heaven et al., 2013; Mezuk, Bohnert, Ratliff, & Zivin, 2011; Wilson, 2012). For example, Thoits (2012) found that among volunteers that provide support for cardiac patients, greater prominence of the volunteer role within one's own identity was associated with better mental and physical health, which was explained through a greater sense of purpose and meaning.

Stress and Coping. Stress and coping models focus on how people respond to events and circumstances that place strain on their current levels of functioning (Wheaton, 1996). Primary constructs include appraisals (e.g., whether a person views an event as a threat), coping styles (e.g., strategies focused on eliminating the stressor itself versus strategies focused on changing emotional responses to a stressor), and contextual factors (e.g., social support and financial resources). Studies of productive engagement have used these concepts to identify conditions under which productive activity leads to negative individual outcomes, such as caregiver stress (Aranda & Knight, 1997) and job strain (Hansson, Robson, & Limas, 2001). Scholars also have used the concept of stress-induced growth to examine how productive activity can lead to individual gains (e.g., Aschbrenner, Greenberg, Allen, & Seltzer, 2010). Musil, Warner, Zauszniewski, Wykle, and Standing (2009) used a stress-and-coping perspective to examine differences in depressive symptoms among US grandmothers raising their grandchildren. Results indicated that while being a custodial grandmother alone was not associated with greater depressive symptoms, being a custodial grandmother and reporting greater strain from the role was associated with higher depressive symptoms. The researchers also found that higher-quality relationships with family and friends protected against the association between greater strain and depressive symptoms.

Empirical Findings

Given the growing availability of individual survey data that ask individuals about their productive activity and potential outcomes, such as health and financial well-being, there is a substantial literature that documents the effects of working, volunteering, and caregiving on participants. Research has demonstrated that working can provide more than an income to older adults; it also has been associated with decreased mortality, better mental health, and better cognitive function (Calvo, 2006; Rohwedder & Willis, 2010). Although the causal effects of working are hard to document because of constraints in study design (Schmitz, 2011), it is proposed that working produces health because of physical, cognitive, and social activity (Glass, De Leon, Bassuk, & Berkman, 2006).

Volunteering also has been associated with a wide range of individual positive health and psychological outcomes (Carlson et al., 2009; Hong & Morrow-Howell, 2010; Li & Ferraro, 2006). Like working, it might be that increased physical, cognitive, and social activity leads to these outcomes (Fried et al., 2004) as well as engagement in purposeful work (Thoits, 2012). Further, volunteering has been associated with higher odds of employment for older adults (Spera, Ghertner, Nerino, & DiTommaso, 2013). Research also has demonstrated that informal volunteering can have positive effects that are similar to formal volunteering, including physical and emotional benefits (Brown et al., 2009; Krause, 2009). All volunteer activities might not produce positive outcomes, as it has been documented that political participation and community activism can be related to higher distress (Berry, Rodgers, & Dear, 2007).

Outcomes of caregiving on the caregiver have probably received the most attention because of the pervasiveness of the role and the health, mental health, and economic costs that have been associated with this largely unpaid work (Talley & Crews, 2007). It has been documented that caregiving can lead to increased stress, lower job satisfaction, poorer physical and mental health, and financial strain (Coughlin, 2010; Feinberg, Reinhard, Houser, & Choula, 2011; Pinquart & Sörensen, 2007).

In contrast, more recent studies have documented better health outcomes for caregivers compared to non-caregivers, including lower mortality rates and better cognitive function (Bertrand et al., 2012; Roth et al., 2013). Caregivers have reported enhanced self-esteem and increased recognition for their work (Zarit, 2012). These positive outcomes might be conditioned by the level of caregiving demand and caregiver strain, especially the presence of advanced dementia. Roth and colleagues (2013) concluded that when caregiving is done willingly, at manageable levels, and for care recipients who are capable of expressing gratitude (which he argues are conditions of most caregiving), it is reasonable to expect health benefits for the caregiver. Fredman, Cauley, Hochberg, Ensrud, and Doros (2010) found that high stress was associated with significantly increased mortality for older female caregivers regardless of caregiving status, lending support to the hypothesis that negative outcomes of caregiving might be attributable to stress, but not caregiving per se. The healthy caregiver hypothesis has also been proposed, suggesting that the positive effects of caregiving might derive from the fact that healthier individuals select into caregiving roles, which might also contribute to the positive outcomes of caregiving observed in recent studies (Bertrand et al., 2012).

Addressing Figure 14.1's attention to how outcomes of productive activity are embedded within individual, community, and societal context, it is clear from the research that individual attributes, as well as the context of the activity, moderate the relationship between productive activity and outcomes for the older adult. For examples, Hispanic caregivers report higher levels of emotional distress than other caregivers (Magana, 2006), and caregivers who live with the care recipient report worse outcomes (NAC & AARP, 2009). There is evidence that volunteers with few personal resources, such as less education and income, benefit more from volunteering (Piliavin & Siegl, 2007). Certain subgroups of workers do not experience the same positive outcomes of employment; the nature of the work, the attributes of the workplace, and the amount of choice involved affect older workers' experience (Calvo, Haverstick, & Sass, 2009; Smith, 2007).

Research also has found that the context of other activities undertaken by the older adults moderates the outcomes of productive engagement. When studying the outcomes of simultaneous caregiving and working, unemployed caregivers have the lowest levels of well-being, especially depression, and employed non-caregivers demonstrate the highest well-being. Employed caregivers have higher levels of physical and mental health than non-employed caregivers, suggesting the positive effects of working for caregivers (Coughlin, 2010). Hao (2008) found that for adults ages 55–66 year olds, working full time and volunteering together produced the best psychological outcomes in contrast to other combinations of productive activities.

In sum, the current understanding is that, in general, the more formalized activities of working and volunteering relate to positive health and mental health outcomes for older adults, but that caregiving has both negative and positive effects. There also is research demonstrating that outcomes depend on individual and broader social contexts.

In addition to the effects of productive engagement on older individuals, Figure 14.1 further indicates that productive engagement in later life can influence outcomes at the societal level, such as by contributing to a more experienced workforce, stronger civic society and more intergenerational exchange within families and communities. However, these associations are largely theoretical at this point. For ethnic minority communities specifically, it is suggested that productive engagement in volunteering and caregiving roles contribute to the passing down of cultural values to younger generations and preservation of language and tradition (Yoshida, Gordon, & Henkin, 2008); to addressing community challenges and enabling ethnic communities to survive (Carlton-LaNey, 2007); and to facilitating family functioning during processes of immigration and acculturation (Min, Moon, & Lubben, 2005). Despite this theorizing, there are few empirical studies documenting these outcomes in terms of public cost savings and effects on organizations and social institutions. Quantitative studies of these associations probably would require longitudinal, multi-level data, inclusive of a wide range of individuals nested within groups (e.g., communities or countries) over time, which are difficult to collect.

CHALLENGES AND FUTURE DIRECTIONS

Since Robert Butler introduced the term "productive aging" more than 30 years ago, gerontologists have moved from an advocacy agenda to confront the stereotypes of later life to a social science agenda. Longitudinal data sets, with large samples representing the older population, have greatly increased the amount and methodological sophistication of research in this area. Moreover, measurement has become more diverse; for example, biomarker and other clinical data have been used to assess predictors (Son & Wilson, 2010) and outcomes (e.g., Burr, Tavares, & Mutchler, 2011; Kim & Ferraro, 2014; Roepke et al., 2011) of productive activity in later life. Further, studies increasingly have considered people's day-by-day and moment-by-moment experiences of productive activity (e.g., Cichy, Stawski, & Almeida, 2013). There also has been growing recognition of historically less recognized types of productive activity, including caregiving and older adults' provision of informal support both within and outside of their families (Greenfield, 2010). Despite these advances, there are several key issues that continue to challenge the advancement of research in this area.

First, this scholarship is fundamentally challenged by the issue of social causation versus social selection. The reciprocal influence of engagement and health has been documented (Thoits & Hewitt, 2001), as well as problems of sample selection in longitudinal analyses (Li & Ferraro, 2005). Providing empirical evidence for simultaneous causation and selection, Wickrama, O'Neal, Kwag, and Lee (2013) used six waves of national US data to document that prior levels of work predicted subsequent changes in cognitive function, physical disability, and depressive symptoms and that changes in health predicted subsequent work. For even stronger evidence of the causal effects of productive activity, there is a need for studies that utilize experimental designs, with randomization into a control group and longitudinal assessment over time (e.g., see Tan, Xue, Li, Carlson, & Fried, 2006).

Another issue is the condition of choice surrounding productive engagement. For example, what choice of jobs is available to older adults, especially those with low levels of education? To what degree does an older adult choose to take on caregiving responsibilities in the context of family constraints? Even volunteer activities, which are often assumed to be done out of a person's free will, could be conducted for other reasons, such as an overwhelming sense of obligation (Wilson, 2000). Previous work has documented that choice indeed matters. Involuntary retirement is associated with negative health and mental health outcomes (Dhaval, Rashad, & Spasojevic, 2008; James & Spiro, 2007), and caregiving outcomes are contingent on the individuals' willingness to assume the role (Roth et al., 2013). The challenge for social scientists is to integrate choice into conceptual frameworks, develop measurement strategies, and systematically include the assessment of choice in data collection. Understanding the role of choice, or stated differently, the role of coercion, is fundamental to ensuring that exploitation and marginalization are minimized, especially for disadvantaged individuals.

Beyond the specific issue of choice, research on understanding the antecedents and outcomes of productive engagement is more broadly challenged by the complexity of multiple and interacting contexts that shape these relationships. Prior research has demonstrated that contextual factors – from the individual to society – influence associations between productive engagement and outcomes, illustrating that simply performing the activity does not produce the most desirable outcomes for all people. For example, discrimination in the workplace affects outcomes of working (Roscigno, 2010), and the type of motivation to volunteer affects health outcomes of volunteering (Konrath, Fuhrel-Forbis, Lou, & Brown, 2011). In fact, there is evidence that older adults who participate in activities associated with work and volunteer roles without psychological engagement might actually experience lower well-being than if they had not been involved at all (Matz-Costa, Besen, James, & Pitt-Catsouphes, 2012).

Also, much of the research on productive engagement in later life has focused on addressing basic social scientific research questions regarding rates of engagement, predictors, and outcomes. While this research is essential for establishing this area of scholarship, additional research that translates this understanding into effective policies and programs is needed. For example, primary data collection with older volunteers has indicated linkages between volunteer management strategies, such as continued support for volunteers and providing a stipend, and volunteer retention (Tang, Morrow-Howell, & Choi, 2010). This type of research enhances an evidence base for supporting nonprofit organizations' adoption of specific practices to recruit and retain older volunteers. More generally, research that focuses on developing and testing theories of contextualized processes that lead from productive engagement to desired outcomes holds the promise of informing the development and of programs to optimize older adults' productive engagement. More consistent inclusion of major social scientific paradigms within studies also will help to connect this area of research to social gerontology more broadly.

Another important translational direction is research that can inform public policy. Despite federal programs to support older adults' productive engagement, there has been very little research to assess the policy that underlies these programs. For example, for every one dollar of public money invested in an infrastructure to support workforce training for older adults, how much public money is saved? To what extent are volunteer programs within public agencies engaging racially/ethnically and socioeconomically diverse elders? Do caregiver support programs funded with federal dollars prevent costly health care utilization? Addressing these questions can help to develop the basic social science of macro-social contexts that influence productive activity, as well as address relevant policy issues at the US national level.

There is further need for more cross-national research, as demographic trends worldwide make productive engagement in later life a relevant topic in both developing and developed countries. For example, in the face of the shrinking number of younger workers, European countries are developing strategies to encourage and enable longer working lives (European Commission, 2011). As the non-governmental sector of China is being established, there is interest in creating volunteer opportunities for older adults, especially because grandparenting duties have decreased as birth rates have dropped (Li, 2012). Most developed countries need to increase long-term care for the growing number of the old-old and are seeking to implement caregiver support policies and programs (Hong & Park, 2012). In short, as cross-national research and education is growing across many areas of gerontology (McCutcheon & Pruchno, 2011), productive engagement represents a very promising topic for global partnerships.

Finally, research has focused mainly on one specific productive activity at a time, largely devoid of attention to advances within research on other activities (Putnam et al., 2013). This fragmentation likely has slowed the pace of overall knowledge advancement. Forging greater connection across research on different types of productive activity, through shared methodologies, similar theoretical frameworks, and patterns across empirical findings, holds the promise of advancing the field more systematically, comprehensively, and expeditiously. Moreover, because older adults oftentimes engage in multiple types of productive activity, greater attention to multiple types of productive activity within single investigations can more holistically reflect older adults' experiences.

In conclusion, although research on older adults' productive engagement has increased in quality and quantity within the past two decades, there remain important substantive, methodological, and theoretical issues to address. Research on the antecedents, forms, and outcomes of productive activity in later life holds exceptional promise as policymakers and program leaders seek evidence to inform decisions concerning key societal challenges, such as enhancing economic security in later life, supporting families' ability to care for the needs of their members, and optimizing individuals' quality of life throughout its entirety. Research in this area can advance a more comprehensive social gerontology, one that simultaneously addresses older adults' strengths and vulnerabilities, while also contributing to efforts toward improving aging societies as a whole.

References

Achenbaum, W. A. (2014). Robert N. Butler, MD (January 21, 1927–July 4, 2010): Visionary leader. *The Gerontologist, 54*, 6–12.

Almeida, D. M., & McDonald, D. A. (2005). The national story: How Americans spend their time on work, family and community. In J. Heymann & C. Beem (Eds.), *Unfinished work: Building equality and democracy in an era of working families* (pp. 180–203). New York, NY: New Press.

Aranda, M. P., & Knight, B. G. (1997). The influence of ethnicity and culture on the caregiver stress and coping process: A sociocultural review and analysis. *The Gerontologist, 37*, 342–354.

Aschbrenner, K. A., Greenberg, J. S., Allen, S. A., & Seltzer, M. M. (2010). Subjective burden and personal gains among older parents of adults with serious mental illness. *Psychiatric Services, 61*, 605–611.

Baker, L. A., & Silverstein, M. (2008). Depressive symptoms among grandparents raising grandchildren: The impact of participation in multiple roles. *Journal of Intergenerational Relationships, 6*, 285–304.

Bass, S., & Caro, F. (2001). Productive aging: A conceptual framework. In N. Morrow-Howell, J. Hinterlong, & M. Sherraden (Eds.), *Productive aging: Concepts and challenges* (pp. 37–81). Baltimore, MD: Johns Hopkins University Press.

Bass, S. A., & Caro, F. G. (1996). Theoretical perspectives in productive aging. In W. H. Crown (Ed.), *Handbook on employment and the elderly* (pp. 265–275). Westport, CT: Greenwood.

Berry, H., Rodgers, B., & Dear, K. (2007). Preliminary development and validation of an Australian community participation questionnaire. *Social Science and Medicine, 64*, 1710–1737.

Bertrand, R. M., Saczynski, J. S., Mezzacappa, C., Hulse, M., Ensrud, K., & Fredman, L. (2012). Caregiving and cognitive function in older women: Evidence for the healthy caregiver hypothesis. *Journal of Aging and Health, 24*, 48–66.

Biddle, B. J. (1986). Recent developments in role theory. *Annual Review of Sociology, 12*, 67–92.

Boerner, K., & Reinhardt, J. P. (2003). Giving while in need: Support provided by disabled older adults. *Journals of Gerontology: Psychological and Social Sciences, 58*, S297–S304.

Bronfenbrenner, U. (1979). *The ecology of human development: Experiments by nature and design*. Cambridge, MA: Harvard University Press.

Bronfenbrenner, U. (1993). Ecological models of human development. In M. Gauvin & M. Cole (Eds.), *Readings on the development of children* (pp. 37–43). New York, NY: Freeman.

Bronfenbrenner, U., & Morris, P. A. (2006). The bioecological model of human development. In W. Damon & R. M. Lerner (Eds.), *Handbook of child psychology, Vol. 1: Theoretical models of human development* (pp. 793–828). New York, NY: John Wiley.

Bronner, S. E. (2011). *Critical theory: A very short introduction*. New York, NY: Oxford University Press.

Brown, S. C., Mason, C. A., Perrino, T., Hirama, I., Verdeja, R., Spokane, A. R., et al. (2009). Longitudinal relationships between neighboring behavior and depressive symptoms in Hispanic older adults in Miami, Florida. *Journal of Community Psychology, 37*, 618–634.

Burr, J. A., Choi, N., Mutchler, J. E., & Caro, F. G. (2005). Caregiving and volunteering: Are private and public helping behaviors linked? *Journal of Gerontology: Social Sciences, 60B*, S247–S256.

Burr, J. A., Mutchler, J. E., & Caro, F. G. (2007). Productive activity clusters among middle-aged and older adults: Intersecting forms and time commitments. *Journal of Gerontology: Social Sciences, 62B*, S267–S275.

Burr, J. A., Tavares, J., & Mutchler, J. E. (2011). Volunteering and hypertension risk in later life. *Journal of Aging & Health, 23*, 24–51.

Butler, R., & Gleason, H. (1985). *Productive aging*. New York, NY: Springer.

Butler, R. N. (1997). Living longer, contributing longer. *The Journal of the American Medical Association, 278*, 1372–1374.

Butrica, B., Smith, K.E., & Steuerle, C.E. (2006). *Working for a good retirement*. Retrieved from: <http://www.urban.org/UploadedPDF/311333_good_retirement.pdf>.

Butrica, B. A., Johnson, R. W., & Zedlewski, S. R. (2009). Volunteer dynamics of older Americans. *Journal of Gerontology: Social Sciences, 64B*, 644–655.

Cahill, K.E., Giandrea, M.D., & Quinn, J.F. (2013). Retirement patterns and the macroeconomy, 1992–2010: Prevalence and determinants of bridge jobs, phased retirement, and re-entry among three recent cohorts of older Americans. *The Gerontologist.* http://dx.doi.org/10.1093/geront/gnt146.

Calvo, E. (2006). *Does working longer make people healthier and happier?* Retrieved from: <http://crr.bc.edu/wp-content/uploads/2006/02/wob_2.pdf>.

Calvo, E., Haverstick, K., & Sass, S. A. (2009). Gradual retirement, sense of control, and retirees' happiness. *Research on Aging, 31*, 112–135.

Carlson, M. C., Erickson, K. I., Kramer, A. F., Voss, M. W., Bolea, N., Mielke, M., et al. (2009). Evidence for neurocognitive plasticity in at-risk older adults: The experience corps program. *Journal of Gerontology, 64*, 1275–1282.

Carlton-LaNey, I. (2007). 'Doing the Lord's work': African American elders' civic engagement. *Generations, 30*, 47–50.

Carstensen, L. L., Isaacowitz, D., & Charles, S. T. (1999). Taking time seriously: A theory of socioemotional selectivity. *American Psychologist, 54*, 165–181.

Centers for Disease Control and Prevention. (2012). *Older employees in the workplace.* Retrieved from: <http://www.cdc.gov/nationalhealthyworksite/docs/Issue_Brief_No_1_Older_Employees_in_the_Workplace_7-12-2012_FINAL(508).pdf>.

Choi, L. H. (2003). Factors affecting volunteerism among older adults. *Journal of Applied Gerontology, 22*, 179–196.

Choi, N. G., Burr, J. A., Mutchler, J. E., & Caro, F. G. (2007). Formal and informal volunteer activity and spousal caregiving among older adults. *Research on Aging, 29*, 99–124.

Cichy, K. E., Stawski, R. S., & Almeida, D. M. (2013). A double-edged sword: Race, daily family support exchanges, and daily well-being. *Journal of Family Issues, 35*, 1–22.

Coleman, J. S. (1988). Social capital in the creation of human capital. *The American Journal of Sociology, 94*, S95–S120.

Coughlin, J. (2010). Estimating the impact of caregiving and employment on well-being. *Outcomes and Insights in Health Management, 2*, 1–7.

Dhaval, D., Rashad, I., & Spasojevic, J. (2008). The effects of retirement on physical and mental health outcomes. *Southern Economic Journal, 75*, 497–523.

Elder, G. H., Jr. (1998). The life course as developmental theory. *Child Development, 69*, 1–12.

Elmore, D. L., & Talley, R. C. (2009). Family caregiving and U.S. federal policy. In S. H. Qualls & S. Zarit (Eds.), *Aging families and caregiving* (pp. 209–231). New York, NY: Wiley.

Employment Benefit Research Institute. (2011). *Labor-force participation rates of the population age 55 and older: What did the recession do to the trends?* Retrieved from: <http://papers.ssrn.com/sol3/papers.cfm?abstract_id=1762707>.

Estes, C. L., & Mahakian, J. L. (2001). The political economy of productive aging. In N. Morrow-Howell, J. Hinterlong, & M. Sherraden (Eds.), *Productive aging: Concepts and challenges* (pp. 197–214). Baltimore, MD: Johns Hopkins University Press.

European Commission. (2011). *Meeting the challenge of Europe's aging workforce: The public employment service response.* Issues paper adopted during the 29th meeting of European Heads of Public Employment Services Warsaw, December 8, 2011.

Feinberg, L., Reinhard, S.C., Houser, A., & Choula, R. (2011). *Valuing the invaluable: 2011 update. The growing contributions and costs of family caregiving.* Retrieved from: <http://hjweinbergfoundation.net/ficsp/documents/10/Caregivers-Save-the-System-Money-With-Uncompensated-Care.pdf>.

Finkelstein, M. A., & Brannick, M. T. (2007). Applying theories of institutional helping to informal volunteering. *Social Behavior & Personality: An International Journal, 35*, 101–114.

Folbre, N. (2012). *For love and money.* New York, NY: Russell Sage Foundation.

Fredman, L., Cauley, J. A., Hochberg, M., Ensrud, K. E., & Doros, G. (2010). Mortality associated with caregiving, general stress, and caregiving-related stress in elderly women. *Journal of the American Geriatrics Society, 58*, 937–943.

Fried, L. P., Carlson, M. C., Freedman, M., Frick, K. D., Glass, T. A., Hill, J., et al. (2004). A social model for health promotion for an aging population: Initial evidence on the Experience Corps model. *Journal of Urban Health, 81*, 64–78.

Fuller-Thomson, E., & Minkler, M. (2007). Mexican American grandparents raising grandchildren. *Families in Society, 88*, 567–574.

Glass, T. A., De Leon, C. F., Bassuk, S. S., & Berkman, L. F. (2006). Social engagement and depressive symptoms in late life: Longitudinal findings. *Journal of Aging and Health, 18*, 604–628.

Goodnow, J. J. (1995). Differentiating among social contexts: By spatial features, forms of participation, and social contracts. In P. Moen, G. H. Elder, & K. Luscher (Eds.), *Examining lives in context: Perspectives on the ecology of human development* (pp. 269–301). Washington, DC: American Psychological Association.

Greene, R. R. (2008). *Human behavior theory and social work practice* (3rd ed.). New Brunswick, NJ: Transaction Publishers.

Greenfield, E. A. (2010). Identifying the boundaries and horizons of the concept of civic engagement for the field of aging. In G. O'Neill & S. F. Wilson (Eds.), *Civic engagement in an older America* (pp. 7–14). Washington, DC: The Gerontological Society of America.

Greenfield, E. A., & Marks, N. F. (2004). Formal volunteering as a protective factor for older adults' psychological well-being. *Journals of Gerontology Series B: Psychological Sciences & Social Sciences, 59B*, S258–S264.

Halvorsen, C., & Emerman, J. (2013). The Encore movement. *Generations, 37*, 33–39.

Hansson, R. O., Robson, S. M., & Limas, M. J. (2001). Stress and coping among older workers. *Work, 17*, 247–256.

Hao, Y. (2008). Productive activities and psychological well-being among older adults: The influence of number of activities and time commitment. *Journal of Gerontology, 63B*, S64–S72.

Heaven, B., Brown, L. J., White, M., Errington, L., Mathers, J. C., & Moffatt, S. (2013). Supporting well-being in retirement through meaningful social roles: Systematic review of intervention studies. *The Milbank Quarterly, 91*, 222–287.

Hendricks, J., & Cutler, S. (2004). Volunteerism and socioemotional selectivity in later life. *Journals of Gerontology, 59B*, S251–S257.

Hinterlong, J. E. (2008). Productive engagement among older Americans: Prevalence, patterns, and implications for public policy. *Journal of Aging & Social Policy, 20*, 141–164.

Hirshorn, B. A., & Settersten, R. A. (2013). Civic involvement across the life course: Moving beyond age-based assumptions. *Advances in Life Course Research, 18*, 199–211.

Holstein, M. (1999). Women and productive aging: Troubling implications. In M. Minkler & C. Estes (Eds.), *Critical gerontology* (pp. 359–373). Amityville, NY: Baywood.

Holstein, M. B., & Minkler, M. (2007). Critical gerontology: Reflections for the 21st century. In M. Bernard & T. Scharf (Eds.), *Critical perspectives on ageing societies* (pp. 12–26). Buckingham, UK: Open University Press.

Hong, S. I., & Morrow-Howell, N. (2010). Health outcomes of Experience Corps: A high-commitment volunteer program. *Social Science & Medicine, 71*, 414–420.

Hong, S. I., & Park, E. (2012). Critical review of older adults' caregiving-related programs and policies in South Korea. In N. Morrow-Howell & A. Mui (Eds.), *Productive engagement in later life* (pp. 113–123). New York, NY: Routledge.

Independent Sector. (2000). *American's senior volunteers*. Retrieved from: <http://www.independentsector.org/publications#sthash.NFk75iCd.dpbs>.

James, J. B., & Spiro, A., III. (2007). The impact of work on the psychological health and well-being of older Americans. In J. B. James & P. Wink (Eds.), *The crown of life. Dynamics of the early postretirement period* (pp. 153–173). New York, NY: Springer Publishing Company.

Johnson, K. J., & Mutchler, J. E. (2014). The emergence of a positive gerontology: From disengagement to social involvement. *The Gerontologist, 54*, 93–100.

Johnson, R.W., & Schaner, S.G. (2005). *The retirement project perspectives on productive aging: Value of unpaid activities by older Americans tops $160 billion per year*. Retrieved from: <http://www.urban.org/UploadedPDF/311227_older_americans.pdf>.

Kim, S., & Ferraro, K. (2014). Do productive activities reduce inflammation in later life? *The Gerontologist, 54*(5), 830–839.

Konrath, S., Fuhrel-Forbis, A., Lou, A., & Brown, S. (2011). Motives for volunteering are associated with mortality risk in older adults. *Health Psychology, 1*, 87–96.

Krause, N. (2009). Meaning in life and mortality. *Journal of Gerontology: Social Sciences, 64B*, 517–527.

Krause, N. (2011). Neighborhood conditions and helping behavior in late life. *Journal of Environmental Psychology, 31*, 62–69. Available from: http://dx.doi.org/10.1016/j.jenvp.2010.11.003.

Lacey, T.A., & Wright, B. (2009). *Occupational employment projections to 2018*. Retrieved from: <http://www.bls.gov/opub/mlr/2009/11/art5full.pdf>.

Lamont, M., & Lareau, A. (1988). Cultural capital: Allusions, gaps and glissandos in recent theoretical developments. *Sociological Theory, 6*, 153–168.

Li, Q. (2012). Senior volunteer services in urban communities: A study in Jinan. In N. Morrow-Howell & A. Mui (Eds.), *Productive engagement in later life* (pp. 167–182). New York, NY: Routledge.

Li, Y., & Ferraro, K. F. (2005). Volunteering and depression in later life: Social benefit or selection processes? *Journal of Health and Social Behavior, 46*, 68–84.

Li, Y., & Ferraro, K. F. (2006). Volunteering in middle and later life: Is health a benefit, barrier or both? *Social Forces, 85*, 497–519.

Livingston, G., & Parker, K. (2010). *Since the start of the Great Repression, more children raised by grandparents*. Retrieved from: <http://www.pewsocialtrends.org/2010/09/09/since-the-start-of-the-great-recession-more-children-raised-by-grandparents/>.

Lum, T. Y., & Lightfoot, E. (2005). The effects of volunteering on the physical and mental health of older people. *Research on Aging, 27*, 31–55.

Luo, Y., LaPierre, T. A., Huges, M. E., & Waite, L. J. (2012). Grandparents providing care to grandchildren: A population-based study of continuity and change. *Journal of Family Issues, 33*, 1143–1167.

Magana, S. (2006). Older Latino family caregivers. In B. Berkman & S. D'Ambruo (Eds.), *Handbook of social work in health and aging* (pp. 371–380). Oxford University Press.

Marks, N. F., Lambert, J. D., & Choi, H. (2002). Transitions to caregiving, gender, and psychological well-being: A prospective U.S. national study. *Journal of Marriage and Family, 64*, 657–667.

Martinson, M., & Minkler, M. (2006). Civic engagement and older adults: A critical perspective. *The Gerontologist, 46*, 318–324.

Matz-Costa, C., Besen, E., James, J. B., & Pitt-Catsouphes, M. (2012). The differential impact of multiple levels of productive activity engagement on psychological well-being in middle and later life. *The Gerontologist, 148*

McCutcheon, M., & Pruchno, R. (2011). Introduction the international spotlight. *The Gerontologist, 51*, 423–424.

McNamara, T. K., & Gonzales, E. (2011). Volunteer transitions among older adults: The role of human, social, and cultural capital in later life. *The Journals of Gerontology, Series B: Psychological Sciences and Social Sciences, 66B*, 490–501.

Mezuk, B., Bohnert, A. S. B., Ratliff, S., & Zivin, K. (2011). Job strain, depressive symptoms, and drinking behavior among older adults: Results from the health and retirement studies. *The Journals of Gerontology, Series B: Psychological Sciences and Social Sciences, 66*, 426–434.

Min, J. W., Moon, A., & Lubben, J. E. (2005). Determinants of psychological distress over time among older Korean immigrants and Non-Hispanic White elders: Evidence from a two-wave panel study. *Aging & Mental Health, 9*, 210–222.

Minkler, M., & Fadem, P. (2002). 'Successful aging:' A disability perspective. *Journal of Disability Policy Studies, 12*, 229–235.

Moen, P., & Flood, S. (2013). Limited engagements? Women's and men's work/volunteer time in the Encore Life Course Stage. *Social Problems, 60*, 206–233.

Morrow-Howell, N., Hingterlong, J., & Sherraden, M. (Eds.), (2001). *Productive aging: Concepts and challenges*. Baltimore, MD: John Hopkins University Press.

Morrow-Howell, N., & Wang, Y. (2013). The productive engagement of older African Americans, Hispanics, Asians, and Native Americans. In K. Whitfield & T. Baker (Eds.), *Handbook of minority aging* (pp. 351–366). New York, NY: Springer Publishing Co.

Mui, A. C., & Shibusawa, T. (2008). *Asian American elders in the 21st century: Key indicators of psychological well-being*. New York, NY: Columbia University Press.

Musick, M. A., & Wilson, J. (2008). *Volunteering: A social profile*. Bloomington, IN: Indiana University Press.

Musil, C., Warner, C., Zauszniewski, J., Wykle, M., & Standing, T. (2009). Grandmother caregiving, family stress and strain, and depressive symptoms. *Western Journal of Nursing Research, 31*, 389–408.

National Alliance for Caregiving (NAC) & AARP. (2005). *Caregiving in the U.S.* Retrieved from: <http://www.caregiving.org/data/04execsumm.pdf>.

National Alliance for Caregiving (NAC) & AARP. (2009). *Caregiving in the U.S. 2009*. Retrieved from: <http://www.caregiving.org/data/Caregiving_in_the_US_2009_full_report.pdf>.

National Association of Area Agencies on Aging. (2014). *The Aging Network volunteer collaborative*. Retrieved from: <http://www.n4a.org/programs/aging-network-volunteer-collaborative/>.

Okun, M. A., & Schultz, A. (2003). Age and motives for volunteering: Testing hypotheses derived from socioemotional selectivity theory. *Psychology and Aging, 18*, 231–239.

Piliavin, J. A., & Siegl, E. (2007). Health benefits of volunteering in the Wisconsin Longitudinal Study. *Journal of Health and Social Behavior, 48*, 450–464.

Pinquart, M., & Sörensen, S. (2007). Correlates of physical health of informal caregivers: A meta-analysis. *The Journals of Gerontology. Series B, Psychological Sciences and Social Sciences, 62*, P126–P137.

Public Law 111–13. (2009). *Serve America Act*. Retrieved from: <http://www.gpo.gov/fdsys/pkg/PLAW-111publ13/pdf/PLAW-111publ13.pdf>.

Putnam, M., Morrow-Howell, N., Inoue, M., Greenfield, J. C., Chen, H., & Lee, Y. (2013). Suitability of public use secondary data sets to study multiple activities. *The Gerontologist*.

Richard, L., Gauvin, L., Kestens, Y., Shatenstein, B., Payette, H., Daniel, M., et al. (2012). Neighborhood resources and social participation among older adults: Results from the VoisiNuage Study. *Journal of Aging and Health, 25*, 296–318.

Roepke, S. K., Mausbach, B. T., Patterson, T. L., Von Kanel, R., Ancoli-Israel, S., Harmell, A. L., et al. (2011). Effects of Alzheimer caregiving on allostatic load. *Journal of Health Psychology, 16*, 58–69.

Rogoff, E. G. (2007). Opportunities for entrepreneurship in later life. *Generations, 31*, 90–95.

Rohwedder, S., & Willis, R. J. (2010). Mental retirement. *Journal of Economic Perspectives, 24*, 119–138.

Roscigno, V. (2010). Ageism in the American workplace. *Contexts, 9*, 16–21.

Roth, D. L., Haley, W. E., Hovater, M., Perkins, M., Wadley, V. G., & Judd, S. (2013). Family caregiving and all-cause mortality: Findings from a population-based propensity-matched analysis. *American Journal of Epidemiology, 178*, 1571–1578.

Rotolo, T., & Wilson, J. (2006). Substitute or complement? Spousal influence on volunteering. *Journal of Marriage & Family, 68*, 305–319.

Sass-Lesser, J., Ghilarducci, T., & Richman, K. (2014). Work and retirement. In K. Whitfield & T. Baker (Eds.), *Handbook of minority aging* (pp. 507–523). New York, NY: Springer.

Schmitz, H. (2011). Why are the unemployed in worse health? The causal effect of unemployment on health. *Labour Economics, 18*, 71–78.

Settersten, R. A., Jr. (2005). Linking the two ends of life: What gerontology can learn from childhood studies. *Journals of Gerontology: Psychological and Social Sciences, 60*, S173–S180.

Sherraden, M., Morrow-Howell, N., Hinterlong, J., & Rozario, P. (2001). Productive aging: Theoretical choices and directions. In N. Morrow-Howell, J. Hinterlong, & M. Sherraden (Eds.), *Productive aging: Concepts and challenges* (pp. 260–284). Baltimore, MD: Johns Hopkins University Press.

Smith, T.W. (2007). *Job satisfaction in America: Trends and socio-demographic correlates*. Retrieved from: <http://www-news.uchicago.edu/releases/07/pdf/070827.jobs.pdf>.

Sok, A. (2010). *Record unemployment among older workers does not keep them out of the job market*. Retrieved from: <http://www.bls.gov/opub/ils/summary_10_04/older_workers.htm>.

Son, J., & Wilson, J. (2010). Genetic variation in volunteerism. *Sociological Quarterly, 51*, 46–64.

Spera, C., Ghertner, R., Nerino, A., & DiTommaso, A. (2013). *Volunteering as a pathway to employment: Does volunteering increase odds of finding a job for the out of work?* Retrieved from: <http://www.nationalservice.gov/sites/default/files/upload/employment_research_report.pdf>.

Szinovacz, M. E., & DeViney, S. (2000). Marital characteristics and retirement decisions. *Research on Aging, 22*, 470–498.

Talley, R. C., & Crews, J. E. (2007). Framing the public health of caregiving. *American Journal of Public Health, 97*, 224–228.

Tan, E. J., Xue, Q. L., Li, T., Carlson, M. C., & Fried, L. P. (2006). Volunteering: A physical activity intervention for older adults – The Experience Corps program in Baltimore. *Journal of Urban Health, 83*, 954–969.

Tang, F. (2006). What resources are needed for volunteerism? A life course perspective. *Journal of Applied Gerontology, 25*, 375–390.

Tang, F., Morrow-Howell, N., & Choi, E. (2010). Why do older adults stop volunteering? *Ageing & Society, 30*, 859–878.

Taniguchi, H. (2012). The determinants of formal and informal volunteering: Evidence from the American Time Use Survey. *Voluntas: International Journal of Voluntary and Nonprofit Organizations, 23*, 920–939.

The Associated Press-NORC Center for Public Affairs and Research (2011). *Working longer: Older Americans' attitudes on work and retirement*. Retrieved from: <http://www.apnorc.org/projects/Pages/working-longer-older-americans-attitudes-on-work-and-retirement.aspx>.

Thoits, P. A. (1983). Multiple identities and psychological well-being: A reformulation and test of the social isolation hypothesis. *American Sociological Review, 48*, 174–187.

Thoits, P. A. (2012). Role-identity salience, purpose and meaning in life, and well-being among volunteers. *Social Psychology Quarterly, 75*, 360–384.

Thoits, P. A., & Hewitt, L. N. (2001). Volunteer work and well-being. *Journal of Health and Social Behavior, 42*, 115–131.

United States Department of Labor: Bureau of Labor Statistics [US BLS]. (2012). *Household data annual averages: Employment status of the civilian noninstitutional population by age, sex, and race*. Retrieved from: <http://www.bls.gov/cps/cpsaat03.pdf>.

United States Department of Labor: Bureau of Labor Statistics [US BLS]. (2013). *Volunteering in the United States*. Retrieved from: <www.bls.gov/news.release/pdf/volun.pdf>.

US Census. (2011). *Facts for features: Grandparents Day 2011: Sept. 11*. Retrieved from: <https://www.census.gov/newsroom/releases/archives/facts_for_features_special_editions/cb11-ff17.html>.

Urban Institute. (2011). *Employment and earnings among 50+ people of color*. Retrieved from: <http://www.urban.org/uploadedpdf/412376-employment-and-earnings.pdf>.

Washko, M. M., Schack, R. W., Goff, B. A., & Pudlin, B. (2011). Title V of the Older Americans Act, the Senior Community Service Employment Program: Participant demographics and service to racially/ethnically diverse populations. *Journal of Aging & Social Policy, 23*, 182–198.

Wheaton, B. (1996). The domains and boundaries of stress concepts. In H. B. Kaplan (Ed.), *From psychosocial stress: Perspectives on structure, theory, life course, and methods* (pp. 29–53). New York, NY: Academic Press.

Whitley, D., Kelley, S., & Sipe, T. (2001). Grandmothers raising grandchildren: Are they at increased risk of health problems? *Health and Social Work, 26*, 105–114.

Wickrama, K. K., O'Neal, C. W., Kwag, K. H., & Lee, T. K. (2013). Is working later in life good or bad for health? An investigation of multiple health outcomes. *Journal of Gerontology, 68*, 807–815.

Wilson, J. (2000). Volunteering. *Annual Review of Sociology, 26*, 215–240.

Wilson, J. (2012). Volunteerism research: A review essay. *Nonprofit and Voluntary Sector Quarterly, 41*, 176–212.

World Health Organization (2007). *Global age-friendly cities: A guide*. Retrieved from: <http://www.who.int/ageing/publications/Global_age_friendly_cities_Guide_English.pdf>.

Yoshida, H., Gordon, D., & Henkin, N. (2008). *Community treasures: Recognizing the contributions of older immigrants and refugees*. Retrieved from: <http://www.projectshine.org/sites/default/files/Community%20Treasures.pdf>.

Zarit, S. H. (2012). Positive aspects of caregiving: More than looking on the bright side. *Aging & Mental Health, 16*, 673–674.

Zedlewski, S.R., & Schaner, S.G. (2006). *The Retirement Project: Perspectives on productive aging*. Retrieved from: <http://www.urban.org/UploadedPDF/311325_older_volunteers.pdf>.

Zimmerman, L., Mitchell, B., Wister, A., & Gutman, G. (2000). Unanticipated consequences: A comparison of expected and actual retirement timing among older women. *Journal of Women & Aging, 12*, 109–128.

CHAPTER

15

Aging, Neighborhoods, and the Built Environment

Carol S. Aneshensel, Frederick Harig, and Richard G. Wight

Department of Community Health Sciences, University of California-Los Angeles, Los Angeles, CA, USA

OUTLINE

Introduction	315
Theoretical Models of Neighborhood	316
The Concept of Neighborhood	316
Contextual and Compositional Neighborhood Effects	317
The Interaction of Person and Environment	318
Neighborhood Stress Process Model	319
Neighborhood Structure and the Health of Older Persons	321
Neighborhood Socioeconomic Disadvantage	321
Racial and Ethnic Segregation	323
Stressors and Resources	324
The Built Environment and the Health of Older Persons	326
Physical Activity and Health	326
The Disability Process	327
Aging in Place	329
Recovery of Mobility	330
Discussion and Directions for Future Research	330
Age and Time	330
Toward Evidence-Based Interventions	331
References	332

INTRODUCTION

A voluminous body of neighborhood research has accumulated over the past two decades demonstrating that the social, economic, demographic, and physical characteristics of neighborhoods are consequential to residents' physical and mental health status, mobility and physical activity, independence, capacity to perform activities of daily living, and ability to participate in social activities (for recent reviews, see Blair, Ross, Gariepy, & Schmitz, 2014; Clarke &

Nieuwenhuijsen, 2009; Hill & Maimon, 2013). Older people generally are considered to be especially vulnerable to adverse neighborhood conditions for several reasons. For example, they tend to have greater exposure to these conditions as a result of spending more time in their neighborhoods than younger adults who leave for work, and exposure is long-term for those who age in place. Daily activities also tend to become increasingly concentrated within a contracting spatial area as a result of age-related declines in health and the onset of mobility disabilities, making older persons increasingly dependent upon their immediate residential communities as sources of social integration, services, and amenities (Clarke & Nieuwenhuijsen, 2009; Glass & Balfour, 2003; Yen, Michael, & Perdue, 2009). In addition, greater sensitivity to neighborhood conditions has been attributed to decreased physical and cognitive functioning and increased fragility (Glass & Balfour, 2003; Yen et al., 2009), and features of the built environment can accelerate the disablement process thereby putting the person at increased risk of developing secondary health conditions (Clarke & George, 2005). However, the positive adaptations that people make to their environments as they grow older are often overlooked and research perspectives that overemphasize detrimental effects of aging have been criticized (Deeg & Fleur-Thomése, 2005). Thus, it is of considerable importance to understand the ways in which neighborhoods and older persons interact with the ultimate objective of creating environments that enable people to attain their goals of continuing to live at home for as long as possible.

This chapter synthesizes information about neighborhoods and the health, emotional well-being, and functioning of older persons. We begin by describing the construct of neighborhood as it is used by social scientists and as it is experienced by the people who live within its environs. The contextual effects of neighborhood are then differentiated from compositional effects, a distinction that is central to the inference that neighborhoods signify more than the aggregation of the characteristics of residents. Theories about neighborhoods and their effects on people are then presented, contrasting "person in environment" models with "person–environment fit" models, and describing the neighborhood stress process model. Research applying these models is reviewed next, focusing first on social structural aspects of neighborhood and then on the built environment. We conclude with a discussion of the policy implications of neighborhood research and directions for future research.

THEORETICAL MODELS OF NEIGHBORHOOD

The Concept of Neighborhood

Neighborhoods refer to aggregations of people living in close proximity to one another within the boundaries of a particular geographical area and are distinguished along *spatial, physical, structural,* and *social* dimensions. The spatial dimension refers to the geopolitical boundaries of the neighborhood that place residents within a specific location. It is most often operationalized by official demarcations of areas, usually Census tracts or block groups – expedient delineations that provide access to official information and statistics about the neighborhood but may not match the perceptions of residents. The physical dimension recently has been elaborated most with respect to the built environment, including features such as density of housing, diversity of land use, and design – the three Ds (Clarke & George, 2005) – especially pedestrian-friendly features that promote walking.

The social structural dimension captures the collective socioeconomic and demographic characteristics of residents that signify the social stratification and racial/ethnic segregation of neighborhoods (Aneshensel, 2010; Aneshensel & Sucoff, 1996). The most researched social structural characteristic is neighborhood socioeconomic disadvantage, which refers to the simultaneous absence of economic, social, and family resources (Ross & Mirowsky, 2001; Wheaton & Clarke, 2003), for instance, percent of households receiving public assistance or percent of households that are single-female-headed. The social aspect refers to the nature of the interactions that occur within the neighborhood, including the formation of collective efficacy and social capital, as well as interactions with institutions, organizations, and agents of social control.

Kemp (2001) differentiates objective environments that are experienced in much the same way by most people from the subjective place that evolves from the lived experience of individuals in a particular environment (cited in Nicotera, 2007). According to Nicotera (2007), these experiences shape the meanings that an individual attaches to an environment and the person's sense of self in relationship to that place. Her analysis of the socially constructed nature of place emphasizes social connections with family and friends and with social institutions and organizations. Similarly, Wahl, Iwarsson, and Oswald (2012) call attention to the cognitive, affective, and behavioral processes that lead to attachment to and identification with environments, transforming "space" into "place," for example, and home. They maintain that environment-related belonging becomes increasingly salient as people approach advanced old age.

Contextual and Compositional Neighborhood Effects

The foundational supposition of neighborhood research is that neighborhoods affect individuals and that these effects can be distinguished from those that flow from the individual's own characteristics. Contextual neighborhood effects refer to the impact of a neighborhood characteristic on an individual outcome that is due to the neighborhood characteristic and is not simply the cumulative result of characteristics of residents that are associated with the outcome, which is known as a compositional effect. Separating these effects can be difficult because although some attributes of neighborhoods clearly pertain to the neighborhood (e.g., acreage in square miles), other attributes are less obviously distinct from resident characteristics because they are measured with aggregated individual-level (e.g., personal income) and neighborhood-level (e.g., median income of neighborhood residents) data. Of particular concern is selection into and out of neighborhoods on characteristics that are related to the outcome of interest.

For example, neighborhood poverty may appear to be associated with depression when it is not because people are sorted into neighborhoods on the basis of housing costs such that residents of impoverished neighborhoods generally are poor and being poor is associated with depression. Alternately, impoverished neighborhoods may affect depression separately from these individual-level processes; for instance, a decaying physical environment may lead to public deviance such as visible drug dealing, which then prompts residents to stay at home and interact primarily with close friends and family, producing a breakdown in social connections within the neighborhood (Massey & Denton, 1993) that is damaging to the mental health of both poor and non-poor residents.

The method of choice for the analytic task of separating contextual and compositional effects is multilevel modeling, also known as hierarchical linear modeling. This method estimates a cross-level effect of the neighborhood-level characteristic on an individual-level outcome while controlling statistically for individual-level characteristics. Some applications also contain cross-level interactions that test the extent to which the effects of individual-level characteristics are contingent upon characteristics of the neighborhood in which the individual lives. For example, Roos, Magoon, Gupta, Chateau, and Veugelers (2004) find that neighborhood socioeconomic disadvantage is not associated with mortality (i.e., absence of a cross-level effect), but the protective impact of personal income is stronger in advantaged than disadvantaged neighborhoods (i.e., presence of a cross-level interaction).

Identifying contextual effects on individual outcomes is one of the most compelling reasons for conducting neighborhood research because these effects attest to the influence of the environment on the individual, including the impact of social stratification and inequality, and that of the built environment. The existence of contextual effects counters the viewpoint that health is solely the responsibility of the individual, which can lead to blaming the individual for factors that are outside his or her control (Heymann & Fischer, 2003). These effects also demonstrate the need for interventions that address these environmental influences because it is more efficient to intervene at the neighborhood level when environmental conditions affect the entire community than it is to treat individuals one-by-one (Gupta, Parkhurst, Ogden, Aggleton, & Mahal, 2008). For depression in particular, Cutrona, Wallace, and Wesner (2006) point out that individuals often do not realize that they are affected by their environments and therefore mistakenly blame themselves for "invisible" stressors over which they have no control – an observation that extends to other stress-related health outcomes.

The Interaction of Person and Environment

Theories about the impact of neighborhood conditions on the people who live within them can be grouped into two broad types (Aneshensel, 2010). "Person in environment" models place the individual within an environment, but do not take into account how the person relates to that environment. Instead neighborhood conditions are implicitly assumed to have a uniform effect on residents. An example is an "advantages of advantaged neighbors" model (Jencks & Mayer, 1990) positing that people benefit from living in proximity to advantaged neighbors and conversely are harmed by living in proximity to disadvantaged neighbors.

In contrast, "person–environment fit" models are based on the premise that neighborhood effects vary from person to person as a function of the interaction between neighborhood conditions and personal attributes and situations. Although some of this variation is idiosyncratic, unique to each individual, these models seek to identify patterns across people with similar capabilities, assets, goals, and so forth.

Gerontological work in this tradition can be traced to Lawton and Nahemow's (1973) seminal person–environment fit model that includes the interaction of the person and the environment. This model predicts optimal outcomes when the individual's competencies correspond to the demands or "press" of the environment. According to Cvitkovich and Wister (2001), criticism of this "competency" model led to the specification of "congruence" models that link positive outcomes to how well the environment meets the needs of the individual. See Cvitkovich and Wister (2001) for the history and additional developments of these models, especially with regard to improving competency and the role that individuals play in changing the environment.

Also influential is Bronfenbrenner's (1979) ecological model, which situates the person within multiple contexts. The "microsystem" refers to the most proximate social groups and institutions, in this instance the home and the neighborhood. The "mesosystem" concerns the intersection of these contexts, including the possibility that their influences on the individual may be synergistic or antagonistic. For example, Wight and colleagues (Wight et al., 2006) find that the proportion of neighborhood residents with a high school education interacts with the person's own educational attainment such that older persons with little formal education are most adversely affected by living in neighborhoods in which other residents also tend to have little formal education. This finding illustrates a synergistic interaction because low neighborhood socioeconomic status (SES) accentuates the adverse effect of low personal SES.

A key feature of this ecological model is the idea that people play an active role in shaping their environments as distinct from being passive recipients of influences from the environment, which corresponds to the concept of agency in the life course perspective. In addition, the "chronosystem" encompasses elements of the life course, calling attention to the sociohistorical context, trajectories and transitions, and the importance of timing, place, and cohort.

However, most quantitative neighborhood research is atheoretical (Yen et al., 2009), although the analytic models employed in these studies place them within the "person in environment" category by default because only the average effect of neighborhood conditions is estimated. This approach calls attention to the ways in which neighborhoods differ from one another and are internally similar, that is, it emphasizes between-neighborhood heterogeneity and within-neighborhood homogeneity.

Although residents of neighborhoods tend to share important characteristics, there usually is substantial variation around these characteristics too, and it is this variation that is emphasized in "person–environment fit" models. This variability means that the same neighborhood may have different effects among residents with dissimilar characteristics and personal situations, amplifying or dampening the extent to which neighborhood conditions affect people.

In other words, "person–environment fit" models examine variation around the average effect of "person in environment" models. An example of this approach is a "disadvantages of advantaged neighbors" model (Jencks & Mayer, 1990), such as the relative deprivation model. This perspective implies that a neighborhood, including an affluent one, may have both desirable and undesirable effects on residents.

Neighborhood Stress Process Model

The neighborhood stress process model attempts to explain why some people in adverse environments are harmed by their surroundings whereas others are seemingly unscathed (Aneshensel, 2010). This integrative model embeds the stress process model (Pearlin & Bierman, 2013; Pearlin, Menaghan, Lieberman, & Mullan, 1981) within the ecological framework in order to trace the impact of social stratification and inequality at multiple layers of the social hierarchy (Wheaton & Clarke, 2003). At issue is the extent to which individual- and neighborhood-level inequality have joint effects on health that are greater than the sum of their separate effects. In this regard, Wheaton and Clarke (2003) identify two potential cross-level interactions: the *compound advantage* model predicts that advantaged neighborhoods benefit personally advantaged residents most, and the *compound disadvantage* model predicts that disadvantaged neighborhoods are most detrimental to personally disadvantaged residents.

The individual-level portion of the stress process model originates with the person's location in the social system as indexed by status characteristics such as SES and race/ethnicity. Low social status is linked to poor health outcomes through greater exposure to stressors, including the process of stress proliferation whereby stressors in one domain of life create stressors in other domains (Pearlin, Aneshensel, & LeBlanc, 1997), in this case neighborhood spillover that creates family conflict, for example. Variations in the impact of stressors on health are a function of the extent to which psychosocial resources, such as social support and mastery, are mobilized and then offset the otherwise deleterious effects of exposure, or instead are depleted and become the means through which stressors exert these harmful effects (Wheaton, 1985). These psychosocial resources also act to buffer the effects of stressors on health, such that the magnitude of these effects decreases as the resource increases.

The ecological portion of the model originates with stratification of neighborhoods by SES and segregation by race/ethnicity. From this perspective, neighborhood-generated health disparities can be explained, at least in part, by neighborhood stratification and segregation creating differences between neighborhoods in features of everyday life that lead to differential exposure to stressors and differential access to psychosocial resources.

Racial segregation has created distinctive ecological environments for African Americans in that most poor African Americans reside in neighborhoods of concentrated poverty (Williams & Collins, 2001). In addition, racial segregation reinforces racial differences in opportunity structures and access to resources, and increases exposure to discrimination (Robert & Ruel, 2006), factors that are likely to elevate stress and psychological distress. Latinos tend to be less segregated and more benign conditions are thought to permeate predominantly Latino neighborhoods, known as "ethnic enclaves." Specifically, as found by Patel and colleagues in a study of older Mexican Americans, these neighborhoods are thought to have high levels of social cohesion that provide considerable social support, including patterns of reciprocity and social exchange, and community institutions (Patel, Eschbach, Rudkin, Peek, & Markides, 2003) – social resources that may mitigate the otherwise damaging health effects of living in socioeconomically disadvantaged neighborhoods. These factors may contribute to the "Latino Paradox," the finding that despite low average SES, Latinos tend to have health outcomes that are as good as or better than non-Hispanic Whites, depending upon nativity, acculturation, and generation (Gallo, Penedo, Espinosa de los Monteros, & Arguelles, 2009). However, any benefits of living among persons of a similar ethnic background may be overridden by the adverse effects of factors associated with segregation (e.g., discrimination) and accompanying socioeconomic disadvantage.

Cross-level effects from neighborhood socioeconomic disadvantage and racial/ethnic segregation are seen as being transmitted through neighborhood disorder, which refers to physical and social signs that social control is lacking – such as the presence of crime, vandalism, unsupervised youth, and vacant buildings – conditions that are experienced as threatening and arouse fear (Ross & Mirowsky, 2001). These conditions are thought to erode collective efficacy – social cohesion among neighbors combined with their willingness to intervene on behalf of the common good (Lochner, Kawachi, & Kennedy, 1999), and social capital – interpersonal trust, norms of reciprocity and social engagement that foster community and social participation (Carpiano, 2006). Neighborhood disorder also indirectly damages health by creating secondary stressors via the process of stress proliferation, described above.

As alluded to above, a key feature of the neighborhood stress process model is the specification of cross-level interactions, specifically interactions that test the hypotheses that the adverse health effects of exposure to stressors are *amplified* by neighborhood socioeconomic disadvantage through a process of compound adversity and *dampened* by high levels of resources through the process of stress buffering (Aneshensel, 2010). Variation in the effectiveness of these resources helps to account for the differential impact of adverse neighborhood conditions.

This line of research is directly relevant to middle-aged and older persons, some of whom find themselves in deteriorating neighborhoods as they and their neighborhoods age in unison. Support networks that are already diminished through death may be further jeopardized if aging persons turn inward, away from neighbors. Krause (1996) contends that dilapidated and crime-ridden neighborhoods project an atmosphere that inhibits social interaction by making older adults suspicious and distrustful of others, leading to social isolation.

NEIGHBORHOOD STRUCTURE AND THE HEALTH OF OLDER PERSONS

Despite the presumed heightened significance of neighborhoods for the older population, relatively few studies focus specifically on this population or examine differences in the effects of neighborhood relative to younger populations (Clarke & Nieuwenhuijsen, 2009). Structural models of neighborhood influences on the health of older persons tend to be atheoretical and descriptive and to apply a "person in environment" model by default insofar as variation by personal characteristics and situations is not explicitly modeled, although these characteristics are controlled statistically. The exception to this generalization is age itself and there is some evidence that age alters the impact of neighborhood on health within the older population.

Neighborhood Socioeconomic Disadvantage

A recent systematic literature review reports that neighborhood SES has the strongest and most consistent relationship with a range of health outcomes within the older population (Yen et al., 2009). For mortality, for instance, Diez Roux, Borrell, Haan, Jackson, and Schultz (2004) find that neighborhood disadvantage is prospectively related to rates of cardiovascular death in older (65+) white persons net of controls for individual-level characteristics, although evidence is equivocal for parallel effects for subclinical cardiovascular disease, a putative etiological pathway (Nordstrom, Diez Roux, Jackson, & Gardin, 2004). Pickett and Pearl (2001) summarize several studies in which neighborhood effects on mortality are modified by age, with neighborhood factors having less of an impact in older age groups, leading the authors to conclude that other factors related to survival are more important in old age, although other studies report no age differences in the impact of neighborhood socioeconomic disadvantage on mortality (e.g., Jaffe, Eisenbach, Neumark, & Manor, 2005).

Robert and Li (2001) examine age variation in the magnitude of the association of neighborhood-level SES with two indicators of physical health status – number of chronic conditions and self-rated health. Holding individual SES constant, the association is not present or is weak during young adulthood, stronger during middle age, strongest at ages 60–69, and weak again at ages 70 and older. The investigators point out that ignoring potential age variations in community effects on physical health may mean that the overall average effect underestimates these effects for some middle-aged or older groups while overestimating these effects among younger and much older groups.

Of the many studies of mental health, principally depressive symptoms, only a few pertain specifically to older persons, and these studies present conflicting evidence of contextual versus compositional effects of neighborhood socioeconomic disadvantage. For instance, one study of persons 65 and older reports a positive association between depressive symptoms and neighborhood poverty when characteristics of the person (e.g., SES) are controlled statistically (Kubzansky et al., 2005). However, a second study reports no association between depressive symptoms and neighborhood disadvantage (as well as neighborhood advantage, age structure, residential stability, and racial/ethnic heterogeneity) in the same age group using similar controls (Hybels et al., 2006). The authors conclude that any association between neighborhood characteristics and symptoms is due to the characteristics of the individuals who reside in the neighborhood rather than the characteristics of the neighborhood, that is, a compositional effect instead of a contextual effect. These discrepant findings may be due to methodological differences, however, in particular the use of distinctly different local samples, New Haven and North Carolina, respectively.

However, a pair of studies using data from the Health and Retirement Study's (HRS) nationally representative sample also yield opposite findings that cannot be attributed to methodological differences: contextual effects are found among persons in late middle age (ages 51–61) (Wight, Ko, & Aneshensel, 2011) and compositional effects are found among older persons (ages 70+) (Aneshensel et al., 2007; Wight, Cummings, Karlamangla, & Aneshensel, 2009). Subsequent analysis of the entire HRS sample of persons ages 52 and older reveal that the negative effect of neighborhood disadvantage on depressive symptoms is strongest among persons under age 65 (Harig, 2012).

Evidence of age variation in the effects of neighborhood socioeconomic disadvantage on older adults raises the question of whether the concentration of older persons in the neighborhood matters to older residents. The tendency for social activities to become increasingly concentrated within a contracting geographical space suggests that a high concentration of older persons in the neighborhood should have a positive effect because it creates a relatively large pool of potential social contacts. However, findings are mixed with some but not all studies showing a positive association between the percentage of the population 65 years of age and older with better mental and physical health among older adults (e.g., Aneshensel et al., 2007; Hybels et al., 2006; Kubzansky et al., 2005). Clarke and Nieuwenhuijsen (2009) suggest that better health outcomes in neighborhoods with a higher proportion of older persons may indicate a greater density of health and social services for older persons.

Neighborhood effects on cognitive functioning are of concern because the occurrence of age-related declines in cognitive functioning is expected to increase enormously in the future (Brunner, 2005). The few existing studies support the idea that neighborhood socioeconomic disadvantage is inversely associated with cognitive functioning among persons who are in late midlife or older. For a national sample of older persons (ages 52 and older) in England, cognitive functioning is inversely associated with neighborhood socioeconomic deprivation for both men and women ages 70 and older and for women but not men under age 70 (Lang et al., 2008). Similarly, cognitive functioning varies inversely with the neighborhood's average level of educational attainment for a US national sample of persons age 70 and older, but this effect is conditional upon the person's own education, being strongest among persons with the least formal education (Wight et al., 2006).

Aneshensel and colleagues (Aneshensel, Ko, Chodosh, & Wight, 2011) examined the impact of neighborhood socioeconomic disadvantage for a somewhat younger US national sample of persons in late middle age (55–65), a stage in the life course when cognitive deficits begin to emerge (Salthouse, 2009). This study also found a synergistic cross-level interaction for (lack of) personal wealth and neighborhood disadvantage. Poor persons living in impoverished neighborhoods exhibit, on average, especially low levels of cognitive functioning, while their neighbors with somewhat more ample financial means appear to be protected from the adverse impact of neighborhood socioeconomic disadvantage. These findings are consistent with the idea of compound disadvantage mentioned above (Wheaton & Clarke, 2003).

In addition to the effects of SES in the absolute, relative social standing also appears to be salient. Deeg and Fleur-Thomése (2005) point out that an income discrepancy relative to neighbors can develop when older persons experience a substantial income decline, for example, through retirement or widowhood, or when the neighborhood deteriorates. They find that these discrepancies adversely affect the physical and mental health of older persons and contribute to loneliness among both low-income residents of high-status neighborhoods – which the investigators attribute to stress of becoming poor late in life and knowing that one will remain poor, and among high-income residents of low-status neighborhoods, including those whose neighborhood has deteriorated over time – which may be due to difficulty resolving the discrepancy between the two statuses.

Although most research focuses on adverse health outcomes, one recent study examined positive mental health in older people (69–78 years), but found no association with area-level deprivation, although perceptions of neighborhood social cohesion and fewer neighborhood problems are associated positively with mental well-being, leading the investigators to conclude that it is how older people feel about their neighborhood that is important for positive mental health outcomes in late life (Gale, Dennison, Cooper, & Sayer, 2011).

Racial and Ethnic Segregation

The second social structural feature of neighborhoods to garner considerable research attention is the racial and ethnic composition of neighborhoods that arises as a result of residential segregation. As discussed above, a recent systematic review of neighborhood and health among older adults concludes that there may be a beneficial effect of living in an "ethnic enclave" for older Latinos in the form of lower morbidity and mortality, fewer depressive symptoms, and better self-rated health (Yen et al., 2009). However, the literature is far from unanimous on this point. For instance, one study finds that living in a neighborhood with a high density of other Mexican Americans is associated with fewer depressive symptoms among older (ages 65+) Mexican Americans (Ostir, Eschbach, Markides, & Goodwin, 2003), but another study (not specific to older persons), finds that neighborhood segregation is associated with higher levels of symptoms of depression and anxiety among both Puerto Ricans and Mexican Americans (Lee, 2009). However, it is difficult to separate the effects of racial or ethnic composition from those of neighborhood socioeconomic disadvantage due to the tendency for most although not all segregated minority neighborhoods to be socioeconomically disadvantaged in comparison to segregated non-Latino White neighborhoods.

In addition, any benefits of living in a predominantly Latino neighborhood may be unique to the Mexican American population. Although not specific to older persons, Lee and Ferraro (2007) find sub-ethnic group variation in the effects of living in highly segregated Latino neighborhoods. Puerto Ricans who live in highly segregated neighborhoods have more health problems (acute physical symptoms and disability) than those who live in communities that are not as segregated. In contrast, residential isolation is beneficial for the health of second- or later-generation Mexican Americans, but not first-generation Mexican Americans. These divergent patterns are not due to neighborhood socioeconomic disadvantage because the Puerto Rican and Mexican American neighborhoods have comparable incomes.

Moreover, the observed effects of residential segregation may mask more complex population dynamics. On the one hand, neighborhood percentage of Mexican Americans is associated with lower all-cause mortality among older

Mexican Americans (ages 65+), leading study investigators to conclude that the sociocultural advantages conferred on Mexican Americans by living in high-density Mexican American neighborhoods outweigh the disadvantages conferred by the high poverty of those neighborhoods (Eschbach, Ostir, Patel, Markides, & Goodwin, 2004). On the other hand, Bond Huie, Hummer, and Rogers (2002) conclude that it is the percentage of persons who are foreign-born that is associated with lower mortality risk among Mexican Americans, Puerto Ricans, and other Latinos (age 45–64), rather than high concentrations of Latinos per se.

Results with regard to cognitive functioning are equally mixed. For example, Mexican Americans living in "barrio" neighborhoods (i.e., low-income, almost exclusively Mexican American neighborhoods) have an elevated risk of cognitive impairment compared to residents of "transitional" and "suburban" neighborhoods, but this study is limited to a single subgroup of the population and a single region of the United States, the Southwest, limiting generalizability (Espino, Lichtenstein, Palmer, & Hazuda, 2001). In contrast, Sheffield and Peek (2009) recently found that the proportion of neighborhood residents who are Mexican American is related to a slower rate of cognitive decline for older Mexican Americans, a potential "ethnic enclave" effect; they also found that neighborhood socioeconomic disadvantage is related to a greater rate of decline over time, although this study has limited generalizability too. Finally, Aneshensel and colleagues (2011) do not find an effect of neighborhood percentage of Hispanics on cognitive functioning among persons in late middle age, although a conditional effect of neighborhood percentage of African Americans is reported (see below).

If there are benefits of residential segregation among the Latino population, these benefits do not accrue to African Americans living in segregated African American neighborhoods (Williams & Collins, 2001; Yen et al., 2009). An isolated exception to this generalization occurs in the study of cognitive functioning among persons in late middle age (55–65) cited above, which finds a significant interaction between the neighborhood concentration of African Americans and the person's own educational attainment (Aneshensel et al., 2011). While education is positively associated with cognition across all neighborhood types, this association is strongest when most if not all residents of the neighborhood are African American. Consequently, segregated African American neighborhoods have the greatest range of cognitive functioning across levels of education. The investigators suggest that the strength of the education effect among this cohort of older African Americans may be indicative of the high levels of ability and motivation that would have been necessary to obtain a higher education in the historical context of legally segregated education, which may outweigh the cognitive costs of racial segregation that is incurred by African Americans overall.

Stressors and Resources

The defining feature of neighborhood research based on the neighborhood stress process model is explicating the social *pathways* that connect the social structural features of neighborhood disadvantage to the health and emotional well-being of its residents (Aneshensel, 2010). As Elliott (2000) points out, most research on the effects of SES on health is conducted at the individual level, ignoring the social context, a practice that assumes implicitly that these effects are invariant across diverse social settings. Wheaton and Clarke (2003) assert that individual-level social class effects on mental health may be misspecified when the intersection of the individual and contextual components of social class is ignored. As described above, cross-level interactions indicative of compound disadvantage have been found for older adults for cognitive functioning (Aneshensel et al., 2011; Wight et al., 2006) and depressive symptoms (Wight et al., 2011).

Neighborhood SES appears to influence health outcomes in part because of its association with individual SES. For instance, Wen, Hawkley, and Cacioppo (2006) find that at late midlife to early old age (50–67), the association between objective neighborhood SES and self-rated health is partly accounted for by individual-level SES. This study also finds that this association is part of a sequential set of pathways that operate through perceived neighborhood quality and several psychosocial factors: loneliness, depression, hostility, and stress. Social support and social networks, however, do not perform this function.

Elements of the stress process model are conditional upon other status characteristics as well. For example, Bierman (2009) contends that the damaging mental health impact of neighborhood disorder is attenuated among the married because marriage tends to provide a predictable social relationship with established role expectations that create a sense of order. He finds that neighborhood disorder is related to increases over time in depressive symptoms among older unmarried persons but not married persons. Moreover, this differential effect is due to a parallel difference in the effect of neighborhood disorder on mastery, which is more likely to decrease over time as a result of exposure to neighborhood disorder among the unmarried than the married.

A cluster of papers from the same study demonstrates that the connections between neighborhood conditions and elements of the stress process are complex: conditional upon how older people compare themselves to their neighbors (Schieman & Pearlin, 2006), specific to particular psychosocial resources and mental health outcomes (Schieman & Meersman, 2004), and contingent on race and gender (Schieman, 2005). For example, Schieman and Meersman (2004) find some evidence that social support and mastery buffer the effects of perceived neighborhood problems on anger for older men but not older women, but this buffering effect is not present for men or women for depression or anxiety.

The impact of some other elements of the stress process also appears to be conditional upon the neighborhood context. The effect of stressful events and circumstances in people's lives may be intensified by living in adverse neighborhood conditions. For example, neighborhood-level economic disadvantage exacerbates the effect of negative life events on the onset of major depression among African American women (Cutrona et al., 2005). Similarly, the impact of at least some psychosocial resources appears to depend in part on neighborhood conditions. As a case in point, social support is beneficial to mental and physical health among midlife and older residents of higher SES neighborhoods, but not similar persons in lower SES neighborhoods (Elliott, 2000). Other studies find that psychosocial resources mediate the effects of neighborhood stressors. For instance, among Hispanic older persons, a positive neighborhood social climate (e.g., supportive acts of neighbors, informal social ties) has a significant indirect relationship with psychological distress that is mediated by social support (Brown, Mason, Spokane, et al., 2009). Based on a recent "realistic" review, the most well-established causal links between neighborhood socioeconomic disadvantage and depression appear to be through heightened exposure to multiple stressors and increased vulnerability to these stressors that lead to perceptions of neighborhood disorder and a sense of powerlessness (Blair et al., 2014).

Most research on the stress process is conducted at the individual level, which carries with it the assumption that the relationships among its components are the same irrespective of the social context. The contingencies just described suggest otherwise. To the extent that these relationships vary across settings, and a recent review concludes that they are conditional on setting (Blair et al., 2014), extant individual-level models of the stress process may well be misspecified.

THE BUILT ENVIRONMENT AND THE HEALTH OF OLDER PERSONS

In recent years, research has proliferated on one aspect of the physical environment in particular – the built environment, which refers to surroundings, spaces, and settings that have been made or modified by people (as distinct from nature) for use in any activity (e.g., work, family life), such as buildings and green spaces, transportation systems, and the physical infrastructure (e.g., roads, sidewalks, water supply) (Clarke, Ailshire, & Lantz, 2009; Renalds, Smith, & Hale, 2010). It encompasses both the internal and external environments, including importantly the place where a person lives and the areas that surround it.

Clarke and Nieuwenhuijsen (2009) maintain that older persons are particularly susceptible to barriers in their surrounding physical environments as a result of declining health and functional status, financial strain and social isolation, especially persons with limitations in their functional abilities and those who need ready access to transportation and other services. Barriers to outdoor mobility that are created by the built environment can negatively impact a person's ability to function independently in the community, for instance to access shops, banks, and health services (Clarke et al., 2009).

Physical Activity and Health

One of the primary avenues through which the built environment affects health among older persons is by facilitating or impeding physical activity, especially walking. For example, neighborhood walkability is related to transport activities and moderate-to-vigorous physical activity, with mobility-impaired older persons (65+) living in walkable neighborhoods reporting transport activity levels that are comparable to less mobility-impaired adults living in less walkable neighborhoods (King et al., 2011). In addition, proximity to local services and amenities is associated with a greater likelihood of maintaining frequent walking over time, demonstrating how aspects of the built environment can act as "buoys" for the continuation of health-promoting behaviors (Gauvin et al., 2012).

A recent systematic literature review concludes that environments that are suitable for leisure- or destination-driven walking contribute to greater physical activity and lower prevalence of overweight, depression, and alcohol abuse (Renalds et al., 2010). Another recent review identified the following as facilitating mobility in older persons: higher street connectivity leading to shorter pedestrian distances; street and traffic conditions such as safety measures; and, proximity to destinations such as retail establishments, parks, and green spaces (Rosso, Auchincloss, & Michael, 2011). For these reasons, Yen and colleagues (2009) conclude that existing research supports an association between accessible neighborhood design and greater levels of walking.

Some aspects of the built environment are especially consequential to the health of older persons. For instance, aspects that are associated with functional deterioration over time include excessive noise, inadequate lighting, and heavy traffic (Balfour & Kaplan, 2002). Also, older persons (75+) living in neighborhoods characterized by high levels of motorized travel have substantially higher odds of developing a mobility disability compared to similar persons living in neighborhoods that are pedestrian-friendly (Clarke et al., 2009). A small-scale exploratory study reports that relative to similar others, disabled older (70+) persons report greater avoidance of some physically challenging features of the environment (e.g., walking long distances, streets with traffic lights) than others (e.g., noisy, busy, crowded, or unfamiliar places), leading the investigators to conclude that some environmental features are more disabling than others (Shumway-Cook et al., 2003).

Some of these effects appear to be transmitted through social processes. For instance, perceived social support is positively associated with architectural features of the front entrance of homes that promote visibility from a building's exterior, such as porches, and negatively associated with window areas that promote visibility from a building's interior, and these features are associated with psychological distress (Brown, Mason, Lombard, et al., 2009). Based on these findings, the investigators conclude that building features that encourage direct, in-person interactions are beneficial for the mental health of older persons.

Along these lines, a recent study examined the extent to which distance from one's residence to community sites that are conducive to social participation is associated with actual participation in social activities among older persons (Richard et al., 2013). The selected sites were public libraries, community centers for older adults, physical activity places, and shopping centers and the most frequently reported social activities were visiting family and friends, going shopping and practicing a hobby outside the home, while taking courses, participating in discussion or self-help groups, doing volunteer work, and attending activities at community or leisure centers were mentioned least often. Results show that objective measures of distance to social locations are associated with social participation even when perceived proximity to local-area services and amenities are controlled, and perceived proximity also is associated with social participation.

Beyond the objective characteristics of the built environment, the subjective assessments of residents about their neighborhoods also are meaningful. According to Clarke and Nieuwenhuijsen (2009), older people experience the following environmental barriers: poor transportation, discontinuous or uneven sidewalks, curbs, noise, and inadequate lighting. Other research indicates that pedestrian-oriented designs (e.g., continuous, barrier-free sidewalks, four-way stop signals, and pedestrian amenities) are associated with physical activity and self-rated health in older adults and inversely related to obesity. Perceived neighborhood problems, social cohesion, and safety are associated with satisfaction with participation in everyday activities (e.g., self-care, mobility, work or social activities) among older adults with chronic health conditions, either directly or indirectly by impacting social support, which in turn increases satisfaction (Hand, Law, Hanna, Elliott, & McColl, 2012).

It should be noted that some studies fail to find health effects of the built environment. For instance, neighborhood walkability does not predict body mass index or risk of obesity over time among older women who were previously nonsedentary (Michael, Gold, Perrin, & Hillier, 2013) or amount of walking among post-menopausal women (Perry et al., 2013). However, the absence of a main effect of the built environment may mask conditional effects insofar as there is strong evidence that these effects are conditional upon the person's functional status (see below). For example, a recent study reports that the impact of walkability on decline over time in dynamic leg strength is conditional upon whether or not the women walked at baseline: walkability was unrelated to declines among older women who were not walking at baseline, but slowed declines among women who did walk at baseline (Michael, Gold, Perrin, & Hillier, 2011). If conditional relationships are ignored, the average effect may be quite misleading.

The Disability Process

Emergent research on the disability process demonstrates that the built environment can smooth or obstruct a person's ability to function independently in the community, holding the disability process in check or accelerating its pace (e.g., Clarke et al., 2009; Clarke & George, 2005). The Disability Process Model describes the dynamic process through which a health condition becomes a disability, emphasizing the mechanisms that speed up this progression or slow it down, including features of the built environment (Clarke & George, 2005). According to Clarke and George, these barriers expand the discrepancy between people's functional capacities and their actualized ability to accomplish the tasks of daily life. The investigators find that declining physical function has less of an effect on independence in performing instrumental activities when people live in areas with greater land-use diversity than in areas that lack this diversity.

Research in this area generally employs a "person–environment fit" model in which the effect of the health problem on disability is conditional on an aspect of the built environment, specifically stronger in environments that pose a greater number of barriers. Stated differently, the impact of the environment is greater among persons with health problems than among those who are healthier. The concept of "fit" is operationalized as an interaction between the person's health condition and a feature of the built environment. Alternately, comparisons are made between people with and without a condition, or between built environments that pose barriers and those that do not.

Consistent with this model, there is good evidence that the impact of the built environment is not uniform throughout the older population, but is more intense among persons with health conditions. In a recent review, Clarke and Nieuwenhuijsen (2009) conclude that older persons are particularly susceptible to physical barriers in their local environments when they experience limitations in their functional abilities and are in need of accessible transportation and services; otherwise, these barriers seem to be relatively unimportant. The authors also identify social groups with heightened vulnerability, including socioeconomically disadvantaged older persons, who need to access social services more often than others, such as community meal programs and senior centers, and; women, minority and low-income older persons who are disproportionately likely to live alone and in substandard housing in disadvantaged neighborhoods.

Clarke and Nieuwenhuijsen (2009) also conclude that residential context regulates the degree to which these limitations in physical function translate into actual difficulty in activities outside of the home, but seems to have little or no effect among those with mild or no impairment. For instance, street conditions do not affect outdoor mobility among midlife and older persons with no physical impairment or only mild impairment, but among those with more severe impairment in neuromuscular and movement-related functions, poor street quality (e.g., cracks, potholes, broken curbs) greatly increases the odds of having severe mobility problems relative to streets that are in good condition (Clarke, Ailshire, Bader, Morenoff, & House, 2008). Compared to those living in less restrictive communities, older people with functional limitations living in communities with mobility barriers and without transportation facilitators felt more limited in doing daily activities; however, they did not perform these daily activities any less frequently (Keysor et al., 2010).

Other studies do not specifically model the conditional relationship implied in the "person–environment fit" model, but instead study only persons who are presumed to have greater competency issues pertaining to the built environment, typically older people with some condition that challenges mobility. Although this design is informative about these particular populations, the results are not informative about the concept of "fit" because there is no comparison to persons with other competencies or needs. However, this design is informative about the group where one expects to see the strongest environmental effect (even if it cannot be shown to be stronger than in other groups).

For instance, a recent qualitative study provides insight into the impact of the built environment on physical activity through the eyes of midlife and older persons with mobility disabilities, predominantly use of canes, walkers, and wheelchairs (Rosenberg, Huang, Simonovich, & Belza, 2013). The key themes that emerged from latent content analysis of in-depth interviews included: availability and condition of ramps, curb ramps, and sidewalks; traffic and crosswalks; paved or smooth walking paths; resting places and shelter on streets; safety; and weather. These barriers to and facilitators of neighborhood-based activity for people with mobility disabilities led the investigators to conclude that changes to the neighborhood built environment are needed in order to foster independence by creating healthy environments for an aging population that uses assistive devices.

In addition to the impact of the built environment, social structural aspects of neighborhoods are implicated in the occurrence of disability among older persons, including "going outside the home disability," such as low neighborhood SES, residential instability, and living in areas with low proportions of foreign-born and high proportions of black residents (Beard et al., 2009).

Aging in Place

The built environment is an important determinant of whether older persons are able to continue living independently in the community (Erickson, Krout, Ewen, & Robison, 2006), which is the preference of most older persons, that is, to "age in place" (Bayer & Harper, 2000). Factors that predict remaining in place among older persons include home equity, financial resources, and strong ties to the community, while moving is predicted by high property taxes and utility costs, changes in family composition, and diminished physical well-being (Sabia, 2008). Although some older persons are "pushed" out of their homes by a crisis, others plan their moves well in advance and are "pulled" into desirable housing (Erickson et al., 2006), and while some stay in declining neighborhoods out of necessity, others remain by choice due to attachment to place that results from long residence in the area and a sense of identify (Brown, Perkins, & Brown, 2003). Attachment to one's neighborhood or community, in turn, is an important correlate of aging in place (Sabia, 2008).

Aspects of the built environment that affect aging in place include housing and transportation, recreational opportunities that encourage physical activity, and places for social interaction and cultural engagement (Wiles, Leibing, Guberman, Reeve, & Allen, 2011) A recent study found that the homes of older persons who moved contain more mobility hazards (e.g., poorly lighted hallways, objects to step over) and fewer accessibility features (e.g., lack of ramp, wheelchair modifications) than homes of those who remain in place (Erickson et al., 2006). Those who moved cited their own poor health as the primary motivation for moving, whereas those considering moving mentioned illness less often than upkeep and maintenance of their homes, anticipating future needs, size of the home, and a desire to be near family. However, the small sample size limits generalizability.

The same study illuminates how older persons perceive the concept of "aging in place" as encompassing not only the physical home, the focus of most research in this area, but also its setting in the surrounding neighborhood and community (Wiles et al., 2011). The study shows that older people see staying where they are as providing a sense of attachment and connection, feelings of security and familiarity, a sense of identity, and independence; whereas the home in its material form was rarely mentioned. Similarly, Sabia (2008) concludes that older homeowners attach meaning to homeownership in terms of financial security, family, and leaving a legacy.

Recovery of Mobility

Although most of the research on the built environment focuses on loss of the ability to move about the neighborhood, assuming that these losses worsen with age overlooks the possibility of reversals. For instance, a recent study examined contraction and expansion of life space evaluated along a continuum ranging from a minimum of the bedroom followed by areas immediately outside the home, in the neighborhood, outside the immediate neighborhood but in the same town or community, to outside the town or community (Shah et al., 2012). Of older persons using the maximum life space at baseline, about half reported incident life space constriction over an average 4-year follow-up; of this group, approximately two-thirds subsequently reported re-expansion of spatial mobility to the largest zone of life space, a transition that is predicted by having a valid driver's license.

Thus models of the disability process should take into consideration the ability to adapt to a health condition or return to previous levels of functioning. Incorporating the life course framework would be beneficial in this regard, especially the concept of agency.

DISCUSSION AND DIRECTIONS FOR FUTURE RESEARCH

Age and Time

Research on neighborhood and the built environment among older persons is predicated on the widely accepted view that these conditions become increasingly relevant as people age; however, some research suggests that this assumption may not be entirely accurate. As reviewed above, there is good, albeit not entirely consistent, evidence of age variation in the impact of neighborhood socioeconomic disadvantage on the health of older persons for at least some health outcomes, with neighborhood effects attenuated at advanced ages. As a result, studies that estimate an average effect of neighborhood disadvantage on health may well be misleading if variation in these effects by age is not taken into consideration (Richardson & Mitchell, 2010).

Along the same lines, longitudinal research on neighborhood SES effects on health among the older persons identifies some important considerations about the timing of neighborhood effects. Consider, for instance, the following finding: Neighborhood socioeconomic disadvantage is associated with worse self-rated health at baseline among adults ages 60 and older, but does not predict subsequent change over time (Yao & Robert, 2008). As noted by the investigators, this finding implies that these neighborhood effects were already in place at baseline and were then perpetuated throughout old age. Moreover, neighborhood effects may be established even earlier in life and sustained over time (Wheaton & Clarke, 2003), giving the appearance of neighborhood effects at older ages. Supporting this idea, an analysis of the impact of a neighborhood's history of unemployment and how it has changed over time suggests that exposure to neighborhood unemployment during earlier stages in the life course may be influential to mental health later in life (Wight, Aneshensel, Barrett, Ko, Chodosh, & Karlamangla 2013). Furthermore, Clarke and Nieuwenhuijsen (2009) point out that contextual effects are cumulative over time.

These considerations are at variance with the idea that neighborhood becomes increasingly salient as people age and the widespread adoption of this premise may have deflected research away from examining age differences in the effects of neighborhood among older persons. The idea that neighborhood effects intensify as people grow older requires further testing because the arguments in favor of it are compelling, but the evidence in support of it is equivocal. The life course perspective would provide a useful lens for considering these issues of age, aging, and possibly cohort.

Toward Evidence-Based Interventions

The upsurge in research on neighborhoods and the built environment over the past two decades has been fueled in large part by the realization that "upstream" interventions are required to remedy the "fundamental causes" of health disparities as distinct from conventional interventions targeting the proximal determinants of poor health through individual behavior change (Link & Phelan, 1995; McKinlay & Marceau, 2000). Although neighborhood socioeconomic disadvantage and racial/ethnic segregation seem intractable, some of their consequences are likely to be more malleable. For example, neighborhood revitalization programs can lead to improvements in psychological well-being (Dalgard & Tambs, 1997), although obstacles are formidable. Changing neighborhood conditions has the potential to enhance the health of all residents, not just a few, which could produce meaningful reductions in neighborhood-generated health disparities (Gupta et al., 2008). Heymann and Fischer (2003) argue that research demonstrating an independent effect of adverse neighborhood conditions beyond individual characteristics: (a) counters public sentiment that individuals are solely responsible for their health, (b) supports public policy holding society partially accountable for improving the health of people living in disadvantaged neighborhoods, and (c) improves the odds that initiatives designed to address neighborhood conditions will be taken seriously.

Neighborhood studies often conclude by advocating for interventions at the community level or changes in the built environment. As a case in point, Renalds et al. (2010) maintain that the adoption of a healthy lifestyle is facilitated by environmental-level changes that embed health-promoting features within the built environment, such as walkable destinations like grocery stores. Along the same lines, Clarke et al. (2009) conclude that simple changes to the built environment, such as home modifications to prevent a fall, may be easier to implement than efforts to change risk factors at the individual level. Similarly, Sabia (2008) contends that aging in place has become a plausible objective to satisfying the housing needs of older persons and therefore recommends that policies designed to reduce the user costs of homeownership be implemented, such as reducing user costs of ownership with property tax cuts and utility subsidies. Finally, Blair and colleagues (2014) identify several proposed modifiable pathways, including exposure to stressors and access to psychosocial resources such as social networks.

However, these recommendations typically are based on observational and cross-sectional research designs that fall short of providing the types of data needed to inform the design of evidence-based interventions and they do not identify the kinds of changes in the built environment that are likely to achieve the desired end. Rather than continuing to demonstrate that harmful neighborhood effects on health exist, it is an opportune time to more systematically probe the causal mechanisms that generate these effects, especially the social psychological processes that link objective elements of the environment to subjective assessments that then influence behavior (cf. Blair et al., 2014). An understanding of these mechanisms is essential to the design of effective remedies (Aneshensel, 2009). This endeavor would benefit from the application of the life course perspective, especially the incorporation of the concept of agency as a counterweight to the tendency to see people as passive reactors to environmental conditions.

In conclusion, the bulk of work in this area focuses on negative socioeconomic aspects of neighborhoods and obstacles in the built environment and employs a model of increasing vulnerability among older people as they age. Although there remains much to learn from these orientations, more information is needed about the positive features of where people live and how these features potentiate the abilities of older persons. Balancing these perspectives is likely to provide a more complete understanding of the ways in which neighborhoods affect older persons with the ultimate objective of creating living spaces that enable people to age actively and to continue to live in the community to the extent that their health allows.

References

Aneshensel, C. S. (2009). Toward explaining mental health disparities. *Journal of Health and Social Behavior, 50,* 377–394.

Aneshensel, C. S. (2010). Neighborhood as a social context of the stress process. In W. R. Avison, C. S. Aneshensel, S. Schieman, & B. Wheaton (Eds.), *Advances in the conceptualization of the stress process: Essays in honor of Leonard I. Pearlin* (pp. 35–52). New York, NY: Springer.

Aneshensel, C. S., Ko, M. J., Chodosh, J., & Wight, R. G. (2011). The urban neighborhood and cognitive functioning in late middle age. *Journal of Health and Social Behavior, 52,* 163–179.

Aneshensel, C. S., & Sucoff, C. A. (1996). The neighborhood context of adolescent mental health. *Journal of Health and Social Behavior, 37,* 293–310.

Aneshensel, C. S., Wight, R. G., Miller-Martinez, D., Botticello, A. L., Karlamangla, A. S., & Seeman, T. E. (2007). Urban neighborhoods and depressive symptoms among older adults. *The Journals of Gerontology Series B: Psychological Sciences and Social Sciences, 62,* S52–S59.

Balfour, J. L., & Kaplan, G. A. (2002). Neighborhood environment and loss of physical function in older adults: Evidence from the Alameda County Study. *American Journal of Epidemiology, 155,* 507–515.

Bayer, A., & Harper, L. (2000). *Fixing to stay: A national survey of housing and home modification issues.* Washington, DC: American Association of Retired Persons.

Beard, J. R., Blaney, S., Cerda, M., Frye, V., Lovasi, G. S., Ompad, D., et al. (2009). Neighborhood characteristics and disability in older adults. *The Journals of Gerontology Series B: Psychological Sciences and Social Sciences, 64B,* 252–257.

Bierman, A. (2009). Marital status as contingency for the effects of neighborhood disorder on older adults' mental health. *Journal of Gerontology: Social Sciences, 64B,* 425–434.

Blair, A., Ross, N. A., Gariepy, G., & Schmitz, N. (2014). How do neighborhoods affect depression outcomes? A realist review and a call for the examination of causal pathways. *Social Psychiatry and Psychiatric Epidemiology, 49,* 873–887.

Bond Huie, S. A., Hummer, R. A., & Rogers, R. G. (2002). Individual and contextual risks of death among race and ethnic groups in the United States. *Journal of Health and Social Behavior, 43,* 359–381.

Bronfenbrenner, U. (1979). *The ecology of human development: Experiments by nature and design.* Cambridge, MA: Harvard University Press.

Brown, B., Perkins, D. D., & Brown, G. (2003). Place attachment in a revitalizing neighborhood: Individual and block levels of analysis. *Journal of Environmental Psychology, 23,* 259–271.

Brown, S. C., Mason, C. A., Lombard, J. L., Martinez, F., Plater-Zyberk, E., Spokane, A. R., et al. (2009). The relationship of built environment to perceived social support and psychological distress in Hispanic elders: The role of "eyes on the street". *The Journals of Gerontology Series B: Psychological Sciences and Social Sciences, 64B,* 234–246.

Brown, S. C., Mason, C. A., Spokane, A. R., Cruza-Guet, M. C., Lopez, B., & Szapocznik, J. (2009). The relationship of neighborhood climate to perceived social support and mental health in older Hispanic immigrants in Miami, Florida. *Journal of Aging and Health, 21,* 431–459.

Brunner, E. J. (2005). Social and biological determinants of cognitive aging. *Neurobiology of Aging Supplement, 1,* 17–20.

Carpiano, R. M. (2006). Toward a neighborhood resource-based theory of social capital for health: Can Bourdieu and sociology help? *Social Science & Medicine, 62,* 165–175.

Clarke, P., Ailshire, J. A., Bader, M., Morenoff, J. D., & House, J. S. (2008). Mobility disability and the urban built environment. *American Journal of Epidemiology, 168,* 506–513.

Clarke, P., Ailshire, J. A., & Lantz, P. (2009). Urban built environment and trajectories of mobility disability: Findings from a national sample of community-dwelling American adults (1986–2001). *Social Science and Medicine, 69,* 964–970.

Clarke, P., & George, L. K. (2005). The role of the built environment in the disablement process. *American Journal of Public Health, 95,* 1933–1939.

Clarke, P., & Nieuwenhuijsen, E. R. (2009). Environments for healthy ageing: A critical review. *Maturitas, 64*, 14–19.

Cutrona, C. E., Russell, D. W., Brown, P. A., Clark, L. A., Hessling, R. M., & Gardner, K. A. (2005). Neighborhood context, personality, and stressful life events as predictors of depression among African American women. *Journal of Abnormal Psychology, 114*, 3–15.

Cutrona, C. E., Wallace, G., & Wesner, K. A. (2006). Neighborhood characteristics and depression: An examination of stress processes. *Current Directions in Psychological Sciences, 15*, 188–192.

Cvitkovich, Y., & Wister, A. (2001). A comparison of four person-environment fit models applied to older adults. *Journal of Housing for the Elderly, 14*, 1–25.

Dalgard, O. S., & Tambs, K. (1997). Urban environment and mental health: A longitudinal study. *The British Journal of Psychiatry, 171*, 530–536.

Deeg, D. J. H., & Fleur-Thomése, G. C. (2005). Discrepancies between personal income and neighbourhood status: Effects on physical and mental health. *European Journal of Ageing, 2*, 98–108.

Diez Roux, A. V., Borrell, L. N., Haan, M., Jackson, S. A., & Schultz, R. (2004). Neighbourhood environments and mortality in an elderly cohort: Results from the Cardiovascular Health Study. *Journal of Epidemiology and Community Health, 58*, 917–923.

Elliott, M. (2000). The stress process in neighborhood context. *Health & Place, 6*, 287–299.

Erickson, M. A., Krout, J., Ewen, H., & Robison, J. (2006). Should I stay or should I go? Moving plans of older adults. *Journal of Housing for the Elderly, 20*, 5–22.

Eschbach, K., Ostir, G. V., Patel, K. V., Markides, K. S., & Goodwin, J. S. (2004). Neighborhood context and mortality among older Mexican Americans: Is there a barrio advantage? *American Journal of Public Health, 94*, 1807–1812.

Espino, D. V., Lichtenstein, M. J., Palmer, R. F., & Hazuda, H. P. (2001). Ethnic differences in Mini-Mental State Examination (MMSE) scores: Where you live makes a difference. *Journal of the American Geriatrics Society, 49*, 538–548.

Gale, C. R., Dennison, E. M., Cooper, C., & Sayer, A. A. (2011). Neighbourhood environment and positive mental health in older people: The Hertfordshire Cohort Study. *Health & Place, 17*, 867–874.

Gallo, L. C., Penedo, F. J., Espinosa de los Monteros, K., & Arguelles, W. (2009). Resiliency in the face of disadvantage: Do Hispanic cultural characteristics protect health outcomes? *Journal of Personality, 77*, 1707–1746.

Gauvin, L., Richard, L., Kestens, Y., Shatenstein, B., Daniel, M., Moore, S., et al. (2012). Living in a well-serviced urban area is associated with maintenance of frequent walking among seniors in the VoisiNuAge Study. *The Journals of Gerontology Series B: Psychological Sciences and Social Sciences, 67B*, 76–88.

Glass, T. A., & Balfour, J. L. (2003). Neighborhoods, aging, and functional limitations. In I. Kawachi & L. F. Berkman (Eds.), *Neighborhoods and health* (pp. 303–334). New York, NY: Oxford University Press.

Gupta, G. R., Parkhurst, J. O., Ogden, J. A., Aggleton, P., & Mahal, A. (2008). Structural approaches to HIV prevention. *The Lancet, 372*, 764–775.

Hand, C., Law, M., Hanna, S., Elliott, S., & McColl, M. A. (2012). Neighbourhood influences on participation in activities among older adults with chronic health conditions. *Health & Place, 18*, 869–876.

Harig, F. A. (2012). *The urban neighborhood, depressive symptoms, and age: Stress and psychosocial resources*. UCLA electronic theses and dissertations, p. 50. <http://escholarship.org/uc/item/67v599gk>. Accessed 16.02.14.

Heymann, J., & Fischer, A. (2003). Neighborhoods, health research, and its relevance to public policy. In I. Kawachi & L. F. Berkman (Eds.), *Neighborhoods and health* (pp. 335–348). Oxford, UK: Oxford University Press.

Hill, T. D., & Maimon, D. (2013). Neighborhood context and mental health. In C. S. Aneshensel, J. C. Phelan, & A. Bierman (Eds.), *Handbook of the sociology of mental health* (pp. 479–501, 2nd ed.). Dordrecht: Springer.

Hybels, C. F., Blazer, D. G., Pieper, C. F., Burchett, B. M., Hays, J. C., Fillenbaum, G. G., et al. (2006). Sociodemographic characteristics of the neighborhood and depressive symptoms in older adults: Using multilevel modeling in geriatric psychiatry. *American Journal of Geriatric Psychiatry, 14*, 498–506.

Jaffe, D. H., Eisenbach, Z., Neumark, Y. D., & Manor, O. (2005). Individual, household and neighborhood socioeconomic status and mortality: A study of absolute and relative deprivation. *Social Science & Medicine, 60*, 989–997.

Jencks, C., & Mayer, S. E. (1990). The social consequences of growing up in a poor neighborhood. In L. E. Lynn Jr. & M. G. H. McGeary (Eds.), *Inner-city poverty in the United States* (pp. 111–185). Washington, DC: National Academy Press.

Kemp, S. P. (2001). Environment through a gendered lens: From person-in-environment to woman-in-environment. *Affilia, 16*, 7–30.

Keysor, J. J., Jette, A. M., LaValley, M. P., Lewis, C. E., Torner, J. C., Nevitt, M. C., et al. (2010). Community environmental factors are associated with disability in older adults with functional limitations: The MOST Study. *The Journals of Gerontology Series A: Biological Sciences and Medical Sciences, 65A*, 393–399.

King, A. C., Sallis, J. F., Frank, L. D., Saelens, B. E., Cain, K., Conway, T. L., et al. (2011). Aging in neighborhoods differing in walkability and income: Associations with physical activity and obesity in older adults. *Social Science & Medicine, 73*, 1525–1533.

Krause, N. (1996). Neighborhood deterioration and self-rated health in later life. *Psychology and Aging, 11*, 342–352.

Kubzansky, L. D., Subramanian, S. V., Kawachi, I., Fay, M. E., Soobader, M. J., & Berkman, L. F. (2005). Neighborhood contextual influences on depressive symptoms in the elderly. *American Journal of Epidemiology, 162*, 253–260.

Lang, I. A., Llewellyn, D. J., Langa, K. M., Wallace, R. B., Huppert, F. A., & Melzer, D. (2008). Neighborhood deprivation, individual socioeconomic status, and cognitive function in older people: Analyses from the English Longitudinal Study of Ageing. *Journal of the American Geriatrics Society, 56*, 191–198.

Lawton, M. P., & Nahemow, L. (1973). An ecological theory of adaptive behavior and aging. In C. Eisdorfer & M. P. Lawton (Eds.), *The psychology of adult development and aging* (pp. 657–667). Washington, DC: American Psychological Association.

Lee, M. A. (2009). Neighborhood residential segregation and mental health: A multilevel analysis on Hispanic Americans in Chicago. *Social Science and Medicine, 68*, 1975–1984.

Lee, M. A., & Ferraro, K. F. (2007). Neighborhood residential segregation and physical health among Hispanic Americans: Good, bad, or benign? *Journal of Health and Social Behavior, 48*, 131–148.

Link, B. G., & Phelan, J. (1995). Social conditions as fundamental causes of disease. *Journal of Health and Social Behavior, 35*(Extra Issue), 80–94.

Lochner, K., Kawachi, I., & Kennedy, B. P. (1999). Social capital: A guide to its measurement. *Health & Place, 5*, 259–270.

Massey, D. S., & Denton, N. A. (1993). *American apartheid: Segregation and the making of the underclass*. Cambridge, MA: Harvard University Press.

McKinlay, J. B., & Marceau, L. D. (2000). To boldly go…. *American Journal of Public Health, 90*, 25–33.

Michael, Y. L., Gold, R., Perrin, N., & Hillier, T. (2011). Built environment and lower-extremity physical performance: Prospective findings from the study of osteoporotic fractures in women. *Journal of Aging and Health, 23*, 1246–1262.

Michael, Y. L., Gold, R., Perrin, N., & Hillier, T. (2013). Built environment and change in body mass index in older women. *Health & Place, 22*, 7–10.

Nicotera, N. (2007). Measuring neighborhood: A conundrum for human services researchers and practitioners. *American Journal of Community Psychology, 40*, 26–51.

Nordstrom, C. K., Diez Roux, A. V., Jackson, S. A., & Gardin, J. M. (2004). The association of personal and neighborhood socioeconomic indicators with subclinical cardiovascular disease in an elderly cohort. The cardiovascular health study. *Social Science & Medicine, 59*, 2139–2147.

Ostir, G. V., Eschbach, K., Markides, K. S., & Goodwin, J. S. (2003). Neighbourhood composition and depressive symptoms among older Mexican Americans. *Journal of Epidemiology and Community Health, 57*, 987–992.

Patel, K. V., Eschbach, K., Rudkin, L. L., Peek, M. K., & Markides, K. S. (2003). Neighborhood context and self-rated health in older Mexican Americans. *Annals of Epidemiology, 13*, 620–628.

Pearlin, L. I., Aneshensel, C. S., & LeBlanc, A. J. (1997). The forms and mechanisms of stress proliferation: The case of AIDS caregivers. *Journal of Health and Social Behavior, 38*, 223–236.

Pearlin, L. I., & Bierman, A. (2013). Current issues and future directions in research into the stress process. In C. S. Aneshensel, J. C. Phelan, & A. Bierman (Eds.), *Handbook of the sociology of mental health* (pp. 325–340 2nd ed.). Dordrecht: Springer.

Pearlin, L. I., Menaghan, E. G., Lieberman, M. A., & Mullan, J. T. (1981). The stress process. *Journal of Health and Social Behavior, 22*, 337–356.

Perry, C. K., Herting, J. R., Berke, E. M., Nguyen, H. Q., Moudon, A. V., Beresford, S. A. A., et al. (2013). Does neighborhood walkability moderate the effects of intrapersonal characteristics on amount of walking in post-menopausal women? *Health & Place, 21*, 39–45.

Pickett, K. E., & Pearl, M. (2001). Multilevel analyses of neighbourhood socioeconomic context and health outcomes: A critical review. *Journal of Epidemiology and Community Health, 55*, 111–122.

Renalds, A., Smith, T. H., & Hale, P. J. (2010). A systematic review of built environment and health. *Family and Community Health, 33*, 68–78.

Richard, L., Gauvin, L., Kestens, Y., Shatenstein, B., Payette, H., Daniel, M., et al. (2013). Neighborhood resources and social participation among older adults: Results from the VoisiNuage Study. *Journal of Aging and Health, 25*, 296–318.

Richardson, E. A., & Mitchell, R. (2010). Gender differences in relationships between urban green space and health in the United Kingdom. *Social Science & Medicine, 71*, 568–575.

Robert, S. A., & Li, L. W. (2001). Age variation in the relationship between community socioeconomic status and adult health. *Research on Aging, 23*, 234–259.

Robert, S. A., & Ruel, E. (2006). Racial segregation and health disparities between black and white older adults. *The Journals of Gerontology Series B: Psychological Sciences and Social Sciences, 61*, S203–S211.

Roos, L. L., Magoon, J., Gupta, S., Chateau, D., & Veugelers, P. J. (2004). Socioeconomic determinants of mortality in two Canadian provinces: Multilevel modelling and neighborhood context. *Social Science & Medicine, 59,* 1435–1447.

Rosenberg, D. E., Huang, D. L., Simonovich, S. D., & Belza, B. (2013). Outdoor built environment barriers and facilitators to activity among midlife and older adults with mobility disabilities. *The Gerontologist, 53,* 268–279.

Ross, C. E., & Mirowsky, J. (2001). Neighborhood disadvantage, disorder, and health. *Journal of Health and Social Behavior, 42,* 258–276.

Rosso, A. L., Auchincloss, A. H., & Michael, Y. L. (2011). The urban built environment and mobility in older adults: A comprehensive review. *Journal of Aging Research* [Article ID 816106]. 1–10.

Sabia, J. J. (2008). There's no place like home: A hazard model analysis of aging in place among older homeowners in the PSID. *Research on Aging, 30,* 3–35.

Salthouse, T. A. (2009). When does age-related cognitive decline begin? *Neurobiology of Aging, 30,* 507–514.

Schieman, S. (2005). Residential stability and the social impact of neighborhood disadvantage: A study of gender- and race-contingent effects. *Social Forces, 83,* 1031–1064.

Schieman, S., & Meersman, S. C. (2004). Neighborhood problems and health among older adults: Received and donated social support and the sense of mastery as effect modifiers. *The Journals of Gerontology Series B: Psychological Sciences and Social Sciences, 59,* S89–S97.

Schieman, S., & Pearlin, L. I. (2006). Neighborhood disadvantage, social comparisons, and the subjective assessment of ambient problems among older adults. *Social Psychology Quarterly, 69,* 253–269.

Shah, R. C., Maitra, K., Barnes, L. L., James, B. D., Leurgans, S., & Bennett, D. A. (2012). Relation of driving status to incident life space constriction in community-dwelling older persons: A prospective cohort study. *The Journals of Gerontology Series A: Biological Sciences and Medical Sciences, 67,* 984–989.

Sheffield, K. M., & Peek, M. K. (2009). Neighborhood context and cognitive decline in older Mexican Americans: Results from the Hispanic established populations for epidemiologic studies of the elderly. *American Journal of Epidemiology, 169,* 1092–1101.

Shumway-Cook, A., Patla, A., Stewart, A., Ferrucci, L., Ciol, M. A., & Guralnik, J. M. (2003). Environmental components of mobility disability in community-living older persons. *Journal of the American Geriatrics Society, 51,* 393–398.

Wahl, H., Iwarsson, S., & Oswald, F. (2012). Aging well and the environment: Toward an integrative model and research agenda for the future. *The Gerontologist, 52,* 306–316.

Wen, M., Hawkley, L. C., & Cacioppo, J. T. (2006). Objective and perceived neighborhood environment, individual SES and psychosocial factors, and self-rated health: An analysis of older adults in Cook County, Illinois. *Social Science & Medicine, 63,* 2575–2590.

Wheaton, B. (1985). Models for the stress-buffering functions of coping resources. *Journal of Health and Social Behavior, 26,* 352–364.

Wheaton, B., & Clarke, P. (2003). Space meets time: Integrating temporal and contextual influences on mental health in early adulthood. *American Sociological Review, 68,* 680–706.

Wight, R. G., Aneshensel, C. S., Barrett, C., Ko, M., Chodosh, J. (2013)., & Karlamangla, A. S. (2013). Urban neighbourhood unemployment history and depressive symptoms over time among late middle age and older adults. *Journal of Epidemiology and Community Health, 67,* 153–158.

Wight, R. G., Aneshensel, C. S., Miller-Martinez, D., Botticello, A., Cummings, J. R., Karlamangla, A. S., et al. (2006). Urban neighborhood context, educational attainment, and cognitive function among older adults. *American Journal of Epidemiology, 163,* 1071–1078.

Wight, R. G., Cummings, J. R., Karlamangla, A. S., Aneshensel, C. S. (2009). Urban neighborhood context and change in depressive symptoms in late life. *The Journals of Gerontology Series B: Psychological Sciences and Social Sciences, 64B,* 247–251.

Wight, R. G., Ko, M. J., & Aneshensel, C. S. (2011). Urban neighborhoods and depressive symptoms in late middle age. *Research on Aging, 33,* 28–50.

Wiles, J. L., Leibing, A., Guberman, N., Reeve, J., & Allen, R. E. S. (2011). The meaning of "ageing in place" to older people. *The Gerontologist, 52,* 357–366.

Williams, D. R., & Collins, C. (2001). Racial residential segregation: A fundamental cause of racial disparities in health. *Public Health Reports, 116,* 404–416.

Yao, L., & Robert, S. A. (2008). The contributions of race, individual socioeconomic status, and neighborhood socioeconomic context on the self-rated health trajectories and mortality of older adults. *Research on Aging, 30,* 251–273.

Yen, I. H., Michael, Y. L., & Perdue, L. (2009). Neighborhood environment in studies of health of older adults: A systematic review. *American Journal of Preventive Medicine, 37,* 455–463.

CHAPTER 16

Abusive Relationships in Late Life

Karen A. Roberto

Center for Gerontology and The Institute for Society, Culture and Environment,
Virginia Tech, Blacksburg, VA, USA

OUTLINE

Introduction	337	Social Interactions and Isolation	345
Prevalence of Elder Abuse	*338*	Perpetrators of Elder Abuse	345
Global Perspectives	*339*	*Spouses/Partners*	*345*
A Socioecological Framework for Understanding Elder Abuse	340	*Adult Children*	*346*
Ecological Theory	*340*	*Other Relatives*	*347*
A Life Course Perspective	*341*	*Trusted Others*	*347*
Models of Social Organization	*341*	Responses to Elder Abuse	348
Feminist Theories	*342*	*Community Perceptions*	*348*
Vulnerabilities and Risk for Elder Abuse	342	*Interventions*	*349*
Age, Gender, Race, and Ethnicity	*343*	*Policy Initiatives*	*350*
Cultural Beliefs and Perceptions	*344*	Future Research	350
Health and Cognitive Abilities	*344*	References	351

INTRODUCTION

Elder abuse knows no boundaries – older adults of all ages, genders, races, incomes, and cultures experience abuse. In the United States, at least 1 in 10 adults aged 60 and older living in their own homes experiences abuse, neglect, or exploitation annually (Acierno et al., 2010). The risk for abuse increases among vulnerable community-dwelling older adults, including individuals who have physical or cognitive impairments or persons who are socially isolated and alone. Some of the harm to victims of elder abuse is obvious, such as physical injuries or the loss of money and valued possessions (Metlife Mature Market Institute, 2009, 2011;

Wiglesworth, Austin, Corona, & Mosqueda, 2009). Not as immediately evident are the less tangible consequences of elder abuse. Long-term outcomes may include emotional or psychological distress (Cisler, Begle, Amstadter, & Acierno, 2011), new or exacerbated health problems and hospitalizations (Dong & Simon, 2013; Thomas, Joshi, Wittenberg, & McCloskey, 2008), premature institutionalization (Rovi, Chen, Johnson, & Mouton, 2009), and a hastened death (Baker et al., 2009; Dong, Simon, & Beck, et al., 2011). Although elder abuse appears more prevalent in community than institutional settings, 1 in 14 complaints regarding care facilities reported to the Long Term Care Ombudsmen in 2010 focused on abuse, neglect, or exploitation (Administration on Aging, 2012). Based on data from the Centers for Medicare and Medicaid for 2000 through 2007, approximately 20% of nursing homes per year received deficiency citations (i.e., violations of Medicare/Medicaid regulations) for resident abuse; in 10% of these nursing homes, the abuse caused harm to at least one resident (Castle, 2011).

First identified in the mid-1970s as "granny bashing" (Baker, 1975), the scientific study of elder abuse emerged in the 1980s and has reached a nascent stage. Comparing findings across this small, albeit growing, body of literature is challenging, however, because within the field of elder abuse there is not an agreed-upon term or uniformly accepted definition used by researchers or the professional community. Terms such as "elder mistreatment" (Bonnie & Wallace, 2003), "elder abuse" (World Health Organization, 2002), and "elder maltreatment" (World Health Organization, 2011) are used interchangeably and the parameters of both the abuse and persons covered vary widely. What the definitions of these terms share in common is the recognition that there are various forms of elder abuse (i.e., physical abuse, sexual abuse, psychological and emotional abuse, financial abuse and exploitation, neglect and abandonment) and that abuse causes "harm or loss to an older victim" (Anetzberger, 2012, p. 12).

This chapter focuses on elder abuse perpetrated by family members and trusted others (e.g., persons known to the elder such as friends, in-home caregivers, legal guardians) in the lives of older adults living in the community. Financial exploitation by strangers (e.g., telemarketing scams, home repair scams, Internet phishing), elder abuse in long-term care settings, and elder self-neglect also are of substantial significance, but beyond the scope of this chapter. The chapter begins with the latest prevalence data about elder abuse available nationally and internationally, followed by a theoretical framework for guiding the study of elder abuse. The next two sections provide an examination of commonly identified vulnerabilities that place older adults at risk for abuse and the characteristics of potential perpetrators of elder abuse. The final sections address the prevention and intervention of elder abuse and priority areas for future research.

Prevalence of Elder Abuse

Discrepancies in definitions, privacy laws, the controversial subject matter, the stigma associated with elder abuse, and underreporting abuse for fear of retribution by family perpetrators makes gathering and interpreting prevalence data on elder abuse challenging and complicated. The scope of investigations (e.g., national, regional, state), design of data sources (e.g., Adult Protective Services (APS) records, crime statistics), type of abuse investigated, and the range of study informants (e.g., older adults, family members, community professionals) contributes to variation in reports of the prevalence, incidence, and understanding of elder abuse (Institute of Medicine & National Research Council, 2013). Despite these limitations, a surge in prevalence research has provided new insights into the magnitude of elder abuse.

Two national studies shed light on the prevalence of elder abuse in the United States. In 2004, as part of the National Social Life, Health and Aging Project, 3 005 community-residing adults aged 57–85 were asked if they had experienced verbal, physical, and financial abuse by a family member in the year prior to the study (Laumann, Leitsch, & Waite, 2008). Nine percent (9%) of the respondents reported verbal abuse, 3.5% financial abuse, and 0.2% physical abuse. The 2008 epidemiological study of prevalence and risk factors for abuse, known as The National Elder Mistreatment Survey (Acierno et al., 2010; Acierno, Hernandez-Tejada, Muzzy, & Steve, 2009), gathered self-reported past-year elder abuse data from a sample of 5 777 community-dwelling elders aged 60–97 years. Overall, 14.1% of respondents experienced abuse and financial exploitation (Government Accountability Office, 2011). Past-year prevalence rates were 4.5% for emotional abuse, 1.5% for physical abuse, 0.6% for sexual abuse, 5.1% for potential neglect, and 5.2% for financial abuse by a family member.

One reason elder abuse appears to be even more common than previously thought is because older victims do not discuss their situations with others or report incidences to the authorities. Factors such as fear of repercussions from involved family members, dependence on perpetrators (e.g., need for care; fear of neglect), and filial norms and responsibilities (e.g., shame associated with raising a child who would abuse a parent; concern about consequences for family perpetrators) contribute to older adults' willingness to endure abusive relationships and keep their situation to themselves. The National Elder Mistreatment Study (Acierno et al., 2009) past-year estimates revealed that only 8% of the 245 older respondents who experienced emotional mistreatment had reported an incident to the police. Comparatively, 16% of the 34 respondents who experienced sexual abuse and 31% of the 86 victims of physical abuse reported the abuse to the authorities. Similarly, a comparison of self-reported data and documented case data collected in 2008 of elder abuse among residents aged 60 and older in New York found self-report rates of elder abuse to be 24 times greater than the number of cases documented by state and community agencies serving victims of elder abuse (Lifespan of Greater Rochester, Inc., 2011). Among these cases, self-reported incidences of elder financial abuse were 44 times higher than the number reported to authorities.

Global Perspectives

With the aging of the population, elder abuse is a global issue recognized as a major public health and policy concern in both developed and developing countries. The number of population-based prevalence studies is growing, particularly in European countries (Sethi et al., 2011). For example, the 2009 Abuse and Health Among the Elderly in Europe study of 4 467 women and men, aged 60–84, living in urban populations in Germany, Greece, Italy, Lithuania, Portugal, Spain, and Sweden reported pooled 1-year prevalence rates of 19.4% for psychological abuse, 2.7% for physical abuse, 0.7% for sexual abuse, and 3.8% for financial abuse (Soares et al., 2010). A national study of 2 111 persons, aged 66 and older, living in private residence households in the United Kingdom in 2006 showed 1-year prevalence rates at 1.1% for neglect, 0.4% for psychological abuse, 0.4% for physical abuse, and 0.2% for sexual abuse (Biggs, Manthorpe, Tinker, Doyle, & Erens, 2009). Other general population studies include the first national prevalence study in Israel conducted during 2004–2005, which relied on interviews with 1 045 community-dwelling Jewish and Arab older adults aged 65 and older, and estimated the prevalence of physical and sexual abuse at 1.8%, financial abuse at 0.5%, verbal abuse at 17.1%, and neglect at 25.6% (Lowenstein, Eisikovits, Band-Winterstein, & Enosh, 2009). In India, 31% of the 5 400 older persons interviewed in 20 cities across the country in 2010 and 2011 reported that they had experienced abuse in the previous year (HelpAge India, 2012). The most common abuse described was disrespect (44%), followed by neglect (30%), and verbal abuse (26%).

With few exceptions, prevalence studies conducted in Asian nations, the Middle East, and Oceania countries tend to be smaller in scope, often limited to a specific geographic area (e.g., Sakarya Province, Turkey, Çevirme, Uğurlu, Çevirme, & Durat, 2012; Western Australia, Boldy, Horner, Crouchley, Davey, & Boylen, 2005) or major city (e.g., Bangkok, Thailand, Chompunud et al., 2010; Seoul, South Korea, Oh, Kim, Martins, & Kim, 2006). There is a paucity of data on elder abuse for much of Africa; however, available literature showed high prevalence of emotional and verbal abuse of older people (Mba, 2007).

Although no country reports with certainty how much elder abuse occurs, rates of psychological abuse (i.e., verbal abuse, emotional abuse, disrespect) and perceived neglect by caregivers are consistently reported as higher than physical and sexual abuse, regardless of national context. As with prevalence studies in the United States, differences in survey methods, demographic characteristics, study settings, and criteria used for defining elder abuse limit understanding of the prevalence and experiences of elder abuse across and often within countries. These limitations pose serious obstacles for universal recognition and understanding of elder abuse, the development of national and international polices concerning elder abuse, and improvements in prevention and treatment strategies to address elder abuse and its aftermath.

A SOCIOECOLOGICAL FRAMEWORK FOR UNDERSTANDING ELDER ABUSE

Although there is not a single explanation or unifying theory of elder abuse, a socioecological approach has emerged as a promising conceptual framework for understanding the complexities of elder abuse (Horsford, Parra-Cardona, Post, & Schiamberg, 2010; Roberto, Teaster, & McPherson, 2015; Teaster, Wangmo, & Vorsky, 2012). Beginning with ecological theory, and drawing upon tenets of the life course perspective, social organization models, and feminist theories, a socioecological framework assesses dynamic and enduring relationships of older individuals within the context of their families, community, and society. It provides guidance for the exploration and examination of the breadth and depth of multiple influences on elder abuse and the experiences of abuse among older adults.

Ecological Theory

Grounded in Bronfenbrenner's ecological systems theory (1979), a socioecological framework focuses attention on characteristics of older adults that may place them at risk for abuse within four influencing concentric and interconnected systems. The microsystem, which includes the older adult and perpetrator, identifies biological and personal factors that influence how individuals behave and risk factors that increase the likelihood of becoming a victim or perpetrator of abuse. The mesosystem focuses on close relationships (e.g., family, friends, neighbors) to explore how these relationships protect or foster abuse. The exosystem identifies the community contexts in which social relationships occur (e.g., neighborhoods) and explores how the characteristics of these settings may affect the victims' well-being and ability to escape the abuse. The macrosystem brings to light broad ideological values, norms, and institutional patterns that help create a climate in which abuse is encouraged or inhibited. The strength of the ecological model as a foundation for the study of elder abuse is that it helps to distinguish between the myriad of influences on elder abuse while at the same time recognizing the interactions between and among these individual and systemic influences.

A Life Course Perspective

The inclusion of life course concepts, including an examination of older individuals' life histories, family relationships, and the social and historical contexts in which they live (Elder, 1977), enhances understanding of micro- and mesosystem influences on older victims of abuse. A central feature of a life course perspective is a focus on time or the temporal context of development. Both positive and adverse ontogenetic (i.e., individual), generational (i.e., family) and societal events, as well as the dynamic interactions among individual time, generational time, and historical time affect life trajectories (Bengtson & Allen, 1993). The timing, sequencing, spacing, density, and duration of events also influence life transitions or turning points, which result in major directional changes or discontinuities in a trajectory (Settersten, 2003). Recent longitudinal findings that early exposure to abuse has lifelong consequences (Greenfield & Marks, 2010; Savla et al., 2013) suggests that the experience of abuse in childhood epitomizes a major life event or turning point that can increase the risk of poor intra- and interpersonal outcomes that are associated with occurrences of abuse throughout the life course (McDonald & Thomas, 2013).

The life course perspective also posits the principle of "linked lives," which emphasizes relational interconnectedness. Individual lives are embedded within and influenced by their relationships with family and others in their social networks. Although relationships may change over time, relationship stories are integral to how older adults make meaning of their lives. Thus the timing of life influences (e.g., abuse in childhood) and the specific characteristics of interactions with others (e.g., physical, emotional, and economic dependence) may increase the risk of abuse in late life (Schiamberg & Gans, 2000).

Models of Social Organization

Tenets of social organization theory inform understanding of the exosystem and macrosystem. Social organization pertains to how people in a community interrelate, cooperate, and provide mutual support. It includes social support norms, social controls that regulate behavior and interaction patterns, and the networks that operate in a community (Mancini & Bowen, 2013). Community members share norms that govern behaviors and expectations that provide for both licit and illicit activities (Furstenberg & Hughes, 1997); different subgroups of individuals may be more tolerant of activities and behaviors considered unacceptable by the general population. These subgroups may be characterized by particular demographic characteristics (i.e., age, ethnicity) and by psychological characteristics, including attitudes about gender, sense of self, and attributes of others.

The variation and complexity of a community is reflected in its structure and the processes or actions engaged in by its members. Community-level networks play a significant role in promoting the physical, psychological, social, and spiritual well-being of individual community members and families (Mancini & Bowen, 2013). Informal relationship networks involve voluntary relationships, such as those with church members, friends, and neighbors, which are characterized by mutual exchanges and reciprocal responsibility. Formal support networks involve obligatory relationships, such as those associated with agencies and organizations. These two networks often interact and are essential for providing support for older adults who experience abuse. Understanding community capacity, or the degree to which network members demonstrate a sense of shared responsibility for the welfare of the community and its individual members (Bowen, Martin, Mancini,

& Nelson, 2000), provides insight about the strengths and weaknesses of networks and the ways in which they coalesce to confront situations that threaten the safety of older adults.

Feminist Theories

Incorporating a feminist lens and concepts regarding gender relations and intersectionality across all layers of the socioecological framework focuses attention on age-related changes in power and control dynamics (e.g., dominance of spouse/partner; reversal of child/parent roles) as well as changes in social positions and financial resources that may contribute to late life experiences of abuse (Straka & Montminy, 2006). A feminist lens calls attention to ways in which gender intersects with age, race, class, and the like. For example, situating microsystem social locations such as old age and gender within the larger macrosystem of inequality focuses attention on the significance of the relationship between abuse and social and economic power (Anderson, 2010). In many cultures women tend to be economically and socially disadvantaged in comparison to men, making gender a risk factor for elder abuse. Without economic or social means, older women are likely to remain in unsafe relationships (Teaster, Roberto, & Dugar, 2006). The established power dynamics within abusive intimate relationships often persist into late life, even after male perpetrators become recipients of care from the female partners they had abused (Koenig, Rinfrette, & Lutz, 2006; Lev-Wiesel & Kleinberg, 2002). Thus, the lack of personal and social resources (e.g., education, skills, income, status, social support) and relationship histories often coalesce to define individual abuse trajectories.

Feminist frameworks also offer an important avenue for examining the complex intersection of social and historical context, age, gender, race, ethnicity, socioeconomics, and sexual orientation in the lives of an increasingly diverse and often marginalized population of victims of elder abuse (Walsh, Olson, Ploeg, Lohfield, & MacMillan, 2011). Because older people are often stigmatized, the intersection of ageism with multiple forms of oppression often exacerbates vulnerabilities and the risk of abuse. For example, LGBT older adults often attribute abuse or neglect by a caregiver to homophobia and resist reporting the abuse because of abuser threats to expose their gender identity or their fear of social prejudice, hostility, and disregard from members of the formal helping community (National Resource Center on LGBT Aging, 2013).

VULNERABILITIES AND RISK FOR ELDER ABUSE

Individual changes associated with aging as well as intrapersonal and interpersonal relationship characteristics can place older persons at risk for abuse. Although investigators have identified many risk factors, there is inconsistency in their influence across studies and different forms of abuse (Johannesen & LoGiudice, 2013). Much of the research on risk factors relies upon small, cross-sectional studies, does not include comparison groups, does not differentiate type of abuse, and does not address interacting factors that contribute to late life vulnerabilities that potentially place older adults at increased risk for elder abuse. Recently, Dong and Simons (2014) constructed a cumulative vulnerability elder abuse index that showed community-dwelling older adults with 3–4 (out of 9) socio-demographic, health-related, and psychosocial risk factors associated with elder abuse had a 3.9-fold greater risk of confirmed abuse than a comparison group of older adults who had not experienced abuse. The risk for confirmed elder abuse increased to 26.7 times as great in older adults with five or more vulnerabilities. Although the new measure is limited in scope and requires further validation, it is an important step toward better risk prognostication.

Age, Gender, Race, and Ethnicity

Although age is one of the most commonly identified risk factors for elder abuse, empirical evidence is mixed. In the two national studies, one found that older adults aged 60–69 were more susceptible to abuse, particularly emotional and physical abuse, than other age groups (Acierno et al., 2009), whereas the other reported that the risk of abuse did not increase with age (Laumann et al., 2008). Smaller investigations focused on specific types of abuse (i.e., sexual, financial) identified adults aged 75 and older as being particularly susceptible to abuse (Burgess, Dowdell, & Prentky, 2000; Metlife Mature Market Institute, 2009, 2011). The association between age and risk of abuse may be linked to a decline in functional health, which often results in a greater dependence on others for care and a higher level of individual vulnerability. With advancing age, older adults also may have decreased contact with others in their social networks whose presence may protect them from abuse (Amstadter, Cisler, et al., 2011). Older individuals tend to judge others' trustworthiness less stringently than younger individuals (Charles & Carstensen, 2010) and show compromises in decision-making capacity that make them more vulnerable to undue influence, a tactic frequently used by individuals perpetrating financial abuse (Wood & Liu, 2012). Other age-related changes in cognitive ability associated with an increased risk for elder abuse include lower levels of episodic memory and slower perceptual speed (Dong, Simon, Rajan, & Evans, 2011).

Researchers frequently identify gender as a risk for elder abuse, with women more likely identified as victims than men. Greater longevity in conjunction with the associated age-related changes may contribute to older women's risk for abuse. In addition, women experience higher rates of family violence throughout the life course, which increases their risk of abuse in late life, particularly for physical and sexual abuse (Acierno et al., 2010). A critique of elder abuse literature identified the potential of differential risk for older men and women and suggested that the focus on individual victim characteristics and family dynamics actually places older men at risk for elder abuse by normalizing abusive experiences (Thompson, Buxton, Gough, & Wahle, 2007). Hence, the abuse of older men has been deemed "invisible," in part due to the failure of men to acknowledge and report abuse (Kosberg, 2014, p. 209).

Belonging to a racial or ethnic minority group also is a frequently identified risk factor for elder abuse; however, race- and ethnicity-based differences in the prevalence of abuse did not hold up in analysis of national prevalence data (Hernandez-Tejada, Amstadter, Muzzy, & Acierno, 2013). The reliance on APS data in which both black and Hispanic elders are often over-represented (Tatara, 1999), and differences in beliefs about what constitutes abuse (DeLiema, Gassoumis, Homeier, & Wilber, 2012; Horsford et al., 2010) may explain differences in the risk of elder abuse. For example, Dakin and Pearlmutter (2009) found that African American and white women with high socioeconomic status (SES), as well as Latina women, did not identify financial abuse as a type of elder abuse. Working-class white women did not identify verbal abuse as mistreatment, whereas working-class African American women included societal maltreatment (i.e., systemic mistreatment by HMOs) and financial abuse in their definitions of elder abuse, but did not include physical abuse. Latina women (all low SES) exhibited higher tolerance for spousal abuse than did women in the other groups. Findings reinforce the premise that macro-level cultural norms, especially for the Latina women who had all migrated to the United States as adults (e.g., traditional gender relations and beliefs about marriage; preserving the family at all costs) may contribute to the risk for elder abuse and the reluctance of both Latino and African American older women (e.g., racial discrimination within the criminal justice system; social inequality) to seek assistance when abuse occurs.

Cultural Beliefs and Perceptions

Cultural influences interact with other parameters of diversity such as race, education, socioeconomics, and gender to influence the risk of elder abuse. Moon and Benton's (2000) examination of cultural norms among different ethnic groups living in an urban area found differences in tolerance of potential elder abuse. Face-to-face interviews with African American, Korean American immigrants, and White older adults revealed that White respondents had significantly higher tolerance for verbal abuse than either African American or Korean American elders. Among the three groups, however, Korean elders were the most distinct. They were the most tolerant of elder abuse overall, particularly financial exploitation. This finding may be an artifact of the traditional practice of Korean parents transferring their wealth and property to children when the parents retire. Adherence to cultural norms regarding family obligations and practices may also explain why the Korean older adults were significantly more likely than other respondents to blame the victim for the occurrence of elder abuse, have significantly more negative attitudes toward involvement of people outside the family, and be less likely to report elder abuse to authorities. In contrast, responses from a mixed-methods study of the ways in which older Korean immigrants define financial abuse revealed some deviation from traditional Korean customs, suggesting the influence of acculturation into American society on perceptions of elder abuse (Lee, Lee, & Eaton, 2012).

Patterns of abuse also may vary in different types of communities. An examination of elder abuse in rural and urban communities found significantly more rural women were victims of physical abuse, emotional abuse, and active caregiver neglect than urban women, while more urban women had experienced more passive caregiver neglect than rural women (Dimah & Dimah, 2003). The structure and culture of rural environments also may inadvertently conceal and consequently facilitate abuse and inhibit prevention and treatment efforts. Close social ties with emergency responders and service providers in "tight knit" southern Appalachian communities, and low levels of education and economic security among older female victims of violence exacerbated the abuse (Teaster et al., 2006). Riddell and colleagues (2009) arrived at similar conclusions about the rural Canadian cultural context, suggesting that strong personal ties to the community, a culture of self-sufficiency, patriarchal views of the family, limited community services, isolation, and economic stressors contribute to and conceal abusive relationships and inhibit help-seeking behaviors.

Health and Cognitive Abilities

Poor health, which requires older adults to seek assistance in order to live independently, increases their risk for abuse. Laumann and colleagues (2008) found that older adults who reported any type of physical vulnerabilities were approximately 13% more likely to report verbal abuse than study participants who indicated they had no physical vulnerabilities. In addition, the worse respondents rated their health, the higher their odds of financial abuse. For elders in South Carolina, the need for assistance with activities of daily living and poor health status were significant correlates of emotional abuse, physical abuse, neglect, and financial abuse (Amstadter, Zajac, et al., 2011).

Cognitive impairment is perhaps the most commonly identified risk factor for abuse, particularly for financial abuse (Metlife Mature

Market Institute, 2011) and abuse by caregivers (Beach et al., 2005). With the onset of cognitive impairment, older adults' financial capacity (i.e., ability to manage financial affairs independently and in a manner consistent with personal self-interest) begins to diminish, thus placing them at greater risk for abuse. Risk of financial abuse stems not only from the diminished judgment that may result in bad decisions, but from the accompanying need for assistance with financial management that may expose the older adults to "helpers" who can easily exploit them (Stiegel, 2012, p. 73). As cognitive abilities decline, the risk of all forms of elder abuse increases significantly (Dong, Simon, Rajan, et al., 2011).

Social Interactions and Isolation

Older adults who are lonely or isolated are much more vulnerable to elder abuse than elders who have strong and actively engaged support systems. Low social support was associated with more than triple the likelihood that mistreatment of any form would be reported by older adults (Acierno et al., 2009). An examination of loneliness as a risk factor for abuse among older Chinese adults showed that each 1-point increase in loneliness scores was associated with a 44% increase in the risk of elder abuse for older women but did not find any significant relationship between loneliness and risk of abuse for older men (Dong, Beck, & Simon, 2009). Although the older adults who reported abuse had lower levels of perceived social support than the older adults not reporting abuse, perceived social support negated the association between loneliness and risk for abuse. Thus, for vulnerable older adults, positive perceptions of social support may reduce the influence of other risk factors of abuse. The structure of older persons' close network ties also may serve as a protective factor for elder abuse. Schafer and Koltai (2014) found that the greater the density (i.e., connections among network members) of older adults' personal networks the less risk of elder mistreatment.

PERPETRATORS OF ELDER ABUSE

The majority of elder abuse incidents are by known perpetrators, usually family members. The relationships between older adults and potential perpetrators of elder abuse are complex and often cited as a contributing factor leading up to abuse. Outsiders often perceive alleged perpetrators as primary sources of support for elders rather than individuals who are causing them harm. Beyond basic descriptive information typically provided by elderly victims, family members, or service providers, limited information derived from empirical research is available about perpetrators' characteristics and their motivations for the abuse.

Spouses/Partners

Information specific to the prevalence of spouse/partner abuse is difficult to ascertain, primarily because the elder abuse literature typically does not single out older couples as a separate study group (Payne, 2008) and the intimate partner violence literature gives little specific attention to older persons (Straka & Montminy, 2006). In addition, while older men, as well as members of same sex couples, experience partner abuse and violence in late life, only a few studies focused on male victim perspectives (Cheung, Leung, & Tsui, 2009; Reeves, Desmarais, Nicholls, & Douglas, 2007; Reid et al., 2008) and no empirical investigations of abuse between older LGBT partners were found.

Spouse/partner abuse in late life often involves various and concurrent forms of abuse and violence including physical harm, sexual assault, psychological humiliation or intimidation, and murder. Abuse by a spouse/partner may manifest as a continuation of longstanding abuse within a single relationship, as engagement in a series of violent intimate relationships over the life course, as violence within a continuing relationship that starts in old age, or as violence that begins with a new relationship in the later years.

Expressions of violence between older couples may change with age. Physical violence tends to decline with victim age, often replaced with new or intensified types of psychological and emotional abuse endured in earlier years (e.g., Daly, Hartz, Stromquist, Peek-Asa, & Jogerst, 2008; Fisher & Regan, 2006; Mezey, Post, & Maxwell, 2002). National prevalence studies support this contention with spouses/partners identified in one-fourth or more of situations involving verbal or emotional abuse experienced by older respondents (Acierno et al., 2009; Laumann et al., 2008). However, a different story emerged from national media reports of intimate partner violence in late life in which murder-suicide, murder, or attempted murder dominated the headlines (Roberto, McCann, & Brossoie, 2013). The intersections of risk factors associated with abuse including social isolation, cognitive impairments, caregiver stress, dependency, and alcohol use, as well as power dynamics frequently surrounding gender-based violence (Straka & Montminy, 2006) contributed to the abuse. While these findings support similar reports of extreme violence in older spouse/partner relationships (Eliason, 2009), they need to be interpreted with caution because of the media tendency to feature sensational incidences of violence (e.g., murder-suicide) over stories about other types of abuse.

Adult Children

The child abuse and family violence literature suggests that for some adults, the propensity for abuse may evolve from experiencing or witnessing abuse/violence within their family of origin (e.g., persons who were maltreated as children in turn abuse their own children; children who see acts of violence within their parent's relationship repeat such behaviors within their adult intimate relationships). No empirical evidence exists to either support or refute the intergenerational transmission of violence hypothesis (e.g., adult children abuse their elderly parents who abused them as young children) in relationship to elder abuse.

Adult children dependency is a contributing factor in the perpetration of elder abuse by adult children. Adult children who are abusive often are reliant upon an elderly parent for housing, finances, and emotional support (Jackson & Hafemeister, 2012; Teaster et al., 2012) and, in some families, the care of their children (Bullock & Thomas, 2007). Underlying factors influencing such dependency on older family members include substance abuse problems (Amstadter, Cisler, et al., 2011; Jogerst, Daly, Galloway, Zheng, & Xu, 2012), a history of mental or emotional illness (Acierno et al., 2009), and chronic unemployment (Jackson & Hafemeister, 2011). These relationships may become abusive or violent when the older adult declines or refuses to provide more money or other types of support. In cases of financial elder abuse, adult children perpetrators often are so desperate for money that they seize upon opportunities with their parents to forge checks, steal credit cards, pilfer bank accounts, and transfer assets (Metlife Mature Market, 2011).

Conversely, when older persons are dependent on an adult child for their care, the potential for abuse may escalate. The overwhelming majority of caregivers provide appropriate care and supportive environments for their older parents; however, changes in roles and the nature of the parent–children relationship affect both caregivers and care recipients. Adult child caregivers often experience distress that can lead to a range of potentially harmful or abusive behaviors, from verbally assaulting their elderly parents to depriving them of daily essentials, care, and services (Amstadter, Cisler, et al., 2011; Beach et al., 2005). Although cognitive impairment increases the risk of mistreatment (Hansberry, Chen, & Gorbien, 2005), national studies of family caregivers that specifically address abuse of relatives with dementia are lacking. Small-scale scholarly investigations found that 30–60% of these caregivers use psychologically and physically aggressive caregiving strategies (Cooper et al., 2009; Wiglesworth et al., 2010).

Elder abuse scholars no longer consider caregiver stress the primary cause of elder abuse (Brandl & Raymond, 2012); most stressed caregivers do not hurt or harm their family care recipients. However, in some families, the amount or level of care the elder requires becomes overwhelming and well-intended caregivers respond by lashing out verbally or physically and the quality of care may degrade (Ramsey-Klawsnik, 2000). Compared to overwhelmed caregivers who often recognize their abusive or neglectful behaviors and seek help to improve the situation, other perpetrators of dependent elders are not so well intended. They are motivated by their own narcissistic and domineering personalities, which leads them to justify their abusive actions (Ramsey-Klawsnik, 2000).

Other Relatives

Few researchers focus specific attention on grandchildren or other relatives who perpetrate elder abuse. Reports from The National Elder Mistreatment Study (Acierno et al., 2009) combine information about children and grandchildren as one family group and other relatives as another; thus, limiting the amount of detail available about the relationship between these family members and their older victims. Among the perpetrators identified, other relatives were responsible for 13% of emotional abuse, 12% of sexual abuse, 9% of physical abuse, and 7% of neglect of their older family members. In the 2008 Metlife national study of elder financial abuse (Metlife Mature Market Institute, 2009), grandchildren (13%), nephews and nieces (15%), and other relatives (i.e., siblings, cousins; 13%) accounted for 41% of family perpetrators ($n = 45$). Similarly, 43% of family perpetrators ($n = 113$) of financial abuse identified by older adults living in New York in 2009 were grandchildren (18%) and other relatives (25%; Peterson et al., 2014). As with adult children, grandchildren, and other relatives who abuse their older family members often provide some assistance to the older adult, but also depend on them for shelter, money, and emotional support.

With the growing numbers of grandparents raising grandchildren, some professionals have expressed concern about the vulnerability for elder abuse among "skip generation" families (Bullock & Thomas, 2007; Kosberg & MacNeil, 2003). Grandparents take on the custodial parenting role due to parental illness, divorce, incarceration, and substance abuse and frequently inherit substantial parenting challenges, particularly if their grandchildren have significant emotional and behavioral problems (Dolbin-MacNab, 2006). Focus groups conducted with custodial grandparents of adolescent grandchildren revealed that physical abuse and financial exploitation of grandparents by grandchildren was common, but often hidden so as not to involve the family's child welfare workers (Brownell, Berman, Nelson, & Fofana, 2003). The grandparents described abusive behaviors of their grandchildren as punching, hitting, throwing objects, stealing money needed for the household, stealing prized possessions, destroying possessions, and threatening them with weapons.

Trusted Others

While most known perpetrators of elder abuse are family members, other individuals with whom older adults have a relationship involving an expectation of trust, including friends, neighbors, and people relied upon for services, also are perpetrators of abuse. Of 580 media stories of elder financial abuse, alleged perpetrators included friends and neighbors (4%), persons who befriended the elders (5%), trusted professionals (e.g., lawyers, investment counselors; 12%), and in-home caregivers (16%) (Metlife Mature Market Institute, 2011). Many acts of elder financial abuse arose from greed, with trust engendered specifically for the intention of obtaining the elders' funds, property, or other possessions as the relationships progressed. In other cases, elders had money, assets, and the like, and an occasion presented itself for the perpetrators to avail themselves of the resource. Perpetrators also believed that, in return for providing assistance and care, whether needed or not by the older person, they were due continual compensation (e.g., money, possessions).

To protect older individuals who become incapable of managing their personal and financial affairs, states provide court-appointed certified guardians, who may be family members, private citizens, or professionals (e.g., case managers, nurses, lawyers, accountants). In some instances guardians entrusted with the welfare of older adults abuse their positions. The Government Accounting Office (GAO) identified hundreds of allegations of physical abuse, neglect, and financial exploitation by guardians in 45 states and the District of Columbia between 1990 and 2010 (GAO, 2010). As part of this study, the GAO conducted an in-depth review of 20 closed cases. Findings suggested that in six cases guardians had a poor track record of managing finances or a criminal background, yet the courts failed to identify these warning signs before appointing them to care for vulnerable older adults; in 12 cases, the courts failed to oversee guardians once they were appointed, allowing the abuse of the older adults and their assets to continue; and in 11 cases, courts and federal agencies did not communicate effectively or at all with each other about abusive guardians, allowing the guardians to continue in their roles.

RESPONSES TO ELDER ABUSE

Despite the growing recognition of elder abuse as a public health problem, very little research focuses on community members' perceptions of and response to elder abuse. Programs to address elder abuse implemented at the local and state level deemed effective, at least anecdotally, often lack stringent evaluations and are mostly invisible in the academic literature. Federal recognition of elder abuse has resulted in several programmatic and policy initiatives to prevent and address elder abuse.

Community Perceptions

A systematic review of the national and international research on public perceptions of elder abuse revealed mixed awareness and understanding of, and tolerance for, elder abuse (Lafferty, 2009). Influences on perceptions of elder abuse included traditional cultural values and norms (e.g., family solidarity, familial privacy, filial piety), their previous experience with elder abuse, the age and gender of study respondents, perceived vulnerability and health of the older person, and characteristics of the perceived victim.

Vignettes and focus groups are commonly used approaches to assess community residents' perspectives on elder abuse. These small-scale investigations point to the significance of personal and contextual influences on beliefs about elder abuse. For college students, the perceived quality of past relationships in the caregiving scenarios presented and their previous exposure to caregiving within their own families were associated with their assessment of elder abuse (Fitzpatrick & Hamill, 2011). Abuse perceived as out of character within the context of a loving relationship was more likely to be seen as an "aberration" and less likely to be reported as the same abuse described within the context of a conflicted relationship. In addition, more students who reported experience with family elder care rated the behavior as more abusive and indicated that they would report behavior to authorities than students without previous exposure to caregiving. The majority of respondents in a study of rural community residents ranging in age from 25 to 65 and older identified scenarios as abusive family situations; their responses differed based on age, sex, and educational level of the respondents and sex of the victim portrayed in the scenarios (Roberto, Teaster, McPherson, Mancini, & Savla, 2015). Older respondents and respondents with more positive perceptions of community cohesion (trust) also were likely to indicate that the older adult described in the scenario would get the necessary help that she or he needed. Among middle-aged and older adults, individuals with lower levels of education and less experience with abuse were less likely to identify elder abuse when it occurred and more likely to have negative emotions (i.e., discomfort, rejection) directed toward abuse victims (Werner, Eisikovits, & Buchbinder, 2005).

Public perceptions of elder abuse are subject to both positive and negative community influences. Through increased attention to elder abuse and legislative changes to law, adult respondents in Alabama showed increased support for criminalization of elder abuse over a 10-year period (Morgan, Johnson, & Sigler, 2006). The endorsement of noncriminal justice sanctions, fines, or misdemeanor penalties in response to elder abuse decreased, and endorsement of felony offense categorizations and longer prison sentences for perpetrators increased. Conversely, reader comments posted to online news items about actual incidents of spouse/partner violence in late life revealed that few posters believed that incidents were episodes of spouse/partner violence (Brossoie, Roberto, & Barrow, 2012). As many posters struggled to make sense of incidents, they attempted to rationalize perpetrator actions by assigning blame elsewhere. Ageist assumptions about abuse and violence in late life, perspectives on relationships and old age, and reporting style of the news items influenced comments posted by other posters.

Interventions

Community-level initiatives to prevent and address elder abuse vary considerably across the United States. Promising programs often incorporate multidisciplinary response teams, inter-agency collaboration, and comprehensive, coordinated community education efforts (Brandl et al., 2007), but lack methodologically sound empirical evidence of their effectiveness (The National Academies Committee on National Statistics, 2010). A recent systematic effort used to evaluate the effectiveness of the Los Angeles County Elder Abuse Forensic Center in prosecuting elder financial exploitation cases (Navarro, Gassoumis, & Wilber, 2013) represents a positive step forward for the development of evidence-based community practices to address elder abuse.

In most states, APS is the principal public agency responsible for investigating elder abuse. When APS receives reports of suspected elder abuse, workers go into the home to investigate, substantiate, and address the situation with legal, medical, psychological, and social services. Unless deemed an emergency, however, APS workers cannot enter private residences to investigate alleged abuse without consent from the older individual or his or her caregiver or legal guardian, a court order, or a search warrant (Roby & Sullivan, 2000). If consent to enter is denied, APS can petition the court for assistance upon showing of probable cause. Although published studies of the effectiveness of APS are few (Ploeg, Fear, Hutchison, MacMillan, & Bolan, 2009), researchers frequently rely on APS data to characterize various aspects of elder abuse (Ernst et al., 2014). While this information is valuable, APS cases represent only a fraction of adults who experience elder abuse.

Policy Initiatives

Although many legislation and policy initiatives have addressed elder abuse at the state (Jirik & Sanders, 2014) and national levels (Teaster, Wangmo, & Anetzberger, 2010), the passage of the Elder Justice Act in 2010 (Public Law 111–148) is considered the most comprehensive federal bill ever passed to combat elder abuse. The purpose of the Elder Justice Act is to increase awareness and knowledge of elder abuse, neglect, and exploitation through training, services, and demonstration programs. While the passing of the Act has significantly increased the nation's focus on elder abuse, Congress has yet to appropriate funds for many of the initiatives authorized, including future research endeavors.

FUTURE RESEARCH

Elder abuse threatens the health, safety, dignity, and overall well-being of far too many older adults. While progress has been made, bringing much needed attention to this global issue, the study of elder abuse demands further attention. To start, greater consensus must emerge about what constitutes abuse. Variability in the nomenclature used to define elder abuse has limited the depth and breadth of generalizable knowledge of all forms of elder abuse, its victims, its perpetrators, and the systemic influences that can serve either as safeguards against or as sources of its occurrence.

While different subtypes of abuse are recognized, current scientific investigations tend to address one or more types of abuse collectively or narrowly focus on one specific subtype of abuse (e.g., psychological abuse, financial abuse). Yet, preliminary evidence embedded within the research literature suggests that cases of elder abuse frequently involve more than one type of abuse or polyvictimization. According to the National Committee for the Prevention of Elder Abuse (NCPEA), polyvictimization occurs when "a person aged 60 or older is harmed through multiple co-occurring or sequential types of elder abuse by one or more perpetrators, or when an older adult experiences one type of abuse perpetrated by multiple others with whom the older adult has a personal, professional or care recipient relationship in which there is a societal expectation of trust" (Ramsey-Klawsnik & Heisler, 2014, p.15). New efforts are underway to illuminate how a polyvictimization perspective may improve the understanding of and response to older adults experiencing multiple victimizations and resultant trauma.

Many factors – individual, relational, cultural, and societal – alone and in combination, place older adults at risk for elder abuse. Yet, we have very little understanding of the discrete contributions of these risk factors or the complex interactions between and among them. Future research must look to disentangle individual and collective influences on the perpetration and consequences of abuse in late life. Although the socioecological framework is receiving increasing attention in the study of elder abuse, additional research is needed to further delineate key variables within each layer of the model in order to begin work on identifying pathways not only of vulnerability for abuse, but protective factors that prevent elder abuse from occurring. To develop this model more fully will require the use of more sophisticated methodologies and statistical approaches to model both risk factors for and consequences of elder abuse. At the same time, rigorous qualitative investigations that go beyond descriptive presentations of the experiences of victims of elder abuse are needed to better understand the dynamics of elder abuse, including gender, race, and cultural variations in definitions of abuse, relationship expectations, and help-seeking behaviors. Researchers must also expand their efforts to learn more about the perpetrators of elder abuse beyond demographic characteristics and descriptions of personal behaviors. The complex interactions between unique facets of different abuse types, dependence, deviance, social isolation, and other relational factors such as gender or family history of abuse appear to play a significant role in patterns of perpetration and merit further exploration.

Seldom empirically validated, the costs and consequences of elder abuse appear to be intersecting, cumulative, and pervasive in the lives of abuse victims. Some costs are material in nature, such as the loss of money and valued possessions; other outcomes can be invisible and may be even more detrimental than readily quantifiable damages. Whereas less tangible long-term consequences of elder abuse have been identified (e.g., psychological distress, social isolation), the full range of the consequences of abuse in the lives of older adults remains unclear.

While sound social science research and theories are powerful tools in understanding and predicting the risks and consequences of elder abuse, an important contribution of theorized research is its potential for informing professional practice and policy development. Given the dearth of intervention research on the efficacy of programs in place to address and prevent elder abuse, researchers must join with practice professionals to gather the comprehensive, evidence-based data needed to determine which strategies and programs work best and to advocate for state and federal support for their implementation.

References

Acierno, R., Hernandez, M. A., Amstadter, A. B., Resnick, H. S., Steve, K., Muzzy, W., et al. (2010). Prevalence and correlates of emotional, physical, sexual, and financial abuse and potential neglect in the United States: The National Elder Mistreatment Study. *American Journal of Public Health*, 100, 292–297.

Acierno, R., Hernandez-Tejada, M., Muzzy, W., & Steve, K. (2009). *National elder mistreatment study*. Rockville, MD: U.S. Department of Justice. Retrieved from: <https://www.ncjrs.gov/pdffiles1/nij/grants/226456.pdf>.

Administration on Aging. (2012). 2010 National ombudsman reporting system. Table A-1: Selected information by State. Retrieved from: <http://www.aoa.gov/aoa_programs/elder_rights/Ombudsman/National_State_Data/2010/Index.aspx>.

Amstadter, A. B., Cisler, J. M., McCauley, J. L., Hernandez, M. A., Muzzy, W., & Acierno, R. (2011). Do incident and perpetrator characteristics of elder mistreatment differ by gender of the victim? Results from the National Elder Mistreatment Study. *Journal of Elder Abuse & Neglect*, 23, 43–57.

Amstadter, A. B., Zajac, K., Strachan, M., Hernandez, M. A., Kilpatrick, D. G., & Acierno, R. (2011). Prevalence and correlates of elder mistreatment in South Carolina: The South Carolina Elder Mistreatment Study. *Journal of Interpersonal Violence*, 26, 2947–2972.

Anderson, K. L. (2010). Conflict, power, and violence in families. *Journal of Marriage and Family*, 72, 726–742.

Anetzberger, G. J. (2012). An update on the nature and scope of elder abuse. *Generations*, 36, 12–20.

Baker, A. A. (1975). Granny bashing. *Modern Geriatrics*, 5, 20–24.

Baker, M. W., LaCroix, A. Z., Wu, C., Cochrane, B. B., Wallace, R. B., & Woods, N. F. (2009). Mortality risk associated with physical and verbal abuse in women aged 50 to 79. *Journal of the American Geriatrics Society*, 57, 1799–1809.

Beach, S. R., Schulz, R., Williamson, G. M., Miller, L. S., Weiner, M. F., & Lance, C. E. (2005). Risk factors for potentially harmful informal caregiver behavior. *Journal of the American Geriatrics Society*, 53, 255–261.

Bengtson, V. L., & Allen, K. A. (1993). The life course perspective applied to families over time. In P. G. Boss, W. J. Doherty, R. LaRossa, W. R. Schumm, & S. K. Steinmertz (Eds.), *Sourcebook of family theories and methods: A contextual approach* (pp. 469–499). New York, NY: Plenum Press.

Biggs, S., Manthorpe, J., Tinker, A., Doyle, M., & Erens, B. (2009). Mistreatment of older people in the United Kingdom: Findings from the first national prevalence study. *Journal of Elder Abuse & Neglect*, 21, 1–14.

Boldy, D., Horner, B., Crouchley, K., Davey, M., & Boylen, S. (2005). Addressing elder abuse: Western Australian case study. *Australian Journal on Ageing*, 24, 3–8.

Bonnie, R. J., & Wallace, R. B. (Eds.). (2003). *Elder mistreatment: Abuse, neglect, and exploitation in an aging America*. Washington, DC: The National Academies Press.

Bowen, G. L., Martin, J. A., Mancini, J. A., & Nelson, J. P. (2000). Community capacity: Antecedents and consequences. *Journal of Community Practice*, 8, 1–21.

Brandl, B., Dyer, C. B., Heisler, C. J., Otto, J. M., Stiegel, L. A., & Thomas, R. W. (2007). *Elder abuse detection and intervention: A collaborative approach*. New York, NY: Springer.

Brandl, B., & Raymond, J. A. (2012). Policy implications of recognizing that caregiver stress is not the primary cause of elder abuse. *Generations, 36*, 32–39.

Bronfenbrenner, U. (1979). *The ecology of human development: Experiments by nature and design*. Cambridge, MA: Harvard University Press.

Brossoie, N., Roberto, K. A., & Barrow, K. M. (2012). Making sense of intimate partner violence in late life: Comments from online news readers. *The Gerontologist, 52*, 792–801.

Brownell, P., Berman, J., Nelson, A., & Fofana, R. C. (2003). Grandparents raising grandchildren: The risks of caregiving. *Journal of Elder Abuse & Neglect, 15*, 5–31.

Bullock, K., & Thomas, R. L. (2007). The vulnerability for elder abuse among a sample of custodial grandfathers: An exploratory study. *Journal of Elder Abuse & Neglect, 19*, 133–150.

Burgess, A. W., Dowdell, E. B., & Prentky, R. A. (2000). Sexual abuse of nursing home residents. *Journal of Psychosocial Nursing and Mental Health Services, 38*, 10–18.

Castle, N. (2011). Nursing home deficiency citations for abuse. *Journal of Applied Gerontology, 30*, 719–743.

Çevirme, A. S., Uğurlu, N., Çevirme, H., & Durat, G. (2012). In Turkish elderly population elder abuse and neglect: A study of prevalence, related risk factors and perceived social support. *HealthMED, 6*, 88–95.

Charles, S. T., & Carstensen, L. L. (2010). Social and emotional aging. *Annual Review of Psychology, 61*, 383–409.

Cheung, M., Leung, P., & Tsui, V. (2009). Asian male domestic violence victims: Services exclusive for men. *Journal of Family Violence, 24*, 447–462.

Chompunud, S., Chounchom, C., Palmer, M., Kanuengnit, P., Thavatchai, V., & Sutthichai, J. (2010). Prevalence, associated factors and predictors of elder abuse in Thailand. *Pacific Rim International Journal of Nursing Research, 14*, 283–296.

Cisler, J. M., Begle, A. M., Amstadter, A. B., & Acierno, R. (2011). Mistreatment and self-reported emotional symptoms: Results from the National Elder Mistreatment Study. *Journal of Elder Abuse & Neglect, 24*, 216–230.

Cooper, C., Selwood, A., Blanchard, M., Walker, Z., Blizard, R., & Livingston, G. (2009). Abuse of people with dementia by family carers: Representative cross sectional survey. *BMJ*, 338.

Dakin, E., & Pearlmutter, S. (2009). Older women's perceptions of elder maltreatment and ethical dilemmas in adult protective services: A cross-cultural, exploratory study. *Journal of Elder Abuse & Neglect, 21*, 15–57.

Daly, J. M., Hartz, A. J., Stromquist, A. M., Peek-Asa, C., & Jogerst, G. J. (2008). Self-reported elder domestic partner violence in one rural Iowa county. *Journal of Emotional Abuse, 7*, 115–134.

DeLiema, M., Gassoumis, Z. D., Homeier, D. C., & Wilber, K. H. (2012). Determining prevalence and correlates of elder abuse using promoters: Low-income immigrant Latinos report high rates of abuse and neglect. *Journal of the American Geriatrics Society, 60*, 1333–1339.

Dimah, K. P., & Dimah, A. (2003). Elder abuse and neglect among rural and urban women. *Journal of Elder Abuse & Neglect, 15*, 75–93.

Dolbin-MacNab, M. L. (2006). Just like raising your own? Grandmothers' perceptions of parenting a second time around. *Family Relations, 55*, 564–575.

Dong, X., Beck, T., & Simon, M. A. (2009). Loneliness and mistreatment of older Chinese women: Does social support matter? *Journal of Women & Aging, 21*, 293–302.

Dong, X., Simon, M., Beck, T., Farran, C., McCann, J., Mendes de Leon, C., et al. (2011). Elder abuse and mortality: The role of psychological and social wellbeing. *Gerontology, 57*, 549–558.

Dong, X., Simon, M., Rajan, K., & Evans, D. A. (2011). Association of cognitive function and risk for elder abuse in a community-dwelling population. *Dementia and Geriatric Cognitive Disorders, 32*, 209–215.

Dong, X., & Simon, M. A. (2013). Elder abuse as a risk factor for hospitalization in older persons. *JAMA Internal Medicine, 173*, 911–917.

Dong, X., & Simon, M. A. (2014). Vulnerability risk index profile for elder abuse in a community-dwelling population. *Journal of the American Geriatrics Society, 62*, 10–15.

Elder, G. H., Jr. (1977). Family history and the life course. *Journal of Family History, 2*, 279–304.

Eliason, S. (2009). Murder-suicide: A review of the recent literature. *Journal of the American Academy of Psychiatry and the Law, 37*, 371–376.

Ernst, J. S., Ramsey-Klawsnik, H., Schillerstrom, J., Dayton, C., Mixson, P., & Counihan, M. (2014). Informing evidence based practice: A review of research analyzing adult protective service data. *Journal of Elder Abuse & Neglect, 26*, 458–494.

Fisher, B. S., & Regan, S. L. (2006). The extent and frequency of abuse in the lives of older women and their relationship with health outcomes. *The Gerontologist, 46*, 200–209.

Fitzpatrick, M. J., & Hamill, S. B. (2011). Elder abuse: Factors related to perceptions of severity and likelihood of reporting. *Journal of Elder Abuse & Neglect, 23*, 1–16.

Furstenberg, F. F., & Hughes, M. E. (1997). The influence of neighborhoods on children's development: A theoretical perspective and research agenda. In J. Brooks-Gunn,

G. J. Duncan, & J. L. Aber (Eds.), *Neighborhood poverty: Policy implications in studying neighborhoods* (pp. 22–47). New York, NY: Russell Sage Foundation.

Government Accounting Office. (2010). *Guardianships: Cases of financial exploitation, neglect, and abuse of seniors* (GAO-10-1046). Retrieved from: <http://www.gao.gov/assets/320/310741.pdf>.

Government Accountability Office. (2011). *Elder justice: Stronger federal leadership could enhance national response to elder abuse* (GAO-11-208). Retrieved from: <http://www.gao.gov/new.items/d11208.pdf>.

Greenfield, E. A., & Marks, N. F. (2010). Identifying experiences of physical and psychological violence in childhood that jeopardize mental health in adulthood. *Child Abuse & Neglect, 34*, 161–171.

Hansberry, M. R., Chen, E., & Gorbien, M. J. (2005). Dementia and elder abuse. *Clinics in Geriatric Medicine, 21*, 315–332.

HelpAge India. (2012). *Elder abuse in India*. Retrieved from: <http://www.helpageindia.org/pdf/Report_Elder-Abuse_India2012.pdf>.

Hernandez-Tejada, M., Amstadter, A., Muzzy, W., & Acierno, R. (2013). The National Elder Mistreatment Study: Race and ethnicity findings. *Journal of Elder Abuse & Neglect, 25*, 281–293.

Horsford, S. R., Parra-Cardona, J. R., Post, L. A., & Schiamberg, L. (2010). Elder abuse and neglect in African American families: Informing practice based on ecological and cultural frameworks. *Journal of Elder Abuse and Neglect, 23*, 75–88.

Institute of Medicine & National Research Council, (2013). *Elder abuse and its prevention: Workshop summary*. Washington, DC: The National Academies Press. Retrieved from: <http://www.nap.edu/catalog.php?record_id=18518>.

Jackson, S. L., & Hafemeister, T. L. (2011). Risk factors associated with elder abuse: The importance of differentiating by type of elder maltreatment. *Violence and Victims, 26*, 738–757.

Jackson, S. L., & Hafemeister, T. L. (2012). Pure financial exploitation vs. hybrid financial exploitation co-occurring with physical abuse and/or neglect of elderly persons. *Psychology of Violence, 2*, 285–296.

Jirik, S., & Sanders, S. (2014). Analysis of elder abuse statutes across the United States, 2011–2012. *Journal of Gerontological Social Work, 57*, 478–497.

Jogerst, G. J., Daly, J. M., Galloway, L. J., Zheng, S., & Xu, Y. (2012). Substance abuse associated with elder abuse in the United States. *The American Journal of Drug and Alcohol Abuse, 38*, 63–69.

Johannesen, M., & LoGiudice, D. (2013). Elder abuse: A systematic review of risk factors in community-dwelling elders. *Age and Ageing, 42*, 292–298.

Koenig, T. S., Rinfrette, E. S., & Lutz, W. A. (2006). Female caregivers' reflections on ethical decision-making: The intersection of domestic violence and elder care. *Clinical Social Work Journal, 34*, 361–372.

Kosberg, J. I. (2014). Rosalie Wolf Memorial Lecture: Reconsidering assumptions regarding men as elder abuse perpetrators and as elder abuse victims. *Journal of Elder Abuse & Neglect, 26*, 207–222.

Kosberg, J. I., & MacNeil, G. (2003). The elder abuse of custodial grandparents: A hidden phenomenon. *Journal of Elder Abuse & Neglect, 15*(3–4), 33–53.

Lafferty, A. (2009). *Public perceptions of elder abuse: A literature review*. National Centre for the Protection of Older People, Dublin. Retrieved from: <http://www.globalaging.org/elderrights/world/2009/elderabuse.pdf>.

Laumann, E. O., Leitsch, S. A., & Waite, L. J. (2008). Elder mistreatment in the United States: Prevalence estimates from a nationally representative study. *The Journals of Gerontology Series B: Psychological Sciences and Social Sciences, 63*, S248–S254.

Lee, H. Y., Lee, S. E., & Eaton, C. K. (2012). Exploring definitions of financial abuse in elderly Korean immigrants: The contribution of traditional cultural values. *Journal of Elder Abuse & Neglect, 24*, 293–311.

Lev-Wiesel, R., & Kleinberg, B. (2002). Elderly battered wives' perceptions of the spousal relationship as reflected in the drawings of the couple. *The Arts in Psychotherapy, 29*, 13–17.

Lifespan of Greater Rochester, Inc. (2011). *Under the radar: New York State elder abuse prevalence study*. Retrieved from: <http://ocfs.ny.gov/main/reports/Under%20the%20Radar%2005%2012%2011%20final%20report.pdf>.

Lowenstein, A., Eisikovits, Z., Band-Winterstein, T., & Enosh, G. (2009). Is elder abuse and neglect a social phenomenon? Data from the first national prevalence survey in Israel. *Journal of Elder Abuse & Neglect, 21*, 253–277.

Mancini, J. A., & Bowen, G. L. (2013). Families and communities: A social organization theory of action and change. In G. W. Peterson & K. R. Bush (Eds.), *Handbook of marriage and the family* (pp. 781–813) (3rd ed.). New York, NY: Springer.

Mba, C. J. (2007). Elder abuse in parts of Africa and the way forward. *Gerontechnology, 6*, 230–235.

McDonald, L., & Thomas, C. (2013). Elder abuse through a life course lens. *International Psychogeriatrics, 25*, 235–243.

Metlife Mature Market Institute. (2009). *Broken trust: Elders, family, and finances*. Retrieved from: <http://www.gerontology.vt.edu/docs/mmi-studies-broken-trust.pdf>.

Metlife Mature Market Institute. (2011). *The MetLife study of elder financial abuse: Crimes of occasion, desperation and predation against America's elders*. Retrieved from: <http://www.metlife.com/assets/cao/mmi/publications/studies/2011/mmi-elder-financial-abuse.pdf>.

Mezey, N. J., Post, L. A., & Maxwell, C. D. (2002). Redefining intimate partner violence: Women's experiences with physical violence and non-physical abuse by age. *International Journal of Sociology and Social Policy, 22,* 122–154.

Moon, A., & Benton, D. (2000). Tolerance of elder abuse and attitudes toward third-party intervention among African American, Korean American, and White elderly. *Journal of Multicultural Social Work, 8,* 283–303.

Morgan, E., Johnson, I., & Sigler, R. (2006). Public definitions and endorsement of the criminalization of elder abuse. *Journal of Criminal Justice, 34,* 275–283.

National Resource Center on LGBT Aging. (2013). A self-help guide for LGBT older adults and their caregivers and loved ones: Preventing, recognizing, and addressing elder abuse. Retrieved from: <http://www.lgbtagingcenter.org/resources/pdfs/SELF-HELP_elderAbuse_Guide.pdf>.

Navarro, A. E., Gassoumis, Z. D., & Wilber, K. H. (2013). Holding abusers accountable: An elder abuse forensic center increases criminal prosecution of financial exploitation. *The Gerontologist, 53,* 303–312.

Oh, J., Kim, H., Martins, D., & Kim, H. (2006). A study of elder abuse in Korea. *International Journal of Nursing Studies, 43,* 203–214.

Payne, B. K. (2008). Training adult protective services workers about domestic violence: Training needs and strategies. *Violence Against Women, 14,* 1119–1123.

Peterson, J. C., Burnes, D. P. R., Caccamise, P. L., Mason, A., Henderson, C. R., Wells, M. T., et al. (2014). Financial exploitation of older adults: A population-based prevalence study. *Journal of General Internal Medicine.*

Ploeg, J., Fear, J., Hutchison, B., MacMillan, H., & Bolan, G. (2009). A systematic review of interventions for elder abuse. *Journal of Elder Abuse & Neglect, 21,* 187–210.

Ramsey-Klawsnik, H. (2000). Elder-abuse offenders: A typology. *Generations, 24,* 17–22.

Ramsey-Klawsnik, H., & Heisler, C. (2014, May–June). Polyvictimization in later life. *Victimization of the Elderly and Disabled, 17,* 3–4, 15–16.

Reeves, K. A., Desmarais, S. L., Nicholls, T. L., & Douglas, K. S. (2007). Intimate partner abuse of older men: Considerations for the assessment of risk. *Journal of Elder Abuse & Neglect, 19,* 7–27.

Reid, R. J., Bonomi, A. E., Rivara, F. P., Anderson, M. L., Fishman, P. A., Carrell, D. S., et al. (2008). Intimate partner violence among men: Prevalence, chronicity, and health effects. *American Journal of Preventive Medicine, 34,* 478–485.

Riddell, T., Ford-Gilboe, M., & Leipert, B. (2009). Strategies used by rural women to stop, avoid, or escape from intimate partner violence. *Health Care for Women International, 30,* 134–159.

Roberto, K. A., McCann, B. R., & Brossoie, N. (2013). Intimate partner violence in late life: An analysis of national news reports. *Journal of Elder Abuse & Neglect, 25,* 230–241.

Roberto, K. A., Teaster, P. B., & McPherson, M. (2015). Abuse in late life: Unsuspecting elders and trusted others. In J. Arditti (Ed.), *Family problems: Stress, risk, and resilience* (pp. 228–248). Hoboken, NJ: John Wiley & Sons, Inc.

Roberto, K. A., Teaster, P. B., McPherson, M., Mancini, J. A., & Savla, J. (2015). A community capacity framework for enhancing a criminal justice response to elder abuse. *Journal of Crime and Justice, 38*(1), 9–26.

Roby, J., & Sullivan, R. (2000). Adult protective service laws: A comparison of state statutes from definition to case closure. *Journal of Elder Abuse & Neglect, 12,* 17–51.

Rovi, S., Chen, P. H., Johnson, M. S., & Mouton, C. P. (2009). Mapping the elder mistreatment iceberg: U.S. hospitalizations with elder abuse and neglect diagnosis. *Journal of Elder Abuse & Neglect, 21,* 346–359.

Savla, J., Roberto, K. A., Jaramillo, A. L., Gambrel, L. E., Karimi, H., & Butner, L. M. (2013). Childhood abuse affects emotional closeness with family in mid and later life. *Child Abuse & Neglect, 37,* 388–399.

Schafer, M. H., & Koltai, J. (2014). Does embeddedness protect? Personal network density and vulnerability to mistreatment among older American adults. *The Journals of Gerontology. Series B: Psychological Sciences and Social Sciences*

Schiamberg, L. B., & Gans, D. (2000). Elder abuse by adult children: An applied ecological framework for understanding contextual risk factors and the intergenerational character of quality of life. *International Journal of Aging and Human Development, 50,* 329–359.

Sethi, D., Wood, S., Mitis, F., Bellis, M., Penhale, B., Marmolejo, I., et al. (2011). European report on preventing elder maltreatment. *WHO European Region.* Retrieved from: <http://www.euro.who.int/__data/assets/pdf_file/0010/144676/e95110.pdf>.

Settersten, R. A. (Ed.). (2003). *Invitation to the life course: Toward new understandings of later life.* Amityville, NY: Baywood.

Soares, J., Barros, H., Torres-Gonzales, F., Toannidi-Kapolou, E., Lamura, G., Lindert, J., et al. (2010). *Abuse and health in Europe.* Lithuanian University of Health Sciences Press. Retrieved from: <http://www.hig.se/download/18.3984f2ed12e6a7b4c3580003555/ABUEL.pdf>.

Stiegel, L. A. (2012). An overview of elder financial exploitation. *Generations, 36,* 73–80.

Straka, S. M., & Montminy, L. (2006). Responding to the needs of older women experiencing domestic violence. *Violence Against Women, 12,* 251–267.

Tatara, T. (1999). Introduction. In T. Tatara (Ed.), *Understanding elder abuse in minority populations* (pp. 1–9). Philadelphia, PA: Taylor & Francis.

Teaster, P. B., Roberto, K. A., & Dugar, T. A. (2006). Intimate partner violence of rural aging women. *Family Relations, 55*, 636–648.

Teaster, P. B., Wangmo, T., & Anetzberger, G. J. (2010). A glass half full: The dubious history of elder abuse policy. *Journal of Elder Abuse & Neglect, 22*, 6–15.

Teaster, P. B., Wangmo, T., & Vorsky, F. (2012). Abuse in aging families. In R. Blieszner & V. Bedford (Eds.), *Handbook of families and aging* (pp. 409–430) (2nd ed.). Denver, CO: Praeger.

The National Academics Committee on National Statistics, (2010). *Meeting on research issues in elder mistreatment and abuse and financial fraud*. Washington, DC: Author. Retrieved from: <http://www.nia.nih.gov/sites/default/files/meeting-report_1.pdf>.

Thomas, K. A., Joshi, M., Wittenberg, E., & McCloskey, L. A. (2008). Intersections of harm and health: A qualitative study of intimate partner violence in women's lives. *Violence Against Women, 14*, 1252–1273.

Thompson, E. H., Buxton, W., Gough, C. P., & Wahle, C. (2007). Gendered policies and practices that increase older men's risk of elder mistreatment. *Journal of Elder Abuse & Neglect, 19*, 129–151.

Walsh, C. A., Olson, J. L., Ploeg, J., Lohfield, L., & MacMillan, H. L. (2011). Elder abuse and oppression: Voices of marginalized elders. *Journal of Elder Abuse & Neglect, 23*, 17–42.

Werner, P., Eisikovits, Z., & Buchbinder, E. (2005). Lay persons' emotional reactions toward an abused elderly person. *Journal of Elder Abuse & Neglect, 17*, 63–76.

Wiglesworth, A., Austin, R., Corona, M., & Mosqueda, L. (2009). *Bruising as a forensic marker of physical elder abuse*. Washington, DC: U.S. Department of Justice.

Wiglesworth, A., Mosqueda, L., Mulnard, R., Liao, S., Gibbs, L., & Fitzgerald, W. (2010). Screening for abuse and neglect of people with dementia. *Journal of the American Geriatrics Society, 58*, 493–500.

Wood, S., & Liu, P. (2012). Undue influence and financial capacity: A clinical perspective. *Generations, 36*, 53–58.

World Health Organization. (2002). *Active ageing: A policy framework*. Retrieved from: <http://whqlibdoc.who.int/hq/2002/who_nmh_nph_02.8.pdf>.

World Health Organization. (2011). *Elder maltreatment*. Retrieved from: <http://www.who.int/mediacentre/factsheets/fs357/en/index.html>.

CHAPTER 17

The Impact of Disasters: Implications for the Well-Being of Older Adults

Lisa M. Brown[1] and Kathryn A. Frahm[2]

[1]Palo Alto University, Palo Alto, CA, USA [2]School of Aging Studies,
University of South Florida, Tampa, FL, USA

OUTLINE

Introduction	357
Types and Definitions of Disasters	358
Influence of Residential Environment on Disaster-Related Activities and Outcomes	359
Group Evacuation versus Individual/Independent Evacuation	360
Temporary Evacuation, Transfer, or Permanent Relocation	361
Age and Vulnerability	362
Stress and Coping	363
Age and Resilience	363
Disaster-Related Physical and Mental Health Issues	365
Social Factors and Disaster Response Outcomes	366
Role of Formal and Informal Social Support and Social Networks	367
Formal Support	*367*
Informal Support and Social Networks	*368*
Future Directions	369
Conclusion	370
References	370

INTRODUCTION

Since 2000, there has been a notable increase in the number of major disasters. Disasters frequently occur, with older adults experiencing a disproportionate number of adverse effects (Hutton, 2008). Disasters may be either natural, such as hurricanes, tornados, and earthquakes, or human-made, such as terrorism or technological accidents. This increase has been accompanied by an upturn in the number of government and nonprofit organizations promulgating detailed instructions describing how to mitigate, prepare for disasters, obtain safe shelter, cope with adverse events, and initiate the recovery process. However, despite these efforts, government officials and researchers consistently report that older adults are at greatest risk for adverse outcomes, but are the least prepared subgroup of the population.

Lack of preparedness and poor post-disaster outcomes can be attributed to a number of personal and system factors. Most policies and plans refer to older adults as though they were a homogeneous subgroup of the population. Not distinguishing between the needs of those who are robustly aging from their less-well-functioning counterparts results in the production and dissemination of generic disaster information. It is well recognized that vulnerability to disasters is a combination of personal, social, environmental, and economic circumstances. Unique differences exist between types of disasters in terms of extent (e.g., timing, duration) and impact (e.g., loss of life, loss of community) (DeWolfe, 2000). The interaction of these factors influences disaster preparation, the experience of a disaster, and the ultimate outcome (Morrow, 1999). Moreover, the same disaster may be perceived and experienced very differently due to race, class, previous experiences, and available resources (Ruscher, 2006). Although most older adults show resilience and are able to effectively cope and recover over time, some develop enduring psychological or physical problems post-disaster.

This chapter describes the key factors that influence whether older adults are resilient or develop problems post-disaster. To set the context for the chapter, definitions of the different phases of disasters and available supports are described to underscore the importance of residential environment on disaster-related activities and outcomes. As post-disaster movement of affected populations is increasingly recognized as influencing outcomes, group (e.g., bussing residents) versus individual evacuation (e.g., driving a private car) and the differences between brief evacuation to a public shelter, temporary transfer to another area, or permanent relocation to a new community are characterized. Next, a review of disaster-related consequences, including physical and mental health is presented. Factors associated with risk and resilience in response to disasters are discussed along with a review of the recent literature on individual, social, and community factors that influence outcomes. Lastly, areas for future direction of disaster-related research are identified and next steps are proposed.

TYPES AND DEFINITIONS OF DISASTERS

Disasters can strike anytime and anywhere. On average, a disaster occurs daily somewhere in the world (Rodriguez, Vos, Below, & Guha-Sapir, 2009) and weekly in the United States (Federal Emergency Management Agency, 2010). In contrast, an emergency, such as a fire that requires a response from the fire department occurs every 22 seconds in the United States (Karter, 2009). Most communities are well prepared for emergencies. Although communities plan and prepare for emergencies, disasters are events of greater magnitude that require more resources to respond and recover than are locally available. Table 17.1 provides a list of disaster types and characteristics.

Emergency managers and planners typically describe disaster management in terms of four phases: mitigation, preparedness, response, and recovery. The four phases are related, affect the impact of an event on a community, and inform allocation of funding and resources. Table 17.2 provides a description of each of the phases that comprise the disaster lifecycle.

TABLE 17.1 Disaster Types and Characteristics

Hazards	An event that has not occurred but has the potential to cause harm to health, life, safety, or the environment and increases vulnerability, although it may not require rapid response
Emergencies	An event that poses an immediate threat to one's health, life, or surroundings and can be adequately managed at the local level by designated responders using local resources (Guha-Sapir, 2000)
Disasters	Events involving 100 or more persons, with ten or more deaths, an official disaster declaration, or an appeal for assistance (Below, Wirtz, & Guha-Sapir, 2009) that requires more resources than are available in the immediate geographic area
Catastrophic events	A sudden and extreme disastrous event, causing an upheaval in the order of communities, which requires an extensive recovery process that fundamentally changes the surrounding environment (Homeland Security, 2008)

TABLE 17.2 Phases Related to Disasters

Phase 1 Mitigation	Activities focus on local vulnerabilities with the goal of minimizing the potential adverse effects of a disaster by not building in a flood zone or by fortifying structures to reduce the threat of wind damage or earthquake
Phase 2 Preparedness	Advance planning that ranges from individual disaster plans and to-go kits to community training, public education, securing equipment and supplies, and developing emergency operation plans
Phase 3 Response	The use of resources and supplies secured during the preparedness phase to minimize the harm resulting from the disaster by operating shelters, evacuating threatened populations, and working to quickly restore critical services (e.g., electricity, telephone, water) after a disaster
Phase 4 Recovery	Actions taken after a disaster that can last from days to years and focus on restoring the community and its residents to their pre-disaster status

INFLUENCE OF RESIDENTIAL ENVIRONMENT ON DISASTER-RELATED ACTIVITIES AND OUTCOMES

The type of residential setting influences the level of disaster preparedness and response. For example, older adults residing in nursing homes and many living in assisted living facilities have health and functional challenges that require assistance with activities of daily living. During the 2004 and 2005 tropical storms and hurricanes affecting the southeastern United States (hurricanes Charley, Frances, Ivan, and Jeanne in 2004 and Katrina and Wilma in 2005), as well as other tropical storms, the level of assistance provided to persons in nursing homes compared to those living in assisted living facilities was pronounced. State regulations for nursing homes specify staff duties and responsibilities, evacuation procedures, and required resource reserves, whereas assisted living facilities were not under the same obligation to assist residents with evacuation and care (Brown, Hyer, & Polivka-West, 2007). Because nursing homes accept Medicaid and Medicare payment for services, they are required to have well-developed disaster plans and provide staff training. It has been recommended that disaster planning efforts account for the wide range of settings in which older adults reside and are specific to the setting (Johnson et al., 2006).

However, the vast majority (78%) of older adults remain independent in the community (Federal Interagency Forum on Aging-Related Statistics, 2010). Those living independently are able to maintain their health and well-being with assistance from family, friends, and home and community-based care provided by the aging services network (e.g., home-delivered meals, home health care). The current managed care climate supports the desire of older adults to remain at home. Approximately 95% receive home delivery of health-related services. After a disaster, many older adults may find themselves in their homes and communities without the usual services and supports that maintain their independence. Interruption of services and supports has been shown to result in a decline in physical and mental health functioning (Laditka, Laditka, Cornman, Davis, & Chandlee, 2008; Laditka et al., 2007). Preparedness planning tends to concentrate on residential care facilities, overlooking the importance of providing emergency services to the majority of older adults who live alone in their own homes or with family.

GROUP EVACUATION VERSUS INDIVIDUAL/INDEPENDENT EVACUATION

In a national survey of people age 50 and older, 13 million reported the need for assistance during evacuation and relocation and nearly half reported they would need help from someone outside of their household (American Association of Retired Persons, 2006). Locating and assisting older residents living at home is complicated, including finding persons in the immediate aftermath, locating sufficient temporary housing, recovering pets, identifying and treating medical problems, obtaining medications, providing food and water, protecting against the natural elements, assisting with clean up, and securing insurance payments, federal compensation, and other assistance. Data collected only after the 2003 and 2004 Louisiana hurricane seasons revealed that approximately 32% of community-dwelling older adults had a disability and 17% needed assistance to evacuate (McGuire, Ford, & Okoro, 2007). While people may be somewhat prepared, specific evacuation plans may be more of a concern (Kusenbach, Simms, & Tobin, 2010). Many frail older adults are at particular risk of transfer trauma, a condition that can occur when older adults are transferred from one living arrangement to another. If people must be relocated, it is advisable to keep them with loved ones, friends, and natural groups (Norris, Friedman, & Watson, 2002).

One way to identify older adults who might be vulnerable during a disaster is to enlist the assistance of the aging services network that provides regular services and supports to older adults in the community (Oriol, 1999). While many older adults in need can be identified in this manner, it is important to recognize that not all older adults who are at risk are currently receiving such supports and services, therefore, additional outreach efforts need to target other potentially vulnerable older adults (Brown, Rothman, & Norris, 2007).

Advance enrollment for special needs shelter services, where a physician has to document medical need, continues to be challenging. Community-dwelling older adults who function independently often do not self-identify as having special needs and register for services prior to a disaster. Loss of electricity frequently results in a sudden influx of need, creating demand for limited resources in public shelters. Another concern is finding new health care providers to address their chronic health problems (Sanders, Bowie, & Bowie, 2003).

For nursing home and assisted living facility residents, those in need of regular assistance with their activities of daily living will need continued aid. During the 2004–2005 Florida hurricanes, not all counties were equally prepared to respond. Across Florida both nursing homes and assisted living facilities had difficulties with evacuation and transportation, as well as extended power loss (Christensen, Brown, & Hyer, 2012). In nursing homes and assisted living facilities, over 70% of residents require some type of assistive mobility device (walker, cane, or wheelchair) and over 50% use a wheelchair, creating challenges to timely evacuation (Clarke, Chan, Santaguida, & Colantonio, 2009). Emergency transport companies were overwhelmed and unable to meet their agreements with nursing homes as hospitals had top priority for services. For facilities that sheltered in place, loss of electrical power was also a significant problem. Not all facilities had working generators, sufficient fuel to operate for extended periods of time, or adequate power. Those that lost electrical power were given priority service restoration, but not all locales recognized priority reconnection for nursing homes, and some facilities providing skilled care went days without power. Decisions about evacuation, whether independent or group, and weighing the advantages of leaving versus staying in place were difficult, and often depended on the circumstances presented. Evacuation was not appropriate in all instances.

A recent study examined the effects of evacuation prior to Hurricane Gustav making landfall on nursing home residents who were cognitively impaired and did not have direct exposure to the actual storm. The dataset included 21 255 residents living in 119 at-risk nursing homes over 3 years of observation. Relative to the 2 years before the storm, there was a 2.8% increase in death at 30 days and a 3.9% increase in death at 90 days for residents with severe dementia who evacuated, controlling for resident demographics and acuity. The findings of this research revealed the deleterious effects of evacuation on residents with severe dementia (Brown, Dosa, et al., 2012).

TEMPORARY EVACUATION, TRANSFER, OR PERMANENT RELOCATION

Evacuation involves residents leaving a location that is under threat of a potential harmful event (Raid & Norris, 1996). Evacuation may take place either in anticipation of or following a disaster and may be short-term, long-term, or even permanent (Kendra, Rozdilsky, & McEntire, 2008). The goal is that people will eventually be able to return to their original setting, but this is not always possible. Evacuation rarely occurs with advanced notice, giving older adults little time to prepare for moving. Because an evacuation is a disruptive event, it may have a negative impact on the psychological and physical well-being of older adults. Evacuation is a significant stressor and may strain coping ability (Geriatric Mental Health Care Foundation, n.d.).

Transfers occur in the process of moving a nursing home or assisted living resident from one health care setting to another, perhaps involuntarily (Hodgson, Freedman, Granger, & Erno, 2004). If the move occurs rapidly, it may also result in adverse resident outcomes (Beirne, Patterson, Galie, & Goodman, 1995). A transfer between facilities disrupts resident patterns and places them in an unfamiliar environment. When residents are moved in the event of an emergency or disaster, they may experience serious declines in health (Administration on Aging, 1995).

For community-dwelling older adults, relocation refers to the impact of a semi-permanent or permanent move to a new location. An aging population, strained health care resources, and a preference toward "aging at home" are contributing to the trend for increasing numbers of older adults with functional and cognitive impairments to receive home-based care (Montz & Tobin, 2005). Moreover, social factors such as changes in family structure, a highly mobile society and a system that discourage communal living arrangements for older adults in favor of independent residency place a growing number of community-dwelling older adults at risk during disasters. Even when relocation is expected, it includes some element of stress as it requires adjustment to a new lifestyle pattern (Aneshensel, Pearlin, Levy-Storms, & Schuler, 2000; Sanders et al., 2003). Relocation may be permanent or temporary depending on the extent of the disaster on the older adults' residential setting, support network, and individual well-being. In all cases, whether a temporary evacuation, transfer, or permanent relocation, the resulting life disruption significantly affects older adults. It is imperative that disaster planning specifically attempts to maximize physical safety and minimize disturbance, confusion, and unfamiliarity for evacuated older adults.

AGE AND VULNERABILITY

By the year 2030, 20% of the American population will be 65 years of age and older (Federal Interagency Forum on Aging-Related Statistics, 2008). The growing number of older people who are likely to have one or more chronic health conditions, such as diabetes, arthritis, osteoporosis, or chronic obstructive pulmonary disease, places them at greater risk for adverse consequences during a disaster (Turcotte & Schellenberg, 2007; US Department of Health and Human Services, 2000). Disruption of daily self-care activities and access to routine medical care can exacerbate existing conditions. Although age is the factor that is most frequently associated with disaster-related morbidity and mortality (Bourque, Siegel, Kano, & Wood, 2007), old age alone does not make a person vulnerable for negative consequences. A variety of interrelated factors, including sensory and cognitive impairment, physical decline, and medical illness, influences vulnerability (Fernandez, Byard, Lin, Benson, & Barbera, 2002).

Early work suggested that older adults receive less warning about impending disasters, were more reluctant to evacuate, more disturbed by disruption in daily life, and were more likely to become physical casualties (Friedsam, 1960, 1961). Older adults were also found to be at greater risk for incurring debt, suffering economic losses, sustaining physical injury, and experiencing disrupted employment relative to younger adults (Kilijanek & Drabek, 1979). It is a constellation of factors that influence post-disaster outcomes for older adults such as trauma severity; social, economic, and cultural factors; health and functional status; psychosocial resources and social support; and gender (Gibbs & Montagnino, n.d.). Regardless of age, people who are dependent or isolated are likely to fare poorly. Simply classifying all old adults as vulnerable does not adequately address how to accommodate their needs or use their talents and abilities to assist others during a disaster.

A variety of economic, structural, political, historical, and cultural conditions affect the ability of people to protect themselves, respond, and recover from disasters. While environmental risk factors, such as housing and location, are important, social risk factors, such as poverty and isolation, also play a critical role in overall vulnerability. For example, during a disaster, a healthy, community-dwelling, older adult with a social network is at risk, along with the rest of the general population, but a frail, isolated, home-bound older adult may be extremely vulnerable. The empirical literature on the effects of disasters on older adults is equivocal.

Some research suggests that older adults tend to be more psychologically resilient than younger adults in response to stressful events, including disasters (Geriatric Mental Health Foundation, n.d.; Huerta & Horton, 1978; Knight, Gatz, Heller, & Bengtson, 2000; Norris et al., 2002). Although most older adults tend to be resilient in the face of life challenges (Boerner, Wortman, & Bonanno, 2005; Bonanno, Wortman, & Nesse, 2004; Norris, 1992; Norris, Byrne, Diaz, & Kaniasty, 2001; Phifer & Norris, 1989; Shalev, 2002), some older adults may be particularly vulnerable (e.g., Holocaust survivors, former prisoners of war, persons exposed to interpersonal violence, persons abused as a child or adult) (Bremner et al., 1992; Breslau et al., 1998; Green et al., 2000; Sigal & Weinfeld, 2001; Yehuda & McFarlane, 1995). Historically, the relatively limited disaster literature on older adults has portrayed this subgroup of the population as either extremely vulnerable or highly resilient to the negative consequences of mass casualty events.

These disparate findings are likely the result of varying research methods. Methods complicating differentiation of age effects and cohort include grouping all adults over a certain age into a single age category, not assessing those who are not independent, community-dwelling older adults, convenience sampling, collection of small sample sizes, and cross-sectional designs with no pre-disaster data. Moreover, most disaster studies have focused exclusively on psychopathology and the negative effects of disasters on well-being.

STRESS AND COPING

In late life, a number of potentially exacerbating and buffering variables can affect outcomes following trauma, such as gender, type of trauma, a history of prior trauma, social support, and age (Breslau et al., 1998; Norris et al., 2002). Stressors that amplify the negative effects of a disaster include decreased social support, declining health, sensory impairment, limited mobility, low literacy, and fewer resources (Brown, Haun, & Peterson, 2014). Prior trauma history sensitizes the individual to new stressors, thus magnifying its impact (Dougall, Heberman, Delahanty, Inslicht, & Baum, 2000). The immediate disaster as well as other current and lifetime stressors all influence peoples' ability to manage post-traumatic stress (Port, Engdahl, Frazier, & Eberly, 2002; Sigal & Weinfeld, 2001; Yehuda & McFarlane, 1995).

Although older adults may endure a severe exposure to disaster, have poorer health, and have fewer social and economic resources, they have a lifetime of learning how to cope with stressful events which promotes adaptation to disasters (Phifer, 1990; Pietrzak, Southwick, Tracy, Galea, & Norris, 2012). Life experiences provide opportunities to make comparisons between current and previous stressors and evaluate the effectiveness of coping strategies. Individual psychological characteristics and personal history influence the use of adaptive or maladaptive coping strategies as a way to overcome adversity in the aftermath of a disaster (King, King, Fairbank, Keane, & Adams, 1998; Phifer, 1990). Examples of adaptive coping strategies include problem solving and planning whereas maladaptive coping strategies involve denial, avoidance, and rumination. Positive cognitive strategies facilitate the ability to maintain a perspective on outcomes and reduce negative reaction to stressful situations and times of extreme stress (Folkman, Lazarus, Dunkel-Schetter, DeLongis, & Gruen, 1986; Janoff-Bulman, 1992). Compared with younger adults, many older adults have learned that they can overcome difficult life events, adapt to adversity, and return to prior levels of psychological functioning (Norris, 1992).

AGE AND RESILIENCE

Early research defined resilience as the ability to adapt quickly or recover from illness, difficult life experiences, misfortune, and traumatic events (Rutter, 1987). More recently, resilience has been conceptualized to include three facets: recovery, resistance, and reconfiguration (Lepore & Revenson, 2006). Recovery is the ability to return to the prior state of functioning after experiencing a disaster. Resistance is the capacity to maintain baseline functioning throughout a stressful or traumatic event. Reconfiguration involves reforming baseline thoughts, beliefs, and behaviors in response to a stressor, which may ultimately enhance ability to endure future traumatic events (Lepore & Revenson, 2006). Using this conceptualization, resilience can be broadly defined as the capacity to adjust to challenges, persevere in the face of hardship, maintain a chosen course of action or pursue desired goals despite the presence of stressors that threaten to disrupt the status quo. Resilience is best conceptualized as adaptability rather than as stability (Handmer & Dovers, 1996; Waller, 2001). In some situations, failure to change or adapt could result in a lack of resilience. However, researchers continue to debate whether resilience is best conceptualized as a personality trait or as a process. Although resilience and its correlates and predictors in older adults after disasters has not been extensively studied, gerontological research examining successful aging, positive adjustment to aging, personality traits across the lifespan, and life satisfaction could be used to inform future research in this area.

Theories that have been proposed to explain resiliency include vulnerability, inoculation, burden, and maturation. *Vulnerability theory* asserts that older adults, compared to their younger counterparts, experience a greater sense of deprivation relative to their losses that results in psychological distress (Kilijanek & Drabek, 1979). Although most research has not supported this theory, older adult survivors of prolonged stress, such as former prisoners of war and Holocaust survivors, are believed to be at greater risk for adverse outcomes because their previous trauma stressors are compounded by age-related losses (Danieli, 1997; Engdahl, Harkness, Eberly, Page, & Bielinski, 1993; Solomon & Prager, 1992). A body of research indicates that prior prolonged stressors are related to increased vulnerability with subsequent trauma (Bremner et al., 1992; Breslau et al., 1998; Green et al., 2000; Nishith, Mechanic, & Resick, 2000; Solomon & Prager, 1992; Yehuda et al., 1995).

Counter to the vulnerability theory, *inoculation theory* posits that earlier traumas foster resilience to subsequent traumas through one of two pathways, direct or cross-tolerance (Eysenck, 1983). Direct tolerance is where a prior traumatic stressor lessens the effects of that particular stressor in the future and cross-tolerance is where prior exposure to a stressor lessens the effects of a different stressor in the future. Earlier research suggested that inoculation theory is supported by older adults who experience intermittent stressors and not by those who have experienced prolonged or extended stress (Norris & Murrell, 1988). In support of this theory, prior earthquake experience was related to lower levels of depression after the 1994 Northridge earthquake (Knight et al., 2000). Inoculation theory is consistent with research that suggests that older adults fare well after disasters because of a lifetime of experience. However, the findings from this study did not support the *maturation theory*, which purports that older adults use a mature coping style that makes them less susceptible to experiencing negative emotional reactions to traumatic events. Advanced age did not moderate the relationship between rumination and disaster exposure.

The premise of the *burden theory* is that middle-aged adults are the subgroup most adversely affected by disasters because of caregiving demands by both younger and older family members and their work-related responsibilities (Thompson, Norris, & Hanacek, 1993). Not surprisingly, no history of major life stressors or traumatic events, as well as no concurrent life stressors or traumatic events are associated with resilience after a disaster (Bonanno, Galea, Bucciarelli, & Vlahov, 2007). While theory is useful to consider when exploring relationships between age and disaster, it is important to remember that simplistic characterizations of older adults as a homogeneous subgroup do not result in an accurate or useful clinical or scientific depiction. People of all ages differ in the amount of psychological distress they experience in response to stressors as well as in the speed in which they are able to recover. Resilience is influenced by a dynamic interaction between external and internal risk and protective factors.

Not surprisingly, people with low stress tolerance are less resilient than those who are adaptive in their response to change and evolving environment or social demands. Personality traits, as measured by the Five Factor Model or the Big Five Inventory, encompass five dimensions: Openness, Conscientiousness, Extraversion, Agreeableness, and Neuroticism. Across the lifespan, Neuroticism, a measure of stress tolerance, is closely related to resilience and adaptive adjustment to change (McCrae & Costa, 2003; Ong, Bergeman, Wallace, & Bisconti, 2006). Personality influences how one relates to others and interacts with society. Social networks can play a key role in supporting adaptive coping and enhancing ability to contend with evolving and difficult situations.

The good news is that a growing body of research indicates that most people who experience a traumatic event will follow a resilient trajectory (Bonanno, 2004; Bonanno et al., 2007; Bonanno, Rennicke, & Dekel, 2005). Hardiness, self-enhancement, repressive coping, and positive emotions have been identified as dimensions related to resilience (Bonanno, 2004). Hardiness was comprised of three dimensions that include: finding meaningful purpose in life, believing one has influence and can effect change, and feeling that it is possible to learn and grow from both negative and positive experiences. Self-enhancement was defined as a somewhat unrealistic and an overly positive self-perception. These attributes may be helpful in facilitating recovery post-event and fostering a sense of well-being. Personality, previous experience successfully recovering from a disaster, adequate social support, and use of adaptive coping strategies all contribute to positive outcomes. However, even without identified factors associated with disaster vulnerability many older adults will benefit from some type of assistance in preparing for, responding to, and recovering from such events.

DISASTER-RELATED PHYSICAL AND MENTAL HEALTH ISSUES

Historically, advanced age is associated with increased disaster-related morbidity and mortality after earthquakes, cyclones, and tsunamis (Doocy, Daniels, Dick, & Kirsch, 2013a, 2013b; Doocy, Daniels, Packer, Dick, & Kirsch, 2013; The Sphere Project, 2011). It is difficult to ascertain the exact number of direct and indirect disaster-related deaths for older adults, as data on age have not been consistently collected in the United States and abroad. Directly related deaths result from the environmental (e.g., wind, water) or direct consequences of environmental force (e.g., structure collapse). Indirect related deaths result from situations where the disaster caused unsafe conditions (e.g., fallen trees) or lead to a loss or disruption of services (e.g., loss of electrical power) (Combs, Quenemoen, Parrish, & Davis, 1998). As warning systems, evacuation procedures, and shelter facilities continue to improve, mortality trends for hurricanes start to shift from pre- to post-event deaths, such as drowning, electrocution, or carbon monoxide poisoning. Disasters disrupt family units, social systems, medical care, access to drugs, and availability of health services (e.g., renal dialysis). A sudden or prolonged disruption can result in a cascade of events that can lead to life-threatening complications.

The 2005 hurricanes along the Gulf Coast of the United States left more than 1 500 dead and 780 000 displaced. While these disasters affected people across the lifespan, older adults were disproportionately affected by Hurricane Katrina (Elmore & Brown, 2007–2008). Half of those who died from Hurricane Katrina were 75 years and older, compared to 2% of those who were 18 years and younger (Bytheway, 2007). The year after Hurricane Katrina, the health of older adult survivors declined by nearly four times when compared to a sample of older adults not affected by the storm (Burton et al., 2009). Morbidity rates increased

12.6% compared with 3.4% nationwide, with increases in the prevalence of cardiac disease, congestive heart failure, and sleep problems. Over the next year there was both an increase in emergency department visits (21%) and hospitalization rates (23%) compared to the pre-Katrina year (Burton et al., 2009). Although many of the research articles reviewed reported descriptive age data, few included age-related injury. Most that did showed decreased risk among children and increased risk among the older adults and with increasing age (Doocy, Daniels, Dick, et al., 2013a, 2013b; Doocy, Daniels, Packer, et al., 2013).

Older adults may be more vulnerable for acute injuries, as dangerous evacuations, unfamiliar surroundings, and altered routines serve as precipitants for adverse outcomes (Brunkard, Namulanda, & Ratard, 2008). Following an earthquake, older adults were more likely to require hospitalization after sustained injuries (Seligson & Shoaf, 2003). Conditions following a disaster have also been shown to increase the likelihood of infectious diseases and to exacerbate existing medical conditions (Rothman & Brown, 2007). As the global population of older adults continues to grow, it is imperative that steps are taken to effectively manage treatment pre- and post-disaster as a way to minimize morbidity and mortality.

After disasters, geriatric health care providers, first responders, and relief workers are called upon to meet not only physical but mental health needs of older adults who have been affected by the event. Like medical first aid, psychological first aid is a broadly used, evidence-informed intervention that appears to be beneficial for younger and older adults alike. Psychological first aid is used in the immediate aftermath of an event to facilitate recovery and promote adaptive functioning. Crisis counseling is offered to those who require more than psychological first aid. People providing post-disaster intervention should possess a solid understanding of how older age and disasters intersect with mental health conditions and disorders, particularly trauma reactions, dementia, and delirium (Rothman & Brown, 2007). Preexisting physical, cognitive, and sensory impairments that impede disaster planning may adversely affect post-disaster access to services and material goods, such as food and water, from disaster assistance centers. People without preexisting cognitive impairment who show signs of cognitive dysfunction in the aftermath of a disaster should be referred for further evaluation to rule out the presence of medical illness, heat stroke, delirium, drug mismanagement, and other conditions that require immediate treatment.

Across the lifespan, health problems and mental health disorders have the potential to exacerbate life's challenges on an ongoing basis. Exposure to a traumatic event, such as a disaster, represents a unique stressor that has influence on an older adult's mental health status. Some disorders with significant cognitive and behavioral symptoms become increasingly likely with age, particularly dementia and delirium. It is important that disaster responders have a basic understanding of each of these issues – trauma, dementia, and delirium – as they present in the older adult population.

After a disaster, many survivors will have to adjust to a new normal. Often personal belongings, homes, and community institutions are destroyed or disappear during a single event. It is not unusual for people to grieve the loss of what was and experience difficulty adjusting to the new normal even if the quality of the post-disaster replacement exceeds the pre-disaster original. In particular, people may experience considerable sadness if irreplaceable family photos and keepsake items are destroyed.

SOCIAL FACTORS AND DISASTER RESPONSE OUTCOMES

In addition to physical and mental health conditions, social isolation, limited financial resources, low education, and low literacy also negatively affect vulnerable adults' ability to remain safe during and after disasters (Brown, Cohen, & Kohlmaier, 2007; Brown, Frahm, & Bongar, 2012; Brown, Hickling, & Frahm, 2010; Institute of Medicine, 2002; Rowel, Sheikhattari, Barber, & Evans-Holland, 2012; Wisner, Blaikie, Cannon, & Davis, 2004). Disaster literacy is defined as an individual's capacity to read, understand, and use information to make informed decisions and follow instructions in the context of preparing for, responding to, and recovering from a disaster (Brown et al., 2014). In general, poor literacy adversely affects the ability of adults to comprehend and act on informational materials (Baker, 2009; Berkman, Sheridan, Donahue, Halpern, & Crotty, 2011; Brown et al., 2014; Institute of Medicine, 2004; Saha, 2006; Sentell & Halpin, 2006). Older adults without sufficient disaster literacy may not adequately prepare for disasters. Inappropriate or lack of preparation further compromises their ability to independently obtain shelter, food, water, or access standard medical care post-disaster (Brown, Rothman, et al., 2007; Rothman & Brown, 2007).

ROLE OF FORMAL AND INFORMAL SOCIAL SUPPORT AND SOCIAL NETWORKS

Formal Support

American Red Cross and Medical Reserve Corp, Community Faith-Based Organizations, Social Agencies

Currently, there is a need for disaster behavioral-health training for other relief workers and disaster responders that includes information about the specific needs and concerns of older adults (Brown et al., 2010). This includes understanding how aspects of the social and physical environment can either exacerbate or reduce stress, particularly for the most vulnerable segment of the older adult population, those who are frail and cognitively impaired (Brown, Frahm, et al., 2012). Understanding the heterogeneity of older adults' functional and cognitive status helps identify what particular services or resources will be the most useful.

Stakeholders who are traditionally included in the disaster planning process are federal, state, and local leaders, emergency management personnel, nonprofit and relief organizations, and policymakers. Many of these same people and organizations are involved during each of the disaster phases of mitigation, preparedness, response, and recovery. The planning process should be wide-ranging, address multiple needs, and identify resources. Efforts to include representatives of populations to be served, including older adults, provide a forum for understanding their perspective of their needs as a way to minimize impact and enhance positive outcomes.

To address disaster mental health concerns, a 1991 Statement of Understanding between the American Psychological Association and the American Red Cross established the Disaster Response Network (DRN) program to manage the local disaster response programs. The American Red Cross is a humanitarian organization that brings relief to survivors of disasters and helps people prepare for and respond to emergencies. The purpose of the DRN is to provide a mechanism for psychologists to provide volunteer assistance to relief workers and survivors after disasters. All DRN psychologists have to complete a series of American Red Cross courses before serving in the field (American Psychological Association, n.d.). Laypeople are also encouraged and eligible to provide psychological first aid after disasters.

The Medical Reserve Corps is a national network of volunteers who are housed within the Office of the Surgeon General, US Department of Health and Human Services. The Medical Reserve Corps was established after 9/11 as a way to recruit, train, and activate medical and health professionals to respond to public health emergencies. Medical Reserve Corps members are dedicated to improving the health, safety, and resiliency of their communities. As the missions of the Medical Reserve Corps and the American Red Cross are complementary, a formal working relationship was established in 2009 to support volunteers who wish to be affiliated with both organizations. Protocols describing how volunteers with both of these organizations would work during disasters of varying magnitude have been developed and published online (Division of the Civilian Volunteer Medical Reserve Corp, 2009).

Faith-based organizations and nonprofit organizations are also integral members of the disaster preparedness and response effort. The National Voluntary Organizations Active in Disaster (VOAD) is a nonprofit, nonpartisan membership organization that includes over 50 national faith-based, community-based and non-governmental organizations as well as 56 State/Territory VOADs. The VOAD was founded after Hurricane Camille, a category-5 storm, struck the Gulf Coast in August 1969, and responders from multiple agencies and faith-based organizations were frustrated by the lack of communication and coordination. To avoid duplication of services and to coordinate an organized response to future events, the national VOAD was formed as a way to share knowledge and manage resources throughout the disaster cycle.

There are a number of new efforts that are currently underway to help older adults during future disasters. One example is the New York Academy of Medicine (NYAM) initiative to have stakeholders create formal and informal support systems for New York's community-dwelling older adults. The initiative was funded after Hurricane Sandy because 24 of the 43 reported deaths in NYC were people over age 60. The goals of the project are to identify best practices, generate recommendations, and mobilize key partners to implement response plans.

Informal Support and Social Networks

Senior Centers, Family, Friends, Neighbors

Most social networks are comprised of family, close friends, neighbors, and community associates (Cantor, 1979). While families provide approximately 70–80% of in-home care for older adults with chronic health conditions, other informal caregivers, such as neighbors and friends, are often first to provide assistance following a disaster and can be instrumental in securing shelter, medical care, food, water, and providing support. Older adults develop social networks of selective family and friends available to provide assistance in needed and anticipated ways (Adams & Blieszner, 1995). During disaster situations, social support systems are critical to the well-being of older adults. The size and closeness of the older adult's social network is related to mental health functioning (Kahn & Antonucci, 1980). These socially protective resources are particularly vulnerable to disruption following a disaster. Significant deterioration of the social support system is likely to result in adverse short- and long-term psychological consequences (Kaniasty & Norris, 1993).

To offset potential health problems or the exacerbation of existing illness, adequate provisions for preparing, responding, and recovering are needed. Friends and close neighbors are often a source of assistance to older adults who need help in accomplishing disaster-related tasks (Crohan & Antonucci, 1989). This support may be critical in assisting older adults recover from disasters. Some older adults are at significant risk for social isolation due to a combination of choice and circumstance. Social factors related to social isolation include a mobile society, living independently, and family structure changes. Personal factors that can contribute to social isolation are the likelihood of living alone, role losses, loss of mobility, financial problems, and poor health (Brown, Gibson, & Elmore, 2012). These factors may result in socially isolated older adults being particularly vulnerable during all phases of a disaster and less resilient in the aftermath.

Although social support is often mobilized after a disaster, assistance is less available when property is damaged or destroyed, electricity or phone communication networks are down, or routines are disrupted (Kaniasty & Norris, 1999). It is likely that many members of an older adult's support network will also be survivors of the same disaster. Social supports may have died, relocated, or be unable to assist because their immediate needs exceed their current resources. Compounding the situation, disruption and destruction of community services diminishes the availability of other formal resources that provide support, resulting in the need for support and services for all disaster survivors surpassing the availability of existing resources. In the ensuing shortage, the needs of older adults are likely to go unmet. While social networks are depleted, temporary formal support from first responders and relief workers can provide some assistance in rebuilding informal systems.

Senior Centers are often used as shelters during disasters and as a clearinghouse for information and resources pre- and post-event. Older adults who are active with their local senior center will typically fare better during a disaster than their less-connected peers. The Administration on Aging and the Aging Network composed of State and Area Agencies on Aging, Native American Tribal Organizations, and other service providers have the legislative mandate to advocate on behalf of older adults and to work in cooperation with other Federal and State programs. The Older Americans Act gives authority and responsibility to the Administration on Aging and the Aging Network to serve older adults in greatest need and provides limited resources to fund disaster response services (Title III, Sec. 310, Disaster Relief Reimbursements). The Aging Network is responsible for basic types of disaster assistance including advocacy and outreach to assure access to needed medical care and assistance in completing and filing applications for financial and other assistance. To help older adults retain independent living, Older Americans Act funds can be used for chore, homemaker, transportation, nutrition, legal, and other temporary or one-time-only expenses.

FUTURE DIRECTIONS

Disaster-related planning and response requires collaboration among all sectors within the community, including those serving older adults informally as well as in more formal settings. A distinct need exists to ensure the unique circumstances of older adults across a variety of settings are fully integrated into current and future disaster preparedness and response efforts. Both community-dwelling older adults and those who reside in residential settings as well as the people who serve as their supports need to be aware and prepared in the event of a disaster. Empowering older adults to take action to become informed for disaster-related situations is critical. Through advance preparation and discussion, some of the morbidity and mortality experienced by older adults in prior disasters may be prevented in future disasters.

Existing disaster educational materials are not adequately tailored to meet the needs of older adults who are at greatest risk for adverse outcomes (Brown et al., 2014). In addition to low disaster literacy and limited resources, personal circumstances that place older adults at higher risk include being dependent on others, isolation, limited social network or support, medical illness, and physical and cognitive impairment (Brown, Rothman, et al., 2007; Hutton, 2008). However, unlike existing funded programs that are dedicated to minimizing negative outcomes on children and teens after traumatic events, at present there is no cohesive, sustained, national infrastructure that supports research, training, and program development for older adults and disasters. In large part, funds that are allocated for older adult disaster initiatives tend to be dispersed post-event by foundations and government funding agencies. However, protecting vulnerable older adults requires continued coordinated action between government agencies, professional associations, and nonprofit organizations. The absence of a multi-sectoral, interagency framework that facilitates coordination and strengthens capacity to effectively respond to this growing and diverse subgroup of our population is a significant gap in our existing disaster system.

CONCLUSION

Since 2001, the United States has allocated billions of dollars to enhance the public health preparedness infrastructure, with the goal of developing a health care system ready to respond effectively and efficiently to a variety of disasters. From fiscal years 2011 to 2013, the federal spending for disaster relief and recovery efforts has been over $136 billion dollars (Center for American Progress, 2013). Despite this fiscal support and subsequent improvements, recent events demonstrate that challenges and gaps remain in the ability to effectively respond to the needs of older adults across settings. Unfortunately, because a majority of the research on disasters and older adults has focused on psychopathology, little is known about the ways in which resilience can be fostered pre-disaster. While the spectrum of residential circumstances varies, from community-dwelling to nursing homes, adequate disaster planning and appropriate response are important to ensure the health and safety of older adults.

References

Adams, R. G., & Blieszner, R. (1995). Aging well with friends and family. *American Behavioral Scientist, 39*(2), 209–224.

Administration in Aging. (1995). *Disaster assistance: Disaster preparedness manual for the aging network*. Retrieved from: <http://www.aoa.gov/aoaroot/preparedness/Resources_Network/manual/disaster_assist_manual.aspx>. Accessed 18.05.10.

American Association of Retired Persons. (2006). *We can do better: Lessons learned for protecting older persons in disasters*. Retrieved from: <http://assets.aarp.org/rgcenter/il/better.pdf>. Accessed 17.03.07.

American Psychological Association. (n.d.). *APA disaster response network: Health communities in times of crisis*. Retrieved from: <http://www.apa.org/practice/programs/drn/brochure.pdf>. Accessed 18.05.10.

Aneshensel, C. S., Pearlin, L. I., Levy-Storms, L., & Schuler, R. H. (2000). The transition from home to nursing home mortality among people with dementia. *Journal of Gerontology, 55B*(3), S152–S162.

Baker, S. M. (2009). Vulnerability and resilience in natural disasters: A marketing and public policy perspective. *Journal of Public Policy and Marketing, 28*, 114–123.

Beirne, N. F., Patterson, M. N., Galie, M., & Goodman, P. (1995). Effects of a fast-track closing on a nursing facility population. *Health and Social Work, 20*, 117–123.

Below, R., Wirtz, A., & Guha-Sapir, D. (2009). *Disaster category classification and peril terminology for operational purposes*. Belguim: Louvain-la-Neuve.

Berkman, N. D., Sheridan, S. L., Donahue, K. E., Halpern, D. J., & Crotty, K. (2011). Low health literacy and health outcomes: An updated systematic review. *Annals of Internal Medicine, 155*, 97–107.

Boerner, K., Wortman, C. B., & Bonanno, G. A. (2005). Resilient or at risk? A 4-year study of older adults who initially showed high or low distress following conjugal loss. *Journal of Gerontology: Psychological Sciences, 60B*, 67–73.

Bonanno, G. A. (2004). Loss, trauma, and human resilience: Have we underestimated the human capacity to thrive after extremely aversive events? *American Psychologist, 59*, 20–28.

Bonanno, G. A., Galea, S., Bucciarelli, A., & Vlahov, D. (2007). What predicts psychological resilience after disaster? The role of demographics, resources, and life stress. *Journal of Consulting and Clinical Psychology, 75*, 671–682.

Bonanno, G. A., Rennicke, C., & Dekel, S. (2005). Self-enhancement among high-exposure survivors of the September 11th terrorist attack: Resilience or social maladjustment? *Journal of Personality and Social Psychology, 88*, 984–998.

… REFERENCES …

Bonanno, G. A., Wortman, C. B., & Nesse, R. M. (2004). Prospective patterns of resilience and maladjustment during widowhood. *Psychology and Aging, 19*, 260–271.

Bourque, L. B., Siegel, J. M., Kano, M., & Wood, M. M. (2007). Morbidity and mortality associated with disasters. In H. Rodriquez, E. L. Quarantelli, & R. R. Dynes (Eds.), *Handbook of disaster research* (pp. 97–112). New York, NY: Springer.

Bremner, J. D., Southwick, S., Brett, E., Fontana, A., Rosenheck, R., & Charney, D. S. (1992). Dissociation and posttraumatic stress disorder in Vietnam combat veterans. *American Journal of Psychiatry, 149*(3), 328–332.

Breslau, N., Kessler, R. C., Chilcoat, H. D., Schultz, L. R., Davis, G. C., & Andreski, P. (1998). Trauma and posttraumatic stress disorder in the community: The 1996 Detroit Area Survey. *Archives of General Psychiatry, 55*, 626–632.

Brown, L. M., Cohen, D., & Kohlmaier, J. R. (2007). Older adults and terrorism. In B. Bongar, L. M. Brown, L. Beutler, J. Breckenridge, & P. Zimbardo (Eds.), *Psychology of terrorism* (pp. 288–310). New York, NY: Oxford University Press.

Brown, L. M., Dosa, D., Thomas, K. S., Hyer, K., Feng, Z., & Mor, V. (2012). The effects of evacuation on nursing home residents with dementia. *American Journal of Alzheimer's Disease and Other Dementias, 27*(6), 406–412.

Brown, L. M., Frahm, K. A., & Bongar, B. (2012). Crisis intervention: Theory and practice. In I. B. Weiner, G. Stricker, & T. A. Widiger (Eds.), *Handbook of psychology, Volume 8: Clinical psychology* (2nd ed.). New York, NY: John Wiley & Sons, Inc.

Brown, L. M., Gibson, M., & Elmore, D. L. (2012). Disaster behavioral health and older adults. In J. L. Framingham & M. L. Teasley (Eds.), *Behavioral response to disasters* (pp. 159–174). Boca Raton, FL: Taylor & Francis.

Brown, L. M., Haun, J. N., & Peterson, L. (2014). A proposed disaster literacy model. *Disaster Medicine and Public Health Preparedness, 8*(3), 267–275.

Brown, L. M., Hickling, E., & Frahm, K. A. (2010). Emergencies, disasters, and catastrophic events: The role of rehabilitation nurses in preparedness, response and recovery. *Rehabilitation Nursing, 35*(6), 236–241.

Brown, L. M., Hyer, K., & Polivka-West, L. (2007). A comparative study of federal and state laws, rules, codes and other influences on nursing homes' disaster preparedness in the gulf coast states. *Behavioral Sciences & Law, 25*, 655–675.

Brown, L. M., Rothman, M., & Norris, F. (2007). Issues in mental health care for older adults after disasters. *Generations, 31*(4), 21–26.

Brunkard, J., Namulanda, G., & Ratard, R. (2008). Hurricane Katrina deaths, Louisiana, 2005. Louisiana Department of Health and Hospitals. Retrieved from: <http://www.dhh.louisiana.gov/offices/reports.asp?ID=192&Detail=207>. Accessed 18.05.10.

Burton, L. C., Skinner, E. A., Uscher-Pines, L., Lieberman, R., Leff, B., Clark, R., et al. (2009). Health of Medicare Advantage plan enrollees at 1 year after Hurricane Katrina. *The American Journal of Managed Care, 15*(1), 13–22.

Bytheway, B. (2007). *The evacuation of older people: The case of Hurricane Katrina*. Retrieved from: <www.understandingkatrina.ssrc.or>. Accessed 10.04.14.

Cantor, M. H. (1979). Neighbors and friends: An overlooked resource in the informal support system. *Research on Aging, 1*(4), 434–463.

Center for American Progress. (2013). *Disastrous spending: Federal disaster-relief expenditures rise amid more extreme weather*. Retrieved from: <http://www.americanprogress.org/wp-content/uploads/2013/04/WeissDisasterSpending-1.pdf>. Accessed 11.04.14.

Christensen, J. J., Brown, L. M., & Hyer, K. (2012). A haven of last resort: The consequences of evacuating Florida nursing home residents to non-clinical buildings. *Geriatric Nursing, 33*(5), 375–383.

Clarke, P., Chan, P., Santaguida, P. L., & Colantonio, A. (2009). The use of mobility devices among institutionalized older adults. *Journal of Aging Health, 21*(4), 611–626.

Combs, D. L., Quenemoen, L. E., Parrish, R. G., & Davis, J. H. (1998). Assessing disaster-attributed mortality: Development and application of a definition and classification matrix. *International Journal of Epidemiology, 28*, 1124–1129.

Crohan, S. E., & Antonucci, T. C. (1989). Friends as a source of social support in old age. In R. G. Adams & R. Blieszner (Eds.), *Older adults friendship: Structure and process* (pp. 129–146). New Bury Park, CA: Sage.

Danieli, Y. (1997). As survivors age: An overview. *Journal of Geriatric Psychiatry, 30*, 9–26.

DeWolfe, D. (2000). *Training manual for mental health and human service workers in major disasters* (2nd ed.). Washington, DC: Department of Health and Human Services.

Division of the Civilian Volunteer Medical Reserve Corp. (2009). *Joint memorandum: American Red Cross and the MRC*. Retrieved from: <https://www.medicalreservecorps.gov/partnerFldr/JointMemoTemplate/ARCMRCJointMemo>. Accessed 07.01.15.

Doocy, S., Daniels, A., Dick, A., & Kirsch, T. D. (2013a). The human impact of tropical cyclones: A historical review of events 1980–2009 and systematic literature review. *PLOS Currents Disasters*. <http://dx.doi.org/10.1371/currents.dis.2664354a5571512063ed29d25ffbce74>.

Doocy, S., Daniels, A., Dick, A., & Kirsch, T. D. (2013b). The human impact of tsunamis: A historical review of events 1900–2009 and systematic literature review.

PLOS Currents Disasters. <http://dx.doi.org/10.1371/currents.dis.40f3c5cf61110a0fef2f9a25908cd795>.

Doocy, S., Daniels, A., Packer, C., Dick, A., & Kirsch, T. D. (2013). The human impact of earthquakes: A historical review of events 1980–2009 and systematic literature review. *PLOS Currents Disasters.* <http://dx.doi.org/10.1371/currents.dis.67bd14fe457f1db0b5433a8ee20fb833>.

Dougall, A. L., Heberman, H. B., Delahanty, D. L., Inslicht, S. S., & Baum, A. (2000). Similarity of prior trauma exposure as a determinant of chronic stress responding to an airline disaster. *Journal of Consulting and Clinical Psychology, 68*, 290–295.

Elmore, D. L., & Brown, L. M. (2007–2008). Emergency preparedness and response: Health and social policy implications for older adults. *Generations, 31*, 66–74.

Engdahl, B. E., Harkness, A. R., Eberly, R. E., Page, W. F., & Bielinski, J. (1993). Structural models of captivity trauma, resilience, and trauma response among former prisoners of war 20 to 40 years after release. *Social Psychiatry and Psychiatric Epidemiology, 28*, 109–115.

Eysenck, H. J. (1983). Stress, disease, and personality: The inoculation effect. In C. L. Cooper (Ed.), *Stress research* (pp. 121–146). New York, NY: John Wiley & Sons.

Federal Emergency Management Agency. (2010). *Federal Emergency Management Agency. Declared disasters by year or state.* Retrieved from: <http://www.fema.gov/news/disaster_totals_annual.fema>. Accessed 27.08.10.

Federal Interagency Forum on Aging-Related Statistics. (2008). *Older Americans 2008: Key indicators of well-being.* Washington, DC: Government Printing Office.

Federal Interagency Forum on Aging-Related Statistics. (2010). *Older Americans 2010: Key indicators of well-being.* Retrieved from: <http://www.agingstats.gov/agingstatsdotnet/Main_Site/Data/2010_Documents/Docs/OA_2010.pdf>. Accessed 15.09.13.

Fernandez, L. S., Byard, D., Lin, C. C., Benson, S., & Barbera, J. A. (2002). Frail elderly as disaster victims: Emergency management strategies. *Prehospital and Disaster Medicine, 17*(2), 67–74.

Folkman, S., Lazarus, R. S., Dunkel-Schetter, C., DeLongis, A., & Gruen, R. (1986). The dynamics of a stressful encounter. *Journal of Personality and Social Psychology, 50*, 992–1003.

Friedsam, H. J. (1960). Older persons as disaster casualties. *Journal of Health and Human Behavior, 1*(4), 269–273.

Friedsam, H. J. (1961). Reactions of older persons to disaster-caused losses: A hypothesis of relative deprivation. *The Gerontologist, 1*, 34–37.

Geriatric Mental Health Care Foundation. (n.d.). *Older adults and disaster: Preparedness and response.* Retrieved from: <http://www.gmhfonline.org/gmhf/consumer/disaster_prprdns.html>. Accessed 20.01.14.

Gibbs, M., & Montagnino, K. (n.d.). *Disasters, a psychological perspective.* Retrieved from: <http://webcache.googleusercontent.com/search?q=cache:tl3MOUGCpNsJ:training.fema.gov/EMIWeb/edu/docs/EMT/GibbsPsychology.doc+&cd=2&hl=en&ct=clnk&gl=us>. Accessed 20.01.14.

Green, B. L., Goodman, L. A., Krupnick, J. L., Corcoran, C. B., Petty, R. M., Stockton, P., et al. (2000). Outcomes of single versus multiple trauma exposure in a screening sample. *Journal of Traumatic Stress, 13*, 271–286.

Guha-Sapir, D. (2000). Disaster preparedness in schools of public health. In L. Y. Landesman (Ed.), *Disaster preparedness in schools of public health: A curriculum for the new century.* Washington, DC: Association of Schools of Public Health (pp. Sect. 1.0–1.1).

Handmer, J., & Dovers, S. (1996). A typology of resilience: Rethinking institutions for sustainable development. *Industrial and Environmental Crisis Quarterly, 9*, 482–511.

Hodgson, N., Freedman, V. A., Granger, D. A., & Erno, A. (2004). Biobehavioral correlates of relocation in the frail elderly: Salivary cortisol, affect, and cognitive function. *Journal of the American Geriatrics Society, 52*, 1856–1862.

Homeland Security. (2008). *US Department of Homeland Security national response framework.* Retrieved from: <http://www.fema.gov/pdf/emergency/nrf/nrf-core.pdf>. Accessed 11.02.10.

Huerta, F., & Horton, R. (1978). Coping behavior of elderly flood victims. *The Gerontologist, 18*, 541–546.

Hutton, D. (2008) *Older people in emergencies: Considerations for action and policy development.* World Health Organization. <http://www.who.int/ageing/publications/Hutton_report_small.pdf>. Accessed 19.01.14.

Institute of Medicine. (2002). *Speaking of health: Assessing health communication strategies for diverse populations.* Board of Neuroscience and Behavioral Health. Washington, DC: The National Academies Press.

Institute of Medicine. (2004). *Health literacy: A prescription to end confusion.* Board of Neuroscience and Behavioral Health. Washington, DC: National Academies Press.

Janoff-Bulman, R. (1992). *Shattered assumptions: Toward a new psychology of trauma.* New York, NY: Free Press.

Johnson, A., Roush, R. E., Jr., et al., Howe, J. L., Sanders, M., McBride, M. R., Sherman, A., et al. (2006). Bioterrorism and emergency preparedness in aging (BTEPA) HRSA-funded GEC collaboration for curricula and training. *Gerontology & Geriatrics Education, 26*(4), 63–86.

Kahn, R. L., & Antonucci, T. C. (1980). Convoys over the life course. Attachment, roles, and social support. In P. B. Baltes & O. G. Brim (Eds.), *Life-span development and behavior* (pp. 253–287). New York, NY: Academic Press.

Kaniasty, K. Z., & Norris, F. H. (1993). A test of the support deterioration model in the context of natural disaster. *Journal of Personality and Social Psychology, 64*, 395–408.

Kaniasty, K. Z., & Norris, F. H. (1999). The experience of disaster: Individuals and communities sharing trauma. In R. Gist & B. Lubin (Eds.), *Response to disaster: Psychosocial, community, and ecological approaches* (pp. 25–61). Philadelphia, PA: Brunner/Mazel.

Karter, M. J. (2009). *Fire loss in the United States 2008.* Quincy, MA: National Fire Protection Association, Fire Analysis and Research Division (p. 44).

Kendra, J., Rozdilsky, J., & McEntire, D. A. (2008). Evacuating large urban areas and challenges for emergency management policies and concepts. *Journal of Homeland Security and Emergency Management, 5*(1), 1–22. [Online article 32].

Kilijanek, T. S., & Drabek, T. E. (1979). Assessing long-term impacts of a natural disaster: A focus on the elderly. *The Gerontologist, 19,* 555–566.

King, L. A., King, D. W., Fairbank, J. A., Keane, T. M., & Adams, G. A. (1998). Resilience-recovery factors in post-traumatic stress disorder among female and male Vietnam veterans: Hardiness, postwar social support and additional stressful life events. *Journal of Personality and Social Psychology, 74,* 420–434.

Knight, B. G., Gatz, M., Heller, K., & Bengtson, V. L. (2000). Age and emotional response to the Northridge earthquake: A longitudinal analysis. *Psychology and Aging, 15,* 624–627.

Kusenbach, M., Simms, J. L., & Tobin, G. A. (2010). Disaster vulnerability and evacuation readiness: Coastal mobile home residents in Florida. *Natural Hazards, 52,* 79–95.

Laditka, S. B., Laditka, J. N., Cornman, C. B., Davis, C. B., & Chandlee, M. J. (2008). Disaster preparedness for vulnerable persons receiving in-home, long-term care in South Carolina. *Prehospital and Disaster Medicine, 23*(2), 133–142.

Laditka, S. B., Laditka, J. N., Xirasagar, S., Cornman, C. B., Davis, C. B., & Richter, J. V. E. (2007). Protecting nursing home residents during emergencies or disasters: An exploratory study from South Carolina. *Prehospital and Disaster Medicine, 22*(1), 42–48.

Lepore, S. J., & Revenson, T. A. (2006). Resilience and post-traumatic growth: Recovery, resistance, and reconfiguration. In L. G. Calhoun & R. G. Tedeschi (Eds.), *Handbook of posttraumatic growth: Research & practice* (pp. 24–46). Mahwah, NJ: Lawrence Erlbaum Associates Publishers.

McCrae, R. R., & Costa, P. T. (2003). *Personality in adulthood: A five-factor theory perspective.* New York, NY: The Guilford Press.

McGuire, L. C., Ford, E. S., & Okoro, C. A. (2007). Natural disasters and older adults with disabilities: Implications for evacuation. *Disasters, 31*(1), 49–56.

Montz, B. E., & Tobin, G. A. (2005). Snowbirds and senior living developments: An analysis of vulnerability associated with Hurricane Charley. Boulder, CO: Natural Hazards Research and Applications Information Center. Quick Response Research Report 177.

Morrow, B. H. (1999). Identifying and mapping community vulnerability. *Disasters, 23,* 1–18.

Nishith, P., Mechanic, M. B., & Resick, P. A. (2000). Prior interpersonal trauma: The contribution to current PTSD symptoms in female rape victims. *Journal of Abnormal Psychology, 109*(1), 20–25.

Norris, F. H. (1992). Epidemiology of trauma: Frequency and impact of different potentially traumatic events on different demographic groups. *Journal of Consulting and Clinical Psychology, 60,* 409–418.

Norris, F.H., Byrne, C.M., Diaz, E., & Kaniasty, K. (2001). The range, magnitude, and duration of effects of natural and human-caused disasters: A review of the empirical literature. *National Center for PTSD Fact Sheet* [Online]. <http://www.ncptsd.org/facts/disasters/fs_range.html>. Accessed 12.11.04.

Norris, F. H., Friedman, M. J., & Watson, P. J. (2002). 60,000 disaster victims speak: Part I. An empirical review of the empirical literature, 1981–2001. *Psychiatry, 65,* 207–239.

Norris, F. H., & Murrell, S. A. (1988). Prior experience as a moderator of disaster impact on anxiety symptoms in older adults. *American Journal of Community Psychology, 16*(5), 665–683.

Ong, A. D., Bergeman, C. S., Wallace, K. A., & Bisconti, T. L. (2006). Psychological resilience, positive emotions, and successful adaptation to the stress in later life. *Journal of Personality and Social and Social Psychology, 91,* 730–749.

Oriol, W. (1999). *Psychological issues for older adults in disasters* (Publication No. ESDRB SMA 99-3323). US Department of Health and Human Services, Substance Abuse and Mental Health Services Administration, Center for Mental Health Services. Retrieved from: <http://mentalhealth.samhsa.gov>. Accessed 23.07.10.

Phifer, J. F. (1990). Psychological distress and somatic symptoms after natural disaster: Differential vulnerability among older adults. *Psychology and Aging, 5*(3), 412–420.

Phifer, J. F., & Norris, F. H. (1989). Psychological symptoms in older adults following natural disaster: Nature, timing, duration, and course. *Journals of Gerontology, 44*(6), S207–S217.

Pietrzak, R. H., Southwick, S. M., Tracy, M., Galea, S., & Norris, F. H. (2012). Posttraumatic stress disorder, depression, and perceived needs for psychological care in older persons affected by Hurricane Ike. *Journal of Affective Disorders, 138*(1), 96–103.

Port, C., Engdahl, B., Frazier, P., & Eberly, R. (2002). Factors related to the long-term course of PTSD in older ex-POWs. *Journal of Clinical Geropsychology, 8,* 203–214.

Raid, J., & Norris, F. (1996). The influence of relocation on the environmental, social, and psychological stress experienced by disaster victims. *Environment and Behavior, 28,* 163–182.

Rodriguez, J., Vos, F., Below, R., & Guha-Sapir, D. (2009). *Annual disaster statistical review 2008: The numbers and trends.* Retrieved from: <http://www.preventionweb.net/files/10251_ADSR20081.pdf>. Accessed 11.02.10.

Rothman, M., & Brown, L. M. (2007). The vulnerable geriatric casualty: Medical needs of frail older adults during disasters. *Generations, 31,* 16–20.

Rowel, R., Sheikhattari, P., Barber, T. M., & Evans-Holland, M. (2012). Introduction of a guide to enhance risk communication among low-income and minority populations: A grassroots community engagement approach. *Health Promotion Practice, 13,* 124–132.

Ruscher, J. B. (2006). Stranded by Katrina: Past and present. *Analyses of Social Issues and Public Policy, 6,* 33–38.

Rutter, M. (1987). Psychosocial resilience and protective mechanisms. *American Journal of Orthopsychiatry, 57,* 316–331.

Saha, S. (2006). Improving literacy as a means to reducing health disparities. *Journal of General Internal Medicine, 21,* 893–895.

Sanders, S., Bowie, S. L., & Bowie, Y. D. (2003). Lessons learned on forced relocation of older adults: The impact of Hurricane Andrew on health, mental health, and social support of public housing residents. *Journal of Gerontological Social Work, 40*(4), 23–35.

Seligson, H. A., & Shoaf, K. I. (2003). Human impacts of earthquake. In W. F. Chen & C. Scawthorn (Eds.), *Earthquake engineering handbook* (pp. 28:21–28:29). Boca Raton, FL: CRC Press.

Sentell, T. L., & Halpin, H. A. (2006). Importance of adult literacy in understanding health disparities. *Journal of General Internal Medicine, 21,* 862–866.

Shalev, A. Y. (2002). Acute stress reactions in adults. *Biological Psychiatry, 51,* 532–543.

Sigal, J. J., & Weinfeld, M. (2001). Do children cope better than adults with potentially traumatic stress? A 40-year follow-up of holocaust survivors. *Psychiatry: Interpersonal and Biological Processes, 64*(1), 69–80.

Solomon, Z., & Prager, E. (1992). Elderly Israeli Holocaust survivors during the Persian Gulf War: A study of psychological distress. *American Journal of Psychiatry, 149,* 1707–1710.

The Sphere Project. (2011). *Humanitarian charter and minimum standards in humanitarian response.* Retrieved from: <http://www.ifrc.org/PageFiles/95530/The-Sphere-Project-Handbook-20111.pdf>. Accessed 10.04.14.

Thompson, M. P., Norris, F. H., & Hanacek, B. (1993). Age differences in the psychological consequences of Hurricane Hugo. *Psychology and Aging, 8,* 606–616.

Turcotte, M., & Schellenberg, G. (2007). *A portrait of seniors in Canada, 2006.* Ottawa, ON: Minister of Industry.

US Department of Health and Human Services. (2000). *Healthy people 2010: With understanding and improving health and objectives for improving health* (2nd ed.). Washington, DC: US Government Printing Office.

Waller, M. (2001). Resilience in ecosystemic context: Evolution of the concept. *American Journal of Orthopsychiatry, 71,* 290–297.

Wisner, B., Blaikie, P., Cannon, T., & Davis, I. (2004). *At risk: Natural hazards, people's vulnerability and disaster* (2nd ed.). London: Routledge.

Yehuda, R., Kahana, B., Schmeidler, J., Southwick, S. M., Wilson, S., & Giller, E. L. (1995). Impact of cumulative lifetime trauma and recent stress on current posttraumatic stress disorder symptoms in Holocaust survivors. *American Journal of Psychiatry, 152,* 1815–1818.

Yehuda, R., & McFarlane, A. C. (1995). Conflict between current knowledge about posttraumatic stress disorder and its original conceptual basis. *American Journal of Psychiatry, 152,* 1705–1713.

CHAPTER 18

End-of-Life Planning and Health Care

Deborah Carr and Elizabeth Luth

Department of Sociology and Institute for Health, Health Care Policy and Aging Research, Rutgers University, New Brunswick, NJ, USA

OUTLINE

Introduction	375	Public Policy Innovations	385
Death and Dying in the United States	376	Physician's Order for Life-Sustaining Treatment (POLST)	385
Demographic and Epidemiologic Contexts	376	Affordable Care Act	386
Cultural Context of Death and Dying	378	Physician-Assisted Suicide	387
Advance Care Planning	380	Conclusion and Future Directions	388
Components and Limitations	380	References	390
ACP Benefits and Consequences	382		
Trends and Differentials	383		

INTRODUCTION

Death is a universal experience, yet the nature of death and dying has changed dramatically throughout the past two centuries; these changes have important implications for how individuals think about and prepare for both their own end-of-life and the final days of their loved ones (Carr, 2012a). Throughout the nineteenth and early twentieth centuries, most deaths occurred with little warning, typically due to short-term infectious diseases (Omran, 1971). In the contemporary United States and most wealthy developed nations, death typically befalls older adults following a long-term chronic illness, often accompanied by physical pain, functional decline, and cognitive impairment (Olshansky & Ault, 1986). Most older adults die in hospitals or nursing homes rather than at home, and many rely on medical technologies and aggressive treatments that may increase the length – although not necessarily the quality – of their lives (Teno et al., 2013). For most older adults, then, it is more accurate to conceptualize the "end of life" as an anticipated and protracted albeit unpredictable process (i.e., dying) rather than a discrete and sudden event (i.e., death) (Carr, 2012a; George, 2002).

These shifts in the timing and cause of death have created a context in which older adults are encouraged to actively prepare and plan for their end of life, conveying to significant others and health care providers their preferences regarding how, where, and under what medical care regimens they would like to die (AMA, 2012; IOM, 2014). Such preparations are widely regarded as an essential step for achieving a "good death" in which physical pain and emotional distress are minimized, and the patient's and family members' treatment preferences are respected (Carr, 2003; Steinhauser et al., 2000; Teno, Gruneir, Schwartz, Nanda, & Wetle, 2007). Yet emerging research identifies psychosocial, economic, structural, and cognitive barriers to effective end-of-life planning (Carr, 2012b,c, 2013; IOM, 2014), leading policy makers to develop new practices such as Physician's Order for Life-Sustaining Treatment (POLST) (National POLST, 2012) and Medicare-reimbursed doctor–patient consultation sessions regarding one's options and preferences for end-of-life care (Belluck, 2014; IOM, 2014; Pear, 2011).

In this chapter, we describe how older adults die in the contemporary United States, and describe the practices that individuals, families, health care providers, and policy makers may use to help increase the chances that dying patients and their families experience a "good death." We begin by describing the changing demographic, technological, and cultural contexts of death and dying throughout the twentieth and twenty-first centuries. We underscore that for current cohorts of older adults, the time period between onset of terminal illness and death may be unpredictable, posing challenges for practitioners and patients in their efforts to prepare for the dying process (Christakis, 2001; Gawande, 2014; George, 2002). We then describe the specific practices that older adults and their families may engage in to prepare for the end of life, with an emphasis on both formal/legal preparations such as the use of advance directives, and informal preparations such as discussing one's general preferences and values with loved ones and health care providers. However, we also point out limitations of formal advance care planning (ACP) that may weaken its effectiveness in promoting a "good death" for dying patients, and we show how both access to and willingness to engage in ACP are powerfully shaped by cultural, religious, and economic factors (Carr, 2011, 2012b,c; Sharp, Carr, & MacDonald, 2012). These disparities in ACP are linked to inequities in end-of-life experiences, including the use of hospice and palliative care, costs related to one's medical care, and family conflicts surrounding the dying process (Carr, 2012b,c; Kramer & Yonker, 2011). We then describe recent innovations and emerging controversies in end-of-life care including debates over POLSTs, Medicare-funded doctor–patient consultations regarding end-of-life care (Belluck, 2014; Pear, 2011), and legalization of physician-assisted suicide (PAS) (Eckholm, 2014). We conclude by suggesting avenues for future research, highlighting areas in which social science research may be particularly effective in complementing and extending findings based primarily on clinical samples and contexts.

DEATH AND DYING IN THE UNITED STATES

Demographic and Epidemiologic Contexts

An "epidemiologic transition" occurred over the past two centuries in which infant and child deaths were replaced by later-life deaths, and infectious diseases were replaced by "lifestyle-related" chronic diseases as the leading causes of death (Olshansky & Ault, 1986; Omran, 1971).

In the nineteenth and early twentieth centuries, deaths occurred primarily due to infectious diseases, such as diphtheria and pneumonia; death occurred relatively quickly after the initial onset of symptoms. Throughout the twentieth century, improved sanitation and nutrition, immunization for communicable diseases, effective treatments for infections, and other medical advances dramatically reduced mortality among younger persons, and increased life expectancy (IOM, 2014). While median life expectancy in 1900 was just 46 years old, it approached 80 years old in 2009 (Arias, 2014). Roughly three-quarters of the 2.4 million deaths in the United States in 2010 were persons ages 65 and older (Federal Interagency Forum on Aging Related Statistics, 2013). The leading causes of death among older adults are chronic and progressive illnesses that can persist for months if not years prior to death, including: heart disease (1 156 deaths per 100 000 people), cancer (982 per 100 000), chronic lower respiratory diseases (291 per 100 000), stroke (264 per 100 000), Alzheimer's disease (184 per 100 000), and diabetes (121 per 100 000) (Federal Interagency Forum on Aging Related Statistics, 2013).

Later-life deaths today rarely occur shortly after the onset of chronic illness, thus the "living–dying interval" (Pattison, 1977) between diagnosis and death is typically marked by compromised quality of life including comorbid conditions, functional impairment, mobility limitations, impaired cognitive functioning, physical discomfort, and the need for assistance with activities of daily living (ADLs) and instrumental activities of daily living (IADLs). In 2009, more than 40% of persons ages 65 and older required assistance with an ADL or IADL (Federal Interagency Forum on Aging Related Statistics, 2013). The number of older Americans with serious cognitive impairment is also high and rising; the number of older adults suffering from Alzheimer's disease and related dementias is expected to grow from 5.5 million in 2010 to 8.7 million in 2030 (HHS/ASPE, 2013).

However, the timing and onset of decline vary widely across disease groups, thus "dying" is a highly heterogeneous (and unpredictable) experience even among older adults with chronic illness. "Dying" is not a medical or diagnostic term, and an individual with a terminal illness could survive anywhere from a few days or weeks to several years (IOM, 2014). Researchers have developed conceptual and empirical models to characterize distinctive patterns of dying (Lynn & Adamson, 2003). Contemporary models have their roots in Glaser and Strauss' (1965) classic writings on the "trajectory of dying." This work was among the first to specify that patients' dying trajectories are based on two core properties: duration, or the time period between illness onset and death; and "shape" of one's trajectory, which may include components such as spikes in symptoms, periods of recovery, and rapidly decreasing levels of functional ability. Researchers generally agree upon three "typical" dying trajectories for patients with progressive chronic disease: a steady progression and a clear terminal phase (e.g., cancer); gradual decline punctuated by episodes of acute deterioration, some recovery, and a seemingly unexpected death (e.g., heart or respiratory failure); and prolonged and gradual decline or "dwindling" (e.g., dementia, frail older adults) (Lynn & Adamson, 2003; Skolnick, 1998). Knowledge of a dying patient's anticipated future trajectory has the potential to guide practitioners as they plan a course of treatment (Murray, Kendall, Boyd, & Sheikh, 2005), to shape patients' preferences regarding the use or rejection of particular treatments (Weeks et al., 1998), to facilitate family members' preparations for their loved one's impending death (Carr, House, Wortman, Nesse, & Kessler, 2001), and to guide decisions regarding the use of and Medicare reimbursement for hospice care (IOM, 2014).

Yet in practice, dying trajectories are fraught with high levels of uncertainty. Many patients' actual trajectories do not conform to "typical" patterns, making it difficult for physicians to offer accurate prognoses (Christakis, 2001). Physicians may be reluctant to share their prognoses with patients and their families, for fear of upsetting or misinforming them (Christakis & Lamont, 2000), especially in cases where dying patients show a strong desire to continue living (Finucane, 2004). To compensate for these concerns, physicians may either fail to provide patients with prognoses, or provide overly optimistic estimates of survival. The closer and more long-standing the physician–patient relationship, the more likely that the physician will make an inaccurate prediction regarding a patient's survival or will shield patients from this potentially distressing information (Christakis & Lamont, 2000). Consequently, dying patients and their kin often report high levels of uncertainty about the patient's survival and symptomatology, which may impede the formation of well-informed preferences regarding end-of-life care (Fried, Bradley, & O'Leary, 2006).

Cultural Context of Death and Dying

Medicalization of Death and Dying

Shifts in the timing and leading causes of death have been accompanied by a cultural transformation: the medicalization of aging and dying. Dying has become an increasingly "medicalized" process, where death is viewed as something to be stopped or delayed, rather than accepted as a natural part of the life cycle (Aries, 1981; McCue, 1995). Throughout the twentieth century, medicine became highly professionalized, with a heightened emphasis on scientific research aimed at finding a "cure," and an increase in the prestige and authority afforded to physicians (Conrad, 1992; Starr, 1982).

Medicalization processes carry two important implications for end-of-life care. First, the location of medical care shifted from the patient's home to a clinical environment, with an increasing reliance on sophisticated technologies, pharmaceuticals, and interventions targeted toward curing symptoms and forestalling death. For example, the development of mechanical ventilation and intensive care units (ICUs) allows patients to use life-sustaining interventions that cannot be easily provided outside a clinical setting. For this reason, the modal place of death shifted from the home to medical institutions during the twentieth century, despite the fact that the majority of Americans say that they would like to die at home (IOM, 2014). In 2009, 25% of older adults died in an acute care hospital, 28% died in a nursing home, and just one in three died at home. Fully one-third of all recent decedents spent time in an ICU in the month prior to death (Teno et al., 2013). Some critics have noted that by "sequestering" dying patients away from their homes and communities, the medical establishment is creating a culture that denies and hides death, heightens death anxiety, and isolates sick and dying persons at precisely the time when they most need interpersonal interaction (Aries, 1981; Elias, 1985).

Second, the increasing professionalization of medicine throughout the twentieth century created a context in which dying patients and their families would cede to physician knowledge and decisions, leading to high levels of reliance on invasive and costly interventions intended to sustain one's life span (IOM, 2014; Starr, 1982). Qualitative research shows that doctors heavily influence both the treatments patients choose, and patients' willingness to seek out or forego life-sustaining technologies (Sudnow, 1967; Timmermans, 2009). Through much of the twentieth century, physicians' training emphasized saving and sustaining lives (Starr, 1982), which partly shaped their reluctance to withhold life-extending treatments (Farber et al., 2006), and their tendency to shield patients from dire prognoses (Christakis & Lamont, 2000).

As a result, patients may passively accept or fail to reject invasive and futile treatments proposed by their physicians. "Futile care" refers to interventions that are unlikely to help patients and that may cause them harm or discomfort (IOM, 2014). One recent study found that critical care clinicians themselves believed almost 20% of their patients received care that was futile (Huynh et al., 2013). Similarly, an estimated one-third of nursing home patients in the final stages of dementia are given feeding tubes, although the practice does not prolong patients' lives and may cause infections (Mitchell, Teno, Roy, Kabumoto, & Mor, 2003). Despite widespread reliance on invasive treatment, attitudinal surveys suggest that older adults would choose to reject treatment if it held no hope for a cure. In 2013, 71% of persons ages 65–74 and 62% of those ages 75 and older say they would "tell their doctor to stop treatment so they could die" if they "had a disease with no hope of improvement and were suffering a great deal of pain" (Pew Research Center, 2013).

For much of the twentieth century, older adults with advanced chronic illnesses "relied almost unquestioningly on their physicians' judgments regarding treatment matters, trusting that physicians would act in their patients' best interests as a matter of professional and personal ethics" (IOM, 2014, pp. 3–5). The use of breathing and feeding tubes and powerful drugs has sustained older adults' lives, although with well-documented emotional, physical, and financial costs. Terminally ill persons (Singer, Martin, & Keltner, 1999), their families (Pierce, 1999), and health care providers (SUPPORT Investigators, 1995) have reported considerable dissatisfaction with end-of-life care, attributing their dissatisfaction to the fact that dying older adults often are non-ambulatory, short of breath, unable to eat, in pain, and unable to recognize family members. The economic costs imposed by end-of-life care also present a substantial threat to older adults, their families, and the federal government. An estimated 13% of the $1.6 trillion spent on health care annually in the United States is for individuals in their last year of life (IOM, 2014).

The Movement toward Patient Autonomy

In response to heightened concerns about costly overtreatment and futile care at the end of life, policy makers and practitioners recently developed initiatives to place greater decision-making latitude in the hands of patients and their families (Daschle, Domenici, Frist, & Rivlin, 2013; IOM, 2014). Most notably, in 1990 the US Congress passed the Patient Self-Determination Act (PSDA), which requires all health care facilities receiving reimbursement from Medicare or Medicaid "to ask patients whether they have advance directives, to provide information about advance directives, and to incorporate advance directives into the medical record" (HHS, 2008), facilitating patients' ability to make decisions about their own medical care.

Another reaction against the highly medicalized context of end-of-life care has been escalating interest in and use of hospice care. Hospice is a comprehensive, socially supportive, pain-reducing, and comforting alternative to technologically elaborate, medically centered interventions (IOM, 2014). The modern hospice movement was founded in the United Kingdom in the 1950s by Dame Cicely Saunders, a physician, nurse, and social worker. The first hospice was established in the United States in 1971 by Florence Wald, the dean of the Yale School of Nursing, who had been inspired by a lecture Saunders delivered. Hospice promotes palliative (or "comfort") care for people with a terminal illness or at high risk of dying in the near future. Hospice has grown rapidly in popularity over the past two decades, with the number of sites increasing at about 3.5% a year during the first decade of the twenty-first century (National Hospice and Palliative Care Organization [NHPCO], 2012). In 1997, 17% of all deaths in the United States occurred under the care of hospice; by 2011, this proportion more than doubled to 45% (NHPCO, 2012). The increase partly reflects attitudinal shifts favoring quality of life over length of life

(Pew Research Center, 2013), and an increased allocation of Medicare funds to hospice services in an effort to reduce the costs associated with high-tech treatments among dying older adults. Medicare beneficiaries who are certified by a physician to have a terminal illness and life expectancy of 6 months or less may elect noncurative medical and support services including hospice (CMS, 2013). Although hospice care typically involves withholding or withdrawing medical treatments that may sustain one's life, it is not a form of euthanasia or PAS. The latter, as we shall describe below, involves proactive steps to hasten death among terminally ill patients (AMA, 2012).

In sum, the movement toward a physician-controlled, highly medicalized death in the mid- to late-twentieth century gave rise to widespread concerns regarding the loss of patient autonomy and reliance on costly, invasive, and often futile treatments at the end of life (Carr, 2012a; IOM, 2014; Steinhauser et al., 2000). Mounting efforts to encourage ACP and to increase both desire for and public awareness of hospice services are indicative of an emerging cultural and social imperative to achieve a "good death," or a death that is "free from avoidable distress and suffering for patients, families, and caregivers; in general accord with patients' and families' wishes; and reasonably consistent with clinical, cultural, and ethical standards" (National Research Council, 1997, p. 24). However, as we will see in the next sections, access to ACP and patient-centered care at the end of life are difficult to achieve, given pervasive socioeconomic and practical barriers (IOM, 2014).

ADVANCE CARE PLANNING

Components and Limitations

Philosophical writings on the "good death" emphasize the central role of patient autonomy (Byock, 1996), while empirical studies show that patient "involvement in decision making" is one of the most frequently mentioned components of a good death (Steinhauser et al., 2000). Bioethicists concur that physicians should share, and in some cases delegate, medical decision-making control to dying patients and their families (President's Council on Bioethics, 2005). In practice, however, many dying persons are unable to convey their preferences for medical treatments because they are incapacitated when the decision is required (IOM, 2014). According to recent estimates, 45–70% of older adults facing end-of-life treatment decisions are incapable of making those decisions themselves (IOM, 2014). As such, difficult decisions about stopping or continuing treatment often fall to family members who may be distressed or may disagree among themselves about an appropriate course of care (Kramer, Boelk, & Auer, 2006). When family members and health care providers cannot agree on a course of action, the default decision is to continue treatments which may be financially and emotionally draining for family members, and physically distressing to the patient (IOM, 2014).

In an effort to prevent problematic, futile, or contested end-of-life care, practitioners encourage adults to express and document their treatment preferences when they are still in good health (AMA, 2012; IOM, 2014). Adults may convey their treatment preferences *formally* through an advance directive, which comprises a living will and/or a durable power of attorney for health care (DPAHC) appointment, or *informally* via discussions with significant others. A living will states the treatments that an individual would want (or not want) at the end of life; such treatments might include ventilators, feeding tubes, or cardiopulmonary resuscitation. A DPAHC is a legal document designating a specific individual (also referred to as a "surrogate" or "proxy") who will make decisions on behalf of the patient in the event that he or she is incapacitated. The vast majority of older adults select a spouse or long-term partner, followed by a child, or other close relative as DPAHC (Carr & Khodyakov, 2007). Spouse proxies tend to be more knowledgeable than adult children regarding a patient's preferences (Parks et al., 2011), with wives more knowledgeable than husbands (Zettel-Watson, Ditto, Danks, & Smucker, 2008).

Despite widespread professional endorsements (AMA, 2012), public awareness and education campaigns (Hammes, 2003), popular books (Gawande, 2014), and public policies (PSDA, 1990) targeted at encouraging ACP, formal ACP has well-documented limitations (Drought & Koenig, 2002; Fagerlin & Schneider, 2004). Criticisms of the living will include: the content or stated preferences may be unclear; the treatment preferences stated may not be relevant to the patient's condition, especially for dying older adults who drafted their living wills years earlier; and physicians may not have access to the document at the critical decision-making moment (Coppola, Ditto, Danks, & Smucker, 2001; Ditto et al., 2001). For example, many advance directives begin with the statement "If I have a terminal condition, then ...." This statement requires a physician to make an evaluation of whether the patient's condition is terminal. Until that determination is made, the content of the advance directive does not hold, despite what the patient and family had hoped. Physicians also may be reluctant to follow the orders stated in the living will for fear of legal liability; in general, physicians believe their liability risk is greater if they do not attempt resuscitation than if they provide it against patient wishes (Burkle, Mueller, Swetz, Hook, & Keegan, 2012). Moreover, family members may not know (or agree with) the document's content, or may not know how to translate vague preferences into specific clinical practices (Ditto et al., 2001).

DPAHC appointments also have practical limitations. Legally appointed proxies are granted decision-making authority, yet some may make decisions that create distress or disagreement among family members (Doukas & Hardwig, 2003; Khodyakov & Carr, 2009). Moreover, surrogate decision makers' knowledge of patient preferences is usually no better than chance (Coppola et al., 2001; Shalowitz, Garrett-Mayer, & Wendler, 2006), and may strongly reflect the surrogate's own preferences (Moorman & Inoue, 2013). As one study concluded, "surrogates are not perfect ambassadors of patient preferences" (Vig, Taylor, Starks, Hopley, & Fryer-Edwards, 2006, p. 1688). Older adults may (erroneously) believe that their loved ones "intuitively" understand their preferences, so they do not see a need to explicitly inform the legal proxy of their views (Coppola et al., 2001).

For some patients, the proxy's limited knowledge is unproblematic; they may prefer that their family members do what they feel is best, rather than abide by the patient's stated preferences (Moorman, 2011). Others may trust their physicians to make decisions for them (Su, 2008). Still, the patient's deference to a specific decision maker's wishes may create distress or conflict for concerned family members who do not hold decision-making power. Family members not designated as decision makers also may trigger distress or disagreement; clinicians often share anecdotes illustrating the "daughter in California" phenomenon, whereby a family member – especially one who resides far away from the dying patient and has had little engagement in a patient's end-of-life care – may enter the family conversation at the patient's final stage of life. These individuals may try to undo, contest, undermine, or alter the decisions made by local family members who had been engaged in the care and decision-making process for a much longer duration (Molloy, Clarnette, Braun, Eisemann, & Sneiderman, 1991). These family disagreements, in turn, may inhibit interdisciplinary health care teams' ability to provide quality end-of-life care (Kramer & Yonker, 2011).

Given these well-documented limitations of formal ACP, some practitioners suggest that informal discussions with significant others and care providers are the most critical component of end-of-life planning (Doukas & Hardwig, 2003). Recent analyses of couple-level data from the Wisconsin Longitudinal Study reveal that discussing end-of-life issues with one's spouse is associated with correctly identifying one's spouse's end-of-life treatment preferences (Moorman & Carr, 2008). In general, discussions can help to facilitate care consistent with the patient's wishes; Winter and Parks (2012, p. 741) find that "those who avoid ... end-of-life conversations are the least likely to have treatment wishes respected, because their proxies are unlikely to know their wishes."

Conversations about an older adult's general values also may be useful because few individuals know precisely how and of what cause they will die, making it difficult to specify particular medical interventions that they would want (or not want) at the end of life, such as feeding tubes or chemotherapy. A general conversation about values (e.g., "I don't want to be a vegetable") and global preferences (e.g., "I don't want to be hooked up to machines") may provide family members a general roadmap for representing their loved one's wishes even in the absence of a formal living will (Doukas & Hardwig, 2003). Discussions also may facilitate decision-making in cases where the patient has not legally appointed a DPAHC. Most states have established default systems for authorizing proxy decision makers. State laws vary, but such lists prioritize the immediate family – starting with spouse, followed by adult child, sibling, and other relatives (American Bar Association, 2009; Kohn & Blumenthal, 2008). Frank conversations about a patient's values may empower and inform state-authorized proxies when making difficult decisions about their loved one's care.

However, research also shows that the timing of discussions is critical, as some discussions may be "too little, too late." Discussions regarding end-of-life issues are typically triggered by a patient's health-related event such as a hospitalization, a period of ineffective mechanical ventilation, a problematic level of forced expiratory volume, or a rapidly deteriorating nutritional status (McGrew, 2001; Pfeifer, Mitchell, & Chamberlain, 2003). When discussions about end-of-life care occur following such "trigger" events, the patient (and family) often is too distressed to make an informed or appropriate decision about imminent care needs (Hoffman, Wenger, Davis, Teno, & Connors, 1997).

Recognition of the importance of timely, in-depth conversations among dying patients, their family members, and health care professionals was partly the impetus for the 2010 House bill that would have reimbursed clinicians for the time spent discussing end-of-life issues with Medicare-beneficiary patients and their families (America's Affordable Health Choices Act of 2009). However, as we will discuss below, unfounded and incendiary rumors surrounding "death panels" led to the deletion of this benefit from the Affordable Care Act (ACA) in 2011 (Belluck, 2014). Other recent initiatives intended to increase the timeliness, rate, and quality of conversations regarding end-of-life care include the development of POLSTs, which we elaborate on later in this chapter.

ACP Benefits and Consequences

Despite concerns regarding the efficacy and effectiveness of formal ACP tools, their use increases the chances of attaining some core components of a "good death," including greater use of hospice or palliative care (Silveira, Kim, & Langa, 2010; Teno et al., 2007); reduced use of invasive or futile treatments such as feeding tubes or ventilators (Mack, Weeks, Wright, Block, & Prigerson, 2010; Nicholas, Langa, Iwashyna, & Weir, 2011; Teno et al., 2007; Wright et al., 2008);

heightened perceptions of patient control over the end-of-life process (Edwards, Pang, Shiu, & Chan, 2010); a greater likelihood of dying at home rather than in an institution (Nicholas et al., 2011; Silveira et al., 2010); and fewer instances of receiving treatments that are discrepant with the patient's wishes (Detering, Hancock, Reade, & Silvester, 2010).

ACP also is associated with superior outcomes for dying patients' family members, including a reduced decision-making burden, and fewer anxiety and depressive symptoms (Detering et al., 2010; Stein et al., 2013). Hospice use is an important pathway linking ACP with survivor well-being; bereaved family members whose loved one used hospice care at the end of life have reduced risks of mortality, depression, and traumatic grief (Iwashyna & Christakis, 2003). However, even when the decedent had an advance directive in place, some family members may still report receiving inadequate support during the dying process (Teno et al., 2007) or may report increased levels of family conflict in cases where the living will was deemed unhelpful or problematic (Khodyakov & Carr, 2009).

Research on the impact of a patient's ACP on end-of-life medical costs is equivocal. In general, studies based on large population-based samples show no significant effect (Kelley et al., 2011; Nicholas et al., 2011), whereas studies focused on specific disease groups, such as advanced cancer patients, suggest that ACP is linked with significantly reduced medical expenditures among older decedents in the last 6 months of life (e.g., Zhang et al., 2009). Nicholas and colleagues (2011) found that median fee-for-service Medicare spending in the last 6 months of life did not differ significantly between those who had versus those who did not have a "treatment-limiting" advance directive; in both subgroups, the median expenditure level was roughly $21 000. By contrast, among patients with advanced cancer, end-of-life medical costs were roughly one-third less for persons who had a treatment-limiting advance directive (Zhang et al., 2009). Intensive care unit (ICU) use is a key pathway linking ACP with reduced medical expenditures; ACP is associated with lower rates of ICU stays, which in turn is linked with lower care costs. One recent study found that the cost of a terminal hospitalization with an ICU stay averaged $38 000, compared with just $13 000 if an ICU was not included (Zilberberg & Shorr, 2012). On the whole, scholars and policy makers agree that standard ACP tools are neither a panacea nor a guarantee of a "good death," although the potential benefits far outweigh the potential costs (IOM, 2014).

Trends and Differentials

Rates of ACP vary widely by age, race, socioeconomic status (SES), and other psychosocial factors. National studies show that only one-third to one-half of all adults in the United States have completed advance directives (HHS, 2008), although rates are as high as 70% among adults age 65+ and persons with terminal illness (Carr & Moorman, 2009; Silveira et al., 2010). ACP rates have increased sharply since 1990; the proportion with a written advance directive more than doubled from 16% in 1990 to 35% in 2013 (Pew Research Center, 2013). This trend is partly attributable to the passage of the Patient Self-Determination Act (1990), high visibility cases of contested end-of-life decisions such as that of Terri Schiavo (Sudore, Landefeld, Pantilat, Noyes, & Schillinger, 2008), public awareness campaigns such as Respecting Choices® (Hammes, 2003) and the "Five Wishes" (Aging with Dignity, 2014), and media programs such as Bill Moyers' PBS Series *On Our Own Terms*. The Five Wishes, for example, is a user-friendly advance directive written in nontechnical language that includes identification of a proxy and preferences for medical and nonmedical treatment and comfort.

Rates of ACP are especially low among Blacks and Latinos, relative to Whites. Estimates vary based on the particular study sample, but most research finds that Whites are two to three times as likely as Blacks and Latinos to have an advance directive, with a much narrower gap for end-of-life discussions (Carr, 2011; Kwak & Haley, 2005; Smith et al., 2008). Explanations for these differentials include: ethnic minorities' lack of access to the medical and legal professionals who may provide assistance in preparing such documents; literacy or language barriers; cultural beliefs that such documents are not needed because family members will make decisions collectively on behalf of the patient; historically rooted distrust of physicians and medical institutions; and adherence to religious beliefs that "God will decide" when it is time for a patient to die (Carr, 2011, 2012b; Morrison, Zayas, Mulvihill, Baskin, & Meier, 1998; West & Hollis, 2012). Some research suggests that Blacks and Latinos believe they don't need a living will, because they tend to desire all possible interventions at the end of life (Barnato, Anthony, Skinner, Gallagher, & Fisher, 2009; Kwak & Haley, 2005; Pew Research Center, 2013), and they believe that living wills limit rather than request treatment (Barnato et al., 2009; Mack, Paulk, Viswanath, & Prigerson, 2010). Given this pervasive misperception that advance directives limit, rather than articulate requests for, particular treatments, one author noted, "advance directives, which are generally accepted in western civilization, hold little or no relevance within the [black and minority ethnic] population" (Cox et al., 2006, p. 20).

Empirical evidence suggests, however, that lack of ACP may prevent minority patients from receiving the treatments they desire. For example, among cancer patients who desire aggressive treatments, Blacks are one-third as likely as Whites to receive treatments that are consistent with their preferences (Loggers et al., 2009). Further research shows a substantial racial gap in end-of-life health care costs, where the average cost of care in the last 6 months of life ranged from $20 166 among Whites, to $26 704 among Blacks, and $31 702 among Latinos in 2001 (Hanchate, Kronman, Young-Xu, Ash, & Emanuel, 2009). Fully 85% of these observed higher costs for Blacks and Hispanics are accounted for by their greater usage of intensive (and costly) invasive treatments. Thus, barriers to ACP among ethnic minority communities are linked to costly intrusive treatments as well as the failure to receive desired treatments.

Research has focused more heavily on racial differences than social class differences in ACP, yet recent work shows that persons with lower levels of education, income, assets, and home ownership rates are significantly less likely than their more advantaged counterparts to do formal ACP, although no differences are found for discussions. Older adults with greater net worth and home owners are nearly twice as likely as those with no or few assets, and non-homeowners to engage in ACP (Carr, 2012c). Older adults with assets to protect are more likely to engage in estate and financial planning (e.g., initiating a signed and witnessed will) than are their less-wealthy counterparts. A visit to one's lawyer to do financial planning often triggers the completion of related documents, including living wills and DPAHC appointments (Carr, 2012c). Educational attainment also is linked indirectly to advance directive completion, as persons with lower levels of literacy are less likely to engage in ACP. One recent study of patients ages 55–74 found that rates of ACP were 12.5%, 25%, and 50% for those with low, marginal, and adequate literacy, respectively (Waite et al., 2013). Limited knowledge about one's specific health condition and the treatments one might receive at the end of life also may impede ACP, as individuals are reluctant to make decisions about treatments they don't understand (Porensky & Carpenter, 2008).

Psychological, religious, and attitudinal factors also may pose obstacles to ACP.

Some scholars argue that this pattern reflects a "death-denying" ethos in contemporary Western society (Kellehear, 1984); empirical research shows that persons with higher levels of death anxiety are less likely to do ACP (Carr & Moorman, 2009). Individuals who have witnessed the painful or prolonged death of loved one are more likely to discuss and make preparations for their own end of life in an effort to avoid the fate experienced by their loved ones (Carr, 2012d). Religious beliefs also affect ACP; those who adhere to Fundamentalist beliefs (Sharp et al., 2012), who believe that the length of their life is in God's hands, who rate religion as "very important," and whose religious beliefs guide their behavior are less likely to do formal ACP (Garrido, Idler, Leventhal, & Carr, 2013; Pew Research Center, 2013). Highly religious persons are likely to desire all treatments possible at the end of life, because they believe that God will either sustain them or let them die when the time is right: "those who believe in God do not have to plan for end-of-life care" (Johnson, Kuchibhatla, & Tulsky, 2008 p. 1956).

ACP is also linked to aspects of one's interpersonal relationships, where persons with closer, less-conflicted relationships are more likely to execute advance directives than those with poor-quality relationships, and married older adults are more likely to appoint their spouse (versus an adult child or someone else) as their DPAHC when their marriage is marked by high levels of warmth and low levels of conflict (Carr, Moorman, & Boerner, 2013). Individuals are also more likely to both do ACP and to engage significant others in the process when they anticipate that their relationship is stable and will persist into the future. For example, on average, cohabiting individuals are significantly less likely than their married counterparts to both engage in ACP and to name their partner as their DPAHC. However, when cohabiting persons are stratified based on whether they intend to marry their partner or whether they expect that the relationship will end, researchers have found that cohabitors who intend to marry their partner are just as likely as married persons to both do ACP and name their partner as their proxy decision maker (Moorman, Carr, & Boerner, 2014).

In sum, research on ACP shows that Whites, and those with the most economic resources, the most supportive and enduring social relationships, and who favor limited versus invasive treatments are most likely to formally state and convey their treatment preferences via the use of advance directives. However, comparable race- and SES-based disparities are either non-existent or considerably narrower with respect to end-of-life discussions – an activity that can be undertaken at no financial cost, and that does not require interactions with health care or legal professionals (Carr, 2011; Carr, 2012c; Wright et al., 2008). As such, economic, informational, and structural barriers may be a more powerful obstacle to ACP among ethnic minorities and poorer adults than are cultural or attitudinal factors. These obstacles are potentially modifiable factors that may be addressed by innovative public policies designed to place decision-making responsibility in the hands of older patients and their families, regardless of their personal resources. We next describe several recent innovations, and highlight strengths, controversies, and limitations in these evolving approaches to end-of-life care.

PUBLIC POLICY INNOVATIONS

Physician's Order for Life-Sustaining Treatment (POLST)

The limitations of living wills and DPAHC appointments are widely documented and led health care providers to develop an alternative approach: the POLST. This one-page document includes standing medical orders about those medical interventions a terminally ill individual wishes to have or forego. POLSTs are advised for dying patients, typically those with less than a year to live. As such, they include preferences for current treatment, whereas living wills may stipulate preferences for treatments that may not be required for many years into the future (Bomba, Kemp, & Black, 2012). Patients complete POLSTs in consultation with their health care providers, providing instructions relevant to those decisions that typically arise in medical crises at the end of life. Because POLSTs are actual doctor's orders, other health professionals are required to follow them (National POLST, 2012).

The medical preferences articulated in POLSTs fall into three main categories: full treatment, comfort measures only, and limited additional interventions. These categories are broader than highly specific treatments named in a living will (e.g., a feeding tube) yet less broad than the general values that might be communicated to family members via discussion (e.g., "don't hook me up to a machine"), so they are particularly effective in conveying patient preferences (Bomba et al., 2012; IOM, 2014). The request for "comfort measures only" indicates that a patient's main goal is maximizing comfort rather than prolonging life span. If adequate comfort cannot be provided in the patient's home, patients are moved to a clinical setting where their needs can be met. The specific types of care provided are those that relieve pain and suffering, such as oxygen. The slightly more involved treatments named under "limited additional interventions" include antibiotics and intravenous fluids. The most intensive category of care, "full treatment," includes measures provided in the other two categories, along with medical interventions, such as mechanical ventilation (National POLST, 2012).

POLSTs are relatively new, yet are growing rapidly in use. The state of Oregon first introduced POLSTs in 1991 as a means for honoring dying patients' end-of-life preferences. As of December 2014, 44 states either used POLSTs or were developing POLST implementation plans (National POLST, 2014). Like all ACP practices, POLSTs have limitations; one concern is that physicians may indicate a patient's preference box without having an in-depth conversation (Bomba et al., 2012). A further concern is that workers on the front lines of care, especially emergency personnel and nursing personnel at long-term care facilities may require training regarding the POLST's content, so that they do not deny treatment for remediable health problems that are not imminently life-threatening (IOM, 2014). However, emerging research generally concludes that POLSTs are effective in helping patients receive care that is concordant with their wishes (Hammes, Rooney, Gundrum, Hickman, & Hager, 2012; Hickman et al., 2011). For this reason, policy makers and health care providers strongly advocate for the use of POLSTs nationwide (IOM, 2014).

Affordable Care Act

An additional way to ensure that older adults make well-informed decisions regarding end-of-life care is to revitalize the original ACA proposal to include one voluntary ACP session as a benefit included in the annual wellness visit for Medicare beneficiaries. In 2009, a bipartisan group of representatives sponsored a provision in the House version of the bill that would have authorized Medicare to reimburse doctors who counsel patients about living wills, advance directives, hospice, and options for end-of-life care (IOM, 2014). This benefit would give all older adults the opportunity to discuss their treatment preferences with a health care provider, regardless of their economic or personal resources (America's Affordable Health Choices Act of 2009). However, political uproar regarding (unsubstantiated) fear of "death panels" contributed to President Barack Obama's deletion of the proposed benefit from ACA in January 2011 (Pear, 2011). For example, House Minority Leader John Boehner and

Representative Thaddeus McCotten (R-MI) incorrectly cautioned that the legislation "may start us down a treacherous path toward government-encouraged euthanasia" (Boehner & McCotter, 2009).

Until this bill passes, doctors may discuss end-of-life concerns with their patients but they cannot bill Medicare for this service, essentially passing the costs along to their patients. Recent data suggest that such conversations in the course of normal care are exceedingly rare; a recent analysis of data from more than 5 000 Medicare beneficiaries who have an ongoing relationship with a primary care physician found that only 1% reported having an end-of-life discussion with their doctor during the course of routine care (Keary & Moorman, 2015). However, those who had such a discussion reported higher rates of ACP, underscoring the potential efficacy of doctor–patient consultations. Debates regarding the ACA bill were revitalized in summer 2014, with advocates recognizing that this simple and relatively low-cost aspect of ACA may be one step toward promoting a better-quality death for financially disadvantaged older adults (Belluck, 2014).

Physician-Assisted Suicide

Discussions of patient autonomy regarding end of life, including the refusal to receive particular treatments, raise debates about active and passive euthanasia. Active euthanasia refers to the "administration of a lethal agent by another person to a patient for the purpose of relieving the patient's intolerable and incurable suffering" (AMA, 2012). Active euthanasia is illegal in all 50 US states. Passive euthanasia, or PAS "occurs when a physician facilitates a patient's death by providing the necessary means and/or information to enable the patient to perform the life-ending act" (AMA, 2012). PAS is fundamentally different from hospice care, which may hasten death by rejecting potentially invasive life-sustaining treatments, whereas PAS actively seeks out medications to hasten death (IOM, 2014). PAS, also referred to as "aid-in-dying," is currently allowed in five states: Oregon (1997), Washington (2008), Montana (2009), Vermont (2013), and New Mexico (2014) although each state has different conditions under which PAS is legal (Pew Research Center, 2013). In all other states, PAS is considered a felony.

Active euthanasia is roundly criticized as an unethical approach to death and dying (AMA, 2012). Assessments of passive euthanasia are much more equivocal; roughly equal proportions of US adults disapprove (49%) versus approve (47%) of PAS (Pew Research Center, 2013). Medical, religious, and political authorities consider PAS a potentially slippery slope, whereby increased acceptance of the practice could lead to subtle pressure for particular individuals to use this option, especially cognitively impaired, oldest old, or other vulnerable populations (Meier, 2010; President's Council on Bioethics, 2005). Data from the state of Oregon indicate that as of mid-2013 1 173 persons had requested prescriptions for barbiturates to end their lives. Two-thirds of those receiving prescriptions ultimately died from taking the prescription, with a median duration of 47 days between the time the drugs were prescribed and the time of death (Oregon Health Authority, 2014).

Further analysis of the Oregon data suggests that those seeking PAS possess both vulnerabilities that may impede and resources that facilitate independent decision-making. On one hand, individuals who seek PAS tend to have higher levels of disability, dependence on others, and poor functioning (Asch, 2005). Yet individuals opting for PAS in Oregon also tend to be white, financially well-off and well-educated; 45% had a college degree, two-thirds had private insurance, and fully 97% were white (Oregon Health Authority, 2014). Debates regarding the legalization of PAS will not be easily resolved, and will continue to scrutinize its purported advantages, such as patient control over the dying process, and stopping the prolongation of patient suffering, as well as its potential disadvantages, including fears that vulnerable populations will be cajoled by family members or care providers into PAS, or that incorrect or overly pessimistic prognoses will lead some patients to opt for PAS, even if they are still far from death (Barone, 2014). These debates intensified in February 2015, when the Supreme Court of Canada struck down a 22-year-long ban on PAS for patients with "grievous and irremediable" medical conditions (Austen, 2015). The reversal of the PAS ban in Canada may inform or guide future debates in the United States, as advocates continue to call for state legislation allowing PAS (Eckholm, 2014).

CONCLUSION AND FUTURE DIRECTIONS

Dying in the twenty-first century is a prolonged process that overwhelmingly strikes older adults suffering from long-term chronic illnesses. This period between diagnosis and death is a time that older adults and their families can spend discussing and preparing for one's impending death, with the hopes that one's preferences will be heeded and respected by health care providers. Yet such preparations require relatively accurate knowledge regarding one's prognosis and future health trajectory, and the means to articulate one's preferences to health care providers who will ultimately carry out those orders. Advance directives and discussions with significant others and health care providers are considered important (albeit imperfect) steps toward achieving a "good death," marked by minimal distress and the receipt of treatment that meshes with one's preferences. Yet policy makers and practitioners continue to develop new tools and practices to ensure a high-quality dying experience for all, including the development of POLSTs and federally funded doctor–patient consultations regarding end-of-life issues (IOM, 2014).

Despite these important strides in end-of-life care, serious problems persist. Compelling evidence shows that the last year of life is still marked by significant levels of pain and other distressing symptoms. One recent study of proxy-reported pain and symptoms among recent decedents found that fully 61% reported that the decedent was in pain in the final year of life, while nearly one-third reported symptoms of depression and confusion prior to death (Singer et al., 2015). These disheartening results raise important questions for future research. The influential IOM (2014) report *Dying in America* highlights key areas for future investigation, focusing primarily on clinical populations and approaches; we suggest here some distinctive contributions that social scientists may make to understanding the context of end-of-life planning and health care, with particular attention to inequalities therein.

Social scientists are particularly well-suited to identify structural, cultural, informational, and interpersonal obstacles to effective end-of-life planning and discussions. First, social scientists should further explore individuals' understanding of the concepts and practices of end-of-life planning, living wills, hospice, and palliative care; an understanding of these perceptions (and misperceptions) may provide insights into how practitioners, policy makers, and even the media can most effectively encourage meaningful preparations. For example, one explanation offered for the relatively low rates of ACP among ethnic minorities and highly religious persons is that they erroneously believe that ACP involves limiting, rather than requesting, treatments. As such, they may view ACP as "irrelevant" given their preferences for life-extending treatments (Cox et al., 2006, p. 20). This pervasive misperception may be partly explained by how information on advance directives is disseminated. Information on end-of-life planning, including living will templates, typically is provided by organizations that promote "death with dignity" such as Aging with Dignity, Americans for Better Care of the Dying, Dying Well, and Last Acts. These templates generally prompt individuals to indicate those treatments that they would *not* want at the end of life, thus promoting the misperception that advance directives limit, rather than request, treatments. High-profile legal cases such as Terri Schiavo also may have led individuals to believe that advance directives are orders to stop life-prolonging treatment, and that in their absence, doctors must ethically and legally continue treatment (Carr & Moorman, 2009; Sudore et al., 2008). As such, reframing messages regarding the general purpose of ACP may be instrumental in increasing rates among those preferring high levels of treatment at the end of life; social scientists could evaluate precisely which messages and frames are most effective, and how these patterns vary based on one's social location.

Second, compelling evidence shows that persons with strong family relationships are particularly likely to engage in ACP, and to be particularly effective in conveying their end-of-life treatment preferences to loved ones. However, we know very little about the ways that persons lacking such social ties approach ACP, and whom they engage in the process. In particular, little is known about the ways that childless persons, those without a long-term romantic partner, or those estranged from family members engage significant others in the end-of-life planning process. Some evidence suggests that unmarried childless persons may turn to a friend or a professional (e.g., doctor, lawyer, or clergy person) as their health care advocate (Carr & Khodyakov, 2007), yet little is known about how frequent, in-depth, or effective their conversations are. Further, it is important to identify obstacles to formally appointing a DPAHC among this subpopulation. As noted earlier, in most states, if a patient does not have a DPAHC, living will, or legally appointed guardian, then the right to make decisions falls to family members in the following order: spouse, adult children, siblings, and other family members. As such, potential proxies who are not a family member may be overlooked, thus creating an additional obstacle to childless or unpartnered persons who would like their treatment preferences heeded.

Third, a well-documented limitation of living wills is that adults may write one at a particular point in time, and fail to update it in responses to changes in one's health, family structure, or other important contextual factors that may guide one's preferences. Life course sociologists, in particular, should explore the ways that one's preferences both for specific treatments and one's selection of a potential DPAHC change as one's illness progresses; as one experiences changes in family structure, such as widowhood, divorce, or remarriage; and as one experiences changes in family functioning – ranging from relocations of adult children, to qualitative shifts in the nature of one's relationships. Understanding the ways that both normative and non-normative changes affect older adults' treatment preferences and preferences for particular family members' engagement in end-of-life decision-making may be particularly useful to practitioners.

Fourth, most research on the end-of-life focuses on coarse subgroup differences, comparing Blacks to Whites, for instance, and paying little attention to intersectionality. For example, even within a particular racial or ethnic group, attitudes and preferences regarding the end-of-life may vary widely based on educational attainment, one's birth cohort, religiosity, and a range of other psychosocial factors. For example, recent research suggests that younger cohorts of African Americans and Latinos are no less likely than their white counterparts to engage in ACP (Carr, 2012b), and that the particular obstacles to ACP among young and midlife African American adults are distinct from those documented among their older counterparts (West & Hollis, 2012). Identifying the specific obstacles to or motivators of effective ACP among particular subgroups may help practitioners to move away from a "one-size-fits-all approach" when discussing and encouraging end-of-life preparations among their patients. By understanding the distinctive motivations, beliefs, fears, and obstacles of particular subgroups of patients, practitioners may be better equipped as they strive to bring a "good death" to all.

References

Aging with Dignity. (2014). *Five Wishes resources.* <http://www.agingwithdignity.org/five-wishes.php>. Accessed 05.12.14.

America's Affordable Health Choices Act of 2009. HR 3200.IH, 111th Congress, 1st sess. <http://www.gpo.gov/fdsys/pkg/BILLS-111hr3200ih/pdf/BILLS-111hr3200ih.pdf>. Accessed 03.12.14.

American Bar Association. (2009). *Default surrogate consent statutes.* <http://www.americanbar.org/content/dam/aba/migrated/aging/PublicDocuments/famcon_2009.authcheckdam.pdf>. Accessed 08.12.14.

American Medical Association (AMA). (2012). *AMA policy on end-of-life care.* <http://www.ama-assn.org/ama/pub/physician-resources/medical-ethics/about-ethics-group/ethics-resource-center/end-of-life-care/ama-statement-end-of-life-care.page>. Accessed 23.11.14.

Arias E. (2014). *United States life tables, 2009. National vital statistics reports* 62 (7). Hyattsville, MD: National Center for Health Statistics.

Aries, P. (1981). *The hour of our death* (H. Weaver, Trans.). New York, NY: Knopf.

Asch, A. (2005). Recognizing death while affirming life: Can end of life reform uphold a disabled person's interest in continued life? *Improving end of life care: Why has it been so difficult? Hastings Center Report Special Report,* 35 S31–S26.

Austen, I. (2015). Canada court strikes down ban on aiding patient suicide. *The New York Times* (February 6, 2015). <http://www.nytimes.com/2015/02/07/world/americas/supreme-court-of-canada-overturns-bans-on-doctor-assisted-suicide.html>. Accessed 08.02.15.

Barnato, A. E., Anthony, D. L., Skinner, J., Gallagher, P. M., & Fisher, E. S. (2009). Racial and ethnic differences in preferences for end-of-life treatment. *Journal of General Internal Medicine,* 24, 695–701.

Barone, E. (2014). See which states allow assisted suicide. *TIME* (November 3, 2014).

Belluck, P. (2014). Coverage for end-of-life talks gaining ground. *New York Times* (August 30, 2014). <http://www.nytimes.com/2014/08/31/health/end-of-life-talks-may-finally-overcome-politics.html>. Accessed 21.11.14.

Boehner, J., & McCotter, T. (2009). *Statement by House GOP leaders Boehner and McCotter on end-of-life treatment counseling in Democrats' health care legislation.* <http://www.speaker.gov/pressrelease/statement-house-gop-leaders-boehner-and-mccotter-end-life-treatment-counseling>. Accessed 06.12.14.

Bomba, P. A., Kemp, M., & Black, J. S. (2012). POLST: An improvement over traditional advance directives. *Cleveland Clinic Journal of Medicine,* 79, 457–464.

Burkle, C. M., Mueller, P. S., Swetz, K. M., Hook, C. C., & Keegan, M. T. (2012). Physician perspectives and compliance with patient advance directives: The role external factors play on physician decision making. *BMC Medical Ethics,* 13, 31.

Byock, I. R. (1996). The nature of suffering and the nature of opportunity at the end of life. *Clinics in Geriatric Medicine,* 12, 237–252.

Carr, D. (2003). A 'good death' for whom? Quality of spouse's death and psychological distress among older widowed persons. *Journal of Health and Social Behavior,* 44, 215–232.

Carr, D. (2011). Racial differences in end-of-life planning: Why don't Blacks and Latinos prepare for the inevitable? *Omega: Journal of Death & Dying,* 63, 1–20.

Carr, D. (2012a). Death and dying in the contemporary United States: What are the psychological implications of anticipated death? *Social and Personality Psychology Compass,* 6(2), 184–195.

Carr, D. (2012b). Racial and ethnic differences in advance care planning: Identifying subgroup patterns and obstacles. *Journal of Aging and Health,* 24, 923–947.

Carr, D. (2012c). The social stratification of older adults' preparations for end-of-life health care. *Journal of Health and Social Behavior,* 53, 297–312.

Carr, D. (2012d). "I don't want to die like that …": The impact of significant others' death quality on advance care planning. *The Gerontologist,* 52, 770–781.

Carr, D., House, J. S., Wortman, C., Nesse, R., & Kessler, R. C. (2001). Psychological adjustment to sudden and anticipated spousal loss among older widowed persons. *Journals of Gerontology: Social Sciences,* 56, S237–S248.

Carr, D., & Khodyakov, D. (2007). Health care proxies in later life: Whom do we choose and why? *Journal of Health and Social Behavior,* 48, 180–194.

Carr, D., & Moorman, S. (2009). End of life treatment preferences among the young-old: An assessment of psychosocial influences. *Sociological Forum,* 24, 754–778.

Carr, D., Moorman, S., & Boerner, K. (2013). End-of-life planning in a family context: Does relationship quality affect whether (and with whom) older adults plan? *Journal of Gerontology: Social Sciences,* 68, 586–592.

Christakis, N. A. (2001). *Death foretold: Prophecy and prognosis in medical care*. Chicago, IL: University of Chicago Press.

Christakis, N. A., & Lamont, E. B. (2000). Extent and determinants of error in physicians' prognoses in terminally ill patients: Prospective cohort study. *Western Journal of Medicine, 172*, 310–313.

CMS. (2013). *Medicare hospice data trends, 1998–2009*. <http://www.cms.gov/Medicare/Medicare-Feefor-Service-Payment/Hospice/Medicare_Hospice_Data.html>. Accessed 25.11.14.

Conrad, P. (1992). Medicalization and social control. *Annual Review of Sociology, 18*, 209–232.

Coppola, K. P., Ditto, P. H., Danks, J. H., & Smucker, W. D. (2001). Accuracy of primary care and hospital-based physicians' predictions of elderly outpatients' treatment preferences with and without advanced directives. *Archives of Internal Medicine, 161*, 431–440.

Cox, C., Cole, E., Reynolds, T., Wandrag, M., Breckenridge, S., & Dingle, M. (2006). Implications of cultural diversity in Do Not Attempt Resuscitation (DNAR) decision making. *Journal of Multicultural Nursing and Health, 12*, 20.

Daschle, T., Domenici, P., Frist, W., & Rivlin, A. (2013). Prescription for patient-centered care and cost containment. *New England Journal of Medicine, 369*, 471–474.

Detering, K. M., Hancock, A. D., Reade, M. C., & Silvester, W. (2010). The impact of advance care planning on end of life care in elderly patients: Randomised controlled trial. *British Medical Journal, 340*, c1345.

Ditto, P. H., Smucker, W. D., Bookwala, J., Coppola, K. M., Dresser, R., Fagerlin, A., et al. (2001). Advance directives as acts of communication. *Annals of Internal Medicine, 161*, 421–430.

Doukas, D. J., & Hardwig, J. (2003). Using the family covenant in planning end-of-life care: Obligations and promises of patients, families, and physicians. *Journal of American Geriatrics Society, 51*, 1155–1158.

Drought, T. S., & Koenig, B. (2002). Choice in end-of-life decision making: Research fact or fiction? *The Gerontologist, 42*(Special Issue), 114–128.

Eckholm, E. (2014). 'Aid in dying' movement takes hold in some states. *New York Times* (February 7, 2014). <http://www.nytimes.com/2014/02/08/us/easing-terminal-patients-path-to-death-legally.html?_r=1>. Accessed 20.11.14.

Edwards, A., Pang, N., Shiu, V., & Chan, C. (2010). Review: The understanding of spirituality and the potential role of spiritual care in end-of-life and palliative care: A meta-study of qualitative research. *Palliative Medicine, 24*, 753–770.

Elias, N. (1985). *The loneliness of the dying*. Oxford: Basil Blackwell.

Fagerlin, A., & Schneider, C. (2004). Enough: The failure of the living will. *Hastings Center Report*, 30–42. (March–April).

Farber, N. J., Simpson, P., Salam, T., Collier, V. U., Weiner, J., & Boyer, E. G. (2006). Physicians' decisions to withhold and withdraw life-sustaining treatment. *Archives of Internal Medicine, 166*, 560–564.

Federal Interagency Forum on Aging Related Statistics, (2013). *Older Americans 2012: Key indicators of well-being*. Washington, DC: Department of Health and Human Services.

Finucane, T. (2004). Preferences and changes in the goals of care. Abstract. *NIH State of the Science Conference on Improving End-of-Life Care* (December 6–8), 23–28.

Fried, T. R., Bradley, E. H., & O'Leary, J. (2006). Changes in prognostic awareness among seriously ill older persons and their caregivers. *Journal of Palliative Medicine, 9*, 61–69.

Garrido, M. M., Idler, E., Leventhal, H., & Carr, D. (2013). Advance care planning: The role of end-of-life values and beliefs about control over the length of life. *The Gerontologist, 53*, 801–816.

Gawande, A. (2014). *Being mortal: Medicine and what matters in the end*. New York, NY: Metropolitan Books.

George, L. K. (2002). Research design in end-of-life research. *The Gerontologist, 42*(Suppl.), 86–98.

Glaser, B. G., & Strauss, A. L. (1965). *Awareness of dying*. New Brunswick, NJ: Aldine.

Hammes, B.J. (2003). Update on Respecting Choices®: Four years on. *Innovations in End-of-Life Care*. <http://www.rpctraining.com.au/module06/mod06_toolkit/Additional%20references/Reference-4-3.pdf>. Accessed 04.12.14.

Hammes, B. J., Rooney, B. L., Gundrum, J. D., Hickman, S. E., & Hager, N. (2012). The POLST program: A retrospective review of the demographics of use and outcomes in one community where advance directives are prevalent. *Journal of Palliative Medicine, 15*, 77–85.

Hanchate, A., Kronman, A. C., Young-Xu, Y., Ash, A. S., & Emanuel, E. (2009). Racial and ethnic differences in end of life costs: Why do minorities cost more than Whites? *Archives of Internal Medicine, 169*, 493–501.

HHS (Department of Health and Human Services). (2008). *Advance directives and advance care planning: Report to Congress*. <http://aspe.hhs.gov/daltcp/reports/2008/adcongrpt.htm>. Accessed 25.11.14.

HHS/ASPE. (Department of Health and Human Services, Assistant Secretary for Planning and Evaluation). (2013). *National plan to address Alzheimer's disease – 2013 update*. <http://aspe.hhs.gov/daltcp/napa/NatlPlan2013.shtml#intro>. Accessed 30.11.14.

Hickman, S. E., Nelson, C. A., Moss, A. H., Tolle, S. W., Perrin, N. A., & Hammes, B. J. (2011). The consistency between treatments provided to nursing facility residents and physician orders for life-sustaining treatment form. *Journal of the American Geriatrics Society, 59*, 2091–2099.

Hoffman, J. C., Wenger, N. S., Davis, R. B., Teno, J., Connors, A. F., Jr., et al. (1997). Patient preferences for communication with physicians about end-of-life decisions. *Annals of Internal Medicine, 127*, 1–12.

Huynh, T. N., Kleerup, E. C., Wiley, J. F., Savitsky, T. D., Guse, D., Garber, B. J., et al. (2013). The frequency and cost of treatment perceived to be futile in critical care. *JAMA Internal Medicine, 173*, 1887–1894.

Institute of Medicine (IOM), (2014). *Dying in America: Improving quality and honoring individual preferences near the end of life*. Washington, DC: The National Academies.

Iwashyna, T. J., & Christakis, N. A. (2003). The health impact of health care on families: A matched cohort study of hospice use by decedents and mortality outcomes in surviving, widowed spouses. *Social Science & Medicine, 57*, 465–475.

Johnson, K. S., Kuchibhatla, M., & Tulsky, J. A. (2008). What explains racial differences in the use of advance directives and attitudes toward hospice care? *Journal of the American Geriatrics Society, 56*, 1953–1958.

Keary, S., & Moorman, S. M. (2015). Patient–physician end-of-life discussions in the routine care of Medicare beneficiaries. *Journal of Aging and Health*.

Kellehear, A. (1984). Are we a 'death-denying' society? A sociological review. *Social Science and Medicine, 18*, 713–723.

Kelley, A. S., Ettner, S. L., Morrison, S., Du, Q., Wenger, N. S., & Sarkisian, C. A. (2011). Determinants of medical expenditures in the last 6 months of life. *Annals of Internal Medicine, 154*, 235–242.

Khodyakov, D., & Carr, D. (2009). The impact of late-life parental death on adult sibling relationships: Do parents' advance directives help or hurt? *Research on Aging, 31*, 495–519.

Kohn, N. A., & Blumenthal, J. A. (2008). Designating health care decision-makers for patients without advance directives: A psychological critique. *Georgia Law Review, 42*, 1–40.

Kramer, B. J., Boelk, A. Z., & Auer, C. (2006). Family conflict at the end of life: Lessons learned in a model program for vulnerable older adults. *Journal of Palliative Medicine, 9*, 79.

Kramer, B. J., & Yonker, J. A. (2011). Perceived success in addressing end-of-life care needs of low-income elders and their families: What's family conflict got to do with it? *Journal of Pain and Symptom Management, 41*, 35–48.

Kwak, J., & Haley, W. E. (2005). Current research findings on end of life decision making among racially or ethnically diverse groups. *The Gerontologist, 45*, 634–641.

Loggers, E. T., Maciejewski, P. K., Paulk, E., DeSanto-Madeva, S., Nilsson, M., Viswanath, K., et al. (2009). Racial differences in predictors of intensive end of life care in advanced cancer patients. *Journal of Clinical Oncology, 27*, 5559–5564.

Lynn, J., & Adamson, D. M. (2003). *Living well at the end of life: Adapting health care to serious chronic illness in old age*. Washington, DC: RAND Health.

Mack, J. W., Paulk, M. E., Viswanath, K., & Prigerson, H. G. (2010). Racial disparities in the outcomes of communication on medical care received near death. *Archives of Internal Medicine, 170*, 1533–1540.

Mack, J. W., Weeks, J. C., Wright, A. A., Block, S. D., & Prigerson, H. G. (2010). End-of-life discussions, goal attainment, and distress at the end of life: Predictors and outcomes of receipt of care consistent with preferences. *Journal of Clinical Oncology, 28*, 1203–1208.

McCue, J. D. (1995). The naturalness of dying. *Journal of the American Medical Association, 273*, 1039–1043.

McGrew, D. M. (2001). Chronic illnesses and the end of life. *Primary Care, 28*, 339–347.

Meier, D. E. (2010). The development, status, and future of palliative care. In D. E. Meier, S. L. Isaacs, & R. Hughes (Eds.), *Palliative care: Transforming the care of serious illness* (pp. 4–76). San Francisco, CA: Jossey-Bass.

Mitchell, S. L., Teno, J. M., Roy, J., Kabumoto, G., & Mor, V. (2003). Clinical and organizational factors associated with feeding tube use among nursing home residents with advanced cognitive impairment. *Journal of the American Medical Association, 290*, 73–80.

Molloy, D. W., Clarnette, R. M., Braun, E., Eisemann, M. R., & Sneiderman, B. (1991). Decision making in the incompetent elderly: "The daughter from California syndrome." *Journal of the American Geriatrics Society, 39*, 396–399.

Moorman, S., Carr, D., & Boerner, K. (2014). The role of relationship biography in advance care planning. *Journal of Aging and Health, 26*, 969–992.

Moorman, S. M. (2011). Older adults' preferences for independent or delegated end-of-life medical decision-making. *Journal of Aging and Health, 23*, 135–157.

Moorman, S. M., & Carr, D. (2008). Spouses' effectiveness as end-of-life health care surrogates: Accuracy, uncertainty, and errors of overtreatment or undertreatment. *The Gerontologist, 48*, 811–819.

Moorman, S. M., & Inoue, M. (2013). Predicting a partner's end of-life preferences, or substituting one's own? *Journal of Marriage and Family, 75*, 734–745.

Morrison, R. S., Zayas, L. H., Mulvihill, M., Baskin, S. A., & Meier, D. E. (1998). Barriers to completion of health care proxies. *Archives of Internal Medicine, 158*, 2493–2497.

Murray, S. A., Kendall, M., Boyd, K., & Sheikh, A. (2005). Illness trajectories and palliative care. *British Medical Journal, 330*, 1007–1011.

National POLST. (2012). *About the National POLST Paradigm*. <http://www.polst.org/about-the-nationalpolstparadigm/>. Accessed 21.11.14.

National POLST. (2014). *Programs in your state*. <http://www.polst.org/programs-in-your-state/>. Accessed 09.12.14.

National Research Council. (1997). *Approaching death: Improving care at the end of life.* Washington, DC: The National Academies Press.

NHPCO. (2012). *NHPCO facts and figures: Hospice care in America.* <http://www.nhpco.org/sites/default/files/public/Statistics_Research/2012_Facts_Figures.pdf>. Accessed 25.11.14.

Nicholas, L. H., Langa, K. M., Iwashyna, T. J., & Weir, D. R. (2011). Regional variation in the association between advance directives and end-of-life Medicare expenditures. *Journal of the American Medical Association, 306,* 1447–1453.

Olshansky, S. J., & Ault, A. B. (1986). The fourth stage of the epidemiologic transition: The age of delayed degenerative diseases. *Millbank Memorial Fund Quarterly, 64,* 355–391.

Omran, A. (1971). The epidemiologic transition: A theory of the epidemiology of population change. *Milbank Memorial Fund Quarterly, 29,* 509–538.

Oregon Health Authority. (2014). *Oregon's Death with Dignity Act, 2013.* <http://public.health.oregon.gov/ProviderPartnerResources/EvaluationResearch/DeathwithDignityAct/Documents/year16.pdf>. Accessed 09.12.14.

Parks, S. M., Winter, L., Santana, A. J., Parker, B., Diamond, J. J., Rose, M., et al. (2011). Family factors in end-of-life decision-making: Family conflict and proxy relationship. *Journal of Palliative Medicine, 14,* 179–184.

Patient Self-Determination Act of 1990. (1990). 554206, 4751 of the Omnibus Reconciliation Act of 1990. Pubs. No. 101-508.

Pattison, E. M. (1977). The living–dying process. In C. A. Garfield (Ed.), *Psychosocial care of the dying patient* (pp. 133–168). New York, NY: McGraw-Hill.

Pear, R. (2011). U.S. alters rule on paying for end-of-life planning. *New York Times* (January 4).

Pew Research Center. (2013). *Views on end-of-life medical treatments* (November 2013). <http://www.pewforum.org/files/2013/11/end-of-life-survey-report-full-pdf.pdf>. Accessed 25.11.14.

Pfeifer, M. P., Mitchell, C. K., & Chamberlain, L. (2003). The value of disease severity in predicting patient readiness to address end-of-life issues. *Archives of Internal Medicine, 163,* 609–612.

Pierce, S. F. (1999). Improving end-of-life care: Gathering suggestions from family members. *Nursing Forum, 34,* 5–14.

Porensky, E. K., & Carpenter, B. D. (2008). Knowledge and perceptions in advance care planning. *Journal of Aging and Health, 20,* 89–106.

President's Council on Bioethics. (2005). *Taking care: Ethical caregiving in our aging society.* Washington, DC: President's Council on Bioethics.

Shalowitz, D. I., Garrett-Mayer, E., & Wendler, D. (2006). The accuracy of surrogate decision makers: A systematic review. *Archives of Internal Medicine, 166,* 493–497.

Sharp, S., Carr, D., & MacDonald, C. (2012). Religion and end-of-life treatment preferences: Assessing the effects of religious denomination and beliefs. *Social Forces, 91,* 275–298.

Silveira, M. J., Kim, S. Y. H., & Langa, K. M. (2010). Advance directives and outcomes of surrogate decision making before death. *New England Journal of Medicine, 362,* 1211–1218.

Singer, A. E., Meeker, D., Teno, J. M., Lynn, J., Lunney, J. R., & Lorenz, K. A. (2015). Symptom trends in the last year of life from 1998 to 2010: A cohort study. *Annals of Internal Medicine, 162,* 175–183.

Singer, P. A., Martin, D. K., & Keltner, M. (1999). Quality end-of-life care: Patients' perspectives. *Journal of the American Medical Association, 281,* 163–168.

Skolnick, A. A. (1998). MediCaring Project to demonstrate, evaluate innovative end-of-life program for chronically ill. *Journal of the American Medical Association, 279,* 1511–1512.

Smith, A. K., McCarthy, E. P., Paulk, E., Balboni, T. A., Maciejewski, P. K., Block, S. D., et al. (2008). Racial and ethnic differences in advance care planning among patients with cancer: Impact of terminal illness acknowledgment, religiousness, and treatment preferences. *Journal of Clinical Oncology, 26,* 4131–4137.

Starr, P. (1982). *The social transformation of American medicine.* New York, NY: Basic Books.

Stein, R. A., Sharpe, L., Bell, M. L., Boyle, F. M., Dunn, S. M., & Clarke, S. J. (2013). Randomized controlled trial of a structured intervention to facilitate end-of-life decision making in patients with advanced cancer. *Journal of Clinical Oncology, 31,* 3403–3410.

Steinhauser, K. E., Clipp, E. C., McNeilly, M., Christakis, N. A., McIntyre, L. M., & Tulsky, J. A. (2000). In search of a good death: Observations of patients, families, and providers. *Annals of Internal Medicine, 132,* 825–832.

Su, Y. (2008). Looking beyond retirement: Patterns and predictors of formal end-of-life planning among retirement age individuals. *Journal of Family and Economic Issues, 29,* 654–673.

Sudnow, D. (1967). *Passing on: The social organization of dying.* Englewood Cliffs, NJ: Prentice Hall.

Sudore, R. L., Landefeld, C. S., Pantilat, S. Z., Noyes, K. M., & Schillinger, D. (2008). Reach and impact of a mass media event among vulnerable patients: The Terri Schiavo Story. *Journal of General Internal Medicine, 233,* 1854–1857.

SUPPORT Principal Investigators. (1995). A controlled trial to improve care for seriously ill hospitalized patients. *Journal of the American Medical Association, 274,* 1591–1598.

Teno, J. M., Gozalo, P. L., Bynum, J. P. W., Leland, N. E., Miller, S. C., Morden, N. E., et al. (2013). Change in end-of-life care for Medicare beneficiaries: Site of death, place of care, and health care transitions in 2000, 2005, and 2009. *Journal of the American Medical Association, 309*, 470–477.

Teno, J. M., Gruneir, A., Schwartz, Z., Nanda, A., & Wetle, T. (2007). Association between advance directives and quality of end-of-life care: A national study. *Journal of the American Geriatrics Society, 55*, 189–194.

Timmermans, S. (2009). Social death as a self-fulfilling prophecy. In P. Conrad (Ed.), *Sociology of Health and Illness* (pp. 370–386) (8th ed.). New York, NY: Worth Publishers.

Vig, E. K., Taylor, J. S., Starks, H., Hopley, E. K., & Fryer-Edwards, K. (2006). Beyond substituted judgment: How surrogates navigate end-of-life decision-making. *Journal of the American Geriatrics Society, 54*, 1688–1693.

Waite, K. R., Federman, A. D., McCarthy, M., Sudore, R., Curtis, L. M., Baker, D. W., et al. (2013). Literacy and race as risk factors for low rates of advance directives in older adults. *Journal of the American Geriatrics Society, 61*, 403–406.

Weeks, J. C., Cook, E., O'Day, S. J., Peterson, L. M., Wenger, N., Reding, D., et al. (1998). Relationship between cancer patients' predictions of prognosis and their treatment preferences. *Journal of the American Medical Association, 279*, 1709–1714.

West, S. K., & Hollis, M. (2012). Barriers to completion of advance care directives among African Americans ages 25–84: A cross-generational study. *Omega, 65*, 125–137.

Winter, L., & Parks, S. M. (2012). Acceptors and rejecters of life-sustaining treatment: Differences in advance care planning characteristics. *Journal of Applied Gerontology, 31*, 734–742.

Wright, A. A., Zhang, B., Ray, A., Mack, J. W., Trice, E., Balboni, T., et al. (2008). Associations between end-of-life discussions, patient mental health, medical care near death, and caregiver bereavement adjustment. *Journal of the American Medical Association, 300*, 1665–1673.

Zettel-Watson, L., Ditto, P. H., Danks, J. H., & Smucker, W. D. (2008). Actual and perceived gender differences in the accuracy of surrogate decisions about life-sustaining medical treatment among older spouses. *Death Studies, 32*, 273–290.

Zhang, B., Wright, A. A., Huskamp, H. A., Nilsson, M. E., Maciejewski, M. L., Earle, C. C., et al. (2009). Health care costs in the last week of life: Associations with end-of-life conversations. *Archives of Internal Medicine, 169*, 480–488.

Zilberberg, M. D., & Shorr, A. F. (2012). Economics at the end of life: Hospital and ICU perspectives. *Seminars in Respiratory and Critical Care Medicine, 33*, 362–369.

PART IV

AGING AND SOCIETY

19 Organization and Financing of Health Care   397
20 Innovations in Long-Term Care   419
21 Politics and Policies of Aging in the United States   441
22 The Future of Retirement Security in Comparative Perspective   461
23 Health Inequalities Among Older Adults in Developed Countries: Reconciling Theories and Policy Approaches   483

CHAPTER 19

Organization and Financing of Health Care

Marilyn Moon

American Institutes for Research, Washington, DC, USA

OUTLINE

Introduction	397
A Brief History of Medicare and Medicaid	399
Moving into an Era of Health Reform	403
Supplementing Medicare with Medicaid and other Insurance	405
Medicaid	407
Employer-Sponsored Plans	408
Medigap Insurance	408
Medicare Advantage	409
No Coverage	410
Major Options for Reform	**411**
A Private Plan Approach	412
More Incremental Approaches	414
The Issues of Financing Medicare and Medicaid	415
References	416

INTRODUCTION

The organization and financing of health care for older Americans has been a rare area of stability for the past decade. While the debate over health care reforms for those under the age of 65 has been a major source of discord and confusion, the Medicare program has retained its strong support from both beneficiaries and the public at large. Yet it too is likely to face potential changes as faith in government and willingness to support the revenues necessary to sustain Medicare and the related Medicaid program once again receive scrutiny. The aging of the Baby Boom population will continue to keep financing concerns near the forefront of debates over the federal budget even though some slowdown in growth has been achieved in recent years. Regardless of their popularity, Medicare and Medicaid will be a key focus of negotiations over how to finance the federal government.

Enacted in 1965 as additions to the Social Security Act, Medicare and Medicaid offer important resources that individuals can rely upon in old age. Without these protections, few Americans could afford to retire, given the costs of health care and the private sector barriers that have traditionally existed to obtaining reasonably priced insurance coverage. But even though these programs are large and growing, Medicare and Medicaid do not cover all the health care needs of the population over 65, leaving gaps that generate substantial risks for low- and moderate-income retirees. Further, declines in retiree health benefits from former employers are beginning to place higher-income retirees at risk as well. These pressures signal a need for expanding, not contracting levels of protection. Thus, Medicare and Medicaid will be caught in a critical public policy battle over how to continue to provide health coverage to older Americans. As the debate takes shape, these programs need to be viewed in the context of both meeting health care needs and offering financial protections to older Americans.

Medicare is a federal health insurance program that serves over 43.5 million people aged 65 and older in the United States (and 8.8 million younger disabled persons), providing basic protection for acute-care needs. Medicaid, which offers help for elderly and other low-income and low-wealth Americans, is also important for older persons because it fills in crucial gaps in Medicare coverage. A joint federal/state program, Medicaid supplements Medicare coverage for about one in every five seniors – approximately 10.2 million people (either in terms of full or limited benefits) (MACPAC, 2014). Medicaid is the only public program that helps cover the long-term care costs of older Americans; about two-thirds of its spending on the over 65 population is for long-term care.

Over the years, as the costs of health care grew rapidly, Medicare became an ever-more important contributor to the well-being of America's oldest citizens. It is the major source of insurance for the elderly (and, since 1972, for individuals who receive federal disability insurance benefits and those with end-stage renal disease). One of Medicare's important accomplishments is that the very old and the very sick have access to the same basic benefits as younger, healthier beneficiaries. While there is certainly room for improvement, Medicare is insurance that is never rescinded because of the poor health of the individual.

Although in many ways Medicare has been one of the most successful federal programs, it has also faced criticism as a result of its rapid expansion. Medicare spending in 2013 stood at $583 billion (Boards of Trustees, 2014). From the late 1970s until 2003, Medicare was a frequent target in efforts to reduce federal spending. Then, in 2003, the addition of a new prescription drug benefit (that began in 2006) represented a major expansion in the program. In 2010, changes in Medicare to reduce costs were included in the Affordable Care Act (ACA) in part to help offset costs of expanding coverage to persons under the age of 65. And now, the new Republican majority in the House of Representatives has viewed slowing growth in Medicare as a major cornerstone of efforts to reduce the overall size of the federal budget (Van de Water, 2012). Medicaid, which enjoys less political popularity, is an even more likely target for cuts.

Growth in the costs of Medicare and Medicaid is attributable to the same factors generating rising costs in the rest of the health care system. To be sure, the size of Medicare and Medicaid means that changes in payment levels or decisions about medical necessity will have an impact on all of health care spending. But these public programs are far from the main drivers of health care costs. The development of new technologies and procedures, increasing rates of utilization of services, and a high rate of inflation in the overall health care arena have greater effects. In the first decade of the twenty-first century, costs rose rapidly in all sectors of health care, particularly for hospital care and prescription drugs, although showing some signs of slowing (Hartman, Martin, Nuccio, Catlin, & National Health Expenditure Account Team, 2010). Since the recent Great Recession, health spending has slowed across the board (Martin, Hartman, Whittie, Catlin, & National Health Expenditure Account Team, 2013). Since growth in Medicare is similar to that for private insurance, it is not only difficult but inappropriate to treat Medicare as separable from the rest of the system.

A BRIEF HISTORY OF MEDICARE AND MEDICAID

Medicare was established by legislation in 1965 as Title XVIII of the Social Security Act and went into effect on July 1, 1966. The overriding goal was to provide mainstream acute health care – hospital, physician, and related services – for persons aged 65 and over. Medicaid began at the same time and filled in the gaps in acute-care coverage that Medicare left for those with very low incomes, and quickly became a program that offered long-term care protection as well. This age group had been under-served by the health care system. Retiree insurance benefits were relatively rare and insurance companies were reluctant to offer coverage even to those who could afford it. The problem was one of both affordability and availability at any price; in 1965, only a little over half of older Americans had any health insurance (Andersen, Lion, & Anderson, 1976).

When Medicare was implemented in 1966, it revolutionized health care coverage for older adults. By 1970, 97% of older Americans were covered, and that proportion has remained about the same ever since (Moon, 2006). Anyone aged 65 or over who is eligible for any type of Social Security benefit (e.g., as a worker or a dependant) receives Part A, Hospital Insurance. Part A covers hospital, skilled nursing, hospice care, and some home health care. It is funded by payroll tax contributions from workers and employers that are earmarked and placed in the Part A trust fund. Supplementary Medical Insurance (Part B) is available to persons age 65 and over on a voluntary basis. It covers physician services, hospital outpatient services, most of home health care, and other ambulatory care and is funded by a combination of general revenues and beneficiary premiums. The 75% share that comes from general revenues means that nearly all eligible seniors enroll in Part B.

Initially, the Part B premium was a flat percentage of Part B costs. It began as 50% and was reduced to 25% as costs of health care began to rise rapidly and Part B premiums began to consume a larger and large share of Social Security benefits. In 2003, an income-related premium was added requiring those with incomes above $85 000 (for individuals and $170 000 for couples) to pay a larger share of Part B costs.

A prescription drug benefit (Part D) was added in 2006 and represented the only major change in coverage since the program began. Part D is voluntary and financed in the same manner as Part B. The drug benefit is provided via private plans and operates as a separate supplemental benefit. As a result, individuals are confronted with complex decisions since many types of plans are offered each year – including variations in the formulary that will be used (i.e., what drugs are covered), the type and amount of cost sharing, what pharmacies participate in each plan, and substantial variations in the premium.

The Part D legislation had two major goals – to cover catastrophic expenses and to offer enough basic coverage to encourage relatively healthy people to sign up (Zhang, Donohue, Newhouse, & Lave, 2009). But the funding level that federal policymakers agreed upon was insufficient to offer a comprehensive benefit. Thus, the benefits put in place were more generous at high and low levels of spending, leaving a gap in coverage in the middle where insurance protection falls to zero for some period. This gap, which has come to be called the "donut hole," means that at some point during any year, high-cost users must pay for all their drug costs until the catastrophic cap begins (Kaiser Family Foundation, 2013).

One promise made in 1965 for Medicare was that it would not interfere with the practice of medicine. Payments were designed to be like standard insurance policies then in place. But costs for the program rose rapidly almost from the beginning, and in the late 1970s the government sought to slow spending growth in Medicare, and chose to do so largely through application of new payment policies. These policies both affected how much would be paid and began to move payments away from per unit pricing.

After initial efforts in the 1970s to limit annual payment increases, reforms went further. Payments for hospital services – the biggest component of Medicare – were modified in the early 1980s to pay on the basis of the patient's primary diagnosis, regardless of length of stay in a hospital or the actual costs of a particular case. This new system encouraged hospitals to be more efficient, although it also resulted in some premature discharges. Over time, however, this payment system has been judged to be relatively successful (Moon, 1996). It has helped to encourage movement away from long inpatient stays and resulted in more care being delivered outside of hospitals for patients of all ages. In general, Medicare deserves credit for helping to improve the efficiency of US hospitals.

Physician reform came later and established payments on the basis of a relative value scale, initially limiting payments to specialists and increasing them for basic primary care (Physician Payment Review Commission, 1987). Many other health care insurers now use Medicare's physician relative value payment system. Both hospital and physician payments require periodic updating, and Medicare is sometimes criticized for falling behind in making adjustments in response to new medical procedures. In particular, the flawed system for updating payments to physicians has come under heavy criticism, generating requirements for cutting fees each year; because of the expense of fixing that flaw, only ad hoc adjustments were made until early 2015 with a permanent change in the update approach. The new system will move away from fee-for-service payments but the details still need to be worked out. The low rates of increase in physician fees for the last 15 years have been a continuing concern. Thus far, physicians have not opted out of Medicare in large numbers but this threat remains a potential problem looming for Medicare even under the new system.

When Medicare came into being in 1966, almost all health care was provided on the basis of fee-for-service – that is, individual providers charged separately for each service, with little effort to coordinate care. This "piece rate" approach still exists today for the majority of Medicare beneficiaries, and is often blamed for encouraging excessive spending. Service providers face very direct incentives to offer more care – more tests, more visits, more procedures. In contrast, younger individuals and families are more likely to be in managed care environments in which the financial incentives are to restrain use of care since the managed care plan is paid a flat annual fee from which it must budget all needed expenses. This approach provides the opposite incentive and is criticized for sometimes skimping on care. Medicare remains caught in the middle of this debate – whether to modify fee-for-service to restrain growth or to rely upon private plans that will take a more activist role in coordinating care. Thus far, fee-for-service remains the dominant approach for "traditional" Medicare, but with an option for beneficiaries to enroll in private managed care plans that has been growing rapidly in recent years.

This private plan option was added to Medicare in 1983, first as a demonstration but soon became a standard option. This option allows beneficiaries to choose to get all of their care from an insurer paid a monthly fee from Medicare to meet the needs of those who enroll. Initially, these were limited to managed care plans – health maintenance organizations (HMOs) – that were touted as the future for health care delivery. The intent of this option was to rely on both innovation by HMOs and competition among private plans to find new ways to keep costs from rising. Over the years, the option has expanded and contracted as the levels of payments offered and the enthusiasm of Congress for such options have gone up and down. And, over time, other less-coordinated plans have also been allowed to participate.

Now called Medicare Advantage (or Part C), enrollment has been rapidly growing. In 2014, over 30% of Medicare beneficiaries signed up for such plans, largely because of extra benefits offered to the enrollees: usually lower premiums, vision and dental care, and more recently more comprehensive drug benefits (Medicare Trustees, 2014). The 2003 drug benefit legislation also established payment levels that effectively offered subsidies to Medicare Advantage plans to encourage greater participation. That subsidy peaked at about 14% above what it cost traditional fee-for-service Medicare to provide similar care (MedPAC, 2009).

When Medicare began, it was dominated by inpatient hospital care, which accounted for about two-thirds of all spending. Indeed, most of the focus of debate before passage was on Part A. But as care has moved out of the inpatient setting, Part B has become a much larger share of the program. Care in hospital outpatient departments and in physicians' offices now replaces many surgeries and treatments formerly performed in inpatient settings. In addition, skilled nursing facility care and home health services – referred to as post-acute care – have also increased in importance as hospital stays have been shortened. When individuals leave a hospital after only a few days, post-acute care is often needed as a transition, either in a nursing facility or at home with visits from nurses or other skilled technicians such as rehabilitation therapists. The financial incentives established by payment to health providers and by coverage of benefits have affected the mix of services used.

Figure 19.1 contrasts the Medicare program at two points in time. In 1967, hospital care (and hence Part A) dominated spending. By 2012, several major changes were notable. Inpatient hospital services declined as a share of the traditional part of the program. And Medicare Advantage – the private option portion – accounted for more than a fifth of all spending.

The situation for Medicaid is quite different because it is controlled by the states, which have considerable latitude in establishing eligibility and coverage. Hence, approaches to holding the line on Medicaid spending can vary substantially across the nation. On the acute-care side, Medicaid serves as a wraparound benefit, filling in gaps in Medicare's coverage both in terms of types of benefits and required cost sharing. To participate, incomes and assets must be very low – often well below 100% of the federal poverty line (FPL, which stands at about $12 000 in annual income for an individual). States vary in terms of these requirements, resulting in considerable variation in who can receive such benefits. For example, eligibility varies from 74% of FPL to 100% for basic benefits (MACPAC, 2014).

Further, states have relied both on keeping payments to providers very low and on encouraging beneficiaries to go into managed care plans. Low payments have sometimes led to access problems when providers of services decide not to take on Medicaid patients. This effectively reduces government costs both through the lower payments and through the reduced volume of services available, although it creates substantial issues for those qualified to receive Medicaid. For example, for persons dually eligible for Medicare and Medicaid,

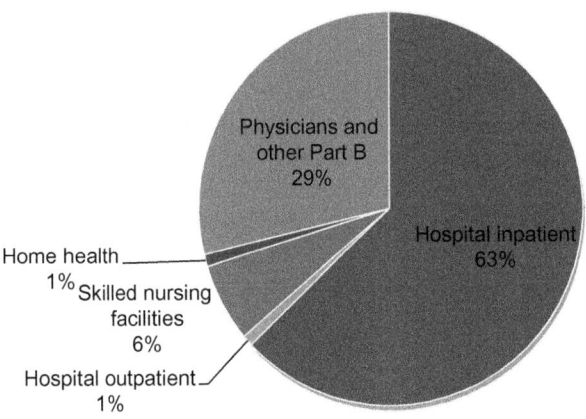

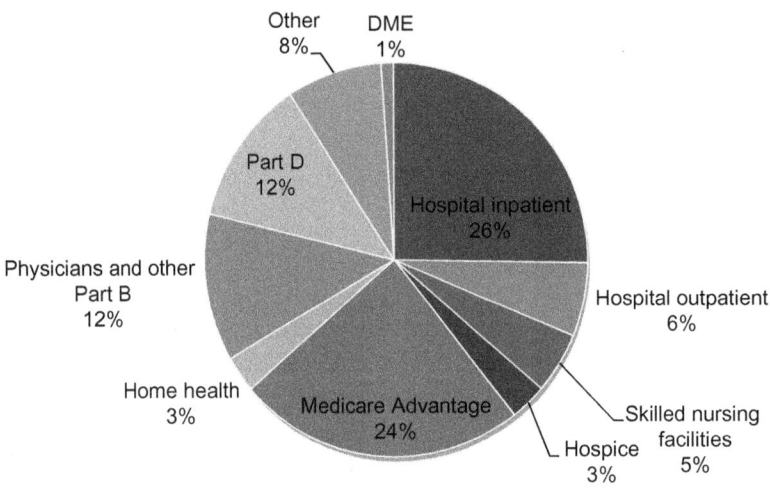

FIGURE 19.1 Medicare expenditures by type of service, 1967 and 2012. Note: DME (durable medical equipment). Data are by calendar year. "Other" includes carrier lab, other carrier, intermediate lab, and other intermediary. *Source: MedPac Data book: Healthcare Spending and the Medicare Program (June 2013).*

a number of states decline to use the federally authorized option of having Medicaid pay for Medicare copayments to physicians, thereby effectively lowering the payments doctors receive. Those who are dually eligible tend to be the sickest and most vulnerable of Medicare's beneficiaries. Consequently, there is periodically discussion about whether care should be split between the two programs or become the sole responsibility of one of them.

Long-term care is another area where Medicaid differs from Medicare. The needs of older Americans for supportive services when they are very frail, disabled, or have cognitive problems are met by Medicaid, not Medicare. And although there are some ways for more modest-income seniors to become eligible, the program remains limited to the poorest seniors. Most generous coverage remains for institutional care, but states are increasingly expanding into care in the home as well.

MOVING INTO AN ERA OF HEALTH REFORM

The passage of the ACA in 2010 and its subsequent rollout has important consequences for care for older Americans for a number of reasons. Although most of that legislation focused on expanding coverage for the under-65 population, a number of changes to Medicare were included. And even more important, the overall health care system is likely to change substantially as a result of the ACA, affecting Americans of all ages. While the full implications of the changes being put into place are still to be determined, the general outline of how health care will change is in place.

A few changes in the ACA improved Medicare benefits by expanding coverage of preventive services and enhancing the prescription drug benefit offered by Medicare. Before 2010, beneficiaries were required to pay the Part B deductible and coinsurance on preventive services, potentially discouraging individuals from using these services (Kaplan, 2011). The ACA eliminated these charges, reducing what beneficiaries must pay for those preventive services deemed to meet a rating of effectiveness by the US Preventive Services Task Force. In addition, the new law adds a yearly wellness visit with no coinsurance requirement. In the past, such visits were only available in the year that the beneficiary entered the Medicare program. More generous coverage of preventive services following implementation of the ACA should lower out-of-pocket costs or premiums for supplemental coverage over time.

The ACA also helped to improve Part D drug protections by gradually closing the "donut hole" where insurance protection stops for some high users. In 2014, the gap in coverage began for beneficiaries with annual expenses over $2 850 (Boards of Trustees, 2014). At that point, insurance protection ceased until affected beneficiaries spent $7 400 on drugs, or a total of $4 550 out-of-pocket. After that, catastrophic protection began. For many Medicare beneficiaries, this is an important gap in coverage. The ACA initially provided each Medicare beneficiary reaching the donut hole a $250 payment to help defray medication costs. Beginning in 2011, further changes were made to gradually eliminate this coverage gap so that people reach the catastrophic protection level with lower actual out-of-pocket spending than in the past. The federal government is gradually increasing the amount covered until the insurance protection reaches 75% of the costs of prescriptions (the same protection now guaranteed before reaching the donut hole); the phase-in period will take 11 years (Kaiser Family Foundation, 2010).

In addition to these expansions in Medicare coverage, changes to Medicare payment served as a major source of financing for the insurance subsidies for the under-65 population in the ACA. Critics characterized these proposals as "starving" the Medicare program and threatening care. In actuality, most of the changes focus on payments to providers of care and not coverage or access. The major concern for beneficiaries is whether cuts will reduce the willingness of providers to continue serving Medicare beneficiaries. These reductions are proportionately lower than reductions that have been made in the program several times over the years (Kaiser Family Foundation, 2009), and – in contrast to past payment reductions – will be offset by improved revenues to providers from expanded coverage for the younger population (Newhouse, 2010).

Most of the savings do not have a direct impact on beneficiaries, with two important exceptions. First, the Medicare Advantage program, where beneficiaries may choose to get their care from a private plan option, is being subjected to substantial reductions in levels of payment that may affect benefits offered. This was in reaction to the overpayments established in the first decade of the century. Payments to private plans are being increased more slowly over time until something close to parity with the costs in traditional Medicare is achieved. Thus far, such plans have found ways to retain at least some extra benefits and continue to attract new enrollees (Newhouse & McGuire, 2014). Managed care plans are demonstrating positive responses that have turned a number of critics into supporters of this option. What is not known, however, is what the reaction of beneficiaries will be in the future as the payment levels to plans are further restricted and extra benefits potentially reduced.

The second change in the ACA that directly affects beneficiaries was the expansion of the income-related portion of the Part B premium that higher-income beneficiaries pay and the addition of a new income-related premium in Part D, increasing their out-of-pocket costs (Kaplan, 2011). Part B had already been changed to require beneficiaries with higher incomes to pay greater premiums. The ACA built on this by increasing the amount that is charged at various levels of income, and extended the requirement to the drug benefit for the first time.

Limitations in scheduled payment increases for many Medicare benefits, including skilled nursing care, home health benefits, and outpatient hospital services helped raise substantial revenue to fund the ACA (Foster, 2010). This generally means that the rate of growth of payments is projected to be slower than the law previously dictated, and is similar to changes made periodically over the past several decades. Critical to evaluating these changes is whether they are consistent with providers' ability to improve efficiency in service delivery or whether they will threaten access to or quality of care, especially relative to non-Medicare patients. But it is also important to consider whether these changes would have been made anyway: would they have been deemed necessary for the health of the Medicare program which continues to be under financial pressure? Will payments be equivalent to what the rest of the health care system pays providers?

Also as part of the ACA, new authority was given to the Centers for Medicare and Medicaid Services (CMS) to establish an Innovation Center to generate reforms in the delivery of care. A broad range of demonstrations are underway that seek to transform our largely fee-for-service health care system affecting both Medicare beneficiaries and the broader private system into one that offers better coordination of care. These reforms include changing the way that providers of services will be paid – offering one comprehensive payment for various bundles of services (such as surgery and all follow-up rehabilitative care) and/or shared savings incentives that will return additional amounts to providers that demonstrate they can deliver care more efficiently at the same or higher quality. Other demonstrations are seeking to give primary care physicians greater responsibilities to coordinate care. Most of these activities show promise of generating savings but are still in their early stages. To give added teeth, the Secretary of HHS has the authority to take these efforts from demonstrations to scale and implement them across Medicare (Shrank, 2013).

One promising area has been the effort to encourage hospitals to find ways to prevent "preventable" readmissions of patients. By generating penalties for hospitals that have high rates of such readmissions, Medicare has spurred hospitals to undertake activities to ensure that when patients are discharged they receive follow-up care in the community to help deter readmissions. Hospitals, however, have complained about being penalized for care over which they have little control.

The goal of these activities is to provide additional discipline in the fee-for-service sector and consequently to achieve a substantial slowdown in the growth of spending both in Medicare and the health care system in general. If successful, Medicare will remain more affordable over time, reducing the pressure for further changes. Thus, the success of these demonstrations will weigh heavily on Medicare's future.

Finally, to assure continued savings in Medicare, the ACA established a 15-member Medicare-independent commission to guarantee lower Medicare spending over time. The Independent Payment Advisory Board (IPAB), is scheduled to begin operations in 2018, and will be tasked with recommending Medicare changes to hold the rate of growth of spending to a formula based on the growth in the economy. Many policymakers have criticized the IPAB, usually arguing that the stringent targets will lead to rationing of care. While it is precluded from changing benefits or cost-sharing requirements, setting absolute growth targets may fail to offer the flexibility needed in future years. Since the allowed growth rate would be tied to Gross Domestic Product (GDP), what would happen in case of another severe recession such as we have been experiencing? The need for health care does not fall just because of a financial crisis. Further, true innovations that can lower costs over time can increase costs in the short run. Finally, establishing constraints only on Medicare may move the program away from the mainstream of care offered to others. Thus, the IPAB represents another unknown in thinking about how Medicare will operate in the future.

Medicaid was also expanded by the ACA, but only for the population under age 65. In fact, the subsidies established in the ACA for younger families and individuals with lower incomes are considerably more generous than what is currently available to the 65-and-older population. No Medicaid improvements were made for seniors in this legislation. Over time, there may be pressure to improve the low-income protections that Medicare and Medicaid now offer to better coordinate with the ACA benefits.

Another way that the ACA will affect health care for seniors is by improving coverage for those a few years from Medicare eligibility. Early retirees, older unemployed individuals and older workers whose employers choose not to offer insurance have traditionally had a difficult time obtaining insurance. They are likely to have health conditions that make purchase in the individual market either expensive or unavailable at any price. The ACA guarantees that older persons can buy insurance, and places limits on the pricing. And for those with incomes below 400% of the federal poverty level, subsidies are available to help make insurance more affordable.

Enhanced coverage of these individuals will ultimately help to moderate the costs of the Medicare and Medicaid programs. Studies have shown that Medicare costs are at least 10% higher for those who were uninsured before becoming eligible. And this is not just an issue of initial pent-up demand for services – this differential persists for at least 10 years (McWilliams, Meara, Zaslavsky, & Ayanian, 2007). Better coverage of the under-65 population and expansion in preventive services in Medicare could contribute further to improvements in health status and reduced need for long-term care over time.

SUPPLEMENTING MEDICARE WITH MEDICAID AND OTHER INSURANCE

Medicare has never been a very generous program, leaving approximately 35% of acute-care spending to be covered by individuals, their families or some type of supplemental insurance (MedPAC, 2013). Crucial gaps in Medicare coverage arise from the high deductibles and copayments required for covered services, and a lack of any upper limit on such payments (as is usually available from most commercial insurance). For example, in 2014, the hospital deductible was $1 216 for each admission in a spell of illness, no matter how short the stay. The Part B deductible is much smaller – just $147 per year in 2014, but coverage of physician and other ambulatory services is limited to 80% of allowed charges. For those who choose Part D, substantial co-pays also exist. Consequently, coverage still lags behind what younger, well-insured families have for insurance. In 2007, the average benefit value of Medicare was only about 87% as generous as the typical large employer plan, largely because of its higher cost-sharing requirements (Yamamoto, Neuman, & Kitchman Strollo, 2008). And since spending is higher for older Americans, the cost-sharing burdens on seniors can be quite substantial. Recent estimates indicate that average out-of-pocket spending, including what individuals spend on insurance premiums, was $4 451 in 2010 (Noel-Miller, 2013).

In addition, Medicare does not cover long-term care costs – nursing home and supportive home benefits for those who need help in daily living. While long-term care is not generally part of insurance for younger families either, it is something that many older Americans need when disabilities and long-term health problems make it difficult to survive independently. The costs of nursing home care can be very high, averaging about $81 000 to $90 000 a year in 2012 (MetLife, 2013). For financial support in this area, seniors turn to the Medicaid program, which is limited to those with very low incomes (or who have depleted their assets to the point where they cannot pay all of the costs of essential long-term care).

The current financing pressures make it unlikely that Medicare will expand in the near future in a manner sufficient to fill in the important gaps in coverage. At present, these are filled by supplemental benefits from both public and private sources. Four kinds of supplemental policies have evolved. As described above, Medicaid, a means-tested public benefit established at the same time as Medicare, subsidizes many poor older persons through several different arrangements. Employer-based retiree insurance and individual supplemental coverage policies (termed Medigap) are provided by private insurers. These three sources of coverage normally are associated with beneficiaries enrolled in the traditional part of the program. The fourth option is MA, in which private health plans that contract with Medicare to provide Medicare-covered services also offer at least some additional supplemental benefits.

These supplemental coverages vary in quality, beneficiaries' ability to access them, and the degree to which they relieve financial burdens. Understanding how these supplemental plans contribute is essential for any analysis of the health care system for older Americans. Figure 19.2 shows the extent and type of supplemental coverage for Medicare beneficiaries.

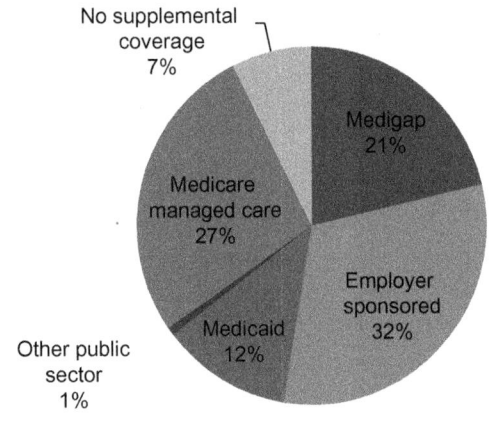

FIGURE 19.2 Sources of supplemental coverage among noninstitutionalized Medicare beneficiaries, 2009. Note: Beneficiaries are assigned to the supplemental coverage category that applied for the most time in 2009. They could have had coverage in other categories during 2009. "Other public sector" includes federal and state programs not included in other categories. Includes only beneficiaries not living in institutions such as nursing homes. Excludes beneficiaries who were not in both Part A and Part B throughout their enrollment in 2009 or who had Medicare as a second payer. *Source: MedPac Data book: Healthcare Spending and the Medicare Program (June 2013).*

Medicaid

Medicaid offers generous fill-in benefits for persons with low incomes and assets. However, because states have latitude in establishing eligibility and coverage, there is considerable variation in the quality and quantity of services provided across the states. For example, Medicaid spending in 2009 on persons aged 65 and above ranged from $5 247 in New Mexico to $24 761 per person in Connecticut (Snyder, Rudowitz, Garfield, & Gordon, 2012). Today, two separate programs provide some benefits under the Medicaid umbrella. First, basic Medicaid coverage is limited to those with the lowest incomes, generally well below the federal poverty level. In addition to paying the Part B premium and relieving beneficiaries of the responsibility for copayments and deductibles in the Medicare program (even if Medicaid fails to compensate providers for these costs), these state-based programs all offer long-term care coverage.

Second, beginning in 1988, some low-income individuals who are not eligible for regular Medicaid receive partial protections through the Medicare Savings Programs. A range of subsidies are available but eligibility ends for anyone above 133% of the FPL (about $16 000 in 2014 for a single individual). Similar protections were contained in the prescription drug benefit legislation, with benefits that extend up to 150% of the poverty line. This is referred to as the low-income subsidy (LIS). This drug legislation removed Medicaid from providing prescription drug coverage to Medicare beneficiaries and many critics have argued that low-income beneficiaries have been treated less well than when they were covered by many state Medicaid plans.

In practice, all of these programs cover only some of the people who qualify. Participation rates remain low because of individuals' reluctance to seek help from a "welfare program," substantial barriers to enrollment, burdensome reporting requirements, and sometimes lack of awareness of potential eligibility. For example, fewer than half of all elderly persons below the poverty line participate in the basic Medicaid program.

In addition, the asset test for determining Medicaid eligibility, which is even more stringent than the income test, results in a number of individuals who would qualify on the basis of income being excluded (Moon, Freidland, & Shirey, 2002). Although income limits have been increased over time, most states have asset tests that have been in place since 1987, often restricting benefits to persons with assets below $2 000 or $4 000 and couples with assets below $3 000 or $6 000. A key policy issue here is how to balance the desire to target benefits to those most in need with the goal of encouraging older Americans to save for their future needs.

The financing issues facing Medicaid are also complicated. Many states complain that Medicaid effectively crowds out spending on other worthy activities. Moreover, states vary in the shares of their programs going to different Medicaid-eligible subgroups – and by amounts greater than would be suggested by population differences alone. Thus, states can decide when to favor the old over the young, or vice versa, within the Medicaid program. Response to hard economic times by states is generally to reduce payments to providers of care rather than cutting benefits or eligibility directly, although these changes can still certainly affect access to care. The federal government can and has supplemented Medicaid to reduce the effects of recessions. For example, this occurred in the fiscal stimulus legislation of 2009, and has been credited with softening the blow of the serious economic downturn of 2008–2009. But the slow recovery from the recession has meant that states have been very reluctant to make Medicaid more generous.

Employer-Sponsored Plans

Employer-based retiree health plans normally offer comprehensive supplemental insurance. Employers usually subsidize retiree premiums and establish benefits comparable to what their working population receives by filling in gaps left by Medicare. A large proportion of these plans, for example, cover prescription drugs. Thus, these plans both reduce out-of-pocket expenses and increase access to services, often without limiting provider choice. Beneficiaries in these plans have among the lowest out-of-pocket costs (Medicare Payment Advisory Commission, 2002), even though they are heavy users of care. They are thus among the best protected of all seniors.

But such plans are limited to workers and dependants whose former employers offers generous retiree benefits. Among current non-institutionalized Medicare beneficiaries, 31.3% have employer-sponsored retiree health plans (MedPAC, 2013). And, these benefits accrue disproportionately to high-income retirees. This privileged group does not need improvements in Medicare to assure them access to care since they are covered very well at present – a fact that is sometimes used to argue against across-the-board improvements in Medicare.

The sense of security felt by those with retiree insurance may decline in the future, however, because employers are beginning to cut back benefits in order to control costs. They are placing more controls on the use of care, raising retiree contributions in the form of premiums or cost sharing, and even changing the benefit package. A study of large firms (with 200 or more workers, the firms most likely to offer such benefits) found the percentage offering retiree health benefits declined from 66% in 1988 to 33% in 2007 and to 28% in 2012 (KFF, 2013). Another approach to reducing employer financial liabilities is to raise the requirements for qualifying for coverage, for example, by adding more years of employment as a condition of participation.

Medigap Insurance

A traditional form of private supplemental coverage, commonly referred to as Medigap, does not lower overall out-of-pocket burdens because the premium is fully paid by the beneficiary and includes substantial administrative and marketing charges, and often, profits for the insurer (US General Accounting Office, 1998). Thus, many beneficiaries have higher, not lower, financial burdens when they buy Medigap. Medigap is most useful for reducing potential catastrophic expenses for those who have high costs in a particular year. It averages out variations in spending over time. The ten standardized plans that insurance companies are allowed to offer under federal law cover a basic package of Medicare's required cost sharing. This form of supplemental insurance provides the least protection for beneficiaries, and participation has declined over time as individuals have gravitated toward Medicare Advantage to lower their out-of-pocket costs.

Medigap premiums rose dramatically over the 1990s (Medicare Payment Advisory Commission, 2000). There have been no recent major studies in this area. However, anecdotal evidence suggests rapid increases in premiums continued, and declines in the number of beneficiaries enrolling in Medigap suggest that it is becoming too expensive for many. Over time, Medigap plans have changed the way they price policies, also contributing to access problems. Medigap providers now usually sell policies that use an "attained age" structure in which policies increase in cost rapidly as people age. This puts greater burdens on beneficiaries just as their incomes are declining. For the unwary buyer at age 65, these plans appear less costly than community-rated options which do not increase by age and hence seem more expensive initially. But switching policies can be difficult at older ages.

Finally, Medigap is often criticized because of its "first dollar" coverage which means that Medigap enrollees do not face any cost-sharing requirements directly. That is, Medigap reduces the cost-sharing that is presumed to discourage overuse of care. Consequently, policy proposals have been advanced to either prevent first dollar coverage or penalize those who do purchase it.

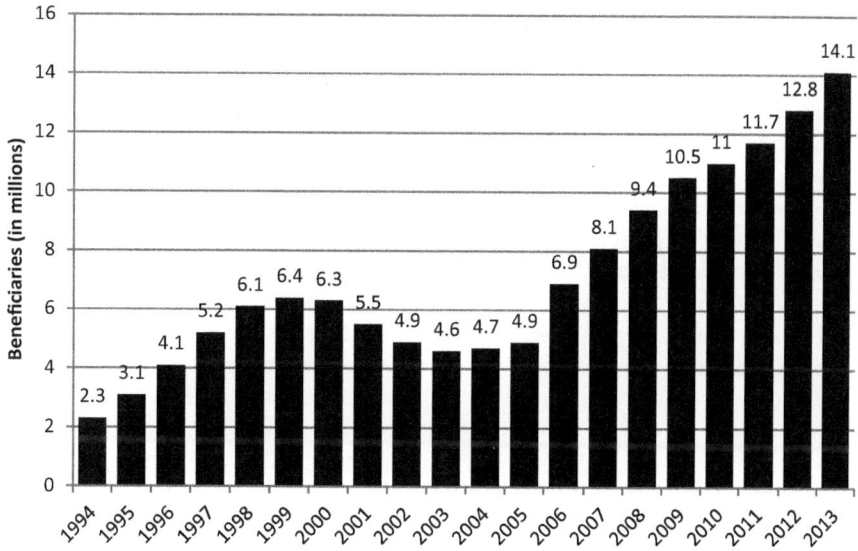

FIGURE 19.3 Enrollment in Medicare Advantage plans, 1994–2013. *Source: MedPac Data book: Healthcare Spending and the Medicare Program (June 2013).*

Medicare Advantage

Beneficiaries also can obtain additional benefits to supplement Medicare's basic package by enrolling in private health plans, termed "Medicare Advantage." In such a case, enrollees agree to get all their coverage from the private plan, rather than from a combination of traditional reimbursements for services from Medicare and private supplemental coverage. For many years in the 1980s and 1990s, private plans that participated in Medicare were mainly HMOs which restrict enrollees to a specific network of doctors and hospitals. Cost sharing was usually lower than for traditional Medicare and some additional benefits were offered for less than the price of a Medigap plan.

After legislative changes in 1997, however, these plans became more expensive for consumers through higher premiums and cost sharing, and many plans withdrew from Medicare altogether. The number of beneficiaries enrolled in these private plans reached as high as 16% in 1999 followed by a decline to about 12% in 2003 (Boards of Trustees, 2009). To counter this retrenchment in private plan participation, the 2003 MMA legislation increased payments from the federal government to plans to encourage them to stay or return to offering coverage. Once again, there was a rapid increase in private plans with Medicare Advantage plans growing ever faster than during the previous expansionary period.

This popularity was due not only to additional payments to plans that allowed them to offer enriched benefits, but also to the rapid rise of less restrictive plans that allow greater choice of doctors or other providers of care. Together these changes led to over 28% of Medicare beneficiaries being enrolled in Medicare Advantage in 2013 up from 14% in 2005 (as shown in Figure 19.3), including a large number of low- and moderate-income beneficiaries who would otherwise do without supplemental coverage.

In 2009, it was estimated that these private plans were still paid, on average, about 14% more than it would cost to provide basic Medicare benefits to the same individuals through the traditional program (MedPAC, 2009). In 2010, changes were made to slow MA payment growth so that, over time, the value of Medicare's contributions would be comparable regardless of whether the beneficiary remains in the traditional program or enrolls in MA. These changes may slow the increase in enrollment. To date, however, enrollment is still rising. The policy changes that have led to these expansions and contractions reflect an active debate over the desirability of relying on the private market as opposed to the traditional Medicare program.

No Coverage

Finally, some beneficiaries cannot afford any supplemental policy. As shown in Figure 19.2, just 7% of Medicare beneficiaries in traditional Medicare had no extra policy to cover what Medicare's benefit package does not. Lack of supplemental coverage is associated with problems of access to care. Those who have no supplemental coverage, for example, are less likely to see a doctor in any given year, less likely to have a usual source of care, and more likely to postpone getting care in a timely fashion. For example, 21% of beneficiaries who relied on traditional Medicare alone reported delaying care due to its cost, while only 5% of beneficiaries with private supplemental coverage reported delaying care (Gluck & Hanson, 2001).

THE AFFORDABILITY AND SUSTAINABILITY QUESTIONS ABOUT THE FUTURE OF MEDICARE AND MEDICAID

Although Medicare has performed well relative to total health care costs since the 1980s, the program nonetheless is often portrayed as a runaway item in the federal budget. The evidence cited for this is the higher overall growth in spending on Medicare as compared to the rest of the budget. Both Medicare and Medicaid have grown as a share of the federal budget and of the GDP. Medicare, however, has not grown faster than the costs of private health insurance (Boccuti & Moon, 2003; MedPAC, 2013). Moreover, contrary to the belief of some policymakers and journalists, it is largely a myth that the senior lobby has been largely responsible for this growth in Medicare (Binstock & Quadagno, 2001). In fact, at various points in time, it has been politically feasible to adopt regulatory and legislative changes that have held down program growth.

Nonetheless, since these public programs are funded with tax dollars in an era of anti-tax sentiment, they get more scrutiny than health expenditures paid for by individuals or by businesses. In the early 1970s, Medicare was only 3.5% of the federal budget and by 1990 accounted for 8.6%. The percent has grown steadily such that Medicare's share totaled 15.6% of the 2013 federal budget (Congressional Budget Office, 2013).

As a result of Medicare's growth, some policymakers believe that it may be crowding out expenditures on other domestic programs. Critics argue that Americans will only accept a certain level of overall public spending, so if Medicare grows rapidly, it hurts other spending even if it has its own revenue sources. In this argument, Medicare gets most of the attention because of the near-term increases in spending that the aging of the Baby Boom is beginning to cause. Already, the number of beneficiaries has grown faster than the population as a whole, reflecting longer life expectancies for the elderly. Looking out over the next decade, Medicare is expected to rise from $583 billion in 2013 to $1099 billion in 2023 if there is no change in policy (Boards of Trustees, 2014).

A second fiscal pressure on Medicare arises from the status of the Part A, Hospital Insurance, trust fund. Current law provides a fixed source of funding, the Federal Insurance Contributions Act payroll tax, for HI. These revenues typically have not grown as fast as spending, creating periodic crises when the date of trust fund exhaustion is close at hand. So far, that day of reckoning has been postponed many times by major cost-cutting efforts and an increase in the wage and salary base subject to taxation. For instance, a strong economy and the slow growth in Medicare spending in 1998 and 1999 pushed that date to 2029, based on the 2002 report of the Medicare trustees. But the date of exhaustion of the Part A trust fund drew closer after the drug benefit was added in 2006 with a predicted exhaustion date of 2019 in the 2008 report (Boards of Trustees, 2008). The exhaustion of the Part A Trust Fund is now projected to occur in 2030 (Boards of Trustees, 2014), reflecting the changes made in the ACA and slower overall spending Medicare.

Despite pressures on financing, the program has key gaps in coverage that could be addressed by policy reforms. When Medicare and its benefit package were created in 1965, medical care needs and insurance looked very different than they do today. For example, many workers had only hospital coverage, in part because health care spending as a share of income was much smaller, and services such as physician care were not inordinately expensive. Today, most good private plans for workers cover almost all aspects of acute care. Yet the basic package for Medicare remained largely unchanged until 2003, when a new drug benefit option was enacted covering about half of drug costs for those who enroll.

The two largest remaining gaps in Medicare coverage are the lack of a limit on how much individuals must pay out-of-pocket for care and the inadequacy of the low-income protections that Medicare offers – even when combined with Medicaid. The lack of "stop loss" protection is a major reason why many people need to supplement Medicare; an individual facing substantial health costs could end up paying thousands of dollars out-of-pocket. Such protections are routinely available for most people through their employer-provided insurance plans. They have not been added to Medicare because of the high projected cost of this protection.

Low-income protections are also particularly inadequate for seniors. The most generous benefit offered is for those who enroll in Part D and have incomes below 150% of the federal poverty level. Other protections are even less generous. These benefits have not improved in many years despite the fact that health care costs have risen faster than income and the burdens on low-income individuals have risen substantially over time. As a consequence, someone with an income just above 150% of the FPL would pay on average 15% of her income for health care needs as compared to 10% in 1988 when the low-income protections were expanded (Moon, 2015). Whether through Medicare or Medicaid, additional protections are particularly crucial for those with modest incomes.

MAJOR OPTIONS FOR REFORM

The financing of the Medicare program is likely to face key challenges in moving forward. Medicare continues to be viewed as a potential source of savings for deficit reduction (Van de Water, 2012). Moreover, critics of the size of the federal government consider Medicare a prime candidate for "entitlement" reform. Whereas entitlements used to be considered popular and relatively untouchable programs, they are now firmly in the sights of those who would like to reduce the size of government in the US Medicare and Medicaid are even more vulnerable than Social Security since they do not provide direct payments to beneficiaries but pay for services delivered. Consequently, it has always been easier to portray changes in the health program as improving efficiency or reducing overpayments to providers rather than as benefit cuts.

The future of Medicare remains politically controversial because this is a large and popular public program that faces projections of rapid growth in the future. As a government program, either new revenues will have to be added to support Medicare, or its growth will have to be curtailed. Some of the problem is driven by the expected increase in the number of persons eligible for Medicare from 52.3 million in 2013 to 81.8 million in 2030 as the Baby Boom generation moves fully into the age of eligibility (Boards of Trustees, 2014). Enrollment in the program will grow from one in every eight Americans to more than one in every five.

In addition, the overall costs of care have been rising substantially over time, but that issue is best tackled by dealing with the entire health care system. If the delivery system reforms being studied by the Innovation Center of CMS are effective, pressure on Medicare will certainly diminish. Already, growth rates have slowed, but many analysts are skeptical that this can continue to the degree needed to alleviate concerns about the size of the program as enrollment increases.

Making changes to Medicare that can improve its viability both in terms of its costs and in how well it serves older and disabled beneficiaries should certainly be pursued. But the challenge remains to contrast what seems like a "simple" solution of turning over Medicare to the private sector and a set of changes that would need to be done to the program through the usual legislative process.

A Private Plan Approach

Medicare could be modified so that beneficiaries choose from an array of private plans. The standard proposal would have the federal government provide oversight and subsidize the premium up to some average level – usually starting at an amount comparable to the current level of per capita spending on the program (now in the range of $11 000 annually). This approach has often been referred to as "premium support." Individuals wishing to buy more comprehensive coverage would have to pay the difference between the federal contribution and the premium charged by a particular plan.

This approach would be quite different from the current system where the traditional program is the default and MA is an option. Instead, everyone would have to choose a plan, likely with higher premiums for those who wish to remain in traditional Medicare. As envisioned by its proponents, there would be less oversight but strong limits on how much government would pay each year. Plans would be left to decide how to restrain costs or could charge beneficiaries higher premiums. The risk of rising costs would be shifted from government to beneficiaries.

This restructuring approach could profoundly affect Medicare's future. In particular, the traditional Medicare program could be priced beyond the means of many beneficiaries, leaving only private plan options from which to choose. Further, if plans begin to sort into two groups of higher-cost and lower-cost plans, the result would be to segregate beneficiaries in plans on the basis of their ability to pay. This would be quite different from today where the basic program treats all beneficiaries alike.

Some supporters of a private approach assume that private plans inherently offer advantages that traditional Medicare cannot achieve. But there is no magic bullet to holding the line on cost growth. Per capita spending rises because of higher use of services, higher prices, or a combination of the two. Medicare's price clout is well known and documented, so it is difficult for private plans to do better in that area. So what about managed care's ability to control use of services? Studies of managed care have concluded that, in the past, most such plans saved money by obtaining price discounts for services and not by changing the practice of health care (Strunk, Ginsburg, & Gabel, 2001). Controlling use of services represents a major challenge for both private insurance and Medicare. Only recently have plans begun to do a better job – and evidence suggests that this is largely in plans that are closely controlled and managed (Newhouse & McGuire, 2014).

A private approach has the *potential* to allow greater flexibility for innovation and change in coverage of benefits. This allows private insurers to respond more quickly than a large government program can to adopt new innovations and to intervene where insurers believe too much care is being delivered (Butler & Moffitt, 1995). But what looks like cost-effectiveness activities from an insurer's perspective may be seen by a beneficiary as the loss of potentially essential care.

Under the traditional Medicare program, beneficiaries do not need to fear loss of coverage when they develop health problems. On the other hand, private insurers are interested in satisfying their own customers and generating profits for stockholders. When the financial incentives they face are very broad (such as receiving fixed monthly payments for providing care), private insurers respond as good business entities should. They seek the easiest ways of holding down costs. Even though many private insurers are willing and able to care for Medicare patients, the easiest way to stay in business as an insurer is to attract healthy enrollees and avoid the sick. Under the current Medicare Advantage program, risk adjustment has improved substantially but has not totally eliminated the incentives to attract healthier patients.

Finally, private insurers will almost surely have higher administrative overhead costs than does Medicare. Private insurance, for example, tends to have administrative costs in the range of 15% (Levit et al., 2004). Insurers need to advertise and promote their plans. They face smaller risk pools than traditional Medicare, requiring them to make more conservative decisions regarding reserves and other protections against losses over time. Private plans expect to return a profit to shareholders. These factors cumulate and work against the likelihood that private companies can perform better than Medicare which has administrative costs totaling about 2% of program spending (Boards of Trustees, 2014).

Reform options stressing competition seek savings not only by relying on private plans but also on competition among those plans. Often this includes allowing premiums paid by beneficiaries to vary, with higher premiums for plans that offer additional benefits for a wide array of providers. The theory is that beneficiaries will become more price-conscious and choose the lower cost plans, rewarding those that are more efficient. But studies on retirees indicate they are less willing to change doctors and learn new insurance rules in order to save a few dollars each month, particularly for those who have health problems (McWilliams, Afendulis, McGuire, & Landon, 2011). This suggests that premium costs may rise as a share of beneficiary incomes over time.

Further, new approaches to the delivery of health care under Medicare may generate a whole new set of problems, including problems in areas where Medicare is now working well. For example, shifting among plans is not necessarily good for patients; it is not only disruptive, it can raise costs of care. Studies have shown that having one physician over a long period of time reduces the costs of care (e.g., Weiss & Blustein, 1996). And if it is only the healthier beneficiaries who choose to switch plans, the sickest and most vulnerable beneficiaries may end up being concentrated in plans that become increasingly expensive over time (Buchmueller, 2000).

Will reforms that lead to a greater reliance on the market still retain the emphasis on equal access to care? Support for a market approach that moves away from a "one-size-fits-all" approach is a prescription for risk-selection problems. If plans have flexibility in tailoring their offerings, they can, for example, raise cost sharing on benefits such as home health care which are disproportionately used by older, sicker beneficiaries. About one in every three Medicare beneficiaries has severe mental or physical health problems. What are the trade-offs from attempts to increase the role of private plans in serving Medicare beneficiaries? Gains in lower costs that come from increased competition and from the flexibility that the private sector enjoys could be offset by increased problems in access to care for the sick and the poor.

More Incremental Approaches

An alternative approach would be to emphasize improvements in *both* the existing private plan options and the traditional Medicare program, basically retaining the current structure in which traditional Medicare is the primary option. But even the traditional plan will change as innovations to improve health care delivery are adopted. This means changing some of the financial incentives in the fee-for-service system (such as encouraging patients to use primary care physicians instead of relying as much on specialists, or on bundling certain groups of services) without going all the way to a formal managed care program. Further, for traditional Medicare to be effective, benefits would need to be expanded to avoid the problem of generating a complex, jerry-rigged system of government and supplemental plans.

In addition, better norms and standards of care are needed if we are to provide quality of care protections to all Americans. Investment in outcomes and effectiveness research, disease management and other techniques that could lead to improvements in treatment of patients require a substantial financial commitment. This cannot be done as well in a proprietary, for-profit environment where innovations may not be shared. Further, innovations in treatment and coordination of care should focus on those with substantial health problems – exactly the population that many private plans avoid.

One area that could benefit from applying knowledge about the effectiveness of care would be with prescription drugs. The Medicare prescription drug benefit will require major efforts to hold down costs over time, including, for example, basing coverage on evidence of the comparative effectiveness and safety of various drugs. For example, higher copayments could be reserved for less-effective drugs or for brandname drugs when equivalent generics are available. Too often, differential copayments are currently established instead on the basis of the price of the drug or on which manufacturer offers the best discounts. Undertaking these studies and evaluations represents a public good and needs to be funded on that basis. Further, resources need to be devoted to disseminating information and educating consumers on what it means to adopt an evidence-based approach to health care delivery. These activities can more efficiently be centralized rather than left to multiple drug plans.

Within the fee-for-service environment, patients and physicians need encouragement to coordinate care. Patients need information and support as well as incentives to become involved. Many caring physicians, who have often resented the low pay in fee-for-service and the controls that managed care imposes upon them, would likely welcome the ability to spend more time with their patients. It has now become popular to talk about a "medical home" in which individuals would rely upon a primary care physician (or team) to help with coordination of care and managing health needs in general. One option would be to give beneficiaries a certificate that spells out the care consultation benefits to which they are entitled and allow them to designate a physician (or team) who will provide those services. In turn, this would trigger a management fee for the participating physician. Such care would likely reduce confusion and unnecessary duplication of services that go on in a fee-for-service environment.

The two dueling approaches to reform suggested here basically differ in terms of how much oversight and active participation in improving care is given to the federal government. If the emphasis is on letting private plans take the lead with little government involvement, the responsibility for holding down costs is shifted to plans and to beneficiaries themselves. The second approach would continue to have the federal government take a major role in seeking innovations and improvements – and also share in the risk of higher costs over time. For Medicaid, this issue is playing out in multiple states as they grapple with how to control costs. The Federal government is participating with states in allowing them to enroll those who qualify for both Medicare and Medicaid in private plans, for example.

THE ISSUES OF FINANCING MEDICARE AND MEDICAID

More resources will likely be needed to finance Medicare and Medicaid over time unless the government sets a strict limit on spending. In actuality, however, even that approach still means that someone pays more – in this case, beneficiaries in the form of higher premiums or lower benefits.

A wide range of mechanisms can be used to explicitly or implicitly require beneficiaries to contribute more to the costs of their care. For example, increased premiums or cost-sharing requirements have been applied over time to shift costs onto those who use the program. In 2005, the Part B deductible was increased. And an income-related premium has been added to both Parts B and D of Medicare.

Efforts to raise the age of eligibility for the program to 67 also implicitly mean cost shifting to beneficiaries. Because Medicare would be eliminating eligibility for its younger, least-expensive older patients on average, the approximately 5% of people disenfranchised if eligibility age goes to 67 would only save Medicare about 2% of its costs (Waidmann, 1998). In 2013, the Congressional Budget Office reiterated that savings from this option would be quite limited (CBO, 2013). And now with the ACA, some of these disenfranchised individuals would become eligible for Federal subsidies.

In terms of Medicaid, it is not feasible to ask beneficiaries – who are by definition lower income – to bear a greater share of the costs of their care. This program will also be important to retirees who need long-term care in the next several decades. Thus, a more severe financing crisis will come to Medicaid when the Baby Boom generation is in its eighties and nineties. So, although less attention is currently focused on Medicaid, it too will need additional financial support over time.

What about relying on greater contributions from taxpayers? The payroll tax rate has not risen since 1987, and technically, new revenues would not have to be raised for a number of years. However, doing so soon would likely increase taxes most on Baby Boomers who will be drawing heavily on Medicare and Medicaid in the future. If broader revenue sources are tapped, they will affect different groups of the population depending upon which sources are used. For example, payroll taxes remain relatively popular with the general public, likely because they know where the revenues raised are supposed to go. But economists criticize these taxes as raising the costs of workers to employers, discouraging employment. Using general revenues (basically the income tax and the other current major source of income for Medicare) would effectively ask higher-income persons to pay more and require that older persons as well as the young contribute. Other taxes – such as those on alcohol and tobacco – often do not bring in enough revenue to resolve the financing issues. Whatever choices are made, raising taxes is likely to be a last resort approach given the political costs often associated with such measures.

Nonetheless, someone will need to pay more to provide care for an aging population. Issues of fairness raise important considerations about how much beneficiaries can be asked to pay and how much should be required from others. The key issue will be how to share that burden, not whether it will increase over time because it surely will.

References

Andersen, R., Lion, J., & Anderson, O. W. (1976). *Two decades of health services: Social survey trends in use and expenditure.* Cambridge, MA: Ballinger Publishing Company.

Binstock, R. H., & Quadagno, J. (2001). Aging and politics. In R. H. Binstock & L. K. George (Eds.), *Handbook of aging and the social sciences* (pp. 333–351) (5th ed.). San Diego, CA: Academic Press.

Boards of Trustees. (2008). *2008 annual report of the Boards of Trustees of the Federal Hospital Insurance and Federal Supplementary Medical Insurance Trust Funds.* Washington, DC: U.S. Government Printing Office.

Boards of Trustees. (2009). *2009 annual report of the Boards of Trustees of the Federal Hospital Insurance and Federal Supplementary Medical Insurance Trust Funds.* Washington, DC: U.S. Government Printing Office.

Boards of Trustees. (2014). *2014 annual report of the Boards of Trustees of the Federal Hospital Insurance and Federal Supplementary Medical Insurance Trust Funds.* Washington, DC: U.S. Government Printing Office.

Boccuti, C., & Moon, M. (2003). Comparing Medicare and private insurance: Growth rates in spending over three decades. *Health Affairs, 22,* 230–237.

Buchmueller, T. (2000). The health plan choices of retirees under managed competition. *Health Affairs, 35,* 949–976.

Butler, S. M., & Moffitt, R. E. (1995). The FEHBP as a model for a new Medicare program. *Health Affairs, 14,* 47–61.

Congressional Budget Office. (2013). *Raising the age of eligibility for Medicare to 67: An updated estimate of the budgetary effects.* Washington, DC: Congressional Budget Office.

Foster, R. (2010). *Estimated financial effects of the "Patient Protection and Affordable Care Act," as passed by the Senate on December 24, 2009.* Office of the Actuary, Centers for Medicare and Medicaid Services, January 8, Memorandum.

Gluck, M., & Hanson, K. (2001). *Medicare chartbook.* Washington, DC: The Henry J. Kaiser Family Foundation.

Hartman, M., Martin, A., Nuccio, O., Catlin, A., & The National Health Expenditure Accounts Team. (2010). Health spending growth at a historic low in 2008. *Health Affairs, 29,* 147–155.

Kaiser Family Foundation. (2009). *Medicare savings in perspective: A comparison of 2009 health reform legislation and other laws in the last 15 years.* Focus on health reform, December.

Kaiser Family Foundation. (2010). *Summary of key changes to Medicare in 2010 health reform law.* Publication 7948-02.

Kaiser Family Foundation. (2013). *Medicare Part D prescription drug plans: The marketplace in 2013 and key trends 2006–2013.* Menlo Park, CA: Kaiser Family Foundation. December 11.

Kaplan, R. (2011). Analyzing the impact of the new health care reform legislation on older Americans. *The Elder Law Journal, 18,* 213–245.

Levit, K., Smith, C., Cowan, C., Sensenig, A., Catlin, A., & Health Accounts Team. (2004). Health spending rebound continues in 2002. *Health Affairs, 23,* 147–159.

Martin, A., Hartman, M., Whittle, L., Catlin, A., & The National Health Expenditure Accounts Team. (2013). National health spending in 2012: Rate of health spending growth remained low for the fourth consecutive year. *Health Affairs, 33,* 67–77.

McWilliams, J. M., Afendulis, C., McGuire, T., & Landon, B. (2011). Complex Medicare Advantage Choices may overwhelm seniors – Especially those with impaired decision making. *Health Affairs, 30,* 1786–1794.

McWilliams, J. M., Meara, E., Zaslavsky, A., & Ayanian, J. (2007). Use of health serviced by previously uninsured Medicare beneficiaries. *New England Journal of Medicine, 357,* 143–153.

Medicaid and CHIP Payment and Access Commission (MACPAC). (2014). *Report to the Congress on Medicaid and CHIP,* Washington, March.

Medicare Payment Advisory Commission (MedPac). (2000). *Report to Congress: Medicare payment policy.* Washington, DC: MedPac.

Medicare Payment Advisory Commission (MedPac). (2002). *Report to Congress: Assessing Medicare benefits.* Washington, DC: MedPac.

Medicare Payment Advisory Commission (MedPac). (2009). *Report to Congress: New approaches in Medicare.* Washington, DC: MedPac.

Medicare Payment Advisory Commission (MedPac). (2013). *A Data Book: Health care spending and the Medicare program.* Washington, DC: MedPac.

MetLife. (2013). *The 2013 MetLife market survey of nursing home, assisted living, adult day services and home care costs.* New York, NY: Mature Market Institute.

Moon, M. (1996). *Medicare now and in the future* (2nd ed.). Washington, DC: Urban Institute Press.

Moon, M. (2006). *Medicare: A policy primer* (2nd ed.). Washington, DC: Urban Institute Press.

Moon, M. (2015). *Addressing gaps in health coverage for low-income seniors*. Washington, DC: AIR Center on Aging Issue Brief. February 2015.

Moon, M., Freidland, R., & Shirey, L. (2002). *Medicare beneficiaries and their assets: Implications for low-income programs*. Menlo Park, CA: Henry J. Kaiser Family Foundation.

Newhouse, J. (2010). Assessing health reform's impact on four key groups of Americans. *Health Affairs, 29*, 1714–1724.

Newhouse, J., & McGuire, T. (2014). How successful is Medicare Advantage? *Milbank Memorial Fund Quarterly, 92*, 351–394.

Noel-Miller, C. (2013). *Medicare beneficiaries' out-of-pocket spending for health care*. Washington, DC: AARP Public Policy Institute.

Physician Payment Review Commission. (1987). *Medicare physician payment: An agenda for reform*. Washington, DC: U.S. Government Printing Office.

Shrank, W. (2013). The Center for Medicare and Medicaid Innovation's blueprint for rapid-cycle evaluation of new care and payment models. *Health Affairs, 32*, 807–812.

Snyder, L., Rudowitz, R., Garfield, R., & Gordon, T. (2012). *Why does Medicaid spending vary across states: A chart book of factors driving state spending. Kaiser Commission on Medicaid and the Uninsured*. Menlo Park, CA: The Henry J. Kaiser Family Foundation.

Strunk, B. C., Ginsburg, P. B., & Gabel, J. R. (2001). Tracking health care costs. *Health Affairs*, W39–W50.

US General Accounting Office. (1998). *Medigap insurance: Compliance with federal standards has increased, GAL/HEHS-98-66*. Washington, DC: U.S. General Accounting Office.

Van de Water, P. (2012). *Medicare in the Ryan budget*. Washington, DC: Center on Budget and Policy Priorities.

Waidmann, T. (1998). Potential effects of raising Medicare's eligibility age. *Health Affairs, 17*, 156–164.

Weiss, L., & Blustein, J. (1996). Faithful patients: The effect of long-term physician-patient relationships on the costs and use of health care by older Americans. *American Journal of Public Health, 86*, 1742–1747.

Yamamoto, D., Neuman, T., & Kitchman Strollo, M. (2008). *How does the benefit value of Medicare compare to the benefit value of typical large employer plans?* Menlo Park, CA: Kaiser Family Foundation.

Zhang, Y., Donohue, J., Newhouse, J., & Lave, J. (2009). The effects of the coverage gap on drug spending: A closer look at Medicare Part D. *Health Affairs, 28*, w317–w325.

CHAPTER 20

Innovations in Long-Term Care

Joseph E. Gaugler

Center on Aging, School of Nursing, University of Minnesota-Twin Cities, Minneapolis, MN, USA

OUTLINE

Introduction	419	Culture Change	430
Defining Long-Term Care	421	Pay-for-Performance	431
		Informal Care	432
A Brief Historical Overview of		Transitional Care/Care Coordination	433
Long-Term Care in the United States	422	Health Information Technology	434
Selected Innovations in Long-Term Care	424	Looking Toward the Future of	
Integrating Acute Care with Long-Term		Long-Term Care	435
Care Services	424	References	436
Rebalancing Long-Term Care Efforts	428		
Consumer-Directed Care Options	429		

INTRODUCTION

Chronic illness poses a, if not *the*, principal challenge to the United States' health care system. Chronic illnesses (also called "chronic conditions") are long-lasting and often lifelong, are typically not curable, place a potentially heavy burden on the life of the person with the illness and family members, and are characterized by a progressive onset with multiple and oftentimes misunderstood causes (Kane, Priester, & Totten, 2005). Chronic illnesses are most prevalent among older adults over the age of 65; approximately 45.4% of US men and 47.4% of US women over the age of 65 had 2–3 chronic illnesses in 2010. A little over 17% of older US men and 14.5% of women had four or more chronic illnesses. In contrast,

28.1% of US women and men 45–64 years of age had 2–3 chronic conditions and 4.7% of men and 6.7% of women in this age group had four or more chronic conditions (Ward & Schiller, 2013).

Prevalent chronic conditions in the United States (which are also among the primary causes of death among older adults) include heart disease, diabetes, stroke, cancer, and Alzheimer's disease. Parallel to the prevalence of chronic diseases in the older population are the extensive health care costs to manage these conditions. Data from the Agency of Healthcare Research and Quality and the Administration on Aging indicated that noninstitutionalized older adults comprised 13% of the US population in 2007 but accounted for 24% of all personal health care expenditures; per capita health care health spending was five times higher for older adults than for those under the age of 65 (Stone & Benson, 2012).

Since the middle of the twentieth century, the infrastructure and financing mechanisms of the US health care system have oriented themselves largely to address acute health conditions (Lubkin & Larsen, 2013). However, such care approaches are not ideally organized to provide effective chronic disease care to older persons. The goal of acute care is to cure the underlying condition, while ideal chronic disease care aims to appropriately manage the underlying condition or conditions. Success in acute care treatment is oriented around immediate improvement, but a longer timeframe is necessary to achieve treatment goals in chronic conditions. The roles of persons with chronic illnesses, their family members, and chronic care providers are different than those in acute care situations; while the latter focuses on patients or families seeking a diagnosis and the care provider administering treatment, older adults with chronic illness live with their conditions around-the-clock and often share the responsibility for care management with the care provider as well as family members (Gaugler, Potter, & Pruinelli, 2014; Kane et al., 2005; Lubkin & Larsen, 2013).

In order to meet the needs of older adults with chronic conditions, a complex array of both formal (paid) and informal community-based and residential sources of care have emerged. This array of services is commonly understood as "long-term care," but what long-term care means and how it is defined varies. For most of the public, the first image that comes to mind when "long-term care" is mentioned is the nursing home (NH) (Reinhard & Young, 2009; White et al., 2013). However, long-term care has developed considerably since the middle of the twentieth century to encompass a mosaic of formal and informal services. To illustrate this point, consider who pays for health care in the United States for older persons. Medicare is federal and is most prominent in paying for hospital/acute, sub-acute, and rehabilitative services for older persons, whereas Medicaid (a federal/state health insurance program) largely pays for long-term care expenditures (Stone & Benson, 2012). However, long-term care is funded by a patchwork of sources beyond Medicare and Medicaid. For example, the Centers for Medicare and Medicaid Services (CMS) in 2004 reported that 49% of all health care spending was covered by Medicare, 16% by private health insurance, 15% by out-of-pocket, 14% by Medicaid, and 6% by other private and public sources. Of the types of services paid for, 57% was for hospital or physician/clinician, while 17% was for NH care and 4% was for home health care. The Lewin Group reported that among those services considered "long-term care," Medicaid and Medicare accounted for 71% of spending in 2010 (Stone & Benson, 2012).

The initial section of this chapter will thus explore various definitions to better ascertain what long-term care is. I will then briefly describe the historical context of long-term care in the United States to determine how we arrived at the current state of long-term care science, practice, and policy. Following this contextual overview, I will summarize what I and others consider the "innovations" of long-term care that have emerged over the past several decades: this core of the chapter will examine innovative models and clinical practices that have aimed to transform and optimize long-term care. There is some argument whether such advances have substantially influenced the quality of long-term care in the United States; personal experiences from renowned researchers and experienced health care providers suggest that these advances have yet to reach down to the day-to-day long-term care experience (Beckett, 2009; Kane & West, 2005). Indeed, many of the innovations described in this chapter have not been widely or consistently implemented throughout the United States, and remain unavailable to many older adults. An important point to remember when reviewing select innovations in long-term care is whether there exist pathways to broad implementation so that the majority of older persons in the United States will eventually have access to these promising programs.

DEFINING LONG-TERM CARE

Several definitions of long-term care exist. Kane and colleagues define long-term care as including: (1) informal, or unpaid, support and care from family and/or friends; (2) formal, or paid care including personal care, meals, homemaking, and custodial/home care services; (3) supportive services in residential care settings such as NHs or assisted living; and (4) assistive technologies/devices and environmental adaptations (Kane et al., 2005). Long-term care principally serves to help chronically impaired older persons accomplish and perform core tasks that allow for everyday function: activities of daily living (ADLs, such as bathing, dressing, grooming, feeding, toileting, and getting in and out of bed or around the home) (Katz, Ford, Moskowitz, Jackson, & Jaffe, 1963) and instrumental activities of daily living (IADLs, or shopping, housework, managing financial affairs, using the telephone) (Lawton & Brody, 1969).

In Kane and colleagues' definition, long-term care is defined separate from *chronic care*, or the health, medical, and supportive services to treat and manage chronic conditions. However, Kane et al. opine that the two overlap significantly and that optimal chronic care cannot take place without appropriately delivered and designed long-term care. Indeed, others have emphasized that chronic care *is* long-term care, but because many persons with chronic conditions can live independently and may not rely on support services or family care, chronic care is sometimes viewed differently from long-term care (the latter of which is usually defined according to services used) (Levine, Halper, Peist, & Gould, 2010).

In the Institute of Medicine (IOM)'s landmark 2001 report *Improving the Quality of Long-Term Care*, long-term care is defined similarly to Kane and colleagues' definitions: "Long-term care covers a diverse array of services provided over a sustained period of time to people of all ages with chronic conditions and functional limitations" (Wunderlich & Kohler, 2001, p. 36). The IOM report emphasizes not only diversity of age in its definition of long-term care but also diversity of need; long-term care can be provided to those with minimal ADL or IADL needs or with total dependency. Moreover, as with the expanded Kane definition, long-term care can be delivered not only in formal residential settings such as NHs but also individuals' homes. The IOM report also emphasizes the temporal dimension, or trajectory, of long-term care use: because of diversity in health needs, preferences, values, socioeconomic status, and geographic location among persons with chronic illness, there is variation in how these services are accessed and used over time. Thus, while long-term care is sometimes conceptualized as existing on a "continuum" (gradually increasing service intensity as one moves across specific settings), in actuality how and which long-term care services are used may be anchored less in place and more on a complex mix of choices, values, accessibility, and other key factors that suggest a sometimes nonlinear long-term care experience (Kane, 2001). For example, an intriguing recent study of transitions into and out of NH for persons with dementia from 2001 to 2008 found that among "long-stay" residents (or those who were admitted for stays in NHs beyond a 90-day interval) only 10.7% did not experience any transition from the NH: 35% experienced a transition back to home, 7.7% to home health care, 44% to a hospital, and 6.3% to death (Callahan et al., 2012). This suggests a decidedly nonlinear, more dynamic trajectory of long-term care use than was previously understood in addition to the potential burdens such transitions may pose to older persons with chronic illnesses and their family members.

Additional definitions of long-term care favor the array of services offered as in the earlier Kane definition, with an eye toward describing how long-term care services have pivoted away from the NH and "rebalanced" toward greater support for community-based options to allow older persons to remain at home for as long as possible (Reinhard & Young, 2009; Stone & Benson, 2012). Others have emphasized that long-term care involves both living environment and treatment, and for this reason it is more complex to judge quality in long-term care than acute care (the latter of which often focuses primarily on the effects of treatment; Binstock & Spector, 1997).

In general, there is agreement on the definition of long-term care: it is an array of informal and formal community-based and residential services offered to those with chronic conditions and/or functional limitations over time. It is important to acknowledge that long-term care use is potentially nonlinear and dynamic, and is subject to interactions between both the services received and the environment (both physical and social) in which such services are delivered.

A BRIEF HISTORICAL OVERVIEW OF LONG-TERM CARE IN THE UNITED STATES

In order to fully appreciate current innovations in long-term care, it is important to explore the history of long-term care in the United States to at least partially explain "how we got here." From the founding of the United States to the early nineteenth century the United States was largely rural and comparatively young when compared to more recent eras. Care for older persons was secured by: (a) the rearing of multiple children who could serve as potential caregivers for parents with chronic or late-life disabilities; or (b) wealth (Stevenson, 2012). The provision of support to older persons was based on communal and reciprocal exchange alongside an attitude of acceptance toward disability. The emphasis on reciprocal exchange meant that older and/or disabled care recipients were to contribute to their caregiving communities when and where possible (Holstein & Cole, 1994, 1996).

In the early 1800s, the era of industrialization rapidly changed American life and more specifically, care for older persons. The eighteenth century saw the advent of "almshouses" to the extent that by the 1820s they became the most prevalent form of long-term care for older adults in need in the United States. The reciprocal nature of care and support present in the preindustrial era was replaced by the notion that the poor or disabled had little to offer society, and thus institutional settings were viewed as places where such individuals could be rehabilitated (Holstein & Cole, 1994, 1996). Moreover, the notion that disability and illness were a natural outcome of old age began to shift toward stigmatizing impairments in old age due to moral failure (particularly for the poor).

The conditions of almshouses were often horrific by today's standards. In the initial half of the nineteenth century, almshouses would commonly house those with various needs together, and this co-housing led to poor outcomes (particularly for frail older persons). Investigations into the conditions of almshouses in the early nineteenth century by Dorothea Dix and others led to several reforms, including establishing precursors to what are now Departments of Welfare and also the segregation of almshouse residents into specialized facilities based on need and age. Because of these initiatives, however, almshouses became *de facto* housing for older adults in the middle of the nineteenth century. Thus, the reform movement that strove to improve quality and standards of residential care for children, the mentally ill, and other groups did not extend to older persons. The additional combination of national financial turmoil and an improving medical system that became less "custodial" also led to almshouses as a primary source of housing for older persons in need (Holstein & Cole, 1994, 1996).

By the end of the nineteenth century, almshouses and their conditions continued to deteriorate as the ongoing tension between providing care for those in need clashed with the desire to "cure" older persons of their illnesses. Since almshouses already included a preponderance of older adults, these institutions became the primary residential locations to provide care and assistance (Cole, 1987; Holstein & Cole, 1994, 1996). Almshouses thus began to transform into more medical environments (including requisite care staff) to help manage the needs of older residents; for these reasons, some have perceived a direct link between the almshouse era and the emergence of NHs in the twentieth century.

By 1935, the Social Security Act was enacted to help reduce the extent of poverty among older persons (such as those in almshouses), and institutional care settings now focused on those with health needs and a lack of available family support. Since Social Security prohibited payments to older persons in public homes until 1950, almshouses became less prevalent in the first half of the twentieth century in favor of proprietary NHs or other residential facilities (Holstein & Cole, 1994; Stevenson, 2012). During the 1940s, partially in response to the 1946 passage of the Hill Burton Act to modernize hospitals, many former hospital spaces were converted to NHs and by the 1950s the US government became the principal payer of NH care. With financing available through Federal Housing Administration loans, the number of for-profit NHs increased substantially during the decade. Federal financing of long-term care in the United States was established by the 1965 passage of the Older Americans Act, Medicare, and Medicaid. Given the quickly escalating costs of Medicare, payments to NHs were slashed, thus creating an incentive for older persons and families to "spend down" personal assets in order to become eligible for Medicaid funding and NH care. An explosion in costs and utilization of NHs had occurred by the middle of the twentieth century (Stevenson, 2012).

Quality of institutional care for older persons has been a concern since the nineteenth century, but a series of scandals in the 1970s related to egregious quality of care and quality of life violations resulted in a number of initiatives to establish regulations to ensure resident safety (Butler, 1975; Lacey, 1999; Moss & Halamandaris, 1977; Vladeck, 1980). The major initiative was the Omnibus Budget Reconciliation Act of 1987 (OBRA), the most sweeping overhaul of federal regulations of NHs in history. Other emerging policies throughout the 1970s and 1980s resulted in more flexible use of Medicaid (via waiver programs to states) to fund community-based long-term care services in favor of institutional care, thus attempting to "rebalance" long-term care (see below). By 1997, Medicare payments to sub-acute units of NHs were further cut via the Balanced Budget Act of 1997, again altering the long-term care landscape.

Throughout the last two decades of the twentieth century, the rapid growth of the assisted living industry (which relies largely on private/out-of-pocket payments), statewide moratoria on new NH bed construction, the culture change movement in NHs, the rebalancing of long-term care services away from NHs toward community-based long-term care services, and the establishment of formal support programs for family caregivers of older persons through Area Agencies on Aging and advocacy foundations (such as the Alzheimer's Association) contribute to the patchwork of long-term care services available to older adults and caregiving families today. These developments have also led to a number of innovations to optimize long-term care for older adults in the United States. The next section highlights several of the most prominent innovations that have emerged in the United States over the past several decades.

SELECTED INNOVATIONS IN LONG-TERM CARE

Table 20.1 summarizes selected innovations in long-term care, including location and date of origination as well as location of current operation (as could be determined). It is important to note that I decided to focus more on innovative long-term care service models, as opposed to other domains such as advances in environmental design.

Integrating Acute Care with Long-Term Care Services

One issue that is apparent in definitions of long-term care is the blurred boundary between long-term and acute care services. For example, more acute care and rehabilitation for older adults has been provided in settings that are outside of traditional hospitals, such as NHs. When such assistance is provided in a setting traditionally oriented toward long-term care provision, the demarcation point between acute and long-term care becomes difficult to identify (Stone, 2000). Alternatively, settings such as hospitals have provided long-term care for some patients. Due to the fragmented funding of services that older adults may require (Medicaid for residential and some community-based long-term care services, Medicare for acute/hospital-based care and rehabilitation), additional blurring of the boundary between long-term care and acute care may also occur. Several innovations have thus attempted to more seamlessly integrate long-term care with acute or "primary" care services in order to reduce the negative outcomes that can occur when older persons transition, often in abrupt fashion, from acute to long-term care (e.g., from hospital to home or from rehabilitative care to a residential long-term care setting such as a NH; Levine et al., 2010). In addition to a number of integrated models that were developed and tested throughout the 1990s, the 2010 Affordable Care Act has provided further funding for the development and testing of approaches to better integrate acute and long-term care services (Stone & Benson, 2012).

Models vary in terms of how "integration" occurs; in some instances, these programs only integrate delivery of services, whereas others integrate both service delivery as well as funding. As noted by Stone (2000, p. 25), the following elements are desirable when integrating acute and long-term care services for older adults in need:

- Broad and flexible benefits;
- Delivery that extends beyond primary care providers or hospitals/post-acute care to encompass community-based long-term care services, care or case management, and other providers;

TABLE 20.1 Examples of Long-Term Care Innovations

Innovation	Origin	Date of origin	Location of current operation
Integrating Acute and Long-Term Care			
Program of All-Inclusive Care for the Elderly	On Lok Health Services, San Francisco, CA	1973	31 states in the United States
Social Health Maintenance Organizations (Thompson, 2002; see also http://endoflifecare.tripod.com/imbeddedlinks/id1.html)	Brooklyn, New York; Portland, Oregon; Long Beach, California; Minneapolis, Minnesota	1985	Portland, Oregon; Long Beach, California; Brooklyn, New York; and Las Vegas, Nevada
Evercare (Polich, Bayard, Jacobson, & Parker, 1990)[a]	Minnesota	1987	More than 350 000 people nationwide as of 2008[b]
Veterans Affairs Home-Based Primary Care Program (Beales & Edes, 2009; Egan, 2010)	Six VA demonstration sites	1972	116 sites serving over 12 000 veterans annually as of 2007
Rebalancing Long-Term Care[c]	Congress added section 1915(c) to the Social Security Act	1983	47 States and the District of Columbia operate at least one 1915(c) waiver
Consumer-Directed Care Options (Nadash & Crisp, 2005)[d,e]	Movement for consumer direction began in the 1960s; states began to offer "consumer-directed" personal care services pursuant to section 1905(a)(24) of the Social Security Act	1990s	Most states offer some form of consumer-directed options, although delivery mechanisms vary widely
Culture Change	National Citizens' Coalition for Nursing Home Reform Consumer Statement of Principles for the Nursing Home Regulatory System	1983	Multiple sites and states
Pay-for-Performance (Minnesota Performance-Based Incentive Program, or PIPP)[f]	Minnesota	2006	As of 2013, 223 nursing facilities that include 119 unique quality improvement projects have participated in PIPP
Informal Care Innovations	Multiple	Multiple	Multiple
Nursing Home Transition Programs (Reinhard, 2010)	Federal grant to 12 states	1998	As of 2009, 3 400 people had moved out of nursing homes
Health Information Technology	Unknown; technology in long-term has had slow adoption compared to other health care settings	Multiple	Multiple

[a]see http://www.unitedhealthgroup.com/About/History.aspx.
[b]As of 2008; as of 2014 EverCare is now operated by Optum Palliative and Hospice Care and numbers are not readily available.
[c]see http://www.medicaid.gov/Medicaid-CHIP-Program-Information/By-Topics/Long-Term-Services-and-Supports/Home-and-Community-Based-Services/Home-and-Community-Based-Services.html.
[d]see http://www.medicaid.gov/Medicaid-CHIP-Program-Information/By-Topics/Delivery-Systems/Self-Directed-Services.html.
[e]see http://www.medicaid.gov/Medicaid-CHIP-Program-Information/By-Topics/Delivery-Systems/Self-Directed-Services.html.
[f]see http://www.dhs.state.mn.us/main/groups/aging/documents/defaultcolumns/dhs16_180213.pdf.

- Approaches that actually integrate care; examples include care management and care planning, interdisciplinary care teams, information sharing and medical record access for all members of the team and even family caregivers where appropriate;
- Flexible, integrated funding to enhance accountability and minimize the tendency to cost shift across acute- or long-term care settings/providers.

The barriers to integrating acute and long-term care services make the design and scientific evaluation of such models difficult and fairly rare. Barriers include fragmented funding sources (Grabowski, 2007), fear of financial risk on the part of providers (and the tendency to cut costs and services), and the dearth of training and knowledge necessary to seamlessly blend acute- and long-term care services (Stone, 2000). Even with these barriers, several demonstrations have attempted to evaluate implementation processes and key outcomes for chronically disabled older adults.

The Program of All-Inclusive Care for the Elderly (PACE) is aimed at older persons who are 55 years of age and over, are certified by their state as requiring NH care, are living at home safely upon enrollment, and live in an area served by a PACE program (Stone & Benson, 2012). The philosophy of PACE is rooted in the principle that older persons with chronic care needs and their families are best served, and health care services are best delivered, in the community (Mui, 2002). Based on the successful On Lok program that served Chinese American older adults in San Francisco, PACE provides: (1) integrated funding and provider financial risks through capitated Medicare and Medicaid reimbursement, (2) a full range of acute and long-term care service delivery through adult day programs, and (3) case management via multidisciplinary assessment teams. As of 2014 there were 104 PACE sites in 31 states; these programs serve anywhere from 14 to 3 813 clients (see http://www.npaonline.org). To date, no randomized controlled evaluation of PACE exists. Eng, Pedulla, Eleazer, McCann, and Fox (1997) compared mortality rates with national samples of NH residents and found that PACE clients had a lower mortality rate per year (138 deaths/1 000 enrollees in PACE versus 186 deaths/1 000 enrollees among NH residents). A number of other investigations have conducted advanced statistical analyses to find that PACE may delay NH admission and reduce costs, but these results are often dependent on various analytical assumptions and again do not include randomly assigned control groups (e.g., Meret-Hanke, 2011; Wieland, Boland, Baskins, & Kinosian, 2010). Limitations of the PACE model are the time and financial commitment necessary to develop adequate sites, the limited attraction of such a model for middle-income older adults who are not eligible for Medicaid and must pay part of the PACE capitation rate privately, and the difficulty in recruiting primary care physicians who are trained geriatricians (Eng et al., 1997). Although PACE is popular and seen as a model community-based program that integrates long-term and acute care services, early evaluations expressed additional concerns about PACE: that it is a "boutique" model that often features slow enrollment rates (suggesting less enthusiasm among clients or families than providers) and that the selection of clients either represents marketing to particular niches, or more skeptically, the skimming of select clients and not those representative of NH residents (Branch, Coulam, & Zimmerman, 1995; Stone, 2000).

Another model that has attempted to integrate acute and long-term care services is the Social Health Maintenance Organization, or SHMO. Beginning in 1985 and implemented throughout the past three decades, SHMOs were seen as a potential breakthrough to integrating services in a seamless fashion. Specifically, SHMOs add community and short-term nursing/rehabilitative care to existing Medicare-HMO plans. Early, nonexperimental evaluations of the first generation of SHMOs were mixed in terms of their benefits, cost savings, and other outcomes (Manton, Newcomer, Vertrees, Lowrimore, & Harrington, 1994; Newcomer, Manton, Harrington, Yordi, & Vertrees, 1995). The second generation of SHMOs established reimbursement rates based on patient's individual illness profiles initially and every year thereafter as well as "geriatric health programs" for all SHMO enrollees, not just those who were long-term care/NH-eligible (Stone, 2000). Evaluation of the SHMO II models indicated no significant benefits to SHMO II enrollees when compared to usual Medicare risk plan care beneficiaries, although follow-up analyses suggested that it took several years for SHMO II models to demonstrate reductions in hospital days and increased use of community-based long-term care services (Newcomer, Harrington, & Kane, 2002; Thompson, 2002).

Additional models have sought to integrate acute and long-term care more effectively in NHs. The Evercare model is designed to offer managed care to NH residents through Medicaid and Medicare waivers. Originally tested in nine states in the early 2000s, Evercare enrolls residents into HMOs with Medicaid or private insurance covering NH costs (Stone, 2000). Geriatric care specialists (geriatricians, nurse practitioners) offer more intensive primary care to residents than what is usually offered in such settings; an emphasis on coordinating these activities with nurses and direct care workers of the facility is another key component of Evercare. As noted above, a core concern to be addressed when integrating acute and long-term care services is to ensure that costs are not shifted from one payer to another (e.g., NHs using hospitals to provide certain services, which then potentially shifts costs from Medicaid to Medicare); Evercare financially covers all care provided to NH residents.

By offering more comprehensive geriatric care to residents, Evercare was designed to reduce hospitalizations or other costly health transitions (Shield, 1996; Stone, 2000). Early evaluations appeared to suggest that Evercare achieved these goals. In addition, NHs were reimbursed for costs associated with caring for residents who would otherwise be hospitalized, and NHs could advertise the improved care their residents were receiving. For example, Evercare enrollees in five states were compared to two groups of NH resident controls in a quasi-experimental post-test design; over a 2-year period incident hospitalizations were twice as high for controls as they were for Evercare residents, which corresponded with the more intensive service days that Evercare provided to residents (Kane, Keckhafer, Flood, Bershadsky, & Siadaty, 2003). However, similar results were not apparent in an Evercare implementation and evaluation in the United Kingdom (Gravelle et al., 2007).

The Veteran's Administration (VA) Home-Based Primary Care program (HBPC) is designed to provide care via a multidisciplinary team to older adults in the home following discharge from a hospital or NH (Beales & Edes, 2009). The HBPC is essentially home-based care management; older persons at-risk are provided care and assistance by a care manager designated from the care team that is available round-the-clock. Another key feature of the program is that older persons require approval from the team for hospitalization and it involves the team in hospital planning prior to admission if hospital care is needed. Care management services provided in addition to coordination include medication management, wound care, tele-health care and monitoring, and blood draws (Stone & Benson, 2012). In 2002, the VA compared service use in the 6 months prior to enrollment in the HBPC to the 6 months following enrollment and found considerable reductions in hospital bed days (62%), NH bed days (88%), and a subsequent 264%

increase in home care visits. The mean total VA cost of care dropped from $38 000 to $29 000. Furthermore, enrollment in HBPC in 2007 was linked to a 59% reduction in hospital bed days and a 89% reduction in NH bed days (Beales & Edes, 2009). Recent efforts are also aimed at embedding mental health providers in HBPC (Karlin & Karel, 2014).

As noted above, the Affordable Care Act provided funds to support a number of other demonstration projects to better integrate acute and long-term care services for older adults. Some of these projects, as described by Stone and Benson (2012, pp. 66–67), include: (a) bundling payments to hospitals, post-acute care providers, and physician groups to reduce hospitalization; (b) provision of additional Medicare payments to accountable care organizations (groups of hospitals, physician group, and community-based care providers) to reduce hospitalizations; (c) the Independence at Home Demonstration that provides a payment based on a spending target to interdisciplinary care teams to integrate primary and long-term care to older adults at home who cannot travel to primary care provider locations; and (d) the Center for Medicaid and Medicare Services' Office of Innovation to develop and evaluate new delivery models that effectively integrate acute- and long-term care services.

Long-term care has often struggled to define care goals and philosophies according to "medical" (identifying and achieving clinical outcomes) and "social" (creating supportive approaches to meet an older person's needs) models. Calls for long-term care improvement suggest that this dichotomy is inappropriate; interdisciplinary collaboration among various care providers is necessary to arrive at a set of shared goals (Kane et al., 2005). While combining funding streams across acute and long-term care providers is important, it will not achieve full integration. Changes in care practice to ensure appropriate coordination are also needed; for example, one limitation is that long-term care is often defined by where care is provided, as opposed to the actual services used. Thus, home health care or adult day services are sometimes seen as distinct from NH care, although the types of services utilized by the older person or their family caregivers across these settings are similar in nature and intensity. With the advent of technologies that can potentially divorce setting from types of care provided, it may be possible to continue to refine and develop new models of integrated acute and long-term care services that also incorporate social and medical components (Kane et al., 2005).

Rebalancing Long-Term Care Efforts

As the payment data cited earlier suggest, long-term care expenditures in the United States are "unbalanced" and heavily weighted toward NH care. Since the 1980s, both federal and state efforts have attempted to shift this bias in favor of serving more persons who live at home via home- and community-based services (HCBS) with the flexible use of Medicaid funds (waivers; Harrington & Kitchener, 2003; Kassner, 2011). In 1999, the *Olmstead Act*, which mandated that older adults and adults with chronic impairments be served in the "least restrictive" settings available, further spurred efforts to rebalance long-term care spending and service delivery efforts (Harrington & Kitchener, 2003; Stone & Benson, 2012).

Four efforts supported by the New Freedom Initiative to states in 2001 were core to more recent rebalancing efforts, including (Stone & Benson, 2012):

(1) Real Choice Systems Change Grants to aid states' expansion of HCBS;
(2) Nursing Facility Transition Grants that supported eligible individuals' moves from NHs to community or other residential care options (see below);
(3) Money Follows the Person Grants, to further expedite transition programs (see below); and

(4) Aging Disability and Resource Centers, which serve as single points of entry for older persons and their families to receive up-to-date information and long-term care services.

Evaluation efforts have suggested that expenditures related to these programs benefit younger developmentally disabled adults more so than older adults. The Affordable Care Act has also offered flexibility to states via the Community First Choice Option to relax income eligibility criteria for Medicaid in order to deliver HCBS more flexibly and also to provide spousal protections (Stone & Benson, 2012).

A recent national survey of long-term care specialists and providers suggested widespread support for rebalancing of long-term care services toward HCBS (Grabowski et al., 2010). An interesting question to consider is that if HCBS increases long-term care expenditures overall (see Grabowski, 2006; Grabowski et al., 2010) then why should rebalancing receive ongoing support? One fairly straightforward reason is that because of the fear many older adults in the United States have toward entering a NH, HCBS should receive additional support based on older adults' and family caregivers' preference for such services. Grabowski and colleagues also call for less of an emphasis on cost savings of HCBS relative to residential long-term care placement and more on value (e.g., cost-effective delivery of these services). One policy-related suggestion is that if HCBS do demonstrate cost savings by replacing NH expenditures, the budget streams for these programs (which are often separate) should be integrated so that savings from reduced NH admissions can be better directed toward increasing HCBS as well as to enhance overall coordination between these two types of programs. More effective engagement of rebalancing efforts with primary care providers (similar to the acute and long-term care integration innovations summarized above) could also enhance and advance the effort to rebalance long-term care in favor of HCBS options (Grabowski et al., 2010).

Consumer-Directed Care Options

Another innovation in long-term care service delivery includes consumer-directed options, which involve disabled older adults paying for family care provision via the use of publicly financed vouchers or similar mechanisms (Doty, Kasper, & Litvak, 1996; Tilly & Wiener, 2001). Borne of advocacy efforts and frustration on the part of intellectually and developmentally disabled adults who demanded expanded community-based options instead of institutional ones, initiatives to increase "consumer" (as opposed to patient) involvement, choice, and control based on preference were developed. These consumer-directed options are designed to enhance choice, preference, and autonomy of older persons with long-term care needs while also allowing older persons to increase their responsibility to manage their own risk (Doty, Mahoney, & Sciegaj, 2010; Stone & Benson, 2012). While questions persist whether older persons with long-term care needs are necessarily able or competent to make what can be fairly complex long-term care decisions, multiple stakeholder organizations and providers have acknowledged that consumer-directed approaches are feasible for at least some older persons with such needs.

One of the most prominent examples of consumer-directed long-term care options is the provision of cash benefits directly to older persons. The Cash and Counseling Demonstration and Evaluation (CCDE) program, in which older persons are provided a budget to purchase community-based services or pay family members to provide care, has been subject to evaluation. While satisfaction among older persons enrolled in consumer-directed programs such as CCDE appears high, CCDE's influence in reducing long-term care costs and delaying NH admission were inconsistent (Dale & Brown, 2006, 2007). Nonetheless, the flexibility featured in consumer-directed options such as the CCDE has led to the development and implementation of a number of these innovations in various states in the United States; studies have shown both feasibility and high satisfaction among older clients who utilize such services (Gaugler, Boldischar, Vujovich, & Yahnke, 2011).

Culture Change

Beginning in the 1980s, the culture change movement attempted to create a fundamental shift about how society views NHs (Rahman & Schnelle, 2008; Weiner & Ronch, 2003). The culture change movement spans consumer advocacy, scientific and demonstration projects, and legal, legislative, and policy efforts. Starting with the work of the National Citizens' Coalition for Nursing Home Reform and its efforts in 1983 to produce the widely endorsed Consumer Statement of Principles for the Nursing Home Regulatory System, key foundations and organizations that support aging-relevant research, policy, and advocacy sought to better define quality in NHs (Koren, 2010). These efforts largely influenced the IOM's committee on how quality of care can be achieved in NHs as well as the sweeping reforms ushered in by OBRA 1987 (see above). Following OBRA 1987, several NH providers in five states began a grassroots movement to transform the prevailing NH model toward smaller "households," resident input, and more engagement about the overall nature of NHs and what they could achieve. In 1997, these vanguard providers along with others interested in culture change in NHs founded the Pioneer Network. The Pioneer Network now collaborates with The Center for Medicare and Medicaid Services to overcome regulatory barriers and to advocate for innovation in NHs (Koren, 2010). As opposed to providing a set of discrete practices and activities, the Pioneer Network has developed a set of principles to guide culture change in NHs. These principles coalesced into a vision of what the "ideal" NH would look like: it would feature resident direction (i.e., activities should be directed by the resident as much as possible); a homelike atmosphere (smaller "households" and elimination of institutional features); close relationships between residents, family members, staff, and the community (e.g., the same aide would care for the same resident); staff empowerment to encourage teamwork; collaborative decision-making to decentralize authority; and continuous quality improvement processes (Koren, 2010; Weiner & Ronch, 2003). One approach rooted in culture change is to alter physical environments to create a more homelike atmosphere that allows residents to direct daily activities, facilitates opportunities for collaborative decision-making (particularly involving direct care staff), changes staffing patterns to encourage closer relationships between staff and residents, and focuses on systematic quality improvement processes (Koren, 2010). In the culture change model, clinical care (while important) is viewed as a complement to overall resident quality of life.

Culture change models that have received the most attention in the United States include the *Eden Alternative*™, which focuses on making facilities more "homelike" with gardens, pets, and interaction with children; the *Green House*™ model, in which individual households of 8–10 residents are cared for by "universal workers" (and represents an evolution of the Eden Alternative); and the *Wellspring*™ model which prioritizes the empowerment of staff to make decisions that impact resident care, implements consistent staff assignments, and focuses on clinical quality improvement. These models emphasize flexibility in scheduling and person-centeredness as central to providing homelike care, in which residents can engage in meaningful activities. Evidence demonstrating the efficacy of culture change initiatives is still developing, and while these initiatives have not resulted in negative outcomes for residents the benefits of such approaches remain mixed (Shier, Khodyakov, Cohen, Zimmerman, & Saliba, 2014; Zimmerman, Shier, & Saliba, 2014).

Tools to facilitate direct engagement with state officials to understand culture change efforts are now available, including case studies and tool kits to initiate culture change (the latter via LeadingAge, which was originally known as the American Association of Homes and Services for the Aging; see http://www.leadingage.org/Culture_Change_Toolkit.aspx). However, comprehensive culture change in NHs has proceeded slowly (Miller, Mor, & Clark, 2010); a 2007 national survey of NHs found that 5% indicated full culture change had occurred at their facilities, while 10% had implemented seven or more culture change activities and a third implemented some culture change (Doty, Koren, & Sturla, 2008).

As summarized in Gaugler, Yu, Davila, and Shippee (2014, p. 655), disparities across NHs may stymie culture change advances. Some have described NH care as a case of the "two tiers," with the lower-tier NHs as including predominantly Medicaid-eligible residents, for-profit, and socioeconomically challenged (Grabowski, Elliot, Leitzell, Cohen, & Zimmerman, 2014; Mor, Zinn, Angelelli, Teno, & Miller, 2004). Culture change appears most likely to take root in NHs that are comparatively "resource-rich," non-profit, and include higher proportions of residents that are not Medicaid-eligible (Grabowski et al., 2014; Miller, Cohen, Lima, & Mor, 2014). Among the factors associated with lack of culture change initiation or sustainability in "low-tier" NHs include leadership turnover and other infrastructural challenges that make such systems' transformation difficult.

As Grabowski et al. (2014) and others emphasize (Institute of Medicine, 2008), there are a number of efforts that can help to promote culture change more fully, including the revision of construction codes, creation of tax credits, or reducing interest rates in order to replace NHs that simply cannot adhere to the environmental upgrades necessary for culture change to flourish. Aligning regulations and financial incentives with culture change principles and providing support to train the health care workforce in person-centered care techniques for older persons are important policy considerations as well. Indeed, policies have been shown to promote culture change efforts. For example, increased state Medicaid payments to NHs appear associated with increased culture change, and innovative policy solutions (such as those in Arkansas that use civil monetary funds to help support the often substantial up-front costs of culture change) are likely required to better traverse the chasm between different tiers of NHs and ensure that culture change is a global long-term care innovation, as opposed to a boutique one.

Pay-for-Performance

A major emphasis of health care reform has been a reorganization of payment systems away from fee-for-service types of models and toward those that provide reimbursement linked to quality outcomes. Such initiatives have fueled the growth of long-term care innovations as well. One such program is the Minnesota Performance-Based Incentive Payment Program (or PIPP). PIPP relies on providers themselves to identify measurable outcomes to improve (Gaugler, Yu, et al., 2014). Specifically, the Minnesota Department of Human Services solicits proposals from providers that specify context, goals, sustainability, and other quality improvement issues. Funds are provided from 1 to 3 years, and if goals are not achieved a 20% reduction in funding occurs. Initial evaluations of PIPP have found increases in quality of care when compared to those facilities that do not participate in PIPP (Arling et al., 2013).

Efforts to implement pay-for-performance models extend beyond PIPP and Minnesota to national initiatives, including the Center for Medicare and Medicaid Services' Quality Assurance Performance Improvement initiative (QAPI; see http://www.cms.gov/Medicare/Provider-Enrollment-and-Certification/QAPI/NHQAPI.html). Data-driven measures and evaluation are at the core of quality improvement efforts in QAPI, but as Arling and others emphasize, the approach PIPP adopts (provider-identified quality improvement goals; the use of grant funds to supplement quality improvement efforts) likely explains the considerable success of PIPP. QAPI will largely depend on existing regulatory mechanisms to facilitate quality improvement and thus may not achieve the same level of success as the PIPP pay-for-performance model (Arling et al., 2013; Gaugler, Yu, et al., 2014).

Informal Care

In 2009, 18.9%, or 43.5 million, Americans cared for someone 50 years of age or over (National Alliance for Caregiving and American Association of Retired Persons, 2009). Sixty-six percent of family caregivers in the United States are women, and the majority of family caregivers are 61 years of age and over. Among caregivers, the 2009 report revealed that the average amount of family care provided on a weekly basis is 20.4 hours. The prevalence and duration of family caregiving in the United States alone illustrates the public health importance of this phenomenon and that families are the "backbone" of the US long-term care system. A considerable segment of the gerontological and geriatric literature has focused on the emotional, psychological, social, economic, and physiological implications of family care provision (Fortinsky, Tennen, Frank, & Affleck, 2007; Gitlin & Schulz, 2012).

In general, interventions designed to support family caregivers aim to ameliorate the negative aspects of caregiving with the goal of improving the health and well-being of family caregivers. Methods used to accomplish this objective include enhancing caregiver strategies to manage disease-related symptoms, bolstering resources through enhanced social support and providing relief/respite from daily care demands. Desired outcomes of these interventions include decreased caregiver stress and depression and delayed NH admission or service utilization of the person with chronic illness or impairment. Characteristics of effective caregiver interventions include programs that are administered over long periods, interventions that approach chronic care as an issue for the entire family, and interventions that train caregivers in the management of behavioral problems (Mittelman, 2013; Zarit, 2009). Multidimensional interventions appear particularly effective. These approaches combine individual consultation, family sessions and support, and ongoing assistance to help caregivers manage changes that occur as a chronic illness progresses. Examples of successful multidimensional interventions are the New York University Caregiver Intervention (Gaugler, Reese, & Mittelman, 2013; Mittelman, Haley, Clay, & Roth, 2006; Mittelman, Roth, Haley, & Zarit, 2004), the Resources for Enhancing Alzheimer's Caregiver Health (REACH) II protocol (Belle et al., 2006; Elliott, Burgio, & Decoster, 2010; Nichols, Martindale-Adams, Burns, Graney, & Zuber, 2011), and the Savvy Caregiver program (Hepburn, Lewis, Tornatore, Sherman, & Bremer, 2007; Hepburn, Tornatore, Center, & Ostwald, 2001).

Although family members often provide the bulk of long-term care to adults in the United States, policy innovations have lagged behind other long-term care reform efforts. This is in part due to the assumption that family care to older relatives is more a moral obligation than a long-term care service proper (Levine et al., 2010). Policy developments include the National Family Caregiver Support Program (established in 2000) to provide fiscal support to Area Agencies on Aging that offer information, referral, and support to family caregivers, and the Lifespan Respite program. It is generally agreed that these programs are underfunded when considering the sheer bulk and economic value of family care occurring in the United States.

Providing information to assist family caregivers to more effectively engage in care planning or decision-making is an important gap in current long-term care research and practice. For example, while intensive multi-component intervention models can exert positive benefits for family caregivers, it is not clear that these types of approaches (with their need for sufficient provider training and time commitment on the part of caregivers) work for all families in every type of situation. Partnering with family caregivers to establish a relationship that results in improved assessment and more effective referral to individualized support services would help to address an important gap in family caregiving practice and science (Gaugler, Potter, et al., 2014).

Transitional Care/Care Coordination

As noted by Levine and others, transitions between care events or settings (such as from hospital to home) are key junction points when acute and long-term care systems join and often clash (Levine et al., 2010). Family caregivers are often in the middle of these transitional events, which can lead to a number of additional challenges (medication mismanagement, misunderstood care regimens, exacerbation of symptoms in the care recipient, etc.). Earlier efforts to improve coordination of care across transitions were provider/professional-centric and often relied on less intensive coordinating support (e.g., telephone follow-up) that did not involve family caregivers. A new series of care coordination models have received attention for the potential to reduce patient care costs, and to a lesser extent, improve family caregiver outcomes. Two of the most prominent examples include the Transitional Care Model and the Care Transitions Program. The Transitional Care Model utilizes transitional care nurses to assess patients prior to discharge and develops a coordinated care plan with other members of the hospital care team; the transitional care nurse visits the patient and family caregiver at home within 1–2 days following discharge and weekly 3 months thereafter. Randomized controlled trials identified a number of benefits, including improved health, functional, and quality of life outcomes and annual cost savings of approximately $4 800 (Levine et al., 2010; Naylor, Brooten, Campbell, Jacobsen, et al., 1999, Naylor, Brooten, Campbell, Maislin, et al., 2004). The Care Transitions Program uses "transition coaches" to follow patients and family members from hospital to home or sub-acute care over a 14-day period to teach families and patients to set goals, recognize potentially adverse events, and manage care; results from randomized controlled evaluations showed lowered hospitalization rates at 30 and 90 days and a 180-day reduction in costs of about $500 (Coleman, Parry, Chalmers, & Min, 2006).

Since the late 1990s, state Medicaid programs have also created NH transition programs which aim to relocate residents into home or community-based settings. A principal goal of these programs is to reduce Medicaid spending by transitioning eligible residents from NH to community. By 2005, the Money Follows the Person federal grants were awarded to 31 states, which eventually succeeded in transitioning 35 000 Medicaid beneficiaries over a 5-year period out of NHs and into community-based or home settings (Reinhard, 2010). Additional policy advances included the creation of community transitions services as well as individuals' flexible deposit of Medicaid funds into security accounts. Ongoing challenges to NH transition programs include the lack of appropriate community housing or care providers/aides in such settings for transitioning residents (Tilley & Boone, 2008).

Combinations of NH transition along with "diversion" strategies (e.g., rebalancing efforts; see above) and training staff to help coordinate and access community options have demonstrated success across multiple states. However, the availability of NH transition programs is a concern (as of 2010, 22 states did not implement these programs via Money Follows the Person or Community Living Programs; Reinhard, 2010). As Reinhard emphasizes, policies that do not require NH residence for transition eligibility (thus avoiding the situation where NH admission is the "required point of entry") eliminate federal barriers (e.g., such as person limits in community residences; improve technical assistance; broaden funding streams for states; resource for staff training), and expand state and local access to transition programs will help to ensure that transition and diversion programs become integrated into each state's set of long-term care programs and policies (Reinhard, 2010).

Health Information Technology

One innovation that can provide more easily accessible information to providers, older persons, and family members is health information technology. In particular, providers are sometimes confronted with too much or too little information to enhance care recommendations and decisions; instead, there is a need for providers to have access to efficient, "just in time" information (Kane et al., 2005). For example, the use of electronic health records can provide useful, searchable fields and can display, in some instances graphically, key information about the course of an older person's chronic illness. Unlike other health care providers, however, long-term care facilities have been relatively slow in adopting health information technologies (Pillemer et al., 2012) and the actual benefits of health information technology adoption in long-term care are not clear. Some studies have actually found potentially adverse events following the adoption of health information technology, including increased confusion related to communication between care providers and patients; an overemphasis on data entry; and inflexible data entry leading to a loss of information suggesting that an overreliance on these technologies may threaten the ideals of resident-centered care (Harrison, Koppel, & Bar-Lev, 2007; Ludwick & Doucette, 2009; Pillemer et al., 2012). The lack of effects of health information technology on NH resident outcomes was also apparent in a recent prospective, quasi-experimental evaluation of 761 NH residents in upper state New York over a 9-month period; no significant effects were found on resident outcomes and in one instance, residents in control facilities demonstrated significant reductions in negative behaviors while residents in treatment facilities indicated no change (Pillemer et al., 2012). Residents were generally positive in terms of using health information technology, but the results of recent controlled research suggest that the overall influence of technology supports for long-term care service users and family caregivers requires greater scientific inquiry (e.g., Bank, Arguelles, Rubert, Eisdorfer, & Czaja, 2006; Czaja & Rubert, 2002; Eisdorfer et al., 2003; Godwin, Mills, Anderson, & Kunik, 2013). As the adoption of health information and other technologies ("smart home" technologies such as remote health monitoring sensors) in long-term care continues to advance rapidly and are touted as "solutions" by care providers and others, studies on processes of implementation as well as controlled studies of outcomes on the part of providers, older persons, and family caregivers is of great import so that these technologies align successfully with resident-centered principles of long-term care (Demiris & Hensel, 2008; Demiris, Hensel, Skubic, & Rantz, 2008).

LOOKING TOWARD THE FUTURE OF LONG-TERM CARE

A survey of 1147 consumer advocates, provider representatives, public officials, and policy experts in the United States sought to identify the opinions of those integral to the policy process to determine how long-term care services can be reformed in the future (Miller, Miller, et al., 2010). Several areas of agreement emerged, and included the following:

- Governmental sources will have to assume more fiscal responsibility for long-term care services than the current private/public arrangement currently allows;
- Rebalancing efforts should continue toward HCBS and in particular toward programs such as PACE and consumer-directed options; interestingly, this was a sentiment that was shared by half of the NH representatives participating in the survey. However, such efforts should not include limiting NH bed supply (reflecting acknowledgment of a greater number of NHs providing sub-acute care for older adults and also the demographic needs to serve the growing aging population);
- Over 65% of respondents were aware of the culture change movement in NHs, and respondents noted that costs, resistance of upper management, and regulatory concerns were among the greatest barriers to implementing culture change in NHs. A follow-up study of this same sample of respondents suggested that culture change efforts in NHs require revisions in regulatory requirements to allow NHs the flexibility to adopt the complex facets required to achieve true, systemic culture change (Miller, Miller, et al., 2010);
- While there was agreement that regulation of NHs and home health care was poorly performed at the federal level, assisted living (which currently is not subject to the same regulations) should receive greater regulatory attention including assessments and inspections;
- Payment incentives were rated as the most promising approach to enhance the quality of long-term care, although there was concern whether current self-reporting mechanisms used to measure NH quality (e.g., CMS' Minimum Data Set/MDS and the Outcome and Assessment Information Set, or OASIS) were sufficient. This suggests that "pay-for-performance" models may be one possible innovation to enhance long-term care quality.

Another issue to ponder is the extent to which innovations in long-term care have been driven, at least in part, by disability rights advocacy groups and how continued alignment with these organizations can continue to advance quality long-term care for older persons. There is now considerable discussion in disability studies about "aging with disabilities," given improved survival among disabled youth and middle-aged adults. This trend will pose unique opportunities for the alignment of interests across groups, as well as challenges regarding long-term care provision. The recent reorganization of various federal agencies into The Administration for Community Living (which has oversight over aging and disabilities programs) – may be a harbinger of what is to come. There is much advocates of older persons and their families can learn from disability studies and affiliated advocacy organizations, as disability advocates often emphasize social integration as opposed to safety and efficiency (the latter of which are often a primary emphasis of long-term care organizations that serve older persons; Kane & Kane, 2001). Empowering older persons, enhancing the quality and availability of personalized information based on older persons' preferences and needs, and aligning advocacy efforts with those of disability rights' groups will help to continue to drive innovations in long-term care.

References

Arling, G., Cooke, V., Lewis, T., Perkins, A., Grabowski, D. C., & Abrahamson, K. (2013). Minnesota's provider-initiated approach yields care quality gains at participating nursing homes. *Health Affairs, 32*, 1631–1638.

Bank, A. L., Arguelles, S., Rubert, M., Eisdorfer, C., & Czaja, S. J. (2006). The value of telephone support groups among ethnically diverse caregivers of persons with dementia. *The Gerontologist, 46*, 134–138.

Beales, J. L., & Edes, T. (2009). Veteran's affairs home based primary care. *Clinics in Geriatric Medicine, 25*, 149–154, viii–ix.

Beckett, J. O. (Ed.). (2009). *Lifting our voices: The journeys into family caregiving of professional social workers*. New York, NY: Columbia University Press.

Belle, S. H., Burgio, L., Burns, R., Coon, D., Czaja, S. J., Gallagher-Thompson, D., & Resources for Enhancing Alzheimer's Caregiver Health (REACH) II Investigators. (2006). Enhancing the quality of life of dementia caregivers from different ethnic or racial groups: A randomized, controlled trial. *Annals of Internal Medicine, 145*, 727–738.

Binstock, R. H., & Spector, W. D. (1997). Five priority areas for research on long-term care. *Health Services Research, 32*, 715–730.

Branch, L. G., Coulam, R. F., & Zimmerman, Y. A. (1995). The PACE evaluation: Initial findings. *The Gerontologist, 35*, 349–359.

Butler, R. (1975). *Why survive? Being old in America*. New York, NY: Harper & Row.

Callahan, C. M., Arling, G., Tu, W., Rosenman, M. B., Counsell, S. R., Stump, T. E., et al. (2012). Transitions in care for older adults with and without dementia. *Journal of the American Geriatrics Society, 60*(5), 813–820.

Cole, T. (1987). Class, culture, and coercion: A historical perspective on long term care. *Generations, 11*, 9–15.

Coleman, E. A., Parry, C., Chalmers, S., & Min, S. J. (2006). The care transitions intervention: Results of a randomized controlled trial. *Archives of Internal Medicine, 166*(17), 1822–1828.

Czaja, S. J., & Rubert, M. P. (2002). Telecommunications technology as an aid to family caregivers of persons with dementia. *Psychosomatic Medicine, 64*(3), 469–476.

Dale, S. B., & Brown, R. (2006). Reducing nursing home use through consumer-directed personal care services. *Medical Care, 44*(8), 760–767.

Dale, S. B., & Brown, R. S. (2007). How does cash and counseling affect costs? *Health Services Research, 42*(1 Pt 2), 488–509.

Demiris, G., & Hensel, B. K. (2008). Technologies for an aging society: A systematic review of "smart home" applications. *Yearbook of Medical Informatics, 47*(Suppl. 1), 33–40.

Demiris, G., Hensel, B. K., Skubic, M., & Rantz, M. (2008). Senior residents' perceived need of and preferences for "smart home" sensor technologies. *International Journal of Technology Assessment in Health Care, 24*(1), 120–124.

Doty, M., Koren, M. J., & Sturla, E. L. (2008). *Culture change in nursing homes: How far have we come? Findings from the Commonwealth Fund 2007 National Survey of Nursing Homes*. New York, NY: Commonwealth Fund.

Doty, P., Kasper, J., & Litvak, S. (1996). Consumer-directed models of personal care: Lessons from Medicaid. *The Milbank Quarterly, 74*, 377–409.

Doty, P., Mahoney, K. J., & Sciegaj, M. (2010). New state strategies to meet long-term care needs. *Health Affairs, 29*, 49–56.

Egan, E. (2010). *VA home based primary care program: A primer and lessons for Medicare*. Washington, DC: American Action Forum.

Eisdorfer, C., Czaja, S. J., Loewenstein, D. A., Rubert, M. P., Arguelles, S., Mitrani, V. B., et al. (2003). The effect of a family therapy and technology-based intervention on caregiver depression. *The Gerontologist, 43*, 521–531.

Elliott, A. F., Burgio, L. D., & Decoster, J. (2010). Enhancing caregiver health: Findings from the resources for enhancing Alzheimer's caregiver health II intervention. *Journal of the American Geriatrics Society, 58*, 30–37.

Eng, C., Pedulla, J., Eleazer, G. P., McCann, R., & Fox, N. (1997). Program of All-Inclusive Care for the Elderly (PACE): An innovative model of integrated geriatric care and financing. *Journal of the American Geriatrics Society, 45*, 223–232.

Fortinsky, R. H., Tennen, H., Frank, N., & Affleck, G. (2007). Health and psychological consequences of caregiving. In C. Aldwin, C. Park, & R. Spiro (Eds.), *Handbook of health psychology and aging* (pp. 227–249). New York, NY: Guilford.

Gaugler, J. E., Boldischar, M., Vujovich, J., & Yahnke, P. (2011). The Minnesota Live Well at Home Project: Screening and client satisfaction. *Home Health Care Services Quarterly, 30*, 63–83.

Gaugler, J. E., Potter, T., & Pruinelli, L. (2014). Partnering with caregivers. *Clinics in Geriatric Medicine, 30*, 493–515.

Gaugler, J. E., Reese, M., & Mittelman, M. S. (2013). Effects of the NYU caregiver intervention-adult child on residential care placement. *The Gerontologist, 53*, 985–997.

Gaugler, J. E., Yu, F., Davila, H. W., & Shippee, T. (2014). Alzheimer's disease and nursing homes. *Health Affairs, 33*, 650–657.

Gitlin, L. N., & Schulz, R. (2012). Family caregiving of older adults. In T. R. Prohaska, L. A. Anderson, & R. H. Binstock (Eds.), *Public health for an aging society* (pp. 181–204). Baltimore, MD: The Johns Hopkins University Press.

Godwin, K. M., Mills, W. L., Anderson, J. A., & Kunik, M. E. (2013). Technology-driven interventions for caregivers of persons with dementia: A systematic review. *American Journal of Alzheimer's Disease and Other Dementias, 28*(3), 216–222.

Grabowski, D. C. (2006). The cost-effectiveness of noninstitutional long-term care services: Review and synthesis of the most recent evidence. *Medical Care Research and Review, 63*, 3–28.

Grabowski, D. C. (2007). Medicare and Medicaid: Conflicting incentives for long-term care. *Milbank Quarterly, 85*, 579–610.

Grabowski, D. C., Cadigan, R. O., Miller, E. A., Stevenson, D. G., Clark, M., & Mor, V. (2010). Supporting home- and community-based care: Views of long-term care specialists. *Medical Care Research and Review, 67*, 82S–101S.

Grabowski, D. C., Elliot, A., Leitzell, B., Cohen, L. W., & Zimmerman, S. (2014). Who are the innovators? Nursing homes implementing culture change. *The Gerontologist, 54*(Suppl. 1), S65–S75.

Gravelle, H., Dusheiko, M., Sheaff, R., Sargent, P., Boaden, R., Pickard, S., et al. (2007). Impact of case management (Evercare) on frail elderly patients: Controlled before and after analysis of quantitative outcome data. *BMJ(Clinical Research Ed.), 334*, 31.

Harrington, C., & Kitchener, M. (2003). *Medicaid long-term care: Changes, innovations, and cost containment*. San Francisco, CA: University of California.

Harrison, M. I., Koppel, R., & Bar-Lev, S. (2007). Unintended consequences of information technologies in health care – An interactive sociotechnical analysis. *Journal of the American Medical Informatics Association, 14*, 542–549.

Hepburn, K., Lewis, M., Tornatore, J., Sherman, C. W., & Bremer, K. L. (2007). The Savvy Caregiver Program: The demonstrated effectiveness of a transportable dementia caregiver psychoeducation program. *Journal of Gerontological Nursing, 33*, 30–36.

Hepburn, K. W., Tornatore, J., Center, B., & Ostwald, S. W. (2001). Dementia family caregiver training: Affecting beliefs about caregiving and caregiver outcomes. *Journal of the American Geriatrics Society, 49*, 450–457.

Holstein, M., & Cole, T. R. (1994). Long-term care: A historical reflection. In L. B. McCullough & N. L. Wilson (Eds.), *Long-term care decisions: Ethical and conceptual dimensions* (pp. 15–34). Baltimore, MD: Johns Hopkins University Press.

Holstein, M., & Cole, T. R. (1996). The evolution of long-term care in America. In R. H. Binstock, L. E. Cluff, & O. van Mering (Eds.), *The future of long-term care: Social and policy issues* (pp. 19–48). Baltimore, MD: Johns Hopkins University Press.

Institute of Medicine (IOM) (2008). *Retooling for an aging America: Building the health care workforce*. Washington, DC: The National Academies Press.

Kane, R. A. (2001). Long-term care and a good quality of life: Bringing them closer together. *The Gerontologist, 41*, 293–304.

Kane, R. L., & Kane, R. A. (2001). What older people want from long-term care, and how they can get it. *Health Affairs, 20*, 114–127.

Kane, R. L., Keckhafer, G., Flood, S., Bershadsky, B., & Siadaty, M. S. (2003). The effect of Evercare on hospital use. *Journal of the American Geriatrics Society, 51*, 1427–1434.

Kane, R. L., Priester, R., & Totten, A. (2005). *Meeting the challenge of chronic illness*. Baltimore, MD: Johns Hopkins University Press.

Kane, R. L., & West, J. C. (2005). *It shouldn't be this way: The failure of long-term care*. Nashville, TN: Vanderbilt University Press.

Karlin, B. E., & Karel, M. J. (2014). National integration of mental health providers in VA home-based primary care: An innovative model for mental health care delivery with older adults. *The Gerontologist, 54*, 868–879.

Kassner, E. (2011). *Home and community-based long-term services and supports for older people*. Washington, DC: AARP Public Policy Institute.

Katz, S., Ford, A. B., Moskowitz, R. W., Jackson, B. A., & Jaffe, M. W. (1963). Studies of illness in the aged. The index of ADL: A standardized measure of biological and psychosocial function. *Journal of the American Medical Association, 185*, 914–919.

Koren, M. J. (2010). Person-centered care for nursing home residents: The culture-change movement. *Health Affairs, 29*, 312–317.

Lacey, D. (1999). The evolution of care. *Journal of Gerontological Social Work, 31*, 101–131.

Lawton, M. P., & Brody, E. M. (1969). Assessment of older people: Self-maintaining and instrumental activities of daily living. *The Gerontologist, 9*, 179–186.

Levine, C., Halper, D., Peist, A., & Gould, D. A. (2010). Bridging troubled waters: Family caregivers, transitions, and long-term care. *Health Affairs, 29*, 116–124.

Lubkin, I. M., & Larsen, P. D. (Eds.). (2013). *Chronic illness: Impact and intervention* (8th ed.). Burlington, MA: Jones and Bartlett Learning, LLC.

Ludwick, D. A., & Doucette, J. (2009). Adopting electronic medical records in primary care: Lessons learned from health information systems implementation experience in seven countries. *International Journal of Medical Informatics, 78*(1), 22–31.

Manton, K. G., Newcomer, R., Vertrees, J. C., Lowrimore, G. R., & Harrington, C. (1994). A method for adjusting capitation payments to managed care plans using multivariate patterns of health and functioning: The experience of social/health maintenance organizations. *Medical Care, 32*(3), 277–297.

Meret-Hanke, L. A. (2011). Effects of the Program of All-Inclusive Care for the Elderly on hospital use. *The Gerontologist, 51*, 774–785.

Miller, E. A., Mor, V., & Clark, M. (2010). Reforming long-term care in the United States: Findings from a national survey of specialists. *The Gerontologist, 50*(2), 238–252.

Miller, S. C., Cohen, N., Lima, J. C., & Mor, V. (2014). Medicaid capital reimbursement policy and environmental artifacts of nursing home culture change. *The Gerontologist, 54*(Suppl. 1), S76–S86.

Miller, S. C., Miller, E. A., Jung, H., Sterns, S., Clark, M., & Mor, V. (2010). Nursing home organizational change: The "Culture Change" movement as viewed by long-term care specialists. *Medical Care Research and Review, 67*, 65S–81S.

Mittelman, M. (2013). Psychosocial interventions to address the emotional needs of caregivers of individuals with Alzheimer's disease. In S. H. Zarit & R. C. Talley (Eds.), *Caregiving for Alzheimer's disease and related disorders* (pp. 17–34). New York, NY: Springer.

Mittelman, M. S., Haley, W. E., Clay, O. J., & Roth, D. L. (2006). Improving caregiver well-being delays nursing home placement of patients with Alzheimer disease. *Neurology, 67*, 1592–1599.

Mittelman, M. S., Roth, D. L., Haley, W. E., & Zarit, S. H. (2004). Effects of a caregiver intervention on negative caregiver appraisals of behavior problems in patients with Alzheimer's disease: Results of a randomized trial. *The Journals of Gerontology. Series B, Psychological Sciences and Social Sciences, 59*, P27–P34.

Mor, V., Zinn, J., Angelelli, J., Teno, J. M., & Miller, S. C. (2004). Driven to tiers: Socioeconomic and racial disparities in the quality of nursing home care. *The Milbank Quarterly, 82*(2), 227–256.

Moss, F. E., & Halamandaris, V. J. (1977). *Too old, too sick, too bad: Nursing homes in America*. Germantown, MD: Aspen Systems Corporation.

Mui, A. C. (2002). The Program of All-Inclusive Care for the Elderly (PACE). *Journal of Aging & Social Policy, 13*, 53–67.

Nadash, P., & Crisp, S. (2005). *Best practices in consumer direction*. Rockville, MD: Westat, Inc.

National Alliance for Caregiving and American Association of Retired Persons. (2009). *Caregiving in the U.S. 2009*. Bethesda, MD: National Alliance of Caregiving.

Naylor, M. D., Brooten, D., Campbell, R., Jacobsen, B. S., Mezey, M. D., Pauly, M. V., et al. (1999). Comprehensive discharge planning and home follow-up of hospitalized elders: A randomized clinical trial. *Journal of the American Medical Association, 281*, 613–620.

Naylor, M. D., Brooten, D. A., Campbell, R. L., Maislin, G., McCauley, K. M., & Schwartz, J. S. (2004). Transitional care of older adults hospitalized with heart failure: A randomized, controlled trial. *Journal of the American Geriatrics Society, 52*, 675–684.

Newcomer, R., Harrington, C., & Kane, R. (2002). Challenges and accomplishments of the second-generation social health maintenance organization. *The Gerontologist, 42*, 843–852.

Newcomer, R., Manton, K., Harrington, C., Yordi, C., & Vertrees, J. (1995). Case mix controlled service use and expenditures in the social/health maintenance organization demonstration. *The Journals of Gerontology. Series A, Biological Sciences and Medical Sciences, 50A*, M35–M44.

Nichols, L. O., Martindale-Adams, J., Burns, R., Graney, M. J., & Zuber, J. (2011). Translation of a dementia caregiver support program in a health care system – REACH VA. *Archives of Internal Medicine, 171*, 353–359.

Pillemer, K., Meador, R. H., Teresi, J. A., Chen, E. K., Henderson, C. R., Lachs, M. S., et al. (2012). Effects of electronic health information technology implementation on nursing home resident outcomes. *Journal of Aging and Health, 24*, 92–112.

Polich, C. L., Bayard, J., Jacobson, R. A., & Parker, M. (1990). A nurse-run business to improve health care for nursing home residents. *Nursing Economics, 8*, 96–101.

Rahman, A. N., & Schnelle, J. F. (2008). The nursing home culture-change movement: Recent past, present, and future directions for research. *The Gerontologist, 48*, 142–148.

Reinhard, S. C. (2010). Diversion, transition programs target nursing homes' status quo. *Health Affairs, 29*, 44–48.

Reinhard, S. C., & Young, H. M. (2009). The nursing workforce in long-term care. *The Nursing Clinics of North America, 44*, 161–168.

Shield, R. R. (1996). Managing the care of nursing home residents: The challenge of integration. *Annual Review of Gerontology and Geriatrics, 16*, 60–77.

Shier, V., Khodyakov, D., Cohen, L. W., Zimmerman, S., & Saliba, D. (2014). What does the evidence really say about culture change in nursing homes? *The Gerontologist, 54*(Suppl 1), S6–S16.

Stevenson, K. (2012). *History of long-term care*. Retrieved from: <http://www.elderweb.com/book/history-long-term-care>.

Stone, R. I. (2000). *Long-term care for the elderly with disabilities: Current policy, emerging trends, and implications for the twenty-first century*. New York, NY: The Milbank Memorial Fund.

Stone, R. I., & Benson, W. F. (2012). Financing and organizing health and long-term care services for older populations. In T. R. Prohaska, L. A. Anderson, & R. H. Binstock (Eds.), *Public health for an aging society* (pp. 53–73). Baltimore, MD: Johns Hopkins University Press.

Thompson, T. (2002). *Evaluation results for the Social/Health Maintenance Organization II Demonstration*. Washington, DC: Department of Health and Human Services.

Tilley, J., & Boone, L. (2008). *Nursing facility transition programs serving older adults with cognitive impairment*. Chicago, IL: The Alzheimer's Association. Retrieved from: <https://www.alz.org/national/documents/issuebrief_112108_transitionporgrams.pdf>.

Tilly, J., & Wiener, J. M. (2001). Consumer-directed home and community services programs in eight states: Policy issues for older people and government. *Journal of Aging & Social Policy, 12*, 1–26.

Vladeck, B. C. (1980). *Unloving care: The nursing home tragedy*. New York, NY: Basic Books.

Ward, B. W., & Schiller, J. S. (2013). Prevalence of multiple chronic conditions among US adults: Estimates from the National Health Interview Survey, 2010. *Preventing Chronic Disease, 10*, E65.

Weiner, A. S., & Ronch, J. L. (2003). *Culture change in long-term care*. New York, NY: Haworth Press, Inc.

White, H. K., Buhr, G., McConnell, E., Sullivan, R. J., Jr., et al., Twersky, J., Colon-Emeric, C., et al. (2013). An advanced course in long term care for geriatric medicine fellows. *Journal of the American Medical Directors Association, 14*, 499–506.

Wieland, D., Boland, R., Baskins, J., & Kinosian, B. (2010). Five-year survival in a Program of All-Inclusive Care for Elderly compared with alternative institutional and home- and community-based care. *The Journals of Gerontology. Series A, Biological Sciences and Medical Sciences, 65*(7), 721–726.

Wunderlich, G. S., & Kohler, P. (Eds.). (2001). *Improving the quality of long-term care (Report of the Institute of Medicine)*. Washington, DC: National Academic Press.

Zarit, S. H. (2009). Empirically supported treatment for family caregivers. In S. H. Qualls & S. H. Zarit (Eds.), *Aging families and caregiving* (pp. 131–154). Hoboken, NJ: John Wiley & Sons, Inc.

Zimmerman, S., Shier, V., & Saliba, D. (2014). Transforming nursing home culture: Evidence for practice and policy. *The Gerontologist, 54*(Suppl. 1), S1–S5.

CHAPTER 21

Politics and Policies of Aging in the United States

Robert B. Hudson

School of Social Work, Boston University, Boston, MA, USA

OUTLINE

Introduction	441	The Newly Conflicted World of Old-Age Policy	448
The Altered Political Perceptions of Older Americans	442	Accounting for Old-Age Policy Enactments	451
Positive Standing and Policy Benefits	442		
Policy Benefits and Political Standing	444	Emerging Issues	455
The Transformation of Seniors' Political Environment	446	References	457
The Shifting Economic Context	447		
New Political Realities	448		

INTRODUCTION

Until some 35 years ago, the politics of aging were relatively benign and policies for the aging were consistently expansionary. With both the public at large and policy elites in support, programs on behalf of older adults helped launch what passes for much of the modern American welfare state. Seniors were near singular beneficiaries of social insurance breakthroughs in the 1930s and the 1960s, enjoying benefits denied to other vulnerable populations. Indeed, they constituted a policy constituency so favorably viewed that seniors periodically served as something of a policy "loss leader" for progressives seeking to broaden eligibility criteria to include younger and politically less sympathetic clienteles.

The more recent period of aging politics and policy has been much more contested. In part, this is a consequence of economic and political forces, emerging in the late 1970s, which affected domestic public policy across the board: slow economic growth, high levels of unemployment and inflation, disaffection with liberalism, and the rise of a rugged new conservatism. Yet, more age-specific forces were at work as well: escalating expenditures for Social Security and Medicare, the prospect of aging Baby Boomers being supported by a far smaller Generation X, and a growing recognition that older Americans were no longer the singularly disadvantaged population of earlier years, thereby rendering old age as an increasingly imperfect proxy for demonstrable need.

This chapter addresses these developments, reviews factors that account for the transformation in understandings of the policy place of older Americans, and identifies a series of policy-related issues that aging politics will need to confront in the years ahead. Central to the discussion are political legitimacy and power, institutional barriers and pathways, public and elite opinion, issue framing, evolving population characteristics and, perhaps most important, the ever-expanding scope of conflict around aging policy. Current events associated with fiscal cliffs, government shutdowns, Tea Party adherents on the right and Occupy Wall Street denizens on the left have, if nothing else, exacerbated recent trends.

THE ALTERED POLITICAL PERCEPTIONS OF OLDER AMERICANS

From an historical perspective, the initial period of old-age policy and politics was built first on a widespread belief in elders' deservingness and on the ability of elite political actors to seize on that positive standing and transform it into concrete legislation. These developments created a second stage with the resulting programs – principally Social Security and Medicare – having heightened the well-being and political identity of older people, thereby generating a potent electoral and policy constituency. Yet, these very policy and political successes led to a third stage wherein a markedly better-off older population came to encounter a new economic and political environment.

Positive Standing and Policy Benefits

Since the birth of US social welfare policies, older people enjoyed, at least in policy terms, a favored existence. Unpleasant as they were, the nineteenth century poor houses that provided sustenance for poor and isolated elders were more accommodating than those established for younger and what were perhaps seen as shiftless populations. Old-age relief and pension programs were always among the first ones established at the state and local levels. Most notable are the pathbreaking initiatives taken on behalf of elders at the federal level in America's "reluctant welfare state" (Wilensky & Lebeaux, 1958, p. xii). The list of "elders first" policies from the mid-1930s through the mid-1970s is impressive: Old Age Assistance as Title I of the original Social Security Act and Old Age Insurance as Title II in 1935; Disability Insurance, initially only for workers over the age of 50, in 1956; Medicare, initially and still primarily for individuals 65 and older in 1965; Age Discrimination and Employment Act for workers over age 40 in 1967; Supplemental Security Income, creating a national minimum income for poor elders and people with disabilities or blindness in 1972: and the Employee Retirement Income Security Act, providing federal insurance protection for defined benefit retirement plans in 1974.

Principally responsible for these largely unique developments was the policy legitimacy that has long been enjoyed by older Americans. In a nation with low levels of class-consciousness and a political ideology centered on individualism and private markets (Lipset, 1997), older people came to enjoy – at least comparatively – a coveted policy spot. Old age stood as a proxy for both need and deservingness, an understanding bestowed on no other population. Haber (1983) notes that from early in American history, the aged were consciously omitted from "the redeemable," that is, those who should be expected to earn their own support. The social insurance pioneer, Isaac Rubinow (1934), spoke of the unquestioned needs of older people as "the final emergency" after what workers might have faced – and not have been publicly protected against – earlier in life. Franklin Roosevelt opined that "poverty in old age should not be regarded as a disgrace or necessarily as a result of lack of thrift or energy … it is a mere by-product of modern industrial life" (Rimlinger, 1971, p. 212). In *Pitied But Not Entitled*, Linda Gordon's (1998) historical analysis of policy toward single mothers early in the twentieth century, she found elders to be "a group quintessentially deserving," much unlike the women whose lives she was investigating. And Marmor's (1970, p. 17) account of Medicare's enactment echoes Haber, stating that the elderly were "one of the few population groupings about whom one could not say that the members should take care of their financial-medical problems by earning or saving more money."

Such data as exist from these earlier periods largely support the needy, if not desperate, situation of many elders. The first issue of *Social Security Bulletin*, the official journal of the US Social Security Administration, featured an article by Shearon (1938) in which she found two-thirds of the older population to be "dependent" on others but found, as well, that one-half of those who were deemed "self-dependent" may have relied on friends of relatives for additional income. Borrowing from Shearon and other sources, Upp (1982) estimated that three-quarters of older people's income prior to Social Security's enactment came from their adult children, a figure that 50 years later had fallen to 1.5%. With less than one-half of older people having health insurance at the time of Medicare's enactment in 1965, Wilbur Cohen, an architect of the plan, noted that the amassing of statistical data to prove that the aged were sicker, poorer, and less insured than other adult groups was "like using a streamroller to crush an ant of opposition" (quoted in Marmor, 1970, p. 17). Programs for protection against chronic illness, inadequate as they may be today, did not exist, the earliest approximation being the Old Age Assistance benefit denied to persons residing in state institutions in order that coverage might be found in the community, as it occasionally was in early "mom and pop nursing homes" (Vladeck, 1980).

In the years leading up to the 1970s, there was little opposition to expanding old-age benefits; indeed, the institution of the automatic cost-of-living adjustment to Social Security benefits in 1972 was seen as a cost-containment measure, designed to preclude continued passage of "Christmas Tree" legislation increasing benefits as Congressional elections approached (Tufte, 1978). This permissive consensus that spawned early age-related programming was based significantly on elders' positive image and unique vulnerability, a political reality well-captured by Binstock's (1983) "compassionate stereotype" label. Using a broader typology of political "target populations," Schneider and Ingram (1993) found older Americans of this earlier period enjoying a "positive social construction," that is a group that "should be the beneficiary of a particular proposal" (p. 335). The great political utility of population legitimacy is captured by Sherman and Kolker (1987) seeing the effects of such construction or legitimacy as far-reaching for "when people agree, power or coercion becomes unnecessary; consensus prevails" (p. 123).

Of course, for purposes of policymaking, this sympathetic standing had to be activated. At a later point, political power became a prime mover in this activation, but in much of the period leading up to the late-1970s, it was a small number of highly placed officials and advocates who advanced the aging policy agenda. Multiple analyses of the enactments enumerated earlier – with the partial exception of Medicare – find elite political actors seizing on supportive views of elders' needs as the driving force behind policy gains. In the 1930s, Roosevelt's Committee on Economic Security worked largely behind closed doors in devising the Old Age Insurance program, and Derthick (1979) found that for the ensuing 40 years a group consisting almost exclusively of program executives and specialty committees within Congress were largely responsible for program governance. In the 1960s, a small coterie of federal officials devised the pull back from earlier national health insurance initiatives put forth by Presidents Roosevelt and Truman to focus exclusively on older Americans.

Enactment of Supplemental Security Income represents the most illuminating example of elite figures employing elders' positive standing for policy gains. Licking their wounds in the wake of defeat of President Nixon's Family Assistance Plan in 1970 – designed to "cash out" Aid to Families with Dependent Children – federal officials quickly redirected the program toward the so-called "adult public assistance categories": the poor aged, blind, and disabled. Within a matter of weeks SSI passed its critical legislative hurdle when the Senate Finance Committee approved the program and its nationwide income guarantee, with Wilber Cohen exclaiming "do you realize what they're doing in there? It's not even controversial, it's not even controversial!" (Burke & Burke, 1974, p. 198). Whether from indifference or support, only seven out of 535 members of Congress even mentioned the guarantee in public testimony. In the then-prevailing constricted world of old-age policymaking, supportive elites were responsible for much of what transpired.

Policy Benefits and Political Standing

Beginning in the mid-1970s, the policy benefits bestowed on older Americans began generating political outcomes. In part, this was due to an increase in the size of the older population, to which the Senate Committee on Aging called attention through a series of reports with the titles telling the story: "Older Americans: One-in-Eleven"; "Older Americans: One-in-Ten"; "Older Americans: One-in-Nine" (Brotman, 1977). More important, however, was seniors finding and acting on a political self-identity that they had not previously held. From an earlier period when Heclo (1988, p. 393) dismissively concluded that "'the elderly' is really a category created by policy analysts, pension officials, and mechanical models of interest group politics," Campbell (2003, p. 36) is later able to introduce policy's expansive effect: "Absent Social Security, senior citizens are a disparate group of people whose common characteristic, age, has little political meaning."

In the earlier years, long after the rise and fall of the Townsend Movement of the 1930s and the old-age political desert of the two decades following World War II that Pratt (1976) describes as "the dismal years," old-age political action was little in evidence. By the mid-1970s, however, political activity among the old had ratcheted up markedly. Beginning with the 1974 off-year elections and the 1976 Presidential election, voting participation among those 65 and older began rising. At roughly the same time, the birth and rise of a growing host of old-age interest groups was underway. Both of these developments speak to the second face of seniors' emerging presence in American politics, that is, the ability to exercise or at least appear to exercise political power.

Two age-based trends speak to the potential power that can be associated with voting and other forms of political participation. First, older people's rate of voting has not only increased

TABLE 21.1 Voting by Age, Selected Presidential Elections

Age	Year	
	1972	2012
18–24	50%	41%
25–44	63%	57%
45–64	71%	68%
65+	63%	72%

Source: US Bureau of the Census (2012a), *Current Population Reports*.

TABLE 21.2 Voting by Age in Recent Off-Year Elections

Age	Year	
	2002	2006
18–29	22%	25%
30–44	42%	43%
45–59	55%	56%
60–74	64%	64%
75+	60%	61%

Source: Barrios, Lopez, Kennedy, and Barr (2008). Young voter registration and turnout trends. Center for Information and Research on Civic Learning and Engagement.

over the past 40 years, but it has done so to the point that since 1988 older Americans vote at a higher rate than any other age group. This is a far cry from the pre-1970s pattern wherein seniors voted less than any other group other than those aged 21–24. Second, while old-age voting was increasing, voting rates among younger populations were actually decreasing (Table 21.1).

Older people's high levels of voting are even more pronounced in off-year elections (and presumably even more so in primaries, where candidates for elective office are chosen). Beginning in 1986, elders voted at a higher rate in off-year elections than any other age groups, voting at more than twice the rate of the youngest voters and considerably more than the middle-aged (Table 21.2).

Beyond voting, older people's commitment to and involvement in the political system has also increased. From a time when Schmidhauser (1970) suggested that older people ranked lowest in sense of political efficacy and among the lowest in "sense of citizen duty," Campbell (2003) finds robust evidence of seniors' active participatory engagement in the system. Seniors have more than quadrupled their rate of campaign contributing since 1952, catching up to the 35–64-year-olds in the 1980s; they have increased their amount of campaign work, now equaling that of the 35–64-year-old group; and they have increasingly been in contact with their Congressional representatives (especially when issues salient to seniors are in question), whereas non-seniors experienced a steady decline from 1973 to the mid-1990s.

A final and perhaps most dramatic indicator of contemporary senior involvement in politics centers on the two most long-standing and strongest predictors of political participation: higher income and greater education. Regarding income, Campbell (2003) finds that elders at all income levels now contribute at a higher rate to political campaigns than younger people, with low-income elders contributing at twice the rate of other low-income groups and high-income elders at three times the rate of younger high-income groups. As for education, Campbell finds the turnout rate at all education levels is higher for seniors than other age groups and also that the turnout rate for the least educated seniors is over twice that at the same education level for people aged 35–49 and a remarkable nine times higher than those aged 18–34. As we will see, Campbell's broader argument centers on the role that Social Security has played in stimulating elder political participation, but, whatever the cause, the activation of low-income and modestly educated seniors is a remarkable finding in the annals of political behavior.

Slightly preceding the heightened participation of older Americans in politics was the birth and growth of a number of organized groups working on behalf of older Americans. The first and best-known, of course, is AARP, which began in 1958 as the National Retired Teachers Association, an insurance company affinity group for those retirees who had had great difficulty in securing life insurance (Morris, 1996). The group later broadened its name and reach, today constituting over 40 million members and an operating budget of several hundred million dollars, mainly devoted to provision of services and contracting for the membership. The other principal mass membership group – today named the Alliance for Retired Americans – began as Senior Citizens for Kennedy in 1960 and later became the National Council of Senior Citizens, where as an offshoot of organized labor it played a major role in the promotion of Medicare legislation in the early 1960s.

A host of additional aging-related groups sprang up around and in the wake of the remarkable watershed 1965–1974 decade of aging policy. Some are independent, others are off-shoots of professional organizations such as nursing and law, and yet others are more directly tied to the implementation of federal and state aging policy, notably the National Association of State Agencies on Aging and Disability and the National Association of Area Agencies on Aging. More recently, this array of groups has quasi-institutionalized itself into the Leadership Council of Aging Organizations, a confederation consisting of no fewer than 64 of these organizations, including, among many others, the American Association for the Blind, the American Society of Consulting Pharmacists, Experience Works, the Medicare Rights Center, the National Association of Home Care and Hospice, and the National Association of Nutrition and Aging Services Programs. These groups advocate for expanded old-age benefits, with their greatest attention concentrated in service and benefit areas where they are planners or providers (Binstock, 1972; Lowi, 1969).

The voting strength and organizational presence of the aging represent a political armamentarium of remarkable potency. They certainly hold the potential for the exercise of political power and, as will be later articulated, recent events suggest that such power is now being exercised. Thus, in something of a two-stage process, elders' political legitimacy and the uses to which it was put by key political actors first generated the major legislation enacted prior to the mid-1970s. In subsequent years, a second stage has been marked by the robust political participation of elders and their advocates in constituting a remarkable bulwark against the trimming of old-age benefits and, occasionally, a base for adding new ones. It is in these ways that the political presence of the aged has evolved for most of the previous century.

THE TRANSFORMATION OF SENIORS' POLITICAL ENVIRONMENT

The resilience of older Americans as a preferred policy constituency and their concomitant emergence as a major force in Washington provided seniors with a privileged position in American social policy, a combination that had them categorized as "advantaged" at the time Schneider and Ingram (1993) were proposing their typology of target populations. Yet, by the mid-1970s emerging forces and factors began to bring new pressures to bear on this well-placed policy domain. Some of these economic and political developments were not directed specifically at older people, but they were to have major impacts on elders' policy domain, introducing new and often hostile actors to aging policy. Heretofore "outsiders" took new interest in the political world of aging because aging-related programs were large and growing, economic growth needed to support those programs was slowing, and support for federal involvement in addressing domestic social problems was waning. And against this more contentious backdrop, evidence mounted that seniors were no longer the singularly disadvantaged group that had been their claim to policy fame over the previous half-century. In Schattschneider's (1960) famous dictum, "the scope of conflict" had expanded.

The Shifting Economic Context

During the immediate post–World War II years and throughout much of the third quarter of the twentieth century, the United States enjoyed a period of unprecedented economic growth. Pent-up consumer demand in the wake of the war and the earlier Depression, the postwar economic struggles of Europe and Japan, and new revenues accruing to the federal government through a groundswell of economic activity contributed directly to the creation and expansion of social policies and expenditures for older people. By the mid-1970s, this economic scene had changed. Oil embargos, high levels of inflation and unemployment, and rising international competition created new challenges for the economy and the resources it could generate. In the ensuing years, spurts of economic growth have periodically appeared, but this time period is bookended by 1970s "stagflation" and the 2008–2009 Great Recession.

Economic data make clear the significant shift in aggregate economic well-being between the two periods and its impact on government revenues. Between 1950 and 1973, growth in the gross domestic product (GDP) averaged 4.25% annually; between 1974 and 2013, that average had fallen to 2.65% annually, more than one-third lower (US Bureau of Economic Analysis, 2013). Increasing federal budget deficits also represented a great divide between the two periods. In constant 2005 dollars, the average annual deficit (including periodic surpluses in the calculation) from 1950 to 1973 was $37 billion, or 0.71% of GDP; for the 1974–2013 period, annual deficits averaged $320 billion (also in constant dollars) and equaled 3.37% of GDP (US Office of Management and Budget, 2013), with the deficit at 9.8% of GDP at the height of the recession in 2009 (US Congressional Budget Office, 2013). The accumulated federal debt level also accelerated during the second period, today totaling $17 trillion.

Whatever the causes of this increase – weak economy, tax policy, growing expenditures – both the trend and the numbers have clear relevance for the state and fate of old-age policy. Social Security and Medicare represent the second and third largest expenditure items in the budget after national defense, Social Security totaling $768 billion and Medicare $551 billion in 2012 (US Congressional Budget Office, 2013). Not all of these dollars are directed to older people – Disability and Survivors Insurance constitute roughly one-third of Social Security spending, and Medicare beneficiaries include people with disabilities and end-stage renal disease – but in much political parlance it is tacitly assumed that older people are virtually the only beneficiaries of these outlays. Calculations of expenditures exclusively on behalf of older people are in any event very large, with a Congressional Budget Office (CBO) analysis (CBO, 2002) finding the amount to be $615 billion in 2000 and more recent calculations by Isaacs placing the figure at $767 billion in 2009 (Isaacs, 2009). In broader perspective, CBO found spending on people aged 65 and above to be 35% of the federal budget in 2000 and projected it to be 42.8% in 2010; Isaacs estimates that such spending may total 47% of the budget by 2050. This contrasts with spending on behalf of elders in the earlier period equaling only 22% of the budget in 1971, or $46 billion, roughly one-twentieth of what the total had risen to by 2010 (CBO, 2002).

New Political Realities

The 1970s also represented the beginning of a marked shift in the political context, again not necessarily directed at older people but affecting them very much nonetheless. The economic retrenchment was accompanied by weak political leadership and an undermining of the liberal impetus that had earlier generated new social programs and expenditures for elders. The latter Nixon and the Ford/Carter years led to a diminished level of trust in government and critiques of 1960s liberalism emerged from a host of new entities and individuals: think tanks (Cato Institute; Heritage Foundation), journals (*The Public Interest, Commentary National Review*), journalists (Robert Samuelson, Charles Krauthammer, William F. Buckley), and academics (Theodore Lowi and Gareth Davies). Former Office of Management and Budget (OMB) Director David Stockman (1975) spoke of "the social pork barrel," and a new sub-field of implementation studies emerged with contributions often implicitly designed to discredit Great Society programs, as in Pressman and Wildavsky's (1973) *Implementation: How Great Expectations in Washington Are Dashed in Oakland*.

Ronald Reagan's election in 1980 solidified and, for at least a generation, institutionalized a new conservative era built on privatization ("getting government off your back"), decentralization ("bring government closer to the people"), and individualization ("promoting traditional family values"). Hacker (2006) captured the totality of this agenda, labeling it "the great risk shift." From a time when government offered protection against selected risks of modern life (old age, disability, illness) and large companies provided guaranteed pensions and health care (defined benefit plans, employee and retiree health insurance), the emergent conservative agenda highlighted the roles of families, communities, and private sector entities in the name of what George W. Bush later labeled "compassionate conservatism."

To Hacker and many liberals, this new agenda favored individual insecurity over social security as a range of social programs were targeted for cutbacks or elimination. Indeed, the broader domestic politics agenda has continued over time to move to the right, as argued by Thrush (2011), saying during the 2011 debt ceiling debate that "the center has shifted to the right" and that in today's politics that "there is no such thing as a liberal victory."

The Newly Conflicted World of Old-Age Policy

Old-age programs became entangled in this new environment, though they fared better than those directed at less well-organized constituencies. An early attempt by the Reagan administration to trim Social Security benefits was withdrawn in a hailstorm of popular and Congressional opposition. A program to extend additional catastrophic care to Medicare was enacted in 1988 but then repealed one year later when seniors objected to an intragenerational financing scheme wherein higher-income seniors would have to pay for their own benefits (Himelfarb, 1995). In the 1990s, President Clinton fended off attacks by the Speaker of the House of Representatives, Newt Gingrich, on Medicare and Medicaid (but did accede to the controversial demise of the Aid to Families with Dependent Children entitlement program in favor of a non-entitlement and largely state-based Temporary Assistance to Needy Families) (Smith, 2002).

Having failed at undermining old-age programs through direct attacks, critics of the programs turned to more indirect strategies. In the case of Social Security, Campbell (2014) highlights three additional strategies employed by program critics: pressing for program privatization (improving non-seniors' rate of return on retirement contributions), alleging generational inequity (resulting in seniors receiving a disproportionate share of federal benefits), and, most recently, highlighting the "entitlement problem" (citing age-related programs as major drivers of federal budget deficits).

George W. Bush made partial privatization of Social Security the main domestic policy initiative of his second term, promoting the presumed improved rate of return, the idea of ownership in which the account would be in one's name, not some amorphous trust fund, and the notion of choice wherein retirement monies could be invested in a manner of one's choosing. Resisted by many elders, virtually all organized senior groups, and burdened by the unanswered question of how current beneficiaries would be supported if one-fourth of contributions intended for the existing trust funds were to be diverted to the new private accounts, the initiative failed. A partial privatization success was, however, found in the case of the Medicare Part D prescription drug program enacted in 2003. In exchange for agreeing to add this new and long-demanded benefit to Medicare, Republicans succeeded in having the benefit provided by private plans, the first time beneficiaries had to sign up with a private insurer to receive benefits (Morgan, 2014). Most recently, Wisconsin Republican Paul Ryan proposed replacing traditional Medicare with government-issued vouchers, a thinly veiled proxy for privatization which, as Hacker would suggest, shifts risk from the government to the beneficiary.

The intergenerational inequity argument has lasted longer and cuts closer to the perception and standing of older people than do other strategies. On both absolute and relative grounds, critics of old-age allocations are, at the very least, outspoken. Books on the topic include Peter Peterson's (1999) *Gray Dawn: How the Coming Age-Wave Will Transform America* and Kotlikoff's (2004) *The Coming Generational Storm – What You Need to Know About America's Economic Future*. MIT economist Lester Thurow (1996) queried: "Will democratic governments be able to cut benefits when the elderly are approaching a voting majority. Universal suffrage ... is going to meet the ultimate test in the elderly." And *Washington Post* columnist Robert Samuelson has been beating the generational equity drum for over 35 years, citing old-age policy costs as the culprit in "The withering freedom to govern," (1978), doubling down in "Why are we in this debt fix? It's the elderly, stupid" (2011), and returning to the theme recently "We need to stop coddling the elderly" (2013).

To argue that older people are receiving more benefits than the young is beyond debate, but to say that such an allocation pattern is disproportionate or inequitable raises a host of other issues. Two faces of this position are relevant to the issues raised in this chapter. In its more moderate version, the intergenerational argument doesn't hold that older people are not deserving of benefits; it's just that they are not seven times as deserving of federal benefits as children (Isaacs, 2009). Here, the legitimacy of elders is joined by the policy legitimacy of children. The latter case is more than reasonable and can be forcefully made (Kristof, 2014; Preston, 1984). It remains unclear, however, if proponents of this dual equity stance are more in favor of increasing benefits to children or cutting benefits to elders. If, say, $50 billion were somehow removed from age-based programs, would those funds be directed to children's issues or to tax cuts or to deficit reduction. The 1996 transition from Aid to Families with Dependent Children (AFDC) to Temporary Assistance to Needy Families (TANF) suggests that replacing or adding an entitlement benefit to children seems highly unlikely. Put more pointedly, to think that such a transfer would occur is not only naïve but cynical (Marmor, Cook, & Scher, 1997).

The second face of the intergenerational equity contention is more absolute, centering on the dramatic improvement in well-being of older Americans experienced in recent decades. The facts in the case are impressive, if not well known. Official poverty among the old has fallen four-fold since 1959, from 39% to 9%. Between 1974 and 2010, average income for older adults (in 2010 dollars) increased 50% from $21 100 to $31 410 (Federal Interagency Forum on Age-Related Statistics, 2012). Median net wealth among older people increased 42% between 1984 and 2009, and older households had 57% greater median equity in their homes than did households headed by younger adults in 1984 (Fry, 2011).

These aggregate figures are not to deny the great variations in economic well-being between older adults of different races, ethnicity, gender, age, and marital status (Gonyea, 2014; Mudrazija & Angel, 2014; US Bureau of the Census, 2012b). Across a series of economic indicators, non-Hispanic Whites are in the range of 50% better-off than non-Hispanic Blacks and Hispanics; differences are almost as pronounced between men and women; and the very old fare far worse than the younger old. The importance and implications of these wide variations and continued vulnerability of millions of older Americans are addressed in this chapter's final section and elsewhere in this volume. However, the *political* point being made here is that the overall improvements over time in elder well-being, while being indisputably positive, have cast aging policy in a harsher new light.

Elders' heightened status over time may be best captured in the codification of this new well-being as "successful aging" or "productive aging." With its origins in the gerontological community, the construct had much of an individual-level focus, urging people to adopt behaviors that would lessen the likelihood of disease and disability, maintain high functional capacity, and be actively engaged in social life (Rowe & Kahn, 1999). The implicit macro-level message was that older people were not nor need be viewed as a drag on society, its economy (the "new old" are active consumers), its workforce (older people's labor force participation is increasing), and its communities (older people volunteer and are otherwise civically engaged).

While no one would argue against a successful or productive old age, the political consequences of this new understanding of old age can be stark. In short, if older people are doing so well, on what grounds should we continue to provide the high levels of monetary and regulatory support that they have come to enjoy over the years? As set forth by Holstein and Minkler (2003, p. 793), through this paradigm policy initiatives for the old "may well suffer at the hands of a populace and a legislature that has bought the stereotypes of a new breed of successfully aging seniors who no longer need much in the way of government support." Critics of today's age-related expenditures are quick to say that they would not do anything to endanger the most vulnerable, but that nonetheless raises the question of whether old age, at least in the policy world, should be defined only by demonstrable need. Thus, welcome as new prosperity and abilities may be on their face, they do call into question many of the underpinnings supporting elders' traditional political legitimacy as set forth earlier.

Finally, those seeking to curtail age-based public expenditures have argued that Social Security and Medicare lie at the heart of America's "entitlement crisis." The programs are large and, being entitlements as the term is understood officially in budgetary terms, expenditures increase automatically rather than being subject to annual Congressional appropriations. Contention is not so much about the outlays, large as they are, but about how to understand the programs' funding mechanisms. Whereas program proponents find that rather modest "tweaking" within the programs' overall frameworks can adequately address long-term financing issues, critics see the trust funds as an accounting slight-of-hand in that they consist of IOUs that the US Treasury owes the US Social Security Administration, obligations that, in the absence of massive new revenues, the Treasury has no ability to meet.

Beyond the population and budgetary numbers, these alternative perspectives have made issue framing a new and contested element in aging policy debates. Unable to directly challenge seniors and their benefits, critics have attempted to recast aging issues in broader terms. Private sector options, generational inequity, and out-of-control spending are each ways of recasting age-based policy debate "beyond aging." Is Social Security a Ponzi scheme in crisis or can long-term viability be achieved by modest adjustments? Are Medicare expenditures on the very old spiraling out of control, or is Medicare caught up in a much larger set of issues around the delivery and financing of health care services in America? If "children are our future," should we spend more on them and less on elders? Is caring for chronically ill elders a responsibility primarily of families, or is it a fault-free risk of modern life that social policy should address? More so than was ever the case, issue framing has become critical to the understanding of and prospects for aging policy.

Significantly, this reframing of roughly the past 20 years re-introduces elite actors back into the aging policy picture. In the more encapsulated world of earlier decades, elite involvement was almost exclusively supportive of aging programs, often to the point of such actors playing pivotal roles in promoting new or expanded programs. Today, there is an ongoing elite-level battle between progressive and conservative analysts around social programs, old-age programs very much included. Most notable on the conservative side is the billionaire investor, Peter Peterson, who has underwritten several individuals and organizations invoking generational and spending concerns. Actions such as his have succeeded in pushing the political spectrum to the right, as in Campbell (2014) noting that what used to be considered conservative is now labeled moderate. More recently, liberal academics and pundits have pushed back, with there now being an ongoing battle on op-ed pages of newspapers and social media nationwide. Liberal participants were instrumental in alerting seniors and others to the dangers they saw in Social Security privatization and helped turn the tide against Bush's initiative. Because there is now evidence that Washington decision-makers pay more attention to the input of elite actors (including economic elites), their ability to sway opinion and votes takes on considerable importance (Cook & Moskowitz, 2014; Gilins, 2012). The emergence of contested elite opinion very much fuels the expanded scope of conflict that marks old-age policy today.

New actors, issues, and understandings have vastly complicated and inflamed the old-age policy arena in recent decades. Once enjoying an almost intimate political life with few prying eyes, these programs are now in the sights of a host of new individuals and organizations. Once primarily about meeting the largely uncontested needs of older people, their overall new-found well-being has raised analytical and jealous eyebrows. Also, an exercise in how to frame or place aging issues has now roiled the environment. Invoking the target population typology put forth by Schneider and Ingram (1993), we can see that old Americans have progressed through three political stages: initially seen as "dependent" (high legitimacy; low power), to then becoming "advantaged" (high legitimacy; high power), and now to encountering concerted opposition and finding themselves and their benefits increasingly rendering them "contenders" (low legitimacy; high power).

ACCOUNTING FOR OLD-AGE POLICY ENACTMENTS

A well-established welfare state literature and recent updates of it are instructive in isolating the factors that have been responsible both for long-standing old-age policy enactments and the more recent efforts at retrenchment.

Two theoretical approaches to welfare state developments speak largely to early developments but are worth brief mention because they help frame the place of older adults in the birth and growth of American social policy. The first approach, under the rubric of *the logic of industrialization*, addresses the origins of welfare states historically, with industrialization both generating the resources to make provision of social benefit possible and creating massive population upheavals which in turn necessitate some form of social provision. These changes especially impacted older people: the status that elders had enjoyed in agrarian economies suffered; factory workers became ill or exhausted, leading to poverty and ill-health in old age; and the smaller families of urban workers undermined the ability of adult children to adequately support their elders. In Wilensky and Lebeaux's (1958, p. 77) words, it was "at this point that all of the things that happen to old people become much more tragic." These factors, now including population aging, made some form of old-age pensions among the first social welfare benefits offered across the industrial world, including the United States.

A nation's *political culture* also plays a fundamental role in promoting or impeding welfare state development. Early European adapters came largely from traditions of strong central governments, autonomous state bureaucracies, corporatist private sectors, marked social stratification, and weak or delayed citizen political involvement. In comparative terms, governments were empowered to act, had the capacity to do so, and had the will to offer social protections out of a sense of privileged obligation or a fear that working-class unions and parties might forcefully press their agendas. Indeed, high levels of working-class consciousness and organization, fueled by emerging left-wing ideologies, led leftist parties to press for benefits and for conservative governments, at least in the cases of Germany and Britain, to seek to co-opt workers by taking the lead in launching social insurance programs.

The American case was largely different and in a manner that speaks to the place of older people on the nation's social policy agenda. A culture of "American exceptionalism," based on individualism, the market economy, and suspicion of government rendered American social policy "late, low, and slow." Working-class consciousness was relatively low and government capacity was relatively weak. As a consequence, the dislocations of industrialization generated little activity on behalf of workers but did come to offer benefits to older people, a population that was not expected to work and had undeniable needs. Beginning with Civil War pensions for Northern veterans, which consumed one quarter of federal spending between 1880 and 1910 (Skocpol, 1992), and then extending to the original Social Security Act, older Americans became the initial, and then the favored, American social policy constituency. That old-age benefits were the leading titles of the Social Security Act, that the United States was the only nation initially to limit national health insurance to the elderly, and that a proposal to provide new cash benefits to low-income single mothers morphed into a guaranteed income largely for the low-income aged speaks both to this American political culture and to the singular legitimacy it afforded the aged.

More proximate to current events is a conventional model of policymaking in the United States, namely, that centered on *individual and group political participation*. Citizens articulate interests, express them through letter-writing, voting, demonstrating, and joining organizations to press their interests in the halls of government. Acknowledging that not everyone participates and that some participants are clearly "more equal" than others, the model itself approximates what democracy is supposed to be all about. And, given the weakness of left-wing ideologies and the lack of powerful unions in the United States, this model most accurately captures the role of political participation and interests in American policy process.

It is also a model that is implicitly invoked when observers speak of the political power of "the elderly," their numbers, their voter turnout, their attendance at political rallies, and their organizations. While elder involvement along these lines is beyond dispute, there are varying opinions about the degree that this presence and activity results in "senior power." Despite the older population's size and growth, it still constitutes only 16% of voters nationwide, and until recently the voting pattern of seniors roughly paralleled that of younger voters. When issues of concern to seniors are in play, seniors' interest rises, but its effect and range remain in question. Aging-oriented interest groups may try to invoke the old-age vote in pressing for policy gains, but there is little evidence that they can deliver such a vote, leading Binstock (1997) to argue that it is largely an "electoral bluff." Moreover, when seniors do appear to be breaking away from other voters, it may or may not be around so-called senior interests. Thus, their disproportionate support for Republicans in the 2010 Congressional elections may have been concern about "Obamacare" and its alleged negative consequences for Medicare, but it may also have been about race, region, and aversion to activist government in general. Public opinion data centered on Tea Party adherents also speaks to the tension between ideology and interest, with older Tea Partiers expressing considerable support for Social Security ("we earned it") while evincing very little support for other government social programs (Galston, 2013).

As power has joined – and may have surpassed – legitimacy as elders' key political resource, recent legislative episodes suggest that presence has increasingly been translated into power. The rapid coming and going of the Medicare Catastrophic Coverage Act (MCCA) is instructive because it revealed the potency of old-age input into the policy process in both the law's enactment and in its repeal. Although a Republican administration secured the support of AARP for its passage, a subsequent grassroots movement by semi-organized older Americans rebelled against the financing scheme – famously pounding House Ways and Means Chairman Dan Rostenkowski's car in opposition – led directly to its repeal (Himelfarb, 1995).

In the late 1990s, responding to sustained pressure from aging interests, President Clinton made government support of seniors' prescription drug costs a major topic of his 2000 State of the Union address, and George W. Bush and Al Gore both endorsed the idea in the 2000 election. With Republicans in control after the election and AARP belatedly endorsing the final legislation, Medicare Part D became law. Republicans did succeed in making the benefit only available through private plans, but a new entitlement was on the books (Morgan, 2014). Of the controversy within the aging community about the wisdom of the plan's design, then-legislative director of AARP John Rother (2005) opined, "we just couldn't see leaving $410 billion on the table" (the 10-year expenditure estimate).

George W. Bush's attempt to partially privatize Social Security represents a third and most telling case of the power older people and their allies can demonstrate when spurred to action. Hoping that a rising generation of Baby Boomers was more wedded to private alternatives than they proved to be, alleging the pending "bankruptcy" of the current system, and aided by the rise and appeal of 401(k) and other private retirement vehicles, Bush hoped to begin the process of weaning Americans from an exclusively federal benefit (Campbell & King, 2010). Instead, current seniors were strongly opposed (though they would not have been affected), the larger public was largely indifferent, and AARP employed its considerable resources to rally opinion against the idea, a welcome political opportunity for it in the wake of the MCCA outcome years earlier.

A final construct is critical to understanding much of the expansion of aging policy and to appreciating its imposing presence in the domestic policy landscape. Variously labeled as "the new institutionalism" or "*path dependency*," this approach sees public policy as a cause as well as a consequence of broader political activity. In an initial stage, policy's enactment generates subsequent political activity, influencing both mass publics and interest groups. This model's importance lies both in its partially reversing the conventional-politics-yields-programs understanding of the policy process and, here, because it can account for much recent age-based policymaking in the United States.

Application of this understanding follows directly from arguments above that much early aging legislation resulted from widespread sympathy for the plight of older people and the ability of highly placed reformers to seize on those feelings to bring forward legislation: early Social Security, Medicare, Supplemental Security Income. The relevance of path dependency becomes clear when one asks what political and policy events transpire in the wake of these developments. As it turns out, a great deal happens, namely, the politicization of heretofore marginally involved older constituents and the creation and expansion of organizations working on their behalf. Campbell (2003) establishes a clear connection between the dramatic expansion of Old Age and Survivors Insurance benefits under Social Security and heightened participation by older Americans in the political process. She juxtaposes Social Security liberalization (involving both eligibility for benefits and benefit increases) against measures of participation (voting, contributing, campaign work, and contacting) and finds increases in each of these types of activity in the wake of Social Security expansion (and more recently to perceived threats to existing Social Security benefits). Based on this evidence, she concludes (p. 77):

> Senior mass membership groups did not create Social Security policy. Rather, the policy helped create the groups. Social Security's effects on individuals ... enhance the likelihood of group membership. Social Security created a constituency for interest group entrepreneurs to organize, just as it defined a group for political parties to mobilize.

And organize they did. AARP was the first and always the largest age-based group, but in its earlier years was not actively involved in politics and, as recently as 1972 opposed as inflationary a 20% increase in Social Security benefits proposed by Congressional Democrats (Pratt, 1976). However, AARP soon became more politically tuned to the interests of seniors and was quickly joined by a host of new aging interest groups, most of these occurrences occurring subsequent to the major aging policy enactments of the mid-1960s. Thus, Walker (1983) determined that more than one-half of the 43 aging-related interest groups in his study came into being after 1965, the breakthrough year of aging policy. In Walker's (1983, p. 403) words:

> In all of these cases, the formation of the new groups was one of the consequences of major new legislation, not one of the causes of its passage. A pressure model of the policymaking process in which an essentially passive legislature responds to petitions from groups of citizens who have spontaneously organized because of common social or economic concerns must yield to a model in which influences for change come as much from inside the government as from beyond its institutional boundaries.

If anything, the evidence of these effects is stronger today than when Walker wrote. A majority of the Leadership Council of Aging Organizations' members are advocacy, provider, educational, and research groups, many of which benefit from policy enactments for jobs, grants, contracts, and dues (Binstock, 1997). They came to constitute what Estes (1979) recognized early on as "the aging enterprise." One can protest these developments as undermining governance (Lowi, 1969) or as being responsible for ever-escalating policy expenditures (Stockman, 1975), but that is a discussion for another day. The point here is that the demand–input model of politics that was absent in 1935–1964 is very active today, created in large part by the policies that singular need, elite involvement, and economic growth of the earlier period made possible (Hudson, 1999).

EMERGING ISSUES

Today's more contentious aging politics raise a number of issues. In keeping with the new centrality of issue framing, the critical overarching question is fundamentally about problem definition. Actors on the right see the age-related policy problem essentially as reining in old-age benefits one way or another, seeing limited resources, recognizing elders' improved well-being, and embodying an antipathy to large governmental benefit programs. Yet, actors on the left use a completely different frame, namely, the ongoing income, health care, retirement, and chronic illness needs facing large numbers of elders must be acknowledged, buttressed by a belief that identifiable and non-fault-based risks should be addressed collectively.

Concerning specific policies, the importance of issue framing is seen most starkly in the case of Medicare. Because of Medicare's popularity and critical contribution to old-age well-being, no actors are willing to go on record as being opposed to its continuation in some form. Thus, both conservatives and progressives each emphatically say that they want to "save" Medicare (White, 2001). Yet, their solutions for saving it are dramatically different. Conservatives seek to largely privatize the system, bring savings through market competition, and reorganize it as a voucher-based defined contribution rather than as a defined benefit program. The left would assure Medicare's future through tighter regulation and curtailing reimbursements to physicians, hospitals, and the pharmaceutical industry (Morgan, 2014). The strategies and possible outcomes are far different in this classic battle between competition and regulation, but both are done in the name of saving the program.

The framing or reframing of Social Security policy centers on how to understand the needs and expectations of current and future beneficiaries. Program critics, citing improved well-being and heightened longevity, call for raising the Social Security retirement age as a straightforward and meaningful way of curtailing program growth. They also are promoting a "chained-CPI" which would lower cost-of-living adjustments over time, the rationale being that the current index does not recognize the substitutions that elders make in the face of rising prices.

Progressives have a very different take on the program and its role. Social Security constitutes the lion's share of income for seniors in the bottom half of the income distribution, and the need is for program maintenance, if not expansion, in light of the income shortfall many Baby Boomers are expected to encounter as they enter retirement. As for population longevity, progressives cite the strained circumstances that long-lived widows and never-married women are already facing. Their principal option for shoring up the program would be raising or eliminating the wage-ceiling on Social Security contributions, set at $117 000 in 2014. Many liberals also favor reining in a number of so-called tax expenditures – such as the exclusion of employer-provided health insurance from both individual income and payroll taxes – which represent upwards of $500 billion in foregone federal revenue annually (Gist, 2014).

A final option that analysts on both the left and right have entertained involves means-testing. At first glance, that is a heretical idea to program defenders, who see Social Security and Medicare as the cornerstones of America's universal social insurance policy, one that is conceptually antithetical to means-testing. Means-testing is a definitional property of public assistance programs, and social insurance is largely about equity. (Social Security's progressive benefit formula is a subtle form of means-testing but has avoided that appellation over the years.) As conventionally employed, means-testing is about cutting or limiting benefits, and some conservative proposals call for curtailing benefits to elders with high levels of non-Social Security income. The Heritage Foundation has proposed that Social Security benefits be eliminated for individuals with more than $110 000 and couples with more than $165 000 in non-Social Security annual income (Howard, 2014).

However, some conservatives and liberals as well have broached the idea of means-testing on the contributions or financing side of the program whereby higher-income individuals would pay more for the same benefit amount. This is already underway in the financing of Medicare Parts B and D, and the possibility has now touched Social Security. Certainly, raising or eliminating the wage-ceiling is high on a list of liberals' reforms, but moderates and conservatives have also expressed interest in this alternative. As well, the Obama administration has introduced two major means-tested revenue enhancements to Medicare Part A through the Affordable Care Act (though they are not labeled as such): individuals with incomes over $200 000 and couples with incomes over $250 000 will pay 2.35% on that income rather than the 1.45% paid on all income below those levels. Many of those same individuals will also pay a new 3.8% tax on their investment income. For liberals these new revenues can be seen as reflecting a long-standing belief in progressive taxation, although program advocates further to the left see this kind of means-testing as starting to erode the political support that so-called universal programs have historically enjoyed (Marmor & Hacker, 1997). As for conservatives, Howard (2014, p. 241) observes that "it appears that conservative officials are anti-tax but not anti-premium."

Beyond the stalemate currently surrounding Social Security and Medicare, there is a final arena instructive for the observations being made here. Public support for those in need of long-term care supports and services has long been an orphan in social policy. Such funding as there is comes overwhelmingly from the Medicaid program, which has long financed institutional rather than community-based services. A philosophical breakthrough in long-term care appeared to occur as part of passage of the Affordable Care Act in 2010. An initiative known as the CLASS Act would have instituted a voluntary insurance program for people with disabilities and, after a period of time, elders suffering from chronic illnesses. Adding long-term care to elders' social insurance protections was hailed by the gerontological and progressive communities (Public Policy & Aging Report, 2010). However, as part of the "fiscal cliff" and government shut-down battles during 2011–2012, Republicans insisted that CLASS be repealed, arguing that it was yet another example of excessive federal reach into a realm more appropriately addressed by families and the private sector. Extending social insurance coverage to long-term care was to be the third and probably final step after the income and acute health care protections of Social Security and Medicare in shielding elders against the vicissitudes of late-life.

The final comment is both political and ethical. Efforts to curtail aging-based programs are usually based on an expressed belief that we should only support elders in need. That was the nation's historic posture toward the old, and even critics of age-related expenditures profess an interest in maintaining those protections, albeit perhaps with new strictures. To the extent that the country might move in the "needs" direction, it could be seen as having the effect of "re-residualizing" elders, that is, redefining old age as a time of weakness and vulnerability (Hudson, 1999). Doing so would negate social policy's success in transforming old age and retirement into a normative stage of life wherein income is maintained and access to health care is assured. Under such a regime some elders would, of course, fare quite well, but most would see their well-being suffer were means-testing of benefits (and even of contributions) widely instituted. The counter-position would be a continuation of what Skocpol (1995) has labeled "targeting within universalism," that is skewing benefits toward lower-income recipients within a universal framework as in the cases of Social Security and Medicare. Even were modifications required due to economic and political pressures, maintenance of this modified social insurance philosophy would continue the preventative social policy that has become the heart of social insurance for older Americans.

References

Barrios, K. M., Lopez, M. H., Kennedy, C., & Barr, K. (2008). *Young voter registration and turnout trends.* Washington, DC: Center for Information and Research on Civic Learning and Engagement.

Binstock, R. H. (1972). Interest group liberalism and the politics of aging. *The Gerontologist, 12,* 65–80.

Binstock, R. H. (1983). The aged as "scapegoat." *The Gerontologist, 23,* 136–143.

Binstock, R. H. (1997). The old-age lobby in a new political era. In R. B. Hudson (Ed.), *The future of age-based public policy* (pp. 56–74). Baltimore, MD: Johns Hopkins University Press.

Brotman, H. (1977). *The graying of every 10th American or every 9th American. Developments in aging (xv–xxiii). U.S. Senate Committee on Aging.* Washington, DC: Government Printing Office.

Burke, V., & Burke, V. (1974). *Nixon's good deed: Welfare reform.* New York, NY: Columbia University Press.

Campbell, A. L. (2003). *How politics makes citizens: Senior political activism and the American welfare state.* Princeton, NJ: Princeton University Press.

Campbell, A. L. (2014). Social Security, the Great Recession, and the entitlements problem. In R. B. Hudson (Ed.), *The new politics of old age policy* (pp. 183–200) (3rd ed.). Baltimore, MD: Johns Hopkins University Press.

Campbell, A. L., & King, R. (2010). Political resilience in the face of conservative strides. In R. B. Hudson (Ed.), *The new politics of old age policy* (pp. 233–253) (2nd ed.). Baltimore, MD: Johns Hopkins University Press.

Cook, F. L., & Moskowitz, R. L. (2014). The great divide: Elite and mass opinion about Social Security. In R. B. Hudson (Ed.), *The new politics of old age policy* (pp. 69–96) (3rd ed). Baltimore, MD: Johns Hopkins University Press.

Derthick, M. (1979). *Policymaking for social security.* Washington, DC: Brookings Institution.

Estes, C. (1979). *The aging enterprise.* San Francisco, CA: Jossey-Bass.

Federal Interagency Forum on Age-Related Statistics. (2012). <http://www.agingstats.gov/agingstatsdotnet/main_site/default.aspx>.

Fry, R. (2011). The rising gap in economic well-being: The old prosper relative to the young. *Pew Research: Social and Demographic Trends*

Galston, W. (2013). The Tea Party and the GOP crackup. *Wall Street Journal* (October 15). <http://online.wsj.com/news/articles/SB10001424052702303376904579135231053555194>.

Gilins, M. (2012). *Affluence and influence: Economic inequality and political power in America.* Princeton, NJ: Princeton University Press.

Gist, J. (2014). Fiscal effects of population aging in the United States. In R. B. Hudson (Ed.), *The new politics of old age policy* (pp. 39–68) (3rd ed). Baltimore, MD: Johns Hopkins University Press.

Gonyea, J. G. (2014). The policy challenges of a larger and more diverse oldest-old population. In R. B. Hudson (Ed.), *The new politics of old age policy* (pp. 155–180) (3rd ed.). Baltimore, MD: Johns Hopkins University Press.

Gordon, L. (1998). *Pitied but not entitled: Single mothers and the history of welfare.* New York, NY: Free Press.

Haber, C. (1983). *Beyond 65: The dilemmas of old age in America's past.* New York, NY: Cambridge University Press.

Hacker, J. (2006). *The great risk shift: The assault on American jobs, families, health care, and retirement.* New York, NY: Oxford University Press.

Heclo, H. (1988). Generational politics. In J. L. Palmer, T. Smeeding, & B. B. Torrey (Eds.), *The vulnerable* (pp. 381–412). Washington, DC: Urban Institute.

Himelfarb, R. (1995). *Catastrophic politics: The rise and fall of the Medicare Catastrophic Coverage Act of 1988.* State College, PA: Pennsylvania State University Press.

Holstein, M., & Minkler, M. (2003). Self, society, and the "new gerontology." *The Gerontologist, 43,* 787–796.

Howard, C. (2014). Means-testing of entitlements: Good policy? Good politics? In R. B. Hudson (Ed.), *The new politics of old age policy* (pp. 236–253) (3rd ed.). Baltimore, MD: Johns Hopkins University Press.

Hudson, R. B. (1999). Conflict in today's aging politics: New population encounters old ideology. *Social Service Review, 53*, 358–379.

Isaacs, S. (2009). How much do we spend on children and the elderly? <http://www.brookings.edu/~/media/Files/rc/reports/2009/1105_spending_childrenisaacs/1_how_much_isaacs.pdf>.

Kotlikoff, L. J. (2004). *The coming generational storm: What you need to know about America's economic future*. Cambridge, MA: MIT Press.

Kristof, N. (2014). Progress in the War on Poverty. *New York Times, A21* (January 9).

Lipset, S. M. (1997). *American exceptionalism: A double-edged sword*. New York, NY: W.W. Norton.

Lowi, T. (1969). *The end of liberalism*. Chicago, IL: W.W. Norton.

Marmor, T. M. (1970). *The politics of Medicare*. Chicago, IL: Aldine.

Marmor, T. R., Cook, F. L., & Scher, S. (1997). Social Security and the conflict between generations: Are we asking the right questions?. In E. R. Kingson & J. H. Schulz (Eds.), *Social security in the 21st century* (pp. 195–207). New York, NY: Oxford University Press.

Marmor, T.R., & Hacker, J. (1997). Ends don't justify the means-test. *Los Angeles Times* (July 14). <http://articles.latimes.com/1997/jul/14/local/me-12570>.

Morgan, K. J. (2014). The Medicare challenge: Clients, cost controls, and Congress. In R. B. Hudson (Ed.), *The new politics of old age policy* (pp. 201–220) (3rd ed.). Baltimore, MD: Johns Hopkins University Press.

Morris, C. R. (1996). *The AARP: America's most powerful lobby and the clash of generations*. New York, NY: Times Books.

Mudrazija, S., & Angel, J. L. (2014). Diversity and the economic security of older Americans. In R. B. Hudson (Ed.), *The new politics of old age policy* (pp. 138–154) (3rd ed.). Baltimore, MD: Johns Hopkins University Press.

Peterson, P. G. (1999). *Gray dawn: How the coming age-wave will transform America – and the world*. New York, NY: Times Books.

Pratt, H. (1976). *The gray lobby*. Chicago, IL: University of Chicago Press.

Pressman, J., & Wildavsky, A. (1973). *Implementation: How great expectations in Washington are dashed in Oakland*. Berkeley, CA: University of California Press.

Preston, S. H. (1984). Children and the elderly: Divergent paths for America's dependents. *Demography, 21*, 435–457.

Public Policy & Aging Report. (2010). Bringing CLASS to long-term care through the Affordable Care Act. *Public Policy & Aging Report, 20*(2), 1–32.

Rimlinger, G. (1971). *Welfare policy and industrialization in Europe, America, and Russia*. New York, NY: Wiley.

Rother, J. (2005). Personal communication.

Rowe, J. W., & Kahn, R. L. (1999). *Successful aging*. New York, NY: Dell.

Rubinow, I. (1934). *The quest for security*. New York, NY: Holt.

Samuelson, R.J. (1978). The withering freedom to govern. *Washington Post* (March 5).

Samuelson, R.J. (2011). Why are we in this debt fix? It's the elderly, stupid! *Washington Post* (July 28). <http://www.washingtonpost.com/opinions/why-are-we-in-this-debt-fix-its-the-elderly-stupid/2011/07/28/gIQA08LtfI_story.html>.

Samuelson, R.J. (2013). We need to stop coddling the elderly. *Washington Post* (November 3). <http://www.washingtonpost.com/opinions/robert-j-samuelson-we-need-to-stop-coddling-the-elderly/2013/11/03/4063ebc0-430f-11e3-a624-41d661b0bb78_story.html>.

Schattschneider, E. E. (1960). *The semi-sovereign people*. New York, NY: Holt, Reinhart, and Winston.

Schmidhauser, J. R. (1970). The elderly and politics. In A. M. Hoffman (Ed.), *The daily needs and interests of older people* (pp. 72–89). Springfield, IL: Charles C Thomas.

Schneider, A., & Ingram, H. (1993). The social construction of target populations. *American Political Science Review, 87*, 334–347.

Shearon, M. (1938). The economic status of the aged. *Social Security Bulletin, 1*, 5–16.

Sherman, A. K., & Kolker, A. (1987). *The social bases of politics*. Belmont, CA: Wadsworth.

Skocpol, T. (1992). *Protecting soldiers and mothers*. Cambridge, MA: Harvard University Press.

Skocpol, T. (1995). Targeting within universalism: Politically viable strategies to combat poverty in the United States. In T. Skocpol (Ed.), *Social policy in the United States* (pp. 250–274). Princeton, NJ: Princeton University Press.

Smith, D. G. (2002). *Entitlement politics: Medicare and Medicaid, 1995–2001*. Chicago, IL: Aldine de Gruyter.

Stockman, D. (1975). The social pork barrel. *Public Interest, 39*, 3–30.

Thrush, G. (2011). The summer of liberal discontent. Politico. <http://dyn.politico.com/printstory.cfm?uuid=9D50ABF4-22CE-4945-B01B-9CAF16F5C9B8>.

Thurow, L. (1996). The birth of a revolutionary class. *New York Times Magazine* (May 19), 47ff.

Tufte, E. R. (1978). *Political control of the economy*. Princeton, NJ: Princeton University Press.

Upp, M. (1982). A look at the economic status of the aged then and now. *Social Security Bulletin, 45*, 16–20.

US Bureau of Economic Analysis. (2013). *U.S. GDP by year since 1929 compared to major events*. Washington, DC: U.S. Department of Commerce.

US Bureau of the Census. (2012a). Reported voting and registration, by race, Hispanic origin, sex, and age, for the United States: November 2012. <http://www.census.

gov/hhes/www/socdemo/voting/publications/p20/2012/tables.html>.

US Bureau of the Census. (2012b). *Current population survey annual social and economic supplement.* <www.census.gov/hhes/www/poverty/publications/pubs-cps.html>.

US Congressional Budget Office. (2002). *Federal spending for the elderly and children.* <http://www.cbo.gov/sites/default/files/cbofiles/ftpdocs/23xx/doc2300/fsec.pdf>.

US Congressional Budget Office. (2013). *The budget and economic outlook, fiscal years 2013 to 2023 and supplementary data.*

US Office of Management and Budget. (2013). *Summary of receipts, outlays, surpluses and deficits.* Table 1.1. <http://www.whitehouse.gov/omb/budget/Historicals>.

Vladeck, B. (1980). *Unloving care: The nursing home tragedy.* New York, NY: Basic Books.

Walker, J. (1983). The origins and maintenance of interest groups in America. *American Political Science Review, 77,* 390–406.

White, J. (2001). *False alarm: Why the greatest threat to Social Security and Medicare is the campaign to "save" them.* Baltimore, MD: Johns Hopkins University Press.

Wilensky, H., & Lebeaux, C. (1958). *Industrialization and social welfare.* New York, NY: Free Press.

CHAPTER

22

The Future of Retirement Security in Comparative Perspective

John B. Williamson[1] and Daniel Béland[2]

[1]Department of Sociology, Boston College, Chestnut Hill, MA, USA [2]Johnson-Shoyama Graduate School of Public Policy, University of Saskatchewan, Saskatoon, Saskatchewan, Canada

OUTLINE

Introduction	462
Social Security in the United States	462
Efforts to Partially Privatize Social Security in the United States	463
Social Security "Parametric" Reform Proposals	464
Employer-Sponsored Pensions in the United States	465
The Rapid Move from DB Pensions to DC Pensions	465
The Emergence and Future of 401(k) Pensions	465
Pros and Cons of 401(k) Plans	466
The Problem of Major Market Corrections	466
The Implicit Assumption of Financial Literacy	466
International Developments and Lessons	467
The Trend toward Partial Privatization	467
Chile: A Very Influential Move from the PAYG-DB to the FDC Model	468
Argentina: An Unsuccessful Latin American Experiment Partial Privatization	469
China: An Innovative New Rural Pension Model	470
The United Kingdom: An Industrial Country Having Noteworthy Problems with its FDC Pillar	473
Canada: A PAYG-DB-Based Scheme with Lessons for the United States	475
Conclusion	476
Acknowledgments	478
References	478

INTRODUCTION

Retirement security is a crucial policy issue in this era of accelerated population aging. This chapter explores recent developments relevant to the future of retirement security, with a focus on public pension policy. The chapter begins by reviewing recent trends in the United States. It then considers developments in five other nations in part to draw lessons for the United States and in part to highlight some innovative new pension models that will be primarily of interest to policy analysts with an interest in recent developments in old-age security policy in developing countries. While the pension burden linked to population aging will be increasing in the United States in the decades ahead, this problem is going to be even more serious in many developing nations, such as China, that are growing old before growing rich.

We give considerable attention to the trend toward the privatization of pension systems in part because the proposal to introduce funded individual accounts as part of the public pension scheme has been hotly debated in the United States and in part because such accounts have been adopted during the past few decades by about 30 countries around the world (James, 2005). To some advocates of movement in this direction, greater privatization is viewed as a means to reduce or eliminate projected government budget deficits linked to population aging and what they consider overly generous benefit promises implicit in existing pension statutes. The goal is generally to reduce spending on pension programs without increasing taxes. To some, more privatization represents a means to shrink the government (welfare state retrenchment) as pensions represent a major component of government spending. But to those who oppose movement in the direction of further privatization, a major concern is that it will increase income inequality, gender inequality, and the fraction of future retirees at an increased risk of poverty or near poverty during their retirement years (Béland & Waddan, 2012; Williamson, 2011).

SOCIAL SECURITY IN THE UNITED STATES

Because so many people (especially those on low and middle incomes) depend on Social Security for a substantial fraction of their retirement income, this program is crucial to any discussion of the future of retirement security in the United States. Social Security is generally understood as a pay-as-you-go-defined-benefit (PAYG-DB) scheme, but it is more accurate to describe it as a "partial-reserve" system (DeWitt, 2010). This is true because the Social Security trust funds always hold at least modest (and currently quite substantial) reserves designed to make it possible to pay benefits during periods when the payroll tax revenues being collected fall below benefits promised under current statutes. Social Security is a PAYG scheme because benefits to current pensioners are paid for largely via payroll taxes on current workers and their employers; and it is a defined benefit (DB) scheme because benefits are based on a wage-indexed average of past covered wages and the number of years as a contributor.

Between the mid-1980s and 2009, Social Security was intentionally taking in more money than it was paying out, but as of 2010 it began to take in less. However, due to the treasury bonds purchased during the surplus years, the size of the trust fund will actually continue to increase until about 2020. But starting then, unless some legislative changes are made, the assets in the Social Security trust funds will rapidly decline and be depleted by 2033 (Board of Trustees, 2014). In recent decades, many commentators have referred to the projected deficit as a crisis (Peterson, 1999; Salsman, 2011). On the other hand, many other analysts have described the situation as a problem that can be dealt with using a combination of parametric reforms, that is, relatively modest and seemingly technical adjustments (Kingson & Williamson, 2001). But it is also the case that analysts of all political persuasions agree that by about 2033 the projected depletion of the trust funds must be dealt with using a combination of payroll tax increases and benefit cuts (National Research Council, 2012).

Efforts to Partially Privatize Social Security in the United States

In debates over pension reform in the United States some commentators support and others oppose the "partial privatization" of Social Security. The terms "privatization" and "partial privatization" tend to be used interchangeably to mean partial privatization (as most of those on both sides of the debate are assuming that at least some version of the traditional PAYG-DB scheme would remain in place even if those pushing for "privatization" were to prevail).

The proposed introduction of funded individual accounts as a formal part of Social Security in the United States is not viewed as a parametric reform; it is viewed as a much more fundamental structure reform. Since the mid-1990s proponents of the partial privatization of Social Security in the United States have typically proposed that we divert a portion of the current Social Security payroll tax into a 401(k)-like personal account while at the same time reducing the size of the DB component of Social Security. The goal, often unstated, is to make the scheme less redistributive and to provide more of an incentive for older workers to remain in the labor force longer than they do at present. The proposed partial privatization of Social Security is not a new idea, as it has been around since the mid-1980s (Ferrara, 1985). For some analysts the proposed partial privatization of Social Security is viewed as a way to bring revenues and benefits into balance over the long run. In debates to date relatively little attention is given to the evidence that during the transition decades spending levels would need to increase to pay promised pension benefits to those already retired while at the same time diverting funds into the new individual accounts (Williamson, 2002a).

Throughout the late 1980s and early 1990s, although it was not viewed as politically feasible by mainstream policy analysts, a small number of analysts and commentators associated with conservative (or libertarian) think tanks such as the Cato Institute proposed that we at least partially privatize Social Security. Due in part to the often-repeated assertion by conservative commentators that Social Security would soon "go bankrupt," the idea of privatization eventually gained some political traction. Although support for the traditional Social Security program remained strong, many Americans became fearful that the funds would not be available to pay their pensions (Reno & Friedland, 1997). Some conservatives used arguments linked to population aging and generational equity to bolster calls for major structural reforms, specifically the partial privatization of Social Security (Béland, 2005; Williamson & Watts-Roy, 2009).

In the United States, in the mid-1990s, the idea of partially privatizing Social Security finally moved onto the mainstream pension policy agenda (Cook, Barabas, & Page, 2002). At the time, many conservative commentators called for the partial privatization of the program, and some clearly viewed something closer to full privatization of this component of the pension system as their ultimate goal (Ferrara & Tanner, 1998; Peterson, 1999). Simultaneously, liberals typically opposed proposals advocating the diversion of some payroll taxes currently used to finance Social Security into funded individual retirement savings accounts (Kingson & Williamson, 1999). Although President Clinton seemed at times willing to consider Social Security reforms that would include funded individual accounts (Gillon, 2008), no such policy changes were made (Béland, 2005).

In 2001, the President's Commission to Strengthen Social Security (2001) was created by George W. Bush, only a few months after he had entered the White House. All members of that pro-privatization commission were selected in part because they favored some form of individual accounts as part of the Social Security system. Three proposals emerged from the commission, each calling for the introduction of individual accounts. When it became clear that along with the individual accounts substantial cuts would materialize in the traditional DB component of Social Security, the proposed reforms looked a lot less attractive to many Americans. With the terrorist attacks of September 11, 2001, and the sharp drop in the stock market between 2000 and 2002, calls for the partial privatization of Social Security all but ended (Béland & Waddan, 2012).

President George W. Bush, however, refused to give up. Soon after he won re-election in 2004, he began a full-court press to sell the American public on the idea of adding funded individual accounts as part of Social Security reform. He highlighted this proposal in his 2005 State of the Union address and then traveled around the country for about 6 months trying to get the American public behind the idea. In an effort to undercut support for Social Security as it is currently structured, he argued that the Social Security Trust Fund was nothing but a bunch of IOUs – his term for the US government bonds held by the Social Security Administration (Hardy & Hazelrigg, 2007). In the end, this argument and his effort to sell partial privatization fell flat; the more the American public learned about the specifics of what would be involved with the partial privatization of Social Security, the less they liked the idea. In the 2006 mid-term elections and in the 2008 Presidential election, many Republicans who were up for re-election went out of their way to avoid discussion of Social Security reform and particularly proposals to partially privatize Social Security (Williamson & Watts-Roy, 2009). During the Obama Administration, Social Security privatization remains off the table, in part because it is clear that the President would veto any such legislation (Béland & Waddan, 2012). However, it is entirely possible that the proposed partial privatization of Social Security will again be given serious consideration, but probably not until Republicans control the House, Senate, and the Presidency.

Social Security "Parametric" Reform Proposals

As Social Security privatization currently remains outside the national policy agenda, a number of parametric changes in US Social Security policy could be enacted to balance projected revenues and pension spending. The list below briefly summarizes some of the parametric proposals likely to be under consideration:

1. shifting to a less generous formula for determining starting benefit levels;
2. shifting to a less generous cost-of-living adjustment (COLA);
3. increasing the age of eligibility for full retirement benefits;
4. increasing the cap on earnings subject to payroll taxation (a proposal supported by President Obama);
5. increasing the number of years of lifetime covered employment used to compute benefits;
6. cutting benefits across the board, in a flat manner, an idea many liberals oppose; and/or
7. increasing the payroll tax rate, an approach strongly opposed by most conservatives.

For more detailed discussions of these and other proposals, see Altman & Kingson (2015), John and Reno (2012), Nuschler (2012), and Sass, Munnell, and Eschtruth (2009).

EMPLOYER-SPONSORED PENSIONS IN THE UNITED STATES

The Rapid Move from DB Pensions to DC Pensions

The United States relies extensively on private, employer-sponsored pensions, which are built on the top of Social Security but remain voluntary, as opposed to the situation prevailing in some European countries like Denmark and the Netherlands. From this perspective, although efforts to partially privatize the federal Social Security program have not been successful, private benefits such as employer-sponsored pensions do play a major role within the American retirement security system (Hacker, 2006).

There are two broad categories of employer-sponsored pensions, DB pensions and defined-contribution (DC) pensions (O'Rand, Ebel, & Isaacs, 2009). Pensions based on the DB model typically provide eligible employees with highly predictable lifetime pensions based on a formula taking into consideration years of service and earnings. Often the size of the benefit is entirely or largely based on earnings during the last few years of employment, and the benefit is sometimes adjusted or partially adjusted for inflation during the payout years. Typically contributions to these DB plans are only made by the employer. In contrast, DC pension plans only specify the amount contributed (by the employee, employer, or both) to the employee's retirement account, not the size of the eventual pension to be paid. The eventual pension benefit is largely a function of the amount contributed, changes in asset allocation made by employees, and market trends over the years.

In the last three decades, the fraction of the private sector wage and salary labor force covered by an employer-based pension scheme has been relatively steady at just under 50%, but the shift from those based on the DB model to those based on the DC model has been dramatic (O'Rand et al., 2009). While coverage by DB schemes has not been declining among public sector workers, the story is quite different in the private sector. In 1980, among private sector workers who were covered by some sort of employer-sponsored pension plan, 60% were covered by DB plans alone, 17% by DC plans alone, and 23% by both. By 2006 only 8% were covered by DB plans alone, while 70% were covered by DC plans alone, and 22% by both (Munnell, Aubry, & Muldoon, 2008). The decline of DB pensions has continued since then. For example, in 2012 only 11 Fortune 100 companies were offering DB plans to new salaried workers, down from 19 in 2009 (Geisel, 2012). In the following discussion, the focus will be on DC schemes due to the rapid shift underway from DB to DC schemes, particularly in the private sector. This trend reflects a form of partial privatization of retirement security currently taking place in the United States, even in the absence of any explicit partial privatization of the Social Security program itself (Hacker, 2006). While Social Security remains a public program, retirement income security in the United States relies extensively on private programs such as 401(k) plans.

The Emergence and Future of 401(k) Pensions

Adopted in 1978, implemented in 1981, and revised on a regular basis (IRS, 2012), 401(k) plans have become the most pervasive employer-sponsored DC pension schemes in the United States (Smith, 2012). Initially, these plans were viewed as savings plans or supplements to employer-sponsored DB plans (Munnell, Golub-Sass, & Muldoon, 2009), that is, as part of the third leg of the proverbial three-legged stool needed to assure adequate retirement income; the other two legs being occupational pensions (often DB plans) and Social Security. However, the 401(k) plans are instead increasingly replacing those DB plans.

Relative to DB plans, the 401(k) plans have a number of benefits for employers, but for employees the picture is more a mixture of pros and cons.

Pros and Cons of 401(k) Plans

For many workers, the shift from DB plans to 401(k) plans has dramatically reduced retirement income security (Boivie, 2011). This is true largely because: (1) despite the recent push for automatic enrollment in the default option (IRS, 2013), enrollment in 401(k) plans frequently remains voluntary and, as a result, many workers (particularly young workers) elect not to participate (Munnell et al., 2009); (2) many do not contribute enough to assure an adequate pension at retirement; (3) many "cash-out" when they move between employers; (4) many do not sufficiently diversify their portfolio, make unwise asset allocation choices, or invest too heavily in their company's stock (Munnell et al., 2009); and (5) their holdings are vulnerable to sharp corrections in financial markets just prior to the time of planned retirement.

Yet, from the perspective of employees, 401(k) plans also have advantages, when compared with DB plans: (1) 401(k) plans make it easier for employees to transfer their account balances when moving to a new employer; (2) 401(k) plans also allow those employees who want to actively manage their pension assets to do so, although this turns out to be a mixed blessing because many of those who actively manage their accounts do a poor job at it (Munnell & Sundén, 2004); and (3), finally, the pension assets are better protected than with many company DB schemes. The assets are administered by an independent third party, such as a mutual fund company, and as a result they are not vulnerable to appropriation in connection with bankruptcy or a merger.

Turning to employers, shifting to DC schemes has a number of clear benefits for them: (1) 401(k) plans reduce the cost of administering the pension system and the burdensome bureaucratic obligations associated with the 1974 ERISA regulations and requirements (Schulz & Borowski, 2006), a particularly important consideration for small businesses; (2) unlike company DB plans, most 401(k) plans are largely employee-funded, with employers often contributing much less than they did to their prior DB schemes; and (3), finally, the risks associated with pension fund financing are shifted from the employer to the employee (Hacker, 2006), including market risks as well as risks associated with increased longevity (of retired workers), corporate downsizing, and pension fund under-funding.

The Problem of Major Market Corrections

Key policy lessons were drawn from the two recent deep stock market corrections, the first between 2000 and 2002 and the second, even deeper, between late 2007 and early 2009. Both were associated with dramatic reductions in 401(k) retirement account assets, and it has become clear to many analysts that pension plans based on the 401(k) model are often less than adequate substitutes for traditional DB schemes, with respect to the predictability and adequacy of the eventual pension. Market risk is only one of many reasons why the 401(k) model is not living up to expectations (Smith, 2012), and various changes to the model have been proposed in an effort to improve the economic security of workers (Ghilarducci, 2008).

The Implicit Assumption of Financial Literacy

Under the 401(k) model, much of the decision making about asset allocation and how much to contribute is shifting from the employer to the employee. Although some workers like the idea of being in control of these decisions, the reality is that financial literacy among many workers is quite low (Schulz & Borowski, 2006). The financial information needed to make well-informed decisions can be hard to find even for those who are interested. More problematic is the evidence suggesting that many workers are unable or unwilling to take the time to acquire the level of financial literacy called for to protect against bad investment decisions. In an effort to deal with this lack of adequate information, a number of recent reforms in 401(k) plans have been proposed which are designed to help protect workers with low levels of financial literacy. As for the increasingly popular idea among pension experts that enhancing the financial literacy of workers would improve their future retirement security, there is a need to be cautious about the alleged benefits of this type of literacy on pension behavior. Although financial literacy initiatives such as the federal web site MyMoney.gov make sense, there is very little evidence to date that such efforts are bringing much by way of benefits to workers and citizens (Olen, 2012; Willis, 2008).

INTERNATIONAL DEVELOPMENTS AND LESSONS

Around the world, the most significant public pension policy trend in recent decades has been the shift toward less reliance on PAYG-DB schemes and to greater dependence on multi-pillar schemes that include a funded defined contribution (FDC) pillar. Typically the FDC pillar is funded by a mandatory tax paid by employees (sometimes supplemented by employer contributions). Each employee has an individual account that is invested in various financial markets, typically by a private sector money management organization. The eventual value of these accounts is a function of the amount contributed, the fees charged to manage the account, asset allocation decisions made by workers or these asset managers, and by trends in financial markets. Because most schemes that include an FDC pillar also include a publicly financed PAYG-DB pillar, it is common to refer to such multi-pillar schemes as being partially privatized. Given this trend and the extent to which the issue has been debated in the United States and is likely to be debated again, it is of interest to discuss how well partial privatization and funded individual accounts more generally are working in other countries.

The Trend toward Partial Privatization

By the early 1980s, many countries with maturing PAYG-DB pension schemes faced large deficits. In 1981, Chile became the first of these nations to confront this problem by introducing an FDC scheme while at the same time starting the process of gradually phasing out its existing PAYG-DB scheme. Its multi-pillar scheme is primarily based on a privately managed FDC individual accounts pillar, but even Chile's public pension system includes other redistributive pillars that are government-financed making its scheme only partially privatized.

By the 1990s, many nations around the world had begun to introduce reforms influenced by the Chilean model (Orenstein, 2008). By the early mid-2000s, approximately 30 countries had at least partially privatized their public pension schemes by adding a mandatory FDC pillar typically managed by private sector organizations, subject to a variety of government regulations and controls (James, 2005). Since then there has been a decline in the rate of additional adoptions and some countries, such as Argentina, have elected to phase out their FDC pillars (Béland & Orenstein, 2013).

Many factors have contributed to the spread of reforms involving reductions in commitments to existing public PAYG-DB pillars in favor of privately managed FDC pillars, the most essential being the influence of the World Bank and the International Monetary Fund

(IMF). These efforts have been reshaping welfare-state-related social policies around the world (Béland & Orenstein, 2013). Of particular note was an influential World Bank (1994) publication, *Averting the Old Age Crisis*. It called for a "three-pillar pension scheme" – with a modest publically financed first pillar in the form of either a universal or a means-tested social pension. The focus of this pillar was on poverty reduction and income redistribution. The second pillar called for a mandatory privately managed and fully funded FDC pillar designed to promote pre-retirement income replacement, and the third was a voluntary private individual savings pillar (World Bank, 1994). By the late 1990s, in an effort to respond to critics both outside (Kay & Sinha, 2008) and inside the World Bank (Béland & Orenstein, 2013), this model was updated to become a five-pillar model that added a PAYG-DB pillar and a "non-financial" pillar that included programs such as housing benefits (Holzmann, Hinz, & Dorfman, 2008).

The discussion below focuses on pension schemes in five countries, each of which has introduced a FDC pillar as an optional or a mandatory pillar (Chile, Argentina, China, the United Kingdom, and Canada). These cases have been selected to illustrate some of the issues and potential outcomes with which pension analysts are dealing. Some have been selected as successful examples others as in some way unsuccessful examples. The case of the United Kingdom and Canada and to a lesser extent those of Chile and Argentina have been selected for their relevance to the past and quite likely future debate over the partial privatization of Social Security in the United States. The case of China has been selected because in the years ahead it is likely to prove to be very influential as a model for many other developing countries around the world. The cases of Chile and Argentina also have lessons for other developing nations concerning the introduction of funded individual accounts.

Chile: A Very Influential Move from the PAYG-DB to the FDC Model

One reason the Chilean FDC scheme has been so influential is that it was the first country to shift from a PAYG-DB model to the FDC model (Kritzer, 2008). Although more accurately described as a partially privatized rather than a fully privatized scheme, it is one of the most privatized pension programs in the world. In addition, it has been in place over 30 years, substantially longer than the FDC schemes introduced in most other countries. Another reason the Chilean scheme has received much attention is that reports on it, particularly during the first 15 years or so, were very favorable and seemed to point to far higher "returns" for covered workers than seemed possible with existing payroll tax-based PAYG-DB schemes in other countries (Williamson, 2005). During the 1990s, it was common for proponents of partial privatization in the United States to point to what was described as a very successful privatized pension scheme in Chile (Borden, 1995). For years pension policy experts from around the world went to Chile to study its pension system, and many left as enthusiastic supporters of what they saw.

However, over time, some of the limitations of the Chilean model began to emerge. For instance, while the system created in the early 1980s has generally benefited affluent workers with steady jobs and at the same time has fostered economic development and particularly the development of the financial sector, it generally did not work out as well for low-wage workers, women, rural workers, and those in the informal sector (Kritzer, Kay, & Sinha, 2011; Williamson, 2005). This evidence has led to a moderation of the enthusiasm for the model both within and outside of Chile (Gill, Packard, & Yermo, 2005; Holzmann & Stiglitz, 2001). Recent reforms in Chile have focused on improving the FDC pillar by increasing coverage and reducing administrative costs. At the same time there has been more government involvement, increasing attention to poverty reduction and efforts to reduce exposure to market risk (Calvo, Bertranou, & Bertranou, 2010). For example, changes were made in 2007 limiting early retirement to those with much larger retirement accounts than had been required in the past, thus ensuring that they have more adequate pension benefits.

A number of major reforms were also introduced in 2008 (Barr & Diamond, 2008; Kritzer et al., 2011). The prior minimum pension guarantee was merged with the prior social assistance program to create a new program with two types of pension benefits. One is a noncontributory social pension benefit added for those of retirement age with incomes in the bottom 60%. The other is a pension top-up for workers who meet the standard pension eligibility criteria, but have low levels of assets in their FDC accounts. Starting in 2009, responsibility for mandatory insurance premiums for disability and survivorship that had been paid by employees was shifted to employers. In addition, a new requirement that the self-employed participate is currently being phased in. Finally, measures have been taken to increase competition among pension fund providers in an effort to lower commission costs. This set of reforms is expected to substantially increase coverage and the adequacy of benefits for women, low-wage workers, rural workers, and those working in the informal sector of the economy more generally (Calvo et al., 2010). In 2014, when Bachelet began a second term as President, she promised to drive down the high cost of administering the pension management funds by introducing a government-managed fund with much lower management fees as one of the available options (Adams, 2013).

One of the most important lessons learned from the example of Chile is that when the goal is to almost entirely privatize a pre-existing PAYG-DB scheme, the result is likely to be inadequate pension benefits for rural workers, informal workers, women, and low-wage workers more generally. Another lesson is that such a scheme can work well, particularly for urban workers with steady jobs with employers that pay well. A related lesson is that a largely FDC-based scheme is likely to require additional pillars that are government-financed, particularly to meet the needs of the economically most vulnerable segments of the population. Finally, the case of Chile illustrates that management fees associated with private management of FDC accounts are likely to be high unless the government takes steps to limit the level of management fees.

Argentina: An Unsuccessful Latin American Experiment in Partial Privatization

After World War II, the Argentine PAYG-DB public pension system expanded rapidly. By the end of the 1980s, it was under significant financial stress, leading to lawsuits over unpaid benefits and to a major restructuring in 1993 (Arza, 2008). The new pension system, which was much influenced by the Chilean scheme, included a mandatory PAYG-DB first pillar partly financed by employer contributions. The mandatory second pillar was financed by employees themselves and offered two options: either (1) a second government-administered PAYG-DB plan or (2) a privately managed FDC plan (Baker & Weisbrot, 2002).

The introduction of the FDC option was motivated by several goals such as: shifting risk from the government to workers, augmenting returns on tax contributions, and reinforcing ownership rights by reducing the political risk that the government would cut previously promised benefits. Unfortunately, the diversion of tax contributions, previously available to pay pensions to current retirees, into the new FDC accounts reduced government revenues far more than anticipated. According to some analysts, the introduction of these new FDC accounts significantly contributed to Argentina's financial meltdown in 2001 (Arza, 2008). The FDC accounts were US-dollar-denominated and at the time the peso was pegged to the US dollar. In an effort to deal with its fiscal crisis the government required that these accounts be shifted from US dollars to pesos. These assets soon lost 60–70% of their value when the peso was allowed to float (Bertranou, Rofman, & Grushka, 2003). In addition, the government was forced to cut benefits associated with the DB component of the public pension system by 13% (Baker & Weisbrot, 2002).

One of the goals of the FDC accounts was to protect pension tax contributions from the political risks of benefit cuts during periods of fiscal crisis, a goal that clearly was not realized in connection with the 2001 crisis. Yet another example of failure in connection with this goal was the reform in 2008, which transformed Argentina into "the first country to reverse the switch to individual savings accounts" (Kritzer et al., 2011, p. 66). The reform required all workers who had previously opted for the FDC second pillar to shift to the government-administered DB option for the second pillar. The assets in their individual accounts were nationalized. At retirement, workers who had been contributing to these FDC accounts have been promised "credit" for the years contributing to the now defunct FDC second pillar (Social Security Administration, 2008). It is not clear that they will ever be fully compensated for the appropriation of the funds in these accounts (Rofman, Fajnzylber, & Herrera, 2008). One of the most important lessons from Argentina for other countries is that there is good reason to question promises or suggestions by FDC proponents that such accounts will be safe from government appropriation because they will be viewed as private property (Kay, 2009).

While Argentina's efforts to reform its pension system in 1993 with the introduction of the option of FDC account for the second pillar of its pension program proved to be unsuccessful, it has more success with recent efforts to increase the proportion of the retirement age population with pension coverage. In 1993 workers were required to contribute for at least 30 years to establish eligibility for a lifetime public pension benefit. This was part of an effort to control pension spending by limiting benefit levels and restricting access. In part due to the pension reforms and in part to the increasing fraction of the labor force working in the informal sector, between 1992 and 2005 the fraction of those of retirement age (65 for men, 60 for women) who were eligible for some sort of retirement benefit decreased from 78% to 62% (Arza, 2012).

In response two major pension policy changes were made. One was to allow early retirement (albeit with a reduced benefit until reaching the retirement age) for those within 5 years of the statutory retirement age, if they had already contributed for 30 or more years. The other, and much more important, was the introduction of a "moratorium" allowing workers with less, in some cases far less, than the statutory 30 years of contribution, to apply for a pension if they had reached retirement age. They may receive lower pension benefits than they would have been eligible for after contributing for 30 years, but it is better than no pension. The impact of these changes was dramatic. Between 2005 and 2010 the fraction of those of retirement age receiving a pension increased from 62% to 85% and the fraction living in a household with at least one pension benefit increased from 77% to 91% (Arza, 2012).

China: An Innovative New Rural Pension Model

There are four main reasons to discuss recent pension developments in China. (1) China has recently introduced a very innovative model making it possible to rapidly expand coverage to a large proportion of the rural population. (2) China's current pension system illustrates both short-term benefits and long-term costs of implementing a highly decentralized scheme. (3) China already has more people covered and more people receiving pensions than any other country in the world. Between 2009 and 2012 the proportion of adults covered by one scheme or another increased from 30% to 55%. Today more than 60% of Chinese adults over age 60 are receiving pensions (HelpAge International, 2013). By some estimates 850 million adults are now covered by one of the Chinese pension schemes (International Social Security Association, 2014). (4) China provides evidence that may be relevant to future debates in the United States over the proposed partial privatization of Social Security.

China introduced its first public pension system in 1951. It was a PAYG-DB scheme (financed entirely by employers) soon covering urban worker in state-owned enterprises (SOEs) and publically owned enterprises more generally. Before 1966 the system was run by the national labor union. In connection with the Cultural Revolution the national labor union was eliminated in 1969 (Oksanen, 2012). The responsibility for funding and administering pensions shifted to the individual SOEs. This legacy led to what has become one of the world's most decentralized pension systems. During the 1980s it became clear that the existing schemes were not going to be sustainable given rapid population aging and the need for many SOEs to lay off large numbers of workers as part of the country's move toward a market economy. During the 1990s, China began to explore pension models that included both a PAYG-DB pillar and a FDC individual account pillar. These early schemes for SOE workers evolved into what is today the Urban Enterprise Pension System (UEPS), the major scheme for urban workers (Pozen, 2013). Currently the UEPS is based on two pillars. One pillar is a PAYG-DB scheme financed by a 20% payroll tax on employers. These funds go to local government to finance the pensions due to current retirees. The second component is an FDC pillar financed by an 8% payroll tax on workers. However, in most provinces much of the 8% is being diverted by local governments, to help pay pensions due to current retirees (Zheng, 2012). Why? Because in many areas the 20% collected from employers is not enough to cover current pension obligations. In addition, when some or all of the 8% employee contribution does make it to a worker's individual FDC account, it is placed in a special category of savings account in a government bank paying an interest rate set by the government that has not been keeping up with wage increases. When some of the FDC contributions are diverted to pay for pensions, a record of these contributions is maintained with the promise that at retirement the pension benefit will be adjusted for the funds that were diverted by local government (Williamson, Shen, & Yang, 2009). By 2013 about 322 million urban residents (including 80 million beneficiaries) were covered by the UEPS program (Ministry of Human Resources and Social Security, 2014). For the average worker the replacement rate was approximately 50% of final earnings (Zheng, 2012).

By 2008 several small-scale rural schemes were in place covering about 10% of the rural population (Fang, Giles, O'Keefe, & Wang, 2012). In 2009 China began to phase in an innovative new rural pension scheme that has the potential to influence rural pension policy in many other low-income nations around the world. This new scheme is voluntary. No rural residents are forced to sign up, but it does include strong incentives for enrolling. The result has been a very rapid increase in pension coverage throughout rural China. The program, called the New Rural Pension Scheme (NRPS), is based on two pillars. One is a contingent social pension. This component is noncontributory, that is, financed entirely by the government (the central and local government split the cost). In some more affluent regions the social pension is augmented by local government. The second component of the NRPS is a voluntary FDC component financed by annual contributions from voluntarily enrolled residents who are not yet of retirement age. The contributor gets to select the level of the contribution over a specified range, typically between ¥100 and ¥500 ($16 and $80) per year and local government adds in an additional ¥30 or more.

The most innovative aspect of this rural pension scheme is that residents who are already of retirement age (60+) become eligible for the contingent social pension of ¥55 per month without having ever contributed to the scheme, but only if their adult children enroll in the "voluntary" contributory scheme. This "family-binding" provision creates a strong incentive for adult children to enroll (Dorfman, Wang, O'Keefe, & Cheng, 2012). A second incentive to enroll is that enrollees who contribute for 15 years or more become eligible for a pension based on two sources, (1) a benefit based on the assets they have accumulated in their individual FDC accounts and (2) a noncontributory social pension financed by the government.

This incentive system is proving very effective in getting rural residents to enroll despite the voluntary nature of the program. However, the size of the actual pension remains an issue as it is very low. In most regions the pension for retirement age parents is currently based entirely on the contingent social pension component of ¥55 ($8) per month. This comes to about 30% of the rural poverty line. By itself this pension is not enough to live on, but it is a welcome supplement to other potential sources of economic security such as continued work, savings, and support from children. By the end of 2013 about 474 million rural residents were covered by the NRPS with 128 million as beneficiaries (Ministry of Human Resources and Social Security, 2014). The income replacement rates for those currently receiving benefits from this scheme are quite low, about 8%, because it is a new scheme and currently most pensioners are only eligible for the ¥55 social pension benefit.

In China's cities many residents are not covered by the UEPS described earlier. To help meet the needs of urban residents not covered by the UEPS, the Urban Resident Pension Scheme (URPS) began to be phased in starting in 2011. This new urban program is in many ways similar to the NRPS. It has the same two-pillar structure with one pillar as a contingent social pension and the other as a contributory FDC pillar. There are, however, some differences. For example, there is no family-binding provision and it is adjusted to fit with the higher cost of living in cities than in rural areas. Residents generally have the option of contributing at a level substantially above the ¥500 maximum level associated with the NRPS. As of 2013 about 24 million urban residents were covered by the URPS including 10 million as pension beneficiaries (Ministry of Human Resources and Social Security, 2014). Currently NRPS and the URPS are in the process of being integrated in many provinces with the eventual goal of full integration of these schemes in all provinces by the end of 2015. After the schemes are integrated, pension assets will become much more portable when rural workers move to urban areas. For those covered by the URPS, the replacement is estimated to be basically the same as for the NRPS.

The three pension schemes discussed to this point all include a funded FDC pillar. The pensions for civil servants, government workers, and military personnel are different. They are entirely funded by the government with no contributions from covered workers. Approximately 43 million Chinese citizens are covered by pensions in this fourth category. The final income replacement rates in this category tend to be high relative to the others, over 90% for the average covered worker (Leckie, 2011).

The evidence from China is a source of lessons for other countries, particularly other low-income ones. For many years policy analysts have commonly assumed that it is not feasible to structure rural pension schemes for low-income nations in a way that a substantial fraction of the rural population will participate in programs that call for contributions either to PAYG-DB or FDC schemes. The evidence from China calls this assumption into question. Pension policymakers in such countries may want to at least consider introducing schemes that draw on China's NRPS. A similar argument can be made for adopting a variant of China's URPS to provide pension coverage for urban workers in the informal economy in many other low-income nations. The family-binding component of the NRPS could be adapted for use in pension schemes in other low-income nations that differ in many ways from China.

There are cautionary lessons from China. The Chinese pension system is made up of the three major pension schemes we have focused on as well as other schemes for government and military workers. There have been some short-term benefits of China's highly fractionalized and decentralized pension system. This decentralization and flexibility have allowed different provinces and cities to modify various pension schemes in an effort to create a better fit to local economic conditions. For example, affluent areas are free to augment the social pensions associated with NRPS and URPS. Similarly, the diversion of funds from the payroll tax paid by employees covered by the UEPS and other adjustments to pension programs have helped reduce local resistance to the introduction of these new schemes. However, the country must now deal with the long-term consequences of this fractionalization (Pozen, 2013). Government policymakers want to move toward much more integration and uniformity in pension policy, but the fractionalization of the current pension system and other consequences of decentralization are hindering such efforts. City-to-city and province-to-province structural differences and differences in eligibility regulations are making it difficult for workers to transfer pension assets and credits when they move from one city to another (Pozen, 2013). While China's focus on keeping pension benefit levels low will make it easier to finance these pensions, the country's rapidly graying age structure is going to make the sustainability of even these modest pensions quite problematic in the decades ahead (Wang & Béland, 2014).

It is rare for those proposing Social Security reforms in the United States to turn to evidence from developing countries for policy ideas. However, a few years ago when the proposed partial privatization of Social Security was being actively debated in Congress, there were frequent references to the "huge success" of Chile's privatization model. If it is reasonable to mention evidence from Chile, it may also be reasonable to consider evidence from other developing countries. As noted earlier, China has introduced individual FDC accounts in connection with both the UEPS and the NRPS. The evidence from China in connection with these schemes illustrates that the government FDC policy has been problematic in two ways. One is the diversion of funds destined for the FDC funds to other uses such as paying pensions to those already retired. The other is the requirement that such funds be placed in government banks paying interest rates that do not keep up with inflation or increasing wage levels (Dorfman et al., 2012). Neither of these issues are directly relevant to the recent privatization debate in the United States, but both illustrate that it cannot be assumed that assets in individual FDC accounts will be free from direct or indirect government manipulation.

The United Kingdom: An Industrial Country Having Noteworthy Problems with its FDC Pillar

The United Kingdom is a highly developed industrial nation with debates between the political right and the left on pension policy that often parallel those in the United States, specifically debates with respect to the role of the state as opposed to the individual in the provision of old-age security. In addition the United Kingdom has already had several years of experience with the partial privatization of its pension system including considerable experience with FDC accounts that are similar to some of those that have in recent years been proposed for the United States. In 1975 legislation was enacted creating an earnings-related pension pillar in the United Kingdom called the State Earnings Related Pension System (SERPS). This same legislation also included an option allowing employers to opt-out of that scheme in favor of their own occupational schemes (Williamson, 2002b). This change transformed the UK pension system into a partially privatized scheme.

More relevant to the debate over the proposed partial privatization of Social Security in the United States was the legislation enacted by Thatcher's government in 1986 creating incentives for workers to opt-out of SERPS (or out of their employer-sponsored alternatives to SERPS) in favor of individually directed FDC schemes called Approved Personal Pensions (hereafter, personal pensions) with freedom to choose from among many privately managed pension funds (Schulz & Borowski, 2006). These pensions were in many ways similar to IRAs in the United States, but they did differ from the IRAs in that they were structured as an alternative place in the private sector to put mandatory pension contributions that would otherwise go to a company pension scheme or to SERPS, the government pension scheme. In contrast, the IRAs in the United States represent an optional add-on pillar.

There are many lessons from the United Kingdom concerning the potential consequences of introducing funded individual accounts. One is that while popular at first, within a few years public support in the United Kingdom for the personal pensions started to weaken. One reason was that workers became more aware of the high fees when buying into or transferring between providers. The high annual management charges were also a problem. The net effect of these fees was low returns for many workers, particularly those with smaller accounts (Waine, 2009). Another problem was the loss of confidence in the scheme due to the misleading financial advice given to potential clients by sales staff working for some of the organizations managing the personal pension plans (Blake, 2004).

An effort was made by the Blair government to respond to some of these problems with the introduction in 2001 of a new scheme call stakeholder pensions. These pensions worked in much the same way as the personal pensions, but they were designed to better serve lower-income employees (Schulz & Borowski, 2006). Employers that did not offer their own occupational pensions were required to offer stakeholder pensions to their workers. The stakeholder pensions were in many ways an improvement over the personal pensions. For example, they reduced the fees that made the personal pensions particularly bad investments for those with small personal accounts, and they eliminated the fees associated with moving from one provider to another (Walker & Foster, 2006). However, relatively few workers signed up for the new stakeholder pensions (Waine, 2009).

In 2002 the State Earnings Related Pension was replaced by the State Second Pension, which over time led to lower state pensions and increased dependence on means-tested benefits for low earners. In the same year a government-appointed Pensions Commission was charged with making a thorough review of the British pension system. It concluded that it was not working well (Pensions Commission, 2005). It was described as the most complex pension system in the world, as contributing to increasing inequality, as failing to provide adequate incentives to save, and as having very high administrative costs (Schulz & Binstock, 2008).

In this context and based largely on recommendations from the Pensions Commission, the Pensions Act of 2008 introduced yet another program designed to deal with the problem of "undersaving" on the part of British workers (Waine, 2009). This legislation mandated that as of 2012 every new employee over age 22 and earning over £8,105 had to be automatically enrolled in either the employer's occupational pension plan or in a new FDC plan called NEST (National Employment Savings Trust) (Williamson, 2011). This new scheme is currently being phased in and by 2018 it will require a contribution of 8% per year split among the employee 4%, the employer 3%, and the government 1% (Sass, 2014). This money is being invested and workers will not have access to the funds until age 55. NEST is an optional add-on pillar like a 401(k) plan in the United States. Employees do not have to contribute, but if they elect not to, they lose out on the employer contribution and a tax break from the government. With these incentives the government is attempting to make the NEST accounts affordable and appealing to many employees, particularly those with lower incomes (BBC News, 2012; Curry, 2008).

The NEST program includes several improvements over earlier individual account schemes in the United Kingdom, of particular note being the automatic enrollment provision that has substantially increased the proportion of workers who are covered as well as the policy of gradually shifting to a more conservative portfolio during a period of about 10 years prior to retirement. The new scheme still puts workers at risk of potentially substantial declines in fund assets due to adverse swings in financial markets. It also puts workers at annuity risk due to the differences in the value of the annuities depending on interest rate fluctuations just prior to retirement. This turns out to be a problem for many low-income workers who are being taken advantage of in the annuity market and as a result ending up with lower pensions (Cumbo, 2013). Some British analysts argue that the long-term real returns for low-wage workers contributing to individual account FDC schemes are generally positive, but others point out that they are often negative. Some emphasize that low returns are particularly likely for women (Ginn & MacIntyre, 2013). These problems and the tendency to foster increased economic inequality are all problems common to the FDC components of partial privatization schemes in the United Kingdom and in the other countries with FDC pillars that we have considered.

Overall, in the United Kingdom since the 1970s, the pension system has witnessed much greater change than its American counterpart, a situation related in part to the high concentration of political power within the British parliamentary system as well as the weakness of the constituencies supporting SERPS compared with those backing Social Security in the United States (Pierson, 1994).

Canada: A PAYG-DB-Based Scheme with Lessons for the United States

Although the trend toward privatization has influenced pension reform in many countries around the world, some, such as Canada, have not moved very far in this direction. We are discussing the Canadian case because it is a country with cultural values similar to the United States, that has to this point been very successful using incremental parametric reforms preserving the basic integrity of its largely PAYG-DB approach adopted decades ago (Jacobs, 2011; Little, 2008).

As created during the post–World War II era, the multi-layered Canadian pension system features three main components: (1) a modest public flat pension financed through general revenues created in the early 1950s (Old Age Security: OAS) and complemented since the late 1960s by an income-tested Guaranteed Income Supplement (GIS); (2) an earnings-related old-age insurance system (the Canada and Quebec Pension Plans: C/QPP), which has a modest replacement rate of 25% of covered wages; and, finally, (3) private yet government-regulated voluntary occupational pensions and personal savings plans (Béland & Myles, 2005). One of the outstanding aspects of this complex pension system is its capacity to reduce poverty among older adults despite relatively limited public spending. For instance, Canada does much better than the United Kingdom and the United States in terms of poverty alleviation. In fact, Canada does nearly as well as Sweden, one of the world leaders in the area (Wiseman & Yčas, 2008). This success is related to the pivotal role of GIS, a program financed through general revenues that covers a larger proportion of the older adult population than its US equivalent, Supplemental Security Income (SSI), which plays a more limited role in old-age security due to SSI's inclusion of both assets and income tests (Berkowitz & DeWitt, 2013; Wiseman & Yčas, 2008).

Reforms enacted in Canada since the 1980s to control costs in a context of population aging and fiscal pressures have largely preserved the integrity of the public pension programs (Béland & Myles, 2005). On one hand, the federal flat pension (OAS) remains formally universal but, since 1989, wealthier Canadians have paid back part or all of this pension through a clawback provision in the income tax system. Moreover, to further control costs, in 2012, the Conservative government of Stephen Harper announced a gradual increase in the OAS entitlement age from 65 to 67 between 2023 and 2029 (Service Canada, 2012). On the other hand, in the mid-1990s, facing the prospect of long-term fiscal unsustainability, federal and provincial officials agreed to reform both the Canada Pension Plan (CPP), which covers workers in nine of the ten provinces and the Quebec Pension Plan (QPP), a similar program covering workers in Quebec. This reform resulted in a significant increase in the payroll tax rate as well as investment of growing CPP surpluses in equity, through the arms-length Canada Pension Plan Investment Board (a similar board has existed in Québec since the mid-1960s). In the end, this reform, which increased the level of advance funding, solved the long-term fiscal challenges facing CPP, which is projected to remain actuarially sound for at least the next 75 years (the QPP faces greater fiscal challenges, which is why another reform was recently enacted in Québec).

Although Canada has found a way to preserve existing public pension programs while enacting incremental changes to preserve them in a context of anticipated fiscal and demographic pressures, it is important to note that this country faces significant changes due to the decline in coverage by DB private pensions (Boychuk & Banting, 2008). From this angle, Canada faces similar challenges as the United States, which are related to broad economic and labor market trends, combined with the intrinsic vulnerability of *voluntary*, employer-provided private pensions. One lesson for the United States from the Canadian experience is that it makes sense to keep any FDC pillar voluntary and in the private sector. However, it is also clear that in the years ahead both countries will need to focus on ways to reform these voluntary FDC schemes so that they better meet the needs of low- and moderate-income workers. Another lesson from Canada is that the PAYG-DB model has not outlived its utility for advanced industrial countries.

CONCLUSION

Our focus has been on recent pension policy developments in the United States with particular attention to the partial privatization of old-age security more generally. As part of this discussion we have presented brief discussions of relevant developments in five other countries selected primarily for their potential relevance to pension policy debates that we anticipate will emerge as we move closer to 2033, the year by which reforms must be made to keep Social Security from becoming insolvent. Depending in part on which party controls Congress and the Presidency when the needed reforms are made, partial privatization may again be getting serious consideration.

The evidence with respect to the consequences of partial privatization is mixed. While a case can be made that the introduction of FDC accounts may contribute to the development of financial markets, particularly in developing countries such as Chile, in many countries partial privatization has not worked as well as had been anticipated, particularly for those at the lower end of the income distribution and those living in rural areas. Chile has had some success in dealing with this problem by increasing spending on noncontributory social pensions designed to meet the needs of those in rural areas and those working in the informal sector more generally. Similarly, both Canada and China have had some success with multi-pillar schemes that have included voluntary FDC pillars.

China is being very pragmatic and is making huge strides in its pension policy efforts. The Chinese consult extensively with international experts and develop models that seem to make sense and then make adjustments when it becomes clear that modifications are needed. This approach is made easier by China's top-down one-party political structure in which policymakers can experiment without as much pushback from public opinion and affected interest groups as would be anticipated in a liberal democracy like the United States. China's current goal is to realize something approaching full coverage for the entire adult population by 2020. Such talk 10 years ago would not have been taken seriously by any international social security experts, but given the progress it has made in recent years, it looks like China may succeed, if not by 2020, very soon thereafter.

The United Kingdom and Canada have been included as examples of countries that are in many ways very similar to the United States that have had very different experiences with FDC pillars. In Canada there is a voluntary FDC pillar that plays a role similar to the 401(k) schemes in the United States, but as in the United States, the core of the Canadian scheme is a PAYG-DB scheme that seems to be working out well with periodic parametric adjustments. In contrast, the United Kingdom has put a great deal more emphasis on its FDC pillar. In the United Kingdom workers can participate in a variety of different FDC schemes, but over the years there have been a number of problems affecting the adequacy of the eventual pensions, particularly for lower-wage workers.

During much of the 1990s and the early 2000s, in the United States, the debate over the future of retirement security focused on the issue of Social Security reform, a topic that has largely been displaced in recent years by the debate over health care reform and the Affordable Care Act. While the debate over the partial privatization of Social Security via the introduction of FDC accounts is currently off the table in the United States, this is not true with respect to the partial privatization of retirement security. The dramatic shift from employer-sponsored DB schemes to 401(k) and similar FDC schemes represents an increase in dependence on the FDC model for retirement security. The question is not whether the United States should consider the partial privatization of retirement security based on the FDC model, as this has already taken place. The discussion now is how to reform the 401(k) plans to better protect the millions of American workers who will have an increasing share of their retirement assets invested in and retirement security linked to these accounts.

To date relatively few countries have had extensive experience with multi-pillar pension schemes that have included FDC pillars. The evidence from the 30 or so countries that have introduced such schemes is mixed. Many of these countries, including Chile, are making changes designed to remedy some of the problems that have emerged in connection with prior versions of their FDC schemes. In the years ahead, these reformed schemes may turn out to work well, and this evidence could bolster the arguments of those who favor partial privatization of the Social Security in the United States. But there will also be evidence to consider from a number of countries in addition to those we have included as case studies. This would include several in Central and Eastern Europe that did not perform at all well in response to the 2008 financial crisis (Drahokoupil & Domonkos, 2012). Evidence from these countries could also influence future privatization debates in the United States.

Acknowledgments

The authors thank the following for their comments on preliminary drafts and other forms of help in connection with the preparation of this chapter: Camila Arza, Esteban Calvo, Lianquan Fang, David Lain, Stephen Kay, Eric Kingson, Steven Sass, Kate Silfen, Alex Waddan, Lijian Wang, Xinmei Wang, April Wu, and Bingwen Zheng. Daniel Béland acknowledges support from the Canada Research Chairs Program.

References

Adams, R. (2013, August 5). Reforming Chile's disputed pension system. *The Santiago Times*. Retrieved from: <http://santiagotimes.cl/>.
Altman, N. J., & Kingson, E. R. (2015). *Social security works: Why social security isn't going broke and how expanding it will help us all*. New York, NY: The New Press.
Arza, C. (2008). The limits of pension privatization: Lessons from Argentine experience. *World Development, 35*, 2696–2712.
Arza, C. (2012). Extending coverage under the Argentinian pension system: Distribution of access and prospect for universal coverage. *International Social Security Review, 65*, 29–49.
Baker, D., & Weisbrot, M. (2002). *The role of social security privatization in Argentina's economic crisis*. Washington, DC: Center for Economic and Policy Research. Retrieved from: <http://www.cepr.net/documents/publications/argentina_2002_04.pdf>.
Barr, N., & Diamond, P. (2008). *Reforming pensions: Principles and policy choices*. New York, NY: Oxford University Press.
BBC News. (2012). Pensions auto-enrolment starts for biggest firms. *BBC News online*. Retrieved from: <http://www.bbc.co.uk/news/business-19760421>.
Béland, D. (2005). *Social Security: History and politics from the New Deal to the privatization debate*. Lawrence, KS: University of Kansas Press.
Béland, D., & Myles, J. (2005). Stasis amidst change: Canadian pension reform in an age of retrenchment. In G. Bonoli & T. Shinkawa (Eds.), *Ageing and pension reform around the world* (pp. 252–272). Cheltenham, UK: Edward Elgar.
Béland, D., & Orenstein, M. A. (2013). International organizations as policy actors: An ideational approach. *Global Social Policy, 13*, 125–143.
Béland, D., & Waddan, A. (2012). *The politics of policy change: Welfare, Medicare, and Social Security reform in the United States*. Washington, DC: Georgetown University Press.
Berkowitz, E. D., & DeWitt, L. (2013). *The other welfare: Supplemental Security Income and U.S. social policy*. Ithaca, NY: Cornell University Press.
Bertranou, F. M., Rofman, R., & Grushka, C. O. (2003). From reform to crisis: Argentina's pension system. *International Social Security Review, 56*, 103–114.
Blake, D. (2004). Contracting out of the state pension system: The British experience of carrots and sticks. In M. Rein & W. Schmal (Eds.), *Rethinking the welfare state: The political economy of pension reform* (pp. 19–55). Cheltenham, UK: Edward Elgar.
Board of Trustees. (2014). *The 2014 annual report of the Board of Trustees of the Federal Old-Age and Survivors Insurance and Federal Disability Insurance Trust Funds*. Washington, DC: Social Security Administration.
Boivie, I. (2011). *Who killed the private sector DB plan?* Washington, DC: National Institute on Retirement Security (NIRS).
Borden, K. J. (1995). *Dismantling the pyramid: The how & why of privatizing Social Security*. Washington, DC: Cato Institute. (Social Security Choice Paper No. 1).
Boychuk, G. W., & Banting, K. G. (2008). The public-private divide: Health insurance and pensions in Canada. In D. Béland & B. Gran (Eds.), *Public and private social policy: Health and pension policies in a new era* (pp. 92–122). Basingstoke, UK: Palgrave Macmillan.
Calvo, E., Bertranou, F. M., & Bertranou, E. (2010). Are old-age pension system reforms moving away from individual retirement accounts in Latin America? *Journal of Social Policy, 39*, 223–234.
Cook, F. L., Barabas, J., & Page, B. I. (2002). Invoking public opinion: Policy elites and Social Security. *Public Opinion Quarterly, 66*, 235–264.
Cumbo, J. (2013, December 16). Millions at risk of a poor pension deal. *Financial Times*. Retrieved from: <http://www.ft.com/home/uk>.

Curry, C. (2008). The introduction of auto-enrolment and personal accounts to the UK in 2012. *Pensions, 13*, 237–245.

DeWitt, L. (2010). The development of Social Security in America. *Social Security Bulletin, 70*(3), 1–26.

Dorfman, M. C., Wang, D., O'Keefe, P., & Cheng, J. (2012). China's pension schemes for rural and urban residents. In R. Hinz, R. Holzmann, D. Tuesta, & N. Takayama (Eds.), *Matching contributions for pensions: A review of international experience* (pp. 217–241). Washington, DC: The World Bank.

Drahokoupil, J., & Domonkos, S. (2012). Averting the funding-gap crisis: Eastern European pension reforms since 2008. *Global Social Policy, 12*, 283–299.

Fang, C., Giles, J., O'Keefe, P., & Wang, D. (2012). *The elderly and old age support in rural China: Challenges and prospects*. Washington, DC: The World Bank.

Ferrara, P. J. (Ed.). (1985). *Social Security: Prospects of real reform*. Washington, DC: Cato Institute.

Ferrara, P. J., & Tanner, M. (1998). *A new deal for Social Security*. Washington, DC: Cato Institute.

Geisel, J. (2012, October 3). Fewer employers offering defined benefit pension plans to new salaried employees. *Workforce* Retrieved from: <http://www.workforce.com/article/20121003/NEWS01/121009976/fewer-employers-offering-defined-benefit-pension-plans-to-new-salaried-employees#>.

Ghilarducci, T. (2008). *When I'm sixty-four: The plot against pensions and the plan to save them*. Princeton, NJ: Princeton University Press.

Gill, I. S., Packard, T., & Yermo, J. (2005). *Keeping the promise of social security in Latin America*. Washington, DC: The World Bank.

Gillon, S. M. (2008). *The pact: Bill Clinton, Newt Gingrich, and the rivalry that defined a generation*. New York, NY: Oxford University Press.

Ginn, J., & MacIntyre, K. (2013). UK pension reforms: Is gender still an issue? *Social Policy and Society, 12*, 91–103.

Hacker, J. (2006). *The great risk shift: The assault on American jobs, families, health care, and retirement and how you can fight back*. New York, NY: Oxford University Press.

Hardy, M., & Hazelrigg, L. (2007). *Pension puzzles: Social Security and the great debate*. New York, NY: Russell Sage Foundation.

HelpAge International. (2013). Pension coverage in China and the expansion of the New Rural Social Pension, *Pension Watch*. (Briefing No. 11). Retrieved from: <www.helpage.org/download/51d564e94677e/>.

Holzmann, R., Hinz, R., & Dorfman, M. (2008). *Pension systems and reform conceptual framework*. Washington, DC: The World Bank. (SP Discussion Paper No. 0824).

Holzmann, R., & Stiglitz, J. E. (Eds.). (2001). *New Ideas about Old Age Security: Toward sustainable pension systems in the 21st century*. Washington, DC: The World Bank.

International Social Security Association. (2014). *Country: China*. Retrieved from: <http://www.issa.int/country-profiles>.

IRS. (2012). *IRS announces 2013 pension plan limitations; taxpayers may contribute up to $17,500 to their 401(k) plans in 2013*. Retrieved from: <http://www.irs.gov/uac/2013-Pension-Plan-Limitations>.

IRS. (2013). *Retirement topics – automatic enrollment*. Retrieved from: <http://www.irs.gov/Retirement-Plans/Plan-Participant,-Employee/Retirement-Topics---Automatic-Enrollment>.

Jacobs, A. M. (2011). *Governing for the long term: Democracy and the politics of investment*. New York, NY: Cambridge University Press.

James, E. (2005). *Reforming social security: Lessons from thirty countries*. Dallas, TX: National Center for Policy Analysis. (NCPA Policy Report No. 277).

John, D., & Reno, V. (2012). *Perspectives: Options for reforming Social Security*. Washington, DC: AARP Public Policy Institute.

Kay, S. J. (2009). Political risk and pension privatization: The case of Argentina (1994–2008). *International Social Security Review, 62*, 1–21.

Kay, S. J., & Sinha, T. (Eds.). (2008). *Lessons from pension reform in the Americas*. New York, NY: Oxford University Press.

Kingson, E. R., & Williamson, J. B. (1999). Why privatizing Social Security is a bad idea. In J. B. Williamson, D. M. Watts-Roy, & E. R. Kingson (Eds.), *The generational equity debate* (pp. 204–219). New York, NY: Columbia University Press.

Kingson, E. R., & Williamson, J. B. (2001). Economic security policies. In R. H. Binstock & L. K. George (Eds.), *Handbook of aging and the social sciences* (pp. 369–386) (5th ed.). San Diego, CA: Academic Press.

Kritzer, B. E. (2008). Chile's next generation pension reform. *Social Security Bulletin, 68*, 69–84.

Kritzer, B. E., Kay, S. J., & Sinha, T. (2011). Next generation of individual account pension reforms in Latin America. *Social Security Bulletin, 71*(1), 35–76.

Leckie, S. H. (2011). *Civil service and military service pensions in China*. Tokyo, Japan: Center for International Studies, Institute of Economic Research, Hitotsubashi University. (PIE/CIS Discussion Paper No. 505).

Little, B. (2008). *Fixing the future: How Canada's usually fractious governments worked together to rescue the Canada pension plan*. Toronto: University of Toronto Press.

Ministry of Human Resources and Social Security. (2014). Retrieved from: <http://www.chinanews.com/gn/2014/05-28/6223421.shtml>.

Munnell, A. H., Aubry, J.-P., & Muldoon, D. (2008). *The financial crisis and private defined benefit plans*. Chestnut Hill, MA: Center for Retirement Research at Boston College. (Issue Brief No. 8–18).

Munnell, A. H., Golub-Sass, F., & Muldoon, D. (2009). *An update on 401(k) plans: Insights from the 2007 SCF.* Chestnut Hill, MA: Center for Retirement Research at Boston College. (Issue Brief No. 95).

Munnell, A. H., & Sundén, A. (2004). *Coming up short: The challenge of 401(k) plans.* Washington, DC: Brookings Institution Press.

National Research Council. (2012). *Aging and the macroeconomy: Long-term implications of an older population.* Committee on the Long-Run Macroeconomic Effects of the Aging U.S. Population. Washington, DC: The National Academies Press.

Nuschler, D. (2012). *Social Security reform: Current issues and legislation.* Washington, DC: Congressional Research Service.

Oksanen, H. (2012). China: Pension reform for an aging economy. In R. Holzmann, E. Palmer, & D. Robalino (Eds.), *Nonfinacial defined contribution pension schemes in a changing pension world: Progress, lessons and implementation* (pp. 213–255). Washington, DC: World Bank.

Olen, H. (2012). *Pound foolish: Exposing the dark side of the personal finance industry.* New York, NY: Penguin.

O'Rand, A. M., Ebel, D., & Isaacs, K. (2009). Private pensions in international perspective. In P. Uhlenberg (Ed.), *International handbook of population aging* (pp. 429–443). New York, NY: Springer.

Orenstein, M. A. (2008). *Privatizing pensions: The transnational campaign for social security reform.* Princeton, NJ: Princeton University Press.

Pensions Commission. (2005). *A new pension settlement for the twenty-first century: Second report of the Pensions Commission.* London: The Stationery Office.

Peterson, P. J. (1999). How will America pay for the retirement of the baby boom generation? In J. B. Williamson, D. M. Watts-Roy, & E. R. Kingson (Eds.), *The generational equity debate* (pp. 41–59). New York, NY: Columbia University Press.

Pierson, Paul. (1994). *Dismantling the welfare state? Reagan, Thatcher and the politics of retrenchment.* New York, NY: Cambridge University Press.

Pozen, R. C. (2013). *Tackling the Chinese pension system. Paulson Policy Memorandum.* Chicago, IL: The Paulson Institute.

President's Commission to Strengthen Social Security. (2001). *Strengthening Social Security and creating personal wealth for all Americans: Final report of the President's Commission.* Washington, DC: President's Commission to Strengthen Social Security.

Reno, V. P., & Friedland, R. B. (1997). Strong support but low confidence: What explains the contradiction? In E. R. Kingson & J. H. Schulz (Eds.), *Social Security in the 21st century* (pp. 178–194). New York, NY: Oxford University Press.

Rofman, R., Fajnzylber, E., & Herrera, G. (2008). *Reforming the pension reforms: The recent initiatives and actions on pensions in Argentina and Chile.* Washington, DC: Social Protection & Labor, The World Bank. (SP Discussion Paper No. 0831).

Salsman, R. M. (2011, September 27). Social Security is much worse than a Ponzi scheme and here's how to end it. *Forbes* Retrieved from: <http://www.forbes.com/sites/richardsalsman/2011/09/27/social-security-is-much-worse-than-a-ponzi-scheme/>.

Sass, S., Munnell, A. H., & Eschtruth, A. (2009). *The Social Security fix-it book* (Rev. ed.). Chestnut Hill, MA: Center for Retirement Research at Boston College.

Sass, S. A. (2014). *The U.K.'s ambitious new retirement savings initiative.* Chestnut Hill, MA: Center for Retirement Research at Boston College. (Issue in Brief 14–5).

Schulz, J. H., & Binstock, R. H. (2008). *Aging nation: The economics and politics of growing older in America.* Baltimore, MD: Johns Hopkins University Press.

Schulz, J. H., & Borowski, A. (2006). Economic security in retirement: Reshaping the public-private pension mix. In R. H. Binstock & L. K. George (Eds.), *Handbook of aging and the social sciences* (pp. 360–379) (6th ed.). San Diego, CA: Academic Press.

Service Canada. (2012). *Age of eligibility for Old Age Security benefits from 2023 to 2029.* Retrieved from: <http://www.servicecanada.gc.ca/eng/isp/oas/changes/oasgis_age_chart.shtml>.

Smith, H. (2012). *Who stole the American dream?* New York, NY: Random House.

Social Security Administration. (2008). *Argentina. International Update, December.* Retrieved from: <http://www.socialsecurity.gov/policy/docs/progdesc/intl_update/2008-12/index.html>.

Waine, B. (2009). New labour and pension reform: Security in retirement? *Social Policy & Administration, 43,* 754–771.

Walker, A., & Foster, L. (2006). Caught between virtue and ideological necessity: A century of pension policies in the UK. *Review of Political Economy, 18,* 427–448.

Wang, L., & Béland, D. (2014). Assessing the financial sustainability of China's rural pension system. *Sustainability, 3*(3), 3271–3290.

Williamson, J. B. (2002a). What's next for Social Security? Partial privatization? *Generations, 26*(2), 34–39.

Williamson, J. B. (2002b). Privatization of social security in the United Kingdom: Warning or exemplar? *Journal of Aging Studies, 16,* 415–430.

Williamson, J. B. (2005). *An Update on Chile's experience with partial privatization and individual accounts.* Washington, DC: AARP. (Issue Paper No. 2005–19).

Williamson, J. B. (2011). The future of retirement security. In R. H. Binstock & L. K. George (Eds.), *Handbook of aging and the social sciences* (pp. 281–294) (7th ed.). San Diego, CA: Academic Press.

REFERENCES

Williamson, J. B., Shen, C., & Yang, Y. (2009). Which pension model holds the most promise for China: A funded defined contribution scheme, a notional defined contribution scheme, or a universal social pension? *Benefits: The Journal of Poverty and Social Justice, 17,* 101–111.

Williamson, J. B., & Watts-Roy, D. M. (2009). Aging boomers, generational equity, and framing the debate over Social Security. In R. B. Hudson (Ed.), *Boomer bust? Economic and political issues of the graying society (Vol. 1): Perspectives on the boomers* (pp. 153–169). Westport, CT: Praeger.

Willis, L. E. (2008). Against financial-literacy education. *Iowa Law Review, 94*(1), 197–285.

Wiseman, M., & Yčas, M. (2008). The Canadian safety net for the elderly. *Social Security Bulletin, 68*(2), 53–67.

World Bank. (1994). *Averting the Old Age Crisis: Policies to protect and promote growth.* New York, NY: Oxford University Press.

Zheng, B. (Ed.). (2012). *China pension report 2012.* China: Economy & Management Publishing House.

CHAPTER

23

Health Inequalities Among Older Adults in Developed Countries: Reconciling Theories and Policy Approaches

Amélie Quesnel-Vallée[1], Andrea Willson[2], and Sandra Reiter-Campeau[3]

[1]Department of Epidemiology, Biostatistics, and Occupational Health; Department of Sociology; Centre for Population Dynamics, McGill University, Montréal, QC, Canada [2]Department of Sociology, Social Science Centre, The University of Western Ontario, London, ON, Canada [3]Faculty of Medicine, Université de Montréal, Montréal, QC, Canada

OUTLINE

Introduction	483	Long-Term Care	492
Theories of Health Inequality in Older Age	484	WHO Health in All Policies	493
Fundamental Cause Theory	485	WHO Age-friendly Environments Programme	494
Life Course Theories of Health Inequality	486		
Welfare States and the Interplay of Social Solidarity and Equity	489	Promising Avenues for Sociological Research	496
Pensions	490	References	498

INTRODUCTION

The extension in high-income countries of systems of income support and access to health care for large, if not all, segments of the elderly population undoubtedly counts as one of major successes of public policy over the past century (OECD, 2013). It is perhaps not surprising then to note that today's elderly find themselves in much better health than their parents and grandparents did in their older age (Crimmins, 2004). Indeed, relative to the past, older populations in developed countries have improved functioning and are afflicted with less disability on average (Wolf, Hunt, & Knickman, 2005); however, these improvements may not have been equally gained across the elderly population (Taylor, 2008). Indeed, the body of literature regarding health inequalities among the elderly is growing as the assumption that the older population is a rather homogeneous group in this regard is increasingly discarded (Grundy & Holt, 2001).

This notably raises the question of the current capacity of social policies to mitigate health inequalities among the elderly. Recent reviews of the impact of the welfare state on health inequalities have been conducted elsewhere (Beckfield & Krieger, 2009; Bergqvist, Åberg Yngwe, & Lundberg, 2013; Brennenstuhl, Quesnel-Vallée, & McDonough, 2012). Among other findings, these show that (a) this field of research is small, but growing; (b) these effects have seldom been studied among the elderly; and (c) most of this discussion draws from political sociology theories (and other political science approaches), but rarely from other sociological theories on health inequalities. Thus, concerns were raised that some of these theoretical frameworks may not apply to the specific study of *health* inequalities (Brennenstuhl et al., 2012), least of all among the elderly.

As such, we propose here to review sociological theories that are most germane to health inequalities in older ages in developed countries, with an eye to their capacity to illuminate processes of social inequality driven by social policies. Then we will critically examine the explicit intent and implicit capacity of current social policies to mitigate these inequalities among the elderly population. As other chapters in this *Handbook* address the question of early life influences on older adult health (see especially, Chapter 5), our focus will be on policy interventions targeted at the elderly.

This critical examination will particularly focus on the contrast between "classical" welfare state strategies of pensions and health insurance and more contemporary policies salient to health inequalities generally and among the elderly specifically, namely the World Health Organization (WHO) Health In All Policies and Age-Friendly Environments programs.

This juxtaposition will show that much could be gained from an intersection of the two areas. Indeed, while sociological theories of health inequalities make room for the impact of social policies, this has not been systematically assessed yet. Conversely, social policies fail to recognize that current inequalities among the elderly are the product of life course trajectories of cumulative dis/advantage (CAD) that may have persistent effects even with receipt of universal benefits. Furthermore, policies are not static, and while some are in need of change to meet current and future needs to avoid further increases in health inequalities [e.g., long-term care (LTC)], others are changing in ways that will need monitoring, lest they contribute to these inequalities (e.g., pensions). Taking stock of these observations, we will end with suggestions for sociological research in this area. Our goal in this chapter is therefore not to provide an exhaustive review of the evidence on social inequalities among the elderly or on the impact of social policies on health inequalities; rather, it is to highlight where sociological research may fruitfully contribute to this research field, and point toward questions that could move this research agenda forward.

THEORIES OF HEALTH INEQUALITY IN OLDER AGE

Studies of health have documented that, similar to other important resources, health and mortality are not randomly distributed across societies. There is considerable variation in health at each life stage, and in health changes experienced by individuals as they age (Ferraro, Shippee, & Schafer, 2009). Like social inequality in other resources, such as income, health inequality refers to systematic and durable differences in health outcomes that are socially determined and unequal. There are a multitude of potential social determinants of health (SDH), and they intersect so that an individual may be advantaged, disadvantaged, or both across multiple dimensions simultaneously. However, an abundance of research across multiple disciplines has now clearly established that socioeconomic resources stratify health at every level of the socioeconomic hierarchy, often referred to as the SEP–health gradient (e.g., Marmot, 2004). Although the association between health and socioeconomic position is clear, the mechanisms that generate this association are not, and a number of theories have emerged to attempt to provide cohesive frameworks to explain these processes. We begin this section with one of the most commonly cited such theories, namely fundamental cause theory, though it has more rarely been associated with investigations of health inequalities among the elderly population, or even over the life course.

Then, we turn to more specific examinations of health inequality and aging, which have generally taken two forms. The first approach draws upon the life course perspective to understand how health inequality unfolds over time and emphasizes the life course processes that culminate in health inequality in old age. Much of the goal of this research lies in understanding the social processes, especially those related to socioeconomic position, that affect health over the life course and lead to health inequality as cohorts age. The second, and relatively less common, approach focuses on the health of older people, and particularly on the elderly as a subpopulation who are at risk of deprivation relative to the working age population. Research in this vein, much of it comparative, examines the role of the welfare state and public policy as major influences on the well-being of the older population in developed countries. The role of the welfare state as a mechanism of social stratification, its potential to reduce inequality produced by markets and redistribute income over the life course, and the effects of trends in welfare state restructuring on levels of inequality are addressed by this literature.

Fundamental Cause Theory

Perhaps the most commonly cited sociological theory currently used to explain the role of social factors in shaping health, *fundamental cause theory*, proposes that efforts to improve population health and reduce health disparities by identifying and targeting individually based risks will be ineffective in the long run because they do not address the social conditions that structure the determinants of disease (Link & Phelan, 1995). According to fundamental cause theory, disparities in health have persisted in developed countries despite the reduction or elimination of risk factors that historically stratified health along socioeconomic lines (e.g., poor sanitation and infectious disease). Indeed, because high socioeconomic position offers a broad range of flexible and multi-purpose resources that can be used to ward off whatever particular threats to health exist at a given time, the elimination of risk factors that are proximally linked to disease does not eliminate the relationship between socioeconomic position and health (Link & Phelan, 1995). The risk factors that mediate the relationship change over time – today high cholesterol has replaced poor sanitation – and greater resources allow individuals to more quickly learn about and protect themselves from new risks as they arise, as well as receive more successful treatment of diseases that do occur, even when those treatments are universally and publicly available.

A few studies have attempted to empirically evaluate the principles of this theory and assess the conditions under which it is valid for both research and policy. For example, Lutfey and Freese (2005) demonstrated that higher-status individuals in the United States tend to have greater access to medical practitioners who also have greater resources and available information for making treatment decisions. Using US mortality data, Phelan, Link and colleagues found that less preventable causes of death, those for which we know little about prevention and treatment, are much more weakly associated with SEP than more preventable causes of death (Phelan, Link, Diez-Roux, Kawachi, & Levin, 2004).

However, the theory's predictive power remains unclear and relatively untested outside the unique health context of the United States. Comparative analysis of the United States and Canada demonstrated that lower levels of SEP lead to a greater likelihood of experiencing a highly preventable disease in the United States, but not in Canada, suggesting that social policies and level of economic inequality may buffer the relationship between socioeconomic resources and the incidence of preventable disease (Willson, 2009). Thus, while fundamental cause theory refers to deep-rooted, systemic, structural mechanisms of stratification, these specific effects have rarely been explicitly tested. As such, the theory is widely cited, but more generally to acknowledge the importance of social conditions as mechanisms generating health inequality, with "piecemeal" investigations of specific social determinants, in particular geographic locations and time periods. More research is needed to determine the generalizability of fundamental cause theory to other welfare state contexts and historical periods.

Life Course Theories of Health Inequality

While fundamental cause theory remains essentially silent on life course processes that produce health inequalities, other theories have emerged from findings that the magnitude of health disparities resulting from SEP varies over the life course and that there is variation in the pace of health change experienced by individuals as they age (see Pavalko & Caputo, 2013, for a review). Thus, many recent studies of health inequality in the social sciences focus on the individual and social patterning of health trajectories over time. Theories of CAD and cumulative inequality (CI) are key frameworks informing studies of health inequality among older people, suggesting that inequality is generated by a process through which initial relative advantage associated with structural location and resources results in systematic divergence in life course processes across individuals or groups over time (Dannefer, 2003; Ferraro & Shippee, 2009; Merton, 1968; O'Rand, 1996).

Cumulative Dis/advantage (CAD)

CAD as a mechanism of stratification suggests that early advantages and disadvantages continue to grow over time. This process occurs through compounding returns to resources such as education, and results in systematic divergence in life course processes across individuals or groups over time (Dannefer, 1987, 2003; Merton, 1968; O'Rand, 1996). Health is a form of life course capital that individuals preserve or deplete at varying rates as they age based on the interaction of structure, human agency, and chance (O'Rand, 2006; O'Rand & Henretta, 1999). Generally, the CAD framework conceptualizes health inequality within a population in later life as the end product of an accumulative process of exposure to risks and opportunities that operates across the individual life course.

Although CAD explains an individual-level process of resource acquisition and risk avoidance, it is a process that has population-level implications for changing levels of inequality with age. However, studying CAD as a mechanism of health inequality presents conceptual and methodological challenges that are related to the unique nature of this particular life course outcome, which was not the original focus of the theory. Thus, the accumulation of advantage slows the rate of health decline, rather than accelerates the rate of growth as it does for other outcomes, such as income and wealth. Furthermore, the health of individuals across positions of both advantage and disadvantage confronts forces of senescence and mortality over time. As such, a remaining question is whether the growing levels of inequality in health that CAD predicts is a process that continues indefinitely or slows upon reaching a particular critical age (Dupre, 2007; Willson, Shuey, & Elder, 2007), or is impacted by other social factors such as welfare state policy and health care institutions (Corna, 2013; Quesnel-Vallée, 2004). This has proven to be a challenge to studies investigating whether or not a systematic divergence in health occurs within cohorts over the life course and has led to conflicting findings.

Indeed, while CAD was substantiated in the United States (Dannefer, 2003; Prus, 2007; Taylor, 2008) as well as across 11 European countries (Huisman et al., 2004), other studies have shown that these health inequalities are stable (Kelley-Moore & Ferraro, 2004; Schöllgen, Huxhold, & Tesch-Römer, 2010), or even that they decrease in later life (Herd, 2006; House, Lantz, & Herd, 2005; Schoeni, Martin, Andreski, & Freedman, 2005). These latter findings led to the "age-as-leveler" hypothesis, and were originally viewed as competing with CAD; however, recent evidence has tended to show that they may not be mutually exclusive (Dupre, 2007). Indeed, the "age-as-leveler" hypothesis may operate through several pathways that mitigate CAD mechanisms, but do not negate them. Thus, decreasing health inequalities in later life could arise out of population-wide processes of biological frailty and senescence, of working conditions no longer having an influential effect on health as people retire, and/or of improvements in policies that support the old, essentially leveling out major effects of SEP differences (Dupre, 2007; Herd, 2006; House et al., 1994, 2005).

Furthermore, this research has highlighted numerous methodological issues with studies of health inequalities among older individuals that may be at the root of these conflicting findings. First, observational studies of general populations tend to exclude institutionalized individuals, and thus the sickest, most isolated and lowest SEP older adults (Robert et al., 2009). In addition, issues of statistical power have historically prevented a focus on the oldest old, where inequalities might become increasingly large again (Herd, 2006).

A second argument against the validity of the "age-as-leveler" findings is whether some of the SEP indicators used to assess inequalities in later life have sufficient predictive power. Some have argued that the bearing of SEP on health may be considerably weaker in the elderly population (Robert et al., 2009), therefore, the choice of indicators used (or in many cases, the availability of certain indicators in a given survey) may condition the extent of inequalities observed. Although indicators such as income, education, and occupation may be accurately predictive earlier in life, they may not be for older adults (Robert & House, 1996). For instance, it has been suggested that education is no longer as relevant as people age and retire, whereas income remains a successful predictor (House et al., 2005; Robert et al., 2009). Yet others have argued that income might also decrease in predictive strength in a similar manner as education, and financial assets might be a more important SEP indicator at older ages (Robert & House, 1996). Finally, these effects may be contextually specific. Indeed, Grundy and Holt (2001) examined the effectiveness of seven SEP indicators including social class, income, and education to study health inequalities among the elderly in the United Kingdom, and found that all were associated with self-reported general health status. However, the best predictive power was attributed to social class or education used in combination with a measure of deprivation.

Finally, selective mortality is often brought forward in the literature as a contributor to the evidence supporting the "age-as-leveler" hypothesis (Dupre, 2007). Cumulative disadvantage over a life course might result in premature death, leaving only the strongest individuals from the disadvantaged groups (be it racial/ethnic groups or low-SEP groups); consequently, when comparing the resilient survivors from the disadvantaged groups to the survivors from more advantaged groups, it appears as though inequalities among the elderly are diminishing relative to earlier in life (Ferraro & Shippee, 2009). Although some studies have found that this indeed contributes to "age-as-leveler" findings, mortality selection cannot fully explain the reduction in health inequalities between low- and high-SEP groups (Herd, 2006; McMunn, Nazroo, & Breeze, 2009).

Cumulative Inequality Theory

Recently, cumulative inequality theory (CI; Ferraro & Shippee, 2009; Ferraro et al., 2009) integrated elements of CAD, life course principles (Elder, 1998), and stress process theories (Pearlin, 1989) to propose formalized axioms that specify the processes leading to accumulated inequalities in health. The synthesis of elements from these theories and the fact that CI was developed specifically for understanding processes generating inequality in *health* outcomes (as opposed to life course inequality more generally, as is CAD) are unique contributions of the theory.

The five axioms of CI theory are covered in depth elsewhere (Ferraro & Shippee, 2009; Ferraro et al., 2009); here we focus on several key aspects of the theory. First, CI theory builds upon the traditional focus on the social antecedents of inequality to emphasize that inequality accumulates over the life course, pointing to the importance of childhood conditions on adult health outcomes (Ferraro & Shippee, 2009). The notion that early life socioeconomic environment initiates pathways or trajectories of health advantage or disadvantage that continue across the life course is not new (for review see Corna, 2013; Graham, 2002), but this emphasis urges researchers of inequality in later life to ask when and how accumulative processes begin and to focus on early life to understand later life outcomes. Life course research has increasingly focused on the effects of timing, duration, and change in childhood circumstances on adult achievement and health (e.g., McDonough, Sacker, & Wiggins, 2005; Wagmiller, Lennon, Kuang, Alberti, & Aber, 2006), and CI theory also formalizes the importance of "magnitude, onset, and duration of exposures to advantage and disadvantage" (Ferraro & Shippee, 2009) in order to better understand how multiple life domains interact and how risk factors accumulate.

Although CI theory heavily emphasizes that inequality in health is structurally generated, it also includes a principle of the life course perspective that is often neglected in research: the proposition that trajectories may change due to human agency and, in particular, the fact that people have the ability to perceive how they are doing, which influences their subsequent actions (Ferraro & Shippee, 2009). CI theory draws upon symbolic interaction and social comparison theories to propose that perceptions of resources may be more important than actual resources in shaping trajectories. It will be interesting to see how this plays out empirically, although research testing this proposition may be difficult to execute due to data limitations. There is some evidence that individuals' perceptions of health are influenced by social factors such as their socioeconomic position and age, but more research is needed to understand the role of perceptions of positions on health and certainly to disentangle the effects of structure and agency on health.

Finally, CI theory tackles the thorny subject of selective mortality and proposes that because

CI in health may result in early mortality, nonrandom selection leads to changes in the composition of cohorts (Ferraro & Shippee, 2009). The loss of the most disadvantaged individuals in a population will give the appearance of decreasing health inequality in later life and leads Ferraro and Shippee (2009, p. 336) to ask, "Stated plainly, should gerontologists study older people only?"

Today, most research on health inequality draws to some extent on the theories discussed above, and views health inequality in old age as an end-product of life-long processes. But life course theories of health inequality generally do not explicitly acknowledge that the social conditions shaping exposures to risks and opportunities shift rather dramatically in old age due to age-based requirements for access to welfare state entitlements such as income supplements and universal health care (particularly, though not exclusively, in the United States). We next discuss broad theories that address the role of the welfare state on inequalities in old age.

WELFARE STATES AND THE INTERPLAY OF SOCIAL SOLIDARITY AND EQUITY

Theories described in the previous section highlight that individual advantages related to SEP in old age are linked to socioeconomic and health advantages across the life course, but this relationship is also mediated by public policy decisions and institutions (Crystal, 2006). Policies that accentuate economic equality may mitigate the effects of an earlier lack of social resources on health. It follows from fundamental cause theory that in societies characterized by less stratification of and competition for resources, health disparities at all ages will be lower than in societies with greater inequality (Willson, 2009).

Early theories of the state's role in the provision of key social transfers and provisions to ensure a level of well-being for its citizens drew upon modernization theory and the "logic of industrialism" and concluded that population aging was a primary motivation for welfare state development (see Quadagno, Kail, & Shekha, 2011, for a review). Industrialization disrupted the ability of families and communities to provide support in case of illness and/or during later life, and, at the same time, contributed to economic growth and a surplus of funds that made it possible for this support to be shifted to governments.

Institution theory (e.g., Scott, 2001) provides a framework for understanding the role of the welfare state in shaping life course policy that affects levels of health inequality in old age. At the level of the state, institutions have the potential to "transform incalculable insecurity into calculable risks and shared meanings" as well as establish opportunity structures and incentives for promoting individual action that leads to collective benefits (Weymann, 2009, p. 114). Thus, welfare states notably attempt to mitigate unexpected and unavoidable risks such as sickness, unemployment, and old age (Briggs, 1961).

The first government-mandated social health insurance program was inaugurated in 1883 by Otto Von Bismarck, first Chancellor of the German Empire, under the moral underpinnings that workers, or perhaps more cynically, their work capacities, were deserving of social insurance (Briggs, 1961). This landmark piece of legislation that was subsequently widely copied and adopted around the world, enacted a law requiring employer (1/3) and employee (2/3) contributions to a sickness fund, ensuring co-subsidizing and pooling of risk (World Health Organization, 2000). Similarly, public pensions were also based on the concept of social insurance and would allow recipients to maintain the standard of living they had in their working years into retirement (Quadagno et al., 2011). Both flagship programs of the welfare states thus originally developed from this view of social solidarity extended to "deserving" workers.

While the first part of the twentieth century is marked by the extension of similar programs to larger segments of the population within countries (workers' dependants, and former workers in the case of health insurance in old age) and to more countries around the world, the second part of this past century saw the emergence of new principles of social protection. Often linked to the seminal British Beveridge report (Beveridge, 1942), reforms that followed broadened the focus of state protection from the worker to all citizens, and "from cradle to grave." This led to a new wave of reforms promoting universal access on the basis of a principle of equity recognizing health and a decent standard of living as fundamental human rights. However, following a period of rapid expansion of these programs, recent decades have seen a retrenchment of these universal welfare state entitlements in many countries (OECD, 2007a). Thus, in most developed countries, Bismarckian and Beveridgian perspectives now co-exist through different programs, but are not necessarily integrated in a coherent framework with clear indications for health inequality, particularly among the elderly.

It is in the context of this framework provided by the welfare state that most of the population makes long-term plans for old age (Weymann, 2009). The social rights associated with welfare state benefits reduce uncertainty and buffer the influence of the market; however, depending on the insurance structure of these entitlements, access to the privileges associated with social rights can mean inclusion for some individuals and groups and exclusion for others; for example, pension benefits that target very specifically defined groups according to work history and earnings (Weymann, 2009). Consequently, welfare states are "mechanisms of social stratification" as social programs can reduce inequality produced by markets and even out income over the life course across class, gender, and racial lines (O'Connor, Orloff, & Shaver, 1999).

The extent to which welfare states redistribute resources can have a significant effect on the well-being of older people and levels of inequality in various outcomes in old age (Myles & Quadagno, 2002). Accordingly, in the next subsections, we will focus on the two flagship social policy programs that welfare states have developed to care for the elderly population, namely pensions and health care systems. First, we will examine recent trends in pensions among Organization for Economic Cooperation and Development (OECD) countries that raise some concerns for rising income (and by extension, health) inequality, if left unchecked. In addition, we will also examine how health systems currently (fail to) respond to the needs for LTC of the elderly population.

Increasingly, however, these is also the recognition that even universal programs may not suffice to stem inequality, particularly when the focus of intervention is on average population effects rather than on the distribution or level of inequality (Frohlich & Potvin, 2008). To respond to this need for targeted intervention, new policy models that are salient for the question of social inequalities in health among the elderly, namely the SDH and Age-Friendly Communities (AFC), have been developed and strongly promoted by the WHO in the last decade. We will review each of these policy models and discuss their potential for limiting health inequalities among the elderly.

Pensions

With population aging leading to a need for increased pension expenditures, pension systems in high-income countries are encountering the common challenge of balancing affordability of pension systems with supplying sufficient pensions to contribute substantially to the retirement income of elderly populations. The OECD monitors the evolution of pension systems among its member countries (34 high- and middle-high-income countries). In its 2013 report, it highlights two major policy trends aiming to achieve this balance: a shift toward funded private pensions and reforms of pay-as-you-go public pension systems. Both of these trends are noted by the OECD as having substantial implications for increases in inequality and, we argue, for potentially compounding existing inequalities as well.

The move toward a greater reliance on private pensions points to the retrenchment of the public pension system in many countries, and most notably in English-speaking countries. In Canada and the United States for instance, public transfers now supply less than 40% of a retiree's income, significantly below the OECD countries' average of 59%. While private pensions may ensure a decent retirement income for their contributors, they typically do not lead to a pooling and redistribution to low-income contributors. Thus, these countries are likely to face an increasing polarization of incomes at retirement between those who rely almost exclusively on public pensions and those who rely on a combination of public and private pensions. Furthermore, as we know from CAD and CI theories, these different statuses in the face of retirement stem from life course trajectories of dis/advantage; thus a greater reliance on private pensions may in fact further compound these trajectories, in particular in the face of shrinking public pensions. Finally, private pensions also expose individual contributors to investment risk and individualize the cost of living longer, rather than mutualize it (OECD, 2012).

At the other end of the policy spectrum, some countries have steered away from private pensions, partly due to public dissatisfaction, and have tended toward the second trend, reforms of pay-as-you-go public pension systems. In these (mostly European continental and Scandinavian) countries, public transfers provide approximately 60% of retirees' incomes. In order to ensure the solvency of the public mutual insurance funds, reforms have included, among other measures, an increase in the retirement age. Accordingly, some countries, like the United Kingdom and the Czech Republic, have already increased the retirement age to 68 or 69 years old, and it is expected that by 2050 normal retirement age will be 67 years old in most OECD countries (OECD, 2013). However, as argued by Lowsky, Olshansky, Bhattacharya, and Goldman (2013), using chronological age as a criterion for program entitlements may increasingly constitute an inaccurate proxy for need in the face of divergent life course trajectories. Again then, a divergence is to be expected, with some easily working to those ages and beyond, and others whose health will fail them well before reaching the age for entitlement. Few countries have enacted policies to address these anticipated disparities, with the notable exception of France. Indeed, one of the main innovations and most hotly debated measures of the pensions reform act voted in by the French Parliament in December 2013 is the *compte personnel de prévention de la pénibilité* [hardship prevention personal account]. These personal accounts will monitor work-related hardships based on ten criteria such as noise levels, night shifts, painful postures, etc. over individuals' work lives. On the basis of the "points" accumulated for work-related hardships, workers will then have access to the following mitigating measures: retraining for another job, a switch to part-time work with full-time pay in later years of their work life, or early retirement (as young as 60 years old) (Service Public, 2014).

Aside from the two trends highlighted above, the OECD (2013) also notes that entitlements have been shrinking and that individuals who do not contribute fully throughout their work lives will increasingly struggle in the future to achieve a decent standard of living. This is true of both public and private pensions, as the latter do not redistribute to lower-incomes groups. Furthermore, this means that future generations of retirees will likely contribute more toward lower pensions, raising the question of intergenerational equity. Among the groups particularly at risk of losing out in these reforms are middle-income groups and women. As there are systems in place to protect workers with the lowest incomes, and those with the highest income have other sources of income and assets to supplement their public pensions, it is those in the middle who are most likely to be lacking retirement income. In addition, as women are less likely to contribute to pensions over full careers, are less likely to have sources of financial support other than public pensions, and are more likely to live longer than men, they are also at greater risk of poverty in older ages. In sum, while pensions have had a significant impact on reducing poverty among the older population, as the OECD (2013) repeatedly highlights, they cannot in and of themselves stem income inequalities, particularly in light of current policy trends. Thus, access to other public services, and particularly to LTC, will prove essential in ensuring living standards among the elderly (OECD, 2013). We turn to this issue next.

Long-Term Care

Most health systems in high-income countries have implemented insurance systems that provide universal access to physician and hospital services for the elderly with minimal or no cost-sharing. However, with the epidemiologic transition to a bulk of chronic and degenerative diseases, this focus on curative care has long been deemed to only partially meet the needs of the elderly population, at least in terms of LTC services. LTC refers to a broad range of services encompassing skilled nursing care, assisted-living facilities, home care, hospice care, respite care, adult day care, and different in-home living arrangements (Sultz & Young, 2011). These services provide a long-term response to chronic increases in disability brought about by functional limitations inherent in the process of individual aging (Feder, Komisar, & Niefeld, 2000).

Even in the absence of comprehensive publicly funded services, the LTC sector has experienced constant growth since the Second World War (Grabowski, 2008). However, while the formal provision of these services was once concentrated in institutions, demand appears to be shifting toward assistance aimed at remaining in the community as long as possible: from 1994 to 2005, the 85+ age group showed a decrease of 9% in the report of at least one Activity of Daily Living (ADL) limitation for those living in institutions, while an increase of 3% took place among those living in the community (Lafortune, Balestat, & Disability Study Expert Group Members, 2007). This suggests that an increasing proportion of those aged 85 and over are in need of LTC services and now remain in the community.

This trend is probably not unrelated to the fact that governments across the board in developed countries have been, since the 1970s, championing a view of seniors as being more independent, more involved in their communities and living as long as possible outside of institutional settings (World Health Organization, 2002). However, this demand and the funding for it so far have been borne mostly by families and informal caregivers.

In fact, to this day, most LT caregiving is still undertaken by informal providers and paid for out-of-pocket, particularly with regards to care in the community. More recently, this informal care network came to be complemented by formal caregiving options, either through scarce and means-tested publicly subsidized services and/or a plethora of private providers (Wiener & Tilly, 2002). Thus, a shift has taken place where a privileged segment of the population can buy into the contemporary conception of individual aging in the private market; indeed, this is where elderly people, as consumers, can spend considerable amounts of money to receive the necessary support which enables them to remain active. Drawbacks to such a loose health sector include lack of information-sharing on available public services, an unstable workforce, little follow-up on elderly people's health and social issues, and greater inequities regarding access to these services (Rantz, Marek, & Zwygart-Stauffacher, 2000). Indeed, the OECD reports that costs associated with greater needs (i.e., 25 hours/week) exceed on average 60% of the disposable income for all but the wealthiest quintile of the elderly population (OECD, 2013).

The reliance on informal caregiving and elderly people as consumers of care generates unmet needs which, at the individual level, accelerate the deterioration of health, and promotes institutionalization which could have been otherwise prevented with efficient and integrated LTC coordination (Keefe, Légaré, & Carrière, 2007). In the United States, a Commonwealth Fund study reported that 58% of the elderly respondents of a 1999 cross-state survey declared having unmet needs (Komisar, Feder, & Kasper, 2005). In Canada, a Statistics Canada study drawing on the 2003 wave of the Canadian Community Health Survey showed that 42% of seniors requiring help with moving about in their home did not receive any help and that even among those who received a mix of formal and informal home care, 19% of seniors expressed unmet home care needs (Carrière, 2006). This sheds light on the fact that even limited and fragmented public provision of LTC services may not be sufficient to ameliorate or even maintain older adults' health status.

Finally, there is substantial evidence that elderly people and their families – especially those from lower socioeconomic classes without an informal caregiving network – do not possess sufficient resources or control over their LTC needs (Feder et al., 2000; Messinger-Rapport, 2009). Indeed, while a vast array of private care options exist, ranging from home care services to high-end retirement communities or elder care centers (Grabowski, 2008), many elderly persons lack the means to consume the variety of LTC products offered by the market and yet fall above the stringent requirements to access publicly covered, means-tested LTC services.

WHO Health in All Policies

According to Marmot and Wilkinson (1999), the SDH encompass individuals' material and psychosocial circumstances, and range from the more micro-level of health behaviors and social support to the more macro-level of unemployment rates and policies regarding food and transportation. Social policies can impact each of these levels, but up until very recently, assessments of the impact of social policies on health inequalities tended to take a "patchwork" approach, reviewing select policy domains as they pertain to major social determinants, such as education, employment, housing, etc. (e.g., Quesnel-Vallée & Jenkins, 2009).

Increasingly, however, the recognition that social inequalities in health stem from multiple and intertwined areas of social intervention has led to the development of a new policy approach to curb social inequalities in health, termed "Health in All Policies." As several chapters in this handbook review some of the most important SDH inequalities among the elderly in great depth, we focus here more specifically on those intersectoral policies and their potential impact on health inequalities among the elderly.

The "Health in All Policies" approach was developed and advocated by the WHO, notably under the impetus of the Finnish Presidency of the European Union in 2006. It favors intersectoral policies that attempt to overcome traditional governmental "silos" and rely on the participation of different levels of government (from the local to the supra-national). Aside from an explicit commitment to reducing health inequalities, the establishment of adequate governance structures allowing for this action is critical (McQueen, Wismar, Lin, Jones, & Davies, 2012). Finally, as it is fundamentally an evidence-based policy approach, it relies quite heavily on Health Impact Assessment (HIA), defined as "a combination of procedures, methods and tools by which a policy, programme, or project may be judged as to its potential effects on the health of a population, and the distribution of those effects within the population" (ECHP, 1999).

Taken together, this integrated policy approach should be more effective at reducing inequalities by offering coordinated mechanisms for action that are articulated around a common goal (McQueen et al., 2012). For instance, if the reduction of health inequalities is explicitly deemed a national priority and provided the appropriate governance structures are in place, an economic policy such as the reduction of income tax for the highest incomes could undergo a HIA, and be reassessed accordingly. Similarly, if the reduction of inequalities in health among the elderly were a focus of intervention, changes to pensions' funding systems might have to undergo an HIA as well.

As yet, few countries have adopted such integrated policies. Accordingly, it is too early for empirical population research to tell whether the countries that adopted such policies fare better than those that did not in terms of the control or reduction of health inequalities. However, we can begin to examine these policies critically in terms of their potential to address social inequalities in health among the elderly.

Australia, Finland, New Zealand, Norway, Sweden, and the United Kingdom stand out as leaders in this policy area. In a recent report, the *Institut national de la santé publique du Québec* (INSPQ, 2014) reviewed the integrated policies of each of these countries. In turn, we searched for information in these policies pertaining to either health inequalities within the elderly population or the elderly as a vulnerable population. All these countries have explicitly targeted early life and childhood as life course periods of intervention, which indicates sensitivity to life course processes producing health inequalities. However, while that perspective focuses on preventing the further development of health inequalities for future cohorts of elderly, it does very little to address health inequalities and the potential divergent trajectories that they both stem from and may further compound among *current* elderly cohorts.

Finland is the only exception in this regard (Ministry of Social Affairs and Health, Finland, 2008). Indeed, in their National Action Plan to Reduce Health Inequalities 2008–2011, Finland adopted the dual perspective of recognizing the elderly both as a population in need of targeted action relative to other groups as well as reducing health inequalities within the elderly population. The twin foci of promoting functional health in the community (notably through the expansion of social, rather than curative, services) and the reduction of health inequalities among the elderly provide, in our opinion, an example of best practices. Finland would therefore constitute a good test case against other countries for testing our theories and monitoring the impact of policies that are mindful of existing and persisting health inequalities among the elderly.

Meanwhile, the WHO made progress on the SDH more generally, with the adoption on May 26, 2012 by the World Health Assembly of resolution 65.8 endorsing the Rio Political Declaration on Social Determinants of Health (WHO, 2014a). This entails that member countries agree that tackling SDH is a fundamental approach to the work of WHO and a priority area for action. It suggests that further policy changes may be afoot in many other countries.

WHO AGE-FRIENDLY ENVIRONMENTS PROGRAMME

The previous sub-section highlighted the dearth of elderly specific policies with regards to health inequalities. We turn next to another integrated policy approach that could be salient in this context as well, which now focuses explicitly on the elderly, namely the WHO Age-friendly Environments Programme (WHO, 2014b). A recurring theme in the age-friendly community literature is the concept of "aging in place," which captures the social and policy goals of age-friendliness (Hanson & Emlet, 2006; Hooyman & Kiyak, 2008; Lui, Everingham, Warburton, Cuthill, & Bartlett, 2009). In a national survey by the AARP concerning housing and home modification issues, over 90% of adults 65 and over indicated that they wished to remain in their homes as long as possible (Bayer, Management, Harper, & Group, 2000). Hooyman and Kiyak (2008) explain that being capable of aging in one's own home is greatly facilitated or hindered by the policies and services that support older individuals in a community who begin to experience cognitive and physical limitations, leading to reduced independence. Furthermore, AFCs enhance older people's quality of life but would also benefit all residents, regardless of age and functional capacity (Alley, Liebig, Pynoos, Banerjee, & Choi, 2007).

Thus, in recognition of these stated needs, the WHO has pursued "healthy" or "active" aging initiatives for a number of years (WHO, 2002). A central component of the WHO Age-friendly Environments Programme is the "Global Age-Friendly Cities Guide," which was developed based on focus groups with lower- and middle-class older adults age 60 years and older, caregivers of older individuals and service providers. The project was partly funded by the Public Health Agency of Canada (PHAC) (Plouffe & Kalache, 2011) and included a total of 35 cities in both developed and developing countries, 33 of which participated in the focus groups (WHO, 2007).

Based on the evidence gathered in these focus groups, the WHO created a checklist of interacting determinants of active aging in a community. These include the essential features of eight domains: outdoor spaces and buildings; transportation; housing; social participation; respect and social inclusion; civic participation and employment; communication and information; and community and health services (WHO, 2007). Programs connected to the WHO Global Network of Age-Friendly Cities and Communities have emerged in Canada, France, Ireland, Portugal, the Russian Federation, Slovenia, Spain, and the United States (WHO, 2012).

The WHO Global Age-Friendly Cities Guide recognizes that the older population is a heterogeneous group in terms of needs and functional capacity, and that inequalities in health and functioning increase with age. Social, economic, and physical features cause differences in life expectancy, functional capacity, and health among older adults (Plouffe & Kalache, 2011) and, therefore, policy should aim to mitigate these disparities (WHO, 2007). Indeed, the physical environment's impact on health and healthy lifestyle choices has been widely explored in recent literature (Menec, Means, Keating, Parkhurst, & Eales, 2011). A study by Dunn (2002) found that disparities in housing conditions played a role in the creation of health inequalities, consistent with the WHO finding that appropriate housing influenced access to services in the community and therefore impacts older adults' quality of life (WHO, 2007). Policy measures and age-friendly services may assist in reducing these inequalities; some have already been put in place, such as in Ponce, Puerto Rico, where free transportation to medical appointments is provided from senior centers (WHO, 2007).

Another example of an age-friendly initiative contributing to shrinking inequalities among the elderly is the Advantage Initiative Model, which was used to assess age-friendliness in a community in western Washington (Hanson & Emlet, 2006). Following the assessment, the community's Senior Information and Assistance program increased the distribution of information and outreach to isolated and low-income older adults; between 2003 and 2004, there was an increase of two and a half times the number of older people seeking help through this program. In addition, the AdvantAge Initiative led to the development of a guide to free or low-cost exercise programs available to the older population as well as to the creation of biweekly walking groups for seniors. However, Hanson and Emlet (2006) emphasize that in the case of the AdvantAge Initiative, assessing livable income would be required as part of the model, because in western Washington it was found that older adults who had incomes under twice the Federal poverty level had poorer outcomes on many different indicators compared to older adults above 200% of the Federal poverty level (Hanson & Emlet, 2006).

In sum, while the Global Age-Friendly Cities Guide was originally developed with middle- and low-income individuals, and presumably also emphasizes the specific needs of these subgroups of elderly, these policies might still have to explicitly assess the level of inequalities in their populations and the differential impact of these population interventions. In the same vein, a lack of certain services in a community can raise inequalities in health. The WHO project found that many cities have nonexistent or poorly organized home care services that are too expensive and too difficult to access due to stringent criteria for eligibility (WHO, 2007). Consequently many elderly in need of these services will not be provided with care and might see their health further deteriorate.

PROMISING AVENUES FOR SOCIOLOGICAL RESEARCH

While the study of social inequalities in health has thrived since the 1970s, it is only in the past two decades that governments have begun enacting policies that explicitly tackle health inequalities and the unequal distribution of their social determinants (Graham & Power, 2004). Thus, public health policy in many countries now has broadened its mandate from a concern with population health to the protection of health equity. This is particularly true in Europe, but can also be found in the United States with the *Healthy People* focus on reducing disparities, or in Canada with its definition of population health as "an approach to health that aims to improve the health of the entire population and to reduce health disparities among population groups" (Health Canada, 2014).

Yet, Graham and Power (2004) contend that these initiatives conflate the (social) determinants of health and the social processes that create an unequal distribution of those determinants in society. This confusion has led to policy perspectives that wrongly assume that addressing health determinants will also de facto reduce inequalities. Thus, while recent decades have seen improvements in determinants of health (e.g., rising living standards, higher average education, lower smoking rates), with concomitant benefits in population health, health inequalities have persisted or even increased. In line with Link and Phelan's (1995) fundamental cause perspective, many have therefore come to recommend that, to truly address health inequalities, policy agendas will have to tackle not only the SDH, but also the determinants of social inequality that shape the myriad ways in which social advantage cumulates over the life course and across generations (Coburn, 2004; Graham & Power, 2004).

Exworthy (2008) identifies seven challenges inherent to a SDH approach to policymaking that compound the conceptual difficulties highlighted by Graham and Power (2004). Among those are the demanding task of enacting policies in the face of multifactorial exposures and complex causal chains that summon concerted initiatives across many governmental departments and beg better data collection. In addition, the life course perspective that underlies much of this research poses the additional problem that the policy cycle is ill-equipped to deal with such long-term processes (since most electoral processes follow timelines measured in years, not decades). While issues relating to the need to rethink policy in the face of new risks across the life course have recently been addressed by policymakers at the request of the OECD (2007a, 2007b), Exworthy (2008) argues that part of the solution to these challenges may come from researchers, through the development of conceptual models and appropriate methodologies that better conform to the subtleties of the SDH approach in a policy setting. It is in recognition of these challenges that we have highlighted the policy initiatives that seem most promising in this regard, namely the WHO Health In All Policies approach and the Age-friendly Environments Programme. As we noted, much remains to be done in evaluating the impact of these programs, and we would argue that a sociological theoretical lens would illuminate important facets of these processes.

However, research evaluating the effects of policy initiatives in mitigating the SDH is fraught with empirical challenges. First is the lack of variation in the population's exposure to universal programs in a given country and point in time. Yet, as we highlighted, natural policy experiments are happening everywhere around the world. These could be exploited empirically to understand how the policy context shapes the structure of opportunity of individuals. Again, all the theories we review, from fundamental cause to CI, would provide fruitful frameworks for the study of these effects. We have highlighted in this chapter how, for the most part, policy reforms such as those pertaining to pensions are still being enacted without much consideration for their ramifications on inequalities (and by extension health inequalities), even though both a policy analysis by the OECD and all the theories we highlighted point to a risk for increasing inequalities as a result of these changes. There are therefore many opportunities for research arising from these policy variations, particularly with methodological approaches drawn from the counterfactual account of causality, such as difference-in-difference models (see Schlotter, Schwerdt, & Woessmann, 2010, for a review of some of these causal approaches to policy evaluation).

Furthermore, the shared issues we have uncovered in this chapter across developed countries mean that we can learn from one another, and that cross-nationally comparative research could prove to be a particularly advantageous strategy in this regard. Here, the sociological imagination can make a distinct contribution. Indeed, the focus in evaluating the impact of those policy interventions (to the extent that there is any evaluation at all) is often toward health outcomes, which obscures the existence, persistence, or even development of social inequalities. Yet, as Foner (2000) reminds us, sociologists are uniquely poised (and tooled) to uncover those processes, particularly as they pertain to aging.

Of course, this line of research comports its own trials, chief among which are data harmonization and the measurement of policy context. Fortunately, headway is being made on both counts. On data harmonization, major cross-national initiatives such as the Survey of Health, Ageing and Retirement (SHARE), as well as the English Longitudinal Study of Ageing (ELSA), are harmonized with the Health and Retirement Study (HRS), and are increasingly bearing fruit (e.g., Banks, Marmot, Oldfield, & Smith, 2006; Crimmins, Kim, & Solé-Auró, 2011). In addition, further resources are being developed to assist in the dissemination of these data initiatives: see the Chicago Core on Biomarkers in Population-Based Aging Research (CCBAR) list of other international longitudinal surveys on aging (CCBAR, 2014), as well as the Integrative Analysis of Longitudinal Studies of Aging (IALSA) network for post-hoc harmonization initiatives of other aging surveys (IALSA, 2014). Other initiatives include the Healthy Ageing across the Life Course (HALCyon) Network (HALCyon, 2014) on British cohort studies.

In terms of policy indicators, one strategy is to focus on particular policies, which has the advantage of seeing a plausible impact on specific outcomes (e.g., Garcia & Crimmins, 2013, for cancer screening policies). Conversely, cross-national assessments of the impact of broader social policies will meet with the challenge of lack of specificity in the outcome, given that these are policies with diffuse effects: where would we expect to see the health effects of changes in pension policies for instance? Lundberg and colleagues (Lundberg et al., 2008) provide a compelling example of research making headway in this area, notably by using specific policy indicators that are congruent with the outcome (i.e., family policies and infant mortality, and pension policies and life expectancy at 65). Thus, the most promising research in this area relies on indicators of the policy context that precisely measure entitlements, such as those provided by initiatives like the Welfare State Entitlements Data Set (Scruggs, Kuitto, & Detlef, 2013) and the Social Citizenship Indicator Program (SCIP, 2014) for general welfare state entitlements, as well as the Comparative Family Policy Database (Gauthier, 2014) for family policies and the Health Insurance Access Database for health insurance policies (Quesnel-Vallée, Renahy, Jenkins, & Cerigo, 2012).

In sum, we have hoped to show in this chapter that much could be gained from an intersection of sociological theories of health inequalities and the study (and evidence-informed development) of social policies. When social policies fail to acknowledge that current inequalities among the elderly are the product of life course trajectories of CAD, even universal benefits may fail to mitigate health inequalities. We highlighted that considerable change occurs in policies that could be fruitfully studied sociologically, using the right theories, methods, micro-data, and contextual data. With this, we hope to have pointed toward questions that could move this research agenda forward.

References

Alley, D., Liebig, P., Pynoos, J., Banerjee, T., & Choi, I. H. (2007). Creating elder-friendly communities. *Journal of Gerontological Social Work, 49,* 1–18.

Banks, J., Marmot, M., Oldfield, Z., & Smith, J. P. (2006). Disease and disadvantage in the United States and in England. *JAMA: The Journal of the American Medical Association, 295,* 2037–2045.

Bayer, A.H., Management, A.K., Harper, L., & Group, A.P.G. (2000). *Fixing to stay: A national survey on housing and home modification issues AARP.* Retrieved from: <http://www.aarp.org/home-garden/housing/info-2000/aresearch-import-783.html>. Accessed 7.12.13.

Beckfield, J., & Krieger, N. (2009). Epi + demos + cracy: Linking political systems and priorities to the magnitude of health inequities – Evidence, gaps, and a research agenda. *Epidemiologic Reviews, 31,* 152–177.

Bergqvist, K., Yngwe, M. Å., & Lundberg, O. (2013). Understanding the role of welfare state characteristics for health and inequalities – An analytical review. *BMC Public Health, 13,* 1234.

Beveridge, S. W. (1942). *Social insurance and allied services.* Presented to Parliament by Command of His Majesty (HMSO CMND 6404). Retrieved May 5, 2015 <http://www.sochealth.co.uk/resources/public-health-and-wellbeing/beveridge-report/>.

Brennenstuhl, S., Quesnel-Vallée, A., & McDonough, P. (2012). Welfare regimes, population health and health inequalities: A research synthesis. *Journal of Epidemiology and Community Health, 66,* 397–409.

Briggs, A. (1961). The Welfare State in Historical Perspective. *European Journal of Sociology, 2,* 221–258.

Carrière, G. (2006). *Seniors' use of home care. Health reports/Statistics Canada, Canadian Centre for Health Information = Rapports Sur La santé/Statistique Canada, Centre Canadien D'information Sur La Santé, 17,* 43.

CCBAR. (2014). *Chicago core on biomarkers in population.* The University of Chicago NORC Center on Aging. Retrieved from: <http://biomarkers.uchicago.edu/chicagocoreonbiomarkers.htm>. Accessed 31.01.14.

Coburn, D. (2004). Beyond the income inequality hypothesis: Class, neo-liberalism, and health inequalities. *Social Science & Medicine, 58,* 41–56.

Corna, L. M. (2013). A life course perspective on socioeconomic inequalities in health: A critical review of conceptual frameworks. *Advances in Life Course Research, 18,* 150–159.

Crimmins, E. M. (2004). Trends in the health of the elderly. *Annual Review of Public Health, 25,* 79–98.

Crimmins, E. M., Kim, J. K., & Solé-Auró, A. (2011). Gender differences in health: Results from SHARE, ELSA and HRS. *The European Journal of Public Health, 21,* 81–91.

Crystal, S. (2006). Dynamics of late-life inequality: Modeling the interplay of health disparities, economic resources, and public policies. In J. Baars, D. Dannefer, C. Phillipson, & A. Walker (Eds.), *Aging, globalization and inequality: The new critical gerontology* (pp. 205–213). Amityville, NY: Baywood.

Dannefer, D. (1987). Aging as intracohort differentiation: Accentuation, the Matthew effect, and the life course. *Sociological Forum, 2,* 211–236.

Dannefer, D. (2003). Cumulative advantage/disadvantage and the life course: Cross-fertilizing age and social science theory. *The Journals of Gerontology. Series B: Psychological Sciences and Social Sciences, 58,* S327–S337.

Dunn, J. R. (2002). Housing and inequalities in health: A study of socioeconomic dimensions of housing and self reported health from a survey of Vancouver residents. *Journal of Epidemiology and Community Health, 56,* 671–681.

Dupre, M. E. (2007). Educational differences in age-related patterns of disease: Reconsidering the cumulative disadvantage and age-as-leveler hypotheses. *Journal of Health and Social Behavior, 48,* 1–15.

ECHP. (1999). *Gothenburg consensus: Health impact assessment* (Report). HIA Gateway, West Midlands Public Health Observatory, <hia@wmpho.org.uk>.

Elder, G. H. (1998). The life course as developmental theory. *Child Development, 69,* 1–12.

Exworthy, M. (2008). Policy to tackle the social determinants of health: Using conceptual models to understand the policy process. *Health Policy and Planning, 23,* 318–327.

Feder, J., Komisar, H. L., & Niefeld, M. (2000). Long-term care in the United States: An overview. *Health Affairs, 19,* 40–56.

Ferraro, K. F., & Shippee, T. P. (2009). Aging and cumulative inequality: How does inequality get under the skin? *The Gerontologist, 49,* 333–343.

Ferraro, K. F., Shippee, T. P., & Schafer, M. H. (2009). Cumulative inequality theory for research on aging and the life course. In V. L. Bergtson, M. Serverstein, N. M. Puney, & D. Gans (Eds.), *Handbook of theories of aging* (pp. 413–433). New York, NY: Springer.

Foner, A. (2000). Age integration or age conflict as society ages? *The Gerontologist, 40,* 272–276.

Frohlich, K. L., & Potvin, L. (2008). Transcending the known in public health practice. *American Journal of Public Health, 98,* 216–221.

Garcia, K., & Crimmins, E. M. (2013). Cancer screening in the U.S. and Europe: Policies, practices, and trends in cancer incidence and mortality. In N. Hoque, M. A. McGehee, & B. S. Bradshaw (Eds.), *Applied demography and public health* (pp. 125–154). Netherlands: Springer.

Gauthier, A. (2014). *The comparative family policy database.* Max Planck Institute for Demographic Research. Retrieved from: <http://www.demogr.mpg.de/cgi-bin/databases/fampoldb/index.plx>. Accessed 31.01.14.

Grabowski, D. C. (2008). The market for long-term care services. *Inquiry, 45,* 58–74.

Graham, H. (2002). Building an inter-disciplinary science of health inequalities: The example of lifecourse research. *Social Science & Medicine, 55,* 2005–2016.

Graham, H., & Power, C. (2004). *Childhood disadvantage and adult health: A lifecourse framework.* London: Health Development Agency. Retrieved from: <http://www.gserve.nice.org.uk/niceMedia/pdf/childhood_disadvantage_health.pdf>. Accessed 31.01.14.

Grundy, E., & Holt, G. (2001). The socioeconomic status of older adults: How should we measure it in studies of health inequalities? *Journal of Epidemiology and Community Health, 55,* 895–904.

HALCyon. (2014). *Healthy ageing across the life course.* HALCyon. Retrieved from: <http://www.halcyon.ac.uk/>. Accessed 30.01.14.

Hanson, D., & Emlet, C. A. (2006). Assessing a community's elder friendliness: A case example of The AdvantAge Initiative. *Family & Community Health, 29,* 266–278.

Health Canada. (2014). *Population health.* Retrieved from: <http://www.hc-sc.gc.ca/ahc-asc/activit/strateg/population-eng.php>. Accessed 30.01.14.

Herd, P. (2006). Do functional health inequalities decrease in old age? Educational Status and functional decline among the 1931–1941 birth cohort. *Research on Aging, 28,* 375–392.

Hooyman, N. R., & Kiyak, H. A. (2008). *Social gerontology: A multidisciplinary perspective.* Boston, MA: Pearson/Allyn & Bacon.

House, J. S., Lantz, P. M., & Herd, P. (2005). Continuity and change in the social stratification of aging and health over the life course: Evidence from a nationally representative longitudinal study from 1986 to 2001/2002 (Americans' Changing Lives Study). *The Journals of Gerontology. Series B: Psychological Sciences and Social Sciences, 60*(Special issue 2), S15–S26.

House, J. S., Lepkowski, J. M., Kinney, A. M., Mero, R. P., Kessler, R. C., & Herzog, A. R. (1994). The social stratification of aging and health. *Journal of Health and Social Behavior, 35,* 213.

Huisman, M., Kunst, A. E., Andersen, O., Bopp, M., Borgan, J.-K., Borrell, C., et al. (2004). Socioeconomic inequalities in mortality among elderly people in 11 European populations. *Journal of Epidemiology and Community Health, 58,* 468–475.

INSPQ. (2014). *Avenues politiques: intervenir pour réduire les inégalités sociales de santé (Policy options: Interventions for reducing social inequalities in health)*. Quebec: INSPQ. Retrieved from: <http://www.inspq.qc.ca/pdf/publications/1822_Avenues_Politiques_Reduire_ISS.pdf>. Accessed 30.09.14.

Integrative Analysis of Longitudinal Studies of Aging (IALSA). (2014). Retrieved from: <www.ialsa.org>. Accessed 05.11.15.

Keefe, J., Légaré, J., & Carrière, Y. (2007). Developing new strategies to support future caregivers of older Canadians with disabilities: Projections of need and their policy implications. *Canadian Public Policy, 33,* S65–S80.

Kelley-Moore, J. A., & Ferraro, K. F. (2004). The black/white disability gap: Persistent inequality in later life? *The Journals of Gerontology. Series B: Psychological Sciences and Social Sciences, 59,* S34–S43.

Komisar, H. L., Feder, J., & Kasper, J. D. (2005). Unmet long-term care needs: An analysis of Medicare-Medicaid dual eligibles. *Inquiry, 42,* 171–182.

Lafortune, G., Balestat, G., & Disability Study Expert Group Members. (2007). *Trends in severe disability among elderly people: Assessing the evidence in 12 OECD countries and the future implications*. Health Working Papers 26.

Link, B. G., & Phelan, J. (1995). Social conditions as fundamental causes of disease. *Journal of Health and Social Behavior,* 80–94.

Lowsky, D. J., Olshansky, S. J., Bhattacharya, J., & Goldman, D. P. (2013). Heterogeneity in healthy aging. *The Journals of Gerontology. Series A: Biological Sciences and Medical Sciences, 69,* 640–649. <http://dx.doi.org/10.1093/gerona/glt162>.

Lui, C.-W., Everingham, J.-A., Warburton, J., Cuthill, M., & Bartlett, H. (2009). What makes a community age-friendly: A review of international literature. *Australasian Journal on Ageing, 28,* 116–121.

Lundberg, O., Åberg Yngwe, M., Kölegård Stjärne, M., Elstad, J. I., Ferrarini, T., Kangas, O., et al. (2008). The role of welfare state principles and generosity in social policy programmes for public health: An international comparative study. *The Lancet, 372,* 1633–1640.

Lutfey, K., & Freese, J. (2005). Toward some fundamentals of fundamental causality: Socioeconomic status and health in the routine clinic visit for diabetes. *American Journal of Sociology, 110,* 1326–1372.

Marmot, M. (2004). Status syndrome. *Significance, 1*(4), 150–154.

Marmot, M., & Wilkinson, R. G. (1999). *Social determinants of health*. Oxford: Oxford University Press.

McDonough, P., Sacker, A., & Wiggins, R. D. (2005). Time on my side? Life course trajectories of poverty and health. *Social Science & Medicine, 61,* 1795–1808.

McMunn, A., Nazroo, J., & Breeze, E. (2009). Inequalities in health at older ages: A longitudinal investigation of the onset of illness and survival effects in England. *Age and Ageing, 38,* 181–187.

McQueen, D., Wismar, M., Lin, V., Jones, C., & Davies, M. (2012). *Intersectoral governance for health in all policies. Structures, actions and experiences*. Retrieved from: <http://www.euro.who.int/en/publications/abstracts/intersectoral-governance-for-health-in-all-policies.-structures,-actions-and-experiences>. Accessed 23.01.14.

Menec, V. H., Means, R., Keating, N., Parkhurst, G., & Eales, J. (2011). Conceptualizing age-friendly communities. *Canadian Journal on Aging/La Revue Canadienne Du Vieillissement, 30,* 479–493.

Merton, R.K. (1968). *Social theory and social structure*. Retrieved from: <http://www.citeulike.org/group/14777/article/8898545>. Accessed 31.01.14.

Messinger-Rapport, B. (2009). Disparities in long-term healthcare. *Nursing Clinics of North America, 44,* 179–185.

Ministry of Social Affairs and Health (Finland). (2008). *National action plan to reduce health inequalities 2008–2011*. Retrieved from: <http://pre20090115.stm.fi/pr1227003636140/passthru.pdf>. Accessed 23.01.14.

Myles, J., & Quadagno, J. (2002). Political theories of the welfare state. *Social Service Review, 76,* 34–57.

O'Connor, J. S., Orloff, A. S., & Shaver, S. (1999). *States, markets, families: Gender, liberalism and social policy in Australia, Canada, Great Britain and the United States*. Cambridge: Cambridge University Press. Retrieved from: <http://journals.cambridge.org/production/action/cjoGetFulltext?fulltextid=6273056>. Accessed 31.01.14.

O'Rand, A. M. (1996). The cumulative stratification of the life course. In R. H. Binstock & L. K. George (Eds.), *Handbook of aging and the social sciences* (pp. 188–207). San Diego, CA: Academic Press.

O'Rand, A. M. (2006). Stratification and the life course: Life course capital, life course risks, and social inequality. In R. H. Binstock & L. K. George (Eds.), *Handbook of aging and the social sciences* (pp. 145–162). San Diego, CA: Academic Press.

O'Rand, A. M., & Henretta, J. C. (1999). *Age and inequality: Diverse pathways through later life*. Boulder, CO, US: Westview Press.

OECD. (2007a). *Life risks, life course and social policy*. Retrieved from: <http://www.oecd.org/els/soc/seminarontheliferiskslifecourseandsocialpolicy.htm>. Accessed 23.01.14.

OECD. (2007b). *Modernising social policy for the new life course*. OECD. Retrieved from: <http://www.oecd.org/social/soc/modernisingsocialpolicyforthenewlifecourse.htm>. Accessed 31.01.14.

OECD. (2012). *OECD pensions outlook 2012*. Retrieved from: <http://www.oecd.org/daf/fin/private-pensions/oecdpensionsoutlook2012.htm>. Accessed 30.01.14.

OECD. (2013). *Pensions at a glance 2013: Retirement-income systems in OECD and G20 countries*. Retrieved from: <http://www.oecd.org/pensions/pensionsataglance.htm>. Accessed 23.01.14.

Pavalko, E. K., & Caputo, J. (2013). Social inequality and health across the life course. *American Behavioral Scientist*.

Pearlin, L. I. (1989). The sociological study of stress. *Journal of Health and Social Behavior, 33*, 241–256.

Phelan, J. C., Link, B. G., Diez-Roux, A., Kawachi, I., & Levin, B. (2004). "Fundamental causes" of social inequalities in mortality: A test of the theory. *Journal of Health and Social Behavior, 45*, 265–285.

Plouffe, L. A., & Kalache, A. (2011). Making communities age friendly: State and municipal initiatives in Canada and other countries. *Gaceta Sanitaria, 25*(Suppl. 2), 131–137.

Prus, S. G. (2007). Age, SES, and health: A population level analysis of health inequalities over the lifecourse. *Sociology of Health & Illness, 29*, 275–296.

Quadagno, J., Kail, B. L., & Shekha, K. R. (2011). Welfare states: Protecting or risking old age. In R. A. Setterson & J. L. Angel (Eds.), *Handbook of Sociology of Aging* (pp. 321–332). New York, NY: Springer.

Quesnel-Vallée, A. (2004). Is it really worse to have public health insurance than to have no insurance at all? Health insurance and adult health in the United States. *Journal of Health and Social Behavior, 45*, 376–392.

Quesnel-Vallée, A., & Jenkins, T. M. (2009). Social policies and health inequalities. In W. C. Cockerham (Ed.), *The new Blackwell companion to medical sociology* (pp. 455–483). Oxford: Blackwell Publishing.

Quesnel-Vallée, A., Renahy, E., Jenkins, T., & Cerigo, H. (2012). Assessing barriers to health insurance and threats to equity in comparative perspective: The Health Insurance Access Database. *BMC Health Services Research, 12*, 107.

Rantz, M. J., Marek, K. D., & Zwygart-Stauffacher, M. (2000). The future of long-term care for the chronically ill. *Nursing Administration Quarterly, 25*, 51–58.

Robert, S., & House, J. S. (1996). SES differentials in health by age and alternative indicators of SES. *Journal of Aging and Health, 8*, 359–388.

Robert, S. A., Cherepanov, D., Palta, M., Dunham, N. C., Feeny, D., & Fryback, D. G. (2009). Socioeconomic status and age variations in health-related quality of life: Results from the national health measurement study. *The Journals of Gerontology. Series B: Psychological Sciences and Social Sciences, 64B*, 378–389.

Schlotter, M., Schwerdt, G., & Woessmann, L. (2010). *Econometric methods for causal evaluation of education policies and practices: A non-technical guide*. Discussion Paper No. 4725. p. 39.

Schoeni, R. F., Martin, L. G., Andreski, P. M., & Freedman, V. A. (2005). Persistent and growing socioeconomic disparities in disability among the elderly: 1982–2002. *American Journal of Public Health, 95*, 2065–2070.

Schöllgen, I., Huxhold, O., & Tesch-Römer, C. (2010). Socioeconomic status and health in the second half of life: Findings from the German Ageing Survey. *European Journal of Ageing, 7*, 17–28.

SCIP. (2014). *Social Citizenship Indicator Program (SCIP)*. DSpace at Stockholm University: Social Citizenship Indicator Program. Retrieved from: <https://dspace.it.su.se/dspace/handle/10102/7>. Accessed 31.01.14.

Scott, A. J. (2001). *Global city-regions: Trends, theory, policy*. Oxford: Oxford University Press.

Scruggs, L., Kuitto, K., & Detlef, J. (2013). *Comparative Welfare Entitlements Dataset 2 (CWED 2)*. CWED. Retrieved from: <http://cwed2.org/>. Accessed 31.01.14.

Service Public. (2014). Réforme des retraites : La loi est publiée au Journal officiel. *Service Public* (Pension reform: The law is published in the Official Journal. Public Service). Retrieved from: <http://www.service-public.fr/actualites/002825.html>. Accessed 30.01.14.

Sultz, H. A., & Young, K. M. (2011). Long-term care. In H. A. Sultz & K. M. Young (Eds.), *In health care, USA: Understanding its organization and delivery* (pp. 279–315). Sudbury: Jones & Bartlett Publishers.

Taylor, M. G. (2008). Timing, accumulation, and the black/white disability gap in later life: A test of weathering. *Research on Aging, 30*, 226–250.

Wagmiller, R. L., Lennon, M. C., Kuang, L., Alberti, P. M., & Aber, J. L. (2006). The dynamics of economic disadvantage and children's life chances. *American Sociological Review, 71*, 847–866.

Weymann, A. (2009). The life course, institutions, and life course policy. In W. R. Heinz, J. Huinink, & A. Weymann (Eds.), *The life course reader: Individuals and societies across time* (pp. 139–158). Frankfurt: Campus Reader.

Wiener, J. M., & Tilly, J. (2002). Population ageing in the United States of America: Implications for public programmes. *International Journal of Epidemiology, 31*, 776–781.

Willson, A. E. (2009). Fundamental causes of health disparities: A comparative analysis of Canada and the United States. *International Sociology, 24*, 93–113.

Willson, A. E., Shuey, K. M., & Elder, G. H., Jr. (2007). Cumulative advantage processes as mechanisms of inequality in life course health. *American Journal of Sociology, 112*, 1886–1924.

Wolf, D. A., Hunt, K., & Knickman, J. (2005). Perspectives on the recent decline in disability at older ages. *Milbank Quarterly, 83*, 365–395.

World Health Organization. (2000). *The world health report 2000: Health systems: improving performance*. Geneva: WHO.

World Health Organization. (2002). *Active ageing: A policy framework*. Geneva: WHO.

World Health Organization. (2007). *Global age-friendly cities: A guide*. Geneva: WHO.

World Health Organization. (2012). *Good health adds life to years: Global brief for World Health Day 2012*. Geneva: WHO.

World Health Organization. (2014a). *Mandate by member states on social determinants of health*. Geneva: WHO.

World Health Organization. (2014b). *WHO age-friendly environments programme*. Geneva: WHO.

Author Index

Note: Page numbers followed by "*f*" and "*t*" refer to figures and tables, respectively.

A

AARP, 296–298, 302–303, 305–306
Abbott, J., 278
Abrams, C., 192
Acevedo-Garcia, D., 132, 154
Achenbaum, W. A., 294
Acierno, R., 337–339, 343–347
Ackland, R., 194
Adair, L. S., 106
Adams, B. N., 212–213
Adams, C., 240
Adams, G. A., 275, 363
Adams, J., 279
Adams, N., 274, 281
Adams, R. G., 368
Adamson, D. M., 377
Adkins, D. E., 130–131, 231–232
Adler, G., 277
Adler, N. E., 192
Administration on Aging, 337–338, 361, 369
Afifi, T. D., 214
Agahi, N. A., 171–172
Aggarwal, R., 84
Aggleton, P., 318
Aging with Dignity, 383, 388–389
Agree, E., 89
Aguila, E., 154–155
Ai, A. L., 257, 266
Ailshire, J. A., 62–63, 326, 328
Ajrouch, K. J., 189, 191–193
Akamigbo, A. B. M., 131–132
Akiyama, H., 187–188, 196, 216
Alarcon, R., 146
Alba, R. D., 145, 148
Albert, I., 210
Alderman, M. H., 131
Aldous, J., 209
Aldwin, C. M., 227–228, 240
Allemand, M., 260

Allen, J. K., 65
Allen, K. A., 341
Allen, K. R., 218–219
Allen, R. E. S., 329
Allen, S. A., 304–305
Alley, D., 60, 65, 91–92
Alley, D. E., 65, 93, 128, 154–155
Allicock, M., 265
Allore, H. G., 93
Allport, G. W., 145
Almeida, D. M., 113–114, 167, 297, 306–307
Amato, P. R., 214, 219
American Association of Retired Persons, 360
American Bar Association, 382
American Medical Association (AMA), 376, 379–381
American Psychological Association, 367
American Red Cross, 235
Amstadter, A. B., 337–338, 343–344, 346–347
Andersen, L. G., 106
Anderson, G., 124–125
Anderson, K. L., 342
Anderson, R. N., 123
Anderson, R. T., 131
Anderson, S., 217–218
Andersson, G., 151
Andersson, L., 173
Andreski, P. M., 87–88
Andress, J., 229–232
Andrisani, P. J., 241
Aneshensel, C. S., 91–92, 316–320, 322–325, 330–332, 361–362
Anetzberger, G. J., 338, 350
Angel, J., 125
Angel, J. L., 91, 124–126, 131–132, 135–136, 146, 152, 154, 254

Angel, R. J., 124–125, 131–132, 135–136, 146, 152, 154–155, 254
Angrist, J. D., 236–241
Antecol, H., 133
Anthony, D. L., 384
Antonucci, T. C., 181–182, 185, 187–189, 191–192, 196, 216, 368–369
Ao, D., 192–193
Appel, H. B., 266
Aranda, M. P., 304–305
Arber, S., 126, 164, 173
Ardelt, M., 237
Argue, A., 253
Arguelles, W., 320
Arias, E., 123–124, 154, 273, 376–377
Aries, P., 378
Armed Forces Health Surveillance Center, 232–233
Arora, K., 175
Arpawong, T., 68
Arvidson, C., 87
Asch, A., 387–388
Aschaffenburg, A., 18
Aschbrenner, K. A., 304–305
Ash, A. S., 384
Ashida, S., 181–182, 186, 188
Asparouhov, T., 34
The Associated Press-NORC Center for Public Affairs and Research, 303
Atchley, R. C., 18, 190–191, 255
Atwood, P. L., 231–232
Auchincloss, A. H., 326
Auer, C., 380
Ault, A. B., 375–377
Austad, S. N., 108–109
Austen, I, 387–388
Austin, R., 337–338
Autor, D. H., 281
Avioli, P. S., 215–216
Avlund, K., 188

Axelrod, A., 236
Aykan, H., 81

B

Bachmeier, J. D., 148–149
Bader, M., 328
Bae, S., 87
Bailey, A. K., 241
Bailey, S. L., 231–232
Baker, A. A., 338
Baker, L. A., 298
Baker, M. W., 337–338
Baker, S. M., 366–367
Baldassar, L., 146
Balfour, J. L., 315–316, 326–327
Balistreri, K., 125
Baltes, M. M., 166
Banaszak-Holl, J., 92
Band-Winterstein, T., 339–340
Bann, D., 106
Barban, N., 24
Barber, T. M., 366–367
Barbera, J. A., 362
Barbos, M., 84
Barker, D. J. P., 104–106, 108
Barna, G., 261
Barnato, A. E., 384
Barnes, L. L., 191–192
Barnes, N. W., 132
Barnett, A. E., 211
Barnett, M. A., 218
Barone, E., 387–388
Barrett, C., 330
Barrett, D. H., 240
Barrow, K. M., 349
Barusch, A. S., 168, 175–176
Baruth, M., 263
Baskin, S. A., 384
Bass, S., 295
Bass, S. A., 295, 299
Bassuk, S. S., 305
Bastida, E., 154, 265–266
Batagelj, V., 183*f*
Batalova, J., 143–144, 146, 148–149, 151–152
Bates, J. E., 214
Bates, L., 154
Batson, C. D., 150–151
Bauer, D. J., 40–41, 46
Baum, A., 363
Bayer, A., 329
Beach, S. R., 344–347
Bean, F. D., 146
Beard, J. R., 329

Beck, T., 345
Becker, S., 167
Beckfield, J., 15
Beckie, T. M., 129–130
Bedard, K., 133, 233, 237
Bedford, V. H., 215–216
Beehr, T., 275
Beggs, J. J., 188
Begle, A. M., 337–338
Begum, N., 193
Beirne, N. F., 361
Belfour, J. L., 15
Bell, S., 117
Bellafaire, J., 239
Bellamy, L., 111
Belluck, P., 376, 382, 387
Below, R., 358, 359*t*
Belsky, D. W., 68
Beltran-Sanchez, H., 59
Belyea, M., 187
Belza, B., 329
Bengtson, V. L., 130–131, 207–208, 210, 221, 341, 362–363
Benjamins, M., 124
Benjamins, M. R., 254
Bennett, J., 189
Bennett, J. M., 91, 125
Bennett, P. R., 233–234, 236–238, 240–241
Ben-Shlomo, Y., 101–107, 113–114, 117
Benson, H., 261
Benson, S., 362
Benton, D., 344
Benzeval, M., 117
Berenson, G. S., 103
Beresford, L., 192–193
Berg, C. A., 221
Bergeman, C. S., 365
Berger, M. C., 239–240
Berger, P. L., 259, 263
Berkman, L. F., 132, 181–182, 185–187, 189, 254–255, 305
Berkman, N. D., 366–367
Berman, J., 347
Berry, H., 305
Bertrand, P., 68
Bertrand, R. M., 305
Besen, E., 307
Bhargava, S. K., 106
Bianchi, S. M., 164, 170–171, 209–210
Biddle, B. J., 304
Bielinski, J., 364
Bienias, J. L., 191–192
Bierman, A., 132, 319, 325

Biggs, S., 339–340
Billari, F. C., 24
Bird, C. E., 172–173
Birditt, K., 207, 210, 221
Birditt, K. S., 196, 210, 213, 216
Birmingham, W., 187
Birnie, K., 115–116
Bisconti, T. L., 365
Bittman, M., 173–176
Bjorck, J., 266
Black, J. S., 385–386
Blackwell, A. D., 114
Blaikie, P., 366–367
Blair, A., 315–316, 325–326, 331–332
Blair, S. N., 263
Blakely, T. A., 15–16
Blalock, C. L., 68
Blandon, A. Y., 189
Blasco, M. A., 111
Blau, D. M., 277, 281
Blazer, D. G., 256
Blieszner, R., 213, 216, 368
Bloch, D. A., 128
Bloemraad, I., 148
Blumenthal, J. A., 382
Board of Trustees of OASDI, 272–274, 278, 281, 286
Boardman, J. D., 68
Boehmer, T. K. C., 240
Boehner, J., 386–387
Boelk, A. Z., 380
Boerner, K., 302, 362–363, 385
Bohnert, A. S. B., 304
Bolan, G., 349–350
Boldy, D., 340
Boll, T., 209, 216
Bollen, K. A., 24, 26, 31, 43
Bolling, S. F., 257
Bolton, C. E., 115
Bolzendahl, C., 219
Bomba, P. A., 385–386
Bonanno, G. A., 362–365
Bonazzo, C., 154
Bond Huie, S. A., 323–324
Bongar, B., 366–367
Bonnie, R. J., 338
Bookwala, J., 220, 239–240
Booth, A., 219
Bootsma-van der Weil, A., 82
Borawski-Clark, E., 192–193
Borjas, G. J., 153
Borrell, L. N., 154–155, 321
Boscarino, J. A., 240
Bosworth, B. P., 272, 283

Bosworth, W., 131
Botticello, A. J., 91
Boughey, C., 18
Bound, J., 65, 129–130, 236–237
Bourdieu, P., 17–18, 164, 173
Bourque, L. B., 362
Bouthillier, D., 68
Bowen, G. L., 341–342
Bowen, K., 187
Bowen, M. E., 91–92
Bowie, S. L., 360
Bowie, Y. D., 360
Boyatzis, C. J., 259
Boyce, T., 107
Boyce, W., 129–130
Boyd, K., 377
Boyle, C., 240
Boyle, M. A., 281
Boylen, S., 340
Braccio, S. M., 154
Brach, J. S., 86–87
Bradley, E. H., 377–378
Brandl, B., 347, 349
Brannick, M. T., 302
Braun, E., 381
Bremner, J. D., 362–364
Brenner, J., 193–194
Breslau, N., 362–364
Brion, S. L., 253
Brissette, I., 189
Britton, A., 117
Brody, E. M., 81–82
Broese van Groenou, M. I., 192–193
Bronfenbrenner, U., 299–300, 318–319, 340–341
Bronner, S. E., 302
Brossoie, N., 346, 349
Brouard, N., 78–79
Brown, B., 329
Brown, G., 329
Brown, L. M., 359–361, 363, 365–370
Brown, M. T., 240–241
Brown, S. C., 305, 307, 325, 327
Brown, S. L., 219–220
Brown, T. H., 91–92, 94, 126, 130–131, 231–232
Brown, W. J., 173
Brownell, P., 347
Browning, C. R., 11
Browning, H. L., 233–234
Bruce, B., 124–125
Bruce, D., 259, 275
Brugha, T., 107
Bruner, E. J., 322

Brunkard, J., 366
Bryk, A. S., 29, 43
Bucciarelli, A., 364–365
Buchbinder, E., 348–349
Buckley, C. J., 154
Bulanda, J. R., 133–134, 219–220
Bull, A. R., 104
Bullard, J., 284
Bullock, K., 346–347
Burbidge, J., 217
Burdette, A. M., 91
Burgard, S. A., 233
Burgess, A. W., 343
Burke, G. L., 93–94
Burkhauser, R. V., 271–273, 281
Burkle, C. M., 381
Burland, D., 241
Burr, J. A., 87, 124, 131–132, 152–154, 298, 302–303, 306–307
Burt, R. S., 184–185
Burtless, G., 272–273, 277–278, 283
Burton, G. J., 105
Burton, L. C., 365–366
Butler, J. S., 236–237
Butler, R. N., 294–295, 297
Butrica, B., 297
Butrica, B. A., 272, 283, 302
Buxton, W., 343
Byard, D., 362
Byrd, R. C., 261
Byock, I. R., 380
Byrne, C. M., 362–363
Bytheway, B., 365–366

C

Cacioppo, J. T., 325
Cagney, K. A., 11
Cahill, K. E., 5, 10, 272, 274–277, 279, 281–282, 284, 303
Cai, L., 57
Calasanti, T. M., 168, 171, 173
Call, V. R. A., 239–240
Calvin, C. M., 107
Calvo, E., 305–306
Camacho, P. R., 231–232
Campbell, D., 235–236
Campbell, L. D., 216–217
Campbell, M. K., 265
Campbell, R. T., 130–131
Cannon, T., 366–367
Cantor, M. H., 368
Cantu, P. A., 154
Cao, G., 84, 86–87
Capps, R., 148–149

Card, J. J., 231–232, 239–240
Carlisle, M., 187
Carlson, E., 229–232
Carlson, M. C., 305, 307
Carlton-LaNey, I., 303, 306
Carney, G., 207–208
Caro, F., 295
Caro, F. G., 295, 298–299
Carpenter, B. D., 384
Carpiano, R. M., 320
Carr, D., 211, 216, 375–377, 380–385, 388–390
Carr, S., 152
Carrasquillo, A. I., 146
Carrasquillo, O., 146
Carreon, D., 191
Carson, V. B., 251–252
Carstensen, L. L., 18, 189–190, 196, 262–263, 301, 343
Casas, J. P., 111
Caspi, A., 68
Castle, N., 337–338
Castora-Binkley, M., 91–92
Cauley, J. A., 305
Cawthon, R. M., 68–69
CDC (Centers for Disease Control and Prevention), 7–8, 11–12
Center for American Progress, 370
Centers for Disease Control and Prevention, 297
Çevirme, A. S., 340
Çevirme, H., 340
Chakravarty, E., 124–125
Chamberlain, L., 382
Chan, A. C. M., 181–182
Chan, C., 382–383
Chan, P., 360–361
Chandlee, M. J., 359–360
Charles, S. T., 189–190, 196, 301, 343
Chateau, D., 317–318
Chatters, L. M., 265–266
Chavez, L. R., 147
Chawla, S., 187
Chen, E., 69, 107, 346–347
Chen, J., 275
Chen, P. H., 337–338
Chen, S. H., 239–240
Chen, X., 93, 207
Cheng, S.-T., 181–182, 188
Cheng, Y. P., 210
Cherlin, A. J., 214, 217, 219
Cherry, K. E., 11–12
Cheung, M., 345
Chevan, A., 87

Chi, I., 153
Chief Medical Officer, 117
Chinese Longitudinal Healthy Longevity Survey, 8
Chiswick, B. R., 148
Chiu, C.-J., 91–93, 125
Chiu, C.-T., 154
Cho, Y., 133
Chodorow, N. J., 18, 210–211, 213
Chodosh, J., 322–323, 330
Choi, E., 307–308
Choi, H., 304
Choi, K. H., 150–151
Choi, L. H., 298
Choi, N. G., 10, 298
Choi, S. H., 154
Chompunud, S., 340
Choula, R., 305
Christakis, N. A., 128, 136, 186–189, 376–378, 383
Christensen, J. J., 360–361
Christensen, K., 68
Cichy, K. E., 167
Cicirelli, V. G., 216
Cisler, J. M., p085, 337–338, 343, 346–347
Clare, L., 220
Clark, D. O., 130–131
Clark, J. L., 93
Clark, M. A., 131–132
Clark, N. M., 133
Clark, R., 281
Clarke, L., 217–218
Clarke, P., 315–317, 319, 321–328, 330–331, 360–361
Clarnette, R. M., 381
Cleveland, W., 93
Cline, K. M., 213
Clipp, E. C., 231–233
Clogg, C. C., 24
Clouston, S. A. P., 116
CMS (Center for Medicare and Medicaid Services), 7, 379–380
Cochran, S. D., 132
Coe, N. B., 175
Cohany, S. R., 239–240
Cohen, B., 61
Cohen, D., 366–367
Cohen, J., 239–240
Cohen, R. D., 15
Cohen, S., 107
Cohn, D., 219
Coile, C., 272, 281–284
Colantonio, A., 360–361

Cole, S. W., 69, 86–87
Coleman, J. S., 188, 301
Colley, J. R. T., 103
Collins, C. A., 131, 320, 324
Collins, J., 18
Collins, L. M., 31, 49–50
Collins, N., 256
Collins, R., 105
Combs, D. L., 365
Conger, R. D., 208, 218
Congressional Budget Office, 273–274
Conley, D., 276
Conlon, C., 207–208
Connidis, I. A., 216–217
Connors, A. F., Jr., 382
Conrad, F., 165–166, 172
Conrad, P., 378
Conroy, S. J., 207, 210
Cook, D. G., 106
Cook, J. M., 187–188
Cooper, C., 323, 346–347
Cooper, R., 102–103, 109–110, 114–116
Cooper, R. S., 84, 86–87
Copeland, C., 274, 278–279
Coppola, K. P., 381
Corder, E. H., 68
Corder, L., 84
Coreil, J., 132–133
Cornman, C. B., 359–360
Cornman, J., 89, 172, 174–176
Cornman, J. C., 87–88, 166
Cornwell, B., 181–182, 184–186, 188–193, 195
Corona, M., 337–338
Corsentino, E. A., 256
Cortina, K. S., 181–182
Coser, L., 213–214, 216–217
Coser, R., 213–214, 216–217
Costa, D., 273, 278
Costa, P. T., 365
Coughlin, J., 296–297, 302, 305–306
Coustasse, A., 87
Covinsky, K. E., 115
Cox, C., 384, 388–389
Cox, E., 148–149
Craig, L., 170
Crews, J. E., 305
Crimmins, E. M., 56–72, 57f, 63f, 66f, 67f, 78–79, 84, 91–92, 124–125, 128, 154–155
Cripps, H. A., 103
Crohan, S. E., 368–369
Crosnoe, R., 4, 146, 231–232
Crotty, K., 366–367

Crouchley, K., 340
Cruickshank, J. K., 132
Csikszentmihalyi, M., 165
Cubanski, J., 286
Cuellar, I., 154
Cummings, J. R., 322
Cunningham, A. S., 154
Curran, P. J., 24, 26, 31, 40–41, 43, 46
Cushman, M., 64–65
Cutler, D. M., 84, 88–89, 124–125
Cutler, S. J., 171–172, 301
Cutrona, C. E., 318, 325
Cvitkovich, U., 318

D

Dakin, E., 343–344
Dalgard, O. S., 331
Daly, J. M., 346
Daly, M. C., 281
Damico, A., 286
Danese, A., 5
Danieli, Y., 364
Daniels, A., 365–366
Daniels, N., 126
Danks, J. H., 380–381
Dannefer, D., 5, 129–131, 134, 147, 191
Danziger, S., 286–287
Darer, J., 124–125
Das, D. A., 128, 136
Das Gupta, P., 87
Daschle, T., 379
Davey, A., 172, 210–211, 216–218
Davey, M., 340
Davey Smith, G., 104, 109, 111–112
Davidian, M., 94
Davidson, K., 126, 164
Davies, L., 216–217
Davignon, J., 68
Davis, C. B., 359–360
Davis, I., 366–367
Davis, J. H., 365
Davis, R. B., 82, 382
Davison, E. H., 234
Daw, J., 187–188
Daxinger, L., 134
Daymont, T. N., 241
De Janvry, A., 147
de Jong, F., 105
de Jong Gierveld, J., 193
de Leeuw, P. W., 115
De Leon, C. F., 305
Dean, E. T., 235
Dean, L., 189
Deane, G., 209

Dear, K., 305
Deary, I. J., 111
Deaton, A., 253
Debats, D. L., 258
Dechter, A. R., 231–234
Deeg, D. J. H., 315–316, 323
Defense Manpower Data Center, 232–233
Dekel, S., 365
Delahanty, D. L., 363
DeLiema, M., 343–344
DeLongis, A., 363
Demissie, S., 68–69
DeNavas-Walt, C., 273
Dennis, J. A., 154–155
Dennison, E. M., 323
Denton, N. A., 317
Deschênes, O., 233, 237
Desmarais, S. L., 345
Detering, K. M., 382–383
DeViney, S., 301–302
DeVinney, L. C., 236
DeVries, M. W., 165
DeWolfe, D., 358
Dhaval, D., 307
Diala, C., 132
Diaz, E., 362–363
Diaz Venegas, C., 154
Dick, A., 365–366
Diehr, P., 93–94
Diewald, M., 147–149
Diez Roux, A. V., 15, 65, 131, 321
Dilley, P. E., 273, 278
Dillon, M., 253
Dilworth-Anderson, P., 136
Dimah, A., 344
Dimah, K. P., 344
DiMatteo, M. R., 188
DiNitto, D. M., 10
DiTommaso, A., 305
Ditto, P. H., 380–381
Division of the Civilian Volunteer Medical Reserve Corp, 367–368
Do, D. P., 126
Dobkin, C., 233, 240
Doblhammer, G., 91
Dobson, A., 193
Dodds, R., 115–116
Dodge, K. A., 214
Dohrenwend, B. P., 240
Dolbin-MacNab, M. L., 347
Domenici, P., 379
Domingue, B. W., 68
Donahue, K. E., 366–367

Donato, K. M., 146, 153
Dong, X., 337–338, 342–345
Donnelly, E. A., 190–191
Doocy, S., 365–366
Doreian, P., 189
Doros, G., 305
Dosa, D., 361
Dougall, A. L., 363
Douglas, J. W. B., 103
Douglas, K. S., 345
Doukas, D. J., 381–382
Dovers, S., 363–364
Dowd, J. J., 130–131
Dowdell, E. B., 343
Doyle, M., 339–340
Drabek, T. E., 362, 364
Dreby, J., 146–147, 150
Drew, L. M., 218
Drought, T. S., 381
Dubos, R., 108–109
Dudbridge, F., 68
Dugar, T. A., 342
Duggleby, S., 104
Dulin, P. L., 263
Dumin, M., 186
Dunkel-Schetter, C., 363
Dunn, T., 275
Durat, G., 340
Durkheim, E., 17, 207
Dykstra, P. A., 190, 212–213, 215–217

E

Easterlin, R. A., 13
Eastham, J., 175–176
Eaton, C. K., 344
Eberly, R., 363
Eberly, R. E., 364
Eberstein, I. W., 132
Eckholm, E., 376, 387–388
Eckstein, S. E., 149–150
The Economist, 12
Edwards, A., 382–383
Edwards, J. D., 91–92
Eggebeen, D. J., 210–211
Eisemann, M. R., 381
Eisenbach, Z., 321
Eisikovits, R. A., 218
Eisikovits, Z., 339–340, 348–349
Elder, G. H., Jr., 4, 6, 23, 130, 146, 149, 208, 217–218, 231–234, 237, 301–302, 341
Elias, N., 378
Eliason, S., 346
Elliott, M., 324–325

Elliott, S., 327
Ellis, R., 219
Ellison, C. G., 254, 258, 260–261, 264
Elmore, D. L., 295, 365–366, 368–369
Elo, I. T., 133, 154–155
ELSA, 8
Elsby, M. W., 277
Emanuel, E., 384
The Emerging Risk Factors Collaboration, 106
Emerman, J., 303
Employee Benefit Research Institute, 279
Employment Benefit Research Institute, 302
Enders, C. K., 47–48, 253
Engdahl, B. E., 363–364
Engelhardt, H., 209
English, N., 11
Enosh, G., 339–340
Enright, R. D., 260
Ensel, W. M., 185–186
Ensrud, K. E., 305
Epel, E., 57–58, 129–130
Epel, E. S., 68–69
Erens, B., 339–340
Erickson, L. B., 194
Erickson, M. A., 329
Eriksen, S., 216–217
Erikson, E., 259
Eriksson, J. G., 106
Erno, A., 361
Ernst, J. S., 349–350
Escarce, J., 154–155
Eschbach, K., 124–125, 131, 133, 143–144, 146, 154, 320, 323–324
Eschtruth, A., 276–277
Espino, D. V., 324
Espinosa de los Monteros, K., 320
Essex, M. J., 113–114
Estes, C. L., 295–296
the Euro-BLCS Study Group, 105
European Commission, 297, 308
Evans, D. A., 191–192
Evans, D. S., 275
Evans-Holland, M., 366–367
Everson, S. A., 260
Ewbank, D. C., 154–155
Ewen, H., 329
Eysenck, H. J., 364

F

Fadem, P., 296
Fagerlin, A., 381

Fairbank, J. A., 363
Falcón, L. M., 136
Fall, C. H. D., 104
Fang, J., 131
Farber, N. J., 378
Farkas, P. A., 194
Farmer, M. M., 126, 130–131
Fast, J. E., 174–175
Faust, K., 181–182, 194–196
Fazio, E. M., 6
Fear, J., 349–350
Federal Emergency Management Agency, 358
Federal Interagency Forum on Aging Related Statistics, 376–377
Federal Interagency Forum on Aging-Related Statistics, 359–360, 362
Federal Reserve Bank of St. Louis, 274
Fehr, R., 260
Feinberg, L., 78, 305
Feng, D., 207
Feng, Z., 131–132
Fennell, M. L., 93, 131–132
Ferguson, B. R., 154
Fernandez, L. S., 362
Ferraro, K. F., 6, 124–126, 129–131, 134, 305–307, 323
Ferring, D., 209
Ferrucci, L., 80, 110, 124–125
Fetzer Institute/National Institute on Aging Working Group, 252
Feussner, J. R., 187
Field, M. J., 80, 89
Filipp, S. H., 209
Finch, B. K., 126
Finch, C. E., 69
Fingerman, K. L., 195, 207, 209–210, 213, 216, 221
Finkelstein, M. A., 302
Finucane, T., 377–378
Fiori, K. L., 181–182, 185, 188–189
Fiscella, K., 15
Fischer, A., 318, 331
Fischer, C. S., 186, 192–193
Fisher, B. S., 346
Fisher, D., 106
Fisher, E. S., 384
Fisher, K., 174–175
Fitting, M., 175–176
Fitzpatrick, M. J., 348–349
Flanders, D., 240
Flap, H., 216–217
Fleur-Thomése, G. C. F., 315–316, 323
Flood, S., 301–302

Flores, G., 131–132
Fofana, R. C., 347
Folbre, N., 294–295
Foley, D. J., 93
Folkman, S., 363
Fomby, P., 214
Foner, A., 178
Ford, E. S., 187, 360
Forsen, T. J., 106
Fowler, J. H., 128, 136, 189
Frahm, K. A., 366–367
Frank, R., 125–126
Frankl, V. E., 259
Franklin, J. C., 276–277
Franks, M. M., 188
Franks, P., 15
Frayling, T. M., 108
Frazier, P., 363
Freathy, R. M., 108
Fredland, J. E., 236–237
Fredman, L., 305
Fredrickson, B. L., 69
Freedman, S., 260
Freedman, V. A., 77–78, 81–89, 93–94, 165–166, 172, 174–176, 361
Freeman, L. C., 184
Freese, J., 68
Fremont, A. M., 173
French, E., 281
Freysinger, V. J., 171–172
Fried, L. P., 124–125, 305, 307
Fried, T. R., 377–378
Friedberg, L., 279, 282
Friedman, E. M., 113–114
Friedman, M. J., 360
Friedrich, N., 112
Friedsam, H. J., 362
Fries, J. F., 124–126, 128
Frieze, I., 239–240
Frisbie, W. P., 133
Frist, W., 379
Frisvold, D., 10
Fronstin, P., 274, 281
Fruhauf, C. A., 218–219
Fry, P. S., 258
Fry, R., 16
Fryer-Edwards, K., 381
Fryers, T., 107
Fuhrel-Forbis, A., 307
Fuller-Rowell, T., 212–213
Fuller-Thomson, E., 210, 214, 218, 303
Fung, H. H., 190
Furstenberg, F. F., 217, 341

G

Gagné, P., 47–48
Gahbauer, E. A., 82–83, 93
Gale, C. R., 111, 323
Gale, W. G., 278–279
Galea, S., 107, 363–365
Galek, K., 264
Galie, M., 361
Gallagher, M., 128
Gallagher, P. M., 384
Gallo, L. C., 112, 320
Galloway, L. J., 346
Gallup, G. H., 261–262
Galobardes, B., 106–107
Galton, F., 251–252
Gamboa, C., 133
Ganea, R. L., 194
Gans, D., 207–208, 210–211, 341
Gardin, J. M., 321
Gardner, M. P., 113
Gardner, P. J., 195
Gariepy, G., 315–316
Garrett-Mayer, E., 381
Garrido, M. M., 384–385
Gary, L. E., 256
Gassoumis, Z. D., 343–344, 349
Gatz, M., 362–363
Gaugler, J. E., 92–93
Gauthier, A. H., 168–169, 171, 173
Gauthier, S., 68
Gauvin, L., 326
Gawande, A., 376, 381
Gee, G. C., 87, 129
Geiser, C., 29–30, 34, 47–48
Geist, C., 219
Gelatt, J., 146, 153
Gelfand, M. J., 260
George, L. K., 14–15, 23, 78, 89, 187, 208, 315–316, 327–328, 375–376
Geriatric Mental Health Care Foundation, 361
Gerin, W., 187
Geronimus, A. T., 65, 128–130
Gershuny, J., 164, 171
Gerst, K., 152–153
Gerstel, N., 212–214, 216–217
Gerstorf, D., 93, 190–191
Geruso, M., 126
Ghertner, R., 305
Ghilarducci, T., 302–303
Giandrea, M. D., 5, 10, 272, 274–276, 282, 303
Giarrusso, R., 207–208, 210, 220
Gibbs, M., 362

Gibson, B. E., 136
Gibson, M., 368–369
Gill, J. M., 65
Gill, T. M., 82–83, 89–90, 93
Gilleskie, D. B., 281
Gilligan, C., 213
Gilligan, M., 206, 208–209, 214, 216
Gilroy, C. L., 241
Gimbel, C., 231–232
Gingras, D. T., 259
Giorgiuli Saucedo, S., 151
Glaser, B. G., 377
Glass, A. P., 210
Glass, R., 15–16
Glass, T. A., 187, 189, 305, 315–316
Gleason, H., 294–295, 297
Gleeson, S., 153
Glei, D. A., 63–64
Glick, J. E., 150–151
Gluckman, P. D., 107–109
Godbey, G., 168
Goddard, M. E., 68
Godfrey, K. M., 107
Goedicke, A., 147–148
Goff, B. A., 295
Golberstein, E., 10
Gold, D. T., 187
Gold, R., 327
Goldbeck, J., 193–194
Goldman, N., 63–64
Goldscheider, F. K., 164
Gong, F., 10
Gonzales, E., 301
Gonzales, R. G., 146, 150, 153
González, H. M., 91–92, 154
Gonzalez, R., 167
González-González, C., 154
Goodkind, D., 86–87
Goodman, L. A., 24
Goodman, P., 361
Goodnow, J. J., 299–300
Goodwin, J. S., 131, 133, 320, 323–324
Goosby, B. J., 6
Gorbien, M. J., 346–347
Gordon, D., 306
Gorin, S. H., 124–125
Gorman, B. K., 128
Gornick, J. C., 164, 169, 171
Gorodnichenko, Y., 272, 284
Gottlieb, B. H., 187–188
Gottschalk, P., 286–287
Gough, C. P., 343
Gough, M., 148
Government Accountability Office, 339

Government Accounting Office, 348
Grace-Martin, K., 275
Granger, D. A., 361
Granovetter, M. S., 184, 186
Graubard, B. I., 92
Gray, G. C., 243–244
Green, B. L., 362–364
Greenberg, J. S., 210, 217, 304–305
Greene, R. R., 299–300
Greenfield, E. A., 213, 296, 304, 306–307, 341
Gresenz, C. R., 15
Grote, N., 239–240
Gruber, J., 272, 277–278, 284–285
Gruen, R., 363
Gruenewald, T., 57–58, 129–130
Gruenewald, T. L., 60, 64–65, 129–130, 187
Gruneir, A., 376
Grusec, J. E., 215
Gu, X., 84
Guberman, N., 329
Gubernskaya, Z., 146, 153, 155
Guha-Sapir, D., 358, 359t
Gundrum, J. D., 386
Gupta, G. R., 318, 331
Gupta, S., 317–318
Guralnik, J. M., 80, 93
Gurven, M., 114
Gustman, A. L., 272, 282–283
Gutman, G., 302

H
Haan, M. N., 115, 154, 321
Haas, S. A., 91–92, 130–131, 134
Haddad, A. A., 207
Hadley, C., 131
Hafemeister, T. L., 346
Hagedorn, A., 78–79
Hager, N., 386
Hagestad, G. O., 164
Haggerty, T., 231–232
Hahn, E. A., 167
Haines, V. A., 188
Hale, P. J., 326
Hales, C. N., 104
Haley, W. E., 167, 384
Hallqvist, J., 102
Halpern, D. J., 366–367
Halpin, H. A., 366–367
Halvorsen, C., 303
Hamill, S. B., 218–219, 348–349
Hamil-Luker, J., 126
Hamilton, P. J. S., 103

Hammes, B. J., 381, 383, 386
Han, L., 91, 93
Hanacek, B., 364–365
Hanchate, A., 384
Hancock, A. D., 382–383
Hancock, G. R., 49
Hand, C., 327
Handmer, J., 363–364
Hank, K., 192–193
Hanna, S., 327
Hansberry, M. R., 346–347
Hanson, M. A., 107
Hansson, R. O., 304–305
Hao, L., 148
Hao, Y., 306
Hardwig, J., 381–382
Hardy, M., 154
Hardy, R., 102–103, 106, 111
Harig, F., 322
Harkness, A. R., 364
Harper, L., 329
Harper, L. V., 134
Harris, J., 60
Hartz, A. J., 346
Hassing, L. B., 93
Hattersley, A. T., 108
Haun, J. N., 363
Haverstick, K., 272, 305–306
Hawkins, R. L., 192
Hawkley, L. C., 325
Hay, E. L., 213, 216
Hayes-Bautista, D. E., 133
Hayslip, B., Jr., 218
Hayward, M. D., 78–79, 124–126, 128, 134, 154–155
Hayward, R. D., 253–254, 256, 258–265
Hazuda, H. P., 324
He, W., 86–87
Heaney, C. A., 181–182, 186, 188
Heaven, B., 304
Heberman, H. B., 363
Heckman, B. D., 91
Heejung, S., 267
Heflin, C. M., 241–242
Heggeness, M. L., 16
Heider, F., 207, 209
Heikkinen, R. L., 189–190
Heisler, C., 350
Heller, K., 207, 362–363
Helman, R., 279, 283, 286
HelpAge India, 339–340
Henderson, A. C., 213
Henderson, C., 209
Henderson, T. L., 11–12

Hendricks, J., 171–172, 301
Henkin, N., 306
Henretta, J. C., 86–87, 130–131
Henry, D., 187
Henry, N. W., 24
Herd, P., 126
Hernandez-Tejada, M., 339, 343–344
Heron, M., 134
Hertzman, C., 103, 107
Herzog, A. R., 83
Hewitt, L. N., 307
Heyman, K. M., 132
Heymann, J., 318, 331
HHS (Department of Health and Human Services), 379, 383
HHS/ASPE, 377
Hiatt, R. A., 133
Hicken, M., 65
Hickling, E., 366–367
Hickman, S. E., 386
Higgins, K. M., 17
Hilber, D., 277
Hill, P. C., 258
Hill, T., 125
Hill, T. D., 91, 254, 315–316
Hillier, T., 327
Himawan, L., 91
Hingorani, A. D., 111
Hinterlong, J. E., 190–191, 294, 298–299
Hipple, S., 275
Hirsch, B. T., 239–241
Hirshorn, B. A., 301–302
Hirst, S., 104
Hobijn, B., 277
Hochberg, M., 305
Hodgson, N., 361
Hofer, S. M., 93, 116
Hoff, A., 173
Hoffman, J., 233, 238
Hoffman, J. C., 382
Hoffmann, R., 91, 130
Hofmeister, H., 220
Hogan, D. P., 131–132, 164, 227–228, 230–232
Hogan, J. W., 94
Hokayem, C., 16
Holdaway, J., 146
Hollis, M., 384, 389–390
Holstein, M. B., 294–296, 302
Holt-Lunstad, J., 186–187
Holtz-Eakin, D., 275–276
Holtzman, R. E., 181–182, 190
Homan, K. J., 259
Homans, G. C., 207, 215

Homeier, D. C., 343–344
Homeland Security, 359t
Hong, S. I., 305, 308
Hood, R. W., 258
Hook, C. C., 381
Hooker, S., 148–149
Hooker, S. P., 263
Hooten, E. G., 265
Hooyman, N. R., 176
Hopley, E. K., 381
Horner, B., 340
Horsford, S. R., 343–344
Horton, R., 362–363
Horwitz, R. I., 64
House, J. S., 62–63, 126, 186–187, 328, 377
Houser, A., 78, 305
Howard, A., 210
Howard, D. L., 131–132
Huang, B., 266
Huang, D. L., 329
Huang, M., 193–194
Huber, M., 108–109
Hubert, H. B., 128
Huerta, F., 362–363
Huges, M. E., 296
Hughes, M. E., 218, 341
Hummer, R. A., 133, 154, 323–324
Hummer, R. M., 124, 132
Hunt, K., 82
Hunt, L. P., 115
Hunter, A. L., 113–114
Hunter, B. D., 255
Hurlbert, J. S., 188
Hurst, L., 116
Hurt, A., 240–242
Hussey, J. R., 263
Hutchens, R. M., 275
Hutchison, B., 349–350
Hutton, D., 357–358, 369–370
Huxley, R., 105, 114
Huynh, T. N., 378–379
Hybels, C. F., 321–322
Hyer, K., 359–361
Hypponen, E., 111

I
ICPSR (Inter-university Consortium for Political and Social Research), 7
Idler, E. L., 254, 384–385
ieker, P. P., 172
Iezzoni, L. I., 77–78, 82
Iida, M., 188
Ikkink, K. K., 189–190

Independent Sector, 297
Ingersoll-Dayton, B., 264
Inoue, M., 381
Inskip, H. M., 107
Inslicht, S. S., 363
Institute for Social Research, 7
Institute of Medicine (IOM), 240, 366–367, 376–381, 383, 386–388
Institute of Medicine & National Research Council, 338
Isaacowitz, D., 301
Israel, S., 68
Iveniuk, J., 11, 220
Ivie, R. L., 231–232
Iwarsson, S., 317
Iwashyna, T. J., 382–383

J
Jackey, L. M. H., 196
Jackson, J. S., 132, 265
Jackson, S. A., 131, 321
Jackson, S. L., 346
Jaffe, D. H., 321
Jager, J., 188
Jain, A. R., 253
Jake, M., 172
James, J. B., 307
James, W., 261
Janevic, M. R., 191–192
Janicki-Deverts, D., 107
Janke, M., 217–218
Janoff-Bulman, R., 363
Jarrott, S. E., 218–219
Jasso, G., 143–144, 154–155
Javaid, M. K., 106
Jencks, C., 318–319
Jette, A. M., 56–57, 80–81, 89
Jirik, S., 350
Jivan, N. A., 279, 282
Jogerst, G. J., 346
Johannesen, M., 342–343
Johansson, B., 93
Johnson, A., 359
Johnson, D. R., 219, 253
Johnson, I., 349
Johnson, K., 209, 214, 216
Johnson, K. J., 294
Johnson, M. K., 4, 146, 178, 231–232
Johnson, M. S., 337–338
Johnson, N. J., 131
Johnson, R. W., 272, 275–277, 294–295, 302
Johnson, T. E., 68
Joinson, A. N., 194

Jones, B. L., 49–50, 94
Jones, J. B., 281
Joshi, M., 337–338
Joulfaian, D., 275
Jovanovic, B., 275
Jung, H. J., 152–153
Jung, T., 39–40
Juster, F. T., 165–167

K

Kabumoto, G., 378–379
Kahana, B., 191
Kahana, E., 191
Kahn, R. L., 185, 368
Kahneman, D., 165–166, 176
Kahramanian, M. I., 133
Kail, B. L., 6
Kajantie, E., 106
Kalish, Y., 194–195
Kalmijn, M., 189–190, 193
Kaminski, P. L., 218
Kamo, Y., 11–12
Kan, M. Y., 166–167
Kandel, W., 149–150
Kandula, N. R., 131
Kang, H. G., 194
Kaniasty, K., 362–363
Kaniasty, K. Z., 368–369
Kano, M., 362
Kantarci, T., 275
Kaplan, G. A., 15, 326–327
Kaplan, H., 114
Kaplan, L., 187
Karlamangla, A. S., 57–58, 63–65, 67, 113–114, 128–130, 154–155, 322
Karoly, L., 281
Karoly, L. A., 275–276, 281
Karp, F., 282
Karter, M. J., 358
Karumanchi, S. A., 111
Kasinitz, P., 146–147, 151–152
Kasl, S. V., 254
Kasper, J. D., 78, 82, 85–86
Katz, S., 81–82
Kawachi, I., 15–16, 64–65, 126, 254–255, 320
Kawachi, J., 276
Kawano, Y., 148
Kaye, H. S., 85–86
Keane, T. M., 363
Keary, S., 387
Keating, N., 186, 216
Keegan, M. T., 381
Keeler, D. R., 17

Keene, D., 65
Keene, J. D., 235–236
Kellehear, A., 384–385
Kelley, A. S., 383
Kelley, S., 303
Kelley-Moore, J. A., 83, 124–125, 129–131, 134, 191
Keltner, M., 379
Kelty, R., 235, 239–241
Kemp, C. L., 216
Kemp, M., 385–386
Kemp, S., 317
Kendall, M., 377
Kendra, J., 361
Kennedy, B. P., 15–16, 320
Kerber, R. A., 68–69
Kessler, R. C., 128, 377
Kestenbaum, B., 154–155
Keyes, C. L. M., 132
Keysor, J. J., 328
Khan, C., 188
Khodyakov, D., 216, 380–381, 383, 389
Kiecolt, J. K., 213–214
Kilijanek, T. S., 362, 364
Kilkey, M., 146
Kim, G., 124
Kim, H., 340
Kim, J., 10, 46, 130–131, 151
Kim, J. E., 220
Kim, J. K., 56–57, 60, 65, 67, 128, 154–155
Kim, N. Y., 148
Kim, S., 209, 306–307
Kim, S. Y. H., 382–383
Kim, W., 152–153
Kim-Cohen, J., 107
King, A. C., 326
King, D. E., 251–252
King, D. W., 233, 363
King, L. A., 233, 363
Kinne, S., 83
Kinsella, K., 69
Kirkwood, T. B., 108–109
Kirsch, T. D., 365
Kiyak, H. A., 176
Klaus, E., 209
Kleban, M. H., 81
Kleiber, D., 172
Klein, D. M., 209
Kleinberg, B., 342
Kleykamp, M., 233, 241
Klumb, P. L., 166
Knickman, J., 82
Knight, B. G., 304–305, 362–364

Ko, M. J., 322–323, 330
Kobor, M. S., 69
Kochanek, K. D., 123
Kochar, R., 171
Koenen, K., 107
Koenig, B., 381
Koenig, H. G., 251–252, 256–257
Koenig, T. S., 342
Koester, S., 233
Kogan, S., 68
Kohli, M., 192–193
Kohlmaier, J. R., 366–367
Kohn, N. A., 382
Kolomer, S., 210
Koltai, J., 196, 345
Kondo, N., 15–16
Konrath, S., 307
Koretz, B., 65
Korkontzelou, C., 94
Korn, E. L., 92
Korporaal, M., 220
Kosberg, J. I., 343, 347
Kramer, B. J., 376, 380–381
Krantz-Kent, R., 168–171, 173
Krause, N., 5, 187–189, 192–193, 196, 252–266, 303, 305, 321
Kreider, R. M., 219
Krekula, C., 126
Kreuter, F., 38
Kroemer, G., 111
Kronenfeld, J. J., 91–92
Kronman, A. C., 384
Krout, J., 329
Krueger, A. B., 165–166, 236–237, 240–241
Krueger, P. M., 134
Kubzansky, L. D., 64–65, 111, 254–255, 321–322
Kudler, T., 264
Kuh, D., 101–107, 109–111, 114–117
Kumari, M., 113
Künemund, H., 192–193
Kuo, W., 185–186
Kuo, Y.-F., 133
Kusenbach, M., 360
Kwag, K., 91
Kwag, K. H., 307
Kwak, J., 384
Kwan, N., 152
Kwon, J., 91

L

Lacey, T. A., 297
Lachman, M. E., 190–191

Ladin, K., 154
Laditka, J. N., 359–360
Laditka, S. B., 265, 359–360
Ladson, H., 154
Lafferty, A., 348
Lahey, J. N., 281
Lambert, J. D., 304
Lamme, S., 190
Lamont, E. B., 377–378
Lamont, M., 301
Lancet, E. A., 154–155
Land, K. C., 33, 44, 134
Landale, N. S., 150–151
Landefeld, C. S., 383
Landes, S. D., 237
Landis, K. R., 186
Landrum, M. B., 124–125
Lang, F. R., 190
Lang, I. A., 322
Langa, K. M., 382–383
Langenberg, C., 106
Lansford, J. E., 196, 214
Lantz, P. M., 126, 326
Lanza, S. T., 31, 49–50
LaPierre, T. A., 186, 218, 296
Lara, M., 133
Lareau, A., 301
Larsen, L. J., 86–87
Larson, R., 188
Larson, R. W., 165, 176
Lash, T. L., 7–8
Lashewicz, B., 216
Laub, J. H., 233–234
Lauderdale, D. S., 151, 155
Laumann, E. O., 181–182, 188, 190–191, 220, 343–344, 346
LaVeist, T. A., 126, 132
Lavery, J., 273–274
Law, C. M., 105
Law, M., 327
Law, R. W., 255–256
Lawler-Row, K. A., 260
Lawlor, D. A., 102–103, 111
Lawton, M. P., 81–82, 318
Layton, J. B., 186
Lazarsfeld, P. F., 24, 215
Lazarus, R. S., 363
Leach, M., 152
Leavell, J., 132
LeBlanc, A. J., 319–320
Lee, C. K. L., 181–182
Lee, G. R., 219–220
Lee, G.-Y., 146, 152
Lee, H. Y., 344

Lee, J., 68, 264
Lee, J.-J., 181–182
Lee, L. C., 91–92
Lee, M. A., 87, 323
Lee, S. E., 344
Lee, T. K., 307
Leibing, A., 329
Leigh, J. P., 126
Lemmon, M., 241–242
Leng, M., 154–155
Leopold, T., 209–211
Lepore, S. J., 363–364
Leung, E. M. F., 181–182
Leung, P., 345
Levenson, M. R., 240
Leventhal, H., 384–385
Levin, J., 266
Levine, M., 68
Levine, M. E., 64–67, 66f
Levine, P. B., 272, 282–284
Levi-Strauss, C., 17
Levkoff, S. E., 127–128
Lev-Wiesel, R., 342
Levy-Storms, L., 361–362
Lewis, B., 124–125
Lewis, E. K., 276
Lewis, K. N., 91
Lewis, M. R., 64–65
Li, L. W., 321
Li, Q., 308
Li, T., 307
Li, X., 281
Li, Y., 305, 307
Liang, J., 91–92, 125, 189
Liao, Y., 84, 86–87
Lichtenstein, M. J., 324
Lieberman, M. A., 319
Lifespan of Greater Rochester, 339
Lightfoot, E., 304
Lightman, S., 113
Lilly, M. B., 175
Limas, M. J., 304–305
Lin, C. C., 362
Lin, I.-F., 210, 219–220
Lin, N., 185–186, 192–193
Lin, Y. H., 63–64
Lindenberger, U., 93
Lindstrom, D., 151
Link, B. G., 15, 134, 331
Lipson, S., 93
Little, R. D., 236–237
Little, R. J. A., 43
Litwin, H., 185–189
Liu, K., 84, 87–88

Liu, M., 49
Liu, P., 343
Liu, X., 172
Livermore, G., 84
Livingston, G., 11, 219, 296–297
Lleras-Muney, A., 88
Lochner, K., 15, 320
Lockenhoff, C. E., 18
Logan, J. R., 148
Loggers, E. T., 384
LoGiudice, D., 342–343
Lohfield, L., 342
Lombard, J. L., 327
London, A. S., 227–228, 231–234, 237, 240–242
Long, J. S., 10
Looney, A., 277
Lopes, A. A. S., 132
Lopez-Otin, C., 111
Lopoo, L., 232–233
Lopreato, S. C., 233–234
Lou, A., 307
Loucks, E. B., 187
Lounds, J., 210
Love, G. D., 69–72
Lowenstein, A., 207–208, 339–340
Lubben, J. E., 185, 306
Lubitz, J., 57
Lucas, M. J., 175–176
Lum, T. Y., 154, 304
Lumme-Sandt, K., 189–190
Lundberg, C. D., 260, 264–265
Lundquist, J. H., 241
Lunney, J. R., 93
Luo, Y., 218, 296, 298, 302
Lüscher, K., 212
Lusher, D., 194–195
Luszcz, M. A., 8
Lutz, A. C., 234–241
Lutz, W. A., 342
Lynch, J. W., 15, 102, 106–107
Lynch, S. M., 10, 33, 43–44, 49–50, 91–92, 94
Lynn, J., 93, 377
Lyyra, A. L., 189–190
Lyyra, T. M., 189–190

M
Ma, R., 69
Maas, I., 18
Mabry, J. B., 210
MacDonald, C., 376
Mack, J. W., 382–384
MacLean, A., 231–239

AUTHOR INDEX

MacMillan, H. L., 342, 349–350
MacNeil, G., 347
Madans, J. H., 78–79
Madden, M., 193–194
Maddox, G. L., 130–131
Madhavan, S., 131
Madrian, B., 281
Maduc, D., 15
Maestas, N., 272, 275, 280–281
Magana, A., 133
Magana, S., 305–306
Maggio, M., 112
Magidson, J., 34
Magoon, J., 317–318
Mahakian, J. L., 295–296
Mahal, A., 318
Maimon, D., 315–316
Mair, C. A., 190
Mancini, J. A., 91, 341–342, 348–349
Mannell, R., 188
Mannheim, K., 147
Manor, O., 321
Manthorpe, J., 339–340
Manton, K. G., 84, 124–125, 132
Marceau, L. D., 331
Mare, R. D., 150–151
Marenco, A., 218
Margetts, B., 104
Margolis, R., 187–188
Marin, T. J., 69
Markides, K. S., 124–125, 131–133, 136, 143–144, 146, 152, 154, 320, 323–324
Marks, N. F., 213, 304, 341
Marmot, M. G., 103
Martin, D. K., 379
Martin, J. A., 341–342
Martin, L. G., 49–50, 78, 84, 87–89, 94
Martin, P. P., 278
Martindale, M., 236–240
Martins, D., 340
Martinson, M., 296, 302
Maselko, J., 254–255
Mason, C. A., 325, 327
Mason, J., 217–218
Massey, D. S., 144, 146–150, 153, 317
Matchar, D. B., 187
Matthews, K. A., 107, 112
Matz-Costa, C., 307
Maughan, B., 107
Maxwell, C. D., 346
May, V., 217–218
Mayer, A. S., 232–233
Mayer, B., 210
Mayer, K. U., 144, 147–148

Mayer, S., 318–319
Mays, V. M., 132
Mazumdar, S., 146, 149–150, 152
Mba, C. J., 340
McCann, B. R., 346
McCarron, P., 115
McCarthy, E. P., 82
McClearn, G., 132
McClellan, J. L., 136
McClintock, M. K., 220
McCloskey, L. A., 337–338
McColl, M. A., 327
McCotter, T., 386–387
McCrae, R. R., 365
McCue, J. D., 378
McCullough, M. E., 253–254, 260
McCutcheon, M., 308
McDade, T. W., 59
McDonald, D. A., 297
McDonald, K. B., 233–234, 236–238, 240–241
McDonald, L., 341
McDonald, M. J., 259
McDonald, S., 190, 192–193
McDonnell, K., 280
McElrath, T. F., 111
McEniery, C. M., 114–115
McEntire, D. A., 361
McEwen, B. S., 5, 57–58, 60, 63–64, 112, 129–130
McEwen, J., 115
McFall, B. H., 11, 272, 283
McFarlane, A. C., 362–363
McGee, D. L., 84, 86–87
McGeehin, M. A., 240
McGrew, D. M., 382
McGuire, L. C., 360
McKinlay, J. B., 331
McLaughlin, D., 193
McNamara, T. K., 169–170, 172, 301
McPherson, M., 187–188, 192–193, 348–349
McVicar, D., 281
Mechanic, M. B., 364
Meersman, S. C., 6, 325
Mehay, S. L., 241
Mehta, C., 153
Mehta, P. D., 44
Meier, D. E., 384, 387
Melchert, F., 90
Meme, K. B., 282
Menaghan, E. G., 319
Mendes de Leon, C. F., 154, 187, 191–192

Meredith, W., 24
Merkin, S. S., 62, 62t, 64–65
Merla, L., 146
Merrill, R. M., 255
Merton, R. K., 5, 215
Merz, E.-M., 213, 215
Metlife Mature Market Institute, 337–338, 343, 347–348
Mettler, S., 231–232, 237
Metzner, H. L., 186
Meyer, M. H., 168
Mezey, N. J., 346
Mezuk, B., 304
Michael, Y. L., 315–316, 326–327
Michelson, W., 168, 175
Midthune, D., 92
Midwinter, R. A., 103
Miech, R. A., 130–131, 233
Miles, T. P., 124–125
Milewski, N., 151
Military Family Resource Center, 241
Milkie, M. A., 164, 258
Miller, G. E., 69
Miller, L., 221
Miller, P. W., 148
Miller, Y. D., 173
Milligan, K. S., 281, 285
Milner, D., 187
Min, J. W., 306
Minkler, M., 218, 294–296, 302–303
Minnes, P., 217
Minnis, H., 113–114
Mirowsky, J., 46, 258, 316–317, 320
Mishra, G., 102–103
Mitchell, B., 302
Mitchell, C. K., 382
Mitchell, R., 330
Mitchell, S. L., 378–379
Miyao, K., 64–65
Modell, J., 231–232
Moen, P., 220, 301–302
Moffett, A., 105
Moffitt, T. E., 68
Mohammed, S. A., 132
Molla, M. T., 78–79
Mollenkopf, J. H., 146–147
Molloy, D. W., 381
Monden, C., 187–188
Monk, C., 272
Monserud, M. A., 218
Montagnino, K., 362
Montez, J. K., 128, 136
Montminy, L., 342, 345–346
Montz, B. E., 361–362

Moon, A., 306, 344
Moorman, S. M., 211, 381–385, 387–389
Mor, V., 93, 131–132, 378–379
Mora, I., 153
Morales, L., 154–155
Morales, L. S., 133
Morenoff, J. D., 328
Morgan, E., 349
Morris, D. C., 256
Morris, M., 189
Morris, P. A., 300
Morrison, R. S., 384
Morrow, B. H., 358
Morrow-Howell, N., 294–295, 299, 305, 307–308
Morton, P. M., 6
Morton, S., 101–102
Mosimann, U. P., 194
Moskos, C. C., Jr., 236–237
Mosqueda, L., 337–338
Mouton, C. P., 337–338
Mrvar, A., 183f
Mueller, M. M., 217–218
Mueller, P. S., 381
Mui, A. C., 302
Mulder, C. H., 210
Mullan, J. T., 319
Mullan, K., 170
Muller, M., 115
Mulvihill, M., 384
Muniz-Terrera, G., 115–116
Munnell, A. H., 272–274, 276–280, 282, 284, 286
Munsch, C. L., 212–213
Müri, R. M., 194
Murray, S. A., 377
Murrell, S. A., 364
Musick, M. A., 260, 298
Musil, C., 304–305
Mussino, E., 151
Mustillo, S. A., 6
Mutchler, J. E., 87, 124, 131–132, 152–154, 294, 298, 306–307
Muthén, B. O., 24, 29–30, 34, 38, 40–41, 47–48
Muthén, L. K., 24, 29–30
Muzzy, W., 339, 343–344
Myers, D., 152
Myers, D. A., 271–272

N

Nag, M., 260
Nagi, S. Z., 81
Nagin, D. S., 24, 47, 49–50, 94
Nahemow, L., 318
Najam, A., 149–150
Nakamura, E., 64–65
Nam, C. B., 132
Nam, Y., 152–154
Namulanda, G., 366
Nanda, A., 376
Naorungroj, S., 93
The National Academics Committee on National Statistics, 349
National Alliance for Caregiving (NAC), 296–298, 302–303, 305–306
National Association of Area Agencies on Aging, 296–297
National Center for Health Statistics, 123, 276
National Center for Veterans Analysis and Statistics, Office of the Actuary, 243f
National Institutes of Health, 24
National POLST, 376, 385–386
National Research Council, 103, 165, 380
National Research Council Committee on Population, 69
National Resource Center on LGBT Aging, 342
Navarro, A. E., 349
Nealey-Moore, J. B., 187
Nee, V., 145
Needham, B. L., 69
Nef, T., 194
Neighbors, H. W., 132
Neil, A., 105
Neimann, S., 188
Nelson, A., 347
Nelson, A. R., 126
Nelson, J. P., 341–342
Neppl, T. K., 218
Nerino, A., 305
Nesse, R., 377
Nesse, R. M., 362–363
Neuman, T., 286
Neumark, Y. D., 321
New York Times, 11, 235
Newman, A. B., 86–87
Newsome, C. A., 105
Ng, J. W., 108
NHPCO, 379–380
Nicholas, L. H., 382–383
Nicholls, T. L., 345
Nicotera, N., 317
Niemi, I., 170
Nies, M. A., 265
Nieuwenhuijsenb, E. R., 315–316, 321–322, 326, 328, 330
Nishith, P., 364
Nomaguchi, K. M., 171
NORC (National Opinion Research Center), 8
Nordstrom, C. K., 321
Norris, F. H., 360–365, 368–369
Norris, J. E., 210, 216–217
North, J., 260–261
Northridge, M. E., 15
Norton, M. C., 255
Norvall, C., 193–194
Noyes, K. M., 383
Noymer, A., 191
Nylund, K. L., 34

O

O'Brien, E., 68–69
Odden, M. C., 115
Odgers, C. L., 24
Oehlert, J. W., 128
Offer, S., 165
Ogden, J. A., 318
Oh, J., 340
Okasha, M., 115
Okoro, C. A., 360
Okun, M. A., 301
Olander, E. B., 207
O'Leary, J., 377–378
Olshansky, S. J., 375–377
Olson, J. L., 342
Olson-Cerny, C., 187
Oman, D., 254
Omran, A., 375–377
Ondrich, J. I., 93–94
O'Neal, C. W., 6, 307
O'Neil, K., 153
O'Neill, G. S., 192
Ong, A. D., 365
Ono, H., 166–167
Ontai, L., 218
O'Rand, A. M., 5, 126–127, 130–131, 164
Orav, E. J., 216
Oregon Health Authority, 387–388
Orel, N. A., 195
Oriol, W., 360
Orrenius, P. M., 153
Orsmond, F. I., 210
Orszag, P., 284
Ory, M. G., 127–128
Osmond, C., 104, 106
Ostir, G. V., 131, 320, 323–324

Oswald, F., 317
Osypuk, T. L., 131
Owen, C. G., 106

P

Pachana, N. A., 193
Packer, C., 365–366
Page, W. F., 364
Palloni, A., 124, 154–155
Palmer, R. F., 324
Palmore, E., 93
Pampel, F. C., 68
Pamuk, E. R., 15
Pang, G., 281
Pang, N., 382–383
Pantilat, S. Z., 383
Papke, L. E., 278–279
Pardo, S., 209
Pargament, K. I., 257
Park, E., 308
Parker, K., 11, 296–297
Parker, M. G., 171–172
Parker, P., 237
Parker, W. M., 168
Parkhurst, J. O., 318
Parks, S. M., 380–382
Parrado, E. A., 151
Parreñas, R. S., 150
Parrish, R. G., 365
Parsons, T., 215–216
Partridge, L., 111
Passel, J. S., 219
Pastor, D. A., 47–48
Patel, K. V., 131, 320, 323–324
Patient Self-Determination Act of 1990, 383
Patterson, M. N., 361
Patterson, R. S., 227–228
Pattison, E. M., 377
Pattison, P., 194–195
Paul, E. L., 216
Paulk, E., 382–383
Pavalko, E. K., 10, 211, 231–232, 237
Payne, B. K., 345
Pear, R., 376, 386–387
Pearl, M., 321
Pearlin, L. I., 6, 258, 319–320, 325, 361–362
Pearlmutter, S., 343–344
Peek, M. K., 125, 154, 192, 320, 324
Peek-Asa, C., 346
Pencavel, J., 165
Penedo, F. J., 320
Penner, R. G., 276–277

Peralta, C. A., 115
Perdue, L., 315–316
Perez, L. M., 257
Perkins, D. D., 329
Perlman, D., 193
Peronto, C. L., 91–92
Perrin, N., 327
Perry, B. L., 188–189
Perry, C. K., 327
Perun, P., 276–277, 279
Perun, P. J., 273, 278
Pescosolido, B. A., 188–189
Peterson, C., 257
Peterson, J. C., 347
Peterson, L., 363
Pettit, G. S., 214
Pew Research Center, 378–380, 383–385, 387
Pfeifer, M. P., 382
Phelan, J. C., 134, 331
Phifer, J. F., 362–363
Phillips, D. I. W., 104
Phillips, R. L., 241
Phipps, P. A., 165–166
Piccinin, A. M., 116
Pickering, T. G., 187
Pickett, K. E., 15–16, 321
Pierce, S. F., 379
Pietrzak, R. H., 363
Piliavin, J. A., 305–306
Pillemer, K. A., 206, 208–216, 220–221
Pinquart, M., 171, 175, 305
Pitt-Catsouphes, M., 307
Plikuhn, M., 209
Ploeg, J., 342, 349–350
Poirier, J., 68
Polivka-West, L., 359
Poloma, M. M., 261–262
Pope, N. D., 210
Population Reference Bureau, 143–144
Porensky, E. K., 384
Port, C., 363
Port, F. K., 132
Portes, A., 145–147, 151–152
Post, L. A., 346
Poston, D. L., 233–234, 236–240
Potischman, N., 111
Potter, J. E., 151
Poulin, M. J., 176
Powell, B., 219
Power, C., 101–102, 111
Prager, E., 364
Prakash, A., 87, 154
Preisser, J. S., 93

Pren, K. A., 147–148, 153
Prenovitz, S., 84
Prentky, R. A., 343
Prescott, C., 68
President's Council on Bioethics, 380, 387
Preston, S. H., 61, 67–68
Prigerson, H. G., 382–383
Prince, M. F., 266
Prochaska, T. R., 127–128
Proctor, B. D., 273
Pruchno, R. A., 175–176, 308
Psaty, B. M., 93–94
Public Law 111–13, 303–304
Pudlin, B., 295
Pudney, S., 166–167
Pudrovska, T., 258
Purcell, P., 271–273
Putnam, M., 308
Putnam, R. D., 171–172

Q

Qian, Z., 150–151
Quenemoen, L. E., 365
Quinn, J. F., 5, 10, 271–278, 282, 303
Quiñones, A. R., 91, 125

R

Raab, M., 209
Rabins, P., 175–176
Racine, E. M., 265
Raghupathy, P., 106
Raid, J., 361
Rajan, K., 343–345
Ram, N., 93
Ramsey-Klawsnik, H., 347, 350
RAND HRS Data, Version M, 24
Rashad, I., 307
Ratard, R., 366
Ratliff, S., 304
Raudenbush, S. W., 29, 43
Ray, L. A., 125, 154
Raymond, J. A., 347
Reade, M. C., 382–383
Redfoot, D., 78
Redstone, I., 144
Reed, D., 254
Reeve, J., 329
Reeves, K. A., 345
Regan, S. L., 346
Rehkopf, D. H., 192
Reid, D. D., 103
Reid, M. C., 220
Reid, R. J., 345

Reinhard, S. C., 305
Reinhardt, J. P., 302
Reinhold, S., 154
Reker, G. T., 259
Renalds, A., 326, 331
Rennicke, C., 365
Repetti, R. L., 107
Resch, N. L., 175–176
Resick, P. A., 364
Resnicow, K., 265
Revenson, T. A., 363–364
Reyes, A. M., 154
Reynolds, S. L., 84
Rhee, M.-K., 153
Richard, L., 303, 327
Richards, M., 102–103
Richardson, E. A., 330
Rich-Edwards, J. W., 111
Richman, K., 302–303
Riddell, T., 344
Riedmann, A., 216
Riffin, C., 212–213, 220
Rigney, D., 5
Riley, J. W., 178
Riley, M. W., 178
Rinfrette, E. S., 342
Riosmena, F., 154–155
Rique, J., 260
Risnes, K. R., 105
Ritzer, G., 17
Rivlin, A., 379
Rix, S. E., 277, 286
Ro, A., 87
Robbins, C., 186
Robert, S. A., 320–321, 330
Roberto, K. A., 11–12, 342, 346, 348–349
Roberts, R. E. L., 207
Robins, G., 194–195
Robinson, C., 281
Robinson, D. E., 126
Robinson, J. P., 164, 168
Robison, J., 329
Robson, S. M., 304–305
Roby, J., 349–350
Röcke, C., 190–191
Rodgers, B., 305
Rodriguez, D. E., 114
Rodriguez, J., 358
Roepke, S. K., 306–307
Roff, L. L., 132
Rogers, I., 105
Rogers, R., 124
Rogers, R. G., 133
Rogers, R. J., 323–324

Rogers, S. J., 219
Rogers, W., 257
Rogoff, E. G., 297
Rogowski, J., 281
Rogowski, J. A., 281
Rohleder, N., 69
Rohlfsen, L., 130–131, 134
Rohlfsen, L. S., 91–92
Rohwedder, S., 305
Rolley, N. C., 132
Rook, K. S., 188, 190–191, 196
Rooney, B. L., 386
Roos, L. L., 317–318
Roscigno, V., 303, 307
Rose, G. A., 103
Rosen, H. S., 275
Rosen, S., 239–240
Rosenberg, D. E., 329
Rosenthal, L., 187
Rosenzweig, M. R., 144
Ross, C. E., 316–317, 320
Ross, N. A., 15, 315–316
Rossi, A. S., 208, 212–213, 215
Rossi, P. H., 208, 212–213, 215
Rosso, A. L., 326
Roth, D. L., 305, 307
Rothman, M., 360, 366–367, 369–370
Rotolo, T., 302
Rovba, L., 272–273, 281
Rovi, S., 337–338
Rowe, J. W., 60, 63–64
Rowel, R., 366–367
Roy, J., 94, 378–379
Rozario, P., 299
Rozdilsky, J., 361
Ruben, J. D., 154
Rubin, D. B., 43
Ruckert, A., 18
Rudenstine, S., 107
Rudkin, L. L., 320
Ruel, E., 320
Ruger, W., 237–240
Ruhm, C. J., 272, 275
Rumbaut, R. G., 145–147, 149
Ruscher, J. B., 358
Rutter, M., 107, 363–364
Ryan, E., 240–241
Ryan, L. H., 167
Ryder, N. B., 9–10, 147
Ryff, C. D., 64–65, 69–72

S

Sabia, J. J., 329–331
Sachs-Ericsson, N., 256

Sackett, P. R., 232–233
Sadoulet, E., 147
Saha, S., 366–367
Sahin, A., 277
Saito, Y., 78–79, 84
Sakamoto, A., 154
Sakkinen, P. A., 64–65
Salthouse, T. A., 322–323
Salva, J., 210–211, 213
Sampson, R. J., 233–234
Sanders, S., 350, 360–362
Santaguida, P. L., 360–361
Sarkisian, C. A., 64
Sarkisian, N., 212–214, 216–217
Sasaki, J. Y., 267
Sass, S. A., 272–273, 276–279, 283, 305–306
Sass-Lesser, J., 302–303
Sasson, I., 92
Savarese, V. W., 233
Savla, J., 217–218, 341, 348–349
Sayer, A. A., 323
Sayer, L. C., 164, 169, 171
Sbarra, D. A., 256
Scaramella, L. V., 218
Schack, R. W., 295
Schafer, M. H., 6, 182–183, 188–190, 194–197, 345
Schaner, S. G., 294–295, 297–298
Scharf, T., 207–208
Schellenberg, G., 362
Schenk, N., 212–213, 215
Schiamberg, L. B., 340–341
Schieman, S., 6, 258, 265–266, 325
Schilling, S., 68
Schillinger, D., 383
Schkade, D. A., 165–166
Schmidt, J. F., 251–252
Schmitz, H., 188, 302, 305
Schmitz, N., 315–316
Schneider, B., 165
Schneider, C., 381
Schnittker, J., 189
Schnurr, P. P., 227–228
Schoeni, R. F., 78, 84, 87–88, 93
Schoepflin, U., 144
Schuengel, C., 213
Schuler, R. H., 361–362
Schultz, A., 301
Schultz, R., 321
Schulze, H.-J., 213
Schumm, L. P., 181–182
Schut, H., 190–191
Schwartz, N., 172

Schwartz, Z., 376
Schwarz, B., 210
Schwarz, N., 165–166, 176
Sears, M. R., 68
Seburn, P. A., 278–279
Sechrist, J., 206–207, 209–210, 213, 215
Seely, E. W., 111
Seeman, M., 65
Seeman, T. E., 57–58, 60, 63–65, 69–72, 107, 113–114, 128–130, 154–155, 181–182, 187, 189–190, 254–255
Segal, D. R., 235, 239–241
Selective Service, 233–234, 238–239
Seligson, H. A., 366
Seltzer, J. A., 209–210
Seltzer, M. M., 210, 217, 304–305
Sembajwe, G., 15–16
Sentell, T. L., 366–367
Seplaki, C. L., 93–94
Serrano, M., 111
Sethi, D., 339–340
Settersten, R. A., Jr., 208, 227–228, 296, 301–302, 341
Shabani, R., 233, 240
Shah, I., 107
Shah, R. C., 330
Shalev, A. Y., 362–363
Shalowitz, D. I., 381
Shanahan, M. J., 231–232
Shapiro, A., 219–220
SHARE, 8
Sharp, S., 376, 384–385
Shaw, B. A., 5, 189–190, 192–193
Shay, J., 235
Shea, S., 146
Sheffield, K. M., 324
Sheikh, A., 377
Sheikhattari, P., 366–367
Sheridan, S. L., 366–367
Sherraden, M., 294, 299
Shi, L., 15
Shibusawa, T., 302
Shiell, A. W., 105
Shiffman, S. S., 165
Shiovitz-Ezra, S., 185, 188
Shipley, M. J., 103
Shippee, T. P., 6, 129, 134
Shiu, V., 382–383
Shoaf, K. I., 366
Shonkoff, J. P., 107, 129–130
Shorr, A. F., 383
Shostak, S., 68
Shrager, S., 65
Shuey, K. M., 130, 164

Shultz, K. S., 272, 275
Shuman, S. K., 93
Shumway-Cook, A., 326–327
Shvydko, T., 277
Shyu, Y. I. L., 91
Siebens, H., 82
Siegel, J. M., 362
Siegl, E., 305–306
Sigad, L. I., 218
Sigal, J. J., 362–363
Sigler, R., 349
Silliman, R. A., 7–8
Silveira, M. J., 382–383
Silverstein, M. D., 207–208, 210–211, 218, 220–221, 298
Silvester, W., 382–383
Simeone, R. S., 185–186
Simmel, G., 209
Simmonds, S. J., 104
Simms, J. L., 360
Simon, M. A, 337–338, 342–345
Simonovich, S. D., 329
Singer, A. E., 388
Singer, B. H., 60, 63–65
Singer, P. A., 379
Singh, G. K., 133
Singh, K. P., 87
Sipe, T., 303
Sisk, B., 146, 153
Sivatchenko, A., 68–69
Skinner, E. A., 258–259
Skinner, J., 384
Skira, M. M., 175
Skolnick, A. A., 377
Sliwinski, M. J., 93
Small, B. J., 91–92
Smedley, B. D., 126
Smeeding, T. M., 168–169, 171, 173
Smith, A. K., 384
Smith, B. W., 257
Smith, D. B., 131–132
Smith, G. D., 106–107, 115
Smith, H., 167
Smith, H. L., 241
Smith, J., 93, 185
Smith, J. C., 273
Smith, J. P., 144
Smith, K. E., 272, 297
Smith, K. P., 186–189
Smith, K. R., 68–69
Smith, M. J., 210, 217
Smith, T. B., 186
Smith, T. H., 326
Smith, T. W., 187, 220–221, 305–306

Smith-Lovin, L., 187–188
Smits, A., 210
Smolensky, E., 286–287
Smucker, W. D., 380–381
Smulders, Y. M., 115
Sneiderman, B., 381
Snijders, T. A. B., 186, 189, 194–195
Snodgrass, J. J., 59
Soares, J., 339–340
Sobolewski, J. M., 214
Sociometrics, 7
Sok, A., 297
Solé-Auró, A., 66f, 67, 67f
Solomon, R. C., 17
Solomon, Z., 364
Sommers, D., 276–277
Son, J., 306–307
Song, C., 148
Song, J., 125, 272
Soobader, M.-J., 132
Sörensen, S., 171, 175, 305
Sorlie, P. D., 131
Southwick, S. M., 363
Spaid, W. M., 168, 175–176
Sparrow, D., 64–65
Spasojevic, J., 307
Spears, M., 114
Spence, M., 195
Spence, N. J., 86–87, 231–232
Spera, C., 305
The Sphere Project, 365
Spilka, B., 258
Spillman, B. C., 83, 85–86, 89, 93–94
Spiro, A. III, 227–228, 240, 243–244, 307
Spiro, C., 276–277
Spitze, G., 209, 216
Spokane, A. R., 325
Stafford, F., 166
Stafford, F. P., 165–167, 172
Stafford, M., 107, 111
Stallard, E., 84, 124–125, 132
Standing, T., 304–305
Stanley, M., 236–238
Star, A. S., 236
Starfield, B., 15
Starks, H., 381
Starr, P., 378
Stawski, R. S., 306–307
Steele, J., 195
Steelman, L. C., 219
Steglich, C. E. G., 194–195
Stehouwer, C. D., 115
Stein, R. A., 383
Steiner, M., 260

Steinhauser, K. E., 376, 380
Steinhour, M., 213
Steinmeier, T. L., 272
Stephens, M. A. P., 188
Stepinsky, J., 17
Sternthal, M., 132
Steuerle, C. E., 297
Steuerle, E., 276–277
Steve, K., 339
Stevens, N. L., 189–190
Stewart, J., 168–171, 173
Stiegel, L. A., 344–345
Stieglitz, J., 114
Stiglitz, J. E., 12–13
Stimpson, J. P., 133
Stith, A. Y., 126
Stoeckel, K. J., 187, 189
Stolyarov, D., 272
Stone, A. A., 165–166
Stouffer, S. A., 236
Straka, S. M., 342, 345–346
Strauss, A. L., 377
Strayley, J., 240–241
Street, D., 233, 238
Stroebe, M., 190–191
Stroebe, W., 190–191
Stromquist, A. M., 346
Strozza, S., 151
Studenski, S., 110
Sturm, R., 15
Su, Y., 381
Suchman, E. A., 236
Sucoff, C. A., 316–317
Sudnow, D., 378
Sudore, R. L., 383
Suitor, J. J., 206, 208–216, 220–221
Sullivan, A. R., 254
Sullivan, R., 349–350
Sun, F., 255
Sundén, A., 274, 278–280
SUPPORT Principal Investigators, 379
Susser, E., 107
Sussman, M. B., 207
Sussman, M. D., 17
Sutherland, M., 87
Suthers, K., 91–92
Suvak, M. K., 233
Swallen, K. C., 154
Swetz, K. M., 381
Swoope, C., 286
Syme, S. L., 181–182, 185–186
Szanton, S. L., 65
Szinovacz, M. E., 301–302
Szydlik, M., 207–208

T

Tabatabai, N., 272
Tait, E. M., 265
Takahashi, K., 216
Talley, R. C., 295, 305
Tamborini, C. R., 219
Tambs, K., 331
Tan, E. J., 307
Tang, F., 302–303, 307–308
Taniguchi, H., 297–298, 302
Tatara, T., 343–344
Taubman, P., 239–240
Tavares, J., 306–307
Taylor, J., 91
Taylor, J. S., 381
Taylor, M. G., 6, 43–44, 49–50, 91–92, 94, 136
Taylor, P., 16
Taylor, R. J., 265
Taylor, S. E., 107, 112
Teachman, J. D., 234, 236–237, 239–242, 245–246
Teaster, P. B., 340, 342, 344, 346, 348–350
Tedrow, L. M., 236–237, 241
Tehranifar, P., 134
Temme, L. V., 239–240
Teno, J. M., 93, 375–376, 378–379, 382–383
Tepperman, L., 168, 175
Theodore, N., 153
Thoits, P. A., 187, 304–305, 307
Thomas, C., 341
Thomas, K. A., 337–338
Thomas, R. L., 346–347
Thomas, W., 145–146
Thompson, E. H., 343
Thompson, J., 276
Thompson, M. P., 364–365
Thomson, C., 174–175
Thornburg, K., 105
Thorpe, R. J., 126
Thorvaldsson, V., 93
Tiedt, A. D., 220
Tienda, M., 152–153
Tighe, L. A., 210
Tiikkainen, P., 189–190
Tilling, K., 102
Timmermans, S., 378
Timmins, C. L., 131–132
Timonen, V., 207–208
Tinker, A., 339–340
Tisak, J., 24
Tobin, G. A., 360–362

Todorova, I., 136
Tofighi, D., 47–48
Tooke, J. E., 108
Toosi, M., 164, 177–178
Toossi, M., 276–277
Topolski, T. D., 83
Torres-Gil, F., 124, 143–144
Toussaint, L. L., 260
Tracy, M., 363
Tracy, R. P., 64–65
Tran, V. C., 147
Treas, J., 143–146, 149–152
Tremblay, R., 47
Trent, K., 216
Trevino, F., 87
Trommsdorff, G., 210
Tseng, M. Y., 91
Tsiatis, A. A., 94
Tsuda, T., 147
Tsui, V., 345
Tucker, K. L., 136
Turcotte, M., 362
Turner, S., 236–237
Turra, C. M., 63–64, 133, 154–155
Tyrrell, J. S., 108

U

Uchino, B. C., 187
Uchino, B. N., 187, 221
Ugurlu, N., 340
Uhlenberg, P., 217–218
Umberson, D. J., 92, 128, 136, 186, 188
United States Department of Labor: Bureau of Labor Statistics [U.S. BLS], 297
University of Wisconsin, 7
Urban Institute, 302–303
US Bureau of Labor Statistics, 11, 164–166, 174, 174t, 277, 284
US Census Bureau, 16, 123, 125, 298
US Code of Federal Regulations, 228, 229f, 236–239
US Department of Commerce, Bureau of Economic Analysis, 274, 279
US Department of Defense, 232–233, 237–238, 240–241, 243–244
US Department of Health and Human Services, 77–78, 362, 367–368
US Department of Veteran Affairs, 235, 241–242
US Department of Veterans Affairs, 234, 241–244

US Social Security Administration, 280, 285–286
Usdansky, M. L., 241

V

Vagenas, D., 193
Vaillant, G. E., 216, 237
Valente, T. W., 189, 195
Valletta, R. G., 277
van Dam, R. M., 15–16
van de Bunt, G. G., 194–195
Van Der Gaag, M., 186
van der Lippe, T., 216–217
van Gaalen, R. I., 210
van Groenou, M. I. B., 190, 220
Van Hook, J., 146, 151–153
Van Houtven, C., 175
Van Soest, A., 275
Van Tilburg, T. G., 189–190, 192–193, 220
Vanderaa, J. P., 154
VanDerhei, J., 278–279, 282
Vanhems, P., 195
Vasunilashorn, S., 56–57, 60, 64–65
Vaupel, J. W., 59, 68–69
Vedhara, K., 113–114
Venkat Narayan, K. M., 154
Venn, S., 126, 164, 168, 171, 173, 177
Verbrugge, L. M., 56–57, 80, 172
Verdery, A., 187–188
Vermunt, J. K., 34
Vernon, M. K., 165–166
Vespa, J., 206
Veugelers, P. J., 317–318
Vig, E. K., 381
Vijayakumar, M., 104
Visscher, P. M., 68
Viswanath, K., 382–383
Vlahov, D., 364–365
Vogt, D. S., 233
Vokonas, P., 243–244
von Oppen, G., 90
von Wachter, T., 273–274
Voorpostel, M., 216–217
Vorsky, F., 340
Vos, F., 358

W

Wachter, K. W., 59, 69
Waddoups, S., 237
Wade, J., 153
Wadsworth, M. E. J., 103, 106
Wagener, D. K., 78–79
Wagner, F., 187–188
Wahl, H., 317

Wahl, P., 64–65
Wahle, C., 343
Waidmann, T. A., 84, 87–88, 129–130
Waite, K. R., 384
Waite, L. J., 128, 136, 181–182, 187, 218, 220, 296
Wakabayashi, C., 154–155
Waldinger, R. J., 216
Wallace, G., 318
Wallace, K. A., 365
Wallace, R. B., 84, 93, 338
Wallace, W. L., 83
Waller, M., 363–364
Walsemann, K. M., 129–131
Walsh, C. A., 342
Wang, H., 210
Wang, L., 231–232
Wang, M., 272, 275
Wang, T. J., 63–64
Wang, W., 219
Wang, Y., 295, 299
Wangmo, T., 340, 350
Ward, R. A., 209, 212–213, 219–220
Warner, C., 304–305
Warner, D. F., 91–92, 94, 126, 154–155
Warnes, T. A., 143–144
Warshawsky, M., 281
Washko, M. M., 295
Wasserman, S., 181–182, 194–196
Waters, D. J., 134
Waters, M. C., 146–147, 151–152
Watkins, I., 193–194
Watson, P. J., 360
Weaver, D. A., 278
Weaver, G. D., 256
Webb, A., 279, 282
Weeks, J. C., 377
Weinfeld, M., 362–363
Weinstein, M., 59, 63–64, 69–72
Weir, D. R., 382–383
Weitzen, M. H. A., 93
Weitzer, B., 281
Weitzman, P. F., 127–128
Wellman, B., 185–186
Wen, M., 325
Wendler, D., 381
Wenger, N. S., 382
Werner, P., 348–349
Wesner, K. A., 318
West, L. A., 86–87
West, S. G., 44
West, S. K., 384, 389–390
Wetle, T., 376
Whalen, D., 84

Wheaton, B., 304–305, 316–317, 319–320, 322–325, 330
Whincup, P. H., 105–106
White, K., 151
White, L. K., 216–217, 253
Whitelaw, E., 134
Whitfield, K. E., 132
Whitley, D., 303
Wickrama, K. A. S., 6, 39–40, 91
Wickrama, K. K., 307
Wight, R. G., 318–319, 322–325, 330
Wiglesworth, A., 337–338, 346–347
Wilber, K. H., 343–344, 349
Wilcox, S., 263
Wilding, R., 146
Wiles, J. L., 329–330
Wilhelm, B., 217–218
Wilkins, R., 281
Wilkinson, L. R., 6
Wilkinson, R. G., 15–16
Williams, D. J., 111
Williams, D. R., 131–132, 260, 266, 320, 324
Williams, I. C., 136
Williams, R. M., Jr., 236
Williams, S., 59
Williamson, J., 93–94
Williamson, J. D., 124–125
Willis, R. J., 305
Wills, A. K., 106, 109, 115, 117
Willson, A. E., 130
Wilmoth, J. M., 227–228, 231–234, 237–238, 240–242
Wilson, J., 298, 302, 304, 306–307
Wilson, P., 113–114
Wilson, P. W., 64
Wilson, S., 237
Wing, C., 232–233
Wingfield, J. C., 112
Wink, P., 253
Winning, A., 111
Winter, L., 382
Winter, P. D., 104
Winters, P., 147, 149–150
Wirtz, A., 359t
Wise, D. A., 272, 277–278, 281, 285
Wisner, B., 366–367
Wister, A., 302, 318
Wittenberg, E., 337–338
Wolf, D. A., 82–83, 85–86, 89–90, 93–94, 175–176, 232–233, 244–245
Wolff, J. L., 78
Wolinsky, F. D., 131–132
Wong, P. T., 259

Wong, R., 154
Wood, M. M., 362
Wood, S., 343
Woodward, A. T., 136
Woolever, C., 259
World Health Organization, 56–57, 79–80, 80f, 303, 338
Worthington, E., 260
Wortley, S., 186
Wortman, C. B., 128, 362–363, 377
Wray, L. A., 91–93, 125
Wray, N. R., 68
Wright, A. A., 382–383, 385
Wright, B., 297
Wu, H. S., 210
Wu, J. H., 124
Wulff, K. M., 264
Wykle, M., 304–305
Wyman, M., 241–242

X

Xie, B., 193–194
Xie, Y., 148, 241
Xu, J., 176, 267
Xu, X., 91, 125
Xu, Y., 346
Xue, Q. L., 94, 307

Y

Yaghootkar, H., 108
Yamagata, Z., 15–16
Yamamoto-Mitani, N., 91
Yang, F. M., 210–211
Yang, Y., 10, 44, 91–92, 124–125, 134
Yao, L., 330
Ye, W., 91, 125
Yehuda, R., 362–364
Yen, I. H., 315–316, 319, 321, 323–324, 326
Yeung, D. Y., 190
Yi, J., 153
Yoder, A., 238–239
Yonker, J. A., 376, 381
Yoo, G. J., 146
York Cornwell, E., 181–182, 187
Yoshida, H., 306
Young-Xu, Y., 384
Yu, T., 214

Z

Zajac, K., 344
Zalli, A., 69
Zarit, S. H., 210, 221, 305
Zauszniewski, J., 304–305
Zavisca, J., 189
Zavodny, M., 153
Zayas, L. H., 384
Zedlewski, S. R., 297–298, 302
Zeki, A. A.-H., 125
Zettel, L. A., 190–191
Zettel-Watson, L., 380–381
Zhang, B., 383
Zhang, H., 64–65
Zhang, W., 148
Zhang, X., 190
Zhang, Z., 133–134
Zheng, H., 14–15
Zheng, S., 346
Zhou, M., 145–147, 151–152
Zhuo, Y., 209
Zickuhr, K., 193–194
Zilberberg, M. D., 383
Zimmer, Z., 49–50, 94
Zimmer-Gembeck, M. J., 258–259
Zimmerman, L., 302
Zinn, J., 131–132
Zissimopoulos, J. M., 275–276
Zivin, K., 304
Znaniecki, F., 145–146
Zuzanek, J., 188

Subject Index

Note: Page numbers followed by "*f*" and "*t*" refer to figures and tables, respectively.

A

AARP, 446, 453–454
Abuse, elder, 337
  future research, 350–351
  global perspectives, 339–340
  perpetrators of, 345–348
    adult children, 346–347
    grandchildren abuse, 347
    spouses/partners, 345–346
    trusted others, 347–348
  prevalence of, 338–339
  responses to, 348–350
    community perceptions, 348–349
    interventions, 349–350
    policy initiatives, 350
  socioecological framework for understanding, 340–342
    ecological theory, 340–341
    feminist theories, 342
    life course perspective, 341
    models of social organization, 341–342
  vulnerabilities and risk for, 342–345
    age, gender, race, and ethnicity, 343–344
    cultural beliefs and perceptions, 344
    health and cognitive abilities, 344–345
    social interactions and isolation, 345
ACA. *See* Affordable Care Act (ACA)
ACL. *See* Americans Changing Lives study (ACL)
ACP. *See* Advance care planning (ACP)
Active euthanasia, 387
Active life expectancy (ALE), 79
Activities of daily living (ADL), 81–83, 377, 421
  disability, 84, 125
Activity limitations, 82–85
Acute care, 420
  integrating with long-term care services, 424–428
Adaptations of behavior, 112
Adaptive coping strategies, 363
ADEA. *See* Age Discrimination and Employment Act (ADEA)
ADL. *See* Activities of daily living (ADL)
The Administration for Community Living, 435
Administration on Aging and the Aging Network, 369
Adrenal glucocorticoid, overactivity of, 112
Adult children, abuse by, 346–347
Adult disease, early life origins of, 104–109
  early adverse environments and stress response, 106–107
  genetic and evolutionary perspective, 107–109
  intrauterine environment, 104–105
  physical growth and development, 106
  postnatal environment, 105–107
Adult life style model, 103
Adult Protective Services (APS), 349–350
Adult public assistance categories, 444
Advance care planning (ACP), 376, 380–385
  benefits and consequences, 382–383
  components and limitations, 380–382
  trends and differentials, 383–385
AFDC. *See* Aid to Families with Dependent Children (AFDC)
Affordable Care Act (ACA), 281, 382, 386–387, 398, 403, 428–429, 456
African Americans, 129–130, 132, 135–136, 219, 241. *See also* Blacks
Age and vulnerability for negative consequences, 362–363
Age Discrimination and Employment Act, 442
"Age-as-leveler" hypothesis, 487–488
Agency, principle of, 4
Aging and social sciences
  contribution, 16–19
  data developments, 6–8
    academic versus government sponsorship, 7
    age ranges and times of measurement, 7
    biomarker, genetic, and physical performance data, 8
    merging survey and administrative data, 7–8
    national versus regional/local samples, 7
    non-U.S. databases, 8
  emerging themes, 9–16
    gradual and/or incremental cultural changes, 12–16
    increased attention to cohort analysis, 9–10
    social and economic disruptions on aging, 10–12
  methods and data, 6–9
  statistical sophistication, 8–9
  theoretical and conceptual developments, 4–6
    cumulative advantage/disadvantage theory (CA/DT), 5–6
    cumulative inequality theory (CIT), 6

"Aging in place,", 329–330
Aging-vector analysis, 46
Aid to Families with Dependent Children (AFDC), 449
Aid-in-dying. *See* Physician-assisted suicide (PAS)
ALE. *See* Active life expectancy (ALE)
Alliance for Retired Americans, 446
Allostatic load, 60, 64–65, 112, 129–130
All-Volunteer Force (AVF), 228, 240
Almshouses, 422–423
ALSA. *See* Australian Longitudinal Study of Aging (ALSA)
Alters, 183
American exceptionalism, 452
American Red Cross, 367–368
American Time Use Study (ATUS), 165–166, 174
Americans Changing Lives study (ACL), 7, 182
"Antagonistic pleiotropy" model of aging, 108–109
APOE. *See* Apolipoprotein E (APOE)
Apolipoprotein E (APOE), 68
APS. *See* Adult Protective Services (APS)
Armed Forces Qualification Test scores, 241
Arthritis, 77–78, 88–89, 125
Asian Americans, 124
ATUS. *See* American Time Use Study (ATUS)
Audacity, 4
Australian Longitudinal Study of Aging (ALSA), 8
AVF. *See* All-Volunteer Force (AVF)

B

"Baby Boom" cohort, 240–241
Baby Boomers, 78, 211, 217, 219–221, 229–230, 240
Balanced Budget Act of 1997, 423–424
Bayes' Rule, 33
Bayesian Information Criterion (BIC), 34
Bengtson's Longitudinal Study of Generations, 208
BESA. *See* Border Epidemiologic Study of Aging (BESA)
BIC. *See* Bayesian Information Criterion (BIC)
Big Five Inventory, 365
"Big questions,", 17

Biodemography, 53
  biological risk, summary indices of, 63–68
  biomarkers
    measuring, in population studies, 59–60
    in social science research, 69–72
    use in assessing population health, 60–63
  expanded biodemographic model of health, 57–59
  genetic markers as new frontier, 68–69
  health change, process of, 56–57
Biological aging, 64, 109
  measurement of, 65–67
Biological risk, 58–59, 58*f*
  summary indices of, 63–68
Biomarkers, 8
  in assessing population health and effectiveness of health care, 60–63
  measuring, in population studies, 59–60
  in social science research, 69–72, 70*t*
Birth cohort, 9–10, 23–25, 44
Birthweight, 115–116
  and blood pressure, 114–115
  and cancer mortality, 105
  and diabetes, 105
Blacks. *See also* African Americans
  activities of daily living (ADL), 125
  life expectancy for, 124
Black–White differences, in aging and health, 124, 130–132
Blood pressure, 60–62, 63*f*, 70*t*, 114–115
BMI. *See* Body mass index (BMI)
Body functions and structure, 79–81
Body mass index (BMI), 24, 106, 115
Border Epidemiologic Study of Aging (BESA), 135
Breastfeeding, 111
Bridging position in the network structure, 184–185
Bronfenbrenner's ecological systems theory, 340–341
Built environment and health of older persons, 326–330
  aging in place, 329–330
  disability process, 327–329
  physical activity and health, 326–327
  recovery of mobility, 330
Burden theory, 364–365

Bush, George W., 449, 453–454
Butler, Robert, 294

C

CA/DT. *See* Cumulative advantage/disadvantage theory (CA/DT)
CAD framework. *See* Cumulative dis/advantage (CAD) framework
Caerphilly Prospective Cohort study, 113
Canada Pension Plan (CPP), 476
CAR. *See* Cortisol awakening response (CAR)
Cardiovascular disease (CVD), 15, 89, 102, 105–106
  and childhood socioeconomic position, 106–107
  intrauterine environment in, 104
  and obesity, 105
  risk in children, 103
Care management, 427–428
Care Transitions Program, 433
Caregiving, 168, 174–176, 174*t*, 296–298, 306–307, 492–493
  measuring caregiving time, 174–176
Cash and Counseling Demonstration and Evaluation (CCDE) program, 429–430
Catastrophic events, 359*t*
CCDE program. *See* Cash and Counseling Demonstration and Evaluation (CCDE) program
Centers for Medicare and Medicaid Services (CMS), 404
Centrality, 184
CHARLS. *See* China Health and Retirement Longitudinal Study (CHARLS)
Child immigrants, 145–146
Childbearing after US immigration, 151
Childhood antisocial behavior, 107
Childhood cognitive ability, 107, 111, 115–116
Childhood obesity, 134–135
Children of the Great Depression (Glen Elder), 149
China Health and Retirement Longitudinal Study (CHARLS), 59–62
Chinese Longitudinal Healthy Longevity Study (CLHLS), 8
Cholesterol, 62, 62*t*

SUBJECT INDEX

Chronic illnesses, 419–420
Church-based social relationships, 262–264
CIT. *See* Cumulative inequality theory (CIT)
CLASS Act, 456
CLHLS. *See* Chinese Longitudinal Healthy Longevity Study (CLHLS)
Closeness centrality, 184
CMS. *See* Centers for Medicare and Medicaid Services (CMS)
Cognitive aging, in adult cohorts, 116
Cohabitation, 220
Cohort analysis, 9–10
Community-level networks, 341–342
Compassionate conservatism, 448
Compensatory reserve, 112
Complexity, of later-life families, 209
Conditional density, 33
Conditional independence assumption, 34
Consumer-directed care options, 429–430
Contemporary immigrant incorporation theories, 147
Contingent exchange theory, 210
Continued work later in life, potential benefits of, 284–287
Continuity theory, 18
Coping strategies, 363
Cortisol awakening response (CAR), 113
CPP. *See* Canada Pension Plan (CPP)
CPS. *See* Current Population Survey (CPS)
C-reactive protein (CRP), 60
Critical perspectives, 302
Crossover effect, 132
CRP. *See* C-reactive protein (CRP)
Culture change models, 430–431
Cumulative advantage/disadvantage theory (CA/DT), 5–6, 129
Cumulative dis/advantage (CAD) framework, 191, 486–488
Cumulative inequality theory (CIT), 6, 488–489
Current Population Survey (CPS), 229–230
CVD. *See* Cardiovascular disease (CVD)
Cystatin C, 60

D

Daily Spiritual Experiences Scale, 252
Data structure and method, in trajectory models, 43–44
Day Reconstruction Method (DRM), 165–166
DC plans. *See* Defined-contribution (DC) plans
Death and dying in US, 376–380
 cultural context, 378–380
 medicalization, 378–379
 movement toward patient autonomy, 379–380
 demographic and epidemiologic contexts, 376–378
Deferred Action for Childhood Arrivals Initiative, 148–149
Defined-contribution (DC) plans, 274
Degree centrality, 184
Dementia, 89–90, 220
Dense networks, 186
Depression, 113–114, 136
Developmental Origins of Health and Disease (DOHaD), 103, 108–109
Developmental plasticity, 108–109
Diabetes, 62, 92–93, 105, 132
Diary data, 168–169, 176–177
Disability and Use of Time (DUST) supplement, 165–166
Disability Insurance, 281, 442
Disability process, 327–329
Disablement, 79–80
Disaster literacy, 366–367
Disaster Response Network (DRN) program, 367
Disasters, 357–358
 age and resilience, 363–365
 age and vulnerability, 362–363
 formal support, 367–368
 future directions, 369–370
 group evacuation versus individual/independent evacuation, 360–361
 influence of residential environment on disaster-related activities and outcomes, 359–360
 informal support and social networks, 368–369
 phases related to, 359t
 physical and mental health issues, disaster-related, 365–366
 social factors and disaster response outcomes, 366–367
 stress and coping, 363
 temporary evacuation, transfer/ permanent relocation, 361–362
 types and definitions of, 358, 359t
Discrimination and racism, 132
Divorce, 214, 218–219
 remarriage after, 219
DOHaD. *See* Developmental Origins of Health and Disease (DOHaD)
Double jeopardy hypothesis, 130, 191
DPAHC. *See* Durable power of attorney for health care (DPAHC)
DRM. *See* Day Reconstruction Method (DRM)
DRN program. *See* Disaster Response Network (DRN) program
Durable power of attorney for health care (DPAHC), 380–381, 385
DUST supplement. *See* Disability and Use of Time (DUST) supplement

E

Earliest Eligibility Age (EEA), 286
Education, 17–18, 87, 126–127
 cardiovascular risk and, 67–68, 67f
 and GI bill, 236–237
EEA. *See* Earliest Eligibility Age (EEA)
Ego, 183
Elder, Glen (*Children of the Great Depression*), 149
Elder abuse. *See* Abuse, elder
Elder maltreatment. *See* Maltreatment, elder
Elder mistreatment. *See* Mistreatment, elder
Electronic networks, 193–194
ELSA. *See* English Longitudinal Study of Ageing (ELSA)
Emergencies, 358, 359t
Emotional abuse, 346
Emotional health, 111
Employee Retirement Income Security Act, 442
Employer-based retiree health plans, 407–408
Employer-provided health insurance, 281
Employment and family role influences on time use, 169–171, 169f

Empty-nest syndrome, 220
Endocrine system, 112–114
End-of-life planning and health care, 375
  advance care planning (ACP), 380–385
    benefits and consequences, 382–383
    components and limitations, 380–382
    trends and differentials, 383–385
  death and dying in US, 376–380
    cultural context, 378–380
    demographic and epidemiologic contexts, 376–378
  future directions, 388–390
  public policy innovations, 385–388
    Affordable Care Act (ACA), 386–387
    physician-assisted suicide, 387–388
    POLSTs (Physician's Order for Life Sustaining Treatment), 385–386
English Longitudinal Study of Ageing (ELSA), 8, 59–60, 497–498
EPESE Program. See Established Populations for Epidemiologic Studies of the Elderly (EPESE) Program
Established Populations for Epidemiologic Studies of the Elderly (EPESE) Program, 7
Ethnic and racial inequalities, in health, 123–125
Ethnic enclaves, 320
Evacuation, 361
  group versus individual, 360–361
Evercare model, 427
Evolutionary biology, for developmental origins of adult disease, 107
Exploitation, financial, 338–339, 349

F
Family care, for impaired older adults, 12
Family Medical Leave Act, 303–304
Family relations, in later life
  grandparents and grandchildren, 217–219
    patterns of support between, 218–219
    relationship quality between, 218
  intergenerational solidarity, 207–208
  life-course perspective in, 208
  marriage in later years, 219–220
  older parents and adult children, relationship quality between, 212–215
    achieved structural characteristics, 213–215
    ascribed characteristics, 212–213
    exchange processes, 215
    social structural characteristics, 212
    value similarity, 215
  siblings, 215–217
    patterns of support between, 216–217
    relationship quality between, 216
  stability, change, and complexity in, 203
  structure, 219
  supportive exchanges between generations, 209–211
  within-family complexity, 209
Family solidarity, components of, 207
Favoritism, parental, 216
FDC. See Funded defined contribution (FDC)
Federal Insurance Contributions Act payroll tax, 410–411
Federal Reserve, 284
Fee-for-service, 400, 414
Fetal Origins of Adult Disease (FOAD), 103–106, 108
Filipino immigrants, 149
Finite mixture model, 33
Five Factor Model, 365
Fixed effects, 29
FOAD. See Fetal Origins of Adult Disease (FOAD)
Forgiveness, religious involvement and, 260–261
Formal support, 367–368
Fortitude, 4
Framingham Risk Score, 64
Functional aging, early life origins of, 109–116
  endocrine system, 112–114
  integrated life course model of aging, 109–111
  life course physical and psychological influences on HPA axis, 113–114
  physical and cognitive capability, 115–116
    lifetime influences on, 115–116
    structural reserve and compensatory mechanisms, 112
  vascular function, 114–115
    lifetime influences on, 114–115
Functional disability, 124
Fundamental cause theory, 485–486
Funded defined contribution (FDC)
  overview, 467
  partial privatization
    Argentina, 469–470
    Canada, 475–476
    Chile, 468–469
    China, 470–473
    trends, 467–468
    United Kingdom, 473–475

G
GAO. See Government Accounting Office (GAO)
GDP. See Gross Domestic Product (GDP)
Gender, 212–213, 219
  and disability, 86–87
  income inequality and, 15
Gender, time use, and aging, 163
  age and gendered time use, 168–169
  caregiving, time use, and well-being, 174–176, 174t
    measuring caregiving time, 174–176
  employment and family role influences on time use, 169–171, 169f
  future directions, 176–178
  measuring time allocation in later life, 165–167
  overview, 163–165
  social versus solitary dimension of time, 173
  work and family roles, gender, and leisure activities, 171–172, 172f
"Generation X" cohort, 229, 240–241
Generations
  ".5 generation", 149
  "1.5 generation", 149
  supportive exchanges between, 209–211
Genetic markers, 68–69
Genome-wide association studies (GWAS), 68
German Reunification, 147–149
Gerontological health research, 134–135

Gerontological Society of America (GSA)-sponsored journals, 230
GI Bill, 233–234, 236–238
Global Age-Friendly Cities Guide, 495–496
Globalization, 143–144
Glycosylated hemoglobin, 62
GMM. *See* Growth mixture modeling (GMM)
God-mediated control beliefs, 256, 258–259
"Good Warriors" cohort, 229–230, 236–238
Government Accounting Office (GAO), 348
Grandchildren, abuse by, 347
Grandparents and grandchildren, 217–219
  patterns of support between, 218–219
  relationship quality between, 218
"Granny bashing,", 338
Great Recession, 11, 272, 274, 276–277, 283–284
Greedy nuclear family, 213, 216–217
Gross Domestic Product (GDP), 13, 405
Group evacuation versus individual/independent evacuation, 360–361
Growth mixture modeling (GMM), 39–43, 91–92
Growth modeling, 25–30, 47
GSA-sponsored journals. *See* Gerontological Society of America (GSA)-sponsored journals
Gulf War Syndrome, 241–242
GWAS. *See* Genome-wide association studies (GWAS)

H

HALCyon. *See* Healthy Ageing across the Life Course (HALCyon)
"Hard Timers" cohort, 228–230, 235
Hazards, 359t
HBPC. *See* Home-Based Primary Care program (HBPC)
HCBS. *See* Home- and community-based services (HCBS)
HCHS. *See* Hispanic Community Health Study (HCHS)
Health
  biodemographic model of, 57–59
  income inequality and, 12–16
  life-course perspectives on, 127–129
  theories of, 129–130
  of older immigrants, 154–155
  outcomes, demographic, socioeconomic, behavioral and biological influences on, 58f
  racial and ethnic inequalities in, 123–125
  racial disparities in, 6
Health and Retirement Study (HRS), 7, 24, 59–60, 68, 82–83, 91–92, 125, 128, 135, 175, 182, 231, 244, 282, 497–498
  family of studies, 71t
Health behaviors, 58f
Health care, 58f
  end-of-life planning and advance care planning (ACP), 380–385
  death and dying in US, 376–380
  future directions, 388–390
  public policy innovations, 385–388
  organization and financing of. *See* Medicare
Health change, process of, 56–57
Health conditions, 79–80
Health Impact Assessment (HIA), 494
Health in All Policies approach, 493–494
Health inequalities among older adults, 483–484
  sociological research, promising avenues for, 496–498
  theories of health inequality, 484–489
    fundamental cause theory, 485–486
    life course theories, 486–489
  welfare states and interplay of social solidarity and equity, 489–494
    long-term care, 492–493
    pensions, 490–492
    WHO Health in All Policies, 493–494
    WHO Age-friendly Environments Programme, 494–496
Health information technology, 434
Health insurance, 281
Health paradox, 154
Healthy Ageing across the Life Course (HALCyon), 113, 115, 497–498
Healthy immigrant effect, 133, 154–155
Healthy People 2020 initiative, 77–78
Healthy warrior/healthy deployer effect, 232–233
H-EPESE. *See* Hispanic Established Populations for the Epidemiologic Study of the Elderly (H-EPESE)
Heraclitis of Ephesus, 3
Heuristic model, 58–59
HIA. *See* Health Impact Assessment (HIA)
Higher education, 126
Hill Burton Act, 423
Hispanic Community Health Study (HCHS), 135–136
Hispanic Established Populations for the Epidemiologic Study of the Elderly (H-EPESE), 125, 128, 135
Hispanic health paradox, 126, 132–133
Hispanic population, 124
  activities of daily living (ADL), 125
  aging and health study for, 132–133
  life expectancy for, 124
  research on, 130
Home- and community-based services (HCBS), 429
Home-Based Primary Care program (HBPC), 427–428
Homeostasis, 108–109
Homophily, 184, 187–188, 211
Hospice care, 379–380, 383
HPA axis. *See* Hypothalamo–pituitary–adrenal (HPA) axis
HRS. *See* Health and Retirement Study (HRS)
Human development and aging, 4
Hurricane Katrina, 11–12, 365–366
Hypertension, 61–62, 105, 108, 111, 132
Hypothalamo–pituitary–adrenal (HPA) axis, 107, 112–114

I

IADL. *See* Instrumental activities of daily living (IADL)
IALSA. *See* Integrative Analysis of Longitudinal Studies of Aging (IALSA)
ICF. *See* International Classification of Functioning, Disability, and Health (ICF)
ICU use. *See* Intensive care unit (ICU) use
IHD. *See* Ischemic heart disease (IHD)
Immigrant adaptation, 145–146
Immigrant families, 146, 150–152
Immigrant health paradox, 154

Immigrant incorporation, 143–144, 146–148, 150
Immigration, 133, 143–144
  agency, principle of, 147
  as life-course experience, 144–146
  childbearing after US immigration, 151
  immigrants and families, 150–152
  life-span development, principle of, 146
  linked lives, principle of, 149–150
  migration routes, to United States, 145f
  older immigrants
    health of, 154–155
    socioeconomic outcomes of, 152–153
  overview, 143–144
  time and place, principle of, 147–148
  timing, principle of, 148–149
Immigration Reform and Control Act of 1986, 147
Income inequality and health, 12–16
Independent Payment Advisory Board (IPAB), 405
Individual and social change, 19
Individual-level disability trajectories, 89–94
  losses due to death, 94
  time to death (TTD), 93–94
Indonesia Family Life Study, 61–62
Industrialization, logic of, 452
Informal care, 432–433
Informal support and social networks, 368–369
Inoculation theory, 364
Institute of Medicine (IOM), 80, 421–422
Institution theory, 489
Instrumental activities of daily living (IADL), 81–83, 377
  disability, 125
Integrated life course model of aging, 109–111
Integrative Analysis of Longitudinal Studies of Aging (IALSA), 497–498
Intensive care unit (ICU) use, 378, 383
Intergenerational solidarity, theoretical perspectives on, 207–208
Internal Revenue Taxation Code, 278–279
International Classification of Functioning, Disability, and Health (ICF), 79, 80f

International migration, 143–144, 148, 157
IOM. *See* Institute of Medicine (IOM)
IPAB. *See* Independent Payment Advisory Board (IPAB)
Ischemic heart disease (IHD), 104–105

J
Journal of Aging and Health, 230

K
Korean War, 237–238, 243–244

L
Labor demand, 277
Labor force participation, 277–281
Labor market reentry, 271–273, 275
LASI. *See* Longitudinal Study of Aging in India (LASI)
Late-life disability
  defined, 77–78
  disability
    conceptualization of, 79–81
    measurement, 81–84
  individual-level disability trajectories, 89–94
    losses due to death, 94
    time to death (TTD), 93–94
  trends-in-prevalence
    between-group differences, 86–88
    overall trends, 84–86
    potential explanations for true change, 88–89
Latent class growth analysis (LCGA) model, 38–40, 39f, 49
Latent class modeling, 31–37
Latent class trajectory models (LCTMs), 91–92
Latent growth curve models (LGCMs), 91–92
Latinos, 124, 135–136, 191–192, 320
LCGA model. *See* Latent class growth analysis (LCGA) model
LCTMs. *See* Latent class trajectory models (LCTMs)
LeadingAge, 431
Leisure activities, of older adults, 171–172, 172f
Lexis diagram of twentieth century cohorts, 228, 229f
LGCMs. *See* Latent growth curve models (LGCMs)
Life course epidemiology, 101–103, 109, 117

Life course research, 4, 117, 190–191
  and gerontological research, 5
  at individual level, 23
Life course theories of health inequality, 486–489
  cumulative dis/advantage (CAD), 486–488
  cumulative inequality (CI) theory, 488–489
Life expectancy, 123–124, 217, 276, 376–377
  and biological risk, 65
  of women and men, 168
Life span development, principle of, 4, 146, 156–157
Life-course experience, immigration as, 144–146
Life-course perspective, 301–302
  on health, 127–129
  in later-life family relationships, 208
Life-course-disruption hypothesis, 233–234
Likelihood function, for latent class model, 32–33
Linked lives, 156, 208, 218, 341
  principle of, 4, 149–150
LIS. *See* Low-income subsidy (LIS)
"Live fast, die young" strategy, 108–109
Living will, 380–381, 384–386, 389
Longitudinal Study of Aging in India (LASI), 59–60
Longitudinal Study of Generations (LSOG), 244
Longitudinal Study on Aging (LSOA), 244
Long-term care, 402–403, 492–493
  definitions, 421–422
  efforts, rebalancing, 428–429
  future of, 435
  historical perspective, 422–424
  innovations in, 420–421, 424–435, 425t
    consumer-directed care options, 429–430
    culture change, 430–431
    health information technology, 434
    informal care, 432–433
    integrating acute care with long-term care services, 424–428
    pay-for-performance, 431–432
    Transitional Care/Care Coordination, 433–434

SUBJECT INDEX

Losses due to death, 94
Low-income subsidy (LIS), 407
LSOA. *See* Longitudinal Study on Aging (LSOA)
LSOA-II. *See* Second Longitudinal Study of Aging (LSOA-II)
LSOG. *See* Longitudinal Study of Generations (LSOG)
"Lucky Few" cohort, 229–230, 238
Lung function, 115

M

MA. *See* Medicare Advantage (MA)
MacArthur Study of Successful Aging, 64–65
Macroeconomic influences, increasing importance of, 282–284
Maltreatment, elder, 338
Mandatory retirement, 273–274, 284–286
MAR. *See* Missing at random (MAR)
Marital quality, in later life, 220
Marital status
  and disability, 86–87
  on parent–adult child relations, 214
Marriage
  in later years, 128, 219–220
  military marriages, 241
  on parent–child relations, 214
Maternal nutrition, 108
Matthew Effect, 5
Maturation theory, 364
MCBS. *See* Medicare Current Beneficiary Survey (MCBS)
MCCA. *See* Medicare Catastrophic Coverage Act (MCCA)
Means-testing, 455–456
Medicaid, 303–304, 397–399, 405, 407, 415, 427–428, 433–434
Medical Reserve Corps, 367–368
Medicare, 7–8, 123, 397–398, 447
  affordability and sustainability, 410–411
  expansion, 399–400
  historical perspective, 399–403
  overview, 397–416
  reform, 403–405, 411–415
    financing revenue, 415–416
    incremental approaches, 414–415
    private plan approach, 412–414
  supplementation
    employer-sponsored plans, 407–408
    Medicaid. *See* Medicaid

Medigap insurance, 408
  prevalence, 410
  private insurance, 409
Medicare Advantage (MA), 401, 403–404, 406, 409–410
Medicare Catastrophic Coverage Act (MCCA), 453
Medicare Current Beneficiary Survey (MCBS), 7, 84
Medicare Modernization Act (MMA), 409
Medigap insurance, 406, 408
Metabolic and respiratory disease, 102, 104
Metabolic syndrome, 64
Mexican Health and Aging Study (MHAS), 59–60, 135
Mexican immigrants, 147, 150–151
Mexican-origin Americans
  data on health of, 128
  research on, 132–133
MHAS. *See* Mexican Health and Aging Study (MHAS)
Midlife in the United States (MIDUS), 59–60, 113–114
MIDUS. *See* Midlife in the United States (MIDUS)
Military service, 227–228
  and aging among specific war cohorts, 234–242
  cohort flow and periods of war, 228–230
  composition of US older adult population, 230
  as "hidden variable" in aging research, 230–232
  influence of, on aging process, 232–234
  studying military service and aging, 242–246
Military-as-turning point hypothesis, 233–234
Millennium Cohort Study, 243–244
Minority groups, health and aging across, 130–133
Minority health and aging, 124, 128, 136
Missing at random (MAR), 43
Mistreatment, elder, 338
Mixed model, 29
MMA. *See* Medicare Modernization Act (MMA)
Morbidity process, 56–57, 57f
Morbidly obese, 36–37, 42

Mortality, 56
  income inequality and, 14–15
Mplus software, 29–30

N

NAS. *See* Normative Aging Study (NAS)
National Committee for the Prevention of Elder Abuse (NCPEA), 350
National Council of Senior Citizens, 446
National Death Index (NDI), 7–8
The National Elder Mistreatment Study, 339, 347
National Health and Aging Trends Study (NHATS), 82–83, 171–172, 244
National Health and Nutrition Examination Survey (NHANES), 69, 135–136
National Health Interview Survey (NHIS), 82, 84
National Institute on Aging (NIA), 135
National Long Term Care Survey (NLTCS), 82–86
National Longitudinal Study of Adolescent to Adult Health (ADD HEALTH), 244
National Longitudinal Study of Youth 1979 (NLSY79), 244
National Retired Teachers Association, 446
National Rural Pension Scheme (NRPS), 471–473
National Social Life, Health, and Aging Project (NSHAP), 8, 11, 59–60, 182, 196–197
National Study of Daily Experiences (NSDE), 113–114, 167
National Survey of Families and Households (NSFH), 244
National Survey of Veterans (NVS), 243–244
National Voluntary Organizations Active in Disaster (VOAD), 368
Native Americans, 124, 135
Naturalization, 145–146, 153
NCPEA. *See* National Committee for the Prevention of Elder Abuse (NCPEA)
NDI. *See* National Death Index (NDI)
Negative interaction in the church, 264
Negative network ties, 196

Neighborhood studies, 315–316
  built environment and health of older persons, 326–330
    aging in place, 329–330
    disability process, 327–329
    physical activity and health, 326–327
    recovery of mobility, 330
  concept of, 316–317
  future research, directions for, 330–332
    age and time, 330–331
    toward evidence-based interventions, 331–332
  poverty, 317
  racial and ethnic segregation, 320, 323–324
  socioeconomic disadvantage, 321–323
  stressors and resources, 324–326
  theoretical models of, 316–321
    contextual and compositional neighborhood effects, 317–318
    interaction of person and environment, 318–319
    neighborhood stress process model, 319–321
Network change
  consequences of, for older adults, 190–191
  theories of, in later life, 189–190
Network composition, 186, 188
Network diffusion processes, 195
Network size, 184
Network structure and composition, elements of, 184–185
Network typology approach, 185
Network-gerontology, 193–196
  electronic networks, 193–194
  negative network ties, 196
  network diffusion processes, 195
  whole networks, 194–195
Network-oriented social gerontologists, 194–195, 197
Network-structural pathways, 188
Never-married sons, 214
"New Boomers" cohort, 229
New Immigrant Survey, 144, 146
"New Worlders" cohort, 228, 230
New York Academy of Medicine (NYAM), 368
NH transition programs, 433–434
NHANES. *See* National Health and Nutrition Examination Survey (NHANES)

NHATS. *See* National Health and Aging Trends Study (NHATS)
NHIS. *See* National Health Interview Survey (NHIS)
NIA. *See* National Institute on Aging (NIA)
NLSY79. *See* National Longitudinal Study of Youth 1979 (NLSY79)
NLTCS. *See* National Long Term Care Survey (NLTCS)
Nonparametric method. *See* Latent class modeling
Normal Retirement Age (NRA), 273–274
Normative Aging Study (NAS), 243–244
NRA. *See* Normal Retirement Age (NRA)
NRPS. *See* National Rural Pension Scheme (NRPS)
NSDE. *See* National Study of Daily Experiences (NSDE)
NSFH. *See* National Survey of Families and Households (NSFH)
NSHAP. *See* National Social Life, Health, and Aging Project (NSHAP)
NVS. *See* National Survey of Veterans (NVS)
NYAM. *See* New York Academy of Medicine (NYAM)

O

OASDI program, 273
  Disability Insurance (DI) component of, 281
Obama, Barack, 148–149, 386–387, 456
Obesity, 34
  disability and, 78
OBRA 1987, 430
Old Age Insurance program, 443–444
Old-Age and Survivors Insurance (OASI) program, 273
Old-age policy enactments, accounting for, 451–455
Older adults, 11–12, 166, 192–193, 366, 368
  "a day in the life" of, 167–173
  age-related transitions, 18
  caregiving to, 174, 174t, 176–177
  community-dwelling, 342–343, 360–362
  consequences of network change for, 190–191

end-of-life planning and health care, 376–377
favorite activity, by age and sex, 172f
financial management, 344–345
higher-SES, 193
immigration of, 150
low-SES, 193
network typology approach, 185
online networks, integration into, 194
social connectedness of, 181–182
Older Americans Act, 369
Older immigrants
  health of, 154–155
  socioeconomic outcomes of, 152–153
Older parents and adult children, relationship quality between, 212–215
  achieved structural characteristics, 213–215
  ascribed characteristics, 212–213
  exchange processes, 215
  social structural characteristics, 212
  value similarity, 215
Older workers, 277, 297, 303
Olmstead Act, 428
Open/empty triad, 184

P

PACE. *See* Program of All-Inclusive Care for the Elderly (PACE)
Paid work, 164, 297
Panel Study of Income Dynamics (PSID), 165–166, 244
PAR. *See* Predictive adaptive response (PAR)
Parental death, on sibling closeness, 216
Parents and adult children
  relationship quality between, 212–215
  supportive exchanges between, 209–211
Partial retirement, 275
PAS. *See* Physician-assisted suicide (PAS)
Patient Self-Determination Act (PSDA) 1990, 379, 383
Pay-as-you-go-defined benefit (PAYG-DB) scheme, 462, 467–469, 471
Pay-for-performance, 431–432, 435
PAYG-DB scheme. *See* Pay-as-you-go-defined benefit (PAYG-DB) scheme

Pension, 490–492
  401 k plan
    emergence and future of, 465–466
    financial literacy assumption, 466–467
    market correction impact, 466
    pros and cons, 466
  defined benefit transition to defined contribution, 465
  partial privatization
    Argentina, 469–470
    Canada, 475–476
    Chile, 468–469
    China, 470–473
    trends, 467–468
    United Kingdom, 473–475
  private, 273–274, 278–279
Performance-Based Incentive Payment Program (PIPP), 431–432
Persistent inequality hypothesis, 130
Personal care, 78
Person–environment fit models, 316, 318–319
Phased retirement, 275
Physical activity and health, 326–327
Physical and cognitive capability, 110–111, 115–116
  lifetime influences on, 115–116
Physical and mental health issues, disaster-related, 365–366
Physical functioning, income inequality and, 15
Physician's Order for Life Sustaining Treatment (POLSTs), 376, 385–386
Physician-assisted suicide (PAS), 387–388
Physiological deterioration, 129–130
PIPP. *See* Performance-Based Incentive Payment Program (PIPP)
Pleiotropic genes, 108–109
Policymaking, 443–444
The Polish Peasant in Europe and America (1918), 145–146
Politics and policies of aging, 441–442
  altered political perceptions, 442–446
    policy benefits and political standing, 444–446
    positive standing and policy benefits, 442–444
  emerging issues, 455–457
  old-age policy enactments, accounting for, 451–455
  transformation of seniors' political environment, 446–451
    newly conflicted world of old-age policy, 448–451
    new political realities, 448
    shifting economic context, 447
POLSTs. *See* Physician's Order for Life Sustaining Treatment (POLSTs)
Polygenic Risk Score (PRS), 68
Polyvictimization, 350
Post-Korean War, 237–238
Poverty status and biological risk, 65
Prayer, 261–262
Predictive adaptive response (PAR), 108–109
Private pensions, 273–274, 278–279
"Productive aging,", 294, 306–307, 450
Productive engagement, 293
  challenges and future directions, 306–309
  conceptual frameworks, 299–301, 300*f*
  conceptual issues, 294–296
    controversies in defining the term, 295–296
    defining the term, 294–295
  current evidence on antecedents of, 301–304
    empirical findings, 302–304
    theoretical perspectives, 301–302
  literature on outcomes of, 304–306
    empirical findings, 305–306
    theoretical perspectives, 304–305
  relevance of, 296–299
    demographic context, 296–297
    prevalence, 297–298
Program of All-Inclusive Care for the Elderly (PACE), 426
Project Talent, 244–245
Pro-religious orientation, 255
PRS. *See* Polygenic Risk Score (PRS)
PSDA 1990. *See* Patient Self-Determination Act (PSDA) 1990
PSID. *See* Panel Study of Income Dynamics (PSID)
Pubertal growth, 115–116
Public policy innovations, 385–388
  Affordable Care Act (ACA), 386–387
  physician-assisted suicide, 387–388
  POLSTs (Physician's Order for Life Sustaining Treatment), 385–386
Pulse wave velocity, 114–115

Q
QAPI. *See* Quality Assurance Performance Improvement Initiative (QAPI)
QPP. *See* Quebec Pension Plan (QPP)
Quality Assurance Performance Improvement Initiative (QAPI), 432
Quebec Pension Plan (QPP), 475–476

R
Race
  differences, in biological age, 65–67, 66*f*
  and disability, 86–87
  and ethnicity
    parent–adult child relationship quality by, 213
  and religion and health, 265–266
Racial and ethnic health disparities, 123–125
  life-course perspectives on health, 127–129
  theories of, 129–130
  organizing framework for explaining, 127*f*
  research on, 130–133
  theoretical perspectives, 125–130
Racial and ethnic segregation, 320, 323–324
Random coefficient model, 29
Reactive ethnicity, 147
"Recareering,", 276
Religion and health, 251
  assessing the dark side of, 264–265
  health-related dimensions of, 256–264
    God-mediated control beliefs, 258–259
    prayer, 261–262
    religion and a sense of meaning in life, 259–260
    religious coping responses, 257–258
    religious involvement and forgiveness, 260–261
    religious services attendance, 256–257
    social relationships in the church, 262–264
  involvement over the life course, 252–254
  physical health, 254–255
  and psychological well-being, 255–256
  race and ethnic differences in, 265–266

Relocation, 361–362
Research on Aging, 230–231
Resilience, 112, 363–365, 446–447
Respiratory diseases, 102, 104
Retirement, 271
    changes to the traditional pillars of retirement income, 277–281
    continued work later in life, potential benefits of, 284–287
    earlier and earlier retirement, beginning and end of, 273–274
    involuntary, 307
    labor force participation, 277–281
    macroeconomic influences, increasing importance of, 282–284
    mandatory, 273–274
    modern era, retirement process in, 275–277
    partial, 275
    phased, 275
Retirement Confidence Survey, 279, 283
Retirement income, 272–273
    changes to the traditional pillars of, 277–281
Retirement security, 462, 477
Return migrants, 147–148
Role theory, 304

S

SAGE. *See* Study on Global Ageing and adult health (SAGE)
Saliva-based markers, 69–72
Salmon bias effect, 133, 154–155
Savings rate, 274, 279
SBP. *See* Systolic blood pressure (SBP)
SEBAS. *See* Social Environment and Biomarkers of Aging Study (SEBAS)
Second Longitudinal Study of Aging (LSOA-II), 7
"Selective acculturation,", 147, 151–152
Self-employment, 275–276
Self-enhancement, 365
Senior Centers, 369
Senior Community Service Employment Program, 295
SEP. *See* Socioeconomic position (SEP)
Sequence analyses, 24
SERPS. *See* State Earnings Related Pension System (SERPS)
SES. *See* Socioeconomic status (SES)
Sexual assault, 241–242

SHARE. *See* Survey of Health, Ageing, and Retirement in Europe (SHARE)
SHMO. *See* Social Health Maintenance Organization (SHMO)
Siblings, 215–217
    patterns of support between, 216–217
    relationship quality between, 216
SIENA. *See* Simulation Investigation for Empirical Network Analysis (SIENA)
Simulation Investigation for Empirical Network Analysis (SIENA), 194–195
Single nucleotide polymorphisms (SNPs), 68
SIPP. *See* Survey of Income and Program Participation (SIPP)
SNAP. *See* Supplemental Nutritional Assistance Program (SNAP)
SNPs. *See* Single nucleotide polymorphisms (SNPs)
Social, human, and cultural capital, 301
Social disparities, 62–63, 302
Social Environment and Biomarkers of Aging Study (SEBAS), 59–60
Social gerontologists, 182–183, 189–191, 194–196
Social Health Maintenance Organization (SHMO), 426–427
Social inequality, 302, 484, 493
Social network composition, 184, 196
Social network indices, 185
Social networks in later life, 144, 181–182, 368
    access to social resources, 185–186
    aging and social network change, 189–191
    consequences of network change for older adults, 190–191
    theories of network change in later life, 189–190
    and immigration, 149–150
    health and well-being, 186–189
    network concepts and definitions, 182–185
    basic social network data concepts, 182–183
    composite network measures, 185
    network structure and composition, elements of, 184–185

    network-gerontology, 193–196
    electronic networks, 193–194
    negative network ties, 196
    network diffusion processes, 195
    whole networks, 194–195
    social networks and stratification, 191–193
    gender, 193
    race/ethnicity, 191–192
    socioeconomic status, 192–193
    and stratification, 191–193
    gender, 193
    race/ethnicity, 191–192
    socioeconomic status, 192–193
Social psychological factors, 58f
Social relationships in the church, 262–264
Social Security, 445–452, 454, 462–465. *See also* Aging and social sciences
    parametric reform proposals, 464
    privatization
    efforts, 463–464
    partial privatization examples. *See* Funded defined contribution (FDC)
Social Security Act of 1935, 273, 278, 286–287, 423
Social Security Amendments of 1983, 278
Social Security program, 278, 280, 465
Social Security's delayed retirement credit (DRC), 273–274
Social stratification, social reproduction of, 17–18, 485
Social support, 136, 185–188, 195, 325, 368–369
    church-based, 263–264, 266–267
Social versus solitary dimension of time, 173
Societal order and stability, 17
Socioeconomic outcomes, of older immigrants, 152–153
Socioeconomic position (SEP), 105, 124, 126, 485
Socioeconomic status (SES), 11, 15, 58, 58f, 65, 69, 125–126, 136, 152, 192–193
Socio-emotional selectivity theory (SST), 18–19, 301
Solidarity theory, 207–208
Spiritual struggles, 264–265
Spouses/partners, abuse by, 345–346
SSI. *See* Supplementary Security Income (SSI)

SST. *See* Socio-emotional selectivity theory (SST)
Stability and societal order, 17
State Earnings Related Pension System (SERPS), 473–474
Status hierarchies, 17–18
Stress, 60, 69, 106, 132, 257–259
   and coping models, 304–305, 363
Stress process model, 319–321
Stress response, early adverse environments and, 106–107
Stressors, 257, 319–320, 325
Stroke, 104
Structural functionalism, 17–18
Structural reserve, 112
Study on Global Ageing and adult health (SAGE), 59–60
"Successful aging", 450
Supplemental Nutritional Assistance Program (SNAP), 152
Supplementary Security Income (SSI), 152, 442, 444
Support, patterns of
  between grandparents and grandchildren, 218–219
  between siblings, 216–217
Supportive exchanges, between parents and adult children, 209 211
Survey of Health, Ageing, and Retirement in Europe (SHARE), 8, 59–60, 182, 497–498
Survey of Income and Program Participation (SIPP), 244
Systolic blood pressure (SBP), 62

T

TANF. *See* Temporary Assistance to Needy Families (TANF)
Tax expenditures, 455
Telomeres, 68–69
Temporary Assistance to Needy Families (TANF), 448–449
Tie strength, 184
Time allocation measurement, in later life, 165–167
Time and place, principle of, 4
Time measurement, in trajectory models, 44–46
Time to death (TTD), as an additional "time" variable, 93–94

Time use, 166–168
  age and gendered, 168–169
  allocation, in later life, 163
  employment and family role influences on, 169–171, 169*f*
Time-varying covariates, 43
Timing, principle of, 4, 148–149
Toxic stress, 129–130
Trajectory modeling, for aging research
  growth mixture modeling (GMM), 39–43
  growth modeling, 25–30
  important issues in implementation of
    data structure and method, 43–44
    extraction of classes and inclusion of covariates, 47–49
    importance of assumptions, 46–47
    measurement of time, 44–46
  latent class growth analysis, 38–39
  latent class modeling, 31–37
Transitional Care model, 433
Trauma, 5
Triad, 184
TTD. *See* Time to death (TTD), as an additional "time" variable
Type II diabetes, 108

U

UEPS. *See* Urban Enterprise Pension Scheme (UEPS)
Unemployment rate, of older adults, 11
"Un-retirement,", 275
Urban Enterprise Pension Scheme (UEPS), 471–473
Urban Resident Pension System (URPS), 472–473
Urine-based markers, 69–72
URPS. *See* Urban Resident Pension System (URPS)
U.S. Bureau of Labor Statistics, 276–277
U.S. Congregational Life Survey data, 262–263

V

VA. *See* Veteran's Administration (VA)
Value similarity, 215

Variance components, 29
Vascular function, 114–115
  lifetime influences on, 114–115
Veteran's Administration (VA), 427–428
Veterans, 231–234, 237–242
Vietnam War, 238–239
VOAD. *See* National Voluntary Organizations Active in Disaster (VOAD)
Volunteering, 297, 305
Vulnerability theory, 364

W

Walkable neighborhoods, 326
War cohorts, 234–242
Weathering hypothesis, 129–130, 155
Web of Science, 10
Wellbeing and capability, 111
White populations, life expectancy for, 124
WHO Age-friendly Environments Programme, 494–496
Whole networks, 194–195
Wisconsin Longitudinal Study (WLS), 7, 382
Within-family complexity, 209
WLS. *See* Wisconsin Longitudinal Study (WLS)
Women
  caregiving responsibilities, 211
  employment, 211
  Hispanic women, 151
  leisure activities, 171
  military participation, 234–235, 237–238, 241
  minority group, 126
  relationships with family, at older ages, 173
  responsibility of, for care work and household work, 164
  retirement and post-retirement status, 171
  time spent alone, 173
Work and family roles, gender, and leisure activities, 171–172, 172*f*
Work and retirement patterns. *See* Retirement
World War I (WWI), 230, 234–236, 242
World War II (WWII), 230, 236–238, 243–244